Acta Numerica 2017

Acta Numerica

Volume 26 2017

CAMBRIDGE
UNIVERSITY PRESS

Published by the Press Syndicate of the University of Cambridge
The Pitt Building, Trumpington Street, Cambridge CB2 1RP
One Liberty Plaza, Floor 20, New York, NY 10006, USA
10 Stamford Road, Oakleigh, Melbourne 3166, Australia

First published 2017

Printed in Great Britain by Bell and Bain Ltd, Glasgow

Library of Congress cataloguing in publication data available

A catalogue record for this book is available from the British Library

ISBN 978 1107 112582
ISSN 0962-4929

Contents

Acta Numerica (2017), pp. 1–94
doi:10.1017/S0962492917000034

The nonlinear eigenvalue problem

Stefan Güttel and Françoise Tisseur*
School of Mathematics,
The University of Manchester,
Oxford Road, Manchester M13 9PL, UK
E-mail: stefan.guettel@manchester.ac.uk,
francoise.tisseur@manchester.ac.uk

Nonlinear eigenvalue problems arise in a variety of science and engineering applications, and in the past ten years there have been numerous breakthroughs in the development of numerical methods. This article surveys nonlinear eigenvalue problems associated with matrix-valued functions which depend nonlinearly on a single scalar parameter, with a particular emphasis on their mathematical properties and available numerical solution techniques. Solvers based on Newton's method, contour integration and sampling via rational interpolation are reviewed. Problems of selecting the appropriate parameters for each of the solver classes are discussed and illustrated with numerical examples. This survey also contains numerous MATLAB code snippets that can be used for interactive exploration of the discussed methods.

CONTENTS

* Supported by EPSRC grant EP/I005293 and by a Royal Society–Wolfson Research Merit Award.

1. Introduction

Nonlinear eigenvalue problems arise in many areas of computational science and engineering, including acoustics, control theory, fluid mechanics and structural engineering. The fundamental formulation of such problems is given in the following definition.

Definition 1.1. Given a non-empty open set $\Omega \subseteq \mathbb{C}$ and a matrix-valued function $F : \Omega \to \mathbb{C}^{n \times n}$, the *nonlinear eigenvalue problem* (NEP) consists of finding scalars $\lambda \in \Omega$ (the *eigenvalues*) and nonzero vectors $v \in \mathbb{C}^n$ and $w \in \mathbb{C}^n$ (right and left *eigenvectors*) such that

$$F(\lambda)v = 0, \quad w^* F(\lambda) = 0^*.$$

Here 0 denotes the zero column vector of $\mathbb{C}^n$, and $(\cdot)^*$ denotes the conjugate transpose of a vector. We refer to (λ, v) as an *eigenpair* of F. The set of all eigenvalues is denoted by $\Lambda(F)$ and referred to as the *spectrum of* F, while $\Omega \setminus \Lambda(F)$ is called the *resolvent set of* F.

Clearly, the eigenvalues λ of F are the solutions of the scalar equation $f(z) = \det F(z) = 0$. The dependence of $F(z)$ on the parameter z is typically nonlinear, giving the 'N' in NEP, but the eigenvectors enter the problem only linearly. Throughout this work we will denote by z the parameter of F as an independent variable, and we will use λ for the eigenvalues.

Three simple but representative examples of NEPs are listed below. We also give links to corresponding problems in the NLEVP collection by Betcke *et al.* (2013). We will use some of these problems later on for numerical illustrations.

(i) *Stability analysis of delay differential equations* (DDEs). Consider the linear first-order homogeneous DDE

$$u'(t) + Au(t) + Bu(t-1) = 0, \quad u(t) \text{ given for } t \in [-1, 0]. \tag{1.1}$$

Solutions of the form $u(t) = \mathrm{e}^{\lambda t} v$ can be obtained from the NEP

$$F(\lambda)v = (\lambda I + A + B\,\mathrm{e}^{-\lambda})v = 0,$$

and the real parts of the eigenvalues λ determine the growth of $\|u\| = \mathrm{e}^{t\,\mathrm{Re}\lambda}\|v\|$. A problem of this type is part of the NLEVP collection under the name `time_delay`. Detailed expositions of the stability theory of time-delay systems can be found in Gu, Kharitonov and Chen (2003) and Michiels and Niculescu (2007).

(ii) *Differential equations with nonlinear boundary conditions.* A minimal example from Solov′ëv (2006) is a boundary value problem on $[0, 1]$,

$$-u''(x) = \lambda u(x), \quad u(0) = 0, \quad -u'(1) = \phi(\lambda)u(1). \tag{1.2}$$

Upon applying a finite element discretization with linear hat functions centred at the equispaced points $x_i = i/n$ $(i = 1, 2, \ldots, n)$, (1.2) can be rewritten as an NEP,

$$F(\lambda)u = (C_1 - \lambda C_2 + \phi(\lambda)C_3)u = 0 \tag{1.3}$$

with $n \times n$ matrices

$$C_1 = n \begin{bmatrix} 2 & -1 & & \\ -1 & \ddots & \ddots & \\ & \ddots & 2 & -1 \\ & & -1 & 1 \end{bmatrix}, \quad C_2 = \frac{1}{6n} \begin{bmatrix} 4 & 1 & & \\ 1 & \ddots & \ddots & \\ & \ddots & 4 & 1 \\ & & 1 & 2 \end{bmatrix}, \quad C_3 = e_n e_n^T,$$

and $e_n = [0, \ldots, 0, 1]^T$. With the rational function $\phi(\lambda) = \lambda/(\lambda - 1)$, this problem is contained in the NLEVP collection under the name `loaded_string`.

A property that is common to many NEPs is that the nonlinearity appears only in a few entries of $F(z)$. Indeed, the nonlinearity in this example affects only a single entry through the rank-1 matrix C_3. This *low-rank structure* is typically worth exploiting.

(iii) *Transparent boundary conditions.* This is a special case of (b), where the right boundary condition is chosen so that (1.2) is satisfied on an interval larger than the discretization interval, *e.g.* for all $x \in [0, +\infty)$. Note that the solutions of (1.2) on $[0, +\infty)$ are of the form

$$u(x) = \alpha \sin(\sqrt{\lambda} x).$$

Therefore, in order to model a 'transparent' boundary, the Dirichlet-to-Neumann map ϕ at $x = 1$ should satisfy

$$\phi(\lambda) = -\frac{\sqrt{\lambda}}{\tan(\sqrt{\lambda})}.$$

Problems with similar types of nonlinearities are encountered, for example, with the modelling of waveguides. The **gun** problem available in the NLEVP collection,

$$F(\lambda)v = \Big(K - \lambda M + \mathrm{i}\sqrt{\lambda - \omega_1^2}\, W_1 + \mathrm{i}\sqrt{\lambda - \omega_2^2}\, W_2\Big)v = 0, \tag{1.4}$$

models a waveguide-loaded accelerator cavity with the values of ω_j corresponding to cut-off wave numbers of modes in two waveguide ports. Here M, K, W_1 and W_2 are real symmetric 9956×9956 matrices arising from the finite element discretization of Maxwell's equation

$$\nabla \times \left(\frac{1}{\mu} \nabla \times E\right) - \lambda \epsilon E = 0$$

for the electric field E on some spatial domain; see Liao, Bai, Lee and

Ko (2010) for details. Also in this example the matrices W_1 and W_2 associated with the waveguide boundaries are of low rank.

For an exposition of other interesting applications where NEPs arise, we refer to the introductory chapters of the PhD theses by Effenberger (2013*a*) and Van Beeumen (2015). Note that, independent of a particular application, NEPs are of mathematical interest as systems of nonlinear equations, or scalar root-finding problems, *e.g.* for $f(z) = \det F(z)$.

There are also problems for which F does not depend on z but depends instead on the eigenvectors or a selection of eigenvectors, with applications in electronic structure calculations (Saad *et al.* 1996), machine learning (Bühler and Hein 2009), and even for the ranking of football teams (Keener 1993). Formally, these eigenvalue problems consist of finding nontrivial solutions of $F(V)V = VD$, where F is an $n \times n$ matrix that depends on an $n \times k$ matrix of eigenvectors V, and D is a $k \times k$ diagonal matrix containing the corresponding eigenvalues. This type of nonlinear dependency will not be considered here. Another interesting class is that of multi-parameter NEPs $F(\lambda, \gamma)v = 0$ which arise, for example, in the stability analysis of parametrized nonlinear wave equations (Beyn and Thümmler 2009) and the delay-independent stability analysis of DDEs (Gu *et al.* 2003, Chapter 4.6). Although we will not discuss multi-parameter NEPs in this survey any further, continuation methods for their solution typically require solving an NEP as in Definition 1.1 at every iteration.

Our aim is to survey NEPs through their interesting mathematical properties and numerical solution techniques. While we mention and list some applications where NEPs arise, these are not our main focus. We hope that this survey will be useful for mathematicians, computational scientists and engineers alike. Those new to the field may want to learn first about the basic mathematical properties of NEPs, such as their solution structure and the sensitivity of eigenvalues, and Section 2 should be a good starting point for that. We also provide an extensive list of references for further exploration, although we do not claim that this is a complete list for this very rich and rapidly developing area. Practitioners who already have a concrete NEP to solve can use this survey to get an overview of available solution techniques. To facilitate choosing the right method, we highlight the guiding principles and various problems that come with each numerical method such as, for example, the selection of suitable parameters.

With practical solution methods being the main focus, and to make this survey more interactive, we include several MATLAB code snippets. Some of the snippets require MATLAB R2016b or later, because they use local functions in a script (Higham and Higham 2017). The codes are also available online at

http://www.maths.manchester.ac.uk/nep/

We encourage the reader to execute them in MATLAB and play with the parameters. Most of the codes also run with little or no modification in GNU Octave (Eaton, Bateman, Hauberg and Wehbring 2016). We emphasize that our code snippets should not be considered as full implementations of robust numerical methods! They merely serve the purpose of illustration.

In addition to the codes, the above website also contains the bibliography of this survey, as well as an up-to-date list of links to available NEP-related software. Alongside this survey, the reader is encouraged to consult the review by Mehrmann and Voss (2004), which discusses many NEP applications and describes some of the methods contained here (but not, for example, the contour-based and rational interpolation-based solvers developed since 2004). Quadratic eigenvalue problems are reviewed in detail by Tisseur and Meerbergen (2001). For quadratic and more generally polynomial eigenvalue problems, the exploitation of special structure in the coefficient matrices is well researched; see for example Mackey, Mackey, Mehl and Mehrmann (2006*a*), Mackey, Mackey and Tisseur (2015) and the references therein. In our discussions of NEPs with special structure we will restrict our attention to the Hermitian case (see Section 3) and problems with low-rank structure. Hermitian NEPs are the most frequently encountered in practice and by far the best understood. The exploitation of low-rank structure in NEPs often leads to considerable computational savings.

The outline of this work is as follows. Section 2 provides an overview of the solution structure of NEPs, including discussions of root functions and invariant pairs, the Smith form and Keldysh's theorem, and generalized Rayleigh functionals, as well as some perturbation results. The short Section 3 is devoted to Hermitian NEPs. Sections 4, 5 and 6 are on numerical methods. Section 4 is about NEP solvers based on Newton's method, with particular attention paid to relating the various existing methods and illustrating their convergence properties with examples. Section 5 is devoted to NEP solvers using contour integrals. Here again our main focus is on summarizing and relating existing algorithms, as well as discussing their dependence on parameters. Section 6 gives an overview of solution methods based on the linearization of rational eigenvalue problems. Here our focus is on the practical and efficient construction of rational interpolants via sampling of the matrix-valued function defining an NEP. Each of the sections on numerical methods ends with a brief overview of existing software, giving links to online resources whenever possible.

Within the sections of this paper we use common counters in the numbering of definitions, theorems, algorithms, figures, *etc.* We believe this simplifies referencing for the reader since, for example, locating Remark 2.2 tells one that a definition with number 2.3 should follow. Subsections and equations are enumerated separately within each section.

2. Solution structure of NEPs

In this section we will shed light on the solution structure of NEPs. As the examples in this section aim to illustrate, the solution properties of general NEPs are quite different from linear eigenvalue problems $Av = \lambda v$, where $A \in \mathbb{C}^{n\times n}$ and $0 \neq v \in \mathbb{C}^n$, although such problems are special cases of NEPs with $F(z) = A - zI$. Nevertheless, we will see that with an appropriate definition of eigenvalue multiplicity and generalized eigenvectors, a rather elegant characterization of NEP solutions can be obtained. Much of the theory we review here can be found in the monograph by Mennicken and Möller (2003), therein developed in more detail and greater generality for operator functions in Banach spaces. The appendix of the monograph by Kozlov and Maz'ja (1999) also contains a good summary of parts of the theory, including a complete proof of Keldysh's theorem. A collection of mathematical facts about NEPs is given by Voss (2014).

2.1. Eigenvalues and eigenvectors

To develop our intuition about the eigenvalue structure of NEPs, it will be instructive to first think of a scalar function $F : \Omega \to \mathbb{C}$. The associated NEP $F(\lambda)v = 0$, $v \neq 0$, is just a scalar root-finding problem. It becomes immediately apparent that an NEP on $\Omega = \mathbb{C}$ may have

- no solutions at all, *e.g.* if $F(z) = \exp(z)$,
- finitely many solutions, *e.g.* if $F(z) = z^3 - 1$,
- infinitely many solutions, *e.g.* if $F(z) = \cos(z)$.

The fact that even an NEP of size 1×1 can have more than one eigenvalue tells us that the eigenvectors belonging to distinct eigenvalues need not be linearly independent: indeed the 'vector' $v = [\,1\,]$ is an eigenvector for all eigenvalues of an 1×1 NEP. An example of size 2×2 is

$$F(z) = \begin{bmatrix} \mathrm{e}^{\mathrm{i}z^2} & 1 \\ 1 & 1 \end{bmatrix}, \tag{2.1}$$

which is a singular matrix exactly at the points $z \in \mathbb{C}$ where $\mathrm{e}^{\mathrm{i}z^2} = 1$. Hence the eigenvalues of F are $\lambda_k = \pm\sqrt{2\pi k}$ $(k = 0, \pm 1, \pm 2, \ldots)$ and $\left[\begin{smallmatrix} 1 \\ -1 \end{smallmatrix}\right]$ is a right and left eigenvector for all of them. It will be useful to keep the function F defined by (2.1) in mind as we will often refer back to it for illustration.

In the following discussions we will assume that $F : \Omega \to \mathbb{C}^{n\times n}$ is holomorphic in a connected open set $\Omega \subseteq \mathbb{C}$ (the *domain*), denoted by $F \in H(\Omega, \mathbb{C}^{n\times n})$. This assumption is satisfied for most NEPs encountered in practice, where the elements of $F(z)$ are usually polynomial, rational, algebraic or exponential functions of $z \in \Omega$, or a combination thereof. As the definition of a matrix determinant involves only finite sums and products of the matrix entries, $F \in H(\Omega, \mathbb{C}^{n\times n})$ also implies that $\det F(z) \in H(\Omega, \mathbb{C})$. Hence the eigenvalues of F are the roots of a scalar holomorphic function in the domain Ω. The following discreteness result is a consequence of that and can be found, for example, in Mennicken and Möller (2003, Theorem 1.3.1).

Theorem 2.1. Let $\Omega \subseteq \mathbb{C}$ be a domain and $F \in H(\Omega, \mathbb{C}^{n\times n})$. Then the resolvent set $\Omega \setminus \Lambda(F)$ is open. If the resolvent set is non-empty, every eigenvalue $\lambda \in \Lambda(F)$ is isolated, that is, there exists an open neighbourhood $\mathcal{U} \subset \Omega$ such that $\mathcal{U} \cap \Lambda(F) = \{\lambda\}$.

The condition that the resolvent set be non-empty is equivalent to saying that $\det F(z)$ does not vanish identically on Ω. In this case we say that F is *regular*. While a regular F can have an infinite number of eigenvalues in Ω, the set of eigenvalues $\Lambda(F)$ does not have accumulation points in Ω. An illustrating example is the scalar function $F(z) = \sin(1/z)$, which is holomorphic in $\Omega = \mathbb{C} \setminus \{0\}$ with an infinite number of roots accumulating at $0 \notin \Omega$.

The *algebraic multiplicity of an eigenvalue* λ is defined as the multiplicity of the root of $\det F(z)$ at $z = \lambda$, that is, the smallest integer $j \geq 1$ such that

$$\frac{\mathrm{d}^j}{\mathrm{d}z^j} \det F(z)\Big|_{z=\lambda} \neq 0.$$

The algebraic multiplicity of an isolated eigenvalue must be finite, but in contrast to linear eigenvalue problems, in the nonlinear case the algebraic multiplicity of an eigenvalue is not necessarily bounded by the problem size n. This is illustrated by the scalar example $F(z) = z^{n+1}$, which has the eigenvalue $\lambda = 0$ of algebraic multiplicity $n + 1$.

By Definition 1.1, the right eigenvectors associated with an eigenvalue λ are the nonzero vectors in the null space of $F(\lambda)$, denoted by $\operatorname{null} F(\lambda)$. The dimension of this null space, $\dim(\operatorname{null} F(\lambda))$, is called the *geometric multiplicity of* λ. An eigenvalue λ is called *semisimple* if its geometric multiplicity is equal to its algebraic multiplicity, and it is called *simple* if its algebraic multiplicity equals one.

Remark 2.2. In some cases it is of interest to consider NEPs with eigenvalues at infinity, for example in the stability analysis of linear time-invariant delay differential-algebraic equations (Du, Linh, Mehrmann and Thuan 2013) or when deflating already computed eigenvalues (see Section 4.3). We say that $F : \Omega \to \mathbb{C}^{n\times n}$ defined on $\Omega \subseteq \overline{\mathbb{C}} := \mathbb{C} \cup \{\infty\}$ with $\infty \in \Omega$ has an eigenvalue at $\lambda = \infty$ if

$$G : z \mapsto F(1/z)$$

has an eigenvalue at $z = 0$. The algebraic and geometric multiplicities of $\lambda = \infty$ in F are simply defined as the corresponding algebraic and geometric multiplicities of $z = 0$ in G. Indeed, if an NEP is considered on the extended complex plane $\overline{\mathbb{C}}$ instead of $\mathbb{C}$, the point $z = \infty$ is not more or less distinguished than any other complex number. In particular, the point at infinity can always be mapped to a finite point via a Möbius transformation. Hence there is no need to further discuss infinite eigenvalues of NEPs separately.

Note that our definition of eigenvalues at infinity is not in contradiction to the common definition of infinite eigenvalues for matrix polynomials $P(z) = \sum_{j=0}^{\ell} z^j C_j$, $C_\ell \neq O$, as the zero eigenvalues of the reversal matrix polynomial $z^\ell P(1/z)$. When considered as an NEP in the sense used here, the function P is actually not well defined at $z = \infty$. Hence the convention is to instead consider the function $F(z) = z^{-\ell} P(z)$, which is well defined (and even holomorphic) at $z = \infty$ and so it makes sense to consider eigenvalues there. Indeed, $G(z) = F(1/z)$ is the reversal of P.

2.2. Root functions and generalized eigenvectors

An elegant definition of generalized eigenvectors is based on root functions, a concept introduced by Trofimov (1968). To motivate the idea of root functions, it is helpful to first consider a linear eigenvalue problem $Av = \lambda v$ for $A \in \mathbb{C}^{n\times n}$ and $0 \neq v \in \mathbb{C}^n$, and to recall from matrix analysis that a sequence $(v_0, v_1, \dots, v_{m-1})$ of nonzero vectors is called a *Jordan chain for the eigenvalue* λ if

$$(A - \lambda I)v_0 = 0, \quad (A - \lambda I)v_1 = v_0, \quad \dots, \quad (A - \lambda I)v_{m-1} = v_{m-2}; \qquad (2.2)$$

see, for example, the monograph by Horn and Johnson (1985, Chapter 3). The vector v_0 is a right eigenvector of A, or equivalently, a nonzero right null vector of $A - \lambda I$, and the $v_0, v_1, \dots, v_{m-1}$ are *generalized eigenvectors of* A. Generalized eigenvectors of a matrix can be easily seen to be linearly independent. The maximal length m of a Jordan chain starting with v_0 is called the *rank of* v_0.

Interestingly, by defining the functions $v(z) = \sum_{j=0}^{m-1} (z-\lambda)^j v_j$ and $F(z) = A - zI$, the m equations in (2.2) can be rewritten as

$$\frac{\mathrm{d}^j}{\mathrm{d}z^j} F(z)v(z)\Big|_{z=\lambda} = j! \sum_{k=0}^{j} \frac{F^{(k)}(\lambda)}{k!} v_{j-k} = 0, \quad j = 0, 1, \dots, m-1, \qquad (2.3)$$

where $F^{(k)}(z) = \frac{\mathrm{d}^k}{\mathrm{d}z^k} F(z)$. In other words, $F(z)v(z)$ has a root of multiplicity at least m at $z = \lambda$. While (2.2) is specific to linear eigenvalue problems, (2.3) extends straightforwardly to the nonlinear case. The main difference for NEPs is that generalized eigenvectors need not necessarily be linearly independent, and even the zero vector is admitted as a generalized eigenvector. We make this precise in the following definition, which has been adopted from Mennicken and Möller (2003, Definition 1.6.1).

Definition 2.3. Let $F \in H(\Omega, \mathbb{C}^{n\times n})$ be regular on a non-empty domain $\Omega \subseteq \mathbb{C}$, and let $\lambda \in \Omega$ be given.

(i) A holomorphic vector-valued function $v \in H(\Omega, \mathbb{C}^n)$ such that $v(\lambda) \neq 0$ and $F(\lambda)v(\lambda) = 0$ is called a *root function for F at λ*. The multiplicity of the root $z = \lambda$ of $F(z)v(z)$ is denoted by $s(v)$.

(ii) A tuple $(v_0, v_1, \ldots, v_{m-1}) \in (\mathbb{C}^n)^m$ with $m \geq 1$ and $v_0 \neq 0$ is called a *Jordan chain for F at λ* if

$$v(z) = \sum_{k=0}^{m-1} (z-\lambda)^k v_k$$

is a root function for F at λ and $s(v) \geq m$.

(iii) For a given vector $v_0 \in \operatorname{null} F(\lambda) \setminus \{0\}$, the number

$$r(v_0) = \max\bigl\{s(v) : v \text{ is a root function for } F \text{ at } \lambda \text{ with } v(\lambda) = v_0\bigr\}$$

is called the *rank of v_0*.

(iv) A system of vectors in $\mathbb{C}^n$,

$$V = (v_k^j : 0 \leq k \leq m_j - 1,\ 1 \leq j \leq d),$$

is a *complete system of Jordan chains for F at λ* if

(a) $d = \dim(\operatorname{null} F(\lambda))$ and $\{v_0^1, v_0^2, \ldots, v_0^d\}$ is a basis of $\operatorname{null} F(\lambda)$,

(b) the tuple $(v_0^j, v_1^j, \ldots, v_{m_j-1}^j)$ is a Jordan chain for F at λ for $j = 1, 2, \ldots, d$,

(c) $m_j = \max\{r(v_0) : v_0 \in \operatorname{null} F(\lambda) \setminus \operatorname{span}\{v_0^\nu : 1 \leq \nu < j\}\}$ for $j = 1, 2, \ldots, d$.

It can be shown (Mennicken and Möller 2003, Proposition 1.6.4) that a complete system of Jordan chains always exists, which essentially requires the verification that for a regular NEP the rank $r(v_0)$ of any eigenvector v_0 is finite. The numbers m_j satisfy $m_1 \geq m_2 \geq \cdots \geq m_d$ by definition and are called the *partial multiplicities of λ*. It can be shown that $\sum_{j=1}^d m_j$ corresponds to the algebraic multiplicity of λ (Mennicken and Möller 2003, Proposition 1.8.5). The number m_1 is called the *index of λ*. If $m_1 = \cdots = m_d = 1$, then λ is a semisimple eigenvalue; if in addition $d = 1$, then λ is a simple eigenvalue.

Example 2.4. Consider the 2×2 function F defined in (2.1). The eigenvalue $\lambda_0 = 0$ has algebraic multiplicity two and geometric multiplicity one. The corresponding right eigenvector v_0 and generalized eigenvector v_1 are both multiples of $\left[\begin{smallmatrix}1\\-1\end{smallmatrix}\right]$, and hence are not linearly independent. The vector-valued function $v(z) = v_0 + zv_1 = \left[\begin{smallmatrix}1+z\\-1-z\end{smallmatrix}\right]$ is a root function for F at

$\lambda_0 = 0$, and the multiplicity of λ_0 as a root of $F(z)v(z)$ can be easily shown to be $s(v) = 2$. Using the MATLAB Symbolic Math Toolbox we find that $\frac{\mathrm{d}^2}{\mathrm{d}z^2} F(z)v(z)$ is the smallest-order derivative which does not vanish at $z = 0$:

```
syms z; order = 2;
F = [ exp(1i*z.^2) 1; 1 1 ]; v = [ 1+z; 1-z ];
subs(diff(F*v, order), z, 0)
ans =
 2i
  0
```

Hence the rank of the right eigenvector v_0 with eigenvalue λ_0 is $r(v_0) \geq 2$, but since λ_0 has algebraic multiplicity two, $r(v_0) = 2$. The pair $\left(\left[\begin{smallmatrix}1\\-1\end{smallmatrix}\right], \left[\begin{smallmatrix}1\\-1\end{smallmatrix}\right]\right)$ is a Jordan chain for F at $\lambda_0 = 0$, and also a complete system of Jordan chains there.

So far we have only considered *right* eigenvectors of F, but clearly left eigenvectors of F can be obtained as the right eigenvectors of F^*. It turns out that there is a natural way to complement a complete system of Jordan chains for F at λ with a complete system of Jordan chains of F^* at λ, and the latter is uniquely determined if certain normalization conditions are enforced. The following result makes this precise (Mennicken and Möller 2003, Theorem 1.6.5).

Theorem 2.5. Let $F \in H(\Omega, \mathbb{C}^{n\times n})$ be regular on a non-empty domain $\Omega \subseteq \mathbb{C}$, and consider an eigenvalue $\lambda \in \Lambda(F)$. Let

$$V = \left(v_k^j : 0 \leq k \leq m_j - 1,\ 1 \leq j \leq d\right)$$

be a complete system of Jordan chains for F at λ. Then there exists a unique complete system of Jordan chains

$$W = \left(w_k^j : 0 \leq k \leq m_j - 1,\ 1 \leq j \leq d\right)$$

for F^* at λ such that each eigenvector w_0^j has rank $r(w_0^j) = m_j$ and

$$\sum_{\alpha=0}^{k} \sum_{\beta=1}^{m_i} w_{k-\alpha}^{j*} \frac{F^{(\alpha+\beta)}(\lambda)}{(\alpha+\beta)!} v_{m_i-\beta}^{i} = \delta_{ij}\delta_{0k}, \quad 0 \leq k \leq m_j - 1,\ 1 \leq i, j \leq d, \tag{2.4}$$

with the Kronecker delta

$$\delta_{ij} = \begin{cases} 0 & \text{if } i \neq j, \\ 1 & \text{if } i = j. \end{cases}$$

Example 2.6. Consider the matrix-valued function F in (2.1). We found in Example 2.4 that the pair

$$V = (v_0, v_1) = \left(\begin{bmatrix} 1 \\ -1 \end{bmatrix}, \begin{bmatrix} 1 \\ -1 \end{bmatrix} \right)$$

is a complete system of Jordan chains for F at $\lambda = 0$. A similar construction shows that

$$W = (w_0, w_1) = \left(\rho_0 \begin{bmatrix} 1 \\ -1 \end{bmatrix}, \rho_1 \begin{bmatrix} 1 \\ -1 \end{bmatrix} \right) \quad \text{with } \rho_0, \rho_1 \in \mathbb{C} \setminus \{0\}$$

is a complete system of Jordan chains for F^* at $\lambda = 0$ satisfying $r(w_0) = 2$. The normalization conditions (2.4) become

$$\begin{aligned} w_0^* F'(0) v_1 + w_0^* \frac{F''(0)}{2} v_0 &= 1, \\ w_1^* F'(0) v_1 + w_1^* \frac{F''(0)}{2} v_0 + w_0^* \frac{F''(0)}{2} v_1 + w_0^* \frac{F'''(0)}{6} v_0 &= 0, \end{aligned}$$

and they are satisfied if we choose $\rho_0 = \mathrm{i}$ and $\rho_1 = -\mathrm{i}$.

While the normalization conditions in Theorem 2.5 look rather messy in general, they simplify if the eigenvalue λ is semisimple. In this case we have $m_1 = \cdots = m_d = 1$ so that (2.4) becomes

$$w_0^{j*} F'(\lambda) v_0^i = \delta_{ij}, \quad 1 \leq i, j \leq d, \tag{2.5}$$

that is, the left and right eigenvectors can be chosen $F'(\lambda)$-biorthonormal.

2.3. Factorizations: Smith form

It is often useful to have a factorization of an NEP about one or multiple eigenvalues, giving rise to *local* or *global* factorizations, respectively. A very useful factorization is the so-called Smith form, which we state in its global version; see Leiterer (1978), Gohberg and Rodman (1981) and Kozlov and Maz'ja (1999, Chapter A.6). It uses the notion of a *unimodular* matrix-valued function $P \in H(\Omega, \mathbb{C}^{n \times n})$, which means that $\det P(z)$ is a nonzero constant over Ω.

Theorem 2.7 (Smith form). Let $F \in H(\Omega, \mathbb{C}^{n \times n})$ be regular on a non-empty domain Ω. Let λ_i $(i = 1, 2, \ldots)$ be the distinct eigenvalues of F in Ω with partial multiplicities $m_{i,1} \geq m_{i,2} \geq \cdots \geq m_{i,d_i}$. Then there exists a *global Smith form*

$$P(z) F(z) Q(z) = D(z), \quad z \in \Omega,$$

with unimodular matrix-valued functions $P, Q \in H(\Omega, \mathbb{C}^{n\times n})$, and

$$D(z) = \operatorname{diag}(\delta_1(z), \delta_2(z), \dots, \delta_n(z)).$$

The diagonal entries δ_j are of the form

$$\delta_j(z) = h_j(z) \prod_{i=1,2,\dots} (z-\lambda_i)^{m_{i,j}}, \quad j = 1,\dots,n,$$

where $m_{i,j} = 0$ if $j > d_i$ and each $h_j \in H(\Omega, \mathbb{C})$ is free of roots on Ω.

From the Smith form it is easy to identify eigenvectors of F for an eigenvalue λ_i. Let Q be partitioned columnwise as $Q(z) = [q_1(z), q_2(z), \dots, q_n(z)]$. By the diagonal structure of D, the canonical unit vectors $e_1, e_2, \dots, e_{d_i}$ form a basis of null $D(\lambda_i)$, and therefore also a basis of $\operatorname{null}(P(\lambda_i)F(\lambda_i)Q(\lambda_i))$. Hence we have found a basis of d_i right eigenvectors of F for λ, namely,

$$q_1(\lambda_i), q_2(\lambda_i), \dots, q_{d_i}(\lambda_i).$$

By the same argument one can identify the first d_i rows of

$$P(z) = [p_1(z), p_2(z), \dots, p_n(z)]^*$$

with left eigenvectors of F for $z = \lambda_i$.

Another application of the Smith form is for the characterization of the singularities of the *resolvent* given by

$$F(z)^{-1} = Q(z)D(z)^{-1}P(z) = \sum_{j=1}^{n} \delta_j(z)^{-1} q_j(z) p_j(z)^*. \tag{2.6}$$

Here the diagonal entries $\delta_j(z)^{-1}$ of $D(z)^{-1}$ are scalar meromorphic functions with poles of multiplicities $m_{i,j}$ at the eigenvalues λ_i,

$$\delta_j(z)^{-1} = \frac{1}{h_j(z)} \prod_{i=1,2,\dots} (z-\lambda_i)^{-m_{i,j}}, \quad j = 1,\dots,n.$$

2.4. Expansions: Keldysh's theorem

Let us consider an eigenvalue $z = \lambda_i$ of F. Then each term $\delta_j(z)^{-1}q_j(z)p_j(z)^*$ in (2.6) can be expanded into a matrix-valued Laurent series (Ahlfors 1953, Chapter 5.1.3) that is valid in some neighbourhood $\mathcal{U} \subseteq \Omega$ about λ_i,

$$\delta_j(z)^{-1}q_j(z)p_j(z)^* = \sum_{k=1}^{m_{i,j}} S_{i,j,k}(z-\lambda_i)^{-k} + R_j(z), \quad z \in \mathcal{U}\setminus\{\lambda_i\},$$

with coefficient matrices $S_{i,j,k} \in \mathbb{C}^{n\times n}$, $S_{i,j,m_{i,j}} \neq O$, and the remainder function $R_j \in H(\mathcal{U}, \mathbb{C}^{n\times n})$. Combining these series for every term in (2.6) into a single expansion at $z = \lambda_i$, we arrive at

$$F(z)^{-1} = \sum_{j=1}^{d_i} \sum_{k=1}^{m_{i,j}} S_{i,j,k}(z - \lambda_i)^{-k} + R(z), \quad z \in \mathcal{U} \setminus \{\lambda_i\}, \tag{2.7}$$

with a remainder function $R \in H(\mathcal{U}, \mathbb{C}^{n\times n})$. This shows that the *resolvent* $F(z)^{-1}$ is meromorphic in $\mathcal{U}$ with a pole of multiplicity $\max_{j=1,\dots,d_i}\{m_{i,j}\} = m_{i,1}$ at λ_i.

Let us now assume that F has only *finitely* many eigenvalues $\lambda_1, \lambda_2, \dots, \lambda_s$ in the domain Ω. Then we can apply the expansion (2.7) recursively at all the eigenvalues as follows: starting with (2.7) about $z = \lambda_1$, we can expand the remainder R into another Laurent series about $z = \lambda_2$, and so forth, giving rise to the global expansion valid on the whole of Ω,

$$F(z)^{-1} = \sum_{i=1}^{s} \sum_{j=1}^{d_i} \sum_{k=1}^{m_{i,j}} S_{i,j,k}(z - \lambda_i)^{-k} + \widetilde{R}(z), \tag{2.8}$$

with $\widetilde{R} \in H(\Omega, \mathbb{C}^{n\times n})$.

It is possible to characterize the matrices $S_{i,j,k}$ in (2.8) in terms of generalized eigenvectors of F, a result that is known as *Keldysh's theorem*. More precisely, for each distinct eigenvalue $\lambda_i \in \Omega$ $(i = 1, 2, \dots, s)$, let

$$\left(v_k^{ij} : 0 \le k \le m_{ij} - 1,\ 1 \le j \le d_i\right), \quad \left(w_k^{ij} : 0 \le k \le m_{ij} - 1,\ 1 \le j \le d_i\right)$$

be pairs of complete systems of Jordan chains for F and F^* at λ_i as defined by Theorem 2.5. Then the matrices $S_{i,j,k}$ in (2.8) are given as

$$S_{i,j,k} = \sum_{\ell=0}^{m_{ij}-k} v_\ell^{ij} w_{m_{ij}-k-\ell}^{ij*}. \tag{2.9}$$

Keldysh's theorem can be found in this form, for example, in Mennicken and Möller (2003, Theorem 1.6.5) and Kozlov and Maz'ja (1999, Theorem A.10.2). A more compact representation can be obtained by defining Jordan blocks

$$J_{ij} = \begin{bmatrix} \lambda_i & 1 & & \\ & \lambda_i & \ddots & \\ & & \ddots & 1 \\ & & & \lambda_i \end{bmatrix} \in \mathbb{C}^{m_{ij}\times m_{ij}}, \tag{2.10}$$

and arranging the generalized eigenvectors in columns of matrices

$$V_{ij} = \left[v_0^{ij}, v_1^{ij}, \dots, v_{m_{ij}-1}^{ij}\right] \quad \text{and} \quad W_{ij} = \left[w_{m_{ij}-1}^{ij}, w_{m_{ij}-2}^{ij}, \dots, w_0^{ij}\right]. \tag{2.11}$$

(Note the descending order of the left generalized eigenvectors in W_{ij}.)

By the Jordan-form definition of a matrix function (Higham 2008, Definition 1.2), we have

$$(zI - J_{ij})^{-1} = \begin{bmatrix} (z-\lambda_i)^{-1} & (z-\lambda_i)^{-2} & \cdots & (z-\lambda_i)^{-m_{ij}} \\ & (z-\lambda_i)^{-1} & \ddots & \vdots \\ & & \ddots & (z-\lambda_i)^{-2} \\ & & & (z-\lambda_i)^{-1} \end{bmatrix},$$

and hence using (2.9),

$$\sum_{k=1}^{m_{i,j}} S_{i,j,k}(z-\lambda_i)^{-k} = V_{ij}(zI - J_{ij})^{-1}W_{ij}^*.$$

We can now rewrite (2.8) and state Keldysh's theorem in the following form.

Theorem 2.8 (Keldysh). Let $F \in H(\Omega, \mathbb{C}^{n\times n})$ be regular on a non-empty domain Ω. Let $\lambda_1, \lambda_2, \ldots, \lambda_s$ be the distinct eigenvalues of F in Ω of partial multiplicities $m_{i,1} \geq m_{i,2} \geq \cdots \geq m_{i,d_i}$, and define

$$\overline{m} = \sum_{i=1}^{s}\sum_{j=1}^{d_i} m_{ij}.$$

Then there are $n \times \overline{m}$ matrices V and W whose columns are generalized eigenvectors, and an $\overline{m} \times \overline{m}$ Jordan matrix J with eigenvalues λ_i of partial multiplicities m_{ij}, such that

$$F(z)^{-1} = V(zI - J)^{-1}W^* + \widetilde{R}(z) \tag{2.12}$$

for some $\widetilde{R} \in H(\Omega, \mathbb{C}^{n\times n})$. With the matrices defined in (2.10)–(2.11),

$$\begin{aligned} J &= \operatorname{diag}(J_1, \ldots, J_s), & J_i &= \operatorname{diag}(J_{i,1}, \ldots, J_{i,d_i}), \\ V &= [V_1, \ldots, V_s], & V_i &= [V_{i,1}, \ldots, V_{i,d_i}], \\ W &= [W_1, \ldots, W_s], & W_i &= [W_{i,1}, \ldots, W_{i,d_i}]. \end{aligned}$$

Let us briefly consider two special cases of Theorem 2.8. Assume that all eigenvalues λ_i of F are semisimple, that is, $m_{i,1} = \cdots = m_{i,d_i} = 1$. Then the matrix J will be diagonal with each eigenvalue λ_i appearing exactly d_i times. Further, if all λ_i of F are simple (*i.e.* $d_1 = \cdots = d_s = 1$), then J is an $s \times s$ diagonal matrix with distinct eigenvalues, and we have

$$F^{-1}(z) = \sum_{i=1}^{s}(z-\lambda_i)^{-1} v_0^i w_0^{i*} + \widetilde{R}(z),$$

where v_0^i and w_0^i are right and left eigenvectors, respectively, satisfying $w_0^{i*}F'(\lambda_i)v_0^i = 1$.

Theorem 2.8 shows that, up to a holomorphic additive term, the behaviour of the resolvent $F^{-1}(z)$ at the eigenvalues λ_i is captured by the inverse of a shifted Jordan matrix. In the linear case $F(z) = zI - A$, we can set $\widetilde{R} \equiv 0$ in (2.12) and choose V as a basis of $n = \overline{m}$ generalized eigenvectors and $W^* = V^{-1}$, resulting in the standard Jordan decomposition $A = VJV^{-1}$. In the nonlinear case, however, V and W can have an arbitrary (but identical) number $\overline{m}$ of columns which need not be linearly independent.

Example 2.9. Consider again the matrix-valued function F in (2.1) and let Ω be the disc of centre 0 and radius 3. Then F has eigenvalues

$$\{0, -\sqrt{2\pi}, \sqrt{2\pi}, -\mathrm{i}\sqrt{2\pi}, \mathrm{i}\sqrt{2\pi}\}$$

inside Ω, all simple except the eigenvalue 0, which has algebraic multiplicity two and geometric multiplicity one. It follows from Examples 2.4 and 2.6 that the matrices J, V, W of Theorem 2.8 are given by

$$J = \operatorname{diag}\left(\begin{bmatrix} 0 & 1 \\ 0 & 0 \end{bmatrix}, -\sqrt{2\pi}, \sqrt{2\pi}, -\mathrm{i}\sqrt{2\pi}, \mathrm{i}\sqrt{2\pi}\right), \tag{2.13}$$

$$V = \begin{bmatrix} 1 & 1 & 1 & 1 & 1 & 1 \\ -1 & -1 & -1 & -1 & -1 & -1 \end{bmatrix}, \tag{2.14}$$

$$W = \frac{1}{2}\begin{bmatrix} -2\mathrm{i} & 2\mathrm{i} & \frac{-\mathrm{i}}{\sqrt{2\pi}} & \frac{\mathrm{i}}{\sqrt{2\pi}} & \frac{1}{\sqrt{2\pi}} & \frac{-1}{\sqrt{2\pi}} \\ 2\mathrm{i} & -2\mathrm{i} & \frac{\mathrm{i}}{\sqrt{2\pi}} & \frac{-\mathrm{i}}{\sqrt{2\pi}} & \frac{-1}{\sqrt{2\pi}} & \frac{1}{\sqrt{2\pi}} \end{bmatrix}.$$

Hence

$$F(z)^{-1} = \begin{bmatrix} -g(z) & g(z) \\ g(z) & -g(z) \end{bmatrix} + \widetilde{R}(z), \quad g(z) = \frac{3\mathrm{i}z^4 - 4\mathrm{i}\pi^2}{z^2(z^4 - 4\pi^2)}.$$

2.5. Invariant pairs

Invariant pairs play an important role in the analysis of NEPs and the derivation of algorithms. Assume again that $F \in H(\Omega, \mathbb{C}^{n\times n})$ on a non-empty domain $\Omega \subseteq \mathbb{C}$. Then by the Cauchy integral formula (Ahlfors 1953, Chapter 4.2) we have the representation

$$F^{(k)}(\mu) = \frac{k!}{2\pi\mathrm{i}} \int_\Gamma \frac{F(z)}{(z-\mu)^{k+1}}\,\mathrm{d}z,$$

where $\Gamma \subset \Omega$ is a contour containing the point μ in its interior. This formula is the basis for the concept of invariant pairs as defined by Beyn, Effenberger and Kressner (2011) and further advocated in Effenberger (2013*a*).

Definition 2.10 (invariant pair). A pair $(V, M) \in \mathbb{C}^{n\times m} \times \mathbb{C}^{m\times m}$ is called an *invariant pair for $F \in H(\Omega, \mathbb{C}^{n\times n})$* if $\mathcal{F}(V, M) = O$, where

$$\mathcal{F}(V, M) := \frac{1}{2\pi\mathrm{i}} \int_\Gamma F(z)V(zI - M)^{-1}\,\mathrm{d}z \in \mathbb{C}^{n\times m}.$$

Here Γ is a contour containing the eigenvalues of M in its interior.

Note that $F \in H(\Omega, \mathbb{C}^{n\times n})$ can always be written in the 'split form'

$$F(z) = f_1(z)C_1 + f_2(z)C_2 + \cdots + f_\ell(z)C_\ell \tag{2.15}$$

with at most $\ell = n^2$ coefficient matrices C_j and scalar functions $f_j \in H(\Omega, \mathbb{C})$ (one for each matrix entry). In many cases, F is naturally given in this form and $\ell \ll n^2$. Clearly, an invariant pair (V, M) for F satisfies

$$\mathcal{F}(V, M) = C_1 V f_1(M) + C_2 V f_2(M) + \cdots + C_\ell V f_\ell(M) = O, \tag{2.16}$$

where each

$$f_j(M) = \frac{1}{2\pi \mathrm{i}} \int_\Gamma f_j(z)(zI - M)^{-1}\,\mathrm{d}z$$

is a matrix function. Hence invariant pairs can be seen as a generalization to NEPs of invariant subspaces for a single matrix, and standard pairs for matrix polynomials (Gohberg, Lancaster and Rodman 2009). The next result shows that invariant pairs can easily be constructed from root functions; see for example Gohberg, Kaashoek and van Schagen (1993).

Lemma 2.11. Let

$$v(z) = \sum_{j=0}^{m-1} (z - \lambda)^j v_j$$

be a root function for $F \in H(\Omega, \mathbb{C}^{n\times n})$ at λ of multiplicity larger than or equal to m. Define

$$V = [v_0, v_1, \ldots, v_{m-1}] \in \mathbb{C}^{n\times m}$$

and let

$$J = \begin{bmatrix} \lambda & 1 & & \\ & \lambda & \ddots & \\ & & \ddots & 1 \\ & & & \lambda \end{bmatrix} \in \mathbb{C}^{m\times m}$$

be the $m \times m$ Jordan block for λ. Then (V, J) is an invariant pair for F.

Proof. By the Jordan-form definition of a matrix function (Higham 2008, Definition 1.2) we have

$$(zI - J)^{-1} = \sum_{k=0}^{m-1} (z - \lambda)^{-(k+1)} E^k,$$

where $J = \lambda I + E$. Therefore,

$$\begin{aligned}\mathcal{F}(V, J) &= \frac{1}{2\pi \mathrm{i}} \int_\Gamma F(z) V (zI - J)^{-1} \,\mathrm{d}z \\ &= \sum_{k=0}^{m-1} \left(\frac{1}{2\pi \mathrm{i}} \int_\Gamma \frac{F(z)}{(z-\lambda)^{k+1}} \,\mathrm{d}z \right) V E^k \\ &= \sum_{k=0}^{m-1} \frac{F^{(k)}(\lambda)}{k!} V E^k.\end{aligned}$$

For $j = 1, 2, \dots, m$, the jth column of $\mathcal{F}(V, J)$ is

$$\sum_{k=0}^{j-1} \frac{F^{(k)}(\lambda)}{k!} v_{j-k},$$

which is equal to the derivative

$$\left(\frac{\mathrm{d}^{j-1}}{\mathrm{d}z^{j-1}} F(z) v(z)|_{z=\lambda} \right) \Big/ (j-1)!;$$

see (2.3). By the definition of a root function this derivative is zero for $j = 1, 2, \dots, m$, hence $\mathcal{F}(V, J) = O$. □

By the block structure of the matrices V and J defined in Theorem 2.8, we immediately find that (V, J) is an invariant pair. An invariant pair (V, M) as defined in Definition 2.10 may contain redundant information as, for example, $([V, V], \mathrm{diag}(M, M))$ is an invariant pair of at least twice the necessary size. Also, any (V, M) with $V = O$ is a trivial invariant pair. The following concept of *minimal pairs* prevents such anomalies.

Definition 2.12 (minimal pair, minimality index). A pair $(V, M) \in \mathbb{C}^{n \times m} \times \mathbb{C}^{m \times m}$ is called *minimal* if there is an integer p such that

$$\operatorname{rank} \mathcal{V}_p(V, M) = m,$$

where

$$\mathcal{V}_p(V, M) = \begin{bmatrix} V \\ VM \\ \vdots \\ VM^{p-1} \end{bmatrix}. \tag{2.17}$$

The smallest such p is called the *minimality index* of the pair (V, M).

For the matrix-valued function $F \in H(\Omega, \mathbb{C}^{2\times 2})$ in (2.1), the pair (V, J) with J, V as in (2.13)–(2.14) is a minimal invariant pair with minimality index 6. Note that the minimality index of an invariant pair (V, M) is generically equal to one when $m \le n$. The following result gives a simple lower and upper bound on the minimality index. The lower bound appears in Beyn (2012, Lemma 5.1) and the upper bound in Kressner (2009).

Lemma 2.13. Let $\Omega \subseteq \mathbb{C}$ be a non-empty domain containing all distinct eigenvalues $\lambda_1, \lambda_2, \ldots, \lambda_s$ of $F \in H(\Omega, \mathbb{C}^{n\times n})$, where each λ_i is of geometric multiplicity d_i and has corresponding partial multiplicities $m_{i,1} \geq m_{i,2} \geq \cdots \geq m_{i,d_i}$. In addition let $(V, M) \in \mathbb{C}^{n\times \overline{m}} \times \mathbb{C}^{\overline{m}\times\overline{m}}$ be an invariant pair with V, M, and $\overline{m}$ as defined in Theorem 2.8. Then the minimality index p of (V, M) satisfies

$$\sum_{i=1}^{s} m_{i,1} \leq p \leq \overline{m}.$$

Let (V, M) be a minimal invariant pair for F with minimality index p, and write F in the form (2.15). If $M = UTU^*$ is the Schur decomposition of M, then on using $f_j(M) = Uf_j(T)U^*$ we have that (2.16) is equivalent to

$$\mathcal{F}(VU, T) = \sum_{j=1}^{\ell} C_j VU f_j(T) = O,$$

that is, (VU, T) is an invariant pair for F and it is also minimal since $\mathcal{V}_p(VU, T) = \mathcal{V}_p(V, M)U$. If we let $v = VUe_1$, then $v \neq 0$ because

$$\mathcal{V}_p(VU, T)e_1 = [1, t_{11}, \ldots, t_{11}^{p-1}]^T \otimes v \neq 0$$

since $\mathcal{V}_p(VU, T)$ is of full column rank. Now, $t_{11} = \lambda$ is an eigenvalue of M and $VUf_j(\lambda)e_1 = f_j(\lambda)VUe_1 = f_j(\lambda)v$ so that

$$\mathcal{F}(VU, T)e_1 = \sum_{j=1}^{\ell} C_j VU f_j(T)e_1 = \sum_{j=1}^{\ell} f_j(\lambda)C_j v = 0,$$

showing that λ is also an eigenvalue of F.

Lemma 2.14. If (V, M) is a minimal pair for $F \in H(\Omega, \mathbb{C}^{n\times n})$, then the eigenvalues of M are eigenvalues of F.

The following theorem summarizes the correspondence between (minimal) invariant pairs and Jordan chains for F at λ. Results of this type have been derived by Gohberg *et al.* (1993, Lemma 2.1), Beyn *et al.* (2011, Proposition 2.4) and Effenberger (2013*a*, Theorem 3.1.13).

Theorem 2.15. Let $\lambda_1, \ldots, \lambda_s$ be distinct eigenvalues of $F \in H(\Omega, \mathbb{C}^{n\times n})$ and consider a matrix $V = [V_1, \ldots, V_s]$ with

$$V_i = \left[V_{i,1}, \ldots, V_{i,\widehat{d}_i}\right] \quad \text{and} \quad V_{ij} = \left[v_0^{ij}, v_1^{ij}, \ldots, v_{\widehat{m}_{ij}-1}^{ij}\right],$$

with all $v_0^{ij} \neq 0$. Then the columns of every V_{ij} form a Jordan chain for F at λ_i if and only if (V, J) is an invariant pair with a block Jordan matrix

$J = \operatorname{diag}(J_1, J_2, \ldots, J_s)$, where $J_i = \operatorname{diag}(J_{i,1}, \ldots, J_{i,\widehat{d}_i})$ and every J_{ij} is an $\widehat{m}_{ij} \times \widehat{m}_{ij}$ Jordan block for λ_i. Moreover, (V, J) is minimal if and only if the vectors $v_0^{i,1}, v_0^{i,2}, \ldots, v_0^{i,\widehat{d}_i}$ are linearly independent for all $i = 1, 2, \ldots, d_i$.

Note that we have not made any explicit assumptions on the integers $\widehat{d}_i$ and $\widehat{m}_{ij}$ in Theorem 2.15. If we assume that $\widehat{d}_i$ equals the geometric multiplicity of each λ_i, then if (V, J) is a minimal pair the columns of each V_i in Theorem 2.15 form a complete system of Jordan chains. In this case the algebraic multiplicities of the eigenvalues of J coincide with the algebraic multiplicities of the corresponding eigenvalues of F. This gives rise to the following definition.

Definition 2.16 (complete invariant pair). An invariant pair (V, M) of $F \in H(\Omega, \mathbb{C}^{n\times n})$ is called *complete* if it is minimal and the algebraic multiplicities of the eigenvalues of M are the same as the algebraic multiplicities of the corresponding eigenvalues of F.

Complete invariant pairs have been introduced by Kressner (2009) under the name *simple invariant pairs*. We prefer to use the term *complete* as it better aligns with the terminology of a complete system of Jordan chains in Definition 2.3.

2.6. Nonlinear Rayleigh functionals

Nonlinear Rayleigh functionals are the generalization of Rayleigh quotients for matrices. Duffin (1955) introduces them for symmetric overdamped quadratic eigenvalue problems. They are generalized to larger classes of Hermitian NEPs in Rogers (1964), Hadeler (1967) and Voss (2009), and to general NEPs in Schreiber (2008) and Schwetlick and Schreiber (2012).

Let λ be an eigenvalue of the matrix-valued function $F \in H(\Omega, \mathbb{C}^{n\times n})$ with corresponding right and left eigenvectors v and w. For some $\rho > 0$ and $\varepsilon < \pi/2$, let

$$\overline{\mathbb{D}}_{\lambda,\rho} = \{z \in \mathbb{C} : \ |z - \lambda| \leq \rho\},$$
$$\overline{\mathbb{K}}_{\varepsilon}(v) = \{x \in \mathbb{C}^n \setminus \{0\} : \angle(x, v) \leq \varepsilon\}$$

be neighbourhoods of λ and v, respectively. Here $\angle(x, v)$ denotes the angle between the nonzero vectors x and v, that is,

$$\angle(x, v) = \cos^{-1} \frac{|v^* x|}{\|x\|_2 \|v\|_2}.$$

A functional

$$p : \overline{\mathbb{K}}_{\varepsilon}(v) \times \overline{\mathbb{K}}_{\varepsilon}(w) \ni (x, y) \mapsto p(x, y) \in \overline{\mathbb{D}}_{\lambda,\rho}$$

is a *two-sided nonlinear Rayleigh functional* if the following conditions hold:

$$p(\alpha x, \beta y) = p(x, y) \text{ for all nonzero } \alpha, \beta \in \mathbb{C}, \tag{2.18}$$
$$y^* F(p(x, y))x = 0, \tag{2.19}$$
$$y^* F'(p(x, y))x \neq 0. \tag{2.20}$$

The condition (2.20) restricts the functional to vectors close to eigenvectors corresponding to a simple eigenvalue.

The next result, which appears in Schwetlick and Schreiber (2012, Theorem 5), concerns the local uniqueness of $p(x, y)$ and bounds the distance of $p(x, y)$ to the exact eigenvalue in terms of the angles between eigenvectors and approximations to eigenvectors.

Theorem 2.17. Let λ be a simple eigenvalue of $F \in H(\Omega, \mathbb{C}^{n\times n})$ with corresponding right and left eigenvectors v and w normalized to have unit 2-norms. Let $\rho > 0$ be such that $\overline{\mathbb{D}}_{\lambda,\rho} \subset \Omega$. Then there exist constants $0 < \rho_0 \leq \rho$, $0 < \varepsilon_0 < \pi/2$ such that for all $(x, y) \in \overline{\mathbb{K}}_{\varepsilon_0}(v) \times \overline{\mathbb{K}}_{\varepsilon_0}(w)$ there exists a unique $p(x, y) \in \overline{\mathbb{D}}_{\lambda,\rho_0}$ satisfying $y^* F(p(x, y))x = 0$ and

$$|p(x, y) - \lambda| \leq \frac{8}{3} \frac{\|F(\lambda)\|_2}{|y^* F'(p(x, y))x|} \tan(\angle(x, v)) \tan(\angle(y, w)).$$

It then follows from Theorem 2.17 that $p(v, w) = \lambda$, that is, $p(v, w)$ is an eigenvalue of F.

Example 2.18. Let F be the matrix-valued function defined in (2.1), and let Ω be the disc with centre 0 and radius 3. If we let $x = \left[\begin{smallmatrix} x_1 \\ x_2 \end{smallmatrix}\right]$ and $y = \left[\begin{smallmatrix} y_1 \\ y_2 \end{smallmatrix}\right]$ then the condition (2.20) is equivalent to $p(x, y) \neq 0$ and $\overline{y}_1 x_1 \neq 0$. Now if $\overline{y}_1 x_1 \neq 0$, (2.19) can be rewritten as

$$\mathrm{e}^{\mathrm{i}p(x,y)^2} = -\frac{\overline{y}_1 x_2 + \overline{y}_2 x_1 + \overline{y}_2 x_2}{\overline{y}_1 x_1}$$

so that

$$\mathrm{i}p(x, y)^2 = \log\left(-\frac{\overline{y}_1 x_2 + \overline{y}_2 x_1 + \overline{y}_2 x_2}{\overline{y}_1 x_1}\right) + 2\pi k\mathrm{i}, \quad k = 0, \pm 1, \pm 2, \ldots$$

for any complex logarithm function log. Hence the set of two-sided nonlinear Rayleigh functionals for F is given by

$$p(x, y) = \sqrt{-\mathrm{i}\log\left(-\frac{\overline{y}_1 x_2 + \overline{y}_2 x_1 + \overline{y}_2 x_2}{\overline{y}_1 x_1}\right) + 2\pi k}, \quad k = 0, \pm 1, \pm 2, \ldots,$$

where for a complex number $z = \rho\,\mathrm{e}^{\mathrm{i}\theta}$, $\sqrt{z} = \pm\sqrt{\rho}\,\mathrm{e}^{\mathrm{i}\theta/2}$. Note that the functional $p(x, y)$ satisfies (2.18).

When $x = v$ and $y = w$ are right and left eigenvectors for F with $v = w = \begin{bmatrix} 1 \\ -1 \end{bmatrix}$, we recover the set of nonzero eigenvalues of F,

$$p(v, w) = \pm\sqrt{2\pi k}, \quad k = \pm 1, \pm 2, \dots,$$

where we excluded $p(v, w) = 0$ corresponding to $k = 0$ since (2.20) implies that $p(v, w) \neq 0$.

Schreiber (2008) shows that the nonlinear Rayleigh functional p is stationary at the eigenvectors (v, w), that is,

$$|p(v + \Delta v, w + \Delta w) - \lambda| = O((\|\Delta v\|_2 + \|\Delta w\|_2)^2),$$

where $\lambda = p(v, w)$. In other words, the first-order terms in a perturbation expansion of p at the eigenvectors (v, w) vanish identically.

2.7. Some perturbation results

It is often of interest, for example in sensitivity analysis and also in numerical computing, to give simple regions in which the eigenvalues of F are located. Probably the most famous result for linear eigenvalue problems is Gershgorin's theorem; see Horn and Johnson (1991, Chapter 6). Bindel and Hood (2013, Theorem 3.1) provide a generalization for NEPs.

Theorem 2.19 (Gershgorin's theorem for NEPs). Let $F(z) = D(z) + E(z)$ with $D, E \in H(\Omega, \mathbb{C}^{n\times n})$ and D diagonal. Then, for any $0 \le \alpha \le 1$,

$$\Lambda(F) \subset \bigcup_{j=1}^{n} \mathbb{G}_j^\alpha,$$

where $\mathbb{G}_j^\alpha$ is the jth generalized Gershgorin region

$$\mathbb{G}_j^\alpha = \{z \in \Omega : |d_{jj}(z)| \le r_j(z)^\alpha c_j(z)^{1-\alpha}\}$$

and $r_j(z)$, $c_j(z)$ are the jth absolute row and column sums of E, that is,

$$r_j(z) = \sum_{k=1}^{n} |e_{jk}(z)|, \quad c_j(z) = \sum_{k=1}^{n} |e_{kj}(z)|.$$

Moreover, if $\mathbb{U}$ is a bounded connected component of $\bigcup_{j=1}^{n} \mathbb{G}_j^\alpha$ such that $\overline{\mathbb{U}} \subset \Omega$, then $\mathbb{U}$ contains the same number of eigenvalues of F and D. Furthermore, if $\mathbb{U}$ includes p connected components of the Gershgorin regions, it must contain at least p eigenvalues of F.

Also of interest is the sensitivity of an eigenvalue λ of $F \in H(\Omega, \mathbb{C}^{n\times n})$ under small perturbations $\Delta F \in H(\Omega, \mathbb{C}^{n\times n})$. To address this point, we assume that λ is a nonzero simple eigenvalue with right eigenvector v and left eigenvector w. We consider F expressed as in (2.15) with the perturbation

$$\Delta F(z) = f_1(z)\Delta C_1 + f_2(z)\Delta C_2 + \cdots + f_\ell(z)\Delta C_\ell \in H(\Omega, \mathbb{C}^{n\times n}). \quad (2.21)$$

To measure the sensitivity of λ, we can use the *normwise condition number*

$$\kappa(\lambda, F) = \limsup_{\varepsilon \to 0} \left\{ \frac{|\Delta\lambda|}{\varepsilon|\lambda|} : \big(F(\lambda + \Delta\lambda) + \Delta F(\lambda + \Delta\lambda)\big)(v + \Delta v) = 0, \right.$$
$$\left. \|\Delta C_j\|_2 \le \varepsilon \alpha_j,\ j = 1, \dots, \ell \right\}, \tag{2.22}$$

where the α_j are nonnegative parameters that allow freedom in how perturbations are measured – for example, in an absolute sense ($\alpha_j = 1$) or in a relative sense ($\alpha_j = \|C_j\|_2$). By setting $\alpha_j = 0$ we can force $\Delta C_j = 0$ and thus keep C_j unperturbed.

Theorem 2.20. The normwise condition number $\kappa(\lambda, F)$ in (2.22) is

$$\kappa(\lambda, F) = \frac{\big(\sum_{j=1}^{\ell} \alpha_j |f_j(\lambda)|\big) \|v\|_2 \|w\|_2}{|\lambda|\, |w^* F'(\lambda) v|}.$$

Proof. By expanding the first constraint in (2.22) and keeping only the first-order terms, we get

$$\Delta\lambda F'(\lambda) v + F(\lambda)\Delta v + \Delta F(\lambda) v = O(\varepsilon^2).$$

Premultiplying by w^* leads to

$$\Delta\lambda w^* F'(\lambda) v + w^* \Delta F(\lambda) v = O(\varepsilon^2).$$

Since λ is a simple eigenvalue, we have from (2.5) that $w^* F'(\lambda) v \ne 0$. Thus

$$\Delta\lambda = -\frac{w^* \Delta F(\lambda) v}{w^* F'(\lambda) v} + O(\varepsilon^2)$$

and so

$$\frac{|\Delta\lambda|}{\varepsilon|\lambda|} \le \frac{\big(\sum_{j=1}^{\ell} \alpha_j |f_j(\lambda)|\big) \|w\|_2 \|v\|_2}{|\lambda|\, |w^* F'(\lambda) v|} + O(\varepsilon).$$

Hence the expression on the right-hand side of $\kappa(\lambda, F)$ in the theorem is an upper bound for the condition number. To show that this bound can be attained we consider the matrix $H = wv^*/(\|w\|_2\|v\|_2)$, for which

$$\|H\|_2 = 1, \quad w^* H v = \|v\|_2 \|w\|_2.$$

Let

$$\Delta C_j = \varepsilon \alpha_j \frac{\overline{f_j(\lambda)}}{|f_j(\lambda)|} H, \quad j = 1, \dots, \ell.$$

Then all the norm inequalities in (2.22) are satisfied as equalities and

$$|w^* \Delta F(\lambda) v| = \varepsilon \left(\sum_{j=1}^{\ell} \alpha_j |f_j(\lambda)| \right) \|w\|_2 \|v\|_2.$$

Dividing by $\varepsilon|\lambda|\,|w^*F'(\lambda)v|$ and taking the limit as $\varepsilon \to 0$ gives the desired equality. □

Pseudospectra are another established tool for gaining insight into the sensitivity of the eigenvalues of a matrix to perturbations; see Trefethen and Embree (2005) and the references therein. The *ε-pseudospectrum* of $F \in H(\Omega, \mathbb{C}^{n\times n})$ is the set

$$\Lambda_\varepsilon(F) = \bigcup_{\substack{\Delta F \in H(\Omega,\mathbb{C}^{n\times n})\\ \|\Delta F\| < \varepsilon}} \Lambda(F + \Delta F), \tag{2.23}$$

where, for example,

$$\|\Delta F\| = \|\Delta F\|_\Omega := \sup_{z\in\Omega} \|\Delta F(z)\|_2. \tag{2.24}$$

If F is expressed as in (2.15), then we can consider perturbations ΔF of the form (2.21) and measure them using

$$\|\Delta F\| = \|\Delta F\|_{\max} := \max_{1\le j\le \ell} \|\Delta F_j\|_2/\alpha_j, \tag{2.25}$$

where the α_j are nonnegative parameters defined as for (2.22). Other measures of the perturbations can be used too; see for example Tisseur and Higham (2001), Higham and Tisseur (2002) and Michiels, Green, Wagenknecht and Niculescu (2006).

The following characterizations of $\Lambda_\varepsilon(F)$ (see Bindel and Hood 2013, Proposition 4.1, and Michiels *et al.* 2006, Theorem 1) provide an easy way to check whether a point $z \in \Omega$ is in the ε-pseudospectrum of F or not.

Theorem 2.21. Let $F \in H(\Omega, \mathbb{C}^{n\times n})$. Then, for $\Lambda_\varepsilon(F)$ defined by (2.23),

$$\Lambda_\varepsilon(F) = \bigl\{z \in \Omega : \|F(z)^{-1}\|_2 > (\varepsilon f(z))^{-1}\bigr\},$$

where $f(z) = 1$ when the perturbations are measured using (2.24), and $f(z) = \sum_{j=1}^{\ell} \alpha_j |f_j(z)|$ when the perturbations are measured using (2.25).

The ε-pseudospectrum is connected to the *backward error* of an approximate eigenvalue $\widetilde{\lambda}$ of F defined by

$$\eta_F(\widetilde{\lambda}) = \min_{\substack{v\in\mathbb{C}^n\\ v\neq 0}} \eta_F(\widetilde{\lambda}, v), \tag{2.26}$$

where

$$\eta_F(\widetilde{\lambda}, v) = \min\bigl\{\varepsilon : (F(\widetilde{\lambda}) + \Delta F(\widetilde{\lambda}))v = 0,\ \|\Delta F\| \le \varepsilon\bigr\}$$

with perturbations measured using, for example, (2.24) or (2.25). Then, by comparing the definitions (2.23) and (2.26), we have that

$$\Lambda_\varepsilon(F) = \{z \in \Omega : \eta_F(z) < \varepsilon\}.$$

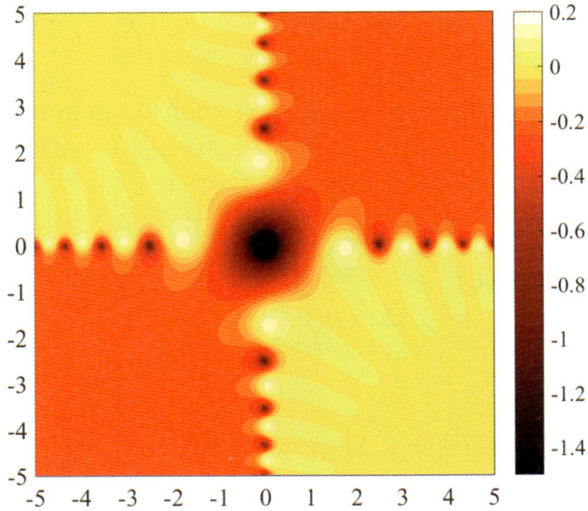

Figure 2.22. Spectral portrait of the matrix-valued function F defined in (2.1). The bar shows the mapping of colours to $\log_{10}$ of the resolvent norm, that is, the mapping $z \mapsto \log_{10} \|F(z)^{-1}\|_2$.

For an approximate eigenpair $(\widetilde{\lambda}, \widetilde{v})$ of F and for perturbations measured using (2.25), a straightforward generalization of Tisseur (2000, Theorem 1 and Lemma 3) leads to explicit expressions for $\eta_F(\widetilde{\lambda}, \widetilde{v})$ and $\eta_F(\widetilde{\lambda})$,

$$\eta_F(\widetilde{\lambda}, \widetilde{v}) = \frac{\|F(\widetilde{\lambda})\widetilde{v}\|_2}{f(\widetilde{\lambda})\,\|\widetilde{v}\|_2} \quad \text{and} \quad \eta_F(\widetilde{\lambda}) = \frac{1}{f(\widetilde{\lambda})\,\|F(\widetilde{\lambda})^{-1}\|_2},$$

where $f(\widetilde{\lambda}) = \sum_{j=1}^{\ell} \alpha_j |f_j(\widetilde{\lambda})|$.

Example 2.23. Let us consider once more the function F from (2.1), now rewritten in the form

$$F(z) = \mathrm{e}^{\mathrm{i}z^2} \begin{bmatrix} 1 & 0 \\ 0 & 0 \end{bmatrix} + \begin{bmatrix} 0 & 1 \\ 1 & 1 \end{bmatrix} =: f_1(z)C_1 + f_2(z)C_2.$$

An easy computation shows that the nonzero eigenvalues of F, $\lambda_k = \pm\sqrt{2\pi k}$, $k = \pm 1, \pm 2, \ldots$, have small condition numbers

$$\kappa(\lambda_k, F) = \frac{1 + \sqrt{1 + (3 + 5^{1/2})/2}}{2\pi|k|},$$

and these condition numbers get smaller as k increases in modulus. This is confirmed by the spectral portrait in Figure 2.22, which shows that nonzero eigenvalues of F move only slightly even under very large perturbations of F. The zero eigenvalue is defective and more sensitive to perturbations.

The largest possible real part of points in the ε-pseudospectrum of F is called the *ε-pseudospectral abscissa* of F, that is,

$$\alpha_\varepsilon(F) = \max\{\mathrm{Re}\,(z) : z \in \Lambda_\varepsilon(F)\}.$$

This quantity is of particular interest for NEPs associated with systems of differential equations such as the DDE (1.1), since perturbations of size ε may cause the system to become unstable whenever $\alpha_\varepsilon(F) \geq 0$. The distance to instability of a stable system associated with F can then be defined as

$$\delta(F) = \inf\{\varepsilon : \alpha_\varepsilon(F) \geq 0\}.$$

Algorithms to compute $\alpha_\varepsilon(F)$ can be found in Michiels and Guglielmi (2012) and Verhees *et al.* (2014).

3. Hermitian NEPs

A matrix-valued function $F(z)$ is said to be Hermitian if $F(\bar{z}) = F(z)^*$ for all $z \in \mathbb{C}$. It is easily seen that the eigenvalues of a Hermitian F are either real or they come in pairs $(\lambda, \bar{\lambda})$. If v and w are right and left eigenvectors of F for the eigenvalue $\lambda \in \mathbb{C}$, then w and v are right and left eigenvectors of F for the eigenvalue $\bar{\lambda}$, whereas if λ is a real eigenvalue of F for the right eigenvector v, then v is also a left eigenvector for the eigenvalue λ.

As for Hermitian linear eigenvalue problems, variational principles for real eigenvalues of Hermitian NEPs exist, as we now discuss. For this we assume that the assumptions (A1)–(A3) below hold.

(A1) The matrix-valued function $F : \mathbb{R} \supseteq \mathbb{I} \to \mathbb{C}^{n\times n}$ is Hermitian and continuously differentiable on the open real interval $\mathbb{I}$.

(A2) For every $x \in \mathbb{C}^n \setminus \{0\}$, the real nonlinear equation

$$x^* F(p(x))x = 0 \tag{3.1}$$

has at most one real solution $p(x) \in \mathbb{I}$.

Note that (3.1) defines implicitly the *one-sided generalized Rayleigh functional* p on some open subset $\mathbb{K}(p) \subseteq \mathbb{C}^n \setminus \{0\}$. When the Rayleigh functional p is defined on the whole space $\mathbb{C}^n \setminus \{0\}$, the NEP $F(\lambda)v = 0$ is called *overdamped*.

(A3) For every $x \in \mathbb{K}(p)$ and any $z \in \mathbb{I}$ such that $z \neq p(x)$,

$$(z - p(x))(x^* F(z)x) > 0, \tag{3.2}$$

that is, $x^* F(z)x$ is increasing at $z = p(x)$.

When F is overdamped, assumption (A3) holds if $x^* F'(p(x))x > 0$ for all $x \in \mathbb{C}^n \setminus \{0\}$. Moreover, if F is overdamped and twice continuously differentiable, and $x^*F''(p(x))x \neq 0$ for all $x \in \mathbb{C}^n \setminus \{0\}$, then (3.2) is equivalent to $x^*F'(p(x))x > 0$ for all $x \in \mathbb{C}^n \setminus \{0\}$ (Szyld and Xue 2016, Proposition 2.4).

Let $\lambda \in \mathbb{I}$ be an eigenvalue of F; then λ is called a *kth eigenvalue of* F if $\mu = 0$ is the kth largest eigenvalue of the Hermitian matrix $F(\lambda)$. The next result, which can be found in Hadeler (1968), Voss and Werner (1982) and Voss (2009), provides a minmax characterization and a maxmin characterization of the real eigenvalues of F.

Theorem 3.1 (nonlinear variational principle). Assume the matrix-valued function F satisfies (A1)–(A3). Then F has at most n eigenvalues in $\mathbb{I}$. Moreover, if λ_k is a kth eigenvalue of F, then

$$\lambda_k = \min_{\substack{V \in \mathbb{S}_k \\ V \cap \mathbb{K}(p) \neq \emptyset}} \max_{\substack{x \in V \cap \mathbb{K}(p) \\ x \neq 0}} p(x) \in \mathbb{I}, \tag{3.3}$$

$$\lambda_k = \max_{\substack{V \in \mathbb{S}_{k-1} \\ V^\perp \cap \mathbb{K}(p) \neq \emptyset}} \min_{\substack{x \in V^\perp \cap \mathbb{K}(p) \\ x \neq 0}} p(x) \in \mathbb{I}, \tag{3.4}$$

where $\mathbb{S}_j$ denotes the set of all subspaces of $\mathbb{C}^n$ of dimension j.

The characterization of the eigenvalues in (3.3) is a generalization of the minmax principle of Poincaré (1890) for linear eigenvalue problems and that in (3.4) is a generalization of the maxmin characterization of Courant (1920), Fischer (1905) and Weyl (1912).

When F satisfies (A1)–(A3) and is overdamped, then there exist exactly n eigenvalues $\lambda_1, \lambda_2, \ldots, \lambda_n$ of F in $\mathbb{I}$, and it follows from Theorem 3.1 that $\lambda_1 \leq \lambda_2 \leq \cdots \leq \lambda_n$, thereby offering an ordering of the real eigenvalues of F in $\mathbb{I}$. The variational principle can be used to derive an interesting property of eigenvalue approximations obtained by projecting F onto low-dimensional spaces; see Szyld and Xue (2016, Theorem 2.7).

Theorem 3.2 (nonlinear Cauchy interlacing theorem). Assume that F satisfies (A1)–(A3) and is overdamped, with eigenvalues $\lambda_1 \leq \lambda_2 \leq \cdots \leq \lambda_n \in \mathbb{I}$. Let $U \in \mathbb{C}^{n \times k}$ be of full column rank. Then the projected matrix-valued function $U^*F(z)U$ has exactly k eigenvalues satisfying the nonlinear variational principle, and if μ_j is the jth eigenvalue of $U^*F(z)U$ then $\lambda_j \leq \mu_j \leq \lambda_{n-k+j}$.

The nonlinear Cauchy interlacing theorem and the variational principle in Theorem 3.1 allow the development of special algorithms for Hermitian NEPs satisfying the assumptions (A1)–(A3) in Section 4.6.

Example 3.3. Let us consider the `loaded_string` problem with F defined in (1.3). The associated quadratic matrix polynomial

$$Q(z) = -(z-1)F(z) = z^2 C_2 - z(C_1 + C_2 + C_3) + C_1$$

is *hyperbolic*, since its leading matrix coefficient C_2 is positive definite and the scalar equation $x^*Q(z)x = 0$ has two distinct real roots (Al-Ammari and Tisseur 2012, Theorem 3.4). As a result, Q has only real eigenvalues. But clearly, if λ is an eigenvalue of F, then it is an eigenvalue of Q. Hence F has only real eigenvalues. For $n = 100$ the first nine eigenvalues of F greater than 1 are given to 10 digits by

$$\begin{aligned} &\lambda_1 = \texttt{4.482176546}, &&\lambda_2 = \texttt{24.22357311}, &&\lambda_3 = \texttt{63.72382114}, \\ &\lambda_4 = \texttt{123.0312211}, &&\lambda_5 = \texttt{202.2008991}, &&\lambda_6 = \texttt{301.3101627}, \\ &\lambda_7 = \texttt{420.4565631}, &&\lambda_8 = \texttt{559.7575863}, &&\lambda_9 = \texttt{719.3506601} \end{aligned} \tag{3.5}$$

(see Solov'ëv 2006), and there is also an eigenvalue smaller than 1,

$$\lambda_0 = \texttt{0.45731848895}. \tag{3.6}$$

Compared to F, Q has $n-1$ extra eigenvalues at $z = 1$.

It is easy to check that property (A1) holds on the interval $\mathbb{I} = (1, +\infty)$. If we let

$$a(x) = x^*C_2x, \quad b(x) = -x^*(C_1 + C_2 + C_3)x, \quad c(x) = x^*C_1x,$$

then

$$p(x) = \frac{-b(x) + \sqrt{b^2(x) - 4a(x)c(x)}}{2a(x)} > 1$$

is well defined for every $x \in \mathbb{C}^n \setminus \{0\}$ since $a(x) > 0$, and it is the only root of $x^*F(p(x))x = 0$ in $\mathbb{I}$. Hence property (A2) holds and F is overdamped on $\mathbb{I}$. Now,

$$x^*F'(z)x = -x^*C_2x - \frac{1}{(z-1)^2}x^*C_3x < 0$$

for all $x \in \mathbb{C}^n \setminus \{0\}$, so that property (A3) holds for $-F(z)$. As a result, the eigenvalues of F in $\mathbb{I}$ satisfy the nonlinear variational principle in Theorem 3.1.

4. Solvers based on Newton's method

Newton's method is a natural approach to compute eigenvalues or eigenpairs of NEPs efficiently and accurately provided that good initial guesses are available. The choice of an initial guess is typically the only crucial parameter of a Newton-type method, which is a great advantage over other NEP solution approaches. As a result, many algorithmic variants have been developed and applied over the years, including the Newton-QR iteration of Kublanovskaya (1970) and its variant in Garrett, Bai and Li (2016), the Newton-trace iteration of Lancaster (1966), nonlinear inverse iteration (Unger 1950), residual inverse iteration (Neumaier 1985),

Rayleigh functional iterations (Lancaster 1961, Schreiber 2008), the block Newton method of Kressner (2009), and for large sparse NEPs, Jacobi–Davidson-type methods (Betcke and Voss 2004, Sleijpen, Booten, Fokkema and van der Vorst 1996).

Generally speaking, there are two broad ways the NEP $F(\lambda)v = 0$ can be tackled by a Newton-type method. First, one can apply Newton's method to a scalar equation $f(z) = 0$ whose roots correspond to the wanted eigenvalues of F, or second, Newton's method can be applied directly to the vector problem $F(\lambda)v = 0$ together with some normalization condition on v. We discuss both approaches.

4.1. Newton's method for scalar functions

For a given initial guess $\lambda^{(0)}$, Newton's method for finding the roots of a scalar equation $f(z) = 0$ is given by

$$\lambda^{(k+1)} = \lambda^{(k)} - \frac{f(\lambda^{(k)})}{f'(\lambda^{(k)})}, \quad k = 0, 1, \ldots . \tag{4.1}$$

This iteration has local quadratic convergence to simple roots (Wilkinson 1965, Section 7.25). In order to apply it to the solution of an NEP $F(\lambda)v = 0$, we only need a scalar function f whose roots are the eigenvalues of F, and its derivative f'. Different methods result from different choices of f.

Probably the most obvious choice is

$$f(z) = \det F(z).$$

The trace theorem (*e.g.* Lancaster 1966, Theorem 5.1), states that if the entries of $F(z)$ are differentiable functions of z, then for any $z \in \mathbb{C}$ for which $f(z) = \det F(z) \neq 0$ we have

$$f'(z) = \det F(z) \operatorname{trace}(F^{-1}(z)F'(z)), \tag{4.2}$$

and so the Newton iteration (4.1) can be rewritten as

$$\lambda^{(k+1)} = \lambda^{(k)} - \frac{1}{\operatorname{trace}(F^{-1}(\lambda^{(k)})F'(\lambda^{(k)}))}, \quad k = 0, 1, \ldots . \tag{4.3}$$

We refer to (4.3) as the *Newton-trace iteration* (Lancaster 1966, Section 5.5); see also Khazanov and Kublanovskaya (1988). Note that we only need the diagonal entries of $F^{-1}(\lambda^{(k)})F'(\lambda^{(k)})$. Nevertheless, with a straightforward implementation each iteration requires the factorization of an $n \times n$ matrix $F(\lambda^{(k)})$ and $2n$ triangular solves, making it a rather expensive method. A basic MATLAB implementation of the Newton-trace iteration (4.2) is provided in Figure 4.1, where we use a simple stopping criterion based on the relative residual $|f(\lambda^{(k)})|/\|F(\lambda^{(k)})\|_F$.

Kublanovskaya (1970) proposes applying Newton's method to

$$f(z) = r_{nn}(z),$$

```
% Newton_trace
F  = @(z) [exp(1i*z.^2) 1; 1 1];
Fp = @(z) [2i*z*exp(1i*z.^2) 0; 0 0];
tol = 1e-8; maxit = 20; lam = 2.2 + 1e-4i;
for k = 0:maxit
    [L,U] = lu(F(lam));
    if abs(prod(diag(U)))/norm(F(lam),'fro')<tol, break, end
    corr = trace(U\(L\Fp(lam)));
    lam = lam - 1/corr;
end
if k < maxit, nbr_iter = k, lambda = lam, end
```

Figure 4.1. Basic MATLAB implementation of the Newton-trace iteration (4.3) for problem (2.1). The NEP parameters F and F' are specified in lines 2–3 and the method's parameters in line 4. Upon convergence, `lambda` is the eigenvalue and `nbr_iter` the number of iterations.

where $r_{nn}(z)$ is the bottom-right entry of the R factor in a rank-revealing QR decomposition of $F(z)$,

$$F(z)\Pi(z) = Q(z)R(z), \tag{4.4}$$

with $\Pi(z)$ being a permutation matrix which ensures that $r_{nn}(z)$ becomes zero before any other diagonal entry does; see also Kublanovskaja (1969). Since $r_{nn}(z) = 0$ is equivalent to $\det F(z) = \det R(z) = 0$, we know that the roots of $f(z)$ are exactly the roots of $\det F(z) = 0$.

The permutation $\Pi(z)$ in (4.4) is not a continuous function of z, but in a small neighbourhood of an approximate eigenvalue $\lambda^{(k)}$ of F the permutation can be kept constant. So if we let

$$F(\lambda^{(k)})\Pi = Q_k R_k = Q_k \begin{bmatrix} R_{11}^{(k)} & r_{12}^{(k)} \\ 0 & r_{nn}^{(k)} \end{bmatrix}, \tag{4.5}$$

then in a small neighbourhood of $\lambda_k^{(k)}$ we can use

$$F(z)\Pi = Q(z)R(z) = Q(z) \begin{bmatrix} R_{11}(z) & r_{12}(z) \\ 0 & r_{nn}(z) \end{bmatrix}$$

with $Q(\lambda^{(k)}) = Q_k$ and $R(\lambda^{(k)}) = R_k$. Garrett *et al.* (2016) show that

$$r_{nn}(z) = r_{nn}^{(k)} + e_n^T Q_k^* F'(\lambda^{(k)})\Pi \begin{bmatrix} -p \\ 1 \end{bmatrix} (z - \lambda^{(k)}) + O(|z - \lambda^{(k)}|^2),$$

```
% Newton_QR
n = 2; F = @(z) [exp(1i*z.^2) 1; 1 1];
Fp = @(z) [2i*z*exp(1i*z.^2) 0; 0 0];
tol = 1e-8; maxit = 20; lam = 2.2 + 1e-4i;
for k = 0:maxit
    [Q,R,P] = qr(F(lam));
    if abs(R(n,n))/norm(F(lam),'fro') < tol, break, end
    p = R(1:n-1,1:n-1)\R(1:n-1,n);
    lam = lam - R(n,n)/(Q(:,n)'*Fp(lam)*P*[-p; 1]);
end
if k < maxit, nbr_iter = k, lambda = lam, end
```

Figure 4.2. Basic MATLAB implementation of the Newton-QR iteration for problem (2.1). The NEP parameters F and F' are specified in lines 2–3 and the method's parameters in line 4. Upon convergence, `lambda` is the eigenvalue and `nbr_iter` the number of iterations.

where $e_n = [0, \dots, 0, 1]^T$ and p solves $R_{11}^{(k)} p = r_{12}^{(k)}$, and therefore

$$r'_{nn}(\lambda^{(k)}) = e_n^T Q_k^* F'(\lambda^{(k)}) \Pi \begin{bmatrix} -p \\ 1 \end{bmatrix}.$$

This leads to the *Newton-QR iteration* for a root of $r_{nn}(z)$,

$$\lambda^{(k+1)} = \lambda^{(k)} - \frac{r_{nn}(\lambda^{(k)})}{r'_{nn}(\lambda^{(k)})} = \lambda^{(k)} - \frac{r_{nn}^{(k)}}{e_n^T Q_k^* F'(\lambda^{(k)}) \Pi \left[\begin{smallmatrix} -p \\ 1 \end{smallmatrix}\right]}. \tag{4.6}$$

At convergence, we can take

$$v = \Pi \begin{bmatrix} -p \\ 1 \end{bmatrix}, \qquad w = Q_k e_n$$

as approximations for the right and left eigenvectors of the converged approximate eigenvalue, respectively.

A basic MATLAB implementation of the Newton-QR iteration (4.6) is given in Figure 4.2. It requires the repeated computation of a rank-revealing QR factorization of $F(\lambda^{(k)})$ and hence is only feasible for problems of moderate size (unless F has some structure that can be exploited; see Section 4.5). However, the iteration (4.6) can be useful in the context of iterative refinement. Yang (1983) and Wobst (1987) consider an approach similar to that of Kublanovskaya by using an LU factorization with column pivoting in place of a rank-revealing QR factorization.

To avoid working with nearly singular matrices $F(\lambda^{(k)})$, Andrew, Chu and Lancaster (1995) propose using in place of $F(z)$ the bordered matrix

$$G(z) = \begin{bmatrix} F(z) & b \\ c^T & 0 \end{bmatrix}, \tag{4.7}$$

where $b, c \in \mathbb{C}^n$ are such that the matrix $G(\lambda)$ is nonsingular and well-conditioned at a simple eigenvalue λ. Now if $c^T v \neq 0$ and $w^T b \neq 0$, where v and w are the right and left eigenvectors of F associated with the eigenvalue λ, then $G(\lambda)$ is nonsingular (Andrew, Chu and Lancaster 1993, Theorem 10.1). For a vector x such that $c^T x = 1$, the authors introduce the linear system

$$\begin{bmatrix} F(z) & b \\ c^T & 0 \end{bmatrix} \begin{bmatrix} x \\ f \end{bmatrix} = \begin{bmatrix} 0 \\ 1 \end{bmatrix}. \tag{4.8}$$

For z near λ, the matrix $G(z)$ is nonsingular and hence $x = x(z)$ and $f = f(z)$ are smooth functions of z. By Cramer's rule we have

$$f(z) = \frac{\det F(z)}{\det G(z)}, \tag{4.9}$$

and therefore $f(z) = 0$ if and only if $\det F(z) = 0$. Differentiating (4.8) with respect to z leads to

$$\begin{bmatrix} F(z) & b \\ c^T & 0 \end{bmatrix} \begin{bmatrix} x'(z) \\ f'(z) \end{bmatrix} = -\begin{bmatrix} F'(z)x(z) \\ 0 \end{bmatrix}. \tag{4.10}$$

The *BDS method* of Andrew *et al.* (1995) and the *implicit determinant method* of Spence and Poulton (2005) consist of applying a Newton iteration to f in (4.9). The latter method is detailed in Algorithm 4.3, with a basic MATLAB implementation given in Figure 4.4.

At convergence, $x^{(k)}$ provides an approximation to the right eigenvector v. Note that the factorization used to solve (4.8) can be re-used to solve (4.10). Andrew *et al.* (1995) propose different choices for the vectors b and c leading

---

Algorithm 4.3: Implicit determinant method

Choose an initial approximate eigenvalue $\lambda^{(0)}$ and vectors $b, c \in \mathbb{C}^n$ such that $G(\lambda^{(0)})$ in (4.7) is nonsingular.

for $k = 0, 1, \ldots$ until convergence **do**

  Solve (4.8) with $z = \lambda^{(k)}$ for $f(\lambda^{(k)})$ and $x^{(k)}$.

  Solve (4.10) with $z = \lambda^{(k)}$ for $f'(\lambda^{(k)})$ using $x(z) = x^{(k)}$ for the right-hand side.

  Perform the Newton update $\lambda^{(k+1)} = \lambda^{(k)} - f(\lambda^{(k)})/f'(\lambda^{(k)})$.

end

```
% Newton_implicit_determinant
n = 2; F = @(z) [exp(1i*z.^2) 1; 1 1];
Fp = @(z) [2i*z*exp(1i*z.^2) 0; 0 0];
tol = 1e-8; maxit = 20; b = [0; 1]; c = b; lam = 2.2 + 1e-4i;
for k = 0:maxit
    [L,U] = lu([F(lam) b; c.' 0]);
    xf = U\(L\[zeros(n,1); 1]);
    if abs(xf(n+1))/norm(F(lam),'fro') < tol, break, end
    xfp = U\(L\[-Fp(lam)*xf(1:n); 0]);
    lam = lam - xf(n+1)/xfp(n+1);
end
if k < maxit, nbr_iter = k, lambda = lam, end
```

Figure 4.4. Basic MATLAB implementation of the implicit determinant method for problem (2.1). The NEP parameters F and F' are specified in lines 2–3 and the method's parameters in line 4. Upon convergence, `lambda` is the eigenvalue and `nbr_iter` the number of iterations.

to the *row/column deletion method* ($b = e_i$, $c = e_j$, where e_k denotes the kth column of the identity matrix), the *column substitution method* ($b \approx v$, $c = e_j$), giving the name BDS (bordered, deletion, substitution) for this class of methods.

Deflation may be necessary when computing several nearby eigenvalues in order to avoid the Newton iteration converging to an already computed eigenvalue. Suppose we have already computed m roots λ_ℓ ($\ell = 1, \dots, m$) of f near λ_0, each of multiplicity m_ℓ. As suggested by Wilkinson (1965, Section 7.48), we can apply Newton's method to

$$\widetilde{f}(z) = \frac{f(z)}{\prod_{\ell=1}^{m}(z-\lambda_\ell)^{m_\ell}},$$

which leads to the iteration

$$\lambda^{(k+1)} = \lambda^{(k)} - \frac{f(\lambda^{(k)})}{f'(\lambda^{(k)}) - f(\lambda^{(k)})\sum_{\ell=1}^{m}(m_\ell/(\lambda^{(k)}-\lambda_\ell))}. \tag{4.11}$$

This strategy has been shown to work well in practice (Garrett *et al.* 2016).

When the derivative of F is not available, it can be replaced by a finite difference approximation, leading to a quasi-Newton method. In particular, the secant method is obtained by using the approximation

$$F'(\lambda^{(k)}) \approx \frac{F(\lambda^{(k)}) - F(\lambda^{(k-1)})}{\lambda^{(k)} - \lambda^{(k-1)}}. \tag{4.12}$$

Example 4.5 (basins of attraction). Newton's method typically requires a good initial guess $\lambda^{(0)}$ for the eigenvalue of interest. Let us illustrate this with the matrix-valued function F defined in (2.1). Recall that F has eigenvalues 0, $\pm\sqrt{2\pi}$ and $\pm\mathrm{i}\sqrt{2\pi}$ in the square

$$\Omega = \{z \in \mathbb{C} : -3 \le \mathrm{Re}\,(z) \le 3,\ -3 \le \mathrm{Im}\,(z) \le 3\}.$$

We generate 10^6 equidistant initial guesses in Ω and for each run the basic MATLAB codes given in Figures 4.1, 4.2 and 4.4 with tolerance `tol = 1e-8` and `maxit = 20`. All codes use as stopping criterion

$$\frac{|f(\lambda^{(k)})|}{\|F(\lambda^{(k)})\|_F} \le \texttt{tol}, \tag{4.13}$$

but $f(z) = \det F(z)$ for the Newton-trace iteration, $f(z) = r_{nn}(z)$ for the Newton-QR iteration, and $f(z) = \det F(z)/\det G(z)$ for the implicit determinant method. The convergence basins of the Newton iteration for these three different scalar functions are illustrated in Figure 4.6. For this particular example and our choice of the parameters b and c for the bordered matrix (4.7), the convergence basins of the Newton-trace iteration and the implicit determinant method are rotated versions of each other (at least to visual accuracy). The basins of convergence for the Newton-QR iteration are quite different and somewhat smaller. The Newton-trace iteration (4.3) with $F'(\lambda^{(k)})$ replaced by the secant approximation (4.12) produces basins of convergence similar to that of Newton's method for $\det F(z) = 0$ but of smaller sizes.

4.2. Newton's method for the vector equation

Newton's method can be applied directly to the NEP $F(\lambda)v = 0$ rather than to a scalar equation $f(z) = 0$ whose roots are eigenvalues of F. For this, a normalization condition on the eigenvector of the form $u^*v = 1$ for some nonzero vector u is added to $F(\lambda)v = 0$ so as to have $n+1$ equations for the $n+1$ unknowns in (λ, v). Note that $u^*v = 1$ ensures that v is not the zero vector. This approach is briefly mentioned in Unger (1950) and discussed further in Ruhe (1973). In order to apply Newton's method to $\mathcal{N}\left[\begin{smallmatrix} v \\ \lambda \end{smallmatrix}\right] = 0$, where

$$\mathcal{N}\begin{bmatrix} v \\ \lambda \end{bmatrix} = \begin{bmatrix} F(\lambda)v \\ u^*v - 1 \end{bmatrix},$$

we need the Jacobian matrix of the $(n+1)$-dimensional operator $\mathcal{N}$, which is readily calculated as

$$J_{\mathcal{N}}\begin{bmatrix} v \\ \lambda \end{bmatrix} = \begin{bmatrix} F(\lambda) & F'(\lambda)v \\ u^* & 0 \end{bmatrix}.$$

Newton's iteration is now given as

$$\begin{bmatrix} v^{(k+1)} \\ \lambda^{(k+1)} \end{bmatrix} = \begin{bmatrix} v^{(k)} \\ \lambda^{(k)} \end{bmatrix} - \left(J_{\mathcal{N}}\begin{bmatrix} v^{(k)} \\ \lambda^{(k)} \end{bmatrix} \right)^{-1} \mathcal{N}\begin{bmatrix} v^{(k)} \\ \lambda^{(k)} \end{bmatrix}. \tag{4.14}$$

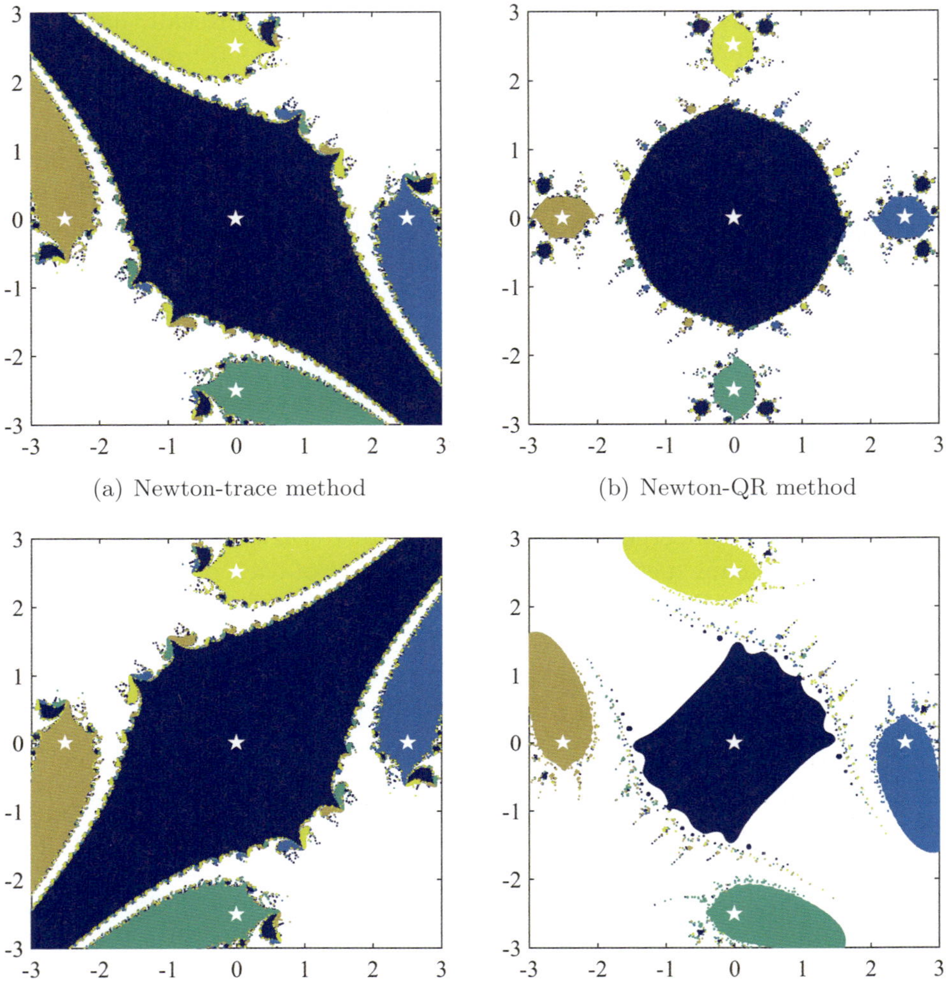

(a) Newton-trace method

(b) Newton-QR method

(c) implicit determinant method

(d) Newton-trace (secant) method

Figure 4.6. Basins of convergence for various Newton-type methods discussed in Example 4.5 for problem (2.1). Each initial point in the square region Ω is coloured according to the eigenvalue a method is converging to, with the five exact eigenvalues of the NEP in Ω indicated by white $\star$ symbols. White areas indicate that no convergence is achieved within 20 iterations or the method converged to an eigenvalue outside Ω.

If we assume that the approximate eigenvector $v^{(k)}$ is normalized such that $u^* v^{(k)} = 1$ for some nonzero vector u, then (4.14) can be rewritten as

$$F(\lambda^{(k)}) v^{(k+1)} = -(\lambda^{(k+1)} - \lambda^{(k)}) F'(\lambda^{(k)}) v^{(k)}, \tag{4.15}$$

$$u^* v^{(k+1)} = 1. \tag{4.16}$$

The Newton iteration (4.15)–(4.16) is equivalent to the *nonlinear inverse iteration* (Unger 1950), which is given in Algorithm 4.7 with a basic MATLAB implementation given in Figure 4.8.

Algorithm 4.7: Nonlinear inverse iteration

Choose an initial pair $(\lambda^{(0)}, v^{(0)})$ with $\|v^{(0)}\| = 1$ and a nonzero vector u.
for $k = 0, 1, \ldots$ until convergence **do**
  Solve $F(\lambda^{(k)}) \widetilde{v}^{(k+1)} = F'(\lambda^{(k)}) v^{(k)}$ for $\widetilde{v}^{(k+1)}$.
  Set $\lambda^{(k+1)} = \lambda^{(k)} - \dfrac{u^* v^{(k)}}{u^* \widetilde{v}^{(k+1)}}$.
  Normalize $v^{(k+1)} = \widetilde{v}^{(k+1)} / \|\widetilde{v}^{(k+1)}\|$.
end

The normalization of $v^{(k+1)}$ is necessary to avoid numerical overflow or underflow and any vector norm can be used. When $\lambda^{(k)}$ is close to an eigenvalue, the matrix $F(\lambda^{(k)})$ of the linear system to be solved is nearly singular. However, standard theory of inverse iteration (see *e.g.* Ipsen 1997, Section 6.3, or Peters and Wilkinson 1979, Section 2) shows that the error in the computed vector $\widetilde{v}^{(k+1)}$ will be almost parallel to the exact solution, that is, the inaccuracy is concentrated in the length of $\widetilde{v}^{(k+1)}$ and not its direction. As it is derived from a Newton iteration, the nonlinear inverse iteration converges locally and quadratically for any simple eigenpair. The vector u in (4.16) and in Algorithm 4.7 can be chosen in a number of ways as discussed below.

- A simple choice is to take u to be the ith unit vector e_i, which corresponds to keeping the ith entry of the vectors $v^{(k)}$ constant.
- To prevent the iteration from converging to previously computed eigenpairs, u can be chosen orthogonal to already computed eigenvectors (Anselone and Rall 1968).
- Choosing $u = F(\lambda^{(k)})^* w^{(k)}$, where $w^{(k)}$ is an approximate left eigenvector, allows us to rewrite the eigenvalue update as

$$\lambda^{(k+1)} = \lambda^{(k)} - \frac{w^{(k)*} F(\lambda^{(k)}) v^{(k)}}{w^{(k)*} F'(\lambda^{(k)}) v^{(k)}},$$

```
% inverse_iteration
n = 2; F = @(z) [exp(1i*z.^2) 1; 1 1];
Fp = @(z) [2i*z*exp(1i*z.^2) 0; 0 0];
tol = 1e-8; maxit = 20; lam = 2.2 + 1e-4i;
v = [1; 1]; v = v/norm(v); u = [1; 0];
for k = 0:maxit-1
    if norm(F(lam)*v) < tol, break; end
    vt = F(lam)\Fp(lam)*v;
    lam = lam - u'*v/(u'*vt)
    v = vt/norm(vt);
end
if k < maxit, nbr_iter = k, lambda = lam, end
```

Figure 4.8. Basic MATLAB implementation of the nonlinear inverse iteration for problem (2.1). The NEP parameters F and F' are specified in lines 2–3 and the method's parameters in lines 4–5. Upon convergence, `lambda` is the eigenvalue and `nbr_iter` the number of iterations.

which is the *generalized Rayleigh quotient iteration* that Lancaster (1966) derived for the polynomial eigenvalue problem.

As mentioned in the third point above, the nonlinear inverse iteration can be combined with the computation of a left eigenvector, and a two-sided Rayleigh functional as defined in Section 2.6, to yield the *two-sided Rayleigh functional iteration* (Schreiber 2008), given in Algorithm 4.9.

Schreiber (2008) shows that Algorithm 4.9 achieves local cubic convergence for a simple eigentriple (λ, v, w). Note that at each iteration, an LU factorization of $F(\lambda^{(k)})$ used to solve the first linear system can be re-used to solve the second linear system involving $F^*(\lambda^{(k)})$.

Algorithm 4.9: Two-sided Rayleigh functional iteration

Choose an initial triple $(\lambda^{(0)}, v^{(0)}, w^{(0)})$ with $\|v^{(0)}\| = \|w^{(0)}\| = 1$.

for $k = 0, 1, \dots$ until convergence **do**

  Solve $F(\lambda^{(k)})\widetilde{v}^{(k+1)} = F'(\lambda^{(k)})v^{(k)}$ for $\widetilde{v}^{(k+1)}$.

  Set $v^{(k+1)} = \widetilde{v}^{(k+1)}/\|\widetilde{v}^{(k+1)}\|$.

  Solve $F(\lambda^{(k)})^*\widetilde{w}^{(k+1)} = F'(\lambda^{(k)})^*w^{(k)}$ for $\widetilde{w}^{(k+1)}$.

  Set $w^{(k+1)} = \widetilde{w}^{(k+1)}/\|\widetilde{w}^{(k+1)}\|$.

  Find the root ρ of the scalar equation $w^{(k+1)*}F(\rho)v^{(k+1)} = 0$ closest to $\lambda^{(k)}$ and set $\lambda^{(k+1)} = \rho$.

end

Each step of the nonlinear inverse iteration, Algorithm 4.7, requires the factorization of $F(\lambda^{(k)})$ to solve the linear system, which cannot be re-used at the next iteration since $\lambda^{(k)}$ varies. We could replace $\lambda^{(k)}$ by a fixed value σ, say $\sigma = \lambda^{(0)}$, but then the resulting iteration would converge to an eigenpair (μ, v) of the linear problem $Av = \mu Bv$, where $A = F(\sigma)$ and $B = F'(\lambda_*)$ for some $\lambda_* \in \mathbb{C}$. If $F(z)$ is linear in z so that $F'(z)$ is a constant matrix, we could easily recover an eigenpair (λ, v) of $F(\lambda)$ from (μ, v), since $F(\lambda)v = 0$ and $Av = \mu Bv$ are directly related eigenvalue problems. This is not the case if $F(z)$ is truly nonlinear in z.

Neumaier (1985) shows that this difficulty can be avoided by considering a variant of the nonlinear inverse iteration based on the use of the residual. If F is twice continuously differentiable, iteration (4.15) can be rewritten as

$$\begin{aligned} v^{(k)} - v^{(k+1)} &= v^{(k)} - (\lambda^{(k+1)} + \lambda^{(k)}) F(\lambda^{(k)})^{-1} F'(\lambda^{(k)}) v^{(k)} \\ &= F(\lambda^{(k)})^{-1} (F(\lambda^{(k)}) + (\lambda^{(k+1)} - \lambda^{(k)}) F'(\lambda^{(k)})) v^{(k)} \\ &= F(\lambda^{(k)})^{-1} F(\lambda^{(k+1)}) v^{(k)} + O(|\lambda^{(k+1)} - \lambda^{(k)}|^2). \end{aligned}$$

Ignoring the second-order term leads to an iteration

$$v^{(k+1)} = v^{(k)} - F(\lambda^{(k)})^{-1} F(\lambda^{(k+1)}) v^{(k)}, \tag{4.17}$$

where the new approximant $\lambda^{(k+1)}$ for the eigenvalue λ needs to be determined beforehand. Neumaier (1985) shows that replacing $\lambda^{(k)}$ by a fixed shift σ in (4.17) does not destroy the convergence of the iteration to the wanted eigenpair. This leads to the *residual inverse iteration*, the pseudocode of which is given in Algorithm 4.10.

The initial vector $v^{(0)}$ can be computed by solving

$$F(\sigma)\widetilde{v}^{(0)} = b, \quad v^{(0)} = \widetilde{v}^{(0)}/(u^*\widetilde{v}^{(0)})$$

Algorithm 4.10: Residual inverse iteration

Choose an initial pair $(\lambda^{(0)}, v^{(0)})$ with $u^*v^{(0)} = 1$ for some nonzero vector u. Set $\sigma = \lambda^{(0)}$.

for $k = 0, 1, \ldots$ until convergence **do**

  Solve for z the scalar equation

$$u^* F(\sigma)^{-1} F'(z) v^{(k)} = 0, \tag{4.18}$$

  and accept as $\lambda^{(k+1)}$ the root z closest to $\lambda^{(k)}$.

  Solve $F(\sigma)x^{(k)} = F(\lambda^{(k+1)})v^{(k)}$ for $x^{(k)}$.

  Set $v^{(k+1)} = \widetilde{v}^{(k+1)}/(u^*\widetilde{v}^{(k+1)})$, where $\widetilde{v}^{(k+1)} = v^{(k)} - x^{(k)}$.

end

---

for some nonzero vector b. As suggested by Wilkinson (1965) for the standard inverse iteration, we can choose $u = e$, where e is the vector of all ones, and $b = Le$, where $F(\sigma) = LU$ is an LU factorization. Then $\widetilde{v}^{(0)} = U^{-1}e$. Note that compared with the nonlinear inverse iteration, Algorithm 4.7, in the residual inverse iteration the eigenvector is updated by solving a linear system whose right-hand side is the NEP residual, giving its name to the iteration. Due to the fixed shift σ, the residual inverse iteration converges only linearly, and in practice, it is advisable to update the shift every now and then, in particular when the convergence is slow.

4.3. Deflation of computed eigenvalues and eigenpairs

The methods we have described so far are directed towards computing one eigenvalue or one eigenpair only. In principle, several runs of the same iteration with different initial guesses could return several eigenpairs, but special care needs to be taken to prevent the iteration from converging to already computed eigenpairs. This is what *deflation* aims to achieve.

A standard deflation technique consists of applying a non-equivalence transformation to the matrix-valued function F that maps the already computed eigenvalues to infinity. This can be done as follows. Suppose we have computed ℓ simple eigenvalues of F, $\lambda_1, \dots, \lambda_\ell$, and let $x_i, y_i \in \mathbb{C}^n$ be such that $y_i^* x_i = 1$, $i = 1, \dots, \ell$. Consider

$$\widetilde{F}(z) = F(z) \prod_{i=1}^{\ell} \left(I - \frac{z - \lambda_i - 1}{z - \lambda_i} y_i x_i^* \right). \tag{4.19}$$

Then it is not difficult to see that

$$\Lambda(\widetilde{F}) = \Lambda(F) \setminus \{\lambda_1, \dots, \lambda_\ell\} \cup \{\infty\}.$$

Indeed, since $y_i^* x_i = 1$,

$$\det \widetilde{F}(z) = \det F(z) \prod_{i=1}^{\ell} \left(\frac{1}{\lambda_i - z} \right),$$

so that λ_i is not a root of $\det \widetilde{F}(z)$ since it is a simple root of $\det F(z)$. With the exception of the eigenvalues λ_i, $i = 1, \dots, \ell$, $\widetilde{F}$ and F have the same finite eigenvalues. Now suppose that F does not have an eigenvalue at infinity, that is, $G(z) = F(1/z)$ does not have an eigenvalue at zero (see Remark 2.2), and let $\widetilde{G}(z) = \widetilde{F}(1/z)$. Then from

$$\det \widetilde{G}(z) = \det G(z) \prod_{i=1}^{\ell} \left(\frac{z}{1 - z\lambda_i} \right)$$

we see that $\widetilde{G}$ has ℓ eigenvalues at zero, and therefore $\widetilde{F}$ has ℓ eigenvalues at infinity. It follows from (4.19) that if $\widetilde{v}$ is an eigenvector of $\widetilde{F}$ with eigenvalue λ, then

$$v = \prod_{i=1}^{\ell} \left(I - \frac{z - \lambda_i - 1}{z - \lambda_i} y_i x_i^* \right) \widetilde{v} \tag{4.20}$$

is an eigenvector of F associated with the eigenvalue λ. To compute the next eigenpair of F, we apply any Newton-based method to $\widetilde{F}(z)$ and recover the eigenvector using (4.20).

The choice of the vectors x_i and y_i affects the basins of convergence of Newton's method. A choice that seems to work well in practice is to set x_i and y_i to be (approximate) right and left eigenvectors corresponding to λ_i. Note that similar deflation strategies have been used in Ferng, Lin, Pierce and Wang (2001) and Huang, Lin and Mehrmann (2016).

Example 4.11. Using the basic MATLAB implementation of the nonlinear inverse iteration given in Figure 4.8, we obtain an approximation $\widehat{\lambda} =$`lam` to the eigenvalue $\lambda = \sqrt{2\pi}$ of the matrix-valued function F in (2.1), which we move to infinity by constructing

$$\widetilde{F}(z) = \begin{bmatrix} e^{iz^2} & 1 \\ 1 & 1 \end{bmatrix} \left(I - \frac{z - \widehat{\lambda} - 1}{z - \widehat{\lambda}} y x^* \right). \tag{4.21}$$

We consider two choices for x and y:

(a) $x = y = \left[\begin{smallmatrix} 1 \\ -1 \end{smallmatrix}\right]$, *i.e.* x and y are right and left eigenvectors of F at λ,

(b) $x = y = \left[\begin{smallmatrix} 0 \\ 1 \end{smallmatrix}\right]$.

As in Example 4.5 we generate 10^6 equidistant initial guesses $\lambda^{(0)}$ in

$$\Omega = \{z \in \mathbb{C} : -3 \le \operatorname{Re}(z) \le 3,\ -3 \le \operatorname{Im}(z) \le 3\},$$

and for each $\lambda^{(0)}$ and $v^{(0)} = \left[\begin{smallmatrix} 1 \\ -1 \end{smallmatrix}\right]$, we run the nonlinear inverse iteration as implemented in Figure 4.8 but with $\widetilde{F}$ in place of F. The basins of convergence are shown in Figure 4.12. The plot on the left corresponds to the above choice (a) for x and y. It shows that the deflation is successful as there is no basin of convergence to the point $\sqrt{2\pi}$. The right plot is obtained with the above choice (b). In this case the point $\sqrt{2\pi}$ still has a basin of convergence even if it is no longer an eigenvalue of $\widetilde{F}$. This unwanted behaviour is likely caused by numerical ill-conditioning of the matrix $\widetilde{F}(z)$ near $\sqrt{2\pi}$. Indeed, even if an eigenvalue λ is exactly deflated from F to obtain $\widetilde{F}$ and hence $\det \widetilde{F}(z)$ is nonzero and holomorphic in a punctured neighbourhood of λ, the matrix $\widetilde{F}(z)$ can have two eigenvalues which converge to zero and infinity, respectively, as $z \to \lambda$. Special care has to be taken when working with such ill-conditioned matrices.

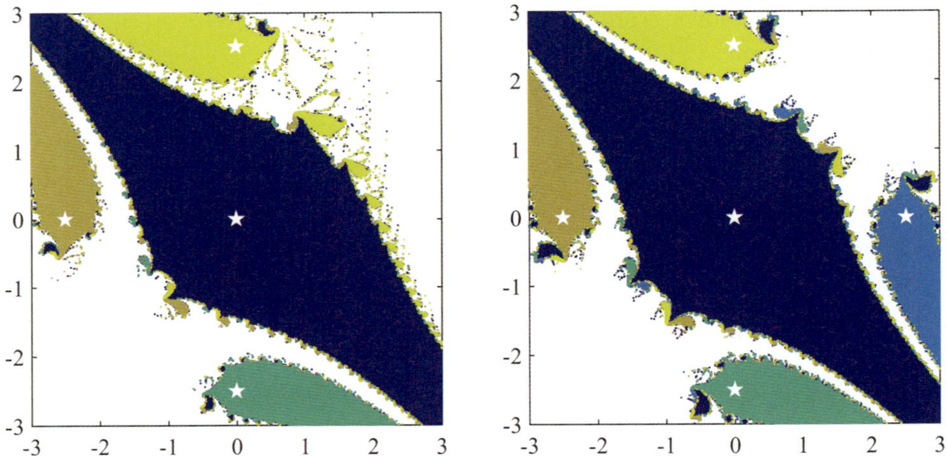

(a) deflation with eigenvectors (b) deflation with non-eigenvectors

Figure 4.12. Basins of convergence for the nonlinear inverse iteration applied to $\widetilde{F}$ in (4.21) for two different choices, (a) and (b), of x and y discussed in Example 4.11. Each initial point in the square region Ω is coloured according to the eigenvalue the method is converging to, with the exact eigenvalues of the NEP in Ω indicated by white $\star$ symbols. White areas indicate that no convergence is achieved within 20 iterations or the method converged to an eigenvalue outside Ω.

Another approach proposed by Effenberger (2013*b*) makes use of minimal invariant pairs as defined in Section 2.5. Effenberger's deflation strategy works as follows. Let $F \in H(\Omega, \mathbb{C}^{n\times n})$ and suppose that we have already computed a minimal invariant pair $(V, M) \in \mathbb{C}^{n\times m} \times \mathbb{C}^{m\times m}$ for F with minimality index p. We want to extend (V, M) into another minimal invariant pair

$$(\widehat{V}, \widehat{M}) = \left(\begin{bmatrix} V & x \end{bmatrix}, \begin{bmatrix} M & b \\ 0 & \lambda \end{bmatrix} \right) \in \mathbb{C}^{n\times(m+1)} \times \mathbb{C}^{(m+1)\times(m+1)}$$

of one size larger. By Definition 2.10, the pair $(\widehat{V}, \widehat{M})$ is invariant if and only if $\mathcal{F}(\widehat{V}, \widehat{M}) = O$. Since

$$(zI - \widehat{M})^{-1} = \begin{bmatrix} (zI - M)^{-1} & (zI - M)^{-1} b (z-\lambda)^{-1} \\ 0 & (z-\lambda)^{-1} \end{bmatrix},$$

we have

$$\mathcal{F}(\widehat{V}, \widehat{M}) = \begin{bmatrix} \mathcal{F}(V, M) & F(\lambda)x + U(\lambda)b \end{bmatrix}$$

with

$$U(\lambda) = \frac{1}{2\pi\mathrm{i}} \int_{\Gamma} F(z) V (zI - M)^{-1} (z-\lambda)^{-1} \,\mathrm{d}z \tag{4.22}$$

and Γ a contour enclosing the eigenvalues of M and λ. Since the pair (V, M) is invariant, $\mathcal{F}(V, M) = O$, and so $(\widehat{V}, \widehat{M})$ is an invariant pair if and only if

$$F(\lambda)x + U(\lambda)b = 0. \tag{4.23}$$

The condition that $(\widehat{V}, \widehat{M})$ be minimal is more involved. Effenberger (2013*b*, Lemma 3.4) shows that $(\widehat{V}, \widehat{M})$ is minimal with minimality index not exceeding $p+1$ if

$$A(\lambda)x + B(\lambda)b = 0, \tag{4.24}$$

where the $m \times n$ matrix-valued function $A(\lambda)$ and the $m \times m$ matrix-valued function $B(\lambda)$ are given by

$$A(\lambda) = \sum_{j=0}^{p} \lambda^j (VM^j)^*, \quad B(\lambda) = \sum_{j=1}^{p} (VM^j)^* V q_j(\lambda)$$

with

$$q_j(\lambda) = \sum_{k=0}^{j-1} \lambda^k M^{j-k-1}.$$

It then follows from (4.23) and (4.24) that $\left(\lambda, \left[\begin{smallmatrix} x \\ b \end{smallmatrix}\right]\right)$ is an eigenpair of the $(n+m) \times (n+m)$ NEP

$$\widetilde{F}(\lambda)\widetilde{v} = 0, \tag{4.25}$$

where

$$\widetilde{F}(\lambda) = \begin{bmatrix} F(\lambda) & U(\lambda) \\ A(\lambda) & B(\lambda) \end{bmatrix}, \quad \widetilde{v} = \begin{bmatrix} x \\ b \end{bmatrix} \neq 0.$$

The matrix-valued function $\widetilde{F}(\lambda)$ is holomorphic since $U(\lambda)$ is holomorphic (Effenberger 2013*b*, Lemma 3.2), and $A(\lambda)$ and $B(\lambda)$ are matrix polynomials. If $F(\lambda)$ is regular, then so is $\widetilde{F}(\lambda)$. Moreover, if $\left(\lambda, \left[\begin{smallmatrix} x \\ b \end{smallmatrix}\right]\right)$ is an eigenpair of $\widetilde{F}(\lambda)$, then $\left([Vx], \left[\begin{smallmatrix} M & b \\ 0 & \lambda \end{smallmatrix}\right]\right)$ is a minimal invariant pair for $F(\lambda)$ (Effenberger 2013*b*, Theorem 3.6). This shows that the pair (V, M) is deflated from the computation when solving the NEP (4.25) in place of $F(\lambda)v = 0$. An eigenpair for the latter can be computed using any of the methods described in Sections 4.1–4.2. But for these, we need to evaluate $U(\lambda)$ in (4.22) and its derivative at a scalar λ. This can be done via numerical integration techniques similar to those described in Section 5.2. Note also that, by using $\lambda I - M = (zI - M) - (z - \lambda)I$, (4.22), and the definition of invariant pairs, we have

$$\begin{aligned} U(\lambda)(\lambda I - M) &= \frac{1}{2\pi \mathrm{i}} \int_\Gamma F(z)V(z-\lambda)^{-1}\,\mathrm{d}z - \frac{1}{2\pi \mathrm{i}} \int_\Gamma F(z)V(zI-M)^{-1}\,\mathrm{d}z \\ &= F(\lambda)V + \mathcal{F}(V, M) = F(\lambda)V; \end{aligned}$$

see Effenberger (2013*b*, Lemma 4.2). Hence, if λ is not an eigenvalue of M, then $U(\lambda) = F(\lambda)V(\lambda I - M)^{-1}$. For large NEPs, Effenberger (2013*b*) combines this deflation strategy with a Jacobi–Davidson method (see Section 4.5) and uses a contour integration method as described in Section 5 to solve the projected NEP.

4.4. Block Newton method

Kressner (2009) presents a block analogue of the Newton iteration on vectors which computes a complete invariant pair $(V, M) \in \mathbb{C}^{n\times m} \times \mathbb{C}^{m\times m}$ for $F \in H(\Omega, \mathbb{C}^{n\times n})$ with minimality index p. Recall from Definition 2.16 that an invariant pair (V, M) is complete if it is minimal and the algebraic multiplicities of the eigenvalues of M are the same as the algebraic multiplicities of the eigenvalues of F.

Let (V, M) be an invariant pair for F so that

$$\mathcal{F}(V, M) = O_{n\times m}, \tag{4.26}$$

where $\mathcal{F}$ is as in Definition 2.10, and assume that (V, M) satisfies the normalization condition

$$\mathcal{N}(V, M) = O_{m\times m}, \tag{4.27}$$

where $\mathcal{N}(V, M) = U^*\mathcal{V}_p(V, M) - I_m$ with $U \in \mathbb{C}^{pn\times m}$ a fixed matrix of full column rank, and $\mathcal{V}_p(V, M)$ is as in (2.17). To apply Newton's method to (4.26)–(4.27), we need the Fréchet derivatives of $\mathcal{F}$ and $\mathcal{N}$ at (V, M), which are given by

$$L_{\mathcal{F}}(\Delta V, \Delta M) = \mathcal{F}(\Delta V, M) + \frac{1}{2\pi \mathrm{i}}\int_{\Gamma} F(z)V(zI - M)^{-1}\Delta M(zI - M)^{-1}\,\mathrm{d}z,$$

$$L_{\mathcal{N}}(\Delta V, \Delta M) = U^*\mathcal{V}_p(\Delta V, M) + \sum_{j=1}^{p-1} U_j^* V\Bigl(\sum_{i=0}^{j} M^i \Delta M M^{j-i-1}\Bigr),$$

where $U^* = [U_0^*, \ldots, U_{p-1}^*]$ with $U_j \in \mathbb{C}^{n\times m}$. Kressner (2009, Theorem 10) shows that the invariant pair (V, M) is complete if and only if the linear matrix operator

$$\begin{aligned}\mathcal{M} : \mathbb{C}^{n\times m} \times \mathbb{C}^{m\times m} &\to \mathbb{C}^{n\times m} \times \mathbb{C}^{m\times m}\\ (\Delta V, \Delta M) &\mapsto \bigl(L_{\mathcal{F}}(\Delta V, \Delta M), L_{\mathcal{N}}(\Delta V, \Delta M)\bigr)\end{aligned}$$

corresponding to the Jacobian of (4.26)–(4.27) at (V, M) is invertible. Then, by the implicit function theorem for holomorphic functions (Krantz 1982), we have that the entries of a complete invariant pair (V, M) vary analytically under analytic changes of F. Hence, complete invariant pairs are well-posed and Newton's method for solving (4.26)–(4.27) converges locally quadratically to a complete invariant pair. Now assuming that $U^*\mathcal{V}_p(V, M) = I_m$, the Newton correction $(\Delta V, \Delta M)$ satisfies

$$\mathcal{M}(\Delta V, \Delta M) = -(\mathcal{F}(V, M), O_{m\times m}).$$

The *block Newton method* for computing an invariant pair $(V, M) \in \mathbb{C}^{n\times m} \times \mathbb{C}^{m\times m}$ of F is given by the pseudocode in Algorithm 4.13.

Algorithm 4.13: Block Newton method for computing an invariant pair

---

Choose an initial pair $(\widetilde{V}^{(0)}, \widetilde{M}^{(0)}) \in \mathbb{C}^{n\times m} \times \mathbb{C}^{m\times m}$ with minimality index p.

for $k = 0, 1, \ldots$ until convergence **do**

  Compute the compact QR factorization $\mathcal{V}_p(\widetilde{V}^{(k)}, \widetilde{M}^{(k)}) = QR$ and let $V^{(k)} = \widetilde{V}^{(k)}R^{-1}$, $M^{(k)} = R\widetilde{M}^{(k)}R^{-1}$.

  Solve the linear matrix equation

$$\mathcal{M}(\Delta V, \Delta M) = -(\mathcal{F}(V^{(k)}, M^{(k)}), O_{m\times m}) \tag{4.28}$$

  for $(\Delta V, \Delta M)$.

  Update $\widetilde{V}^{(k+1)} = V^{(k)} + \Delta V$ and $\widetilde{M}^{(k+1)} = M^{(k)} + \Delta M$.

end

The initial pair $(\widetilde{V}^{(0)}, \widetilde{M}^{(0)})$ can be constructed with a block variant of inverse iteration (Kressner 2009, Algorithm 2). The solution to the linear matrix equation (4.28) is the most expensive part of the iteration. Kressner (2009) shows how to do that efficiently using ideas from Beyn and Thümmler (2009), and provides a MATLAB function `nlevp_newtonstep` that implements these ideas and returns the pair $(\Delta V, \Delta M)$.

Example 4.14. Let us consider the matrix-valued function F in (2.1),

$$F(z) = \begin{bmatrix} 0 & 1 \\ 1 & 1 \end{bmatrix} + \mathrm{e}^{\mathrm{i}z^2} \begin{bmatrix} 1 & 0 \\ 0 & 0 \end{bmatrix},$$

and aim to compute approximations for the eigenvalues $\pm\sqrt{2\pi}$ and $\pm\mathrm{i}\sqrt{2\pi}$. As initial pair we use a random perturbation of the exact invariant pair

$$V = \begin{bmatrix} 1 & 1 & 1 & 1 \\ -1 & -1 & -1 & -1 \end{bmatrix}, \quad M = \mathrm{diag}(\sqrt{2\pi}, -\sqrt{2\pi}, \mathrm{i}\sqrt{2\pi}, -\mathrm{i}\sqrt{2\pi}).$$

Using the MATLAB code given in Figure 4.15, we consistently obtain all four eigenvalues with a relative error of order 10^{-15} or less. No convergence takes place if we perturb V and M with random perturbations of order 10^{-1}.

```
% block_Newton
A(:,:,1) = [0 1; 1 1]; A(:,:,2) = [1 0; 0 0];
n = 2; k = 4; ell = k; maxit = 30; tol = n*eps;
% construct initial pair (V,M)
a = sqrt(2*pi); % eigenvalues are [a, -a, 1i*a, -1i*a]
d = [a -a a*1i -a*1i]+1e-2*(randn(1,k)+1i*randn(1,k)); M = diag(d);
V = diag([1 -1])*ones(n,k) + 1e-2*(randn(n,k)+1i*randn(n,k));
for iter = 0:maxit
  Z = zeros(n*ell,k); Z(1:n,:) = V;
  for j = 2:ell, Z((j-1)*n+1:j*n,:) = Z((j-2)*n+1:(j-1)*n,:)*M; end
  [Q,R] = qr(Z); R = R(1:k,1:k); V = V/R; M = R*(M/R);
  W(:,:,1) = V; for j = 2:ell, W(:,:,j) = W(:,:,j-1)*M; end
  Res = A(:,:,1)*V*f(1,M) + A(:,:,2)*V*f(2,M);
  if norm(Res,'fro') < tol, break, end
  [DV,DM] = nlevp_newtonstep(A,@f,V,M,W,Res,zeros(k));
  V = V - DV; M = M - DM;
end
if k < maxit, nbr_iter = k, evs = eig(M), end

function X = f(j,M)
if j == 1, X = eye(size(M)); end
if j == 2, X = expm(1i*M*M); end
end
```

Figure 4.15. Basic MATLAB implementation of the block Newton method for problem (2.1). The lines preceding the outer `for` loop specify the problem and the method's parameters. When the iteration converges, it returns the number of iterations and the eigenvalues associated with the computed invariant pair (V, M).

We ran the same code but with lines 3–7 replaced by

```
n = 2; k = 6; ell = k; maxit = 100; tol = n*eps;
% construct initial pair (V,M)
a = sqrt(2*pi); % eigenvalues are [a, -a, 1i*a, -1i*a, 0, 0]
d = [a -a a*1i -a*1i 0 0]+1e-10*(randn(1,k)+1i*randn(1,k));
M = diag(d);
V = diag([1 -1])*ones(n,k)+1e-10*(randn(n,k)+1i*randn(n,k));
```

so as to find an invariant pair containing all the eigenvalues inside the circle $\{z \in \mathbb{C} : |z| = 3\}$. The result varies from run to run due to the random perturbations of V and M, but most often after `maxit` iterations we obtain the simple eigenvalues to about 10 digits and the double eigenvalue $\lambda = 0$ to about 7 digits, which shows that the defective eigenvalue 0 affects the accuracy of the other non-defective eigenvalues. We refer to Szyld and Xue (2013*b*) for a discussion of the sensitivity of invariant pairs. The iterations fail to converge if we apply larger perturbations to V and M when constructing the initial pair.

4.5. Large sparse NEPs

Most of the methods we have described so far require the solution of linear systems. For large sparse matrices, there are efficient direct methods implementing Gaussian elimination with some form of pivoting that make clever use of the sparsity structure to avoid fill-in and save storage. These include HSL (2016), MUMPS (2016), PARDISO (Schenk and Gärtner 2004) and UMFPACK (Davis 2004). In place of direct methods to solve the sparse linear systems in Algorithms 4.7–4.10, we can also use iterative methods, for example Krylov subspace methods such as the generalized minimal residual method (GMRES) of Saad and Schultz (1986), the biconjugate gradient method in its stabilized form (BICGSTAB) by van der Vorst (1992), or the quasi-minimal residual method (QMR) of Freund and Nachtigal (1996). Solving the linear systems iteratively leads to inexact versions of nonlinear inverse iteration, residual inverse iteration, and Rayleigh functional iteration. Let us consider in particular the inexact nonlinear inverse iteration, Algorithm 4.7, with the exact solve replaced by an inexact solve. Let $\tau^{(k)}$ be a tolerance such that the approximate solution $\widetilde{v}^{(k+1)}$ returned by the iterative solver at step k satisfies

$$\text{Res} = \|F'(\lambda^{(k)})v^{(k)} - F(\lambda^{(k)})\widetilde{v}^{(k+1)}\| \leq \tau^{(k)}\|F'(\lambda^{(k)})v^{(k)}\|,$$

and let

$$e^{(k)} = \begin{bmatrix} v^{(k)} \\ \lambda^{(k)} \end{bmatrix} - \begin{bmatrix} v \\ \lambda \end{bmatrix}$$

denote the error at iteration k. Szyld and Xue (2013*a*, Theorem 6) show that if, at each iteration,

$$\tau^{(k)} \leq c\ \|e^{(k)}\|$$

is ensured with a constant c independent of k, then the inexact nonlinear inverse iteration converges at least quadratically. In a similar way, they show that for inexact versions of the residual inverse iteration and the two-sided Rayleigh functional iteration, the same order of convergence as for the exact iterations can be achieved if an appropriate sequence of tolerances is used for the inner solves.

Garrett *et al.* (2016) propose a variation of Kublanovskaya's Newton-QR method for either large banded NEPs with narrow bandwidths relative to the size n of the NEP, or large sparse NEPs that can be reduced to such banded NEPs. They make use of a special data structure for storing the matrices to keep memory and computational costs low. They replace the rank-revealing QR factorization of $F(\lambda^{(k)})$ in (4.5) by a banded QR factorization followed by a rank-revealing QR factorization of the R factor. MATLAB and C++ implementations of this approach including the deflation strategy (4.11) are publicly available (Garrett and Li 2013).

Example 4.16. Let us consider the `loaded_string` problem defined by (1.3) with $n = 100$. This NEP has real eigenvalues, and we are interested in the five eigenvalues in the interval $[4, 296]$, all of which are given in (3.5). The MATLAB code in Figure 4.17 calls the function `NQR4UB` from Garrett and Li (2013) implementing the Newton-QR method (*i.e.* Kublanovskaya's method) for unstructurally banded matrix-valued functions. With initial guess `lambda0 = 4.0`, it returns

```
evs =
    4.4822   63.7238  123.0312  202.2009  719.3507
```

which, compared with (3.5), are excellent approximations to the eigenvalues $\lambda_1, \lambda_3, \lambda_4, \lambda_5$, and to another eigenvalue λ_9 outside the interval of interest. Note that the eigenvalue λ_2 is missed. The parameter `h` returned by `NQR4UB` is a cell array containing the values of the residual in (4.13) at each iteration (these are used as a stopping criterion). The left and right normalized residuals,

$$\eta_F(\lambda, v) = \frac{\|F(\lambda)v\|_2}{\|F(\lambda)\|_F\,\|v\|_2} \quad \text{and} \quad \eta_F(\lambda, w^*) = \frac{\|w^*F(\lambda)\|_2}{\|F(\lambda)\|_F\,\|w\|_2}, \tag{4.29}$$

are reported in Figure 4.18. The eigenvalue λ_2 is found by the code in Figure 4.17 if we replace `lambda0 = 4.0` by `lambda0 = 20.0`, but this is the only eigenvalue of F found with this choice of initial guess.

For large sparse NEPs, the iterations described in Sections 4.1–4.2 can be used as part of an iterative projection method in the following way. Suppose we are given a matrix $U \in \mathbb{C}^{n\times k}$ with $k \ll n$ having orthonormal columns that span a subspace (the *search space*) containing an eigenvector of interest. Then instead of solving $F(\lambda)v = 0$ we can solve the $k \times k$ projected NEP

$$Q^*F(\vartheta)Ux = 0, \tag{4.30}$$

where $Q \in \mathbb{C}^{n\times k}$ is some matrix with orthonormal columns spanning the *test space*. Now let (ϑ, x) be an eigenpair of the projected problem (4.30), selected so that ϑ is as close as possible to the target eigenvalue. The pair (ϑ, v) with $v = Ux$ is the corresponding *Ritz eigenpair* for F, and if the residual $\|F(\vartheta)v\|$ is small enough, we can accept (ϑ, v) as an approximate eigenpair for F. If the residual $\|F(\vartheta)v\|$ is too large, then the search space $\operatorname{span}\{U\}$ can be extended by one step of the Newton iteration with initial guess (ϑ, v) and v normalized to have unit 2-norm. The Newton step

```
% Newton_QR_banded
n = 100; lambda0 = 4.0;
ind(1,:) = [1 (1:n-1)]; ind(2,:) = [(2:n) n]; ind(3,:) = ind(2,:);
opts.tol = 5.0e-10; opts.maxitn = 20; opts.rr = 0; opts.nevs = 0;
fun = @(z) f_loaded_string(z,n);
for k = 1:5
    [lam,v,w,h] = NQR4UB(n,fun,ind,lambda0,opts);
    opts.evs(k) = lam; opts.nevs = opts.nevs+1;
end
evs = opts.evs

function [F Fp] = f_loaded_string(z,n)
% Return F(z) and derivative F'(z) for the loaded_string problem.
% Both F(z) and F'(z) are compactly stored as required by NQR4UB.
F(1,:) = [2*n-2*z/3/n (-n-z/6/n)*ones(1,n-1)];
F(2,:) = [-n-z/(6*n)  (2*n-2*z/3/n)*ones(1,n-2) n-z/3/n+z/(z-1)];
F(3,:) = [0 (-n-z/6/n)*ones(1,n-2) 0];
Fp(1,:) = [-2/3/n (-1/6/n)*ones(1,n-1)];
Fp(2,:) = [-1/(6*n)  (-2/3/n)*ones(1,n-2) -1/3/n-1/((z-1)^2)];
Fp(3,:) = [0 (-1/6/n)*ones(1,n-2) 0];
F = sparse(F); Fp = sparse(Fp);
end
```

Figure 4.17. Basic MATLAB M-file calling the Newton-QR method implemented in the Unstructurally Banded Nonlinear Eigenvalue Software as `NQR4UB`. Lines 2–5 define the input parameters for `NQR4UB`. The M-file requires the function `f_loaded_string`.

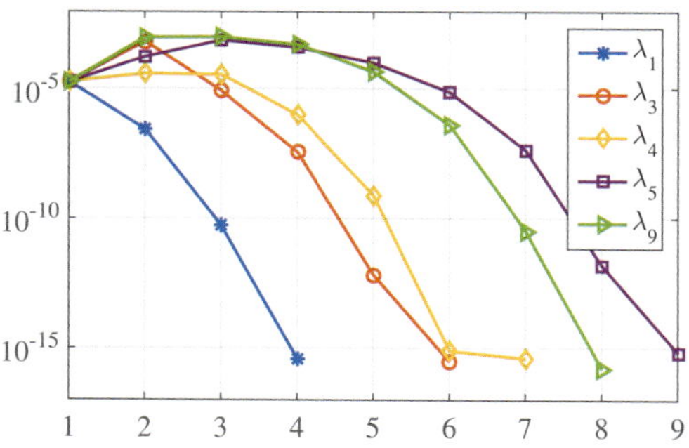

(a)

i	λ_i	$\eta_F(\lambda_i, v_i)$	$\eta_F(\lambda_i, w_i^*)$
1	4.482	3.5e-17	3.7e-16
3	63.72	3.1e-17	2.1e-16
4	123.0	2.7e-17	9.2e-17
5	202.2	4.9e-17	5.4e-16
9	719.4	3.5e-17	1.8e-16

(b)

Figure 4.18. Eigenvalue computation for the `loaded_string` problem using basic MATLAB calls to `NQR4UB` as in Figure 4.17. (a) Residuals $|r_{nn}(\lambda)|/\|F(\lambda)\|_F$ at each iteration for the found eigenvalues λ. (b) Final residuals as defined in (4.29).

in (4.14) is rewritten as

$$\begin{cases} F(\vartheta)\Delta v &= -F(\vartheta)v - \Delta\vartheta\, F'(\vartheta)v, \\ v^*\Delta v &= 0, \end{cases} \tag{4.31}$$

where $\Delta\vartheta$ and Δv define the Newton correction (we omit the index k). Since we only need Δv to extend the search space, we premultiply the first equation in (4.31) by the oblique projector $I_n - F'(\vartheta)vq^*$, where $q \in \mathbb{C}^n$ is such that $q^*F(\vartheta)v = 0$ and normalized such that $q^*F'(\vartheta)v = 1$. This yields

$$\begin{cases} (I_n - F'(\vartheta)vq^*)F(\vartheta)\Delta v &= -F(\vartheta)v, \\ v^*\Delta v &= 0. \end{cases} \tag{4.32}$$

Because of the orthogonality condition $v^*\Delta v = 0$, we can rewrite the first equation in (4.32) as

$$(I_n - F'(\vartheta)vq^*)F(\vartheta)(I_n - vv^*)\Delta v = -F(\vartheta)v, \tag{4.33}$$

which has the form of a *Jacobi–Davidson correction equation.* We can use $\operatorname{span}\{U, \Delta v\}$ as the new search space and $\operatorname{span}\{Q, F(\vartheta)v\}$ as the new test space. This process is repeated until we obtain a Ritz eigenpair with a small residual. We can expect quadratic convergence of this process if the correction equation (4.33) is solved exactly. As for linear eigenproblems (Sleijpen and van der Vorst 1996), the correction equation does not need to be solved accurately to preserve convergence.

Pseudocode is presented in Algorithm 4.19. Usually only a few steps of a preconditioned Krylov subspace method such as GMRES (Saad and Schultz 1986) are necessary to solve the correction equation. If P is a preconditioner for $F(\vartheta)$ so that $Py = r$ is easy to solve, the preconditioner P should be modified to

$$\widetilde{P} = (I_n - F'(\vartheta)vq^*)P(I_n - vv^*)$$

for the correction equation (4.33). Davidson (1975) proposes using $P = \operatorname{diag} F(\vartheta)$ for linear problems.

Variations of the Jacobi–Davidson approach are proposed by Sleijpen *et al.* (1996) for polynomial eigenvalue problems, Betcke and Voss (2004) for symmetric NEPs, and Voss (2007) and Effenberger (2013*b*) for nonsymmetric NEPs. Variants of a two-sided Jacobi–Davidson method are discussed in Hochstenbach and Sleijpen (2003). The nonlinear Arnoldi method of Voss (2004*a*) corresponds to using one step of residual inverse iteration in place of the Jacobi–Davidson correction equation, that is, (4.33) is replaced by $F(\sigma)\Delta v = F(\vartheta)v$. The linear system is then solved approximately, $\Delta v \approx MF(\vartheta)v$, using an approximation M to $F(\sigma)^{-1}$.

Algorithm 4.19: Nonlinear two-sided Jacobi–Davidson method

Choose initial bases U_0 and Q_0 such that $U_0^*U_0 = Q_0^*Q_0 = I$.
for $k = 0, 1, \ldots$ until convergence **do**
  Compute the eigenpair (ϑ, x) of the projected matrix-valued function $Q_k^*F(z)U_k$ with ϑ closest to the wanted eigenvalue and $\|x\|_2 = 1$.
  Compute the Ritz vector $v = U_k x$ and the residual $r = F(\vartheta)v$.
  if $\|r\|_2 \leq$ tol, accept (ϑ, v) as approximate eigenpair, **stop**
  Approximately solve the correction equation

$$(I_n - F'(\vartheta)vq^*)F(\vartheta)(I_n - vv^*)\Delta v = -r$$

  for Δv orthogonal to v.
  Orthogonalize Δv against U_k, normalize $\Delta v \leftarrow \frac{\Delta v}{\|\Delta v\|}$ and expand search space $U_{k+1} = [U_k, \Delta v]$.
  Orthogonalize r against Q_k, normalize $r \leftarrow \frac{r}{\|r\|}$ and expand test space $Q_{k+1} = [Q_k, r]$.
end

4.6. Hermitian NEPs

Assume that the matrix-valued function $F(z)$ is Hermitian, that is, $F(\bar{z}) = F(z)^*$ for all $z \in \mathbb{C}$. We can take advantage of this property in the algorithms described in Sections 4.1–4.2 when factorizing the Hermitian matrix $F(\lambda^{(k)})$ using a block LDL* factorization. Also, when λ is real, the solution to the nonlinear scalar equation in (4.18) in the first step of the residual inverse iteration can be replaced by finding z (*i.e.* the Rayleigh functional) such that $v^{(k)*}F(z)v^{(k)} = 0$. The resulting residual inverse iteration converges quadratically (Neumaier 1985).

Special algorithms can be designed when the Hermitian matrix-valued function F satisfies properties (A1)–(A3) in Section 3 on some real interval $\mathbb{I}$. It follows from Theorem 3.1 that the eigenvalues of F can be characterized as minmax and maxmin values of the Rayleigh functional $p(x)$, which is the only root of $x^*F(p(x))x = 0$ in $\mathbb{I}$. Indeed, Theorem 3.1 asserts that if λ_k is a kth eigenvalue of $F(z)$ (that is, $\mu = 0$ is the kth largest eigenvalue of the Hermitian matrix $F(\lambda_k)$), then

$$\lambda_k = \min_{\substack{V \in \mathbb{S}_k \\ V \cap \mathbb{K}(p) \neq \emptyset}} \max_{\substack{x \in V \cap \mathbb{K}(p) \\ x \neq 0}} p(x) \in \mathbb{I}, \tag{4.34}$$

where $\mathbb{S}_k$ denotes the set of all subspaces of $\mathbb{C}^n$ of dimension k, and $\mathbb{K}_p$ is a subspace of $\mathbb{C}^n$ onto which the Rayleigh functional p is defined. The minimum in (4.34) is attained for an invariant subspace of the Hermitian matrix $F(\lambda_k)$ corresponding to its k largest eigenvalues and the maximum is attained for some $x \in \operatorname{null} F(\lambda_k)$. This suggests the method in Algorithm 4.20 for computing the jth eigenvalue of F, called *safeguarded iteration* (Werner 1970).

Algorithm 4.20: Safeguarded iteration for the jth eigenvalue of F

Choose an initial approximation $\lambda^{(0)}$ to the jth eigenvalue of F.

for $k = 0, 1, \ldots$ until convergence **do**

  Compute an eigenvector $x^{(k)}$ corresponding to the jth largest eigenvalue of the Hermitian matrix $F(\lambda^{(k)})$.

  Compute the real root ρ of $x^{(k)*}F(\rho)x^{(k)} = 0$ closest to $\lambda^{(k)}$ and set $\lambda^{(k+1)} = \rho$.

end

Niendorf and Voss (2010) show that if λ_j is a simple eigenvalue of F, then the safeguarded iteration converges locally and quadratically to λ_j. For large sparse problems, the safeguarded iteration can be embedded into an iterative projection method such as the Jacobi–Davidson scheme or the nonlinear Arnoldi scheme described in Section 4.5; see Voss (2004*b*) and Niendorf and Voss (2010).

For large sparse NEPs satisfying properties (A1)–(A3) defined in Section 3, Szyld and Xue (2015) describe several variants of a preconditioned conjugate gradient method, which make use of the nonlinear variational principle (Theorem 3.1) and the nonlinear Cauchy interlacing theorem (Theorem 3.2).

Example 4.21. As shown in Example 3.3, the `loaded_string` NEP defined with F in (1.3) satisfies assumptions (A1)–(A3) in Section 3 on the open real interval $(1, +\infty)$. Hence we can use the safeguarded iteration to compute the five smallest eigenvalues of F in that interval. A basic MATLAB implementation is given in Figure 4.22. The inner loop implements the safeguarded iteration. Once an eigenpair is found, we increase the index `j` of the eigenvalue and restart the safeguarded iteration with the previous converged eigenvalue as initial guess. Using this code we obtain

```
eigenvalue  #iter   residual
 4.48e+00     4     7.43e-17
 2.42e+01     2     1.02e-16
 6.37e+01     2     8.28e-17
 1.23e+02     2     8.44e-17
 2.02e+02     2     9.80e-17
```

```
% safeguarded_iteration
n = 100; C1 = n*gallery('tridiag',n); C1(end) = C1(end)/2;
C2 = (abs(gallery('tridiag',n)) + 2*speye(n))/(6*n);
C2(end) = C2(end)/2; C3 = sparse(n,n); C3(n,n) = 1;
F = @(z) C1 - z*C2 + C3*z/(z-1);
lam = 1.1; maxit = 5; nb_evs = 5; tol = n*eps;
fprintf('eigenvalue  #iter   residual\n')
for j = 1:nb_evs
    for k = 0:maxit
        [X,E] = eigs(F(lam),nb_evs+0,'sa'); v = X(:,j);
        res = norm(F(lam)*v)/norm(F(lam),'fro');
        if res < tol, break, end
        c1 = v'*C1*v; c2 = v'*C2*v; c3 = v'*C3*v;
        f = @(z) c1-c2*z+c3*z/(z-1); lam = fzero(f,lam);
    end
    fprintf('%9.2e %5.0f %12.2e\n',lam,k,res)
end
```

Figure 4.22. Basic MATLAB implementation of the safeguarded iteration to compute the five smallest eigenvalues of the `loaded_string` problem (1.3). Lines 2–5 define the NEP F and line 6 specifies the parameters for the iteration.

Note that only a small number of iterations per eigenvalue are required for this problem.

4.7. Additional comments and software

While Newton's method is widely used, for example, as a final refinement step in the Chebyshev interpolation-based method of Effenberger and Kressner (2012) (MATLAB code available at http://anchp.epfl.ch/files/content/sites/anchp/files/software/chebapprox.tar.gz), there appears to be very little software beyond mere research code available. Gander, Gander and Kwok (2014, Section 8.3.4) provide MATLAB codes for Newton's method to compute the eigenvalues of a quadratic eigenvalue problem. They show how to use algorithmic differentiation of $f(z) = \det F(z)$ to obtain the Newton update. Indeed, algorithmic differentiation is a powerful tool to compute derivative information when these are not available explicitly (Arbenz and Gander 1986, Griewank and Walther 2008).

MATLAB code for a nonlinear Jacobi–Davidson algorithm with deflation by Effenberger (2013*b*) is available at http://anchp.epfl.ch/nonlinjd. As stated by the author, this is research code and not intended for production use.

The most comprehensive suite of high-performance solvers for large-scale linear and nonlinear eigenvalue problems is SLEPc (Hernandez, Roman and Vidal 2005), available at http://slepc.upv.es/. SLEPc builds on top of PETSc (Balay *et al.* 2016), which means that PETSc must be installed first in order to use SLEPc. SLEPc provides several Newton-based solvers for NEPs, namely the residual inverse iteration of Neumaier (1985) (Algorithm 4.10), the method of successive linear problems by Ruhe (1973), and the nonlinear Arnoldi method of Voss (2004*a*). All of these methods are implemented without deflation, that is, for computing a single eigenpair near an initial guess. However, SLEPc implements Effenberger's method in Section 4.3 for the *iterative refinement* of approximations computed by some other method. The implementation is described in Campos and Roman (2016*a*) for polynomial eigenvalue problems, but it can be used for NEPs as well.

5. Solvers based on contour integration

Keldysh's theorem (Theorem 2.8) provides the main tool for a class of NEP solvers based on contour integration. These methods have been developed in a series of works by Asakura *et al.* (2009), Beyn (2012) and Yokota and Sakurai (2013), and they have become popular due to their relatively straightforward implementation and great potential for parallelization.

Let $F \in H(\Omega, \mathbb{C}^{n\times n})$ be a regular matrix-valued function with finitely many eigenvalues $\lambda_1, \lambda_2, \ldots, \lambda_s$ in the domain Ω. Further, let $\Gamma \subset \Omega$ be a contour which encloses all eigenvalues, and let $f \in H(\Omega, \mathbb{C})$ be a scalar holomorphic function. Then with the notation of Theorem 2.8 and the Cauchy-integral definition of a matrix function (Higham 2008, Definition 1.11), we have

$$\frac{1}{2\pi \mathrm{i}} \int_\Gamma f(z) F(z)^{-1}\, \mathrm{d}z = V f(J) W^*. \tag{5.1}$$

Here, following Theorem 2.8, V and W are $n \times \overline{m}$ matrices whose columns are right and left generalized eigenvectors, respectively, and J is an $\overline{m} \times \overline{m}$ block-diagonal Jordan matrix. The idea common to many contour-based methods is to 'probe' the matrix decomposition (5.1) from the right (and left) to infer information about J and V (and W).

5.1. Beyn's 'Integral algorithm 1'

Let us start with the simple case in which there are only a small number $\overline{m} \le n$ of generalized eigenvectors and they are all linearly independent, that is, both V and W are of full column rank $\overline{m}$. In this case it suffices to use polynomials f of degree zero and one in (5.1), also called the zeroth and first-order *moments*. For simplicity we choose the monomials 1 and z

for f, but the derivation extends to any choice of two linearly independent polynomials of degree at most one.

Let $R \in \mathbb{C}^{n\times r}$ ($\overline{m} \leq r \leq n$) be a probing matrix (typically chosen at random) such that W^*R is of maximal rank $\overline{m}$. Then the pair of $n \times r$ matrices

$$A_0 = \frac{1}{2\pi \mathrm{i}} \int_\Gamma F(z)^{-1} R \,\mathrm{d}z = VW^*R, \tag{5.2}$$

$$A_1 = \frac{1}{2\pi \mathrm{i}} \int_\Gamma zF(z)^{-1} R \,\mathrm{d}z = VJW^*R \tag{5.3}$$

can be manipulated to infer J and V as follows.

(1) Compute an economy-size singular value decomposition (SVD) of

$$A_0 = V_0 \Sigma_0 W_0^*, \tag{5.4}$$

where $V_0 \in \mathbb{C}^{n\times\overline{m}}$ and $W_0 \in \mathbb{C}^{r\times\overline{m}}$ have orthonormal columns, and $\Sigma_0 = \mathrm{diag}(\sigma_1, \dots, \sigma_{\overline{m}})$ is invertible. This SVD exists since $\mathrm{rank}(A_0) = \overline{m}$ by the above rank conditions on V, W and R.

(2) Since $\mathrm{range}(V) = \mathrm{range}(V_0)$, there exists an invertible matrix $X \in \mathbb{C}^{\overline{m}\times\overline{m}}$ such that $V = V_0X$. From (5.2) and (5.4) we find

$$W^*R = X^{-1}\Sigma_0 W_0^*.$$

This relation can be used to eliminate W^*R from $A_1 = VJW^*R$,

$$A_1 = (V_0X)JX^{-1}\Sigma_0 W_0^*.$$

(3) We have thus arrived at

$$M := V_0^* A_1 W_0 \Sigma_0^{-1} = XJX^{-1},$$

showing that XJX^{-1} is a Jordan decomposition form of the matrix M. Hence the eigenvalues of the NEP can be obtained from the eigenvalues of M, and the columns of $V = V_0X$ contain the corresponding right generalized eigenvectors of the NEP.

By this simple three-step procedure we have effectively reduced an NEP with $\overline{m}$ eigenvalues inside Γ to a linear eigenproblem of size $\overline{m}\times\overline{m}$. By construction, this procedure returns all the eigenvalues λ_i of F inside Γ, together with corresponding right generalized eigenvectors v_i which we assumed to be linearly independent; see also Beyn (2012, Theorem 3.3). However, we need a number of further ingredients to make this method practical.

First of all, we should be aware that this method effectively amounts to the computation of a Jordan decomposition of the matrix M, which is known to be highly sensitive to perturbations. Indeed, Beyn (2012) first derives his method for the case that all eigenvalues inside Γ are simple. In this case, and even in the semisimple case, M will be diagonalizable and a full set of linearly independent eigenvectors of M can be computed reliably by the QR algorithm.

Second, we typically do not know the number $\overline{m}$ of linearly independent eigenvectors in advance. However, as long as r is chosen large enough (*i.e.* $r \geq \overline{m}$), we can detect $\overline{m}$ as a *numerical rank of* A_0, for example by counting the number of singular values of A_0 that are above a user-defined tolerance `tol`.

The third ingredient, the numerical approximation of the contour integrals in (5.2) and (5.3) by appropriate quadrature rules, will be discussed in the following section.

5.2. *Quadrature rules and rational filters*

Let us assume that Γ is a piecewise regular contour with parametrization $\gamma : [0, 2\pi] \to \Gamma$. Then by substituting $z = \gamma(t)$ in (5.1) we have

$$A_f := \frac{1}{2\pi \mathrm{i}} \int_\Gamma f(z)F(z)^{-1}R \, \mathrm{d}z = \frac{1}{2\pi \mathrm{i}} \int_0^{2\pi} f(\gamma(t))\gamma'(t)F(\gamma(t))^{-1}R \, \mathrm{d}t. \quad (5.5)$$

The next step is to apply a quadrature rule with, say, $n_c \geq 1$ points to the integral on the right.

To first consider the simplest case, assume that $\gamma(t) = \mathrm{e}^{\mathrm{i}t}$ parametrizes the unit circle $\Gamma = \{z \in \mathbb{C} : |z| = 1\}$. Then the most natural quadrature rule, the *trapezoid rule*, amounts to using equispaced points $t_\ell = 2\pi\ell/n_c \in [0, 2\pi]$ for $\ell = 1, 2, \ldots, n_c$, resulting in the approximation

$$A_{f,n_c} := \sum_{\ell=1}^{n_c} \omega_\ell F(z_\ell)^{-1}R \approx A_f, \quad (5.6)$$

with quadrature weights and nodes

$$\omega_\ell = \frac{f(z_\ell)z_\ell}{n_c}, \qquad z_\ell = \mathrm{e}^{2\pi \mathrm{i}\ell/n_c}.$$

Note that only the weights ω_ℓ depend on f, hence the quadrature of both A_0 and A_1 in (5.2)–(5.3) requires only n_c evaluations of $F(z_\ell)^{-1}R$, not $2n_c$. A further reduction in the number of evaluations is possible if the $F(z_\ell)^{-1}R$ in (5.6) appear in complex conjugate pairs, as is the case when F is a Hermitian matrix-valued function and R is a matrix with real entries. The evaluations of $F(z_\ell)^{-1}R$ are typically the computationally most expensive part of a contour-based eigensolver, but they are completely decoupled and can hence be assigned to different processing units on a parallel computer. This potential for coarse-grain parallelism is one reason for the popularity of these methods.

Beyn (2012) studies the quality of the quadrature approximation (5.6). Assume that each element of $f(z)F^{-1}$ is holomorphic and bounded on an annulus of modulus ρ,

$$\mathbb{A}_\rho = \{z \in \mathbb{C} : \rho^{-1} < |z| < \rho\}, \quad \rho > 1.$$

Then each element of $f(z)F(z)^{-1}R$ is a bounded holomorphic function on that same annulus and we can apply standard results on the convergence of the trapezoidal rule for holomorphic functions elementwise; see, for example, Trefethen and Weideman (2014, Theorem 2.2). This results in

$$\|A_f - A_{f,n_c}\| = O(\rho^{-n_c}),$$

that is, the trapezoid rule yields exponential convergence in the number of quadrature nodes. The convergence factor ρ^{-1} is determined by a singularity of $F(z)^{-1}$ (an eigenvalue of F) closest to the unit circle (where 'closeness' is measured in terms of the level lines $\partial\mathbb{A}_\rho$).

Care has to be taken with contours that are not circles, as the exponential accuracy of the trapezoidal rule may deteriorate or be completely lost if the rule is not transplanted conformally. As an example, consider the square contour

$$\Gamma := \{z \in \mathbb{C} : \max\{|\mathrm{Re}(z)|, |\mathrm{Im}(z)|\} = 1\}.$$

It seems tempting to simply apply the trapezoidal rule on each of the four sides of the square separately and, assuming that n_c is divisible by four, discretizing each side by $n_c/4+1$ equispaced quadrature nodes including the endpoints. A schematic view is given in Figure 5.1(a), with the red pluses corresponding to the trapezoid quadrature nodes on the square. As the function to be integrated, $(2\pi\mathrm{i})^{-1}f(z)F(z)^{-1}$, is generally not periodic on each of the four sides, this approach only gives convergence of order 2, $\|A_f - A_{f,n_c}\| = O(n_c^{-2})$, in line with the standard error bounds for the trapezoidal rule for non-periodic functions (with continuous second derivative); see for example Davis and Rabinowitz (2007, equation (2.1.11)).

A better approach to obtaining a quadrature rule on a non-circular contour Γ is to use a conformal map Ψ from the annulus $\mathbb{A}_\rho$ (the w-domain) onto a doubly connected domain Ω with continuous inner and outer boundary (the z-domain), and then to define the contour Γ as the image of the unit circle under Ψ, $\Gamma = \{z = \Psi(w) : |w| = 1\}$. (Conversely, by the Riemann mapping theorem (Ahlfors 1953, p. 255), any bounded domain Ω whose boundary consists of two disjoint continua can be identified conformally with an annulus.) Let $\gamma(t) = \Psi(\mathrm{e}^{\mathrm{i}t})$, $t \in [0, 2\pi]$, be a parametrization of Γ. Then, by applying the trapezoid rule to (5.5), we obtain the quadrature approximation

$$A_{f,n_c} := \sum_{\ell=1}^{n_c} \omega_\ell F(z_\ell)^{-1} R \approx A_f,$$

with quadrature weights and nodes

$$\omega_\ell = \frac{f(z_\ell)\Psi'(e^{2\pi i\ell/n_c})\, e^{2\pi i\ell/n_c}}{n_c}, \qquad z_\ell = \Psi(e^{2\pi i\ell/n_c}).$$

We can interpret $F(z)$ as a function $F(\Psi(w))$ which is holomorphic and bounded on the annulus $\mathbb{A}_\rho$, hence this quadrature rule will again converge exponentially with convergence factor ρ^{-1}. This is illustrated in Figure 5.1.

For comparison we show the quadrature error obtained when applying Gauss–Legendre quadrature with $n_c/4$ nodes on each of the four sides of the square. Such an approach will also lead to exponential convergence, with the convergence factor determined by the largest Bernstein ellipse enclosing each of the four sides such that $F(z)^{-1}$ is holomorphic in its interior. More precisely, the *Bernstein ellipse region of elliptical radius* $\rho > 1$ *associated with the interval* $[-1, 1]$ is

$$\mathbb{E}_\rho = \bigl\{z = \cos(t + i \ln r) : t \in [0, 2\pi],\ r \in [1, \rho)\bigr\}, \tag{5.7}$$

and the corresponding Bernstein ellipse region associated with a (possibly complex) interval $[a, b]$ is the image $\varphi_{[a,b]}(\mathbb{E}_\rho)$ with $\varphi_{[a,b]}(z) = (a - b)z/2 + (a + b)/2$. The boundary $\partial\mathbb{E}_\rho$ is referred to as the *Bernstein ellipse*. For our example in Figure 5.1, the most restricted Bernstein ellipse is the one of radius $\rho = 1.105$ associated with the upper side $[a, b] = [-1 + i, 1 + i]$ of the square. The Gauss–Legendre rule associated with this side has $n_c/4$ nodes, hence the quadrature error reduces with a convergence factor of $\rho^{-2n_c/4}$. The overall quadrature error behaves like $\|A_f - A_{f,n}\| = O(\rho^{-n_c/2})$, which in this example is worse than the convergence achieved with the trapezoid rule.

Conformally mapped quadrature rules have been applied, for example, to the problem of matrix function approximation (Hale, Higham and Trefethen 2008). An alternative interpretation of quadrature rules in terms of filter functions has been discussed by Van Barel and Kravanja (2016). This interpretation also motivated the numerical approach to designing a rational filter for linear and nonlinear eigenvalue problems presented in Van Barel (2016).

5.3. Higher-order moments and the Sakurai–Sugiura method

Let us now consider the general case where the number $\overline{m}$ of generalized eigenvectors is not necessarily smaller than or equal to the problem size n, and that the generalized eigenvectors are not necessarily linearly independent. In this case we need to use higher-order moments in (5.1), that is, employ polynomials f of degree larger than one. Eigenvalue techniques based on higher-order moments have been applied successfully by Sakurai and coauthors in a number of papers on the generalized eigenvalue problem,

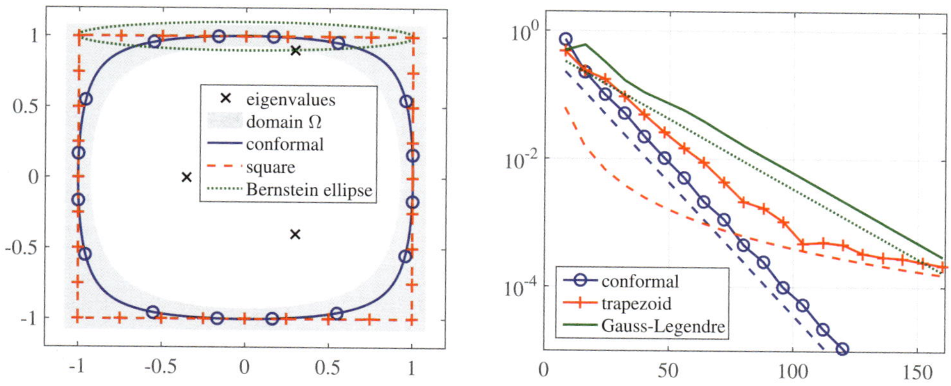

(a) (b)

Figure 5.1. Comparison of three quadrature rules for approximating the integral in (5.1). (a) A portion of the complex plane, showing three eigenvalues of a 3×3 NEP $F(\lambda)v = 0$ (black $\times$). The doubly connected region Ω (shown in grey) is the conformal image of an annulus with modulus $\rho = 1.1$, and the blue contour is the image of the unit circle with mapped quadrature nodes shown as blue circles. The red crosses on the square contour correspond to the quadrature nodes of trapezoidal rules applied to each side. The green curve is the largest Bernstein ellipse associated with the upper side of the square in which $F(z)^{-1}$ is analytic. (b) Quadrature errors $\|A_f - A_{f,n_c}\|$ as the number of quadrature nodes n_c increases (solid curves), together with their predicted error behaviours as discussed in Section 5.2.

and for the nonlinear case by Asakura *et al.* (2009) and Yokota and Sakurai (2013). Beyn (2012) also considers higher-order moments in his 'Integral algorithm 2'. In the following we will use a general formulation of these methods and then discuss how the methods available in the literature relate to this formulation.

Let $L \in \mathbb{C}^{n\times\ell}$ and $R \in \mathbb{C}^{n\times r}$ be given left and right probing matrices, respectively, and let $\overline{p} \geq 0$ be a given integer. For any $p = 0, 1, \ldots, \overline{p}$ we define

$$A_p = \frac{1}{2\pi \mathrm{i}} \int_\Gamma z^p L^* F(z)^{-1} R \,\mathrm{d}z = L^* V J^p W^* R \in \mathbb{C}^{\ell\times r} \tag{5.8}$$

and arrange these matrices in $\overline{p}\ell \times \overline{p}r$ block-Hankel matrices as follows:

$$B_0 = \begin{bmatrix} A_0 & \cdots & A_{\overline{p}-1} \\ \vdots & & \vdots \\ A_{\overline{p}-1} & \cdots & A_{2\overline{p}-2} \end{bmatrix} \quad \text{and} \quad B_1 = \begin{bmatrix} A_1 & \cdots & A_{\overline{p}} \\ \vdots & & \vdots \\ A_{\overline{p}} & \cdots & A_{2\overline{p}-1} \end{bmatrix}. \tag{5.9}$$

Further, let us define the matrices

$$V_{[\overline{p}]} = \begin{bmatrix} L^*V \\ \vdots \\ L^*VJ^{\overline{p}-1} \end{bmatrix} \in \mathbb{C}^{\overline{p}\ell\times\overline{m}} \quad \text{and} \quad W^*_{[\overline{p}]} = \left[W^*R, \ldots, J^{\overline{p}-1}W^*R\right] \in \mathbb{C}^{\overline{m}\times\overline{p}r}. \tag{5.10}$$

Then by (5.8) we have factorizations

$$B_0 = V_{[\overline{p}]}W^*_{[\overline{p}]} \quad \text{and} \quad B_1 = V_{[\overline{p}]}JW^*_{[\overline{p}]}.$$

Let us assume that L, R and $\overline{p}$ have been chosen so that $V_{[\overline{p}]}$ and $W^*_{[\overline{p}]}$ are of maximal rank, that is,

$$\operatorname{rank}(V_{[\overline{p}]}) = \operatorname{rank}(W^*_{[\overline{p}]}) = \overline{m}. \tag{5.11}$$

Then, using the same derivation as in the three-step procedure given in Section 5.1, it is again possible to infer J from the pair (B_0, B_1). (A sufficient condition for (5.11) is given by Lemma 2.13, namely $\overline{p} \geq \sum_{i=1}^{s} m_{i,1}$.) The following theorem, given by Beyn (2012, Theorem 5.2) for the case $L = I_n$, makes this precise.

Theorem 5.2. Suppose that $F \in H(\Omega, \mathbb{C}^{n\times n})$ has no eigenvalues on the contour $\Gamma \subset \Omega$ and finitely many pairwise distinct eigenvalues $\lambda_1, \lambda_2, \ldots, \lambda_s$ inside Γ. For each eigenvalue λ_i let d_i denote its geometric multiplicity with partial multiplicities $m_{i,1} \geq \cdots \geq m_{i,d_s}$. Further assume that the matrices $V_{[\overline{p}]}$ and $W^*_{[\overline{p}]}$ defined in (5.10) satisfy the rank condition (5.11). Let

$$V_{[\overline{p}]}W^*_{[\overline{p}]} = V_0\Sigma_0W_0^*$$

be an economy-size SVD, where $V_0 \in \mathbb{C}^{\overline{p}\ell\times\overline{m}}$ and $W_0 \in \mathbb{C}^{\overline{p}r\times\overline{m}}$ have orthonormal columns and $\Sigma_0 = \operatorname{diag}(\sigma_1, \ldots, \sigma_{\overline{m}})$ is invertible. Then the matrix $X = V_0^*V_{[\overline{p}]}$ is nonsingular and the matrix

$$M := V_0^*B_1W_0\Sigma_0^{-1} = XJX^{-1}$$

has a Jordan normal form J and hence the same eigenvalues λ_i with identical partial multiplicities $m_{i,j}$ as F.

In conjunction with numerical quadrature to approximate the matrices A_p defined by (5.8), the above theorem translates directly into a numerical method; see Figure 5.3 for a basic MATLAB implementation. Let us discuss some special cases of this method.

(i) If $\ell = r$, the numerical method indicated by Theorem 5.2 corresponds to the 'block-SS method' described by Asakura *et al.* (2009). This method has been used successfully in applications, *e.g.* for solving NEPs arising in acoustics (Leblanc and Lavie 2013).

(ii) If $\ell = r = 1$ and $F(\lambda) = A - \lambda B$ with an $n \times n$ matrix pencil (A, B), we recover the method of Sakurai and Sugiura (2003) for generalized eigenvalue problems, which is sometimes referred to as the 'SS method'. In this case the block matrices B_0, B_1 defined in (5.9) reduce to standard square Hankel matrices of size $(2\overline{p} - 1) \times (2\overline{p} - 1)$.

(iii) If $\overline{p} = 1$, $\overline{m} \leq r \leq n$ and $L = I_n$, the method reduces to the 'Integral algorithm 1' of Beyn (2012) described in Section 5.1. This method has been applied successfully, *e.g.* for the solution of NEPs arising from resonance problems related to fluid–solid interaction (Kimeswenger, Steinbach and Unger 2014) and Maxwell eigenvalue problems (Wieners and Xin 2013). The latter reference also contains numerical comparisons of the integral approach with a Newton method.

(iv) For $\overline{p} \geq 1$ and $L = I_n$ we obtain the 'Integral algorithm 2' of Beyn (2012).

(v) Another closely related projection method was given by Yokota and Sakurai (2013). Therein the idea is not to use the matrix M defined in Theorem 5.2 to extract eigenvalues and eigenvectors, but to use the matrices A_p defined in (5.8) with $L = I_n$ to compute an orthonormal basis V_0 of $\operatorname{span}(V)$ from an economy-size SVD of $N := [A_0, A_1, \dots, A_{\overline{p}-1}] = V_0 \Sigma_0 W_0$. Note that by (5.8) we have

$$N = V[J^0 W^* R, J^1 W^* R, \dots, J^{\overline{p}-1} W^* R],$$

and hence $\operatorname{span}(N) = \operatorname{span}(V)$ is guaranteed by (5.11). The method of Yokota and Sakurai (2013) then amounts to solving the projected NEP $\widehat{F}(\lambda)\widehat{x} = 0$ with $\widehat{F}(\lambda) = V_0^* F(\lambda) V_0$. The authors recommend using one of Beyn's integral methods for the solution of this lower-dimensional NEP. The reported results indicate that this explicit projection approach may give better accuracy than the methods of Asakura *et al.* (2009) and Beyn (2012) applied directly to the original problem.

5.4. Numerical illustration

Let us consider the matrix-valued function F defined in (2.1). We choose L and R to be random vectors ($\ell = r = 1$) and look for the six eigenvalues of F inside the circle $\Gamma = \{z \in \mathbb{C} : |z| = 3\}$. In line with Lemma 2.13 we choose $\overline{p} = 6$. MATLAB code is given in Figure 5.3. For numerical stability we have scaled the moment functions z^p to have unit norm on Γ.

```
% contour_eigensolver
n = 2; F = @(z) [exp(1i*z.^2), 1; 1, 1];
ell = 1; r = 1; L = rand(n,ell); R = rand(n,r); % probing matrices
gam = 0; rad = 3; nc = 200; % circle centre & radius, nbr of nodes
w = exp(2i*pi*(1:nc)/nc); z = gam+rad*w; % unit roots and quad pts
pbar = 6; A = zeros(ell,r,2*pbar);        % matrices of moments
for k = 1:nc
    Fz = L'*(F(z(k))\R);
    for j = 0:2*pbar-1
        A(:,:,j+1) = A(:,:,j+1) + (w(k)^j*rad*w(k)/nc)*Fz;
    end
end
A = A(:,:); B0 = zeros(pbar*ell,pbar*r); B1 = B0;
for j = 0:pbar-1
    B0(1+j*ell:(j+1)*ell,:) = A(:,1+j*r:pbar*r+j*r);
    B1(1+j*ell:(j+1)*ell,:) = A(:,1+(j+1)*r:pbar*r+(j+1)*r);
end
[V,Sig,W] = svd(B0); mbar = find(diag(Sig)/Sig(1)>1e-12,1,'last');
V0 = V(:,1:mbar); Sig0 = Sig(1:mbar,1:mbar); W0 = W(:,1:mbar);
M = (V0'*B1*W0)/Sig0; evs = eig(M); evs = gam+rad*evs(abs(evs)<1)
```

Figure 5.3. Basic MATLAB implementation of an integral method for NEPs with a circular contour. The method's parameters are specified in lines 2–4, and the variable evs computed in the final line contains the eigenvalue approximations.

Using a trapezoidal rule with $n_c = 200$ nodes, we obtain the eigenvalue approximations

```
evs =
 -2.506628274630944 - 0.000000000000013i
 -0.000000000000028 + 2.506628274631054i
  0.000000000000045 - 2.506628274631075i
  2.506628274630940 - 0.000000000000047i
 -0.000000202602049 + 0.000001229470211i
  0.000000202602004 - 0.000001229470420i
```

The correct digits are underlined. The result will vary slightly from run to run if another random initialization of the probing matrices L and R is used, however, we observe consistently that the simple eigenvalues are accurate to about 12 digits, whereas the double eigenvalue $\lambda = 0$ is only accurate to about 6 digits. This accuracy reduction is not surprising given the higher sensitivity of $\lambda = 0$; see in particular the spectral portrait in Figure 2.22.

5.5. Eigenvalue localization

Theorem 5.2 effectively provides a method for counting the number of eigenvalues $\overline{m}$ of F inside the contour Γ by computing the (numerical) rank of the matrix B_0 defined in (5.9). In the MATLAB code in Figure 5.3, $\overline{m}$ is provided in the variable `mbar`. A more direct approach for counting the eigenvalues is to make use of integrals of the form

$$N_p(f, \Gamma) = \frac{1}{2\pi \mathrm{i}} \int_{\Gamma} z^p \frac{f'(z)}{f(z)} \, \mathrm{d}z. \tag{5.12}$$

By the argument principle (Ahlfors 1953, Section 5.2), the number $N_0(f, \Gamma)$ corresponds to the number of roots (counting multiplicity) of a holomorphic function f enclosed by Γ. By choosing $f(z) = \det F(z)$ and using the trace relation (4.2), we have

$$N_0(\det F, \Gamma) = \frac{1}{2\pi \mathrm{i}} \int_{\Gamma} \operatorname{trace}(F^{-1}(z) F'(z)) \, \mathrm{d}z. \tag{5.13}$$

We can apply the same quadrature techniques as explained in Section 5.2 to approximate this integral. The domain Ω on which to count the eigenvalues of F need not be simply connected. For example, if Γ_1 is a contour inside another contour Γ_2, then F has

$$N_0(\det F, \Gamma_2) - N_0(\det F, \Gamma_1)$$

eigenvalues in the doubly connected region bounded by Γ_1 and Γ_2.

Example 5.4. Let us determine the number of eigenvalues of F defined in (2.1) inside the annulus Ω bounded by the contours $\Gamma_j = \{z \in \mathbb{C} : |z| = \rho_j\}$, $j = 1, 2$, where $\rho_1 = 4$ and $\rho_2 = 5.25$. For this we apply the trapezoidal rule to the integral (5.13). This is implemented in the basic MATLAB code in Figure 5.5 with the parameters set up to compute an approximation to $N_0(\det F, \Gamma_1)$ using $n_c = 90$ quadrature nodes. The code returns

```
nbr_evs =
   9.999998431959657 - 0.000000000000001i
```

suggesting that $N_0(\det F, \Gamma_1) = 10$. Then replacing `rad = 4` with `rad = 5.25` produces

```
nbr_evs =
  18.000958027557175 - 0.000000000000001i
```

suggesting that $N_0(\det F, \Gamma_2) = 18$. Indeed, F has exactly $18 - 10 = 8$ eigenvalues in Ω.

5.6. Further remarks and available software

Contour-based methods for linear and nonlinear eigenproblems are closely related to earlier techniques developed for finding roots of holomorphic

```
% contour_count
F = @(z) [exp(1i*z.^2) 1; 1 1];
Fp = @(z) [2i*z*exp(1i*z.^2) 0; 0 0];
gam = 0; rad = 4; nc = 90; % centre, radius, nr of nodes on circle
w = exp(2i*pi*(1:nc)/nc); z = gam+rad*w; % unit roots and quad pts
nr = 0;
for k = 1:nc
    nr = nr + rad*w(k)*trace(F(z(k))\Fp(z(k)));
end
nbr_evs = nr/nc % number of eigenvalues
```

Figure 5.5. Basic MATLAB code implementing the trapezoidal rule approximation of (5.13) for computing the number of eigenvalues of F inside the disc with centre `gam` and radius `rad`. The NEP parameters are specified in lines 2–3.

functions f; see the review by Ioakimidis (1987). The pioneering work by Delves and Lyness (1967) uses the contour integrals (5.12) for $p \geq 0$, which can be related to Newton sums of the unknown roots of f enclosed by Γ. However, this relation is prone to ill-conditioning, in particular when the number of unknown roots surrounded by Γ is large. In order to address this issue, Delves and Lyness (1967) propose a subdivision technique. A FORTRAN 77 implementation of their method is provided in Botten, Craig and McPhedran (1983).

Kravanja, Sakurai and Van Barel (1999*a*) present an alternative root-finding approach with better numerical stability based on formal orthogonal polynomials and the solution of a generalized eigenvalue problem. A Fortran 90 implementation is given in Kravanja *et al.* (2000). Similar ideas can be used to compute roots and poles of a meromorphic function f inside Γ (Kravanja, Van Barel and Haegemans 1999*b*); see also Austin, Kravanja and Trefethen (2014), who discuss connections of contour integration with (rational) interpolation and provide several MATLAB code snippets. An NEP solver based on rational interpolation and resolvent sampling is presented and applied in Xiao, Zhang, Huang and Sakurai (2016*a*) and Xiao, Zhou, Zhang and Xu (2016*b*).

The function f in (5.1) can be interpreted as a filter acting on the eigenvalues of F; see for example Van Barel and Kravanja (2016). While we have chosen monomials z^p in (5.8) (for notational simplicity) and scaled-and-shifted monomials $((z-\sigma)/\rho)^p$ in Figure 5.3 (for numerical stability), one is free to use other sets of linearly independent filter functions f which are in a linear relation to each other. The linearity is required in order to derive a factorization of the form (5.10) from which J can be inferred. Interestingly, if F is a meromorphic function with a finite number of poles in Ω, represented as $F(z) = G(z)/q(z)$ with $G \in H(\Omega, \mathbb{C}^{n\times n})$ and some polynomial q, then $F(z)^{-1} = q(z)G(z)^{-1}$ and (5.1) can be rewritten as

$$\frac{1}{2\pi \mathrm{i}} \int_\Gamma f(z)F(z)^{-1}\,\mathrm{d}z = \frac{1}{2\pi \mathrm{i}} \int_\Gamma \widetilde{f}(z)G(z)^{-1}\,\mathrm{d}z$$

with $\widetilde{f} = fq$. Therefore any contour-based method applied to the meromorphic function F with filter functions f is equivalent to applying the same method to the holomorphic function G with modified filters $\widetilde{f}$. This means that contour-based methods should be able to handle meromorphic NEPs without modification.[1] Indeed, the reader can easily test numerically that the algorithm in Figure 5.3 also works fine if the second line is replaced with the definition of the `loaded_string` problem (using lines 2–5 in Figure 4.22). This is a meromorphic (in fact, rational) NEP but the contour solver in Figure 5.3 still computes good eigenvalue approximations even when the contour Γ contains the pole $z = 1$ in its interior.

A high-performance implementation of a contour-based eigensolver is the Contour Integral Spectrum Slicing (CISS) method in SLEPc (Maeda, Sakurai and Roman 2016): http://slepc.upv.es/. CISS is based on the method of Sakurai and Sugiura (2003), and can solve generalized linear and nonlinear eigenvalue problems; see also http://zpares.cs.tsukuba.ac.jp/?page_id=56.

6. Methods based on linearization

Instead of solving an NEP with $F \in H(\Omega, \mathbb{C}^{n\times n})$ directly, we might first replace it with a 'proxy' NEP of a simpler structure. In this section we shall focus on proxies that can be obtained via polynomial or, more generally, rational approximation of F. More specifically, we will consider NEPs $R_m(\lambda)v = 0$ with

$$R_m(z) = b_0(z)D_0 + b_1(z)D_1 + \cdots + b_m(z)D_m, \tag{6.1}$$

where the $D_j \in \mathbb{C}^{n\times n}$ are constant coefficient matrices and the b_j are rational functions of type (m, m), that is, quotients of polynomials of degree at most m. Note that polynomial eigenvalue problems are special cases of (6.1) when all the functions b_j are polynomials.

It is crucial that $R_m \approx F$ in some sense, for otherwise the eigenpairs of F and R_m are not necessarily related to each other. For example, if $\Sigma \subset \Omega$ is a compact set, we may want to impose that

$$\|F - R_m\|_\Sigma := \max_{z\in\Sigma} \|F(z) - R_m(z)\|_2 \le \varepsilon,$$

because then the eigenvalues of R_m in Σ can be guaranteed to be approximations to some of the eigenvalues of F. This can be seen as follows.

[1] This fact was pointed out to us by Wolf-Jürgen Beyn in private communication.

Assume that (λ, v) with $\lambda \in \Sigma$ and $\|v\|_2 = 1$ is an eigenpair of R_m, that is, $R_m(\lambda)v = 0$. Then from

$$\|F(\lambda)v\|_2 = \|(F(\lambda) - R_m(\lambda))v\|_2 \leq \|F(\lambda) - R_m(\lambda)\|_2 \leq \varepsilon$$

we find that (λ, v) has a bounded residual for the original NEP $F(\lambda)v = 0$. Conversely, if $\mu \in \Sigma$ is *not* an eigenvalue of F, that is, $F(\mu)$ is nonsingular, then a sufficiently accurate approximant R_m is also nonsingular at μ, which can be argued as follows. Assume that $\|F(\mu) - R_m(\mu)\|_2 < \|F(\mu)^{-1}\|_2^{-1}$. Then

$$\|I - F(\mu)^{-1}R_m(\mu)\|_2 \leq \|F(\mu)^{-1}\|_2\|F(\mu) - R_m(\mu)\|_2 < 1.$$

Hence all eigenvalues of $I - F(\mu)^{-1}R_m(\mu)$ are strictly smaller in modulus than 1. As a consequence, $F(\mu)^{-1}R_m(\mu)$ and hence $R_m(\mu)$ are nonsingular. Ideally, R_m does not have any eigenvalues in Σ which are in the resolvent set of F. In this case we say that R_m is free of *spurious* eigenvalues on Σ.

In the following Section 6.1 we discuss how to obtain an effective approximant R_m via interpolation of F and how to quantify its approximation error. In Section 6.2 we explain and demonstrate some practical approaches for the construction of R_m. If the functions b_j in (6.1) satisfy a linear recurrence relation, then we can linearize R_m to obtain an equivalent generalized eigenvalue problem. This linearization technique will be the subject of Section 6.3. In Section 6.4 we review some solution techniques for the linear problem, and Section 6.5 lists related work and available software.

6.1. Polynomial and linear rational interpolation

Assume that we are given a function $F \in H(\Omega, \mathbb{C}^{n\times n})$ on a domain $\Omega \subseteq \mathbb{C}$, a compact and connected *target set* $\Sigma \subset \Omega$, and two disjoint sequences of *interpolation nodes* $(\sigma_j)_{j=0}^m \subset \Sigma$ and *poles* $(\xi_j)_{j=1}^m \subset \overline{\mathbb{C}} \setminus \Omega$. Let the associated *nodal (rational) function* be defined as

$$s_m(z) = \prod_{j=0}^{m}(z - \sigma_j) \Big/ \prod_{\substack{j=1 \\ \xi_j \neq \infty}}^{m} (z - \xi_j). \tag{6.2}$$

Further, let $\Gamma \subset \Omega$ be a contour which encloses Σ, and hence contains all nodes σ_j in its interior. Then, by the Walsh–Hermite integral formula (see *e.g.* Walsh 1935, Chapter VIII, Theorem 2),

$$R_m(z) = \frac{1}{2\pi \mathrm{i}} \int_\Gamma \left(1 - \frac{s_m(z)}{s_m(\zeta)}\right) \frac{F(\zeta)}{\zeta - z}\, \mathrm{d}\zeta$$

is the unique matrix-valued rational function of type (m, m) with preassigned poles ξ_j that interpolates F in the Hermite sense (that is, counting multiplicities) at the nodes σ_j. Note that the pole parameters ξ_j are fixed and hence we are dealing here with *linearized rational interpolants*, which in contrast to nonlinear rational interpolants with free poles have existence and uniqueness properties very similar to polynomial interpolants. We refer to Stahl (1996) for an overview. As a consequence, if F itself is a rational matrix-valued function of type (m, m) with poles $\xi_1, \dots, \xi_m$, then the interpolant R_m with these poles will be exact for any choice of sampling points σ_j away from the poles, that is, $R_m \equiv F$.

Different choices of the nodal function in (6.2) give rise to different types of interpolants. For example, if we set all poles $\xi_j = \infty$, then R_m reduces to a matrix polynomial of degree m. Two popular instances of polynomial interpolants are as follows.

(i) If all $\xi_j = \infty$ and all $\sigma_j = \sigma$ are identical, then $s_m(z) = (z - \sigma)^{m+1}$ and R_m is the degree m truncated Taylor expansion of F at $z = \sigma$,

$$R_m(z) = F(\sigma) + F'(\sigma)(z - \sigma) + \cdots + F^{(m)}(\sigma)\frac{(z - \sigma)^m}{m!}. \tag{6.3}$$

As $m \to \infty$, this Taylor series converges geometrically in concentric discs about the expansion point σ in which F is holomorphic. More precisely, let $\mathbb{D}_{\sigma,\rho} = \{z \in \mathbb{C} : |z - \sigma| < \rho\}$ with $\rho > 0$ be a disc in which F is holomorphic and bounded (by which we mean that $\|F(z)\|_2$ is bounded for all $z \in \mathbb{D}_{\sigma,\rho}$). Further, let $\Sigma = \overline{\mathbb{D}}_{\sigma,\rho_0}$ be a smaller closed disc with $0 < \rho_0 < \rho$. Then there exists a constant c depending only on F, σ, ρ, ρ_0 such that

$$\|F - R_m\|_\Sigma \le c\left(\frac{\rho_0}{\rho}\right)^m. \tag{6.4}$$

Clearly, this interpolation type is appropriate if the wanted eigenvalues of F lie inside a circular region about the target point σ.

(ii) If all $\xi_j = \infty$, $\Sigma = [-1, 1]$, and the σ_j are chosen to be Chebyshev points of the first kind, that is,

$$\sigma_j = \cos\left(\frac{j + 1/2}{m + 1}\pi\right), \quad j = 0, 1, \dots, m,$$

then $s_m(z) = T_{m+1}(z)$ is a Chebyshev polynomial defined by the recursion

$$T_0(z) = 1, \quad T_1(z) = z, \quad T_{m+1}(z) = 2zT_m(z) - T_{m-1}(z). \tag{6.5}$$

For $m \to \infty$, the polynomial interpolant R_m converges inside Bernstein ellipse regions (5.7) in which F is holomorphic. More precisely, let $\mathbb{E}_\rho$ with $\rho > 1$ be a Bernstein ellipse region in which F is holomorphic and bounded, and let $\Sigma = \overline{\mathbb{E}}_{\rho_0}$ be a smaller closed Bernstein ellipse region with $1 < \rho_0 < \rho$. Then there exists a constant c depending only on

F, ρ, ρ_0 such that

$$\|F - R_m\|_{\Sigma} \leq c\left(\frac{\rho_0}{\rho}\right)^m;$$

see Effenberger and Kressner (2012, Proposition 3.1). The same convergence rate is achieved with interpolation at Chebyshev points of the second kind, that is,

$$\sigma_j = \cos(j\pi/m), \quad j = 0, 1, \ldots, m.$$

Chebyshev interpolation is most appropriate if the wanted eigenvalues of F lie in or near the interval $[-1, 1]$. Via linear mapping it is possible to interpolate F on (complex) intervals $[a, b]$ other than $[-1, 1]$, and via parametrization $t \mapsto F(\gamma(t))$ on arbitrary smooth curves in the complex plane.

These two examples indicate that a 'good' choice of the nodes σ_j for polynomial interpolation is dictated by the target set Σ in which the wanted eigenvalues of F are located, and that the rate of convergence is dictated by the location of the singularities of F relative to Σ.

More generally, in linear rational interpolation we also have the freedom to select the pole parameters ξ_j. An informed choice of these parameters can be made by inspecting the Walsh–Hermite formula for the error

$$F(z) - R_m(z) = \frac{1}{2\pi \mathrm{i}} \int_{\Gamma} \frac{s_m(z)}{s_m(\zeta)} \frac{F(\zeta)}{\zeta - z} \, \mathrm{d}\zeta.$$

By standard estimation of integrals we have

$$\|F(z) - R_m(z)\|_2 \leq c \frac{|s_m(z)|}{\min_{\zeta \in \Gamma} |s_m(\zeta)|}, \quad \text{for all } z \in \Sigma, \tag{6.6}$$

with a constant c that only depends on F, Σ and Γ. The pair (Σ, Γ) is called a *condenser* (Bagby 1967, Gonchar 1969) and in view of (6.6) our goal must be to construct a rational function s_m which is uniformly small on Σ and large on Γ. A greedy approach to selecting parameters $\sigma_j \in \Sigma$ and $\xi_j \in \Gamma$ that achieve this goal is as follows. Start with an arbitrary $\sigma_0 \in \Sigma$, and then define the nodes $\sigma_j \in \Sigma$ and poles $\xi_j \in \Gamma$ recursively such that the following conditions are satisfied:

$$\max_{z \in \Sigma} |s_j(z)| = |s_j(\sigma_{j+1})| \quad \text{and} \quad \min_{z \in \Gamma} |s_j(z)| = |s_j(\xi_{j+1})|, \quad j = 0, 1, \ldots.$$

The resulting points are called *Leja–Bagby points* for (Σ, Γ) (Walsh 1932, Bagby 1969). One can show (see *e.g.* Levin and Saff 2006, Theorem 3.5) that there exists a number $\text{cap}(\Sigma, \Gamma) > 0$, called the *condenser capacity*, such that any sequence of rational functions (s_m) constructed with the Leja–Bagby procedure satisfies

$$\lim_{m\to\infty}\left(\frac{\max_{z\in\Sigma}|s_m(z)|}{\min_{z\in\Gamma}|s_m(z)|}\right)^{1/m} = \exp(-1/\operatorname{cap}(\Sigma,\Gamma)). \tag{6.7}$$

By the maximum modulus principle for holomorphic functions, the points σ_j lie on $\partial\Sigma$, the boundary of Σ, and Γ can be replaced by its closed exterior, say Ξ, without changing the condenser capacity, that is, $\operatorname{cap}(\Sigma,\Gamma) = \operatorname{cap}(\Sigma,\Xi)$. Combining the inequality (6.6) and (6.7), we arrive at the asymptotic convergence result

$$\limsup_{m\to\infty}\|F - R_m\|_\Sigma^{1/m} \le \exp(-1/\operatorname{cap}(\Sigma,\Xi))$$

for rational interpolation at Leja–Bagby points. The convergence is thus asymptotically exponential with a rate depending on the target set Σ and the poles on Ξ, which should stay a positive distance away from Σ.

The determination of the numerical value $\operatorname{cap}(\Sigma,\Xi)$ is difficult for general condensers (Σ,Ξ). However, in some cases there are known closed formulas derived from conformal maps of doubly connected domains; see Nehari (1975, Chapter VII). For example, if $\Xi = (-\infty,\alpha]$ is a real interval and $\Sigma = \overline{\mathbb{D}}_{\sigma,\rho}$ is the closed disc of radius $\rho > 0$ centred at $\sigma > \alpha + \rho$, then one can show that

$$\operatorname{cap}(\Sigma,\Xi) = \frac{4}{\pi}\frac{K(\kappa)}{K(\sqrt{1-\kappa^2})}, \quad \kappa = \left(\frac{\sigma-\alpha}{\rho} - \sqrt{\left(\frac{\sigma-\alpha}{\rho}\right)^2 - 1}\right)^2, \tag{6.8}$$

where

$$K(\kappa) = \int_0^1 \frac{1}{\sqrt{(1-t^2)(1-\kappa^2 t^2)}}\,\mathrm{d}t$$

is the complete elliptic integral of the first kind;[2] see Nehari (1975, pp. 293–294). For a list of some other special condensers and formulas for their capacities see Güttel (2013) and the references therein.

6.2. *Sampling approaches*

To make the interpolation process described in the previous section constructive, we need to agree on some basis functions b_j for (6.1). For practical purposes it is usually advantageous to choose basis functions that are in a linear relation with one another. Let us list some examples.

[2] The definition of $K(\kappa)$ varies in the literature. We stick to the definition used by Nehari (1975, Chapter VI). In MATLAB the value $K(\kappa)$ is obtained with `ellipke(kappa^2)`.

- Shifted and scaled monomials satisfy the linear relation

$$b_0(z) = \frac{1}{\beta_0}, \quad b_{j+1}(z) = \frac{z - \sigma}{\beta_{j+1}} b_j(z) \tag{6.9}$$

  with some nonzero scaling factors β_j. An interpolant R_m with this basis corresponds to the degree m truncated Taylor expansion of F at $z = \sigma$ given in (6.3).

- Orthogonal polynomials satisfy a three-term recurrence

$$b_0(z) = \frac{1}{\beta_0}, \quad b_{j+1}(z) = \frac{zb_j(z) + \alpha_j b_j(z) + \gamma_j b_{j-1}(z)}{\beta_{j+1}}. \tag{6.10}$$

  Both the monomials (6.9) and the Chebyshev polynomials $b_j = T_j$ in (6.5) are special cases of orthogonal polynomial sequences.

- Given nodes $\sigma_0, \sigma_1, \ldots, \sigma_m$, the (scaled) Newton polynomials are defined by the linear relation

$$b_0(z) = \frac{1}{\beta_0}, \quad b_{j+1}(z) = \frac{z - \sigma_j}{\beta_{j+1}} b_j(z). \tag{6.11}$$

- Given distinct nodes $\sigma_0, \sigma_1, \ldots, \sigma_m$, the (scaled) Lagrange polynomials

$$b_j(z) = \frac{1}{\beta_j} \prod_{\substack{i=0 \\ i \neq j}}^{m} (z - \sigma_i) \tag{6.12}$$

  satisfy the linear relation

$$\beta_j(z - \sigma_j) b_j(z) = \beta_k(z - \sigma_k) b_k(z).$$

  In contrast to the above listed polynomials b_j whose degrees coincide with their indices j, the Lagrange polynomials are not degree-graded.

From an approximation point of view it does not matter in which basis an interpolant $R_m \approx F$ is represented, and indeed the results in Section 6.1 are independent of the particular representation of R_m. For practical computations, however, the choice of basis functions b_j is important. In this section we will focus on degree-graded *rational Newton basis functions*, which give rise to rational interpolants that can be constructed in a greedy fashion, that is, one (distinct) interpolation node at a time, and by only knowing the values of F at these nodes.

Given sequences of interpolation nodes $(\sigma_j)_{j=0}^m \in \Sigma$ and nonzero[3] poles $(\xi_j)_{j=1}^m \subset \overline{\mathbb{C}} \setminus \Sigma$, we define rational basis functions by the recursion

$$b_0(z) = \frac{1}{\beta_0}, \quad b_{j+1}(z) = \frac{z - \sigma_j}{\beta_{j+1}(1 - z/\xi_{j+1})} b_j(z). \tag{6.13}$$

Each rational function b_{j+1} has roots at the interpolation nodes $\sigma_0, \sigma_1, \ldots, \sigma_j$ and poles at $\xi_1, \ldots, \xi_j$. If all poles ξ_j are chosen at infinity, (6.13) reduces to the polynomial Newton recursion (6.11).

If all nodes $\sigma_0, \sigma_1, \ldots, \sigma_m$ are pairwise distinct, we can compute the coefficient matrices D_j of the interpolant (6.1) in a straightforward manner: by the interpolation condition $F(\sigma_0) = R_m(\sigma_0)$ and the formula for b_0 we have $D_0 = \beta_0 F(\sigma_0)$. From the interpolation conditions $F(\sigma_j) = R_m(\sigma_j)$ we then find recursively

$$D_j = \frac{F(\sigma_j) - b_0(\sigma_j)D_0 - \cdots - b_{j-1}(\sigma_j)D_{j-1}}{b_j(\sigma_j)}, \quad j = 1, \ldots, m. \tag{6.14}$$

The matrix-valued numerator of D_j can be evaluated via the Horner scheme starting with the coefficient matrix D_{j-1}, using the fact that each b_{j-1} divides b_j. Computing the matrices D_j this way is mathematically equivalent to computing the diagonal entries of a divided-difference tableau with matrix entries; see also Güttel, Van Beeumen, Meerbergen and Michiels (2014).

In the confluent case, where some of the interpolation nodes coincide, derivatives of F will enter the formulas. If F is given in the form

$$F(z) = f_1(z)C_1 + f_2(z)C_2 + \cdots + f_\ell(z)C_\ell \tag{6.15}$$

with constant coefficient matrices $C_j \in \mathbb{C}^{n \times n}$, we can compute

$$F^{(k)}(z) = f_1^{(k)}(z)C_1 + f_2^{(k)}(z)C_2 + \cdots + f_\ell^{(k)}(z)C_\ell$$

simply by calculating derivatives of scalar functions and use a straightforwardly modified version of the sampling formula (6.14). There is, however, a more convenient approach to computing the interpolant R_m which does not require any sampling at all. To explain this, let us consider again the basis recursion (6.13) and write it in linearized form as

$$\beta_{j+1}(1 - z/\xi_{j+1})b_{j+1}(z) = (z - \sigma_j)b_j(z), \quad j = 0, 1, \ldots, m-1.$$

[3] The exclusion of poles at the origin is merely for ease of notation and can easily be remedied by replacing all ξ_j by $\xi_j + \tau$ and $F(z)$ by $F(z + \tau)$. Alternatively, one can use the more general recursion

$$b_{j+1}(z) = \frac{z - \sigma_j}{\beta_{j+1}(\nu_{j+1}z - \mu_{j+1})} b_j(z)$$

with $\xi_{j+1} = \mu_{j+1}/\nu_{j+1}$ at the cost of dealing with two parameters (μ_{j+1}, ν_{j+1}) instead of ξ_{j+1} only.

With the help of a z-dependent vector $b(z) = [b_0(z), b_1(z), \ldots, b_m(z)]^T$, we can combine these relations in matrix form as

$$zb^T(z)\underline{K_m} = b^T(z)\underline{H_m}, \tag{6.16}$$

where

$$\underline{K_m} = \left[\begin{array}{cccc} 1 & & & \\ \frac{\beta_1}{\xi_1} & 1 & & \\ & \ddots & \ddots & \\ & & \frac{\beta_{m-1}}{\xi_{m-1}} & 1 \\ \hline & & & \frac{\beta_m}{\xi_m} \end{array}\right], \quad \underline{H_m} = \left[\begin{array}{cccc} \sigma_0 & & & \\ \beta_1 & \sigma_1 & & \\ & \ddots & \ddots & \\ & & \beta_{m-1} & \sigma_{m-1} \\ \hline & & & \beta_m \end{array}\right] \tag{6.17}$$

are $(m+1) \times m$-matrices. The underscores in our notation of $\underline{K_m}$ and $\underline{H_m}$ indicate the additional rows below the horizontal line, and we will denote by (H_m, K_m) the $m \times m$ matrix pencil obtained by omitting this row.

Linear relations of the form (6.16) also exist for Taylor, orthogonal polynomial and Lagrange basis functions. It is not difficult to construct the corresponding matrices $\underline{K_m}$ and $\underline{H_m}$ using the basis recursions (6.9)–(6.12).

The decomposition (6.16) is called *rational Krylov decomposition*, and there is a close connection between matrix functions associated with the pencil (H_m, K_m) and rational divided differences. This connection was first observed by Opitz (1964) for the case of polynomial interpolation, and later extended to the rational case; see for example Güttel (2013, Section 3.4). The following theorem summarizes this approach; *cf.* Güttel *et al.* (2014, Theorem 2.1).

Theorem 6.1. Given a matrix-valued function F in the split form (6.15) and rational basis functions b_j as in (6.13), define the matrices $D_j = \sum_{i=1}^{\ell} d_{i,j} C_i$ for $j = 0, 1, \ldots, m$, where

$$\begin{bmatrix} d_{i,0} \\ d_{i,1} \\ \vdots \\ d_{i,m} \end{bmatrix} = \beta_0 f_i(H_{m+1} K_{m+1}^{-1}) e_1, \quad i = 1, 2, \ldots, \ell,$$

with K_{m+1} and H_{m+1} as in (6.17) (with m replaced by $m+1$ and last row removed), and $e_1 = [1, 0, \ldots, 0]^T \in \mathbb{R}^{m+1}$. Then

$$R_m(z) = b_0(z) D_0 + b_1(z) D_1 + \cdots + b_m(z) D_m$$

is the rational matrix-valued function of type (m, m) with poles at the points $\xi_1, \ldots, \xi_m$ that interpolates F in the Hermite sense at the nodes $\sigma_0, \sigma_1, \ldots, \sigma_m$.

The following two examples compare the performance of polynomial and rational interpolation, with the first example also demonstrating that if F itself is a rational matrix-valued function of type (m, m) with poles $\xi_1, \ldots, \xi_m$, then its rational interpolant R_m with these poles will be exact.

Example 6.2. Let us reconsider the `loaded_string` NEP (1.3) of size $n = 100$. Recall that this is in fact a rational NEP of type $(2, 1)$ which could be transformed into a polynomial one via multiplication by $z - 1$; however, this would introduce a spurious eigenvalue $\lambda = 1$ of multiplicity $n-1$, and we prefer not to do this here. We are interested in the eigenvalues of F located in the interval $[4, 296]$ and use the sampling routine implemented in the MATLAB Rational Krylov Toolbox (RKToolbox) by Berljafa and Güttel (2014). A code example is shown in Figure 6.3. The function `util_nleigs` returns a so-called RKFUN object `Rm` which represents the rational interpolant R_m. This object can be evaluated at any point $z \in \mathbb{C}$ by typing `Rm(z)`, and its poles are listed with the command `poles(Rm)`; see Berljafa and Güttel (2015) for implementation details.

The uniform interpolation errors

$$\|F - R_m\|_{[4,296]} = \max_{z\in[4,296]} \|F(z) - R_m(z)\|_2,$$

resulting from NLEIGS sampling with three alternatives for the sampling nodes σ_j and the poles ξ_j, are plotted as functions of the degree m in Figure 6.5(a) (see page 75), together with the predicted convergence factors shown as dashed lines. Figure 6.5(c) shows the eigenvalues of the different interpolants R_m.

```
% NLEIGS_sampling
n = 100; C1 = n*gallery('tridiag',n); C1(end) = C1(end)/2;
C2 = (abs(gallery('tridiag',n)) + 2*speye(n))/(6*n);
C2(end) = C2(end)/2; C3 = sparse(n,n); C3(n,n) = 1;
F = @(z) C1 - z*C2 + C3*z/(z-1); tol = 0; mmax = 50;
Sigma = 150+146*exp(2i*pi*(0:99)/100); Xi = inf; % Leja on circle
Sigma = [4,296]; Xi = inf;                        % Leja on interval
Sigma = [4,296]; Xi = [1,inf]; mmax = 2;          % Leja-Bagby
Rm = util_nleigs(F, Sigma, Xi, tol, mmax);        % NLEIGS sampling
Lm = linearize(Rm); [Am,Bm] = Lm.get_matrices(); % linearization
```

Figure 6.3. Basic MATLAB calls of the Leja–Bagby sampling routine implemented in the RKToolbox. The first four lines define the `loaded_string` problem. The three lines starting with '`Sigma = ...`' correspond to the three different choices of Leja–Bagby points discussed in Example 6.2. The interpolant R_m computed by `util_nleigs` is represented by an object `Rm`, and can be evaluated at any $z \in \mathbb{C}$ by typing `Rm(z)`. The last line generates a matrix pencil $\boldsymbol{L}_m(z) = \boldsymbol{A}_m - z\boldsymbol{B}_m$ which corresponds to a linearization of R_m. This will be discussed in Section 6.3.

(i) Choosing the sampling points σ_j on a circle of radius $\rho_0 = 146$ centred about the point $z = 150$, while keeping all $\xi_j = \infty$, results in a sequence of Leja interpolants (the polynomial special case of Leja–Bagby interpolants) whose errors over $[4, 296]$ tend to reduce by a factor $\rho_0/\rho = 146/149$ as the degree m increases; see the grey curve with pluses and the dashed line in Figure 6.5(a). Here, $\rho = 149$ is the radius of the largest possible disc centred at $z = 150$ in which F is holomorphic. Recall from (6.4) that this is exactly the same convergence factor that would be achieved by truncated Taylor expansions of F about $z = 150$; however, the 'sweeping out' (or, in the language of potential theory, *balayage*) of the interpolation nodes to the boundary of a disc centred at $z = 150$ allows for derivative-free sampling at the same convergence rate.

Figure 6.5(c) shows some of the eigenvalues of the matrix polynomial R_{50} as grey pluses. Here are the first five eigenvalues of R_{50} whose imaginary part is smallest in modulus, listed in order of increasing real part:

$$
\begin{array}{r}
\texttt{\underline{4}.5086126318959 + \underline{0.00}78529409700i}\\
\texttt{\underline{123.0}253244697823 + \underline{0.00}61650205038i}\\
\texttt{\underline{202}.1974982022052 + \underline{0.00}19912027330i}\\
\texttt{301.3093923260613 - 0.0166126006570i}\\
\texttt{356.3668306213224 - 0.0114103053079i}
\end{array}
$$

The first three of these eigenvalues are inside the convergence disc, and seem to approach some of the exact eigenvalues of F given in (3.5). We have underlined the correct digits of these three eigenvalues of R_{50} as approximations to their closest counterparts of F. The other eigenvalues of R_{50}, however, cannot be trusted, as they are not inside the convergence disc. Interestingly, many of these 'spurious' eigenvalues tend to align on the boundary of that disc (see again the grey pluses in Figure 6.5(c)). This phenomenon is observed frequently in the literature; see for example Jarlebring and Güttel (2014, Section 4.3). While there appears to be no analysis of this effect, it may plausibly be related to a similar behaviour of roots of scalar truncated Taylor expansions. In particular, Jentzsch (1916) showed that every point of the circle of convergence for a scalar power series is an accumulation point for the roots of its partial sums. More general results of this type can be made for polynomial best approximants on compact sets; see for example Blatt, Saff and Simkani (1988). With $\mathrm{Taylor}_{\sigma,m}[\,\cdot\,]$ being the operator that returns the degree m truncated Taylor expansion about σ of its argument, it is easy to verify that

$$\det(\mathrm{Taylor}_{\sigma,m}[F(z)]) = \mathrm{Taylor}_{\sigma,m}[\det F(z)] + O((z-\sigma)^{m+1}).$$

This may be a possible starting point for an analysis of the limiting behaviour of spurious eigenvalues in NEP linearizations.

(ii) Choosing the sampling points σ_j on the interval $[4, 296]$, while keeping all $\xi_j = \infty$, results in a sequence of Leja interpolants whose errors over $[4, 296]$ tend to reduce by a factor $\rho_0^{-1} = 0.817$ as the degree m increases. Here, ρ_0 is the elliptical radius of the largest Bernstein ellipse region associated with $[4, 296]$ in which F is holomorphic. This is exactly the same convergence factor that would be obtained with interpolation at Chebyshev points on the interval Σ; however, the greedy choice of Leja–Bagby nodes allows for the degree m to be adaptively increased while keeping previously chosen nodes σ_j fixed. Computationally this is often advantageous, as the degree m required for a desired sampling accuracy is typically unknown in advance.

The small blue dots in Figure 6.5(c) correspond to the eigenvalues of the matrix polynomial R_{50}. Among them are some very good approximations to the eigenvalues of F in the interval $[4, 296]$ (correct digits are underlined):

```
  4.4821710017932 + 0.000000000000000i
 24.2235759507902 + 0.000000000000000i
 63.7238218842619 + 0.000000000000000i
123.0312214021378 + 0.000000000000000i
202.2009012596192 + 0.000000000000000i
```

Again some of the spurious eigenvalues align on the boundary of the convergence region, which in this case is a Bernstein ellipse.

(iii) Finally, we choose the 'exact' poles $\xi_1 = 1$, $\xi_2 = \infty$ of F, and arbitrary sampling points away from the singularity $z = 1$. As expected, the resulting type $(2, 1)$ rational interpolant R_2 coincides exactly with F, which is indicated by the sharp drop in the interpolation error; see the solid red curve with circles in Figure 6.5(a). Note that, in contrast to the previous two polynomial interpolants, $R_2 \equiv F$ is exact not only in the sampling region, but throughout the complex plane. As a consequence, R_2 also discovers the eigenvalue `0.45731848895` of F; see the red circles in Figure 6.5(c).

Example 6.4 (gun). We consider the NEP (1.4). The complex square root $\sqrt{\cdot}$ corresponds to the principal branch, and the parameters are chosen as $\omega_1 = 0$ and $\omega_2 = 108.8774$. The target set Σ is the closed disc $\overline{\mathbb{D}}_{\sigma,\rho_0}$ of radius $\rho_0 = 5 \times 10^4$ centred at $\sigma = 250^2$. There is numerical evidence that F has 21 eigenvalues in Σ. The singularity set $\Xi = (-\infty, \omega_2^2]$ corresponds to the union of branch cuts of the square roots in (1.4).

For this example we try two different sets of Leja–Bagby points and plot the convergence of the resulting interpolants together with their predicted convergence factors (plotted as dashed lines) in Figure 6.5(b), with the eigenvalues of the interpolants shown in Figure 6.5(d).

(i) Choosing Leja points σ_j on the boundary of Σ while keeping all $\xi_j = \infty$ results in a sequence of polynomial interpolants R_m whose errors over Σ reduce with a factor $\rho_0/\rho \approx 0.987$ as m increases; see the grey curve with pluses and the dashed line in Figure 6.5(b). Here, $\rho = \sigma - \omega_2^2$ is the radius of the largest possible disc centred at σ so that F is holomorphic in its interior. This is the same convergence factor that would be obtained with a truncated Taylor expansion of F about σ, but choosing the nodes σ_j on the boundary of Σ avoids the need for derivatives of F.

In Figure 6.5(d) we show the eigenvalues λ of R_{100} with a relative residual $\|F(\lambda)v\|_2/\|v\|_2$ below 10^{-3} as grey pluses. Inside the target set Σ there are 13 such eigenvalues.

(ii) We now choose Leja–Bagby nodes on the condenser (Σ, Ξ), resulting in a rational interpolant with a geometric convergence factor

$$\exp(-1/\operatorname{cap}(\Sigma, \Xi)) \approx 0.465,$$

where $\operatorname{cap}(\Sigma, \Xi)$ is given by (6.8). This convergence is significantly faster than for the polynomial interpolant. Some of the poles of R_{38} are shown as magenta dots in Figure 6.5(d), and some of the eigenvalues of R_{38} are shown as red circles. All 21 eigenvalues of F in Σ are very well approximated by eigenvalues of R_{38}. Moreover, R_{38} appears to be free of spurious eigenvalues on Σ, and there are even some eigenvalues outside Σ which have a small relative residual $\|F(\lambda)v\|_2/\|v\|_2$.

Looking at the examples in Figure 6.5 we conclude that rational interpolation can significantly outperform polynomial interpolation. This is in particular the case when F has singularities near the target set Σ.

It remains to discuss the choice of scaling parameters β_j for the rational functions b_j defined in (6.13). Perhaps the most natural approach, also advocated in Güttel *et al.* (2014, Section 5.1), is to select each β_j such that $\|b_j\|_\Sigma = 1$. In this case we have

$$\|F - R_m\|_\Sigma \leq \|D_{m+1}\|_2 + \|D_{m+2}\|_2 + \cdots,$$

so if the norms of the coefficient matrices D_j decay quickly we can use $\|D_{m+1}\|_2$ (or $\|D_{m+1}\|_F$) as a cheap upper estimate for the interpolation error $\|F - R_m\|_\Sigma$. (As will be seen in the following section, this scaling also leads to well-balanced block entries in the eigenvectors of the linearization pencil $\boldsymbol{L}_m(z)$ of $R_m(z)$.) In practice, the normalization of each b_j can be achieved by setting $\beta_j = 1/\max_{\partial\Sigma_p} |\widehat{b}_j(z)|$, where $\partial\Sigma_p$ is a fine discretization of the boundary of Σ with p points (typically $p = 1000$ is enough) and $\widehat{b}_j$ is

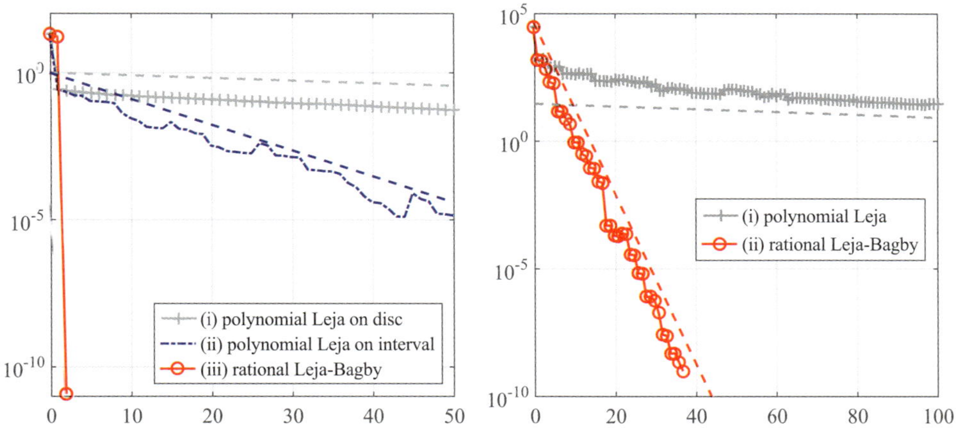

(a) loaded_string: approximation error

(b) gun: approximation error

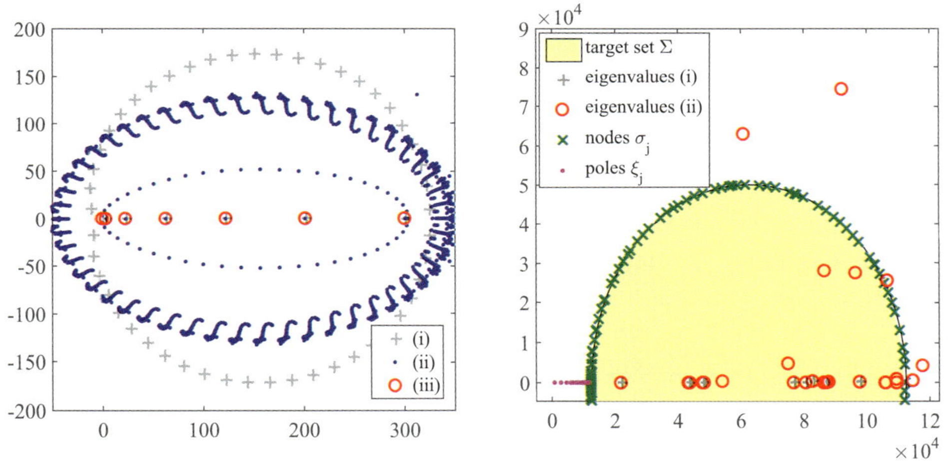

(c) loaded_string: eigenvalues

(d) gun: eigenvalues

Figure 6.5. (a, b) Approximation errors $\|F - R_m\|_\Sigma$ of Leja–Bagby interpolants R_m for different choices of Leja–Bagby points as the degree $m \geq 0$ increases, together with the predicted error decay rates (dashed lines). (a) Sampling of the loaded_string problem as discussed in Example 6.2 with $\Sigma = [4, 296]$. (b) Polynomial versus rational sampling of the gun problem as discussed in Example 6.4 with Σ chosen to be the closed disc of radius 5×10^4 centred at 6.25×10^4. (c, d) Eigenvalues of the different Leja–Bagby interpolants R_m for both the loaded_string problem (c) and the gun problem (d). For the gun problem in the case of rational Leja–Bagby interpolation, we also show the interpolation nodes σ_j and poles ξ_j. Part of the lower half of the target set Σ is outside the visible region, but this part is free of eigenvalues.

the basis function prior to scaling. As this normalization procedure only involves evaluations of scalar functions, its computational cost is typically negligible.

While we have focused in this section on rational Newton interpolation for F due to its greedy approach, the approximation R_m in (6.1) via Lagrange interpolation is easy to construct since $D_j = F(\sigma_j)$. For Chebyshev interpolation, Effenberger and Kressner (2012) propose using a sequence of inverse discrete cosine transforms whose type depends on the choice of interpolation nodes.

6.3. Linearization

The polynomial or rational eigenvalue problem $R_m(\lambda)v = 0$ with R_m of the form (6.1) can easily be converted into a linear eigenvalue problem when the basis functions b_j are in a linear relation with one another, that is, when there exist matrices $\underline{K_m}, \underline{H_m} \in \mathbb{C}^{(m+1)\times m}$ satisfying a relation of the form (6.16). Using the linear relation between b_m and b_{m-1}, $R_m(z)$ is rewritten in the form

$$g(z)R_m(z) = \sum_{j=0}^{m-1} b_j(z)(A_j - zB_j) \tag{6.18}$$

for some function $g(z)$. For example, using the Newton-type basis functions defined in (6.13), we get

$$R_m(z) = b_0(z)D_0 + \cdots + b_{m-1}(z)D_{m-1} + \underbrace{\frac{z - \sigma_{m-1}}{\beta_m(1 - z/\xi_m)} b_{m-1}(z)}_{b_m(z)} D_m,$$

and by multiplying the above equation by $g(z) = (1 - z/\xi_m)$, (6.18) is obtained with

$$\begin{aligned} A_j &= D_j, \quad B_j = D_j/\xi_m, \quad j = 0, 1, \ldots, m-2, \\ A_{m-1} &= D_{m-1} - \frac{\sigma_{m-1}}{\beta_m} D_m, \quad B_{m-1} = \frac{D_{m-1}}{\xi_m} - \frac{D_m}{\beta_m}. \end{aligned}$$

If R_m is represented in the Taylor, orthogonal polynomial or Lagrange basis, then R_m in (6.1) is a matrix polynomial and $g(z) = 1$ in (6.18). For completeness, the conversion rules relating the coefficient matrices D_j and the (A_j, B_j) used in (6.18) are provided in Table 6.6; see also Van Beeumen, Meerbergen and Michiels (2015*a*, Table 1).

Now (6.18) and the first m equations in (6.16) can be rewritten as

$$\boldsymbol{L}_m(z)(b(z) \otimes I_n) = (g(z)e_1 \otimes I_n)R_m(z), \tag{6.19}$$

Table 6.6. Conversion of R_m in the form (6.1) into the form (6.18). The ranges of the variable j indicated in the column for B_j are the same for A_j along each row.

Basis b_j	A_j	B_j
Taylor (6.9)	$\begin{cases} D_j \\ D_{m-1} - \dfrac{\sigma}{\beta_m} D_m \end{cases}$	$\begin{cases} O, & j < m-1 \\ -\dfrac{D_m}{\beta_m}, & j = m-1 \end{cases}$
Orthogonal (6.10)	$\begin{cases} D_j \\ D_{m-2} + \dfrac{\gamma_{m-1}}{\beta_m} D_m \\ D_{m-1} + \dfrac{\alpha_{m-1}}{\beta_m} D_m \end{cases}$	$\begin{cases} O, & j < m-2 \\ O, & j = m-2 \\ -\dfrac{D_m}{\beta_m}, & j = m-1 \end{cases}$
Lagrange (6.12)	$\begin{cases} \sigma_m D_j \\ \sigma_m D_{m-1} + \dfrac{\beta_{m-1}\sigma_m}{\beta_m} D_m \end{cases}$	$\begin{cases} D_j, & j < m-1 \\ D_{m-1} + \dfrac{\beta_{m-1}}{\beta_m} D_m, & j = m-1 \end{cases}$
RatNewton (6.13)	$\begin{cases} D_j \\ D_{m-1} - \dfrac{\sigma_{m-1}}{\beta_m} D_m \end{cases}$	$\begin{cases} O, & j < m-1 \\ -\dfrac{D_m}{\beta_m}, & j = m-1 \end{cases}$

where $\boldsymbol{L}_m(z) = \boldsymbol{A}_m - z\boldsymbol{B}_m$ is an $mn \times nm$ pencil with

$$\boldsymbol{A}_m = \left[\begin{array}{cccc} A_0 & A_1 & \cdots & A_{m-1} \\ \hline \multicolumn{4}{c}{\underline{H_{m-1}}^T \otimes I_n} \end{array}\right], \quad \boldsymbol{B}_m = \left[\begin{array}{cccc} B_0 & B_1 & \cdots & B_{m-1} \\ \hline \multicolumn{4}{c}{\underline{K_{m-1}}^T \otimes I_n} \end{array}\right] \tag{6.20}$$

and

$$b(z) = \begin{bmatrix} b_0(z) & b_1(z) & \cdots & b_{m-1}(z) \end{bmatrix}^T.$$

Using a block (permuted) UL decomposition of $\boldsymbol{L}_m(z_0)$ for $z_0 \in \mathbb{C}$, Van Beeumen *et al.* (2015*a*, Corollary 2.4) show that if $\underline{H_{m-1}}^T - z\underline{K_{m-1}}^T$ is of rank $m-1$ for all z, then R_m regular implies that $\boldsymbol{L}_m$ is regular. While the spectra of R_m and $\boldsymbol{L}_m$ are not necessarily identical when R_m is a rational matrix-valued function, Grammont, Higham and Tisseur (2011, Theorem 3.1) show that the one-sided factorization in (6.19) implies useful relations between the eigensystem of R_m and that of $\boldsymbol{L}_m$.

Theorem 6.7. Let $R_m(\lambda)$ and $\boldsymbol{L}_m(z)$ be matrix functions of dimensions $n \times n$ and $mn \times mn$, respectively. Assume that (6.19) holds at $z = \lambda \in \mathbb{C}$ with $g(\lambda) \neq 0$ and $b(\lambda) \neq 0$. Then

(i) $b(\lambda) \otimes v$ is a right eigenvector of $\boldsymbol{L}_m$ with eigenvalue λ if and only if v is a right eigenvector of R_m with eigenvalue λ,

(ii) if $\boldsymbol{w} \in \mathbb{C}^{mn}$ is a left eigenvector of $\boldsymbol{L}_m$ with eigenvalue λ, then $(g(z)e_1 \otimes I_n)^* \boldsymbol{w}$ is a left eigenvector of R_m with eigenvalue λ provided that it is nonzero.

The pencil $\boldsymbol{A}_m - z\boldsymbol{B}_m$ with $\boldsymbol{A}_m, \boldsymbol{B}_m$ of the form (6.20) is referred to as *CORK linearization*. Such pencils have been used in Güttel *et al.* (2014) and Van Beeumen *et al.* (2015*a*). They are a generalization of the polynomial Newton-type linearizations in Van Beeumen, Meerbergen and Michiels (2013). While these linearizations are most naturally constructed by interpolatory sampling as explained in the previous section, they may also result from non-interpolatory approximation procedures. The following example illustrates this.

Example 6.8. Consider again the `loaded_string` problem (1.3),

$$F(z) = C_1 - zC_2 + \frac{z}{z-1}C_3.$$

Using the basis functions $b_0(z) = 1$, $b_1(z) = z/(1-z)$ and $b_2(z) = -z$, we can rewrite F in the form (6.1), namely

$$R_2(z) = b_0(z)D_0 + b_1(z)D_1 + b_2(z)D_2$$

with $D_0 = C_1$, $D_1 = -C_3$ and $D_2 = C_2$. By choosing the basis functions b_j, we have implicitly specified the parameters $\sigma_0 = 0$, $\sigma_1 = 1$, $\xi_1 = 1$, $\xi_2 = \infty$ and $\beta_0 = \beta_1 = \beta_2 = 1$. Note that $\sigma_1 = 1$ would be an invalid choice as a sampling point as F has a pole there. Nevertheless, F and R_2 are identical. By the rational Krylov decomposition

$$z[b_0(z), b_1(z), b_2(z)] \underbrace{\begin{bmatrix} 1 & 0 \\ 1 & 1 \\ 0 & 0 \end{bmatrix}}_{\underline{K_2}} = [b_0(z), b_1(z), b_2(z)] \underbrace{\begin{bmatrix} 0 & 0 \\ 1 & 1 \\ 0 & 1 \end{bmatrix}}_{\underline{H_2}},$$

Table 6.6, and (6.20), we arrive at the $2n \times 2n$ CORK linearization

$$\begin{aligned} \boldsymbol{L}_m(z) &= \left[\begin{array}{cc} D_0 & D_1 - D_2 \\ \hline \multicolumn{2}{c}{\underline{H_1}^T \otimes I_n} \end{array}\right] - z \left[\begin{array}{cc} O & -D_2 \\ \hline \multicolumn{2}{c}{\underline{K_1}^T \otimes I_n} \end{array}\right] \\ &= \left[\begin{array}{cc} D_0 & D_1 - D_2 \\ \hline O & I_n \end{array}\right] - z \left[\begin{array}{cc} O & -D_2 \\ \hline I_n & I_n \end{array}\right]. \end{aligned}$$

If some of the matrix pairs (A_j, B_j) in (6.18) are of low rank, one can further reduce the size of the linearization. In the following example we illustrate this so-called *trimmed linearization* approach.

Example 6.9. In the `loaded_string` problem (1.3),

$$F(z) = C_1 - zC_2 + \frac{z}{z-1}C_3,$$

the matrix $C_3 = e_n e_n^T$ is of rank 1. Using the basis functions $b_0(z) = 1$ and $b_1(z) = 1/(1-z)$, we can rewrite F in the form (6.18) as

$$R_2(z) = b_0(z)(A_0 - zB_0) + b_1(z)(A_1 - zB_1)$$

with $A_0 = C_1$, $B_0 = C_2$, $A_1 = O$ and $B_1 = C_3$. Note that $b_1(z)$ is now a rational function of subdiagonal type $(0,1)$, referred to as 'proper form' in Su and Bai (2011). The $2n \times 2n$ CORK linearization is

$$\boldsymbol{L}_2(z) = \left[\begin{array}{cc} C_1 & O \\ \hline -I_n & I_n \end{array}\right] - z \left[\begin{array}{cc} C_2 & e_n e_n^T \\ \hline O & I_n \end{array}\right].$$

Using the structure of the eigenvector $\boldsymbol{v} = [b_0(\lambda)v^T, b_1(\lambda)v^T]^T$, we find that $\boldsymbol{L}_2(\lambda)\boldsymbol{v} = \boldsymbol{0}$ is equivalent to the equations

$$\begin{aligned} C_1 b_0(\lambda)v - \lambda C_2 b_0(\lambda)v - \lambda e_n e_n^T b_1(\lambda)v &= 0, \\ -b_0(\lambda)v + b_1(\lambda)v - \lambda b_1(\lambda)v &= 0. \end{aligned}$$

The last of the two equations specifies the linear relation between $b_0(z)$ and $b_1(z)$, and it does so even if we multiply it from the left by e_n^T (as long as $e_n^T v \neq 0$). Hence we can rewrite the reduced relations as a trimmed $(n+1) \times (n+1)$ linearization $\widehat{\boldsymbol{L}}_2(\lambda)\widehat{\boldsymbol{v}} = \boldsymbol{0}$, where

$$\widehat{\boldsymbol{L}}_2(z) = \left[\begin{array}{cc} C_1 & 0 \\ \hline -e_n^T & 1 \end{array}\right] - z \left[\begin{array}{cc} C_2 & e_n \\ \hline 0^T & 1 \end{array}\right], \quad \widehat{\boldsymbol{v}} = \begin{bmatrix} b_0(z)v \\ b_1(z)e_n^T v \end{bmatrix}.$$

Trimmed linearizations appeared in the context of polynomial eigenvalue problems with singular coefficient matrices in Byers, Mehrmann and Xu (2008) and for rational eigenvalue problems in Su and Bai (2011) and Alam and Behera (2016). The approach has been extended to NEPs involving several low-rank terms in a series of works, for example Van Beeumen *et al.* (2013), Güttel *et al.* (2014), Van Beeumen, Jarlebring and Michiels (2016*a*) and Lu, Huang, Bai and Su (2015).

6.4. Solving the linearized problem

Theorem 6.7 establishes a one-to-one correspondence between right eigenpairs of R_m and structured right eigenpairs of the linear $mn \times mn$ pencil

$\boldsymbol{L}_m(z) = \boldsymbol{A}_m - z\boldsymbol{B}_m$, and all that remains is to solve this generalized eigenvalue problem. If mn is moderate, all eigenpairs of the linearization can be found via the QZ algorithm in $O((mn)^3)$ floating point operations. For many problems arising in applications, however, mn is too large for QZ, and iterative techniques for large-scale eigenproblems are required.

In principle, any available iterative algorithm for generalized eigenproblems can be used; see for example Saad (2011) and Bai *et al.* (2000) for comprehensive overviews and Hernandez, Roman, Tomas and Vidal (2009) for a survey of available software. In the context of NEPs, one of the most popular approaches to finding a few eigenpairs of the pencil $\boldsymbol{L}_m$ uses the rational Arnoldi algorithm by Ruhe (1998). Given a nonzero starting vector $\boldsymbol{v} \in \mathbb{C}^{mn}$ and a sequence of shift parameters $\tau_1, \dots, \tau_k \in \mathbb{C} \setminus \Lambda(\boldsymbol{L}_m)$, this algorithm attempts to compute an orthonormal basis $\boldsymbol{V}_{k+1} = [\boldsymbol{v}_1, \dots, \boldsymbol{v}_{k+1}] \in \mathbb{C}^{mn \times (k+1)}$ of a *rational Krylov space* defined by Algorithm 6.10.

Algorithm 6.10: Rational Arnoldi algorithm

Given $\{\boldsymbol{A}_m, \boldsymbol{B}_m\} \subset \mathbb{C}^{mn \times mn}$, $\boldsymbol{v} \in \mathbb{C}^{mm} \setminus \{\boldsymbol{0}\}$, shifts $(\tau_j)_{j=1}^{k} \subset \mathbb{C}$.
Set $\boldsymbol{v}_1 := \boldsymbol{v}/\|\boldsymbol{v}\|_2$.
for $j = 1, 2, \dots, k$ **do**
  Compute $\boldsymbol{w} := (\boldsymbol{A}_m - \tau_j \boldsymbol{B}_m)^{-1} \boldsymbol{B}_m \boldsymbol{v}_j$.
  Orthogonalize $\widehat{\boldsymbol{w}} := \boldsymbol{w} - \sum_{i=1}^{j} \mu_{i,j} \boldsymbol{v}_i$, where $\mu_{i,j} = \boldsymbol{v}_i^* \boldsymbol{w}$.
  Set $\mu_{j+1,j} = \|\widehat{\boldsymbol{w}}\|_2$ and normalize $\boldsymbol{v}_{j+1} := \widehat{\boldsymbol{w}}/\mu_{j+1,j}$.
end

If this algorithm completes without early termination, which is the generic case when all $\mu_{j+1,j} \neq 0$, the computed quantities can be combined into a rational Arnoldi decomposition

$$\boldsymbol{A}_m \boldsymbol{V}_{k+1} \underline{M_k} = \boldsymbol{B}_m \boldsymbol{V}_{k+1} \underline{N_k}, \tag{6.21}$$

where $(\underline{N_k}, \underline{M_k})$ is a $(k+1) \times k$ upper Hessenberg pencil with $\underline{M_k} = [\mu_{i,j}]$ and $\underline{N_k} = \underline{I_k} + \underline{M_k}\,\mathrm{diag}(\tau_1, \dots, \tau_k)$. The rational Arnoldi decomposition (6.21) can then be used to extract Ritz pairs $(\vartheta_i, \boldsymbol{w}_i = \boldsymbol{V}_{k+1}\underline{N_k} s_i)$, where (ϑ_i, s_i) are solutions of the generalized eigenproblem

$$N_k s_i = \vartheta_i M_k s_i, \quad s_i \neq 0,$$

involving the upper $k \times k$ part of the pencil $(\underline{N_k}, \underline{M_k})$. Typically, the Ritz pairs are expected to be good approximations to some of the eigenpairs of the linearization $\boldsymbol{L}_m$, in particular, close to the shifts τ_j. In practice, one may distribute the shift parameters inside the target region Σ, run a few rational Arnoldi iterations k, and then compute the residuals of the extracted Ritz pairs. If the residuals are not satisfactory, one can extend the rational Arnoldi decomposition to a larger value of k by continuing the rational Arnoldi iteration.

The most expensive part in Algorithm 6.10 is the solution of a shifted linear system

$$\boldsymbol{L}_m(\tau_j)\boldsymbol{w} = (\boldsymbol{A}_m - \tau_j \boldsymbol{B}_m)\boldsymbol{w} = \boldsymbol{B}_m \boldsymbol{v}_j$$

for $\boldsymbol{w}$, involving the $mn \times mn$ matrices $\boldsymbol{A}_m$ and $\boldsymbol{B}_m$. It turns out that these systems can be solved very efficiently if the underlying Kronecker structure in the lower block part of $\boldsymbol{A}_m - \tau_j \boldsymbol{B}_m$ is exploited; see Theorem 6.7. Let us illustrate this pictorially at a degree $m = 4$ linearization arising from sampling with a (rational) Newton basis. In this case $\boldsymbol{A}_4 - \tau_j \boldsymbol{B}_4$ has the block sparsity pattern shown on the left of the following equation:

$$\left[\begin{array}{cccc} \times & \times & \times & \times \\ \hline + & + & & \\ & + & + & \\ & & + & + \end{array}\right] = \left[\begin{array}{cccc} \otimes & \times & \times & \times \\ \hline & + & & \\ & & + & \\ & & & + \end{array}\right] \left[\begin{array}{cccc} I & & & \\ \hline + & I & & \\ & + & I & \\ & & + & I \end{array}\right].$$

Here, the first block row contains coefficient matrices of an NEP associated with the Newton basis (symbolized by $\times$), while all blocks below the horizontal line are multiples of the identity matrix (symbolized by $+$). Provided that the block diagonal entries of the matrix on the left are nonzero (which is guaranteed as long as τ_j does not coincide with any of the poles of R_m), we can compute a block UL decomposition as shown on the right of the equation. Hence a linear system solve reduces to a forward and backsubstitution with the matrix factors on the right-hand side of the equation. Luckily, the only matrix to be inverted is the upper left block entry in the left factor of size $n \times n$, symbolized by $\otimes$. Hence, for each distinct shift τ_j, only a single factorization of an $n \times n$ matrix is required. The exploitation of this structure in $\boldsymbol{L}_m(\tau_j)$ is crucial for the efficient solution of the linearized problem, and it has been done, for example, in Effenberger and Kressner (2012), Jarlebring, Meerbergen and Michiels (2012*b*), Van Beeumen *et al.* (2013) and Güttel *et al.* (2014).

Further considerable savings in arithmetic operations for the orthogonalization and storage of the orthonormal rational Krylov basis are possible by exploiting a *compact representation* of the basis vectors

$$\boldsymbol{V}_{k+1} = (I_m \otimes Q)\,\boldsymbol{U}, \tag{6.22}$$

where $Q \in \mathbb{C}^{n \times r}$ and $\boldsymbol{U} \in \mathbb{C}^{mr \times (k+1)}$ have orthonormal columns. The rank r is bounded by $m + k + 1$ and typically much smaller than $m(k+1)$. All operations required in the rational Arnoldi algorithm (matrix-vector products, linear system solves with $\boldsymbol{L}_m(\tau_j)$, and inner products) can be implemented efficiently using this representation. The stability of the compact representation within the two-level orthogonal Arnoldi procedure (TOAR) for quadratic eigenvalue problems is investigated in Lu, Su and Bai (2016). TOAR has been extended to (polynomial) NEPs in a number of works, for example Kressner and Roman (2014) and Van Beeumen *et al.* (2015*a*).

Finally, let us clarify that some algorithmic details are still missing in our simplistic presentation starting from Algorithm 6.10. However, most of these details appear in identical form for the Arnoldi solution of linear eigenvalue problems, including the use of reorthogonalization to avoid the loss of orthogonality in the rational Krylov basis $\boldsymbol{V}_{k+1}$, the use of inexact solves (Lehoucq and Meerbergen 1998), and Krylov–Schur restarting to further reduce storage and orthogonalization costs (Stewart 2002). For further reading we refer to the ARPACK users' guide by Lehoucq, Sorensen and Yang (1998), and to Ruhe (1998).

6.5. Related work and software

Solution approaches based on approximation or interpolation of the NEP are frequently employed in the literature, in particular by the boundary element method (BEM) community; see Kamiya, Andoh and Nogae (1993) for a review. Theoretical aspects of such approximations were investigated by Karma (1996*a*, 1996*b*).

For a given matrix polynomial or rational matrix-valued function R_m there are many (in fact infinitely many) possible linearizations (see Mackey *et al.* 2015 and references therein). Our aim in Section 6.3 was to present a unified and practical approach. For other examples of linearizations based on degree-graded polynomial bases, Bernstein, Lagrange and Hermite bases, see for example Corless (2004), Mackey, Mackey, Mehl and Mehrmann (2006*b*), Higham, Mackey, Mackey and Tisseur (2006), Amiraslani, Corless and Lancaster (2009), Van Beeumen, Michiels and Meerbergen (2015*b*), Mackey and Perović (2016) and Noferini and Pérez (2016). For constructions of linearizations of rational matrix-valued functions, we refer to Su and Bai (2011) and Alam and Behera (2016).

Van Beeumen *et al.* (2013, Section 4.5) made a connection with rational Krylov techniques for solving the linearized problem with Newton's method, using a link between the rational Arnoldi and Jacobi–Davidson algorithms pointed out by Ruhe (1998), and the interpretation of a Jacobi–Davidson iteration as a Newton update (Sleijpen and van der Vorst 1996). A connection between rational Krylov methods for (nonlinear) eigenvalue problems and a basic contour-based method is discussed in Van Beeumen, Meerbergen and Michiels (2016*b*).

A practical advancement of linearization-based methods is the so-called *infinite Arnoldi method*, which in its original form by Jarlebring *et al.* (2012*b*) uses Taylor approximation and has been applied successfully to various NEPs, for example those associated with delay differential equations; see also Jarlebring and Güttel (2014) and Jarlebring, Meerbergen and Michiels (2012*a*, 2014). The main idea of the infinite Arnoldi method is to increase the degree m of the linearization $\boldsymbol{L}_m$ dynamically with every outer rational Krylov iteration k for finding its eigenvalues, that is, $k = m$. This is possible if the starting vector $\boldsymbol{v}$ in Algorithm 6.10 has a particular structure and the interpolation nodes σ_j are chosen identically to the shifts τ_j for all j. The advantage of this approach over the first-sample-then-solve approach is that the degree m of the linearization does not need to be determined in advance. However, the restriction on the sampling points $\sigma_j = \tau_j$ may enforce the use of suboptimal interpolants for the linearization, and the accuracy of this linearization (degree m) and the convergence of the outer rational Krylov iteration for finding its eigenvalues (index k) are not necessarily synchronous.

The following MATLAB implementations of the infinite Arnoldi method are available online.

- A rank-exploiting variant is described in Van Beeumen *et al.* (2016*a*):

  https://people.kth.se/~eliasj/src/lowranknep

- A tensor version applied to a waveguide eigenvalue problem is described in Jarlebring, Mele and Runborg (2017):

  http://www.math.kth.se/~gmele/waveguide

- A bi-Lanczos method is described in Gaaf and Jarlebring (2016):

  http://www.math.kth.se/~eliasj/src/infbilanczos/

Other NEP solvers based on polynomial interpolation are as follows.

- A method using empirical interpolation to solve nonlinear Helmholtz eigenvalue problems is given in Botchev, Sleijpen and Sopaheluwakan (2009).
- The Chebyshev interpolation approach in Effenberger and Kressner (2012) focuses on problems arising from the BEM discretization of 3D elliptic PDE eigenvalue problems and comes with MATLAB code available at

http://anchp.epfl.ch/files/content/sites/anchp/files/software/chebapprox.tar.gz

- A linearization approach of Lagrange and Hermite interpolating matrix polynomials is described in Van Beeumen *et al.* (2015*b*):

  http://twr.cs.kuleuven.be/research/software/nleps/lin-lagr.php

MATLAB codes making use of rational Leja–Bagby sampling combined with a Krylov solution of the linearization include the following.

- The NLEIGS method described in Güttel *et al.* (2014), which also supports exploitation of low-rank structure in the NEP:

  http://twr.cs.kuleuven.be/research/software/nleps/nleigs.php

- The CORK method described in Van Beeumen *et al.* (2015*a*), which implements the compact representation of Krylov basis vectors (6.22), exploitation of low-rank terms, and implicit restarting:

  http://twr.cs.kuleuven.be/research/software/nleps/cork.php

- The `util_nleigs` function in the Rational Krylov Toolbox, which we demonstrated in Figure 6.3, as well as the `rat_krylov` function which implements the (parallel) rational Arnoldi algorithm described in Berljafa and Güttel (2017):

  http://rktoolbox.org/

The `NEP` module of the SLEPc package (Hernandez *et al.* 2005) provides various linearization-based solvers for NEPs: http://slepc.upv.es/. (There is also a specialized module `PEP` for polynomial eigenvalue problems.) In particular, it implements Chebyshev interpolation on an interval and the NLEIGS method using rational Leja–Bagby sampling. The eigenpairs of the resulting linearization are computed by a Krylov–Schur implementation that fully exploits the compact representation (6.22); see Campos and Roman (2016*b*) for implementation details in the case of matrix polynomials.

SLEPc also provides matrix function routines to compute rational interpolants R_m of NEPs given in split form, as suggested by Theorem 6.1.

Acknowledgements

We are grateful to Wolf-Jürgen Beyn, Nick Higham, Daniel Kressner, Volker Mehrmann and Jose E. Roman for their insightful comments and further references which significantly improved this review. We also thank Mary Aprahamian, Steven Elsworth and Jennifer Lau for a number of useful remarks. We are thankful to Arieh Iserles for his patience with handling our manuscript, and to Glennis Starling for the copy-editing.

REFERENCES[4]

L. V. Ahlfors (1953), *Complex Analysis: An Introduction to the Theory of Analytic Functions of One Complex Variable*, McGraw-Hill.

M. Al-Ammari and F. Tisseur (2012), 'Hermitian matrix polynomials with real eigenvalues of definite type I: Classification', *Linear Algebra Appl.* **436**, 3954–3973.

R. Alam and N. Behera (2016), 'Linearizations for rational matrix functions and Rosenbrock system polynomials', *SIAM J. Matrix Anal. Appl.* **37**, 354–380.

A. Amiraslani, R. M. Corless and P. Lancaster (2009), 'Linearization of matrix polynomials expressed in polynomial bases', *IMA J. Numer. Anal.* **29**, 141–157.

A. L. Andrew, K. E. Chu and P. Lancaster (1993), 'Derivatives of eigenvalues and eigenvectors of matrix functions', *SIAM J. Matrix Anal. Appl.* **14**, 903–926.

A. L. Andrew, K. E. Chu and P. Lancaster (1995), 'On the numerical solution of nonlinear eigenvalue problems', *Computing* **55**, 91–111.

P. M. Anselone and L. B. Rall (1968), 'The solution of characteristic value-vector problems by Newton's method', *Numer. Math.* **11**, 38–45.

P. Arbenz and W. Gander (1986), 'Solving nonlinear eigenvalue problems by algorithmic differentiation', *Computing* **36**, 205–215.

J. Asakura, T. Sakurai, H. Tadano, T. Ikegami and K. Kimura (2009), 'A numerical method for nonlinear eigenvalue problems using contour integral', *JSIAM Letters* **1**, 52–55.

A. P. Austin, P. Kravanja and L. N. Trefethen (2014), 'Numerical algorithms based on analytic function values at roots of unity', *SIAM J. Numer. Anal.* **52**, 1795–1821.

T. Bagby (1967), 'The modulus of a plane condenser', *J. Math. Mech.* **17**, 315–329.

T. Bagby (1969), 'On interpolation by rational functions', *Duke Math. J.* **36**, 95–104.

Z. Bai, J. W. Demmel, J. J. Dongarra, A. Ruhe and H. A. van der Vorst, eds (2000), *Templates for the Solution of Algebraic Eigenvalue Problems: A Practical Guide*, SIAM.

S. Balay, S. Abhyankar, M. F. Adams, J. Brown, P. Brune, K. Buschelman, L. Dalcin, V. Eijkhout, W. D. Gropp, D. Kaushik, M. G. Knepley, L. C. McInnes, K. Rupp, B. F. Smith, S. Zampini, H. Zhang and H. Zhang (2016), PETSc users' manual. Technical report ANL-95/11, revision 3.7, Argonne National Laboratory.

M. Berljafa and S. Güttel (2014), A Rational Krylov Toolbox for MATLAB. MIMS EPrint 2014.56, Manchester Institute for Mathematical Sciences, The University of Manchester, UK. http://rktoolbox.org/

M. Berljafa and S. Güttel (2015), 'Generalized rational Krylov decompositions with an application to rational approximation', *SIAM J. Matrix Anal. Appl.* **36**, 894–916.

[4] The URLs cited in this work were correct at the time of going to press, but the publisher and the authors make no undertaking that the citations remain live or are accurate or appropriate.

M. Berljafa and S. Güttel (2017), Parallelization of the rational Arnoldi algorithm. MIMS EPrint 2016.32, Manchester Institute for Mathematical Sciences, The University of Manchester, UK. *SIAM J. Sci. Comput.*, to appear.

T. Betcke and H. Voss (2004), 'A Jacobi–Davidson-type projection method for nonlinear eigenvalue problems', *Future Gener. Comput. Syst.* **20**, 363–372.

T. Betcke, N. J. Higham, V. Mehrmann, C. Schröder and F. Tisseur (2013), 'NLEVP: A collection of nonlinear eigenvalue problems', *ACM Trans. Math. Software* **39**, 7:1–7:28.

W.-J. Beyn (2012), 'An integral method for solving nonlinear eigenvalue problems', *Linear Algebra Appl.* **436**, 3839–3863.

W.-J. Beyn and V. Thümmler (2009), 'Continuation of invariant subspaces for parameterized quadratic eigenvalue problems', *SIAM J. Matrix Anal. Appl.* **31**, 1361–1381.

W.-J. Beyn, C. Effenberger and D. Kressner (2011), 'Continuation of eigenvalues and invariant pairs for parameterized nonlinear eigenvalue problems', *Numer. Math.* **119**, 489–516.

D. Bindel and A. Hood (2013), 'Localization theorems for nonlinear eigenvalue problems', *SIAM J. Matrix Anal. Appl.* **34**, 1728–1749.

H.-P. Blatt, E. B. Saff and M. Simkani (1988), 'Jentzsch–Szegö type theorems for the zeros of best approximants', *J. London Math. Soc.* **2**, 307–316.

M. A. Botchev, G. L. G. Sleijpen and A. Sopaheluwakan (2009), 'An SVD-approach to Jacobi–Davidson solution of nonlinear Helmholtz eigenvalue problems', *Linear Algebra Appl.* **431**, 427–440.

L. Botten, M. Craig and R. McPhedran (1983), 'Complex zeros of analytic functions', *Comput. Phys. Commun.* **29**, 245–259.

T. Bühler and M. Hein (2009), Spectral clustering based on the graph p-Laplacian. In *26th Annual International Conference on Machine Learning*, ACM, pp. 81–88.

R. Byers, V. Mehrmann and H. Xu (2008), 'Trimmed linearizations for structured matrix polynomials', *Linear Algebra Appl.* **429**, 2373–2400.

C. Campos and J. E. Roman (2016*a*), 'Parallel iterative refinement in polynomial eigenvalue problems', *Numer. Linear Algebra Appl.* **23**, 730–745.

C. Campos and J. E. Roman (2016*b*), 'Parallel Krylov solvers for the polynomial eigenvalue problem in SLEPc', *SIAM J. Sci. Comput.* **38**, S385–S411.

R. M. Corless (2004), Generalized companion matrices in the Lagrange basis. In *Proceedings EACA* (L. Gonzalez-Vega and T. Recio, eds), SIAM, pp. 317–322.

R. Courant (1920), 'Über die Eigenwerte bei den Differentialgleichungen der mathematischen Physik', *Math. Z.* **7**, 1–57.

E. R. Davidson (1975), 'The iterative calculation of a few of the lowest eigenvalues and corresponding eigenvectors of large real-symmetric matrices', *J. Comput. Phys.* **17**, 87–94.

P. J. Davis and P. Rabinowitz (2007), *Methods of Numerical Integration*, Courier Corporation.

T. A. Davis (2004), 'Algorithm 832: UMFPACK V4.3: An unsymmetric-pattern multifrontal method', *ACM Trans. Math. Software* **30**, 196–199.

L. Delves and J. Lyness (1967), 'A numerical method for locating the zeros of an analytic function', *Math. Comp.* **21**(100), 543–560.

N. H. Du, V. H. Linh, V. Mehrmann and D. D. Thuan (2013), 'Stability and robust stability of linear time-invariant delay differential-algebraic equations', *SIAM J. Matrix Anal. Appl.* **34**, 1631–1654.

R. J. Duffin (1955), 'A minimax theory for overdamped networks', *J. Rat. Mech. Anal.* **4**, 221–233.

J. W. Eaton, D. Bateman, S. Hauberg and R. Wehbring (2016), *GNU Octave Version 4.2.0 Manual: A High-Level Interactive Language for Numerical Computations.* http://www.gnu.org/software/octave/doc/interpreter

C. Effenberger (2013*a*), Robust solution methods for nonlinear eigenvalue problems. PhD thesis, EPFL, Lausanne.

C. Effenberger (2013*b*), 'Robust successive computation of eigenpairs for nonlinear eigenvalue problems', *SIAM J. Matrix Anal. Appl.* **34**, 1231–1256.

C. Effenberger and D. Kressner (2012), 'Chebyshev interpolation for nonlinear eigenvalue problems', *BIT* **52**, 933–951.

W. R. Ferng, W.-W. Lin, D. J. Pierce and C.-S. Wang (2001), 'Nonequivalence transformation of λ-matrix eigenproblems and model embedding approach to model tuning', *Numer. Linear Algebra Appl.* **8**, 53–70.

E. Fischer (1905), 'Über quadratische Formen mit reellen Koeffizienten', *Monatshefte für Mathematik und Physik* **16**, 234–249.

R. W. Freund and N. M. Nachtigal (1996), 'QMRPACK: A package of QMR algorithms', *ACM Trans. Math. Software* **22**, 46–77.

S. W. Gaaf and E. Jarlebring (2016), The infinite bi-Lanczos method for nonlinear eigenvalue problems. arXiv:1607.03454

W. Gander, M. J. Gander and F. Kwok (2014), *Scientific Computing: An Introduction using Maple and MATLAB*, Springer.

C. K. Garrett and R.-C. Li (2013), Unstructurally banded nonlinear eigenvalue solver software. http://www.csm.ornl.gov/newsite/software.html

C. K. Garrett, Z. Bai and R.-C. Li (2016), 'A nonlinear QR algorithm for banded nonlinear eigenvalue problems', *ACM Trans. Math. Software* **43**, 4:1–4:19.

I. Gohberg and L. Rodman (1981), 'Analytic matrix functions with prescribed local data', *J. Analyse Math.* **40**, 90–128.

I. Gohberg, M. Kaashoek and F. van Schagen (1993), 'On the local theory of regular analytic matrix functions', *Linear Algebra Appl.* **182**, 9–25.

I. Gohberg, P. Lancaster and L. Rodman (2009), *Matrix Polynomials*, SIAM. Unabridged republication of book first published by Academic Press in 1982.

A. A. Gonchar (1969), 'Zolotarev problems connected with rational functions', *Math. USSR Sb.* **7**, 623–635.

L. Grammont, N. J. Higham and F. Tisseur (2011), 'A framework for analyzing nonlinear eigenproblems and parametrized linear systems', *Linear Algebra Appl.* **435**, 623–640.

A. Griewank and A. Walther (2008), *Evaluating Derivatives: Principles and Techniques of Algorithmic Differentiation*, SIAM.

K. Gu, V. L. Kharitonov and J. Chen (2003), *Stability of Time-Delay Systems*, Springer Science & Business Media.

S. Güttel (2013), 'Rational Krylov approximation of matrix functions: Numerical methods and optimal pole selection', *GAMM-Mitt.* **36**, 8–31.

S. Güttel, R. Van Beeumen, K. Meerbergen and W. Michiels (2014), 'NLEIGS: A class of fully rational Krylov methods for nonlinear eigenvalue problems', *SIAM J. Sci. Comput.* **36**, A2842–A2864.

K. P. Hadeler (1967), 'Mehrparametrige und nichtlineare Eigenwertaufgaben', *Arch. Rational Mech. Anal.* **27**, 306–328.

K. P. Hadeler (1968), 'Variationsprinzipien bei nichtlinearen Eigenwertaufgaben', *Arch. Rational Mech. Anal.* **30**, 297–307.

N. Hale, N. J. Higham and L. N. Trefethen (2008), 'Computing A^α, $\log(A)$, and related matrix functions by contour integrals', *SIAM J. Numer. Anal.* **46**, 2505–2523.

V. Hernandez, J. E. Roman and V. Vidal (2005), 'SLEPc: A scalable and flexible toolkit for the solution of eigenvalue problems', *ACM Trans. Math. Software* **31**, 351–362.

V. Hernandez, J. E. Roman, A. Tomas and V. Vidal (2009), A survey of software for sparse eigenvalue problems. SLEPc Technical Report STR-6, Universidad Politécnica de Valencia.

D. J. Higham and N. J. Higham (2017), *MATLAB Guide*, third edition, SIAM.

N. J. Higham (2008), *Functions of Matrices: Theory and Computation*, SIAM.

N. J. Higham and F. Tisseur (2002), 'More on pseudospectra for polynomial eigenvalue problems and applications in control theory', *Linear Algebra Appl.* **351–352**, 435–453.

N. J. Higham, D. S. Mackey, N. Mackey and F. Tisseur (2006), 'Symmetric linearizations for matrix polynomials', *SIAM J. Matrix Anal. Appl.* **29**, 143–159.

M. E. Hochstenbach and G. L. G. Sleijpen (2003), 'Two-sided and alternating Jacobi–Davidson', *Linear Algebra Appl.* **358**, 145–172.

R. A. Horn and C. R. Johnson (1985), *Matrix Analysis*, Cambridge University Press.

R. A. Horn and C. R. Johnson (1991), *Topics in Matrix Analysis*, Cambridge University Press.

HSL (2016), A collection of Fortran codes for large-scale scientific computation. http://www.hsl.rl.ac.uk/

T.-M. Huang, W.-W. Lin and V. Mehrmann (2016), 'A Newton-type method with nonequivalence deflation for nonlinear eigenvalue problems arising in photonic crystal modeling', *SIAM J. Sci. Comput.* **38**, B191–B218.

N. I. Ioakimidis (1987), Quadrature methods for the determination of zeros of transcendental functions: A review. In *Numerical Integration: Recent Developments, Software and Applications*, Springer, pp. 61–82.

I. C. F. Ipsen (1997), 'Computing an eigenvector with inverse iteration', *SIAM Rev.* **39**, 254–291.

E. Jarlebring and S. Güttel (2014), 'A spatially adaptive iterative method for a class of nonlinear operator eigenproblems', *Electron. Trans. Numer. Anal.* **41**, 21–41.

E. Jarlebring, K. Meerbergen and W. Michiels (2012*a*), The infinite Arnoldi method and an application to time-delay systems with distributed delay. In *Time Delay Systems: Methods, Applications and New Trends* (R. Sipahi, T. Vyhlídal, S.-I. Niculescu and P. Pepe, eds), Vol. 423 of Lecture Notes in Control and Information Sciences, Springer, pp. 229–239.

E. Jarlebring, K. Meerbergen and W. Michiels (2012*b*), 'A linear eigenvalue algorithm for the nonlinear eigenvalue problem', *Numer. Math.* **122**, 169–195.

E. Jarlebring, K. Meerbergen and W. Michiels (2014), 'Computing a partial Schur factorization of nonlinear eigenvalue problems using the infinite Arnoldi method', *SIAM J. Matrix Anal. Appl.* **35**, 411–436.

E. Jarlebring, G. Mele and O. Runborg (2017), 'The waveguide eigenvalue problem and the tensor infinite Arnoldi method', *SIAM J. Sci. Comput.*, to appear.

R. Jentzsch (1916), 'Untersuchungen zur Theorie der Folgen analytischer Funktionen', *Acta Math.* **41**, 219–251.

N. Kamiya, E. Andoh and K. Nogae (1993), 'Eigenvalue analysis by the boundary element method: New developments', *Eng. Anal. Bound. Elem.* **12**, 151–162.

O. Karma (1996*a*), 'Approximation in eigenvalue problems for holomorphic Fredholm operator functions I', *Numer. Funct. Anal. Optim.* **17**, 365–387.

O. Karma (1996*b*), 'Approximation in eigenvalue problems for holomorphic Fredholm operator functions II: Convergence rate', *Numer. Funct. Anal. Optim.* **17**, 389–408.

J. P. Keener (1993), 'The Perron–Frobenius theorem and the ranking of football teams', *SIAM Rev.* **35**, 80–93.

V. B. Khazanov and V. N. Kublanovskaya (1988), 'Spectral problems for matrix pencils: Methods and algorithms II', *Sov. J. Numer. Anal. Math. Modelling* **3**, 467–485.

A. Kimeswenger, O. Steinbach and G. Unger (2014), 'Coupled finite and boundary element methods for fluid–solid interaction eigenvalue problems', *SIAM J. Numer. Anal.* **52**, 2400–2414.

V. Kozlov and V. G. Maz'ja (1999), *Differential Equations with Operator Coefficients*, Springer Monographs in Mathematics, Springer.

S. G. Krantz (1982), *Function Theory of Several Complex Variables*, Wiley.

P. Kravanja, T. Sakurai and M. Van Barel (1999*a*), 'On locating clusters of zeros of analytic functions', *BIT* **39**, 646–682.

P. Kravanja, M. Van Barel and A. Haegemans (1999*b*), On computing zeros and poles of meromorphic functions. In *Third Conference on Computational Methods and Function Theory 1997*, World Scientific, pp. 359–370.

P. Kravanja, M. Van Barel, O. Ragos, M. Vrahatis and F. Zafiropoulos (2000), 'ZEAL: A mathematical software package for computing zeros of analytic functions', *Comput. Phys. Commun.* **124**, 212–232.

D. Kressner (2009), 'A block Newton method for nonlinear eigenvalue problems', *Numer. Math.* **114**, 355–372.

D. Kressner and J. E. Roman (2014), 'Memory-efficient Arnoldi algorithms for linearizations of matrix polynomials in Chebyshev basis', *Numer. Linear Algebra Appl.* **21**, 569–588.

V. N. Kublanovskaja (1969), 'On an application of Newton's method to the determination of eigenvalues of λ-matrices', *Soviet Math. Dokl.* **10**, 1240–1241.

V. N. Kublanovskaya (1970), 'On an approach to the solution of the generalized latent value problem for λ-matrices', *SIAM J. Numer. Anal.* **7**, 532–537.

P. Lancaster (1961), 'A generalised Rayleigh quotient iteration for lambda-matrices', *Arch. Rational Mech. Anal.* **8**, 309–322.

P. Lancaster (1966), *Lambda-Matrices and Vibrating Systems*, Pergamon Press. Reprinted by Dover, 2002.

A. Leblanc and A. Lavie (2013), 'Solving acoustic nonlinear eigenvalue problems with a contour integral method', *Engrg Anal. Bound. Elem.* **37**, 162–166.

R. B. Lehoucq and K. Meerbergen (1998), 'Using generalized Cayley transformations within an inexact rational Krylov sequence method', *SIAM J. Matrix Anal. Appl.* **20**, 131–148.

R. B. Lehoucq, D. C. Sorensen and C. Yang (1998), *ARPACK Users' Guide: Solution of Large-Scale Eigenvalue Problems with Implicitly Restarted Arnoldi Methods*, SIAM.

J. Leiterer (1978), 'Local and global equivalence of meromorphic operator functions II', *Math. Nachr.* **84**, 145–170.

E. Levin and E. B. Saff (2006), Potential theoretic tools in polynomial and rational approximation. In *Harmonic Analysis and Rational Approximation* (J.-D. Fournier *et al.*, eds), Vol. 327 of Lecture Notes in Control and Information Sciences, Springer, pp. 71–94.

B.-S. Liao, Z. Bai, L.-Q. Lee and K. Ko (2010), 'Nonlinear Rayleigh–Ritz iterative method for solving large scale nonlinear eigenvalue problems', *Taiwan. J. Math.* **14**, 869–883.

D. Lu, X. Huang, Z. Bai and Y. Su (2015), 'A Padé approximate linearization algorithm for solving the quadratic eigenvalue problem with low-rank damping', *Internat. J. Numer. Methods Engrg* **103**, 840–858.

D. Lu, Y. Su and Z. Bai (2016), 'Stability analysis of the two-level orthogonal Arnoldi procedure', *SIAM J. Matrix Anal. Appl.* **37**, 195–214.

D. S. Mackey and V. Perović (2016), 'Linearizations of matrix polynomials in Bernstein bases', *Linear Algebra Appl.* **501**, 162–197.

D. S. Mackey, N. Mackey and F. Tisseur (2015), Polynomial eigenvalue problems: Theory, computation, and structure. In *Numerical Algebra, Matrix Theory, Differential-Algebraic Equations and Control Theory* (P. Benner *et al.*, eds), Springer, pp. 319–348.

D. S. Mackey, N. Mackey, C. Mehl and V. Mehrmann (2006*a*), 'Structured polynomial eigenvalue problems: Good vibrations from good linearizations', *SIAM J. Matrix Anal. Appl.* **28**, 1029–1051.

D. S. Mackey, N. Mackey, C. Mehl and V. Mehrmann (2006*b*), 'Vector spaces of linearizations for matrix polynomials', *SIAM J. Matrix Anal. Appl.* **28**, 971–1004.

Y. Maeda, T. Sakurai and J. Roman (2016), Contour integral spectrum slicing method in SLEPc. SLEPc Technical Report STR-11, Universidad Politécnica de Valencia.

V. Mehrmann and H. Voss (2004), 'Nonlinear eigenvalue problems: A challenge for modern eigenvalue methods', *GAMM-Mitt.* **27**, 121–152.

R. Mennicken and M. Möller (2003), *Non-Self-Adjoint Boundary Eigenvalue Problems*, Vol. 192 of North-Holland Mathematics Studies, Elsevier Science.

W. Michiels and N. Guglielmi (2012), 'An iterative method for computing the pseudospectral abscissa for a class of nonlinear eigenvalue problems', *SIAM J. Sci. Comput.* **34**, A2366–A2393.

W. Michiels and S.-I. Niculescu (2007), *Stability and Stabilization of Time-Delay Systems: An Eigenvalue-Based Approach*, SIAM.

W. Michiels, K. Green, T. Wagenknecht and S.-I. Niculescu (2006), 'Pseudospectra and stability radii for analytic matrix functions with application to time-delay systems', *Linear Algebra Appl.* **418**, 315–335.

MUMPS (2016), MUltifrontal Massively Parallel Solver: Users' guide. http://mumps.enseeiht.fr/.

Z. Nehari (1975), *Conformal Mapping*, Dover. Unabridged and unaltered republication of book originally published by McGraw-Hill in 1952.

A. Neumaier (1985), 'Residual inverse iteration for the nonlinear eigenvalue problem', *SIAM J. Numer. Anal.* **22**, 914–923.

V. Niendorf and H. Voss (2010), 'Detecting hyperbolic and definite matrix polynomials', *Linear Algebra Appl.* **432**, 1017–1035.

V. Noferini and J. Pérez (2016), 'Fiedler-comrade and Fiedler–Chebyshev pencils', *SIAM J. Matrix Anal. Appl.* **37**, 1600–1624.

G. Opitz (1964), 'Steigungsmatrizen', *Z. Angew. Math. Mech.* **44**, T52–T54.

G. Peters and J. H. Wilkinson (1979), 'Inverse iteration, ill-conditioned equations and Newton's method', *SIAM Rev.* **21**, 339–360.

H. Poincaré (1890), 'Sur les équations aux dérivées partielles de la physique mathématique', *Amer. J. Math.* pp. 211–294.

E. H. Rogers (1964), 'A minimax theory for overdamped systems', *Arch. Rational Mech. Anal.* **16**, 89–96.

A. Ruhe (1973), 'Algorithms for the nonlinear eigenvalue problem', *SIAM J. Numer. Anal.* **10**, 674–689.

A. Ruhe (1998), 'Rational Krylov: A practical algorithm for large sparse nonsymmetric matrix pencils', *SIAM J. Sci. Comput.* **19**, 1535–1551.

Y. Saad (2011), *Numerical Methods for Large Eigenvalue Problems*, revised edition, SIAM.

Y. Saad and M. H. Schultz (1986), 'GMRES: A generalized minimal residual algorithm for solving nonsymmetric linear systems', *SIAM J. Sci. Statist. Comput.* **7**, 856–869.

Y. Saad, A. Stathopoulos, J. Chelikowsky, K. Wu and S. Öğüt (1996), 'Solution of large eigenvalue problems in electronic structure calculations', *BIT* **36**, 563–578.

T. Sakurai and H. Sugiura (2003), 'A projection method for generalized eigenvalue problems using numerical integration', *J. Comput. Appl. Math.* **159**, 119–128.

O. Schenk and K. Gärtner (2004), 'Solving unsymmetric sparse systems of linear equations with PARDISO', *Future Gener. Comput. Syst.* **20**, 475–487. http://www.pardiso-project.org/

K. Schreiber (2008), Nonlinear eigenvalue problems: Newton-type methods and nonlinear Rayleigh functionals. PhD thesis, Technische Universität Berlin. http://www.math.tu-berlin.de/~schreibe/

H. Schwetlick and K. Schreiber (2012), 'Nonlinear Rayleigh functionals', *Linear Algebra Appl.* **436**, 3991–4016.

G. L. G. Sleijpen and H. A. van der Vorst (1996), 'A Jacobi–Davidson iteration method for linear eigenvalue problems', *SIAM J. Matrix Anal. Appl.* **17**, 401–425.

G. L. G. Sleijpen, A. G. L. Booten, D. R. Fokkema and H. A. van der Vorst (1996), 'Jacobi–Davidson type methods for generalized eigenproblems and polynomial eigenproblems', *BIT* **36**, 595–633.

S. I. Solov′ëv (2006), 'Preconditioned iterative methods for a class of nonlinear eigenvalue problems', *Linear Algebra Appl.* **415**, 210–229.

A. Spence and C. Poulton (2005), 'Photonic band structure calculations using nonlinear eigenvalue techniques', *J. Comput. Phys.* **204**, 65–81.

H. Stahl (1996), 'Convergence of rational interpolants', *Bull. Belg. Math. Soc. Simon Stevin* **3**, 11–32.

G. W. Stewart (2002), 'A Krylov–Schur algorithm for large eigenproblems', *SIAM J. Matrix Anal. Appl.* **23**, 601–614.

Y. Su and Z. Bai (2011), 'Solving rational eigenvalue problems via linearization', *SIAM J. Matrix Anal. Appl.* **32**, 201–216.

D. B. Szyld and F. Xue (2013*a*), 'Local convergence analysis of several inexact Newton-type algorithms for general nonlinear eigenvalue problems', *Numer. Math.* **123**, 333–362.

D. B. Szyld and F. Xue (2013*b*), 'Several properties of invariant pairs of nonlinear algebraic eigenvalue problems', *IMA J. Numer. Anal.* **34**, 921–954.

D. B. Szyld and F. Xue (2015), 'Local convergence of Newton-like methods for degenerate eigenvalues of nonlinear eigenproblems I: Classical algorithms', *Numer. Math.* **129**, 353–381.

D. B. Szyld and F. Xue (2016), 'Preconditioned eigensolvers for large-scale nonlinear Hermitian eigenproblems with variational characterizations I: Extreme eigenvalues', *Math. Comp.* **85**(302), 2887–2918.

F. Tisseur (2000), 'Backward error and condition of polynomial eigenvalue problems', *Linear Algebra Appl.* **309**, 339–361.

F. Tisseur and N. J. Higham (2001), 'Structured pseudospectra for polynomial eigenvalue problems, with applications', *SIAM J. Matrix Anal. Appl.* **23**, 187–208.

F. Tisseur and K. Meerbergen (2001), 'The quadratic eigenvalue problem', *SIAM Rev.* **43**, 235–286.

L. N. Trefethen and M. Embree (2005), *Spectra and Pseudospectra: The Behavior of Nonnormal Matrices and Operators*, Princeton University Press.

L. N. Trefethen and J. Weideman (2014), 'The exponentially convergent trapezoidal rule', *SIAM Rev.* **56**, 385–458.

V. Trofimov (1968), 'The root subspaces of operators that depend analytically on a parameter', *Mat. Issled* **3**, 117–125.

H. Unger (1950), 'Nichtlineare Behandlung von Eigenwertaufgaben', *Z. Angew. Math. Mech.* **30**, 281–282.

M. Van Barel (2016), 'Designing rational filter functions for solving eigenvalue problems by contour integration', *Linear Algebra Appl.* **502**, 346–365.

M. Van Barel and P. Kravanja (2016), 'Nonlinear eigenvalue problems and contour integrals', *J. Comput. Appl. Math.* **292**, 526–540.

R. Van Beeumen (2015), Rational Krylov methods for nonlinear eigenvalue problems. PhD thesis, KU Leuven.

R. Van Beeumen, E. Jarlebring and W. Michiels (2016*a*), 'A rank-exploiting infinite Arnoldi algorithm for nonlinear eigenvalue problems', *Numer. Linear Algebra Appl.* **23**, 607–628.

R. Van Beeumen, K. Meerbergen and W. Michiels (2013), 'A rational Krylov method based on Hermite interpolation for nonlinear eigenvalue problems', *SIAM J. Sci. Comput.* **35**, A327–A350.

R. Van Beeumen, K. Meerbergen and W. Michiels (2015*a*), 'Compact rational Krylov methods for nonlinear eigenvalue problems', *SIAM J. Matrix Anal. Appl.* **36**, 820–838.

R. Van Beeumen, K. Meerbergen and W. Michiels (2016*b*), Connections between contour integration and rational Krylov methods for eigenvalue problems. Technical report TW673, Department of Computer Science, KU Leuven.

R. Van Beeumen, W. Michiels and K. Meerbergen (2015*b*), 'Linearization of Lagrange and Hermite interpolating matrix polynomials', *IMA J. Numer. Anal.* **35**, 909–930.

H. A. van der Vorst (1992), 'Bi-CGSTAB: A fast and smoothly converging variant of Bi-CG for the solution of nonsymmetric linear systems', *SIAM J. Sci. Statist. Comput.* **13**, 631–644.

D. Verhees, R. Van Beeumen, K. Meerbergen, N. Guglielmi and W. Michiels (2014), 'Fast algorithms for computing the distance to instability of nonlinear eigenvalue problems, with application to time-delay systems', *Int. J. Dynam. Control* **2**, 133–142.

H. Voss (2004*a*), 'An Arnoldi method for nonlinear eigenvalue problems', *BIT* **44**, 387–401.

H. Voss (2004*b*), Eigenvibrations of a plate with elastically attached loads. In *European Congress on Computational Methods in Applied Sciences and Engineering: ECCOMAS 2004* (P. Neittaanmäki *et al.*, eds). http://www.mit.jyu.fi/eccomas2004/proceedings/proceed.html

H. Voss (2007), 'A Jacobi–Davidson method for nonlinear and nonsymmetric eigenproblems', *Comput. Struct.* **85**, 1284–1292.

H. Voss (2009), 'A minmax principle for nonlinear eigenproblems depending continuously on the eigenparameter', *Numer. Linear Algebra Appl.* **16**, 899–913.

H. Voss (2014), Nonlinear eigenvalue problems. In *Handbook of Linear Algebra* (L. Hogben, ed.), second edition, Chapman and Hall/CRC, pp. 115:1–115:24.

H. Voss and B. Werner (1982), 'A minimax principle for nonlinear eigenvalue problems with applications to nonoverdamped systems', *Math. Methods Appl. Sci.* **4**, 415–424.

J. L. Walsh (1932), 'On interpolation and approximation by rational functions with preassigned poles', *Trans. Amer. Math. Soc.* **34**, 22–74.

J. L. Walsh (1935), *Interpolation and Approximation by Rational Functions in the Complex Domain*, Vol. 20 of American Mathematical Society Colloquium Publications, AMS.

B. Werner (1970), Das Spektrum von Operatorenscharen mit verallgemeinerten Rayleighquotienten. PhD thesis, Universität Hamburg.

H. Weyl (1912), 'Das asymptotische Verteilungsgesetz der Eigenwerte linearer partieller Differentialgleichungen (mit einer Anwendung auf die Theorie der Hohlraumstrahlung)', *Math. Ann.* **71**, 441–479.

C. Wieners and J. Xin (2013), 'Boundary element approximation for Maxwell's eigenvalue problem', *Math. Methods Appl. Sci.* **36**, 2524–2539.

J. H. Wilkinson (1965), *The Algebraic Eigenvalue Problem*, Oxford University Press.

R. Wobst (1987), 'The generalized eigenvalue problem and acoustic surface wave computations', *Computing* **39**, 57–69.

J. Xiao, C. Zhang, T.-M. Huang and T. Sakurai (2016*a*), 'Solving large-scale nonlinear eigenvalue problems by rational interpolation and resolvent sampling based Rayleigh–Ritz method'. *Internat. J. Numer. Methods Engrg.* https://doi.org/10.1002/nme.5441

J. Xiao, H. Zhou, C. Zhang and C. Xu (2016*b*), 'Solving large-scale finite element nonlinear eigenvalue problems by resolvent sampling based Rayleigh–Ritz method', *Comput. Mech.* pp. 1–18.

W. H. Yang (1983), 'A method for eigenvalues of sparse λ-matrices', *Internat. J. Numer. Methods Engrg* **19**, 943–948.

S. Yokota and T. Sakurai (2013), 'A projection method for nonlinear eigenvalue problems using contour integrals', *JSIAM Letters* **5**, 41–44.

Acta Numerica (2017), pp. 95–135
doi:10.1017/S0962492917000058

Randomized algorithms in numerical linear algebra

Ravindran Kannan
Microsoft Research Labs, Bangalore, Karnataka 560001, India
E-mail: kannan@microsoft.com

Santosh Vempala
Georgia Institute of Technology, North Avenue NW, Atlanta, GA 30332, USA
E-mail: vempala@gatech.edu

This survey provides an introduction to the use of randomization in the design of fast algorithms for numerical linear algebra. These algorithms typically examine only a subset of the input to solve basic problems approximately, including matrix multiplication, regression and low-rank approximation. The survey describes the key ideas and gives complete proofs of the main results in the field. A central unifying idea is sampling the columns (or rows) of a matrix according to their squared lengths.

CONTENTS

1. Introduction

Algorithms for matrix multiplication, low-rank approximations, singular value decomposition, dimensionality reduction and other compressed representations of matrices, linear regression, *etc.*, are widely used. For modern data sets, these computations take too much time and space to perform on the entire input matrix. Instead one can pick a random subset of columns (or rows) of the input matrix. If s (for sample size) is the number of columns we are willing to work with, we execute s statistically independent identical trials, each selecting a column of the matrix. Sampling uniformly at random (u.a.r.) is not always good, for example when only a few columns are significant.

Using sampling probabilities proportional to squared lengths of columns (henceforth called 'length-squared sampling') leads to many provable error bounds. If the input matrix A has n columns, we define[1]

$$p_j = \frac{|A(:,j)|^2}{\|A\|_F^2}, \quad \text{for } j = 1, 2, \ldots, n,$$

and in each trial, pick a random $X \in \{1, 2, \ldots, n\}$, with $\Pr(X = j) = p_j$.

We will prove error bounds of the form $\varepsilon\|A\|_F^2$, provided s grows as a function of $1/\varepsilon$ (s is independent of the size of the matrix) for *all* matrices. So, the guarantees are *worst-case* bounds rather than average-case bounds. They are most useful when $\|A\|_2^2/\|A\|_F^2$ is not too small, as is indeed the case for the important topic of principal component analysis (PCA). The algorithms are randomized (*i.e.*, they use a random number generator) and hence errors are random variables. We bound the expectations or tail probabilities of the errors. In this paper, we strike a compromise between readability and comprehensive coverage by presenting full proofs of conceptually central theorems and stating stronger results without proofs.

1.1. Overview of the paper

The first problem we consider is computing the matrix product AA^T. Given as input an $m \times n$ matrix A, we select a (random) subset of s columns of A (in s independent identical trials). We then scale the selected columns and form an $m \times s$ matrix C. We wish to satisfy two conditions: (i) (each entry of) CC^T is an unbiased estimator of (the corresponding entry of) AA^T, and (ii) the sum of the variances of all entries of CC^T (which we refer to as

[1] All matrices in this paper have real entries. We use the usual norms for matrices: the Frobenius norm ($\|\cdot\|_F$) and the spectral norm ($\|\cdot\|_2$). We also use standard MATLAB colon notation, *i.e.*, $A(:,j)$ is the jth column of A; see Golub and Van Loan (1996, Section 1.1.8).

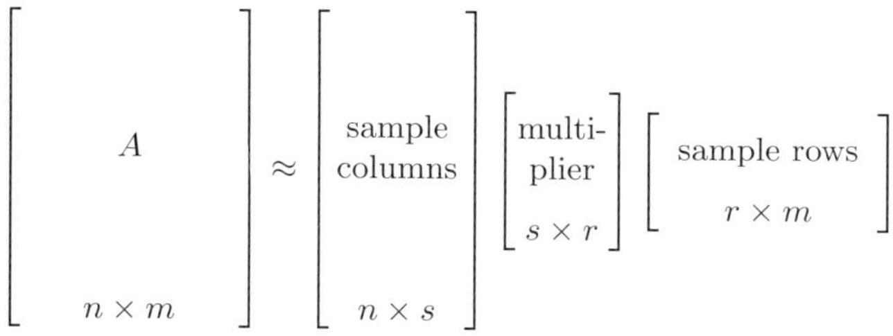

Figure 1.1. Approximating A by a sample of s columns and r rows.

the 'variance') is at most $\varepsilon^2\|A\|_F^4$. Formally,

$$E(CC^T) = AA^T, \quad E(\|AA^T - CC^T\|_F^2) \leq \varepsilon^2\|A\|_F^4. \tag{1.1}$$

Note that (1.1) implies $E(\|AA^T - CC^T\|_F) \leq \varepsilon\|A\|_F^2$ by Jensen's inequality.

The starting point of sampling-based matrix algorithms was the discovery of length-squared sampling by Frieze, Kannan and Vempala (1998, 2004), motivated by low-rank approximation. We start with two properties of length-squared sampling, which will be proved in Theorem 2.1.

- Length-squared sampling minimizes variance among all unbiased estimators.
- A sample of size $s = 1/\varepsilon^2$ suffices to ensure (1.1). In particular, s is independent of m and n.

Length-squared sampling achieves similar error bounds for general matrix multiplication, matrix sketches, low-rank approximation, *etc.* The main result in matrix sketching is as follows. For any $m \times n$ input matrix A, form an $m \times s$ sub-matrix C of the columns of A and an $r \times n$ sub-matrix R of the rows of A, both picked using length-squared sampling. We can then compute an $s \times r$ matrix U so that

$$E(\|A - CUR\|_2^2) \leq \varepsilon\|A\|_F^2$$

provided $r \geq c/\varepsilon^2, s \geq c/\varepsilon^3$. This is proved in Theorem 2.5. A schematic diagram of C, U and R is shown in Figure 1.1.

Section 2.3 shows a result based purely on matrix perturbation theory (no probabilities involved) which in words states: If C and A are any two matrices with $CC^T \approx AA^T$ (C, A may be of different dimensions), then the restriction of A to the space spanned by the top k singular vectors of C is a good approximation to A. Used in conjunction with (1.1), this reduces the problem of low-rank approximation of A to computing the singular value decomposition (SVD) of a submatrix C.

In Section 3 we apply length-squared sampling to tensors (higher-dimensional arrays) to obtain good low-rank tensor approximation. Unlike matrices, finding the best low-rank approximation for tensors is NP-hard.

Low-rank tensor approximation has several applications, and we show that there is a natural class of tensors (which includes dense hypergraphs and generalizations of metrics) for which, after scaling, $\|A\|_2 = \Omega(\|A\|_F)$. Low-rank approximation based on length-squared sampling yields good results.

In Sections 4, 5 and 6 we turn to more refined error bounds for matrices. Can we improve the right-hand side of (1.1) if we want to bound only the spectral norm instead of the Frobenius norm? Using deep results from functional analysis, Rudelson and Vershynin (2007) showed that if we use length-squared sampling,

$$\|CC^T - AA^T\|_2 \leq \varepsilon\|A\|_2\|A\|_F \text{ holds with high probability}$$
$$\text{provided } s \geq \frac{c\ln(1/\varepsilon)}{\varepsilon^2}. \tag{1.2}$$

While the change from Frobenius to spectral norm was first considered for its mathematical appeal, it is also suitable for the Hoeffding–Chernoff-type inequality for matrix-valued random variables, proved by Ahlswede and Winter (2002). We present this theorem and proof (Theorem 4.1), since it is of independent interest. In Section 4 we prove the result of Rudelson and Vershynin.

A central question in probability is to determine how many i.i.d. samples from a probability distribution suffice to make the empirical estimate of the variance close to the true variance in every direction. Mathematically, there is an $n \times m$ matrix A (with $m > n$, m possibly infinite) such that AA^T is the covariance matrix, and for any unit vector $\mathbf{x} \in \mathbb{R}^n$, the variance of the distribution in direction $\mathbf{x}$ is $\mathbf{x}^T AA^T \mathbf{x} = |\mathbf{x}^T A|^2$. The problem is to find a (small) sample of s columns of A to form an $n \times s$ matrix C so that the variance is (approximately) preserved, to relative error ε, in every direction $\mathbf{x}$. That is, we wish to satisfy, for all $\mathbf{x}$,

$$\|\mathbf{x}^T C\|^2 \in \big[(1-\varepsilon)\|\mathbf{x}^T A\|^2,\ (1+\varepsilon)\|\mathbf{x}^T A\|^2\big], \text{ denoted } \|\mathbf{x}^T C\|^2 \cong_\varepsilon \|\mathbf{x}^T A\|^2. \tag{1.3}$$

It is important that (1.3) holds simultaneously for all $\mathbf{x}$. It turns out that, in general, if we sample columns of A with probability proportional to the squared length not of its own columns but of the columns of A^+A, where A^+ is the pseudo-inverse of A, then (1.3) follows from (1.2). We call this 'preconditioned' length-squared sampling (since multiplying A by A^+ can be thought of as preconditioning).

There is another, seemingly unrelated context, where exactly the same question (1.3) arises, namely, graph sparsification, considered by Spielman and Srivastava (2011). Here, A is the node–edge signed incidence matrix of a graph, and the goal is to find a subset of edges satisfying (1.3). Spielman and Srivastava (2011) showed in this case that the squared length of the columns of A^+A are proportional to the (weighted) electrical resistances and can also be computed in linear time (in the number of edges of the graph).

In theoretical computer science, preconditioned length-squared sampling (also called leverage score sampling) arose from a different motivation: Can the additive error guarantee (1.1) be improved to a relative error guarantee, perhaps with more sophisticated sampling methods? Several answers have been given to this question. The first is simply to iterate length-squared sampling on the residual matrix in the space orthogonal to the span of the sample chosen so far (Deshpande, Rademacher, Vempala and Wang 2006). Another is to use preconditioned length-squared sampling, where the preconditioning effectively makes the sampling probabilities proportional to the *leverage scores* (Drineas, Mahoney and Muthukrishnan 2008). A third is *volume sampling*, which picks a subset of k columns with probabilities proportional to the square of the volume of the k-simplex they span, together with the origin (Deshpande and Vempala 2006, Deshpande and Rademacher 2010, Anari, Gharan and Rezaei 2016). We discuss preconditioned length-squared sampling in Section 5.

In Section 6 we consider an approach pioneered by Clarkson and Woodruff (2009, 2013) for obtaining similar relative-error approximations but in input sparsity time, *i.e.*, time that is asymptotically linear in the number of non-zeros of the input matrix. This is possible via a method known as subspace embedding, which can be performed in linear time using the sparsity of the matrix. We first discuss an inefficient method using random projection, then an efficient method due to Clarkson and Woodruff (2013) based on a sparse projection.

As the title of this article indicates, we focus here on randomized algorithms with linear algebra as the important application area. For treatments from a linear algebra perspective (with randomized algorithms as one of the tools), the reader might consult Halko, Martinsson and Tropp (2011), Woodruff (2014) and references therein.

2. Basic algorithms

2.1. Matrix multiplication using sampling

Suppose A is an $m \times n$ matrix and B is an $n \times p$ matrix, and the product AB is desired. We can use sampling to get an approximate product faster than the traditional multiplication. Let $A(:,k)$ denote the kth column of A. $A(:,k)$ is an $m \times 1$ matrix. Let $B(k,:)$ be the kth row of B. $B(k,:)$ is a $1 \times n$ matrix. We have

$$AB = \sum_{k=1}^{n} A(:,k)B(k,:).$$

Can we estimate the sum over all k by summing over a sub-sample? Let $p_1, p_2, \ldots, p_n$ be non-negative reals summing to 1 (to be determined later). Let z be a random variable that takes values in $\{1, 2, \ldots, n\}$ with $\Pr(z = j) = p_j$. Define an associated matrix random variable X such that

$$\Pr\left(X = \frac{1}{p_k} A(:, k)B(k, :)\right) = p_k. \tag{2.1}$$

Let $E(X)$ denote the entry-wise expectation, that is,

$$E(X) = \sum_{k=1}^{n} \Pr(z = k)\frac{1}{p_k} A(:, k)B(k, :) = \sum_{k=1}^{n} A(:, k)B(k, :) = AB.$$

The scaling by $1/p_k$ makes X an unbiased estimator of AB. We will be interested in $E(\|AB - X\|_F^2)$, which is just the sum of the variances of all entries of X,[2] since

$$\begin{aligned} E(\|AB - X\|_F^2) &= \sum_{i=1}^{m}\sum_{j=1}^{p} \mathrm{Var}(x_{ij}) = \sum_{ij} E(x_{ij}^2) - E(x_{ij})^2 \\ &= \left(\sum_{ij}\sum_{k} p_k \frac{1}{p_k^2} a_{ik}^2 b_{kj}^2\right) - \|AB\|_F^2. \end{aligned}$$

We can ignore the $\|AB\|_F^2$ term since it does not depend on the p_k. Now

$$\begin{aligned} \sum_{ij}\sum_{k} p_k \frac{1}{p_k^2} a_{ik}^2 b_{kj}^2 &= \sum_{k} \frac{1}{p_k} \left(\sum_{i} a_{ik}^2\right)\left(\sum_{j} b_{kj}^2\right) \\ &= \sum_{k} \frac{1}{p_k} |A(:, k)|^2 |B(k, :)|^2. \end{aligned}$$

It can be seen by calculus[3] that the minimizing p_k must be proportional to $|A(:, k)||B(k, :)|$. In the important special case when $B = A^T$, this means picking columns of A with probabilities proportional to the squared length of the columns. In fact, even in the general case when $B \neq A^T$, doing so simplifies the bounds, so we will use it. If p_k is proportional to $|A(:, k)|^2$, that is,

$$p_k = \frac{|A(:, k)|^2}{\|A\|_F^2},$$

then

$$E(\|AB - X\|_F^2) = \mathrm{Var}(X) \le \|A\|_F^2 \sum_{k} |B(k, :)|^2 = \|A\|_F^2 \|B\|_F^2.$$

[2] We use a_{ij} to denote entries of matrix A.

[3] For any non-negative c_k, minimizing $\sum_k c_k p_k^{-1}$ subject to $\sum_k p_k = 1$ via Lagrange multipliers implies that p_k is proportional to $\sqrt{c_k}$.

To reduce the variance, we can take the average of s independent trials. Each trial i, $i = 1, 2, \ldots, s$ yields a matrix X_i as in (2.1). We take

$$\frac{1}{s}\sum_{i=1}^{s} X_i$$

as our estimate of AB. Since the variance of the sum of independent random variables is the sum of the variances, we find that

$$\mathrm{Var}\left(\frac{1}{s}\sum_{i=1}^{s} X_i\right) = \frac{1}{s}\mathrm{Var}(X) \leq \frac{1}{s}\|A\|_F^2\|B\|_F^2.$$

Let $k_1, \ldots, k_s$ be the k chosen in each trial. Expanding this, we obtain

$$\frac{1}{s}\sum_{i=1}^{s} X_i = \frac{1}{s}\left(\frac{A(:,k_1)B(k_1,:)}{p_{k_1}} + \frac{A(:,k_2)B(k_2,:)}{p_{k_2}} + \cdots + \frac{A(:,k_s)B(k_s,:)}{p_{k_s}}\right). \tag{2.2}$$

We write this as the product of an $m \times s$ matrix with an $s \times p$ matrix as follows. Let C be the $m \times s$ matrix consisting of the following columns, which are scaled versions of the chosen columns of A:

$$\frac{A(:,k_1)}{\sqrt{sp_{k_1}}}, \quad \frac{A(:,k_2)}{\sqrt{sp_{k_2}}}, \ldots, \frac{A(:,k_s)}{\sqrt{sp_{k_s}}}.$$

Note that this scaling has a nice property, which we leave to the reader to verify:

$$E(CC^T) = AA^T. \tag{2.3}$$

Define R to be the $s \times p$ matrix with the corresponding rows of B similarly scaled, namely, R has rows

$$\frac{B(k_1,:)}{\sqrt{sp_{k_1}}}, \quad \frac{B(k_2,:)}{\sqrt{sp_{k_2}}}, \ldots, \frac{B(k_s,:)}{\sqrt{sp_{k_s}}}.$$

The reader may also verify that

$$E(R^T R) = A^T A. \tag{2.4}$$

From (2.2), we see that

$$\frac{1}{s}\sum_{i=1}^{s} X_i = CR.$$

This is represented in Figure 2.1. We summarize our discussion in Theorem 2.1.

Theorem 2.1. Suppose A is an $m \times n$ matrix and B is an $n \times p$ matrix. The product AB can be estimated by CR, where C is an $m \times s$ matrix consisting of s columns of A picked according to length-squared distribution and scaled

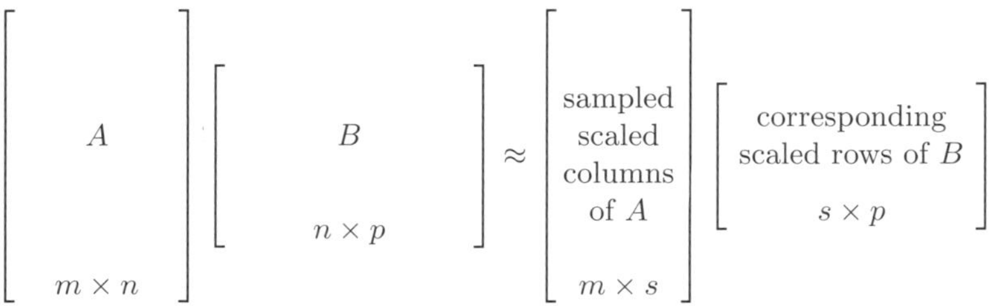

Figure 2.1. Approximate matrix multiplication using sampling.

to satisfy (2.3), and R is the $s \times p$ matrix consisting of the corresponding rows of B scaled to satisfy (2.4). The error is bounded by

$$E(\|AB - CR\|_F^2) \leq \frac{\|A\|_F^2 \, \|B\|_F^2}{s}.$$

Thus, to ensure that

$$E(\|AB - CR\|_F^2) \leq \varepsilon^2 \|A\|_F^2 \|B\|_F^2,$$

it suffices to choose $s \geq 1/\varepsilon^2$. If now $\varepsilon = \Omega(1)$ (and so $s = O(1)$), then the multiplication CR can be carried out in time $O(mp)$.

When is this the error bound useful? Let us focus on the case $B = A^T$ so that we have just one matrix to consider. If A is the identity matrix, then the guarantee is *not* very good. In this case, $\|AA^T\|_F^2 = n$, but the right-hand side of the inequality is n^2/s. So we would need $s > n$ for the bound to be any better than approximating the product with the zero matrix.

More generally, the trivial estimate of the zero matrix for AA^T makes an error in the Frobenius norm of $\|AA^T\|_F$. If $\sigma_1, \sigma_2, \ldots$ are the singular values of A, then the singular values of AA^T are $\sigma_1^2, \sigma_2^2, \ldots$, and we have

$$\|A\|_F^2 = \sum_t \sigma_t^2 \quad \text{and} \quad \|AA^T\|_F^2 = \sum_t \sigma_t^4.$$

So from the theorem we can assert

$$E(\|AA^T - CR\|_F^2) \leq \|AA^T\|_F^2$$

provided that

$$s \geq \frac{(\sigma_1^2 + \sigma_2^2 + \cdots)^2}{\sigma_1^4 + \sigma_2^4 + \cdots}.$$

If $\text{rank}(A) = r$, then there are r non-zero σ_t and the best general upper bound on the ratio $(\sigma_1^2 + \sigma_2^2 + \cdots)^2/(\sigma_1^4 + \sigma_2^4 + \cdots)$ is r, so in general, s needs to be at least r. If A is of full rank, this means that sampling will not give us any gain over taking the whole matrix!

However, if there is a constant c and a small integer p such that

$$\sigma_1^2 + \sigma_2^2 + \cdots + \sigma_p^2 \geq c(\sigma_1^2 + \sigma_2^2 + \cdots + \sigma_r^2), \tag{2.5}$$

then

$$\frac{(\sigma_1^2 + \sigma_2^2 + \cdots + \sigma_r^2)^2}{\sigma_1^4 + \sigma_2^4 + \cdots + \sigma_r^4} \leq \frac{1}{c^2} \frac{(\sigma_1^2 + \sigma_2^2 + \cdots + \sigma_p^2)^2}{\sigma_1^4 + \sigma_2^4 + \cdots + \sigma_p^2} \leq \frac{p}{c^2},$$

and so $s \geq p/c^2$ gives us a better estimate than the zero matrix. Further increasing s by a factor decreases the error by the same factor. The condition (2.5) (in words, the top p singular values make up a constant fraction of the spectrum) is indeed the hypothesis of the subject of principal component analysis, and there are many situations when the data matrix does satisfy the condition and so sampling algorithms are useful.

2.1.1. Implementing length-squared sampling in two passes

Traditional matrix algorithms often assume that the input matrix is in random access memory (RAM) and so any particular entry of the matrix can be accessed in unit time. For massive matrices, RAM may be too small to hold the entire matrix, but may be able to hold and compute with the sampled columns/rows.

Let us consider a high-level model where the input matrix or matrices have to be read from 'external memory' using a 'pass'. In one pass, we can read sequentially all entries of the matrix in some order. We may do some 'sampling on the fly' as the pass is going on.

It is easy to see that two passes suffice to draw a sample of columns of A according to length-squared probabilities, even if the matrix is not in row order or column order and entries are presented as a linked list (as in sparse representations). In the first pass, we just compute the squared length of each column and store this information in RAM. The squared lengths can be computed as running sums. Then, we use a random number generator in RAM to figure out the columns to be sampled (according to length-squared probabilities). Then, we make a second pass in which we pick out the columns to be sampled.

What if the matrix is already presented in external memory in column order? In this case, one pass will do, based on a primitive using rejection sampling.

The primitive is as follows. We are given a stream (*i.e.* a read-once only input sequence) of positive real numbers $a_1, a_2, \ldots, a_n$. We seek to have a random $i \in \{1, 2, \ldots, n\}$ at the end, with the property that the probability of choosing i is exactly equal to $a_i / \sum_{j=1}^n a_j$, for all i. This is solved as follows. After having read $a_1, a_2, \ldots, a_i$, suppose we have (i) $\sum_{j=1}^i a_j$ and (ii) a sample $a_j, j \leq i$, picked with probability $a_j / \sum_{k=1}^i a_k$. On reading a_{i+1}, we update the sum and, with the correct probability, reject the earlier sample and replace it with a_{i+1}. If we need s independent identical samples, we just run s such processes in parallel.

2.2. Sketch of a large matrix

The main result of this section is that for any matrix, a sample of columns and rows, each picked according to length-squared distribution, provides a good sketch of the matrix in a formal sense that will be described briefly. Let A be an $m \times n$ matrix. Pick s columns of A according to length-squared distribution. Let C be the $m \times s$ matrix containing the picked columns scaled so as to satisfy (2.3), that is, if $A(:,k)$ is picked, it is scaled by $1/\sqrt{sp_k}$. Similarly, pick r rows of A according to length-squared distribution on the rows of A. Let R be the $r \times n$ matrix of the picked rows, scaled as follows. If row k of A is picked, it is scaled by $1/\sqrt{rp_k}$. We then have $E(R^T R) = A^T A$. From C and R, we can find a matrix U so that $A \approx CUR$.

One may recall that the top k singular vectors give a similar picture. But the SVD takes more time to compute, requires all of A to be stored in RAM, and does not have the property that the singular vectors, the basis of the reduced space, are directly from A The last property – that the approximation involves actual rows/columns of the matrix rather than linear combinations – is called an *interpolative approximation*, and is useful in many contexts. Some structural results of such approximations are found in the work of Stewart (Stewart 1999, Stewart 2004, Berry, Pulatova and Stewart 2004) and Goreinov, Tyrtyshnikov and Zamarashkin (Goreinov, Tyrtyshnikov and Zamarashkin 1997, Goreinov and Tyrtyshnikov 2001).

We briefly mention two motivations for such a sketch. Suppose A is the document-term matrix of a large collection of documents. We are to 'read' the collection at the outset and store a sketch so that later, when a query represented by a vector with one entry per term arrives, we can find its similarity to each document in the collection Similarity is defined by the dot product. In Figure 1.1 it is clear that the matrix–vector product of a query with the right-hand side can be done in time $O(ns + sr + rm)$, which would be linear in n and m if s and r are $O(1)$. To bound errors for this process, we need to show that the difference between A and the sketch of A has small 2-norm. The fact that the sketch is an interpolative approximation means that our approximation essentially consists a subset of documents and a subset of terms, which may be thought of as a representative set of documents and terms. Moreover, if A is *sparse* in its rows and columns – each document contains only a small fraction of the terms and each term is in only a small fraction of the documents – then this property will be preserved in C and R, unlike with the SVD.

A second motivation comes from recommendation systems. Here A would be a customer–product matrix whose (i, j)th entry is the preference of customer i for product j. The objective is to collect a few sample entries of A and, based on these, get an approximation to A so that we can make future recommendations. A few sampled rows of A (all preferences of a few customers) and a few sampled columns (all customer preferences for a few products) give a good approximation to A provided that the samples are drawn according to the length-squared distribution.

It now remains to describe how to find U from C and R. There is an $n \times n$ matrix P of the form $P = QR$ which acts as the identity on the space spanned by the rows of R and zeros out all vectors orthogonal to this space. The matrix Q is just the pseudo-inverse of R.

Lemma 2.2. If $\text{rank}(R) = r'$ and $R = \sum_{t=1}^{r'} \sigma_t \mathbf{u}_t \mathbf{v}_t^T$ is the SVD of R, then the matrix $P = \left(\sum_{t=1}^{r'} {\sigma_t}^{-1} \mathbf{v}_t \mathbf{u}_t^T\right) R$ satisfies:

(i) $P\mathbf{x} = \mathbf{x}$ for every vector $\mathbf{x}$ of the form $\mathbf{x} = R^T \mathbf{y}$,

(ii) if $\mathbf{x}$ is orthogonal to the row space of R, then $P\mathbf{x} = \mathbf{0}$,

(iii) $P = R^T \left(\sum_{t=1}^{r'} {\sigma_t}^{-2} \mathbf{u}_t \mathbf{u}_t^T\right) R$.

We begin with some intuition. In particular, we first present a simpler idea that does not work but will then motivate the idea that *does* work. Write A as AI, where I is the $n \times n$ identity matrix. Now, let us approximate the product AI using the algorithm of Theorem 2.1 from the previous section, that is, by sampling s columns of A according to their squared lengths. Then, as in the last section, write $AI \approx CW$, where W consists of a scaled version of the s rows of I corresponding to the s columns of A that were picked. Theorem 2.1 bounds the error $\|A - CW\|_F^2$ by

$$\|A\|_F^2 \|I\|_F^2 / s = \|A\|_F^2 \frac{n}{s}.$$

But we would like the error to be a small fraction of $\|A\|_F^2$ which would require $s \geq n$, which clearly is of no use since this would pick at least as many columns as the whole of A.

Instead, let us use the identity-like matrix P instead of I in the above discussion. Using the fact that R is picked according to the squared length, we will show the following proposition later.

Proposition 2.3. $A \approx AP$ with error $E(\|A - AP\|_2^2) \leq \|A\|_F^2 / \sqrt{r}$.

We then use Theorem 2.1 to argue that instead of doing the multiplication AP, we can use the sampled columns of A and the corresponding rows of P. The sampled s columns of A form C. We have to take the corresponding s rows of the matrix

$$P = R^T \left(\sum_{t=1}^{r'} \frac{1}{\sigma_t^2} \mathbf{u}_t \mathbf{u}_t^T \right) R.$$

This is the same as taking the corresponding s rows of R^T and multiplying this by $\left(\sum_{t=1}^{r'} \sigma_t{}^{-2} \mathbf{u}_t \mathbf{u}_t^T\right) R$. It is easy to check that this leads to an expression of the form CUR. Moreover, by Theorem 2.1, the error is bounded by

$$E(\|AP - CUR\|_2^2) \le E(\|AP - CUR\|_F^2) \le \frac{\|A\|_F^2 \|P\|_F^2}{s} \le \frac{r}{s} \|A\|_F^2, \quad (2.6)$$

since we will deduce the following inequality.

Proposition 2.4. $\|P\|_F^2 \le r$.

Combining (2.6) and Proposition 2.3, and using the fact that by the triangle inequality we have

$$\|A - CUR\|_2 \le \|A - AP\|_2 + \|AP - CUR\|_2,$$

which in turn implies that

$$\|A - CUR\|_2^2 \le 2\|A - AP\|_2^2 + 2\|AP - CUR\|_2^2,$$

we obtain the main result.

Theorem 2.5. Suppose A is any $m \times n$ matrix and r and s are positive integers. Suppose C is an $m \times s$ matrix of s columns of A picked according to length-squared sampling, and similarly R is a matrix of r rows of A picked according to length-squared sampling. Then we can find from C, R an $s \times r$ matrix U such that

$$E(\|A - CUR\|_2^2) \le \|A\|_F^2 \left(\frac{2}{\sqrt{r}} + \frac{2r}{s} \right).$$

We see that if s is fixed, the error is minimized when $r = (s/2)^{2/3}$. Choosing $s = O(r/\varepsilon)$ and $r = 1/\varepsilon^2$, the bound becomes $O(\varepsilon)\|A\|_F^2$.

Now we prove Proposition 2.3. First,

$$\|A - AP\|_2^2 = \max_{\{\mathbf{x}:|\mathbf{x}|=1\}} |(A - AP)\mathbf{x}|^2.$$

Let us first suppose that $\mathbf{x}$ is in the row space V of R. We have $P\mathbf{x} = \mathbf{x}$, so for $\mathbf{x} \in V$ we have $(A - AP)\mathbf{x} = \mathbf{0}$. Now, since every vector can be written as the sum of a vector in V plus a vector orthogonal to V, this implies that the maximum must therefore occur at some $\mathbf{x} \in V^\perp$. For such $\mathbf{x}$, we have $(A - AP)\mathbf{x} = A\mathbf{x}$. Thus, the question now becomes: For unit-length $\mathbf{x} \in V^\perp$, how large can $\|A\mathbf{x}\|^2$ be? To analyse this, we can write

$$\begin{aligned} \|A\mathbf{x}\|^2 &= \mathbf{x}^T A^T A \mathbf{x} = \mathbf{x}^T (A^T A - R^T R)\mathbf{x} \\ &\le \|A^T A - R^T R\|_2 |\mathbf{x}|^2 \le \|A^T A - R^T R\|_2. \end{aligned}$$

This implies that we get $\|A - AP\|_2^2 \le \|A^T A - R^T R\|_2$. So, it suffices to prove that

$$E(\|A^T A - R^T R\|_2^2) \le \|A\|_F^4 / r,$$

which follows directly from Theorem 2.1, since we can think of $R^T R$ as a way of estimating $A^T A$ by picking (according to length-squared distribution) columns of A^T, that is, rows of A. By Jensen's inequality, this implies that

$$E(\|A^T A - R^T R\|_2) \leq \frac{\|A\|_F^2}{\sqrt{r}}$$

and proves Proposition 2.3.

Proposition 2.4 is easy to see. Since, by Lemma 2.2, P is the identity on the space V spanned by the rows of R, and $P\mathbf{x} = 0$ for $\mathbf{x}$ perpendicular to the rows of R, we have that $\|P\|_F^2$ is the sum of its singular values squared, which is at most r as claimed.

Finally, to bound the time needed to compute U, the only step involved in computing U is to find the SVD of R. But note that RR^T is an $r \times r$ matrix, and since r is much smaller than n, m, this is fast.

2.3. Low-rank approximations and the SVD

Singular value decomposition yields the best approximation of given rank to a matrix in both spectral norm and Frobenius norm. Here, we show that a near-optimal low-rank approximation (LRA) of a matrix A can be found by computing the SVD of a smaller matrix R. The main result is that for *any* matrix R with

$$A^T A \approx R^T R,$$

the restriction of A to the space spanned by the top few right singular vectors of R is a good approximation to A. This is purely based on matrix perturbation theory: there is no probability involved.

If we can find an R with smaller dimensions than A, with $R^T R \approx A^T A$, then computing the SVD for R would be faster. Indeed, we have already seen one way of getting such an R: pick a set of r rows of A in r i.i.d. trials according to the length-squared distribution and scale them so that $E(R^T R) = A^T A$. We saw in Theorem 2.1 that $R^T R \approx A^T A$.

The Hoffman–Wielandt inequality (stated below) implies that if $\|R^T R - A^T A\|_F$ is small, then the singular *values* of A and R are close. If the top singular vectors of R and A were close, the desired result would follow easily. However, as is well known, multiple singular values are points of discontinuity of singular vectors. But it is not difficult to see intuitively that LRA is not discontinuous: using a singular vector of a slightly worse singular value does not induce gross errors in LRA, and this intuition underlies the result proved formally here.

2.3.1. Restriction to the SVD subspace

If $\mathbf{w}_1, \mathbf{w}_2, \ldots, \mathbf{w}_k$ is a basis for the vector space V, the restriction of A to V is

$$\bar{A} = A \sum_{t=1}^{k} \mathbf{w}_t \mathbf{w}_t^T \text{ satisfies } \bar{A}\mathbf{x} = \begin{cases} A\mathbf{x} & \text{if } x \in V, \\ 0 & \text{if } \mathbf{x} \perp V. \end{cases} \tag{2.7}$$

Theorem 2.6. For *any* $r \times n$ matrix R and *any* $m \times n$ matrix A, if $\mathbf{v}_1, \mathbf{v}_2, \ldots, \mathbf{v}_k$ are the top k right singular vectors of R and $A(k)$ is the best rank-k approximation to A, then

$$\left\| A - A \sum_{t=1}^{k} \mathbf{v}_t \mathbf{v}_t^T \right\|_F^2 \leq \|A - A(k)\|_F^2 + 2\sqrt{k} \|R^T R - A^T A\|_F, \tag{2.8}$$

$$\left\| A - A \sum_{t=1}^{k} \mathbf{v}_t \mathbf{v}_t^T \right\|_2^2 \leq \|A - A(k)\|_2^2 + 2\|R^T R - A^T A\|_2. \tag{2.9}$$

In (2.8), the first term $\|A - A(k)\|_F^2$ is the best possible error we can make with exact SVD of A. We cannot avoid that. The second term is the penalty we pay for computing with R instead of A, and similarly for (2.9). The theorem is for any r including the cases $r = 0$ and $r = n$. Central to the proof is the Hoffman–Wielandt inequality, stated next without proof.

Lemma 2.7. If P, Q are two real symmetric $n \times n$ matrices and $\lambda_1, \lambda_2, \ldots$ denote eigenvalues in non-increasing order, then

$$\sum_{t=1}^{n} (\lambda_t(P) - \lambda_t(Q))^2 \leq \|P - Q\|_F^2.$$

We can now prove the low-rank approximation guarantee.

Proof of Theorem 2.6. Complete $\mathbf{v}_1, \mathbf{v}_2, \ldots, \mathbf{v}_k$ to a basis $\mathbf{v}_1, \mathbf{v}_2, \ldots, \mathbf{v}_n$ of $\mathbb{R}^n$. Then

$$\begin{aligned}
&\left\| A - A \sum_{t=1}^{k} \mathbf{v}_t \mathbf{v}_t^T \right\|_F^2 - \|A - A(k)\|_F^2 \\
&\quad = \|A\|_F^2 - \left\| A \sum_{t=1}^{k} \mathbf{v}_t \mathbf{v}_t^T \right\|_F^2 - (\|A\|_F^2 - \|A(k)\|_F^2) \\
&\quad = \|A(k)\|_F^2 - \left\| A \sum_{t=1}^{k} \mathbf{v}_t \mathbf{v}_t^T \right\|_F^2 \\
&\quad = \sum_{t=1}^{k} \sigma_t^2(A) - \sum_{t=1}^{k} \|A\mathbf{v}_t\|^2
\end{aligned}$$

$$= \sum_{t=1}^{k} \sigma_t^2(A) - \sum_{t=1}^{k} \mathbf{v}_t^T(A^TA - R^TR)\mathbf{v}_t - \sum_{t=1}^{k} \mathbf{v}_t^T R^T R \mathbf{v}_t$$

$$= \sum_{t=1}^{k} (\sigma_t^2(A) - \sigma_t^2(R)) - \sum_{t=1}^{k} \mathbf{v}_t^T(A^TA - R^TR)\mathbf{v}_t.$$

We can now deduce that

$$\left|\sum_{t=1}^{k} \mathbf{v}_t^T(A^TA - R^TR)\mathbf{v}_t\right| \leq k\|A^TA - R^TR\|_2,$$

but we want $\sqrt{k}$ instead of k. For this, we use the Cauchy–Schwarz inequality to assert

$$\begin{aligned}\left|\sum_{t=1}^{k} \mathbf{v}_t^T(A^TA - R^TR)\mathbf{v}_t\right| &\leq \sqrt{k}\left(\sum_{t=1}^{k} (\mathbf{v}_t^T(A^TA - R^TR)\mathbf{v}_t)^2\right)^{1/2} \\ &\leq \sqrt{k}\|A^TA - R^TR\|_F,\end{aligned}$$

since the Frobenius norm is invariant under change of basis and, in a basis containing $\mathbf{v}_1, \mathbf{v}_2, \ldots, \mathbf{v}_k$,

$$\sum_{t=1}^{k} (\mathbf{v}_t^T(A^TA - R^TR)\mathbf{v}_t)^2$$

is the sum of squares of the first k diagonal entries of $A^TA - R^TR$. We still have to bound

$$\sum_{t=1}^{k} (\sigma_t^2(A) - \sigma_t^2(R)),$$

for which we use the Hoffman–Wielandt inequality, after another use of Cauchy–Schwarz:

$$\sum_{t=1}^{k} (\sigma_t^2(A) - \sigma_t^2(R)) \leq \sqrt{k}\left(\sum_{t=1}^{k} (\sigma_t^2(A) - \sigma_t^2(R))^2\right)^{1/2}.$$

Now $\sigma_t^2(A) = \lambda_t(A^TA)$ and $\sigma_t^2(R) = \lambda_t(R^TR)$, and so

$$\sum_{t=1}^{k} (\sigma_t^2(A) - \sigma_t^2(R))^2 \leq \|A^TA - R^TR\|_F^2.$$

Plugging these in, we obtain (2.8).

For (2.9), first note that the top singular vector $\mathbf{u}$ of $A - A\sum_{t=1}^{k} \mathbf{v}_t\mathbf{v}_t^T$ must be orthogonal to the span of $\mathbf{v}_1, \mathbf{v}_2, \ldots, \mathbf{v}_k$. Hence

$$\left(A - A\sum_{t=1}^{k} \mathbf{v}_t\mathbf{v}_t^T\right)\mathbf{u} = A\mathbf{u}$$

by (2.7). Thus

$$\begin{aligned}
&\left\| A - A\sum_{t=1}^{k} \mathbf{v}_t\mathbf{v}_t^T \right\|_2^2 \\
&\quad = \|A\mathbf{u}\|^2 \\
&\quad = \mathbf{u}^T A^T A\mathbf{u} \\
&\quad = \mathbf{u}^T(A^TA - R^TR)\mathbf{u} + \mathbf{u}^T R^T R\mathbf{u} \leq \|A^TA - R^TR\|_2 + \sigma_{k+1}^2(R) \\
&\quad = \|A^TA - R^TR\|_2 + (\sigma_{k+1}^2(R) - \sigma_{k+1}^2(A)) + \sigma_{k+1}^2(A) \\
&\quad \leq 2\|A^TA - R^TR\|_2 + \sigma_{k+1}^2(A),
\end{aligned}$$

by Weyl's inequality (Horn and Johnson 2012, Section 4.3), which states that

$$\lambda_{k+1}(R^TR) \in [\lambda_{k+1}(A^TA) - \|A^TA - R^TR\|_2,\ \lambda_{k+1}(A^TA) + \|A^TA - R^TR\|_2].$$

□

To summarize, here is the algorithm for LRA of A after one final note: we need the right singular vectors of R. For this, we can compute the SVD of RR^T (an $r \times r$ matrix) to find the left singular vectors of R, from which we can get the right singular vectors.

(1) Pick r rows of A by length-squared sampling, and scale them so that for the resulting $r \times n$ matrix R, $E(R^TR) = A^TA$.

(2) Find RR^T. Find the left singular vectors of R by computing the SVD of the $r \times r$ matrix RR^T. Premultiply R by the left singular vectors to get the right singular vectors $\mathbf{v}_1, \mathbf{v}_2, \ldots, \mathbf{v}_k$ of R.

(3) Return $A\sum_{t=1} \mathbf{v}_t\mathbf{v}_t^T$ as the implicit LRA (or if required, multiply out and return the matrix).

Let p be the maximum number of non-zero entries in any row of A ($p \leq n$). The first step can be done in two passes from external memory. For the second step, RR^T can be found in $O(r^2p)$ time. The spectral decomposition of RR^T can be done in time $O(r^3)$ (or better). Finally, multiplying the k left singular values by R can be done in time $O(krp)$.

The running time has been improved by Drineas, Kannan and Mahoney (2006) and further by Clarkson and Woodruff (see Woodruff 2014).

3. Tensor approximation via length-squared sampling

An r-tensor A is an r-dimensional array with real entries $A_{i_1,i_2,\ldots,i_r}$, for $i_1, i_2, \ldots, i_r \in \{1, 2, \ldots, n\}$. Tensors encode mutual information about subsets of size three or higher (matrix entries are pairwise information) and arise naturally in a number of applications.

In analogy with matrices, we may define a rank-1 tensor to be the *outer product* of r vectors $\mathbf{x}^{(1)}, \mathbf{x}^{(2)}, \ldots, \mathbf{x}^{(r)}$ denoted $\mathbf{x}^{(1)} \otimes \mathbf{x}^{(2)} \otimes \cdots \otimes \mathbf{x}^{(r)}$, with $(i_1, i_2, \ldots, i_r)$th entry $x^{(1)}_{i_1} x^{(2)}_{i_2} \cdots x^{(r)}_{i_r}$. We will show the following.

(1) For any r-tensor A, there exists a good approximation by the sum of a small number of rank-1 tensors (Lemma 3.2).

(2) We can algorithmically find such an approximation (Theorem 3.3).

In the case of matrices, traditional linear algebra algorithms find optimal approximations in polynomial time. Unfortunately, there is no such theory (or algorithm) for r-dimensional arrays when $r > 2$. Indeed, there are computational hardness results (Hillar and Lim 2013). But our focus here is what we can do, not hardness results. We assume throughout that r is a fixed number, whereas n grows to infinity. We will develop polynomial-time algorithms for finding low-rank approximations. The algorithms make crucial use of length-squared sampling.

Definition 3.1. Corresponding to an r-tensor A, there is an r-linear form defined as follows: for vectors $\mathbf{x}^{(1)}, \mathbf{x}^{(2)}, \ldots, \mathbf{x}^{(r)}$,

$$A(\mathbf{x}^{(1)}, \mathbf{x}^{(2)}, \ldots, \mathbf{x}^{(r)}) = \sum_{i_1,i_2,\ldots,i_r} A_{i_1,i_2,\ldots,i_r} x^{(1)}_{i_1} x^{(2)}_{i_2} \cdots x^{(r)}_{i_r}. \tag{3.1}$$

We use the following two norms of r-dimensional arrays corresponding to the Frobenius and spectral norms for matrices:

$$\|A\|_F = \Big(\sum A^2_{i_1,i_2,\ldots,i_r}\Big)^{1/2}, \tag{3.2}$$

$$\|A\|_2 = \max_{\mathbf{x}^{(1)},\mathbf{x}^{(2)},\ldots,\mathbf{x}^{(r)}} \frac{A(\mathbf{x}^{(1)}, \mathbf{x}^{(2)}, \ldots, \mathbf{x}^{(r)})}{\|\mathbf{x}^{(1)}\|\|\mathbf{x}^{(2)}\| \cdots \|\mathbf{x}^{(r)}\|}. \tag{3.3}$$

Lemma 3.2. For any tensor A and any $\epsilon > 0$, there exist $k \leq 1/\epsilon^2$ rank-1 tensors $B_1, B_2, \ldots, B_k$ such that

$$\|A - (B_1 + B_2 + \cdots + B_k)\|_2 \leq \epsilon \|A\|_F.$$

Theorem 3.3. For any tensor A and any $\epsilon > 0$, we can find $k \leq 4/\epsilon^2$ rank-1 tensors $B_1, B_2, \ldots, B_k$, using a randomized algorithm in time $(n/\epsilon)^{O(1/\epsilon^4)}$, such that with high probability (over coin tosses of the algorithm), we have

$$\|A - (B_1 + B_2 + \cdots + B_k)\|_2 \leq \epsilon \|A\|_F.$$

Proof of Lemma 3.2. If $\|A\|_2 \le \epsilon\|A\|_F$, then we are done. If not, there are unit vectors $\mathbf{x}^{(1)}, \mathbf{x}^{(2)}, \ldots, \mathbf{x}^{(r)}$ such that $A(\mathbf{x}^{(1)}, \mathbf{x}^{(2)}, \ldots, \mathbf{x}^{(r)}) \ge \epsilon\|A\|_F$. Now consider the r-dimensional array

$$B = A - (A(\mathbf{x}^{(1)}, \mathbf{x}^{(2)}, \ldots, \mathbf{x}^{(r)}))\mathbf{x}^{(1)} \otimes \mathbf{x}^{(2)} \otimes \cdots \otimes \mathbf{x}^{(r)}.$$

It is easy to see that $\|B\|_F^2 \le \|A\|_F^2(1-\epsilon^2)$. We may repeat for B: either $\|B\|_2 \le \epsilon\|A\|_F$ and we may stop, or there exist $\mathbf{y}^{(1)}, \mathbf{y}^{(2)}, \ldots, \mathbf{y}^{(r)}$ with $B(\mathbf{y}^{(1)}, \mathbf{y}^{(2)}, \ldots, \mathbf{y}^{(r)}) \ge \epsilon\|A\|_F$. Let

$$C = B - B(\mathbf{y}^{(1)}, \mathbf{y}^{(2)}, \ldots, \mathbf{y}^{(r)})\,(\mathbf{y}^{(1)} \otimes \mathbf{y}^{(2)} \otimes \cdots \otimes \mathbf{y}^{(r)}).$$

Each time the Frobenius norm squared falls by $\epsilon^2\|A\|_F^2$, so this process will only continue for at most $1/\epsilon^2$ steps. □

The algorithm that proves Theorem 3.3 will take up the rest of this section. First, from the proof of Lemma 3.2, it suffices to find $\mathbf{x}^{(1)}, \mathbf{x}^{(2)}, \ldots, \mathbf{x}^{(r)}$ all of length 1, maximizing $A(\mathbf{x}^{(1)}, \mathbf{x}^{(2)}, \ldots, \mathbf{x}^{(r)})$ to within *additive error* $\epsilon\|A\|_F/2$. We will give an algorithm to solve this problem. We need a bit more notation. For any $r-1$ vectors $\mathbf{z}^{(1)}, \mathbf{z}^{(2)}, \ldots, \mathbf{z}^{(r-1)}$, we define a vector with ith component

$$A(\mathbf{z}^{(1)}, \mathbf{z}^{(2)}, \ldots, \mathbf{z}^{(r-1)}, \cdot)_i = \sum_{i_1, i_2, \ldots, i_{r-1}} A_{i_1, i_2, \ldots, i_{r-1}, i} z_{i_1}^{(1)} z_{i_2}^{(2)} \cdots z_{i_{r-1}}^{(r-1)}. \quad (3.4)$$

Here is the idea behind the algorithm. Suppose $\mathbf{z}^{(1)}, \mathbf{z}^{(2)}, \ldots, \mathbf{z}^{(r)}$ are the unknown unit vectors that maximize $A(\mathbf{x}^{(1)}, \mathbf{x}^{(2)}, \ldots)$. Since

$$A(\mathbf{z}^{(1)}, \mathbf{z}^{(2)}, \ldots, \mathbf{z}^{(r-1)}, \mathbf{z}^{(r)}) = \mathbf{z}^{(r)^T} A(\mathbf{z}^{(1)}, \mathbf{z}^{(2)}, \ldots, \mathbf{z}^{(r-1)}, \cdot),$$

we have

$$\mathbf{z}^{(r)} = \frac{A(\mathbf{z}^{(1)}, \mathbf{z}^{(2)}, \ldots, \mathbf{z}^{(r-1)}, \cdot)}{|A(\mathbf{z}^{(1)}, \mathbf{z}^{(2)}, \ldots, \mathbf{z}^{(r-1)}, \cdot)|}.$$

Thus, if we knew $\mathbf{z}^{(1)}, \mathbf{z}^{(2)}, \ldots, \mathbf{z}^{(r-1)}$, then we could compute $\mathbf{z}^{(r)}$. In fact, each component of $\mathbf{z}^{(r)}$ is the sum of n^{r-1} terms as in (3.4). We can hope to estimate the components of $\mathbf{z}^{(r)}$ with a sample of fewer than n^{r-1} terms. In fact, the main point is that we will show that if we pick a sample I of $s = O(1/\epsilon^2)$ elements $(i_1, i_2, \ldots, i_{r-1})$, with probabilities proportional to $\sum_i A_{i_1, i_2, \ldots, i_{r-1}, i}^2$, which is the squared length of the 'line' or 'column' $(i_1, i_2, \ldots, i_{r-1})$, the sums are well estimated. In more detail, let $f((i_1, i_2, \ldots, i_{r-1})) = z_{i_1}^{(1)} z_{i_2}^{(2)} \cdots z_{i_{r-1}}^{(r-1)}$. We will show that

$$\sum_{(i_1, i_2, \ldots, i_{r-1}) \in I} A_{i_1, i_2, \ldots, i_{r-1}, i} f((i_1, i_2, \ldots, i_{r-1})) \approx c z_i^{(r)},$$

where c is a scalar (independent of i).

Now we still need to compute the $s(r-1)$ real numbers $f((i_1, i_2, \ldots, i_{r-1}))$, for all $(i_1, i_2, \ldots, i_{r-1}) \in I$. We just enumerate all possibilities for the $s(r-1)$-tuple of values in steps of a certain size η, and for each possibility, we get a candidate $\mathbf{z}^{(r)}$. One of these candidates will be (close to) the optimal $\mathbf{z}^{(r)}$, but we do not know which one. To solve this puzzle, we just turn the problem on its head: for each candidate $\mathbf{z}^{(r)}$, we can define an $r-1$ tensor $A(\cdot, \ldots, \cdot, \mathbf{z}^{(r)})$ similar to (3.4), and recursively find the $\mathbf{y}^{(1)}, \mathbf{y}^{(2)}, \ldots, \mathbf{y}^{(r-1)}$ (approximately) maximizing $A(\mathbf{y}^{(1)}, \mathbf{y}^{(2)}, \ldots, \mathbf{y}^{(r-1)}, \mathbf{z}^{(r)})$. The best of these will satisfy the theorem. One subtle point: each enumerated $s(r-1)$-tuple need not really come from $\mathbf{z}^{(1)}, \mathbf{z}^{(2)}, \ldots, \mathbf{z}^{(r-1)}$, since all we need for this argument is that one of the enumerated $s(r-1)$-tuples of values gets close to the true $\{z_{i_1}^{(1)} z_{i_2}^{(2)} \cdots z_{i_{r-1}}^{(r-1)} : (i_1, i_2, \ldots, i_{r-1}) \in I\}$.

Algorithm 3.4 (tensor decomposition). Set

$$\eta = \frac{\epsilon^2}{100r\sqrt{n}} \quad \text{and} \quad s = \frac{10^5 r}{\epsilon^2}.$$

(1) Pick s random $(r-1)$-tuples $(i_1, i_2, \ldots, i_{r-1})$ with probabilities proportional to the sum of squared entries on the corresponding line, that is,

$$p(i_1, i_2, \ldots, i_{r-1}) = \frac{\sum_i A^2_{i_1, i_2, \ldots, i_{r-1}, i}}{\|A\|_F^2}.$$

Let I be the set of s $(r-1)$tuples so chosen.

(2) Enumerate all functions

$$f : I \to \{-1, -1+\eta, -1+2\eta, \ldots, 0, \ldots, 1-\eta, 1\}.$$

(a) For each of the $((2/\eta)+1)^{s(r-1)}$ functions so enumerated, find a vector $\mathbf{y}$ defined by

$$y_i = \sum_{(i_1, i_2, \ldots, i_{r-1}) \in I} \frac{A_{i_1, i_2, \ldots, i_{r-1}, i} f((i_1, i_2, \ldots, i_{r-1}))}{p(i_1, i_2, \ldots, i_{r-1})}.$$

Replace $\mathbf{y}$ by $\mathbf{y}/\|\mathbf{y}\|$.

(b) Consider the $(r-1)$-dimensional array $A(\mathbf{y})$ defined by

$$(A(\mathbf{y}))_{i_1, i_2, \ldots, i_{r-1}} = \sum_i A_{i_1, i_2, i_3 \ldots, i_{r-1}, i}\, y_i$$

and apply the algorithm recursively to find the approximate maximum

$$A(\mathbf{y})(\mathbf{x}^{(1)}, \mathbf{x}^{(2)}, \ldots, \mathbf{x}^{(r-1)}) \text{ subject to } \|\mathbf{x}^{(1)}\| = \cdots = \|\mathbf{x}^{(r-1)}\| = 1,$$

to within additive error $\epsilon\|A(\mathbf{y})\|_F/2$. Note that $\|A(\mathbf{y})\|_F \le \|A\|_F$ by Cauchy–Schwarz.

(3) Output the set of vectors that gives the maximum among all these candidates.

We now analyse Algorithm 3.4 and prove Theorem 3.3. We begin by showing that the discretization does not cause any significant loss.

Lemma 3.5. Let $\mathbf{z}^{(1)}, \mathbf{z}^{(2)}, \ldots, \mathbf{z}^{(r-1)}$ be the optimal unit vectors. Suppose $\mathbf{w}^{(1)}, \mathbf{w}^{(2)}, \ldots, \mathbf{w}^{(r-1)}$ are obtained from the $\mathbf{z}^{(t)}$ by rounding each coordinate down to the nearest integer multiple of η. Then

$$\|A(\mathbf{z}^{(1)}, \mathbf{z}^{(2)}, \ldots, \mathbf{z}^{(r-1)}, \cdot) - A(\mathbf{w}^{(1)}, \mathbf{w}^{(2)}, \ldots, \mathbf{w}^{(r-1)}, \cdot)\| \le \frac{\epsilon^2}{100}\|A\|_F.$$

Proof. We write

$$\begin{aligned}
&|A(\mathbf{z}^{(1)}, \mathbf{z}^{(2)}, \ldots, \mathbf{z}^{(r-1)}, \cdot) - A(\mathbf{w}^{(1)}, \mathbf{w}^{(2)}, \ldots, \mathbf{w}^{(r-1)}, \cdot)| \\
&\quad \le |A(\mathbf{z}^{(1)}, \mathbf{z}^{(2)}, \ldots, \mathbf{z}^{(r-1)}, \cdot) - A(\mathbf{w}^{(1)}, \mathbf{z}^{(2)}, \ldots, \mathbf{z}^{(r-1)}, \cdot)| \\
&\qquad + |A(\mathbf{w}^{(1)}, \mathbf{z}^{(2)}, \ldots, \mathbf{z}^{(r-1)}, \cdot) - A(\mathbf{w}^{(1)}, \mathbf{w}^{(2)}, \mathbf{z}^{(3)}, \ldots, \mathbf{z}^{(r-1)}, \cdot)| + \cdots.
\end{aligned}$$

A typical term above is

$$\begin{aligned}
&\big|A(\mathbf{w}^{(1)}, \mathbf{w}^{(2)}, \ldots, \mathbf{w}^{(t)}, \mathbf{z}^{(t+1)}, \ldots, \mathbf{z}^{(r-1)}, \cdot) \\
&\quad - A(\mathbf{w}^{(1)}, \mathbf{w}^{(2)}, \ldots, \mathbf{w}^{(t)}, \mathbf{w}^{(t+1)}, \mathbf{z}^{(t+2)}, \ldots, \mathbf{z}^{(r-1)}, \cdot)\big|.
\end{aligned}$$

Define B to be the matrix with components

$$B_{ij} = \sum_{j_1, j_2, \ldots, j_t, j_{t+2}, \ldots, j_{r-1}} A_{j_1, j_2, \ldots, j_t, i, j_{t+2}, \ldots, j_{r-1}, j} w_{j_1}^{(1)} \cdots w_{j_t}^{(t)} z_{j_{t+2}}^{(t+2)} \cdots z_{j_{r-1}}^{(r-1)}$$

Then the term above is bounded by

$$|B(\mathbf{z}^{(t+1)} - \mathbf{w}^{(t+1)})| \le \|B\|_2 |\mathbf{z}^{(t+1)} - \mathbf{w}^{(t+1)}| \le \|B\|_F \eta \sqrt{n} \le \|A\|_F \eta \sqrt{n}.$$

The claim follows from our choice of η. □

Next, we analyse the error incurred by sampling. Consider the $(r-1)$-tuple $(i_1, i_2, \ldots, i_{r-1}) \in I$, and define the random variables X_i by

$$X_i = \frac{A_{i_1, i_2, \ldots, i_{r-1}, i} w_{i_1}^{(1)} w_{i_2}^{(2)} \cdots w_{i_{r-1}}^{(r-1)}}{p(i_1, i_2, \ldots, i_{r-1})}.$$

It follows that

$$E(X_i) = A(\mathbf{w}^{(1)}, \mathbf{w}^{(2)}, \ldots, \mathbf{w}^{(r-1)}, \cdot)_i.$$

We estimate the variance using a calculation similar to the matrix case:

$$\begin{aligned}\sum_i \mathrm{Var}(X_i) &\le \sum_i \sum_{i_1,i_2,\ldots,i_{r-1}} \frac{A^2_{i_1,i_2,\ldots,i_{r-1},i}(w^{(1)}_{i_1}\cdots w^{(r-1)}_{i_{r-1}})^2}{p(i_1,i_2,\ldots,i_{r-1})}\\ &= \sum_{i_1,i_2,\ldots,i_{r-1}} \frac{(w^{(1)}_{i_1}w^{(2)}_{i_2}\cdots w^{(r-1)}_{i_{r-1}})^2}{p(i_1,i_2,\ldots,i_{r-1})} \sum_i A^2_{i_1,i_2,\ldots,i_{r-1},i}\\ &= \|A\|_F^2 \sum_{i_1,i_2,\ldots,i_{r-1}} (w^{(1)}_{i_1}w^{(2)}_{i_2}\cdots w^{(r-1)}_{i_{r-1}})^2 = \|A\|_F^2.\end{aligned}$$

Consider the y_i computed by the algorithm when all $\hat{z}^{(t)}_{i_t}$ are set to $w^{(t)}_{i_t}$. This will clearly happen at some point during the enumeration. This $\mathbf{y}_i$ is just the sum of s i.i.d. copies of X_i, one for each element of I. Thus we have

$$E(\mathbf{y}) = sA(\mathbf{w}^{(1)},\mathbf{w}^{(2)},\ldots,\mathbf{w}^{(r-1)},\cdot), \quad \mathrm{Var}(\mathbf{y}) = E(\|\mathbf{y}-E(\mathbf{y})\|^2) \le s\|A\|_F^2.$$

Let

$$\zeta = sA(\mathbf{z}^{(1)},\mathbf{z}^{(2)},\ldots,\mathbf{z}^{(r-1)}).$$

Since we only want to find the maximum to within an additive error of $\epsilon s\|A\|_F/2$, without loss of generality, we may assume $\|\zeta\| \ge \epsilon\|A\|_F/2$. By Chebyshev's inequality, it follows that with high probability $\|\mathbf{y}-\zeta\| \le c\sqrt{s}\|A\|_F$. One can now show that

$$\left\|\frac{\mathbf{y}}{\|\mathbf{y}\|} - \frac{\zeta}{\|\zeta\|}\right\| \le c\epsilon.$$

From this we obtain

$$\left\|A\left(\frac{\mathbf{y}}{\|\mathbf{y}\|}\right) - A\left(\frac{\zeta}{\|\zeta\|}\right)\right\|_F \le \frac{\epsilon}{10}\|A\|_F.$$

Thus, for any $r-1$ unit vectors $\mathbf{a}^{(1)},\mathbf{a}^{(2)},\ldots,\mathbf{a}^{(r-1)}$, we have

$$\left\|A\left(\mathbf{a}^{(1)},\mathbf{a}^{(2)},\ldots,\mathbf{a}^{(r-1)},\frac{\mathbf{y}}{\|\mathbf{y}\|}\right) - A\left(\mathbf{a}^{(1)},\mathbf{a}^{(2)},\ldots,\mathbf{a}^{(r-1)},\frac{\zeta}{\|\zeta\|}\right)\right\| \le \frac{\epsilon}{10}\|A\|_F.$$

This implies that the optimal set of vectors for $A(\mathbf{y}/\|\mathbf{y}\|)$ are nearly optimal for $A(\zeta/\|\zeta\|)$. Since $\mathbf{z}^{(r)} = \zeta/\|\zeta\|$, the optimal vectors for the latter problem are $\mathbf{z}^{(1)},\ldots,\mathbf{z}^{(r-1)}$. The error bound follows using Lemma 3.5.

Finally, the running time of Algorithm 3.4 is dominated by the number of candidates we enumerate, and is given by

$$\mathrm{poly}(n)\left(\frac{1}{\eta}\right)^{s^2 r} = \left(\frac{n}{\epsilon}\right)^{O(1/\epsilon^4)}.$$

3.1. Symmetric non-negative tensors with good low-rank approximations

As we mentioned earlier, a small set of samples cannot hope to deal with every matrix. But we saw that sampling yields good results for numerically low-rank matrices (matrices where, say, $\sigma_1^2(A)+\sigma_2^2(A)+\cdots+\sigma_p^2(A) \geq c\|A\|_F^2$ for a small p). Here, we will discuss instances of tensors with a similar property.

Property X. The spectral norm of the tensor is a constant fraction of the Frobenius norm after we scale the tensor, once for each dimension.

For a matrix, the permitted scaling is to scale the rows, scale the columns or both. In this section we only deal with tensors with non-negative real entries which are also symmetric, that is, for any permutation τ of $\{1, 2, \ldots, r\}$, we have

$$A_{i_1,i_2,\ldots,i_r} = A_{i_{\tau(1)},i_{\tau(2)},\ldots,i_{\tau(r)}}.$$

We will show that two natural classes of tensors have Property X. We begin by discussing these classes for matrices. The first class of matrices are adjacency matrices of dense graphs, that is, $n \times n$ matrices with entries $\{0, 1\}$ with $\Omega(n)$ non-zeros in each row. These arise in many applications. They are easily seen to satisfy Property X with no scaling, since

$$\|A\|_F = O(n), \quad \|A\|_2 \geq \frac{1}{n}\sum_{i,j} A_{ij} \in \Omega(n).$$

Another important class of matrices represent metrics, where A_{ij} is the distance between points i and j in a metric space, that is, distances satisfy the triangle inequality. Let $D_i = \sum_{j=1}^n A_{ij}$ be the total distance of point i to other points. Then, one can prove the following result.

Lemma 3.6 (local density of metrics). A_{ij} is at most $(D_i + D_j)/n$.

Proof. By the triangle inequality, $d(x_i, x_j) \leq d(x_i, x_k) + d(x_k, x_j)$, for all k. Summing, for $k = 1, \ldots, n$, we obtain

$$n\, d(x_i, x_j) \leq \sum_{k=1}^{n} d(x_i, x_k) + \sum_{k=1}^{n} d(x_k, x_j) = D_i + D_j,$$

whence $d(x_i, x_j) \leq (D_i + D_j)/n$. □

We call the lemma the 'local density property' since, for the previous example of dense graphs, each entry of the adjacency matrix is indeed at most a constant times the row and column average. Metrics also have this property for each entry.

Using this, one can show that a row and column scaled version of A satisfies Property X. Let

$$\bar{D} = \frac{1}{n}\sum_{i=1}^{n} D_i.$$

The scaled matrix B is defined by

$$B_{ij} = \frac{A_{ij}}{\sqrt{(D_i + \bar{D})(D_j + \bar{D})}}.$$

We will not prove here that B has Property X, but this will follow from the general tensor case below. In fact, we will see that for tensors, a global averaged version of the density lemma holds, and implies Property X.

Let A be any symmetric non-negative tensor. Let

$$D_i = \sum_{i_2,i_3,\ldots,i_r} A_{i,i_2,i_3,\ldots,i_r}, \quad \bar{D} = \frac{1}{n}\sum_i D_i.$$

Definition 3.7. The *density* of a tensor A is defined by

$$\gamma(A) = \left(\sum_{i=1}^{n} D_i\right)^{r-2} \sum_{i_1,i_2,\ldots,i_r} \frac{A^2_{i_1,i_2,\ldots,i_r}}{\prod_{t=1}^{r}(D_{i_t} + \bar{D})}.$$

Lemma 3.8. Let A be an r-dimensional tensor satisfying the following local density condition:

$$A_{i_1,\ldots,i_r} \le \frac{c}{rn^{r-1}} \sum_{j=1}^{r} D_{i_j}, \quad \text{for all } i_1,\ldots,i_r \in V,$$

where c is a constant. Then A has local density at most c.

Remark. Examples of graphs that satisfy the local density condition above include graphs with total number of edges at least cn^2, metrics and quasi-metrics where there is some $\gamma > 0$ with $A_{ij} = (\text{distance between } i \text{ and } j)^\gamma$.

Proof. We need to bound the density of A. To this end,

$$\begin{aligned}
&\sum_{i_1,i_2,\ldots,i_r\in V} \frac{A^2_{i_1,\ldots,i_r}}{\prod_{j=1}^{r}(D_{i_j} + \bar{D})} \\
&\quad \le \frac{c}{rn^{r-1}} \sum_{i_1,i_2,\ldots,i_r\in V} \frac{A_{i_1,\ldots,i_r}\sum_{j=1}^{r} D_{i_j}}{\prod_{j=1}^{r}(D_{i_j} + \bar{D})} \\
&\quad \le \frac{c}{rn^{r-1}} \sum_{i_1,i_2,\ldots,i_r\in V} A_{i_1,\ldots,i_r} \sum_{j=1}^{r} \frac{1}{\prod_{k\in\{1,\ldots,r\}\setminus j}(D_{i_k} + \bar{D})}
\end{aligned}$$

$$\leq \frac{c}{rn^{r-1}}\left(\sum_{i_1,i_2,\ldots,i_r\in E} A_{i_1,\ldots,i_r}\right)\frac{r}{\bar{D}^{r-1}}$$
$$= \frac{c}{(\sum_{i=1}^n D_i)^{r-2}}.$$

Thus, the density is at most

$$\left(\sum_{i=1}^n D_i\right)^{r-2} \sum_{i_1,i_2,\ldots,i_r\in E} \frac{A^2_{i_1,\ldots,i_r}}{\prod_{j=1}^r (D_{i_j}+\bar{D})} \leq c. \qquad \square$$

Let B be a scaled version of A:

$$B_{i_1,i_2,\ldots,i_r} = \frac{A_{i_1,i_2,\ldots,i_r}}{\prod_{t=1}^r \alpha_{i_t}}, \quad \text{where } \alpha_i = \sqrt{D_i+\bar{D}}.$$

Lemma 3.9. $\|B\|_F \leq c\|B\|_2$.

Proof. We have

$$\|B\|_F^2 = \frac{\gamma(A)}{(n\bar{D})^{r-2}},$$

while, if $\alpha = (\alpha_1,\ldots,\alpha_r)^T$, then

$$\|B\|_2 \geq \frac{B(\alpha,\alpha,\ldots,\alpha)}{\|\alpha\|^r}$$
$$= \frac{1}{\|\alpha\|^r}\sum_{i_1,i_2,\ldots,i_r} B_{i_1,i_2,\ldots,i_r}\alpha_{i_1}\alpha_{i_2}\cdots\alpha_{i_r}$$
$$= \frac{1}{\|\alpha\|^r}\sum_{i_1,i_2,\ldots,i_r} A_{i_1,i_2,\ldots,i_r}$$
$$= \frac{n\bar{D}}{(2n\bar{D})^{r/2}}.$$

Hence

$$\|B\|_F^2 \leq \gamma(A)2^r\|B\|_2^2. \qquad \square$$

4. Spectral norm error for matrices

In this section we prove the result of Rudelson and Vershynin (2007) summarized earlier in (1.2). The proof involves the Hoeffding–Chernoff inequality for matrix-valued random variables, and we prove this first. The Hoeffding–Chernoff inequality for real-valued independent identically distributed random variables $X_1, X_2, \ldots, X_s$ can be stated as follows. For any positive real numbers a, t, we have

$$\Pr\left(\sum_{i=1}^{s} X_i > a\right) \le \mathrm{e}^{-ta}\,(E(\mathrm{e}^{tX_1}))^s.$$

Also, recall that the matrix exponential

$$\mathrm{e}^U = \sum_{t=0}^{\infty} \frac{U^t}{t!}$$

exists for any square matrix U. For a real symmetric matrix U with eigenvalues $\lambda_i(U)$, e^U has eigenvalues $\mathrm{e}^{\lambda_i(U)}$ and the same eigenvectors as U. Therefore e^U is always positive semidefinite.

4.1. Hoeffding–Chernoff inequality for matrix-valued random variables

Theorem 4.1. Let X be a random variable taking values which are real symmetric $d \times d$ matrices. Suppose $X_1, X_2, \dots, X_s$ are i.i.d. draws of X. For any positive real numbers a, t, we have

$$\Pr\left(\lambda_{\max}\left(\sum_{i=1}^{s} X_i\right) \ge a\right) \le d\,\mathrm{e}^{-ta}\|E\,\mathrm{e}^{tX}\|_2^s, \tag{4.1}$$

$$\Pr\left(\left\|\sum_{i=1}^{s} X_i\right\|_2 \ge a\right) \le d\,\mathrm{e}^{-ta}\big(\|E\,\mathrm{e}^{tX}\|_2^s + \|E\,\mathrm{e}^{-tX}\|_2^s\big). \tag{4.2}$$

Remark. $\lambda_{\max}$ is the largest eigenvalue. Note that having the expectation inside the norm is better than having it outside on the right-hand side, since, by the convexity of the norm function and Jensen's inequality, it follows that for a matrix-valued random variable B, we have $\|E(B)\| \le E(\|B\|)$, and it can be much less. Also, it is easy to see that applying the real-valued Hoeffding–Chernoff inequality to each entry of $\sum_{i=1}^{s} X_i$ would not yield the theorem.

Proof. Inequality (4.2) follows from (4.1), since

$$\left\|\sum_i X_i\right\|_2 = \max\left(\lambda_{\max}\left(\sum_i X_i\right),\ \lambda_{\max}\left(\sum_i (-X_i)\right)\right),$$

and we can apply the first inequality twice: once with X_i and once with $-X_i$. So we prove only the first inequality. Let $S = X_1 + X_2 + \cdots + X_s$. Then

$$\lambda_{\max}(S) \ge a \iff \lambda_{\max}(tS) \ge ta \iff \lambda_{\max}(\mathrm{e}^{tS}) \ge \mathrm{e}^{ta} \implies \mathrm{Tr}[\mathrm{e}^{tS}] \ge \mathrm{e}^{ta}.$$

Now $\mathrm{Tr}[\mathrm{e}^{tS}]$ is a non-negative real-valued random variable, so by Markov's inequality, we get that

$$\Pr(\mathrm{Tr}[\mathrm{e}^{tS}] \ge \mathrm{e}^{ta}) \le \mathrm{e}^{-ta} E(\mathrm{Tr}[\mathrm{e}^{tS}]).$$

We will upper-bound $E(\mathrm{Tr}[e^{tS}])$. To this end, we first use the Golden–Thomson inequality (Bhatia 1996), which asserts that for Hermitian matrices U, V,

$$\mathrm{Tr}[e^{U+V}] \leq \mathrm{Tr}[e^U\, e^V].$$

(The proof is not given here. Note that this is false for three matrices.) We will use this with $U = t(X_1 + X_2 + \cdots + X_{s-1})$ and $V = tX_s$. Also note that Tr and E commute. Then

$$\begin{aligned}
E(\mathrm{Tr}[e^{tS}]) &\leq E(\mathrm{Tr}[e^U\, e^V]) \\
&= \mathrm{Tr}[E(e^U\, e^V)] \\
&= \mathrm{Tr}\big[E_{X_1,X_2,\dots,X_{s-1}}(e^U\, E_{X_s}(e^V))\big] \qquad\qquad \text{(independence)} \\
&\leq \|E(e^{tX})\|_2\, \mathrm{Tr}\big[E_{X_1,X_2,\dots,X_{s-1}}(e^U)\big] \\
&\qquad\qquad (\mathrm{Tr}[BC] \leq \mathrm{Tr}[B]\, \|C\|_2 \text{ for positive semidefinite } B, C) \\
&\leq \|E(e^{tX})\|_2^{s-1} \mathrm{Tr}\big[E(e^{tX})\big] \\
&\leq d\|E(e^{tX})\|_2^s,
\end{aligned}$$

where, in the penultimate inequality, we have peeled off an X_i $s-2$ times. The factor of d arises because $\mathrm{Tr}[e^{tX}] \leq d\lambda_{\max}(e^{tX})$. No direct way of bounding $\lambda_{\max}$ without going to the trace is known. □

Notation. For two $n \times n$ real symmetric matrices B, C, we write $B \preceq C$ when $C - B$ is positive semidefinite.

Lemma 4.2. If B is a real symmetric matrix for which $\|B\|_2 \leq 1$, then $e^B \preceq I + B + B^2$.

Proof. The inequality $e^{\lambda_i} \leq 1 + \lambda_i + \lambda_i^2$, for $|\lambda_i| \leq 1$, implies

$$e^{\lambda_i} \mathbf{v}_i \mathbf{v}_i^T \preceq (1 + \lambda_i + \lambda_i^2) \mathbf{v}_i \mathbf{v}_i^T,$$

whence

$$e^B = \sum_{i=1}^d e^{\lambda_i} \mathbf{v}_i \mathbf{v}_i^T \preceq \sum_{i=1}^d (1 + \lambda_i + \lambda_i^2) \mathbf{v}_i \mathbf{v}_i^T = I + B + B^2. \qquad □$$

4.2. Applying Hoeffding–Chernoff to length-squared sampling

Suppose we use length-squared sampling on a matrix A to draw s columns in s i.i.d. trials. Let

$$p_j = \frac{|A(:, j)|^2}{\|A\|_F^2}$$

be the probability of picking column j. Define the real symmetric matrix-valued random variable Y satisfying

$$\Pr\left(Y = \frac{1}{s\, p_j}\, A(:,j)A(:,j)^T\right) = p_j. \tag{4.3}$$

We saw that $EY = AA^T/s$. Then the random variable

$$X = Y - EY \quad \text{satisfies } EX = 0. \tag{4.4}$$

Also, we have $E(X^2) \preceq E(Y^2)$; this is proved along the same lines as for real random variables, where the variance of a random variable is at most the second moment. Therefore

$$E(X^2) \preceq E(Y^2) = \sum_j p_j \frac{A(:,j)A(:,j)^T A(:,j)A(:,j)^T}{s^2\, p_j^2} = AA^T \|A\|_F^2 \frac{1}{s^2}, \tag{4.5}$$

which implies

$$\|E(X^2)\|_2 \le \frac{1}{s^2}\|A\|_2^2\|A\|_F^2. \tag{4.6}$$

We will also need an absolute upper bound on $\|X\|_2$. For this, note that

$$\begin{aligned} \|X\|_2 &\le \frac{1}{s}\max(\|A(:,j)A(:,j)^T/p_j\|_2, \|AA^T\|_2) \qquad (4.7)\\ &\le \max\left(\frac{\|A(:,j)A(:,j)^T\|_2}{s\|A(:,j)\|^2}\|A\|_F^2, \frac{\|A\|_F^2}{s}\right) = \frac{1}{s}\|A\|_F^2. \end{aligned}$$

Proposition 4.3. If t is a positive real number such that $\|tX\|_2 \le 1$ for all possible values of X, then

$$\|E\,\mathrm{e}^{\pm tX}\|_2 \le 1 + \frac{t^2}{s^2}\|A\|_2^2\|A\|_F^2 \le \mathrm{e}^{t^2\|A\|_2^2\|A\|_F^2/s^2}$$

Proof. We have $E(\mathrm{e}^{tX}) \preceq E(I + tX + t^2X^2) = I + t^2E(X^2)$, since $EX = 0$. Thus we have the proposition using (4.5). □

Note. It was important to argue that $\mathrm{e}^B \preceq I + B + B^2$ in the lemma. A weaker inequality such as $\|\mathrm{e}^B\|_2 \le \|I + B + B^2\|_2$ does not suffice. We are ready to prove (1.2).

Theorem 4.4. Let A be any $m \times n$ matrix and let C be an $m \times s$ matrix obtained by length-squared sampling and scaling to have $E(CC^T) = AA^T$. (C consists of columns $Y_1, Y_2, \ldots, Y_s$, which are i.i.d. copies of Y defined in (4.3).) Then, for all $\varepsilon \in [0, \|A\|_2/\|A\|_F]$,[4] we have

$$\Pr\big(\|CC^T - AA^T\|_2 \ge \varepsilon\|A\|_2\|A\|_F\big) \le 2n\,\mathrm{e}^{-\varepsilon^2 s/4}.$$

[4] If $\varepsilon \ge \|A\|_2/\|A\|_F$, then the zero matrix is a good enough approximation to AA^T.

Hence, for $s \geq (c \ln n)/\varepsilon^2$, with high probability we have

$$\|CC^T - AA^T\|_2 \leq \varepsilon \|A\|_2 \|A\|_F. \tag{4.8}$$

Proof. It is easy to see that

$$CC^T - AA^T = \sum_{i=1}^{s} X_i,$$

where the X_i are i.i.d. copies of X defined in (4.4). Then

$$\begin{aligned}
&\Pr\left(\left\|\sum_{i=1}^{s} X_i\right\|_2 \geq \varepsilon \|A\|_2 \|A\|_F\right) \\
&\quad \leq n\, \mathrm{e}^{-t\varepsilon \|A\|_2 \|A\|_F} \left(\|E(\mathrm{e}^{tX_1})\|_2^s + \|E(\mathrm{e}^{-tX_1})\|_2^s\right) \\
&\qquad\qquad\qquad\qquad\qquad \text{(for any } t > 0\text{, by Theorem 4.1)} \\
&\quad \leq 2n\, \mathrm{e}^{-t\varepsilon \|A\|_2 \|A\|_F} \mathrm{e}^{t^2 \|A\|_2^2 \|A\|_F^2 / s} \\
&\qquad\qquad \text{(provided } t \leq s/\|A\|_F^2\text{, by Proposition 4.3 and (4.7))} \\
&\quad \leq 2n\, \mathrm{e}^{-\varepsilon^2 s/4},
\end{aligned}$$

setting

$$t = \frac{\varepsilon s}{2\|A\|_F \|A\|_2} \leq \frac{s}{\|A\|_F^2}.$$

So we see that for $s \geq (c \ln n)/\varepsilon^2$ and a large enough constant c, with high probability,

$$\|CC^T - AA^T\|_2 \leq \varepsilon \|A\|_2 \|A\|_F. \qquad \square$$

5. Preconditioned length-squared sampling

In this section we show how to solve problem (1.3). We first discuss a principal motivation.

5.1. Relative error in the residual

We have seen that if we sample a set of columns of A and scale them to form an $m \times s$ matrix C, the restriction of A to the column space of the top k right singular vectors of C is a good approximation to A, in the sense of Theorem 2.1. But the error is in terms of $\|A\|_F$. In this section we see how to get errors of the form $O(\|A - A_k\|_F)$, where A_k denotes the best rank-k approximation to A. Of course, we can get such errors by using the SVD of A. The point here is that we will see that a random sample corresponding to preconditioned length-squared sampling on A_k (instead of A) gives us a subset of columns of A in whose span we find an $O(\|A - A_k\|_F)$-type error.

There are at least three approaches to getting such a relative error bound. The first is to use length-squared sampling *iteratively* (Deshpande *et al.* 2006). In this approach, after performing length-squared sampling on the initial matrix A to obtain a sample S_1, the columns of A are sampled again, but this time with probability proportional to their squared lengths in the residual, that is, after orthogonal projection onto the span of S_1. This process is repeated, and results in a geometrically decaying error factor with each round. Such an approach was also used by Li, Miller and Peng (2013). More recent papers analyse a similar approach with uniform sampling (Cohen *et al.* 2015, Cohen, Musco and Musco 2017).

The second approach is *volume sampling*, which extends length-squared sampling by picking subsets of k columns jointly (Deshpande and Vempala 2006). The probability of picking a k-subset is proportional to the squared volume of the k-dimensional simplex induced by the columns along with the origin, a generalization of length-squared sampling. This single sample approach has expected squared error $(k+1)\|A-A_k\|_F^2$, the best possible bound using k columns of the matrix. This can be improved by one round to length-squared sampling in the residual to $(1+\epsilon)$ (relative error) using $O(k/\epsilon)$ samples.

The third approach (Drineas *et al.* 2008), which we present here, uses only one round of length-squared sampling, but the lengths are according to a different matrix, which can be viewed as a preconditioned version of the original matrix, and corresponds to sampling columns according to their *leverage scores*. Before we go to the main proof of the preconditioned length-squared sampling method, we discuss its application to low-rank approximation in more detail.

Let S be the $n \times s$ column selector (no scaling of columns) matrix, so that $C = AS$. Let V_k be the $n \times k$ matrix with top k right singular vectors of A as columns. The probability of picking column j of A to include in C is according to the squared length of column j of $A_k^+ A_k = V_k V_k^T$, where A_k^+ is the pseudo-inverse of A. Since V_k has orthonormal columns, this is the same as the squared length of column j of V_k^T. Then the approximation to A in the column space of C is

$$X = C(V_k^T S)^+ V_k^T.$$

The error is bounded as follows:

$$\begin{aligned}
A - X &= A - AS(V_k^T S)^+ V_k^T \\
&= A - (AV_k V_k^T + (A - A_k))S(V_k^T S)^+ V_k^T \\
&= A - AV_k V_k^T - (A - A_k)S(V_k^T S)^+ V_k^T \\
&= \underbrace{(A - A_k)}_{X_1} - \underbrace{(A - A_k)S(V_k^T S)^+ V_k^T}_{X_2}.
\end{aligned}$$

Then

$$\|X_1\|_F = \|A - A_k\|_F,$$
$$\|X_2\|_F \le \|(A - A_k)S\|_F \underbrace{\|(V_k^T S)^+ V_k^T\|_2}_{X_3}.$$

Hence $E(\|(A - A_k)S\|_F) \le \|A - A_k\|_F$. Now, $X_3 = 1/\sigma_{\min}(V_k^T S)$ (the last V_k^T can be omitted since it has orthonormal rows). We can bound this as follows:

$$X_3 = \frac{1}{\sqrt{\lambda_{\min}(V_k^T S S^T V_k)}}.$$

Noting that $V_k V_k^T = I_k$, we can view this as selecting columns (as in S) according to length-squared sampling on V_k to multiply V_k^T by V_k. By the guarantee for approximate matrix multiplication using length-squared sampling, the error is at most $(2 + \varepsilon)\|A - A_k\|_F$. This can be improved to $(1 + \epsilon)\|A - A_k\|_F$.

5.2. Leverage score sampling

Here we present the main guarantee for sampling according to the leverage scores, that is, the preconditioned squared lengths.

Theorem 5.1. Suppose A is any $d \times n$ matrix with rank r and SVD $A = \sum_{t=1}^r \sigma_t \mathbf{u}_t \mathbf{v}_t^T$, and W is an $n \times d$ symmetric matrix $\sum_{t=1}^r \sigma_t{}^{-1} \mathbf{v}_t \mathbf{u}_t^T$.[5] Let $p_j, j = 1, 2, \ldots, n$ be defined by[6]

$$p_j = \frac{|(WA)(:, j)|^2}{\|WA\|_F^2}.$$

If $\varepsilon \in [0, 1]$ and $s \ge (cr \ln n)/\varepsilon^2$, and we draw s columns of A in i.i.d. trials, with probabilities $\{p_j\}$, scale the chosen column j of A by $1/\sqrt{sp_j}$, and form a $d \times s$ matrix C with these scaled columns, then with high probability we have

$$\|\mathbf{x}^T C\|^2 \in \big[(1 - \varepsilon)\|\mathbf{x}^T A\|^2,\ (1 + \varepsilon)\|\mathbf{x}^T A\|^2\big], \quad \text{for all } \mathbf{x}.$$

Remark. The p_j are proportional to the squared lengths of columns of WA, but the actual sampling picks corresponding columns of A.

Proof. Let V be the space spanned by columns of A. Write vector $\mathbf{x}$ as $\mathbf{x} = \mathbf{z} + \mathbf{y}$, with $\mathbf{z} \in V$ and $\mathbf{y}$ orthogonal to V. Clearly, $\|\mathbf{x}^T A\| = \|\mathbf{z}^T A\|$ and $\|\mathbf{x}^T C\| = \|\mathbf{z}^T C\|$. So it suffices to prove the theorem assuming $\mathbf{x} \in V$.

[5] W is the pseudo-inverse of A.

[6] We call p_j the 'preconditioned length-squared' probabilities.

For any $\mathbf{x} \in V$, there is a $\mathbf{y} \in \text{span}(\mathbf{v}_1, \mathbf{v}_2, \ldots, \mathbf{v}_r)$ such that $\mathbf{x}^T = \mathbf{y}^T W$. Note that

$$WA = \sum_{t=1}^{r} \mathbf{v}_t \mathbf{v}_t^T,$$

so $\|WA\|_2^2 = 1$ and $\|WA\|_F^2 = r$. Apply Theorem 4.4 to WA with the ε of that theorem set to the ε here divided by $\sqrt{r}$. Since $s \geq r \ln n/\varepsilon^2$, the hypothesis of Theorem 4.4 is satisfied. Thus

$$\|WC(WC)^T - WA(WA)^T\|_2 \leq \varepsilon,$$

which implies

$$|\|\mathbf{y}^T WC\|^2 - \|\mathbf{y}^T WA\|^2| \leq \varepsilon \|\mathbf{y}\|^2 = \varepsilon \|\mathbf{y}^T (WA)\|^2, \quad \text{for all } \mathbf{y},$$

since, for $\mathbf{y} \in \text{span}(\mathbf{v}_1, \mathbf{v}_2, \ldots, \mathbf{v}_r)$,

$$\|\mathbf{y}^T WA\| = \|\mathbf{y}\|.$$

This implies the theorem. □

6. Subspace embeddings and applications

The methods of the previous section allow us to get relative errors in the residual, but can be more expensive computationally. They can be implemented in time dominated by the number of non-zeros of the input matrix times k/ϵ for a rank-k approximation. In this section we will see methods to improve this to only the order of the number of non-zeros in A (*i.e.* input sparsity time) plus lower-order terms that depend only on one of the dimensions of A. The guarantee is slightly weaker, but suffices for applications.

6.1. Very sparse subspace embeddings

Suppose A is an $n \times d$ given data matrix. For this section we assume that n is much larger than d, so A is a tall skinny matrix. For positive reals a, b, we say $a \cong_\varepsilon b$ if $a \in ((1-\varepsilon)b, (1+\varepsilon)b)$.

Definition 6.1. A $t \times n$ matrix S is a subspace embedding (for A) with error ε if

$$\|SA\mathbf{x}\| \cong_\varepsilon \|A\mathbf{x}\|, \quad \text{for all } \mathbf{x} \in \mathbb{R}^d.$$

We aim to bound t by a function of only d and ε. The point of subspace embedding is as follows: we can compute with SA instead of A, and if $t \ll n$, the matrix SA, which is $t \times d$, is much smaller than A. For example, if we want to find the top singular value (vector) of A within relative error ε, we can work on SA instead of A. The next theorem is from Clarkson and Woodruff (2013).

Theorem 6.2 (Clarkson and Woodruff). For a matrix A of rank r, the following matrix S is a subspace embedding with high probability provided $t \geq \text{poly}(r/\varepsilon)$. For each $j \in \{1, 2, \ldots, n\}$, pick an $i \in \{1, 2, \ldots, d\}$ uniformly at random, and set $S_{ij} = \pm 1$ with probability 1/2. Then, with high probability,

$$\|SA\mathbf{x}\| \cong_\varepsilon \|A\mathbf{x}\|, \quad \text{for all } \mathbf{x} \in \mathbb{R}^d.$$

Remarks.

(i) Note that the theorem states that, with high probability, $\cong_\varepsilon$ *for all* x. It is a much stronger statement than saying 'for each x, $\cong_\varepsilon$ holds, with high probability'.

(ii) Note that the construction of S is 'oblivious', *i.e.*, independent of A.

(iii) Since S has only one non-zero entry per column, SA can be computed in time equal to a constant times the number of non-zero entries of A (often denoted $\text{nnz}(A)$), using a sparse representation of the input matrix.

Subspace embeddings as presented here are due to Clarkson and Woodruff (2013) using the work of Dasgupta, Kumar and Sarlós (2010). For further improvements and applications, see Woodruff (2014).

6.2. *Elementary subspace embedding via Johnson–Lindenstrauss*

First, we prove a simpler theorem using the Johnson–Lindenstrauss random projection theorem. It says that we can project high-dimensional vectors to a lower-dimensional space and still preserve lengths approximately.

The projection $f : \mathbb{R}^n \to \mathbb{R}^k$ that we will examine (in fact, many related projections are known to work as well) is as follows. Pick k vectors $\mathbf{u}_1, \mathbf{u}_2, \ldots, \mathbf{u}_k$ in $\mathbb{R}^n$, by choosing all the nk coordinates independently, each from the Gaussian distribution

$$\frac{1}{(2\pi)^{n/2}} \exp(-\|\mathbf{x}\|^2/2).$$

Then, for any vector $\mathbf{v}$ in $\mathbb{R}^n$, we define the projection $f(\mathbf{v})$ by

$$f(\mathbf{v}) = (\mathbf{u}_1 \cdot \mathbf{v}, \mathbf{u}_2 \cdot \mathbf{v}, \ldots, \mathbf{u}_k \cdot \mathbf{v}). \tag{6.1}$$

We will show that with high probability $\|f(\mathbf{v})\| \approx \sqrt{k}\|\mathbf{v}\|$, so if we have to find, say, $\|\mathbf{v}\|$, it suffices to find $\|f(\mathbf{v})\|$ (since the factor of $\sqrt{k}$ is known). The original proof was to project $\mathbf{v}$ onto a random k-dimensional subspace of $\mathbb{R}^n$. The proof is more complicated for that situation, since projecting onto a random subspace is not equivalent to picking k vectors independently at random and taking dot products. The bound below for a Gaussian random projection is simpler to prove, and still has the same set of applications

(Indyk and Motwani 1998, Arriaga and Vempala 1999, Dasgupta and Gupta 2003, Arriaga and Vempala 2006, Vempala 2004).

Theorem 6.3 (Johnson–Lindenstrauss). Let $\mathbf{v}$ be a fixed vector in $\mathbb{R}^n$ and let f be defined by (6.1). Then, for $\varepsilon \in (0,1)$,

$$\Pr\bigl(\bigl|\|f(\mathbf{v})\| - \sqrt{k}\|\mathbf{v}\|\bigr| \geq \varepsilon\sqrt{k}\|\mathbf{v}\|\bigr) \leq 2\,\mathrm{e}^{-(\epsilon^2-\epsilon^3)k/4},$$

where the probability is taken over the random draws of vectors $\mathbf{u}_i$ used to construct f.

Proof. See Dasgupta and Gupta (2003, Theorem 2.1). □

The theorem deals with one vector. However, subspace embeddings have to work for infinitely many vectors. We cannot do a simple union bound because the number of vectors is infinite. Instead, we use what is called an ε-net.

Definition 6.4. A set N of unit vectors is an *ε-net* for the unit sphere $S = \{\mathbf{x} \in \mathbb{R}^d : \|\mathbf{x}\| = 1\}$ if, for any $\mathbf{x} \in S$, there exists $\mathbf{y} \in N$ such that $\|\mathbf{x} - \mathbf{y}\| \leq \varepsilon$.

The next lemma bounds the size of the ϵ-net for a sphere.

Lemma 6.5. There is an ε-net N with $|N| \leq (1 + 2/\varepsilon)^d$.

Proof. The proof is by taking a maximal set N of points that are pairwise ϵ-separated and using a volume argument (see *e.g.* Lemma 4.10 of Pisier 1989), that is, balls of radii $\epsilon/2$ centred at each point of the net are disjoint and the union of all these balls is contained in a ball of radius $1 + (\epsilon/2)$. Hence

$$|N| \leq \left(\frac{1+\epsilon/2}{\epsilon/2}\right)^d = \left(1 + \frac{2}{\varepsilon}\right)^d. \qquad \square$$

Now we may apply Theorem 6.3 as follows. Suppose V is a d-dimensional subspace of $\mathbb{R}^n$, for example, $V = \{A\mathbf{x}\}$ for an $n \times d$ matrix A. The set of unit vectors in V has an ε-net N of size at most $\mathrm{e}^{cd\ln(1/\varepsilon)}$. Choose $k \geq cd\ln(1/\varepsilon)/\varepsilon^2$ so that, for a single $\mathbf{x} \in V$, the probability that $\|M\mathbf{x}\| \not\cong_\varepsilon \sqrt{k}\|\mathbf{x}\|$ is at most $1/|N|^2$. Then, just by union bounds, we get

$$\Pr\bigl(\text{for all } \mathbf{x} \in N,\ \|M\mathbf{x}\| \cong_\varepsilon \sqrt{k}\|\mathbf{x}\|\bigr) \geq 1 - \frac{1}{|N|}.$$

Now $f(\mathbf{v})$ can be written as a matrix product, $f(\mathbf{v}) = M\mathbf{v}$, where M is a $k \times n$ matrix with $\mathbf{u}_1, \mathbf{u}_2, \dots, \mathbf{u}_k$ as its rows. We claim that the above suffices to prove that $\|M\mathbf{x}\| \cong_{3\varepsilon} \sqrt{k}\|\mathbf{x}\|$ for *every* $x \in V$.

Here is a first attempt to prove the claim. For any $\mathbf{x} \in V$, there is some $\mathbf{y} \in N$ with $\|\mathbf{x} - \mathbf{y}\| \leq \varepsilon$. So, $\|M\mathbf{x}\| = \|M\mathbf{y}\| + \delta$, where $|\delta| \leq \|M(\mathbf{x} - \mathbf{y})\|$. But bounding $\|M(\mathbf{x} - \mathbf{y})\|$ needs a good bound on $\|M\|_2$ which we do not yet have. Instead, we will see that by repeated use of the ε-net, we can write $\mathbf{x}$ as a convergent series of linear combinations of elements in N.

First, there exists a $\mathbf{y}^{(1)} \in N$ with $\|\mathbf{x} - \mathbf{y}^{(1)}\| = \theta_1 \leq \varepsilon$. Now there exists a $\mathbf{y}^{(2)} \in N$ with

$$\left\| \mathbf{y}^{(2)} - \frac{\mathbf{x} - \mathbf{y}^{(1)}}{\theta_1} \right\| = \theta_2 \leq \varepsilon.$$

Continuing, we get that $\mathbf{x} = \mathbf{y}^{(1)} + \theta_1 \mathbf{y}^{(2)} + \theta_1 \theta_2 \mathbf{y}^{(3)} + \cdots$, where $\theta_i \in [0, \varepsilon]$. Therefore,

$$\begin{aligned}
\frac{1}{\sqrt{k}} \|M\mathbf{x}\| &= \frac{1}{\sqrt{k}} \|M(\mathbf{y}^{(1)} + \theta_1 \mathbf{y}^{(2)} + \cdots)\| \\
&\geq \frac{1}{\sqrt{k}} (\|M\mathbf{y}^{(1)}\| - \theta_1 \|M\mathbf{y}^{(2)}\| - \theta_1\theta_2 \|M\mathbf{y}^{(3)}\| - \cdots) \\
&\geq (1 - \varepsilon) - (1 + \varepsilon)(\theta_1 + \theta_1\theta_2 + \theta_1\theta_2\theta_3 + \cdots) \\
&\geq 1 - 3\varepsilon.
\end{aligned}$$

A similar argument also proves an upper bound of $1 + 3\varepsilon$ on $\|M\mathbf{x}\|$.

Now we return to the situation where A is an $n \times d$ matrix of rank r. The set $\{A\mathbf{x} : \mathbf{x} \in \mathbb{R}^d\}$ is an r-dimensional subspace of $\mathbb{R}^n$. Hence, by the above, as long as $k \geq \mathrm{poly}(r/\varepsilon)$, we deduce that M is a subspace embedding for A. But M is dense, and computing MA is costly (and in general, superlinear).

6.3. Sparse embeddings

Now consider the sparse matrix S defined in Theorem 6.2. Let V be the subspace spanned by the columns of A. Unfortunately, we cannot carry out the same sort of argument as above, because it is not true that for every (individual) $\mathbf{x} \in V$ we have

$$\Pr(\|S\mathbf{x}\| \not\cong_\varepsilon \|\mathbf{x}\|) \leq \exp(-cd).$$

Counter-examples are 'lumpy' $\mathbf{x}$ with some 'big' coordinates: for example, if $x_1 = x_2 = 1/\sqrt{2}$, with probability $\mathrm{poly}(1/t)$, we could have $S_{11}, S_{12} \neq 0$, whence $\|S\mathbf{x}\| \not\cong_\varepsilon |\mathbf{x}|$. On the other hand, 'non-lumpy' $\mathbf{x}$ – with, say, all $\|x_i\| \leq c\|\mathbf{x}\|/\sqrt{n}$ – can be shown to satisfy $\Pr(\|S\mathbf{x}\| \not\cong_\varepsilon \|\mathbf{x}\|) \leq \exp(-cd)$. We do not do this since it is not required for our proof.

Here is a simple and crucial point.

Lemma 6.6. Suppose $\alpha \in [0, 1]$. There is a subset H of $\{1, 2, \ldots, n\}$ with $|H| \leq d/\alpha$ such that

$$|(A\mathbf{x})_i| \leq \sqrt{\alpha}\|A\mathbf{x}\|, \quad \text{for all } \mathbf{x} \text{ and all } i \notin H.$$

The lemma asserts that only coordinates in H (for heavy) can be bigger than $\sqrt{\alpha}\|A\mathbf{x}\|$ in absolute value. Note that H is independent of $\mathbf{x}$.

Proof. Let $r = \text{rank}(A)$. Form an $n \times r$ matrix U with the left singular vectors of A as its columns. Since $\|U\|_F^2 = r$, we have that[7]

$$H = \{i : \|U(i,:)\| \geq \sqrt{\alpha}\} \text{ satisfies } |H| \leq r/\alpha \leq d/\alpha. \tag{6.2}$$

We assert that this H satisfies the conclusion of the lemma. For any $\mathbf{x}$, we have that $A\mathbf{x}$ is a linear combination of the columns of U, and so $A\mathbf{x} = U\mathbf{z}$ for some $\mathbf{z}$. Since the columns of U are orthonormal, we have $\|\mathbf{z}\| = \|A\mathbf{x}\|$. Also $(A\mathbf{x})_i = U(i,:)^T\mathbf{z}$, and so for $i \notin H$,

$$|(A\mathbf{x})_i| \leq \|U(i,:)\|\|\mathbf{z}\| \leq \sqrt{\alpha}\|A\mathbf{x}\|. \qquad \square$$

Let $[n] \setminus H = L$ (L for light). We can write any $\mathbf{y}$ in the column space of A as the sum of two vectors $\mathbf{y}^H, \mathbf{y}^L$, where $y_i^H = y_i$ for $i \in H$ and $y_i^H = 0$ for $i \in L$, and similarly $\mathbf{y}^L$ is $\mathbf{y}$ restricted to L. Then

$$\|S\mathbf{y}\|^2 = \|S\mathbf{y}^H\|^2 + \|S\mathbf{y}^L\|^2 + 2(S\mathbf{y}^H)^T(S\mathbf{y}^L). \tag{6.3}$$

We deal with each of these three terms. For each $j \in \{1,2,\ldots,n\}$, we picked an $i \in \{1,2,\ldots,t\}$ at random and set $S_{i,j} = \pm 1$. It will be convenient to call the i we picked the 'hash of j'. Since $n \gg t$, $j \in \{1,2,\ldots,n\}$ and $i \in \{1,2,\ldots,t\}$, it resembles a random hash function.

Lemma 6.7. With probability at least $1 - (|H|^2/t)$, we have $\|S\mathbf{y}^H\| = \|\mathbf{y}^H\|$ when $\mathbf{y} = A\mathbf{x}$.

Proof. The probability that the hashes of two different $j \in H$ are the same is at most $|H|^2/t$, and we will choose $t > c|H|^2$, so the probability of this hash collision is small. (Recall the 'birthday paradox'.) Note that this event has nothing to do with a particular $\mathbf{y}$. If there is no hash collision, then the submatrix of S in the columns corresponding to H is just a permutation matrix (with signs), and the lemma follows. $\square$

For bounding the other two terms of (6.3), the following theorem due to Dasgupta *et al.* (2010) (which we do not prove here) is useful. Call a vector $\mathbf{y} = A\mathbf{x}$ non-lumpy if no coordinate of $\mathbf{y}$ is too big in absolute value. The theorem asserts that with high probability, for all non-lumpy vectors simultaneously, we have bounds on the difference $\|S\mathbf{y}\| - \|\mathbf{y}\|$. Since we have already taken care of the heavy coordinates, this will help us take care of the rest.

Theorem 6.8 (Theorem 2 of Dasgupta *et al.* 2010). Suppose $\varepsilon, \delta \in (0,1)$ and $t \geq 12\log(1/\delta)/\varepsilon^2$, and let S be a $t \times n$ matrix chosen as in Theorem 6.2. Suppose $\mathbf{y}$ is any vector in $\mathbb{R}^n$ with $|y_i| \leq \sqrt{\alpha}$, where $1/\alpha = 16\log(1/\delta)\log^2(t/\delta)/\varepsilon$. Then

$$\Pr\big(\big|\|S\mathbf{y}\| - \|\mathbf{y}\|\big| \geq \varepsilon\big) \leq 3\delta.$$

[7] $U(i,:)$ is the ith row of U.

Remark. The theorem is for one vector $\mathbf{y}$. It is a sparse version of Johnson–Lindenstrauss, and finds other uses as well.

To prove that the probability result holds simultaneously for all $\mathbf{y}$, we apply Theorem 6.8 as follows.

(i) We examine the construction of an ε-net to see that if we want to cover only non-lumpy $\mathbf{y}$ (each $|y_i| \le \sqrt{\alpha}$) in a d-dimensional subspace of $\mathbb{R}^n$ (like $\{A\mathbf{x}\}$), then we can get the same size net with all coordinates of vectors in the net having absolute value at most $2\sqrt{\alpha}$.

(ii) We choose $\delta \le \exp(-cd \ln d/\varepsilon^2)$ and apply Theorem 6.8 to show that, for each non-lumpy $\mathbf{y}$ *in the net*, the probability of failure is at most δ. Then use the union bound over all elements of the net.

(iii) Finally, we use the argument of Section 6.2 to write any non-lumpy $\mathbf{y}$ as a convergent series of linear combinations of net elements, and use this to conclude that the following holds.

Theorem 6.9. Suppose $\varepsilon \in (0, 1)$ and $t = \text{poly}(d/\varepsilon)$ and $\alpha = \text{poly}(\varepsilon/d)$ satisfying[8]

$$t \ge \frac{d}{\alpha\varepsilon^2}, \quad \frac{1}{\alpha} \ge c(\log^2 t)(d^3(\ln d)^3/\varepsilon^6). \tag{6.4}$$

Let S be a $t \times n$ matrix chosen as in Theorem 6.2. Then,

$$\Pr\bigl(\text{for all } \mathbf{y}, \text{ with } |y_i| \le \sqrt{\alpha},\ \|S\mathbf{y}\| \cong_\varepsilon \|\mathbf{y}\|\bigr) \ge 1 - \varepsilon.$$

Theorem 6.9 can be used to imply immediately that

$$|\|S\mathbf{y}^L\| - \|\mathbf{y}^L\|| \le c\varepsilon, \quad \text{for all } \mathbf{y} \in \{Ax\}.$$

This takes care of the second term on the right-hand side of (6.3).

The third term requires a further argument. Let

$$L' = \{j \in L : \text{the hash of } j = \text{the hash of some } j' \in H\}.$$

Note that L' depends only upon the choices made for S, that is, it does not depend on any particular y. We make three assertions, all holding with high probability:

$$|(S\mathbf{y}^H)^T(S\mathbf{y}^L)| = |(S\mathbf{y}^H)^T(S\mathbf{y}^{L'})| \le \|S\mathbf{y}^H\|\,\|S\mathbf{y}^{L'}\| \le \|S\mathbf{y}^{L'}\|, \tag{6.5}$$

$$\|S\mathbf{y}^{L'}\| \le O(\varepsilon) + \|\mathbf{y}^{L'}\|, \tag{6.6}$$

$$\|\mathbf{y}^{L'}\| \le \varepsilon. \tag{6.7}$$

Statement (6.5) is easy to check. For (6.6), note that for a $j \in L'$ conditioned on its hash being one of the $|H|$ hashes of $j' \in H$, it is uniform random. So, S restricted to columns in L' is an $|H| \times |L'|$ matrix constructed the

[8] It is easy to see that t, α satisfying these inequalities exist.

same way as the whole of S: pick a random row for each column and set the entry in that row and column to ± 1 with probability 1/2 each. So we may apply Theorem 6.9 to this situation to get (6.6). If the actual $|H|$ is much smaller than the upper bound in (6.2), we can just augment it to get to size $cd \ln d/\varepsilon^2$. Statement (6.7) will follow using a version of Bernstein's inequality (often called Freedman's inequality) from probability theory, as follows.

Lemma 6.10. Suppose $Y_1, Y_2, \ldots, Y_n$ are independent real-valued random variables with $|Y_i| \leq M$. Let $b = \mathrm{Var}(\sum_{i=1}^n Y_i)$, and let a be any positive real number. Then

$$\Pr\left(\left|\sum_{i=1}^n Y_i\right| \geq a\right) \leq 2\exp\left(-\frac{a^2}{2(aM+b)}\right).$$

Proof. See Freedman (1975). □

To prove (6.7), consider a single $\mathbf{y}$ first. Since for $j \in L$, $y_j^{L'} = y_j \mathbf{1}(j \in L')$ (where $\mathbf{1}$ is an indicator random variable), we obtain

$$\mathrm{Var}((y_j^{L'})^2) \leq y_j^4 \frac{|H|}{t} \leq y_j^4 \frac{d}{t\alpha} \leq \varepsilon^2 y_j^4,$$

using (6.2) and (6.4). Thus

$$\mathrm{Var}\left(\sum_{j\in L} (y_j^{L'})^2\right) \leq \varepsilon^2 \sum_j y_j^4 \leq \varepsilon^2 |\mathbf{y}|^2 \alpha \leq \alpha\varepsilon^2.$$

Hence Lemma 6.10 with $a = \varepsilon^2$ implies that

$$\Pr(\|\mathbf{y}^{L'}\|^2 \geq \varepsilon^2) \leq 2\exp(-c\varepsilon^4/(\varepsilon^2\alpha + \varepsilon^2\alpha)) \leq 2\exp(-cd\ln d/\varepsilon^2).$$

This probability is small enough that we can take a union bound over an ε-net and then extend to all vectors again by expressing the vector as a convergent series of net vectors.

7. Conclusion

The goal of this survey is to convey some core ideas of using random sampling to obtain fast algorithms for matrix problems in numerical linear algebra. These algorithms work for arbitrary input matrices (and tensors where appropriate). They are randomized approximation algorithms with a probability of failure that can be controlled by using a larger sample.

The problems and their randomized algorithms have many applications, including machine learning, combinatorial optimization, solving linear systems and graph decompositions. For a more extensive discussion of these applications we refer the reader to Kannan and Vempala (2009), Mahoney (2011), Vishnoi (2013) and Woodruff (2014).

Randomized numerical linear algebra is now a large and active field. The basic theorems presented here have also been extended in many ways. We mention some of them briefly.

Achlioptas and McSherry (2007) gave a different method for additive error low-rank approximation based on sparsifying the original matrix by sampling each entry independently. The CUR approximation has been improved by Boutsidis and Woodruff (2014). For multiplicative error low-rank approximation, the methods based on iterative length-squared sampling, volume sampling and preconditioned length-squared sampling all require two or more passes over the data. Sarlós (2006) gave the first algorithm that uses only one pass over the data and gives a multiplicative error. Deshpande and Rademacher (2010) showed that volume sampling can be implemented efficiently with random projections, while Anari *et al.* (2016) gave a Markov chain method to sample efficiently from the volume sampling distribution. Drineas, Magdon-Ismail, Mahoney and Woodruff (2012) improved the complexity of preconditioned length-squared sampling (leverage score sampling). While the original method of Spielman and Srivastava for graph sparsification was a polynomial-time algorithm establishing an asymptotically optimal bound on the size of the sparse graph, Lee and Sun (2015) have improved the running time to almost linear. For sparse embeddings, the work of Clarkson and Woodruff described here has been improved by Meng and Mahoney (2013), Nelson and Nguyên (2013) and further by Cohen (2016). In related work, Tropp (2011) and Kane and Nelson (2014) have improved the sparsity of random projection matrices for Euclidean length preservation.

REFERENCES

D. Achlioptas and F. McSherry (2007), 'Fast computation of low-rank matrix approximations', *J. Assoc. Comput. Mach.* **54**, 9.

R. Ahlswede and A. Winter (2002), 'Strong converse for identification via quantum channels', *IEEE Trans. Inform. Theory* **48**, 568–579.

N. Anari, S. Gharan and A. Rezaei (2016), Monte Carlo Markov chain algorithms for sampling strongly Rayleigh distributions and determinantal point processes. In *COLT 2016: 29th Conference on Learning Theory*, pp. 103–115.

R. Arriaga and S. Vempala (1999), An algorithmic theory of learning: Robust concepts and random projection. In *FOCS 1999: 40th Annual Symposium on Foundations of Computer Science*, pp. 616–623.

R. Arriaga and S. Vempala (2006), 'An algorithmic theory of learning: Robust concepts and random projection', *Machine Learning* **63**, 161–182.

M. Berry, S. Pulatova and G. Stewart (2004), Computing sparse reduced-rank approximations to sparse matrices. Technical report, UMIACS, University of Maryland.

R. Bhatia (1996), *Matrix Analysis*, Vol. 169 of Graduate Texts in Mathematics, Springer.

C. Boutsidis and D. Woodruff (2014), Optimal CUR matrix decompositions. In *STOC 2014: Symposium on Theory of Computing*, pp. 353–362.
K. Clarkson and D. Woodruff (2009), Numerical linear algebra in the streaming model. In *STOC 2009: 41st Annual ACM Symposium on Theory of Computing*, pp. 205–214.
K. Clarkson and D. Woodruff (2013), Low rank approximation and regression in input sparsity time. In *STOC 2013: Symposium on Theory of Computing Conference*, pp. 81–90.
M. Cohen (2016), Nearly tight oblivious subspace embeddings by trace inequalities. In *SODA 2016: 27th Annual ACM–SIAM Symposium on Discrete Algorithms*, pp. 278–287.
M. Cohen, Y. Lee, C. Musco, C. Musco, R. Peng and A. Sidford (2015), Uniform sampling for matrix approximation. In *ITCS 2015: Conference on Innovations in Theoretical Computer Science*, pp. 181–190.
M. Cohen, C. Musco and C. Musco (2017), Ridge leverage scores for low-rank approximation. In *SODA 2017: 27th Annual ACM–SIAM Symposium on Discrete Algorithms*, pp. 1758–1777.
A. Dasgupta, R. Kumar and T. Sarlós (2010), A sparse Johnson–Lindenstrauss transform. In *STOC 2010: 42nd ACM Symposium on Theory of Computing*, pp. 341–350.
S. Dasgupta and A. Gupta (2003), 'An elementary proof of a theorem of Johnson and Lindenstrauss', *Random Struct. Alg.* **22**, 60–65.
A. Deshpande and L. Rademacher (2010), Efficient volume sampling for row/column subset selection. In *FOCS 2010: 51th Annual IEEE Symposium on Foundations of Computer Science*, pp. 329–338.
A. Deshpande and S. Vempala (2006), Adaptive sampling and fast low-rank matrix approximation. In *APPROX–RANDOM 2006*, Vol. 4110 of Lecture Notes in Computer Science, Springer, pp. 292–303.
A. Deshpande, L. Rademacher, S. Vempala and G. Wang (2006), 'Matrix approximation and projective clustering via volume sampling', *Theory of Computing* **2**, 225–247.
P. Drineas, R. Kannan and M. Mahoney (2006), 'Fast Monte Carlo algorithms for matrices II: Computing a low-rank approximation to a matrix', *SIAM J. Comput.* **36**, 158–183.
P. Drineas, M. Magdon-Ismail, M. Mahoney and D. Woodruff (2012), 'Fast approximation of matrix coherence and statistical leverage', *J. Mach. Learn. Res.* **13**, 3475–3506.
P. Drineas, M. Mahoney and S. Muthukrishnan (2008), 'Relative-error CUR matrix decompositions', *SIAM J. Matrix Anal. Appl.* **30**, 844–881.
D. Freedman (1975), 'On tail probabilities for martingales', *Ann. Probab.* **3**, 100–118.
A. Frieze, R. Kannan and S. Vempala (1998), Fast Monte-Carlo algorithms for finding low-rank approximations. In *FOCS 1998: 39th Annual Symposium on Foundations of Computer Science*, pp. 370–378.
A. Frieze, R. Kannan and S. Vempala (2004), 'Fast Monte-Carlo algorithms for finding low-rank approximations', *J. Assoc. Comput. Mach.* **51**, 1025–1041.

G. Golub and C. Van Loan (1996), *Matrix Computations*, third edition, Johns Hopkins University Press.

S. Goreinov and E. Tyrtyshnikov (2001), 'The maximum-volume concept in approximation by low-rank matrices', *Contemp. Math.* **280**, 47–51.

S. Goreinov, E. Tyrtyshnikov and N. Zamarashkin (1997), 'A theory of pseudoskeleton approximations', *Linear Algebra Appl.* **261**, 1–21.

N. Halko, P. Martinsson and J. Tropp (2011), 'Finding structure with randomness: Probabilistic algorithms for constructing approximate matrix decompositions', *SIAM Review* **53**, 217–288.

C. Hillar and L. Lim (2013), 'Most tensor problems are NP-hard', *J. Assoc. Comput. Mach.* **60**, 45.

R. Horn and C. Johnson (2012), *Matrix Analysis*, second edition, Cambridge University Press.

P. Indyk and R. Motwani (1998), Approximate nearest neighbors: Towards removing the curse of dimensionality. In *STOC 1998: 30th Annual ACM Symposium on Theory of Computing*, pp. 604–613.

D. Kane and J. Nelson (2014), 'Sparser Johnson–Lindenstrauss transforms', *J. Assoc. Comput. Mach.* **61**, 4.

R. Kannan and S. Vempala (2009), 'Spectral algorithms', *Found. Trends Theoret. Comput. Sci.* **4**, 157–288.

Y. Lee and H. Sun (2015), Constructing linear-sized spectral sparsification in almost-linear time. In *FOCS 2015: IEEE 56th Annual Symposium on Foundations of Computer Science*, pp. 250–269.

M. Li, G. Miller and R. Peng (2013), Iterative row sampling. In *FOCS 2013: 54th Annual IEEE Symposium on Foundations of Computer Science*, pp. 127–136.

M. Mahoney (2011), 'Randomized algorithms for matrices and data', *Found. Trends Mach. Learning* **3**, 123–224.

X. Meng and M. Mahoney (2013), Low-distortion subspace embeddings in input-sparsity time and applications to robust linear regression. In *STOC 2013: Symposium on Theory of Computing Conference*, pp. 91–100.

J. Nelson and H. Nguyên (2013), OSNAP: Faster numerical linear algebra algorithms via sparser subspace embeddings. In *FOCS 2013: IEEE 54th Annual Symposium on Foundations of Computer Science*, pp. 117–126.

G. Pisier (1989), *The Volume of Convex Bodies and Banach Space Geometry*, Cambridge University Press.

M. Rudelson and R. Vershynin (2007), 'Sampling from large matrices: An approach through geometric functional analysis', *J. Assoc. Comput. Mach.* **54**, 21.

T. Sarlós (2006), Improved approximation algorithms for large matrices via random projections. In *FOCS 2006: 47th Annual IEEE Symposium on Foundations of Computer Science*, pp. 143–152.

D. Spielman and N. Srivastava (2011), 'Graph sparsification by effective resistances', *SIAM J. Comput.* **40**, 1913–1926.

G. Stewart (1999), 'Four algorithms for the efficient computation of truncated QR approximations to a sparse matrix', *Numer. Math.* **83**, 313–323.

G. Stewart (2004), Error analysis of the quasi-Gram–Schmidt algorithm. Technical report, UMIACS, University of Maryland.

J. Tropp (2011), 'Improved analysis of the subsampled randomized Hadamard transform', *Adv. Adapt. Data Anal.* **3**, 115–126.

S. Vempala (2004), *The Random Projection Method*, Vol. 65 of DIMACS Series in Discrete Mathematics and Theoretical Computer Science, DIMACS/AMS.

N. Vishnoi (2013), 'Lx = b', *Found. Trends Theoret. Comput. Sci.* **8**, 1–141.

D. Woodruff (2014), 'Sketching as a tool for numerical linear algebra', *Found. Trends Theoret. Comput. Sci.* **10**, 1–157.

Acta Numerica (2017), pp. 137–303
doi:10.1017/S0962492917000071

Numerical analysis of strongly nonlinear PDEs

Michael Neilan
Department of Mathematics,
University of Pittsburgh, Pittsburgh, PA, USA
E-mail: neilan@pitt.edu

Abner J. Salgado
Department of Mathematics,
The University of Tennessee, Knoxville, TN, USA
E-mail: asalgad1@utk.edu

Wujun Zhang
Department of Mathematics,
Rutgers University, Piscataway, NJ, USA
E-mail: wujun@math.rutgers.edu

We review the construction and analysis of numerical methods for strongly nonlinear PDEs, with an emphasis on convex and non-convex fully nonlinear equations and the convergence to viscosity solutions. We begin by describing a fundamental result in this area which states that stable, consistent and monotone schemes converge as the discretization parameter tends to zero. We review methodologies to construct finite difference, finite element and semi-Lagrangian schemes that satisfy these criteria, and, in addition, discuss some rather novel tools that have paved the way to derive rates of convergence within this framework.

CONTENTS

1. Introduction

Все счастливые семьи похожи друг на друга, каждая несчастливая семья несчастлива по-своему.

Leo Tolstoy

The quote above from Leo Tolstoy (1998), which roughly translates to 'All happy families resemble each other, but each unhappy one is unhappy in its own way', was used by Iserles (2009, Section 8.1) to point out that the numerical approximation of partial differential equations substantially differs from that of ordinary differential equations. The same quote was used in the preface of Holden and Risebro (2015) to say that the theory for nonlinear equations is very different to that for linear problems, and that each nonlinearity needs to be treated in its own way. For these reasons we feel compelled to begin our discussion with the same quote, since we have chosen the unhappiest of all possible families for numerical approximation: fully nonlinear equations.

The goal of this paper is to summarize recent advances and trends in the numerical approximation and theory of strongly second-order nonlinear PDEs with an emphasis on fully nonlinear convex and non-convex PDEs. Such PDEs appear in diverse applications such as weather and climate modelling, stochastic optimal control, determining the initial shape of the universe, optimal reflector design, differential geometry, optimal transport, mathematical finance, image processing and mesh generation. Despite their importance in these application areas, and in contrast to the PDE and solution theory, the study of numerical methods for fully nonlinear problems is still an emerging field in numerical analysis. The reasons for the delayed development are plentiful. As well as the strong nonlinearity, the fundamental difficulties in computing solutions of fully nonlinear problems are the lack of regularity of solutions, the conditional uniqueness of solutions and, most importantly, the notion of viscosity solutions. Similar to weak solutions for PDEs in divergence form, the viscosity solution concept relaxes the pointwise meaning of the PDE, and while doing so, broadens the class of admissible functions in which to seek a solution. However, unlike the weak solution framework, the definition of viscosity solutions is not based on variational principles, but rather comparison principles. While viscosity solution theory and the PDE theory of fully nonlinear problems have made incredible progress during the last 25 years, numerical results for such problems have been slow to catch up due to the pointwise nature and non-variational structure found in the theory.

A breakthrough occurred in Barles and Souganidis (1991), which, roughly speaking, asserts that a consistent, stable and *monotone* numerical method (or general approximation scheme) converges to the viscosity solution as the discretization or regularization parameter tends to zero. The first two conditions in this framework – consistency and stability – are expected; they are the cornerstone of any convergence theory of numerical PDEs and are recognized as the basis of the Lax–Richtmyer equivalence theorem. While arguably less well known, monotonicity of numerical methods is also a long-established area of study; for example, its importance in the context of linear finite difference schemes was realized (at least) 80 years ago (see *e.g.* Gerschgorin 1930, Motzkin and Wasow 1953). On the other hand, the construction of numerical methods that satisfy all three criteria, at least for fully nonlinear problems, is not immediately obvious.

Around the same time, and complementary to the Barles–Souganidis framework, Kuo and Trudinger (1990, 1992) gave a methodology to construct consistent, stable and monotone finite difference schemes for uniformly elliptic fully nonlinear operators. In addition, they showed that such discrete schemes satisfy properties found in viscosity solution theory, including Alexandrov–Bakelman–Pucci (ABP) maximum principles, Harnack inequalities and Hölder estimates. While these results provided a fairly practical guide to computing viscosity solutions, and the theory paved the way for future advances, the fundamental issue of convergence *rates* was explicitly stated as an open and elusive problem.

For the next 15 years, progress in numerical fully nonlinear second-order PDEs was relatively limited and mostly constrained to the theory and convergence rates for *convex* PDEs, in particular the Hamilton–Jacobi–Bellman equation. In this direction, Krylov (1997) introduced the groundbreaking idea of 'shaking the coefficients' to obtain sufficiently smooth subsolutions, which, along with comparison principles, yield rates of convergence even for degenerate problems. These techniques were later refined under various scenarios and assumptions of the problem and discretization (*e.g.* Barles and Jakobsen 2005, Barles and Jakobsen 2002, Krylov 2000), although convexity of the equation always played an essential role in the analysis.

The last ten years have seen an explosion of results for numerical nonlinear PDEs, including a variety of discretization types and convergence results. These include the construction of relatively simple and practical wide-stencil finite difference schemes tailored to specific PDEs (Oberman 2008, Froese and Oberman 2011, Benamou, Froese and Oberman 2014, Froese 2016), and the emergence of Galerkin methods for fully nonlinear problems (Feng and Neilan 2011, Feng and Neilan 2009, Feng and Lewis 2014, Dean and Glowinski 2006*a*, Dean and Glowinski 2006*b*). With regard to rates of convergence, Caffarelli and Souganidis (2008), using intricate regularity results, derived algebraic rates of convergence for finite difference approximations for non-convex PDEs with constant coefficients, and these results were quickly extended by Krylov (2015) and Turanova (2015) to problems with variable coefficients and lower-order terms. On the Galerkin front, Nochetto and Zhang (2017*a*) extended the Kuo–Trudinger theory to finite element methods and derived ABP maximum principles for linear elliptic problems. These results were soon extended in several directions, including rates of convergence for a discrete Monge–Ampère equation (Nochetto and Zhang 2017*b*) and wide-stencil finite difference schemes (Nochetto, Ntogakas and Zhang 2017) and the construction and analysis of finite element methods for non-convex fully nonlinear problems (Salgado and Zhang 2016).

The intention of this survey is to summarize the 25 years of development of fully nonlinear numerical PDEs. Let us describe the organization and the problems we consider in the paper. After setting the notation and stating some instances of fully nonlinear problems, we review some of the basic theory and analysis of elliptic PDEs in Section 2. Here different notions of solutions are introduced and the underlying properties of the solutions and operators are discussed. Besides being of independent interest, these fundamental results motivate both the construction and analysis of the numerical methods. We develop a general framework to compute second-order elliptic problems in Section 3. Basic properties of numerical methods are given, namely consistency, stability and monotonicity, which will lay the groundwork for future developments in the paper. Special attention will be on discrete ABP maximum principles in both the finite difference and finite element setting. Section 4 concerns finite element approximations for linear problems in non-divergence form and with non-smooth coefficients. In Section 5 we combine the ideas of the previous sections and consider finite element and finite difference approximations for fully nonlinear convex PDEs. Alongside the construction and convergence of the schemes, this section will focus on the rates of convergence and the techniques to obtain these results. In Section 6 these results are extended to a particular convex PDE, the Monge–Ampère equation. In Section 7 we discuss recent results of the numerical approximations for fully nonlinear non-convex PDEs. Finally, in Section 8, we give our concluding remarks and state some open problems.

Before starting our discussion let us first briefly outline the derivation of some fully nonlinear PDEs that we focus on in the paper and illustrate their connection with some applications and other areas of mathematics.

1.1. *Convex PDEs*

This section obtains an instance of the Hamilton–Jacobi–Bellman (HJB) equation, a prototypical fully nonlinear second-order *convex* PDE, and shows how such problems arise in stochastic optimal control problems. In addition, we show below that *every* uniformly elliptic, convex operator with bounded gradient is an implicit HJB problem.

Following Øksendal (2003, Chapter 11) and Fleming and Soner (2006), we consider a stochastic process X_τ governed by the differential equation

$$\begin{cases} \mathrm{d}X_\tau = \ell(X_\tau, \tau, \alpha_\tau)\,\mathrm{d}\tau + \sigma(X_\tau, \tau, \alpha_\tau)\,\mathrm{d}W_\tau & \tau > 0, \\ X_0 = x_0 \in \mathbb{R}^d. \end{cases} \tag{1.1}$$

Here, W_τ is a Brownian motion of dimension d, σ is a $d \times d$ matrix-valued function, $\alpha_\tau \in \mathcal{A}$ is a (Markov) control and $\mathcal{A}$ is the control space. Problem (1.1) describes a dynamical system driven by additive white noise with diffusion coefficient (or volatility) σ and non-stochastic drift ℓ. Under appropriate smoothness and growth conditions on ℓ and σ, and for a fixed $\alpha(\cdot) \in \mathcal{A}$, there is a path-wise unique solution to (1.1). Associated with problem (1.1) is the family of stochastic processes that satisfy (1.1) but with initial time $t \geq 0$:

$$\begin{cases} \mathrm{d}X_\tau^{x,t} = \ell(X_\tau^{x,t}, \tau, \alpha_\tau)\,\mathrm{d}\tau + \sigma(X_\tau^{x,t}, \tau, \alpha_\tau)\,\mathrm{d}W_\tau & \tau > t, \\ X_t^{x,t} = x \in \mathbb{R}^d. \end{cases} \tag{1.2}$$

Now let $\Omega \subset \mathbb{R}^d$ be an open bounded domain, and set $Q = \Omega \times (0, T)$ for some $T \in (0, \infty]$. Let $\widehat{T}$ denote the first exit time for the process $X_\tau^{x,t}$ satisfying (1.2), that is,

$$\widehat{T} = \widehat{T}^{x,t} = \inf\{\tau > t; (X_\tau^{x,t}, \tau) \notin Q\}.$$

We then define the performance function (or cost functional)

$$J(x, t, \alpha) = \mathbb{E}_{x,t}\bigg\{\int_t^{\widehat{T}} f(X_\tau^{x,t}, \tau, \alpha_\tau)\,\mathrm{d}\tau + \chi_{\widehat{T}<\infty}\psi(X_{\widehat{T}}^{x,t}, \widehat{T})\bigg\}$$

over all $\alpha \in \mathcal{A}$ under the constraint (1.2). Here, $f : \mathbb{R}^d \times \mathbb{R} \times \mathcal{A} \to \mathbb{R}$ is the profit rate function, $\psi : \mathbb{R}^d \times \mathbb{R} \to \mathbb{R}$ is the bequest function, $\chi_{\widehat{T}<\infty}$ is the indicator function, which equals one if $\widehat{T} < \infty$ and zero otherwise, and $\mathbb{E}_{x,t}$ represents the expectation conditional on $X_t^{x,t} = x$. We then consider the problem of finding the value function $u : \bar{Q} \to \mathbb{R}$ and optimal control $\alpha^* \in \mathcal{A}$ such that

$$u(x,t) = \sup_{\alpha \in \mathcal{A}} J(x,t,\alpha) = J(x,t,\alpha^*).$$

Let us now give a heuristic derivation of the Hamilton–Jacobi–Bellman equation based on Bellman's principle of optimality and Itô's lemma. First, Bellman's principle of optimality states that (Bellman 2010):

Whatever the initial state $\{X_s^{x,t}\}_{s<\tau}$ and initial decision $\{\alpha_s\}_{s<\tau}$ are, the remaining controls $\{\alpha_s\}_{s>\tau}$ must constitute an optimal policy with regard to the state $X_\tau^{x,t}$ resulting from the initial decision.

More precisely, this principle equates to

$$u(x,t) = \sup_{\alpha \in \mathcal{A}} \mathbb{E}_{x,t}\left\{ u(X_{t+\delta t}^{x,t}, t+\delta t) + \int_t^{t+\delta t} f(X_\tau^{x,t}, \tau, \alpha_\tau)\,\mathrm{d}\tau \right\}. \tag{1.3}$$

Next, recalling that $X_t^{x,t} = x$, by Itô's lemma we obtain

$$\begin{aligned} u(X_{t+\delta t}^{x,t}, t+\delta t) = u(x,t) &+ \int_t^{t+\delta t} \frac{\partial u}{\partial t}\,\mathrm{d}\tau + \int_t^{t+\delta t} Du \cdot \mathrm{d}X_\tau^{x,t} \\ &+ \frac{1}{2}\int_t^{t+\delta t} \mathrm{d}X_\tau^{x,t} \cdot D^2 u\, \mathrm{d}X_\tau^{x,t}. \end{aligned}$$

Since the state variable $X_\tau^{x,t}$ is governed by the stochastic equation (1.2), and by using the identity $\mathrm{d}W_\tau \otimes \mathrm{d}W_\tau = I\,\mathrm{d}\tau$, we obtain

$$\begin{aligned} &u(X_{t+\delta t}^{x,t}, t+\delta t) - u(x,t) \\ &\quad = \int_t^{t+\delta t} \left(\frac{\partial u}{\partial t} + Du \cdot \ell + \frac{1}{2}(\sigma\sigma^\intercal) : D^2 u \right) \mathrm{d}\tau + \int_t^{t+\delta t} (\sigma^\intercal Du) \cdot \mathrm{d}W_\tau. \end{aligned}$$

Therefore, from principle of optimality (1.3),

$$\sup_{\alpha \in \mathcal{A}} \mathbb{E}_{x,t}\left\{ u(X_{t+\delta t}^{x,t}, t+\delta t) - u(x,t) + \int_t^{t+\delta t} f(X_\tau^{x,t}, \tau, \alpha)\,\mathrm{d}\tau \right\} = 0,$$

and Itô's formula above, we obtain

$$\sup_{\alpha \in \mathcal{A}} \mathbb{E}_{x,t}\left\{ \int_t^{t+\delta t} \left(\frac{\partial u}{\partial t} + Du \cdot \ell + \frac{1}{2}(\sigma\sigma^\intercal) : D^2 u + f \right) \mathrm{d}\tau \right\} = 0,$$

where we used that

$$\mathbb{E}_{x,t}\left\{\int_t^{t+\delta t}(\sigma^\intercal Du)\cdot \mathrm{d}W_\tau\right\}=0.$$

Dividing by δt and formally taking the limit $\delta t \to 0$, we then obtain a deterministic equation, namely the Hamilton–Jacobi–Bellman equation

$$\frac{\partial u}{\partial t}(x,t)+\sup_{\alpha\in\mathcal{A}}(\mathcal{L}^\alpha u(x,t)+f(x,t,\alpha))=0,$$

where

$$\mathcal{L}^\alpha u(x,t)=Du(x,t)\cdot \ell(x,t,\alpha)+\frac{1}{2}\sigma(x,t,\alpha)\sigma(x,t,\alpha)^\intercal : D^2u(x,t).$$

To derive the equation, we assumed that the value function $u(x,t)$ has continuous second-order derivatives. However, the PDE theory reveals that the solution of the HJB equation in general does not satisfy this regularity assumption. To justify the derivation above, the concept of viscosity solutions was introduced in Crandall and Lions (1983) for first-order Hamilton–Jacobi equations and later generalized to second-order Hamilton–Jacobi–Bellman equations (Lions 1982, 1983). Viscosity solutions – an essential concept in the PDE theory – will also play an important role in this paper.

1.2. Non-convex PDEs

The Isaacs equation, a prototypical non-convex PDE, describes zero sum stochastic games with two players. Each player has one control and they have opposite objectives. The first player chooses a control to maximize the expected payoff, whereas the second player chooses a control to minimize it. Stochastic game theory has wide applications in engineering and mathematical finance. Here, we follow the ideas in Fleming and Souganidis (1989) to derive the equation.

The dynamics of the stochastic differential game we investigate are given by the controlled stochastic differential equation

$$\begin{cases}\mathrm{d}X_\tau^{x,t}=\ell(X_\tau^{x,t},\tau,\alpha_\tau,\beta_\tau)\,\mathrm{d}\tau+\sigma(X_\tau^{x,t},\tau,\alpha_\tau,\beta_\tau)\,\mathrm{d}W_\tau & \tau>t,\\ X_t^{x,t}=x\in\mathbb{R}^d,\end{cases}\tag{1.4}$$

and the expected payoff is

$$J(x,t,\alpha,\beta)=\mathbb{E}_{x,t}\left\{\int_t^{\widehat{T}} f(X_\tau^{x,t},\tau,\alpha_\tau,\beta_\tau)\,\mathrm{d}\tau+\chi_{\widehat{T}<\infty}\psi(X_{\widehat{T}}^{x,t},\widehat{T})\right\},$$

where $X_\tau^{x,t}$ satisfies the differential equation (1.4) for $\tau > t$ and has initial condition $X_t^{x,t}=x$.

The Isaacs equation can be derived in a similar fashion to the HJB equation. If, at time t, the first player chooses a control α to maximize the expected payoff J, and the second player, based on the decision of the first player, chooses the control β to minimize it, then we set

$$u^+(x,t) = \inf_{\beta \in \mathcal{B}} \sup_{\alpha \in \mathcal{A}} \mathbb{E}_{x,t}\{J(x,t,\alpha,\beta)\}$$

and call this the upper value function. On the other hand, if the second player makes the decision first and the first player reacts accordingly, then we set the lower value function as

$$u^-(x,t) = \sup_{\alpha \in \mathcal{A}} \inf_{\beta \in \mathcal{B}} \mathbb{E}_{x,t}\{J(x,t,\alpha,\beta)\}.$$

By the principle of optimality we have

$$u^+(x,t) = \inf_{\beta \in \mathcal{B}} \sup_{\alpha \in \mathcal{A}} \mathbb{E}_{x,t}\left\{ u^+(X^{x,t}_{t+\delta t}, t+\delta t) + \int_t^{t+\delta t} f(X^{x,t}_\tau, \tau, \alpha, \beta)\,\mathrm{d}\tau \right\}.$$

Using a derivation similar to the HJB equation, we obtain that for all $(x,t) \in Q$ the upper value function must satisfy

$$\frac{\partial u^+}{\partial t}(x,t) + H^+(x,t,Du^+,D^2u^+) = 0,$$

where

$$H^+(x,t,\mathbf{p},M) = \inf_{\beta \in \mathcal{B}} \sup_{\alpha \in \mathcal{A}} \left\{ \frac{1}{2}\sigma\sigma^\intercal : M + \ell(x,t,\alpha,\beta)\cdot \mathbf{p} + f(x,t,\alpha,\beta) \right\}.$$

Similarly we obtain, for $(x,t) \in Q$,

$$\frac{\partial u^-}{\partial t}(x,t) + H^-(x,t,Du^-,D^2u^-) = 0,$$

with

$$H^-(x,t,\mathbf{p},M) = \sup_{\alpha \in \mathcal{A}} \inf_{\beta \in \mathcal{B}} \left\{ \frac{1}{2}\sigma\sigma^\intercal : M + \ell(x,t,\alpha,\beta)\cdot \mathbf{p} + f(x,t,\alpha,\beta) \right\}.$$

We finally comment that if the Isaacs' condition holds, that is, we have that for all $x, t, \mathbf{p}, M$,

$$H^+(x,t,\mathbf{p},M) = H^-(x,t,\mathbf{p},M),$$

then we can conclude that $u^+(x,t) = u^-(x,t)$ for all $(x,t) \in Q$. In this case, we say that an optimal policy exists.

1.3. Characterizations of elliptic PDEs

Let us show that there is no loss in generality in confining our considerations to these two equations by following the construction proposed by Evans (1980, Lemma 2.2). Let $\Omega \subset \mathbb{R}^d$ be a bounded domain and let $F : \Omega \times \mathbb{S}^d \to \mathbb{R}$ be continuously differentiable, non-decreasing with respect to its second argument and with bounded gradient. Here, $\mathbb{S}^d$ denotes the space of symmetric $d \times d$ matrices. We comment that, as we will see below (see Definition 2.10), these conditions guarantee that the operator F is elliptic. We have the following representation: for every $x \in \Omega$ and $M \in \mathbb{S}^d$, we have

$$F(x, M) = \inf_{\beta \in \mathbb{S}^d} \sup_{\alpha \in \mathbb{S}^d} \left[\int_0^1 \frac{\partial F}{\partial M}(x, (1-t)\beta + t\alpha) : (M - \beta) + F(x, \beta) \right]. \quad (1.5)$$

To see this, let *IS* denote the right-hand side of (1.5), and note that by setting $\beta = M$ we obtain

$$\begin{aligned} IS &\leq \sup_{\alpha \in \mathbb{S}^d} \left[\int_0^1 \frac{\partial F}{\partial M}(x, (1-t)M + t\alpha) : (M - M) + F(x, M) \right] \\ &= F(x, M). \end{aligned}$$

On the other hand, setting $\alpha = M$ we obtain that

$$\begin{aligned} &\int_0^1 \frac{\partial F}{\partial M}(x, (1-t)\beta + tM) : (M - \beta) + F(x, \beta) \\ &\qquad \leq \sup_{\alpha \in \mathbb{S}^d} \left[\int_0^1 \frac{\partial F}{\partial M}(x, (1-t)\beta + t\alpha) : (M - \beta) + F(x, \beta) \right]. \end{aligned}$$

Under the given assumptions on F, the left-hand side of this inequality can be rewritten as

$$\begin{aligned} &\int_0^1 \frac{\partial F}{\partial M}(x, (1-t)\beta + tM) : (M - \beta) + F(x, \beta) \\ &\quad = \int_0^1 \frac{\mathrm{d}}{\mathrm{d}t} F(x, (1-t)\beta + tM) + F(x, \beta) = F(x, M) - F(x, \beta) + F(x, \beta), \end{aligned}$$

and consequently

$$F(x, M) \leq \sup_{\alpha \in \mathbb{S}^d} \left[\int_0^1 \frac{\partial F}{\partial M}(x, (1-t)\beta + t\alpha) : (M - \beta) + F(x, \beta) \right],$$

or $F(x, M) \leq IS$.

With representation (1.5) at hand we define

$$\begin{aligned} A^{\alpha,\beta}(x) &= \int_0^1 \frac{\partial F}{\partial M}(x, (1-t)\beta + t\alpha), \\ f^{\alpha,\beta}(x) &= A^{\alpha,\beta}(x) : \beta - F(x, \beta), \\ \mathcal{A} &= \mathcal{B} = \mathbb{S}^d, \end{aligned}$$

and note that, since $F(x, \cdot)$ is non-decreasing, $A^{\alpha,\beta}(x) \geq 0$ for all $\alpha \in \mathcal{A}$, $\beta \in \mathcal{B}$ and $x \in \Omega$. In conclusion, $F(x, \cdot)$ can be represented as the inf-sup of a family of affine maps, that is,

$$F(x, M) = \inf_{\beta \in \mathcal{B}} \sup_{\alpha \in \mathcal{A}} \big[A^{\alpha,\beta}(x) : M - f^{\alpha,\beta}(x)\big].$$

If we assume in addition that F is convex in its second argument, then we have

$$F(x, M) - F(x, \alpha) \geq \frac{\partial F}{\partial M}(x, \alpha) : (M - \alpha), \quad \text{for all } \alpha \in \mathbb{S}^d.$$

Setting $\beta = \alpha$ in (1.5) yields

$$F(x, M) \leq \sup_{\alpha \in \mathbb{S}^d} \left[\frac{\partial F}{\partial M}(x, \alpha) : (M - \alpha) + F(x, \alpha)\right],$$

so that

$$F(x, M) = \sup_{\alpha \in \mathcal{A}} \big[A^{\alpha}(x) : M - f^{\alpha}(x)\big],$$

with $\mathcal{A} = \mathbb{S}^d$ and

$$A^{\alpha}(x) = \frac{\partial F}{\partial M}(x, \alpha) \geq 0, \quad f^{\alpha}(x) = \frac{\partial F}{\partial M}(x, \alpha) : \alpha - F(x, \alpha).$$

We conclude by remarking that more general types of dependence can also be reduced to a similar inf-sup form; see, for instance, Kuo and Trudinger (1992, Section 2.1), Katsoulakis (1995) and Koike and Święch (2013).

Nomenclature

A_z	adjacent set of node z
$\{\!\{\cdot\}\!\}$	average operator over faces F
B_r	ball with radius r centred at the origin
$\delta_{y,h}$	centred difference operator in the direction y with step h
$\partial\Omega$	boundary of Ω
$\delta^-_{y,h}$	backward difference operator in the direction y with step h
$A \Subset B$	the set A is compactly contained in B
C	a generic positive constant
$\mathcal{C}^-(v)$	lower contact set of v
$C^k(\bar{\Omega})$	space of k-times continuously differentiable functions defined in $\bar{\Omega}$, $k \in \mathbb{N}$
conv A	convex hull of the set A
$C^\infty_0(\Omega)$	infinitely differentiable functions whose support is compactly contained in Ω
$\mathcal{C}^-_h(v_h)$	discrete lower contact set of v_h

$\det M$	determinant of the matrix M
I_h	abstract canonical interpolant
d	spatial dimension
$\Gamma_h(u_h)$	discrete convex envelope of u_h
$\operatorname{dist}(x, A)$	distance between the point x and the set A
$D\cdot$	divergence operator
Ω	domain in d-dimensional space $\mathbb{R}^d$
Ω_h^B	abstract discrete boundary domain
Ω_h	abstract discrete domain
Ω_h^I	abstract discrete interior domain
$\cdot$	inner/dot product in $\mathbb{R}^d$
X_h	abstract approximation space with parameter $h > 0$
K	elements in a mesh $\mathcal{T}_h$
$\lvert\mathbf{q}\rvert_M$	norm in $\mathbb{R}^d$ induced by the matrix $M > 0$, that is, $(\mathbf{q} \cdot M\mathbf{q})^{1/2}$
$\Gamma(u)$	convex envelope of function u
F	faces in a mesh $\mathcal{T}_h$
$\mathcal{F}_h$	set of faces
$\mathcal{F}_h^I$	interior faces
$\delta_{y,h}^+$	forward difference operator in the direction y with step h
Δ_h	finite element Laplacian
X_h^l	finite element space of piecewise linear continuous functions
X_h^{cd}	H^2-conforming finite element space
X_h^{dg}	discontinuous finite element space
X_h^{cg}	H^1-conforming finite element space with polynomials of degree $k \in \mathbb{N}$
$X_{0,h}^l$	functions in X_h^l that vanish at the boundary
$\lvert\cdot\rvert$	ℓ^2-norm/Fröbenius norm of a matrix
D	gradient operator
$\{\phi_i\}$	normalized hat basis functions of X_h^l
$\{\tilde{\phi}_i\}$	canonical hat basis functions of X_h^l
D^2u	Hessian of the function u

$C^{k,\alpha}(\bar{\Omega})$	space of functions in $C^k(\bar{\Omega})$ whose kth derivative is Hölder-continuous with exponent $\alpha \in (0,1]$
I_h^{ep}	elliptic projection onto X_h^l
I_h^{fd}	canonical nodal interpolant onto X_h^{fd}
I_h^{fe}	canonical nodal interpolant onto X_h^l
u_τ^-	inf-convolution of the function u
$\fint_\Omega v$	integral of v over the domain Ω
$[[\cdot]]$	jump operator on faces F
$L^p(\Omega)$	space of Lebesgue-integrable functions with exponent $p \in [1,\infty]$
$\lvert \cdot \rvert_\infty$	ℓ^∞-norm
$LSC(\Omega)$	space of lower semi-continuous functions
$\mathbb{S}^d$	space of symmetric $d \times d$ matrices
$\lvert U \rvert$	measure of the set U
$v^- = -\min\{v,0\}$	negative part of v
X_h^{fd}	space of nodal functions
$\mathbb{SO}^d$	special orthogonal group
$\partial v/\partial u$	partial derivative of the function v with respect to its argument u
$\{\omega_i\}_{i=1}^N$	arbitrary (not necessarily simplicial) partition of $\bar{\Omega}$
ω_z	patch of vertex z
$\mathbb{P}_k$	space of polynomials of degree $k \geq 0$
$v^+ = \max\{v,0\}$	positive part of v
$\delta_{y,h}^2$	second difference operator in the direction y with step h
$W^{s,p}(\Omega)$	Sobolev spaces of functions whose kth derivatives belong to $L^p(\Omega)$, $s \geq 0$
$W_0^{s,p}(\Omega)$	closure of $C_0^\infty(\Omega)$ in $W^{s,p}(\Omega)$, $s \geq 0$
$H_0^2(\Omega)$	Sobolev space $W_0^{s,2}(\Omega)$
$H^s(\Omega)$	Sobolev space $W^{s,2}(\Omega)$
$W^{-s,p'}(\Omega)$	dual of $W_0^{s,p}(\Omega)$ $s \geq 0$
u_h	discrete solution of PDEs
$\mathbb{S}_+^d$	symmetric non-negative definite $d \times d$ matrices
S	stencil
m	stencil size

$\partial v(x)$	subdifferential of the function v at the point x
u_τ^+	sup-convolution of the function u
$\bar{\partial} u(x)$	superdifferential of u at x
$D_T \cdot$	tangential divergence operator
D_T	tangential gradient operator
Δ_T	tangential Laplacian
$\operatorname{tr} M$	trace of the matrix M
$\mathbb{S}_1^d$	set of matrices M in $\mathbb{S}_+^d$ for which $\operatorname{tr} M = 1$
$\mathbb{D}_1^d$	set of diagonal positive definite matrices with unit trace
Tv_h	translates of the function v_h
$M^\intercal$	transpose of the matrix M
$\mathcal{T}_h$	triangulation
$USC(\Omega)$	space of upper semi-continuous functions
z	vertices/nodes of mesh $\mathcal{T}_h$
$VMO(\Omega)$	space of functions with vanishing mean oscillation

2. Elements of the theory of strongly nonlinear elliptic PDEs

In order to get an idea of how to discretize strongly nonlinear partial differential equations, we must first understand their underlying structure and the main ideas that are at the basis of their theory and analysis. In this introductory section we collect all the relevant information that will later serve as a guide in the construction and analysis of numerical schemes. We will describe the fundamental properties that define an elliptic equation, even in the case of strong nonlinearities, and thereby define various suitable notions of solutions and their properties. We will provide existence and non-existence results, as well as a review of the available regularity results. While it is not our intention to provide a thorough exposition of the theory, which can be found in textbooks such as those of Evans (2010), Han and Lin (2011), Gilbarg and Trudinger (2001), Krylov (1996, 2008) and Maugeri, Palagachev and Softova (2000), we believe it is fundamental to understand this if one wishes to provide a rigorous analysis of approximation schemes.

2.1. Two defining consequences of ellipticity

We begin our description by providing two fundamental properties that lie at the heart of much of the theory for elliptic PDEs, namely energy considerations and comparison principles. We will see how these give rise to various concepts of solutions and how much of the existence and regularity theory stems from these two simple ideas.

To make matters precise, we set $\Omega \subset \mathbb{R}^d$ with $d \geq 1$ to be a bounded domain with Lipschitz boundary. If further smoothness of the domain becomes necessary we will specify this at every stage. For simplicity, and because these ideas are better motivated in this case, let us consider *the Laplacian* which, for a function $u \in C^2(\Omega)$, is defined by

$$\Delta u = \sum_{i=1}^{d} \frac{\partial^2 u}{\partial x_i^2}.$$

The first fundamental property that can be observed for this operator is a *maximum principle.*

Theorem 2.1 (maximum principle). Let $u \in C^2(\Omega) \cap C(\bar{\Omega})$ be such that $\Delta u \geq 0$. Then

$$\sup_{x \in \bar{\Omega}} u(x) = \sup_{x \in \partial\Omega} u(x).$$

While we will not provide a detailed proof here, we wish to provide some intuition into this fact. Namely, if we assume that a strict global maximum is attained at an interior point $z \in \Omega$, then elementary considerations from calculus will give us that

$$Du(z) = 0, \quad D^2 u(z) < 0,$$

where Du denotes the gradient and $D^2 u$ the Hessian of u, respectively. Since it can be easily seen that $\Delta u = \operatorname{tr} D^2 u$, a contradiction ensues.

An important consequence of this result is the following *comparison principle.*

Corollary 2.2 (comparison principle). Let $u, v \in C^2(\Omega) \cap C(\bar{\Omega})$ be such that $u \leq v$ in $\partial\Omega$ and $\Delta u \geq \Delta v$ in Ω. Then $u \leq v$ in Ω.

This result easily follows by setting $w = u - v$ and observing that $w \leq 0$ on $\partial\Omega$ and $\Delta w \geq 0$ in Ω. An application of the maximum principle then allows us to conclude the result.

The comparison principle is one of the fundamental properties of an elliptic operator, namely that an ordering on the boundary and a (reverse) ordering of the operators implies an order of the underlying functions. Throughout this survey the application of a similar principle will be a recurring feature.

Having understood comparison principles we now proceed to describe *energy* considerations. Consider the equation

$$\Delta u = f \tag{2.1}$$

in Ω and multiply it by a sufficiently smooth function φ that vanishes on

$\partial\Omega$. An application of Green's identity reveals that

$$\int_\Omega Du \cdot D\varphi = -\int_\Omega f\varphi, \tag{2.2}$$

which we immediately recognize as the Euler–Lagrange equation for the minimization of the *energy* functional

$$J(v) = \int_\Omega \left(\frac{1}{2}|Dv|^2 + fv\right)$$

subject to the condition $v = u$ on $\partial\Omega$.

It is important to notice that, as opposed to (2.1), identity (2.2) only requires the existence of square-integrable first derivatives. Note also that the second variation of J is non-negative:

$$\partial^2 J(v)[w_1, w_2] = \int_\Omega Dw_1 \cdot Dw_2, \quad \partial^2 J(v)[w, w] = \int_\Omega |Dw|^2 \geq 0, \tag{2.3}$$

which shows its positivity in a certain sense.

On the basis of this observation, we now introduce our first definition of ellipticity (Gilbarg and Trudinger 2001, Chapter 3). In order to do so, in what follows we let $\mathbb{S}^d$ denote the space of symmetric $d \times d$ matrices. We endow $\mathbb{S}^d$ with the usual partial order

$$M, N \in \mathbb{S}^d: \quad M \leq N \quad \Longleftrightarrow \quad \boldsymbol{\xi} \cdot M\boldsymbol{\xi} \leq \boldsymbol{\xi} \cdot N\boldsymbol{\xi}, \quad \text{for all } \boldsymbol{\xi} \in \mathbb{R}^d.$$

We denote the identity matrix by $I \in \mathbb{S}^d$.

Definition 2.3 (elliptic operator in divergence form). Let $A : \Omega \to \mathbb{S}^d$. We say that the operator

$$Lu(x) = -D \cdot (A(x)Du(x)) \tag{2.4}$$

is *elliptic* at $x \in \Omega$ if $0 < \lambda(x)I \leq A(x) \leq \Lambda(x)I$, *strictly* elliptic if $\lambda(x) \geq \lambda_0 > 0$ for all $x \in \Omega$, and *uniformly* elliptic if $\Lambda(x)/\lambda(x)$ is bounded in Ω.

Example 2.4 (lower-order terms). The concept of elliptic operators in divergence form can be extended to operators having lower-order terms. For instance, consider the operator

$$\tilde{L}u(x) = -D \cdot (A(x)Du(x) + \mathbf{b}(x)u(x)) + \mathbf{c}(x) \cdot Du(x) + e(x)u(x),$$

where A is as in Definition 2.3, and the functions $\mathbf{b}, \mathbf{c} : \Omega \to \mathbb{R}^d$ and $e : \Omega \to \mathbb{R}$ are assumed to be measurable.

Example 2.5 (quasi-linear operators). Given a differentiable vector-valued function $\Omega \times \mathbb{R} \times \mathbb{R}^d \ni (x, z, \mathbf{p}) \mapsto \mathbf{a}(x, z, \mathbf{p}) \in \mathbb{R}^d$ and a scalar function $\Omega \times \mathbb{R} \times \mathbb{R}^d \ni (x, z, \mathbf{p}) \mapsto b(x, z, \mathbf{p}) \in \mathbb{R}$, we consider the *quasi-linear* operator

$$Qu(x) = -D \cdot \mathbf{a}(x, u, Du) + b(x, u, Du)$$

defined for $u \in C^2(\Omega)$. We say that this operator is *variational* if it is the Euler–Lagrange operator of the energy functional

$$\int_\Omega E(x, u, Du),$$

that is, $\mathbf{a}(x, z, \mathbf{p}) = D_{\mathbf{p}} E(x, z, \mathbf{p})$ and $b(x, z, \mathbf{p}) = D_z E(x, z, \mathbf{p})$. Following Definition 2.3 we realize that the ellipticity of Q is equivalent to the strict convexity of E with respect to the $\mathbf{p}$ variables. This immediately hints at the fact that tools from calculus of variations will be essential in the study of equations with this type of operator. Examples of quasi-linear operators can be given by choosing appropriate energies E. For instance, setting

$$E = E(\mathbf{p}) = (1 + |\mathbf{p}|^2)^{s/2}$$

(Gilbarg and Trudinger 2001, Chapter 10) for $s > 1$, we obtain a family of uniformly elliptic quasi-linear operators.

On the other hand, many problems cannot be cast in this form. The prototypical example is that given by the operator

$$\mathcal{L}u(x) = A(x) : D^2 u(x), \tag{2.5}$$

where, for $M, N \in \mathbb{S}^d$, $M : N$ denotes the Fröbenius inner product:

$$M : N = \sum_{i,j=1}^{d} M_{i,j} N_{i,j}.$$

The Fröbenius norm of a matrix M will be denoted by $|M| := \sqrt{M : M}$.

If A is sufficiently smooth, the operator (2.5) can be recast in divergence form and $-\mathcal{L}$ can be understood as an elliptic operator in the sense of Definition 2.3. However, this is not always possible, and consequently we must extend the notion of ellipticity to non-divergence-form operators (Gilbarg and Trudinger 2001, Chapter 3).

Definition 2.6 (non-divergence elliptic operator). We say that the operator (2.5) is *elliptic* at $x \in \Omega$ if $0 < \lambda(x) I \leq A(x) \leq \Lambda(x) I$, *strictly* elliptic if $\lambda(x) \geq \lambda_0 > 0$ for all $x \in \Omega$, and *uniformly* elliptic if $\Lambda(x)/\lambda(x)$ is bounded in Ω.

The reader is encouraged to verify that, for an operator that is strictly elliptic in the sense of Definition 2.6, variants of Theorem 2.1 and Corollary 2.2 hold.

Example 2.7 (lower-order terms). As in the divergence-form case, the notion of elliptic operators extend to those with lower-order terms, for example,

$$\tilde{\mathcal{L}}u(x) = A(x) : D^2 u(x) + \mathbf{b}(x) \cdot Du(x) + c(x) u(x),$$

where A is as in Definition 2.6, and the lower-order coefficients $\mathbf{b}$ and c are (vector- and scalar-valued) functions defined on Ω.

Example 2.8 (linear operators in divergence form). Consider the operator L of Definition 2.3 and assume that the coefficient matrix A is differentiable. One can then rewrite $Lu(x)$ as

$$Lu(x) = -A(x) : D^2u(x) - (D \cdot A(x)) \cdot Du(x),$$

where the divergence operator acts on A column-wise. Consequently, the operator $-L$ is of the form $\tilde{\mathcal{L}}$ of Example 2.7.

Example 2.9 (quasi-linear operators in non-divergence form). For a function $u \in C^2(\Omega)$ we define the *quasi-linear* operator

$$\mathcal{Q}u(x) = A(x, u, Du) : D^2u(x) + b(x, u, Du)$$

where $\Omega \times \mathbb{R} \times \mathbb{R}^d \ni (x, z, \mathbf{p}) \mapsto A(x, z, \mathbf{p}) \in \mathbb{S}^d$ and $\Omega \times \mathbb{R} \times \mathbb{R}^d \ni (x, z, \mathbf{p}) \mapsto b(x, z, \mathbf{p}) \in \mathbb{R}$. We say that $\mathcal{Q}$ is elliptic at the function u if $A(x, u, Du)$ satisfies the positivity conditions of Definition 2.6.

The previous definitions and examples entailed linear and quasi-linear operators. While, by linearization, one could extend Definitions 2.3 and 2.6 to more general nonlinear problems, we shall instead give a general definition of ellipticity, one that preserves the fundamental concept of comparison (Crandall, Ishii and Lions 1992, Caffarelli and Cabré 1995).

Definition 2.10 (elliptic operator). Let $F \in C(\Omega \times \mathbb{R} \times \mathbb{R}^d \times \mathbb{S}^d)$. We say that F is *elliptic* in Ω if F satisfies the following *monotonicity condition*. If $r, s \in \mathbb{R}$ and $M, N \in \mathbb{S}^d$ with $r \geq s$ and $M \leq N$ then

$$F(x, r, \mathbf{p}, M) \leq F(x, s, \mathbf{p}, N).$$

We will say, moreover, that F is *uniformly elliptic* if there are constants $0 < \lambda \leq \Lambda$ such that for all $M \in \mathbb{S}^d$ and $r \geq s$ we have

$$\lambda |N| \leq F(x, r, \mathbf{p}, M + N) - F(x, s, \mathbf{p}, M) \leq \Lambda |N|, \quad \text{for all } N \geq 0.$$

Example 2.11 (linear and quasi-linear equations). As a first example, let us show that the linear operator in non-divergence form $\mathcal{L}$ of (2.5) is elliptic in the sense of Definition 2.10. By doing so and following the considerations of the examples previously given, we see that all the other cases also fit into this framework. Define

$$F(x, r, \mathbf{p}, M) = \operatorname{tr}(A(x)M).$$

Since, for symmetric matrices, $A : B = \operatorname{tr}(AB)$, we see that $\mathcal{L}u(x) = F(x, u(x), Du(x), D^2u(x))$. Moreover, the positivity of A implies the monotonicity of F.

We now present several examples of *fully nonlinear* equations that fit into Definition 2.10.

Example 2.12 (Hamilton–Jacobi–Bellman operator). Let $\mathcal{A}$ be any compact set and assume that for every $\alpha \in \mathcal{A}$ we are given a uniformly elliptic linear operator

$$\mathcal{L}^\alpha u(x) = A^\alpha(x) : D^2 u(x)$$

and a function $f^\alpha \in C(\Omega)$. Define

$$F(x, r, \mathbf{p}, M) = \sup_{\alpha \in \mathcal{A}} \big[\mathrm{tr}(A^\alpha(x)M) - f^\alpha(x)\big].$$

Note immediately that

$$F(x, u(x), Du(x), D^2 u(x)) = \sup_{\alpha \in \mathcal{A}} \big[\mathcal{L}^\alpha u(x) - f^\alpha(x)\big].$$

Moreover, since for every $\alpha \in \mathcal{A}$, $x \in \Omega$ and $M, N \in \mathbb{S}^d$ we have that $A^\alpha(x)M \leq A^\alpha(x)N$ whenever $M \leq N$, we immediately conclude that the operator F is monotone and thus elliptic in the sense of Definition 2.10. A similar argument shows that F is uniformly elliptic whenever the family of linear operators $\{\mathcal{L}^\alpha\}_{\alpha \in \mathcal{A}}$ is uniformly elliptic. More importantly, we note that the function F is convex with respect to M.

Example 2.13 (Isaacs operator). The previous example can be generalized as follows. Assume now that we have two index sets $\mathcal{A}$ and $\mathcal{B}$ and, for each $(\alpha, \beta) \in \mathcal{A} \times \mathcal{B}$, we have a uniformly elliptic linear operator

$$\mathcal{L}^{\alpha,\beta} u(x) = A^{\alpha,\beta}(x) : D^2 u(x).$$

Define

$$F(x, r, \mathbf{p}, M) = \inf_{\beta \in \mathcal{B}} \sup_{\alpha \in \mathcal{A}} \big[\mathrm{tr}(A^{\alpha,\beta}(x)M) - f^{\alpha,\beta}(x)\big],$$

and note that

$$F(x, u(x), Du(x), D^2 u(x)) = \inf_{\beta \in \mathcal{B}} \sup_{\alpha \in \mathcal{A}} \big[\mathcal{L}^{\alpha,\beta} u(x) - f^{\alpha,\beta}(x)\big].$$

One more time, the uniform ellipticity of the operators $\mathcal{L}^{\alpha,\beta}$ yields the uniform ellipticity of F. Note that F is neither convex nor concave with respect to M.

Example 2.14 (Monge–Ampère operator). As a final example, consider the operator

$$F(x, r, \mathbf{p}, M) = \det M - f(x).$$

Note that, in general, this operator does not satisfy the monotonicity condition of Definition 2.10. However, if we restrict it to *positive definite* matrices, then this operator is elliptic. Consequently, for a positive f and a *strictly convex* function $u \in C^2(\Omega)$, we define the Monge–Ampère operator as

$$F(x, u(x), Du(x), D^2u(x)) = \det D^2u(x) - f(x).$$

With Definition 2.10 at hand, we turn our attention to boundary value problems for elliptic operators. In other words, for an elliptic operator F we consider the following problem: find $u : \bar{\Omega} \to \mathbb{R}$ such that

$$F(x, u, Du, D^2u) = 0 \text{ in } \Omega, \quad u = g \text{ on } \partial\Omega. \tag{2.6}$$

The sense in which (2.6) is satisfied will give rise to the various existing concepts of solutions.

2.2. Classical solutions

The first, and obvious, notion of solution is when identity (2.6) is understood in a pointwise sense. This gives rise to so-called classical solutions.

Definition 2.15 (classical solution). Let $F \in C(\Omega \times \mathbb{R} \times \mathbb{R}^d \times \mathbb{S}^d)$. Then the function $u \in C^2(\Omega) \cap C(\bar{\Omega})$ is said to be a *classical* solution of (2.6) if this identity holds for every $x \in \bar{\Omega}$.

An immediate consequence of ellipticity is that classical solutions are unique.

Theorem 2.16 (uniqueness). Let F be elliptic in the sense of Definition 2.10. If F is strictly decreasing in the r variable or uniformly elliptic, then problem (2.6) cannot have more than one classical solution.

Proof. Let us prove this result under the assumption that the map F is strictly decreasing in the r variable. The remaining case can be found, for instance, in Gilbarg and Trudinger (2001, Corollary 17.2). Assume that u and v are classical solutions to (2.6) and set $w = u - v$. Note that $w = 0$ on $\partial\Omega$, and that if w attains a (positive) maximum at $x_0 \in \Omega$, then $Dw(x_0) = 0$ and $D^2w(x_0) \leq 0$. Therefore, if $u(x_0) > v(x_0)$, ellipticity and the fact that the map is strictly decreasing imply

$$0 = F(x_0, u(x_0), Du(x_0), D^2u(x_0)) < F(x_0, v(x_0), Dv(x_0), D^2v(x_0)),$$

which is a contradiction. Similarly, the function w cannot attain a negative minimum in Ω, and consequently $w \equiv 0$. □

As an example, we mention that Theorem 2.16 holds for the operator $\tilde{\mathcal{L}}$ of Example 2.7 whenever the zero order coefficient $c \leq 0$.

In the linear case of Example 2.11, the Dirichlet problem (2.6) reads as follows: find $u \in C^2(\Omega) \cap C(\bar{\Omega})$ such that

$$\mathcal{L}u = f \text{ in } \Omega, \quad u = g \text{ on } \partial\Omega. \tag{2.7}$$

In this case, the existence of classical solutions is guaranteed by what is known as Schauder estimates which, simply put, boil down to freezing the coefficients and a continuity argument; see Gilbarg and Trudinger (2001, Theorems 6.13–6.14).

Theorem 2.17 (existence). Let Ω satisfy an exterior sphere condition at every boundary point. Assume the operator (2.5) is strictly elliptic in the sense of Definition 2.6. If $g \in C(\partial\Omega)$ and, for some $\alpha \in (0,1)$, f and the coefficients of $\mathcal{L}$ are bounded and belong to $C^\alpha(\Omega)$, then the Dirichlet problem (2.7) has a unique classical solution $u \in C^{2,\alpha}(\Omega) \cap C(\bar{\Omega})$. If, in addition, we assume that $\partial\Omega \in C^{2,\alpha}$, that f and the coefficients of $\mathcal{L}$ belong to $C^\alpha(\bar{\Omega})$; and $g \in C^{2,\alpha}(\bar{\Omega})$, then $u \in C^{2,\alpha}(\bar{\Omega})$ and

$$\|u\|_{C^{2,\alpha}(\bar{\Omega})} \leq C\big(\|f\|_{C^\alpha(\bar{\Omega})} + \|g\|_{C^{2,\alpha}(\bar{\Omega})}\big),$$

where the constant C is independent of u, f and g.

At this point, the reader may wonder if Hölder continuity is indeed necessary for these results. Example 2.23 below will show us that this is the case.

If more regularity is assumed on the domain and problem data, it can be shown that the (unique) classical solution is also more regular (Gilbarg and Trudinger 2001, Theorem 6.19).

Theorem 2.18 (regularity). In the setting of Theorem 2.17, assume additionally that, for some $k \geq 0$ we have $\partial\Omega \in C^{k+2,\alpha}$, $g \in C^{k+2,\alpha}(\bar{\Omega})$ and that f and the coefficients of $\mathcal{L}$ belong to $C^{k,\alpha}(\bar{\Omega})$. Then $u \in C^{k+2,\alpha}(\bar{\Omega})$.

While Theorems 2.17 and 2.18 provide a satisfactory and conclusive answer for a linear operator with smooth coefficients, it does not cover rough coefficients or nonlinear problems, in which a classical solution may not exist. This is why we must depart from classical solutions and consider *weakened* or *generalized* concepts of solutions.

2.3. Weak (variational) solutions

We now turn our attention to the case of divergence-form operators as in Definition 2.3 and consider, for this particular operator, the Dirichlet problem (2.6). The natural solution concept in this case is called weak solution, and follows from an integration-by-parts argument, and an integral identity similar to (2.2).

Definition 2.19 (weak solutions of linear equations). A function $u \in H^1(\Omega)$ is said to be a *weak* solution of the Dirichlet problem

$$Lu = f \quad \text{in } \Omega, \qquad u = g \quad \text{on } \partial\Omega \tag{2.8}$$

if $u - g \in H^1_0(\Omega)$ and, for every $\varphi \in H^1_0(\Omega)$, we have

$$\int_\Omega (ADu) \cdot D\varphi = \int_\Omega f\varphi. \tag{2.9}$$

Existence and uniqueness follow from the classical Lax–Milgram lemma or, more generally, from so-called inf-sup conditions.

Theorem 2.20 (existence and uniqueness). Let Ω be bounded, and let L be uniformly elliptic in the sense of Definition 2.3 and such that its coefficients belong to $L^\infty(\Omega)$. If $f \in H^{-1}(\Omega)$ and $g \in H^1(\Omega)$, then problem (2.8) has a unique weak solution $u \in H^1(\Omega)$.

Again, under additional smoothness assumptions on the domain and problem data, one can assert further differentiability of the solution. This is the content of the following result (Gilbarg and Trudinger 2001, Grisvard 2011).

Theorem 2.21 (regularity). Assume, in addition to the conditions of Theorem 2.20, that $\partial\Omega \in C^2$ or that Ω is convex. If $A \in C^{0,1}(\bar{\Omega}, \mathbb{S}^d)$, $f \in L^2(\Omega)$ and $g \in H^2(\Omega)$, then we have $u \in H^2(\Omega) \cap H^1(\Omega)$ and

$$\|u\|_{H^2(\Omega)} \leq C\big(\|u\|_{L^2(\Omega)} + \|f\|_{L^2(\Omega)} + \|g\|_{H^2(\Omega)}\big),$$

where the constant C is independent of u, f and g.

We also wish to mention the remarkable result by E. De Giorgi concerning the Hölder regularity of weak solutions.

Theorem 2.22 (De Giorgi I). Let $u \in H^1(\Omega)$ be a weak solution to (2.9) with $g = 0$, $f \in L^q(\Omega)$ with $q > d/2$. Then there exists $\alpha \in (0,1)$ for which $u \in C^\alpha_{loc}(\Omega)$. If, in addition, $q > d$ and $A \in C^\beta(\bar{\Omega}, \mathbb{S}^d)$, with $\beta = 1 - d/q$, then $u \in C^{1,\beta}_{loc}(\Omega)$.

In light of the second part of the previous result, it is natural to ask whether $f \in L^\infty(\Omega)$ with appropriate assumptions on the boundary data g and the coefficients of L would yield that $Du \in C^1_{loc}(\Omega)$. The following example shows that this is generally false (Han and Lin 2011, Section 3.4).

Example 2.23 (second derivatives are not continuous). For $R < 1$ let $\Omega = \{x \in \mathbb{R}^d : |x| < R\}$ and consider

$$f(x) = \begin{cases} 0 & x = 0, \\ \dfrac{x_2^2 - x_1^2}{2|x|^2}\left(\dfrac{d+2}{\sqrt{-\ln|x|}} + \dfrac{1}{2(-\ln|x|)^{3/2}}\right) & x \neq 0. \end{cases}$$

Note that $f \in C(\bar{\Omega})$ and that the function

$$u(x) = (x_1^2 - x_2^2)\sqrt{-\ln|x|} \in C(\bar{\Omega}) \cap C^\infty(\bar{\Omega} \setminus \{0\})$$

satisfies $\Delta u = f$ with boundary conditions

$$g = \sqrt{-\ln R}(x_1^2 - x_2^2).$$

However, this function cannot be a classical solution since

$$\lim_{|x|\to 0} \frac{\partial^2 u(x)}{\partial x_1^2} = \infty$$

so that $u \notin C^2(\Omega)$. In fact, although the problem has a weak solution, it does not have a classical one. This example also shows that, in the classical solution theory given in Theorem 2.17, mere continuity of the data is not sufficient, thus justifying the need for Hölder continuity.

Let us now focus our attention on the quasi-linear operator Q of Example 2.5 and consider the Dirichlet problem

$$Qu = f \;\text{ in } \Omega, \quad u = g \;\text{ on } \partial\Omega. \tag{2.10}$$

The definition of weak solution is as follows.

Definition 2.24 (weak solutions of quasi-linear equations). A function $u \in W^{1,p}(\Omega)$ $(1 < p < \infty)$ is called a *weak* solution of (2.10) if $u - g \in W_0^{1,p}(\Omega)$ and

$$\int_\Omega (\mathbf{a}(x, u, Du) \cdot Dv + b(x, u, Du)v) = \int_\Omega fv$$

for all $v \in W_0^{1,p}(\Omega)$.

Note that the equation that defines weak solutions to (2.10) are the Euler–Lagrange equations of the functional

$$I(u) = \int_\Omega (E(x, u, Du) - fu)$$

over the set of functions $v \in W^{1,p}(\Omega)$ such that $u - g \in W_0^{1,p}(\Omega)$. Consequently, the existence of weak solutions is tightly bound with the calculus of variations.

Theorem 2.25 (existence and uniqueness). Assume that there is a $p \in (1, \infty)$ for which the function E satisfies the coercivity condition: there are constants $C_1 > 0$, $C_2 \geq 0$ such that, for every $x \in \Omega$, $z \in \mathbb{R}$ and $\mathbf{p} \in \mathbb{R}^d$, we have

$$E(x, z, \mathbf{p}) \geq C_1|\mathbf{p}|^p - C_2.$$

Assume, in addition, that E is convex in the $\mathbf{p}$ variable. Then, for $f \in L^{p'}(\Omega)$, the functional I has a minimizer $u \in W^{1,p}(\Omega)$ such that $u - g \in W_0^{1,p}(\Omega)$. Finally, if E does not depend on z and is uniformly convex, then this minimizer is unique.

With this theorem at hand it can be readily shown that, in this setting, the (unique) minimizer u of I is a weak solution of (2.10) in the sense of Definition 2.24.

We can also establish, under additional assumptions on E, further differentiability of minimizers. To shorten the exposition we confine ourselves to the case where E is independent of x and z, it is coercive with $p = 2$, satisfies the *growth condition*

$$|D_{\mathbf{p}}E(\mathbf{p})| \leq C(|\mathbf{p}| + 1), \quad \text{for all } \mathbf{p} \in \mathbb{R}^d, \tag{2.11}$$

and

$$|D^2E(\mathbf{p})| \leq C, \quad \text{for all } \mathbf{p} \in \mathbb{R}^d. \tag{2.12}$$

With these additional assumptions we have the following regularity result (Evans 2010, Theorem 8.3.1).

Theorem 2.26 (regularity). In the setting of Theorem 2.25 assume, in addition, that E depends only on $\mathbf{p}$ and satisfies (2.11) and (2.12). If $g = 0$, $f \in L^2(\Omega)$ and $\partial\Omega \in C^2$, we have that $u \in H^2(\Omega)$ with the estimate

$$\|u\|_{H^2(\Omega)} \leq C\|f\|_{L^2(\Omega)}.$$

What is more interesting and remarkable is that the results of De Giorgi presented in Theorem 2.22 can be extended to this case as well.

Theorem 2.27 (De Giorgi II). Let $u \in W^{1,p}(\Omega)$ be a minimizer of I. If E satisfies the growth and monotonicity conditions

$$|D_{\mathbf{p}}E(x,z,\mathbf{p})| \leq C_1(1 + |\mathbf{p}|^{p-1}), \quad D_{\mathbf{p}}E(x,z,\mathbf{p}).\mathbf{p} \geq C_2|\mathbf{p}|^p - C_3,$$

then there exists $\alpha \in (0,1)$ for which $u \in C^{\alpha}_{loc}(\Omega)$.

Under suitable assumptions, local Hölder continuity of the gradients of the minimizers can also be established. For further regularity results for quasilinear problems the reader is referred, for instance, to Mingione (2006).

While, in this setting, we have a sufficiently rich theory, it only applies to divergence-form operators. Below, in Section 2.5, we will describe the right generalization of the notion of solutions for more general problems.

2.4. Strong solutions

We now describe a solution concept that, in a sense, lies in between classical and weak solutions, and that can also be applied to non-divergence-form operators such as (2.5) and that of Example 2.9. These solutions are called strong.

Definition 2.28 (strong solutions). The function $u \in W^{2,p}(\Omega)$ $(1 < p < \infty)$ is a *strong* solution of the boundary value problem (2.6) if the equation and boundary conditions hold almost everywhere in Ω and $\partial\Omega$, respectively.

We immediately remark that every classical solution is a strong solution. Moreover, an integration-by-parts and density argument shows that a sufficiently regular weak solution (see Theorems 2.21 and 2.26) is also a strong solution. Therefore, strong solutions for the divergence-form equations (2.8) and (2.10) can be obtained from regularity considerations.

Let us now turn our attention to the non-divergence-form problem (2.7) and study the existence of strong solutions. In this case we have the following result.

Theorem 2.29 (existence). Let Ω be a $C^{1,1}$ domain and the coefficients of the operator L belong to $C(\bar{\Omega})$. If $f \in L^p(\Omega)$ and $g \in W^{2,p}(\Omega)$ ($1 < p < \infty$), then the Dirichlet problem (2.7) has a unique strong solution $u \in W^{2,p}(\Omega)$, and moreover

$$\|u\|_{W^{2,p}(\Omega)} \leq C\big(\|f\|_{L^p(\Omega)} + \|g\|_{W^{2,p}(\Omega)}\big),$$

where C is independent of u, f and g but depends on $\|A\|_{C(\bar{\Omega},\mathbb{S}^d)}$, the dimension d and the exponent p.

We must comment on the technique of proof for this result. First, for $A = I$ and $p = 2$, this follows from the regularity result of Theorem 2.21. An interpolation result, in conjunction with the celebrated Calderón–Zygmund decomposition technique (Calderón and Zygmund 1952), yields the result for any p. Using the continuity of A, the result can be extended to a general $\mathcal{L}$.

Remark 2.30 (Hölder regularity). Let us briefly describe the results of Krylov and Safonov; see Gilbarg and Trudinger (2001, Section 9.8) and Theorem 2.65 below. To do so, we assume that $f \in L^d(\Omega)$, $g \in C^\beta(\bar{\Omega})$ for some $\beta \in (0,1)$ and $\partial\Omega$ satisfies a uniform exterior cone condition. Then, given $\omega \Subset \Omega$, there is a constant $\alpha \in (0,1)$ such that

$$|u|_{C^\alpha(\omega)} \leq C.$$

The constants α and C depend, in particular, on the dimension d and the ratio Λ/λ that defines the ellipticity of $\mathcal{L}$. A natural question to ask is whether a similar estimate for the gradient Du (possibly under stricter smoothness assumptions) is possible. A result by Nirenberg (see Theorem 2.57) showed that this is the case for $d = 2$. For higher dimensions, however, this turns out to be false. Safonov (1987) showed that in $B_1 \subset \mathbb{R}^3$, the unit ball, for every $\alpha \in (0,1]$ there are:

- a bounded function $v \in C^\infty(\mathbb{R}^3 \setminus \{0\})$,
- a constant $\nu \in (0,1)$,

- a family $\{A_\varepsilon\}_{\varepsilon>0} \subset C^\infty(\bar{B}_1, \mathbb{S}^d)$ such that the associated non-divergence operators $\mathcal{L}_\varepsilon$ are uniformly elliptic with $\Lambda/\lambda = 1/\nu^2$.

With these objects at hand, he showed that the solution to the problem

$$\mathcal{L}_\varepsilon u_\varepsilon = 0 \quad \text{in } B_1, \qquad u_\varepsilon = v \quad \text{on } \partial B_1,$$

satisfies $u_\varepsilon \in C^\infty(\bar{B}_1)$, $\|u_\varepsilon\|_{L^\infty(B_1)} = 1$ but

$$\lim_{\varepsilon \downarrow 0} |u_\varepsilon|_{C^\alpha(B_{1/2})} = \infty.$$

From this it immediately follows that Hölder estimates on the derivatives are not possible.

Let us point out now that, in Theorem 2.29, the assumption that $A \in C(\bar{\Omega}, \mathbb{S}^d)$ cannot generally be weakened. The following example is due to Pucci.

Example 2.31 (non-uniqueness). Let us show, following Maugeri *et al.* (2000, Section 1.1), that for $d \geq 3$ there is a bounded measurable matrix A such that problem (2.7) with $f = 0$ and $g = 0$ has more than one strong solution in $H^2(\Omega) \cap H_0^1(\Omega)$. Let Ω be the unit ball of $\mathbb{R}^d$, and define

$$A(x) = I + b\frac{xx^\intercal}{|x|^2}, \qquad b = \frac{d-2+\lambda}{1-\lambda}, \qquad \max\{2 - d/2, 0\} < \lambda < 1.$$

Obviously

$$|\boldsymbol{\xi}|^2 \leq \boldsymbol{\xi} \cdot A\boldsymbol{\xi} \leq (1+b)|\boldsymbol{\xi}|^2,$$

so that A is bounded and the associated operators $\mathcal{L}$ are uniformly elliptic. Define $u(x) = |x|^\lambda - 1$ and note that

$$D^2u(x) = \lambda(\lambda - 2)|x|^{\lambda-4}xx^\intercal + \lambda|x|^{\lambda-2}I,$$

which, since $\lambda > 2 - d/2$, shows that $u \in H^2(\Omega) \cap H_0^1(\Omega)$. Moreover, due to the choice of b, we have

$$\mathcal{L}u(x) = A : D^2u(x) = \lambda|x|^{\lambda-2}\big[(1+b)\lambda + d - b - 2\big] = 0.$$

Since this will be important in subsequent developments, we now focus on conditions weaker than continuity that allow for the existence and uniqueness of a strong solution for (2.7).

2.4.1. The Cordes condition

Since, as Example 2.31 shows, mere boundedness of the coefficients in the operator of (2.7) does not suffice to ensure uniqueness of strong solutions, here we study the so-called Cordes condition for linear operators in non-divergence form. The idea behind it and the theory that follows is to reformulate the operator in such a way that the result is ‘close’ to one in divergence form, in particular, the Poisson equation. This reformulation allows us to apply classical tools in functional analysis to study the existence, uniqueness and *a priori* estimates for problem (2.5).

To motivate and derive this condition, consider the following problem: given $x \in \Omega$, find $\gamma(x) \in \mathbb{R}$ that minimizes the quadratic function

$$\tau \mapsto |\tau A(x) - I|^2.$$

Simple arguments show that the minimum is attained at

$$\gamma(x) = \frac{\operatorname{tr} A(x)}{|A(x)|^2} \quad \text{and} \quad |\gamma(x)A(x) - I|^2 = d - \frac{(\operatorname{tr} A(x))^2}{|A(x)|^2}. \tag{2.13}$$

In particular, this simple calculation shows that

$$\begin{aligned} |\gamma(x)\mathcal{L}v(x) - \Delta v(x)|^2 &= |(\gamma(x)A(x) - I) : D^2 v(x)|^2 \\ &\leq \left(d - \frac{\operatorname{tr} A(x)^2}{|A(x)|^2}\right)|D^2 v(x)|^2. \end{aligned} \tag{2.14}$$

The Cordes condition ensures that the multiplicative constant on the right-hand side of (2.14) is less than one.

Definition 2.32 (Cordes condition). A positive definite matrix $A \in L^\infty(\Omega, \mathbb{S}^d)$ satisfies the *Cordes condition* provided that there exists an $\epsilon \in (0, 1]$ such that

$$\frac{|A|^2}{(\operatorname{tr} A)^2} \leq \frac{1}{d - 1 + \epsilon} \quad \text{a.e. } \Omega. \tag{2.15}$$

Note that the Cordes condition ensures that there exists $\gamma > 0$ such that, for all $v \in H^2(\Omega)$, we have

$$\|\gamma \mathcal{L}v - \Delta v\|_{L^2(\Omega)} \leq \sqrt{1 - \epsilon}\|D^2 v\|_{L^2(\Omega)}. \tag{2.16}$$

Remark 2.33 (the Cordes condition in spectral terms). Since, by assumption, for a.e. $x \in \Omega$ we have that $A(x) \in \mathbb{S}^d$ and that it is positive definite, it is diagonalizable and all its eigenvalues $\{\lambda_i\}_{i=1}^d = \{\lambda_i(x)\}_{i=1}^d$ are positive. Using the well-known identities

$$|A|^2 = \sum_{i=1}^d \lambda_i^2, \quad \operatorname{tr} A = \sum_{i=1}^d \lambda_i \quad \text{and} \quad \left(\sum_{i=1}^d \lambda_i\right)^2 \leq d \sum_{i=1}^d \lambda_i^2,$$

condition (2.15) can be recast in terms of the eigenvalues of A as follows:

$$\frac{1}{d} \leq \frac{\sum_{i=1}^d \lambda_i^2}{\left(\sum_{i=1}^d \lambda_i\right)^2} \leq \frac{1}{d - 1 + \epsilon} \quad \text{a.e. } \Omega.$$

In other words, (2.15) is an anisotropy condition on A that becomes more stringent in higher dimensions.

The considerations in Remark 2.4.1 show that the Cordes condition is always satisfied in two dimensions with $\epsilon = \inf_{x\in\Omega} 2\lambda_1\lambda_2/(\lambda_1^2+\lambda_2^2) \in (0,1]$. On the other hand, there exist symmetric positive definite matrices in three dimensions (and higher) that do not satisfy (2.15).

Example 2.34 (three dimensions). Consider the matrix

$$A = \begin{pmatrix} 1 & 0 & c \\ 0 & 1 & b \\ c & b & 4 \end{pmatrix}$$

with $b^2 + c^2 < 4$ so that $\det(A) = 4 - (b^2 + c^2) > 0$. Sylvester's criterion ensures that the matrix is positive definite, and a straightforward calculation shows that

$$\frac{|A|^2}{(\operatorname{tr} A)^2} = \frac{18 + 2(b^2 + c^2)}{6^2} \geq \frac{1}{2}.$$

Thus A does not satisfy the Cordes condition.

Example 2.35 (the example of Pucci). As another example consider the matrix of Example 2.31 for $d = 3$. Note first of all that in this case we have

$$\frac{1}{2} < \lambda < 1, \quad b = \frac{3 - 2 + \lambda}{1 - \lambda} > 3.$$

Simple calculations then yield

$$|A|^2 = d + 2b + b^2,$$
$$(\operatorname{tr} A)^2 = (d + b)^2,$$

and therefore

$$|A|^2 - \frac{1}{d-1}(\operatorname{tr} A)^2 = d + 2b + b^2 - \frac{(d+b)^2}{d-1} > 0$$

Thus one concludes that A does not satisfy the Cordes condition.

The Cordes condition is a key assumption to establish the well-posedness of the elliptic problem (2.5) with discontinuous coefficients. Another crucial ingredient is the Miranda–Talenti estimate, which is summarized in the next lemma.

Lemma 2.36 (Miranda–Talenti estimate). Let $\Omega \subset \mathbb{R}^d$ be a bounded convex domain. Then, for any $v \in H^2(\Omega) \cap H_0^1(\Omega)$,

$$|v|_{H^2(\Omega)} \leq \|\Delta v\|_{L^2(\Omega)}. \tag{2.17}$$

While this result can be understood as a regularity estimate in the spirit of Theorem 2.21, we remark that it can be obtained without appealing to this theory; we refer the reader to Maugeri *et al.* (2000, Lemma 1.2.2) for a proof. Moreover, while the aforementioned regularity results yield that, for functions in $H^2(\Omega) \cap H_0^1(\Omega)$, the norm $v \mapsto \|\Delta v\|_{L^2(\Omega)}$ is equivalent to the $H^2(\Omega)$-norm, the important feature of estimate (2.17) is that the equivalence constant is exactly one on convex domains.

Remark 2.37 (polygonal domains). In two dimensions the Miranda–Talenti estimate (2.17) holds for a polygonal domain Ω *without the convexity assumption.* Indeed, assuming that $u \in C^\infty(\bar{\Omega})$, we have

$$|D^2u|^2 = |\Delta u|^2 + 2(|\partial_{12}u|^2 - \partial_{11}u\partial_{22}u),$$

where we explicitly used that we are in two dimensions. In addition, integration by parts and some algebraic manipulations show that

$$\int_\Omega (|\partial_{12}u|^2 - \partial_{11}u\partial_{22}u) = \int_{\partial\Omega} (-\partial_{\tau\tau}u\partial_n u + \partial_{n\tau}u\partial_\tau u)$$

(see Ciarlet 2002, equation (1.2.9)), where $\boldsymbol{\tau}$ is the unit tangential vector along the boundary $\partial\Omega$, ∂_τ is the derivative in its direction and ∂_n denotes the normal derivative. Now, if $u = 0$ on $\partial\Omega$ then we have that $\partial_\tau u = 0$, so that the second term on the right-hand side of this expression vanishes. If, in addition, $\partial\Omega$ is polygonal, this also implies that $\partial_{\tau\tau}u = 0$, which allows us to obtain (2.17). By density, the same result holds for every $u \in H^2(\Omega) \cap H_0^1(\Omega)$.

Identities (2.13) and (2.17) motivate the introduction of the bilinear form

$$a(\cdot,\cdot) : (H^2(\Omega) \cap H_0^1(\Omega))^2 \ni (v,w) \mapsto a(v,w) = \int_\Omega \gamma\mathcal{L}v\Delta w \in \mathbb{R}.$$

The properties of a are as follows.

Lemma 2.38 (properties of a). Assume that the coefficient A of the operator $\mathcal{L}$ satisfies the Cordes condition (2.15). If Ω is convex, then the bilinear form a is bounded and coercive on $H^2(\Omega) \cap H_0^1(\Omega)$.

Proof. Since γ is bounded, the continuity immediately follows.

If the Cordes condition (2.15) is satisfied and Ω is convex, then from the Miranda–Talenti estimate (2.17) and the Cauchy–Schwarz inequality we obtain

$$\begin{aligned} a(v,v) &= \|\Delta v\|_{L^2(\Omega)}^2 + \int_\Omega (\gamma\mathcal{L}v - \Delta v)\Delta v \\ &\geq \|\Delta v\|_{L^2(\Omega)}^2 - \sqrt{1-\epsilon}|v|_{H^2(\Omega)}\|\Delta v\|_{L^2(\Omega)} \\ &\geq (1-\sqrt{1-\epsilon})\|\Delta v\|_{L^2(\Omega)}^2. \end{aligned} \tag{2.18}$$

In conclusion, a is coercive on $H^2(\Omega) \cap H_0^1(\Omega)$. □

The coercivity estimate of Lemma 2.38 allows us to show the existence and uniqueness of strong solutions under the Cordes condition.

Theorem 2.39 (existence and uniqueness). Assume that the coefficient A of the operator $\mathcal{L}$ satisfies the Cordes condition (2.15). If Ω is convex, then the Dirichlet problem (2.7) with $f \in L^2(\Omega)$ and $g = 0$ has a unique strong solution $u \in H^2(\Omega) \cap H_0^1(\Omega)$. Moreover, we have

$$\|u\|_{H^2(\Omega)} \leq C \frac{\|\gamma\|_{L^\infty(\Omega)}}{1-\sqrt{1-\epsilon}} \|f\|_{L^2(\Omega)},$$

where the constant C is independent of u and f.

Proof. From Lemma 2.38 and the Lax–Milgram lemma, there exists a unique $u \in H^2(\Omega) \cap H_0^1(\Omega)$ satisfying

$$a(u,v) = \int_\Omega \gamma f \Delta v, \quad \text{for all } v \in H^2(\Omega) \cap H_0^1(\Omega).$$

Since $u \in H^2(\Omega)$ and the Laplace operator $\Delta : H^2(\Omega) \cap H_0^1(\Omega) \to L^2(\Omega)$ is surjective on convex domains, standard arguments show that u satisfies $\mathcal{L}u = f$ almost everywhere, that is, it is a strong solution to the elliptic problem (2.7). The coercivity condition (2.18) also implies the *a priori* estimate

$$\|u\|_{H^2(\Omega)} \leq C\|\Delta u\|_{L^2(\Omega)} \leq C \frac{\|\gamma\|_{L^\infty(\Omega)}}{1-\sqrt{1-\epsilon}} \|f\|_{L^2(\Omega)}. \qquad \square$$

Remark 2.40 (inf-sup conditions). Since the Laplace operator is surjective from $H^2(\Omega) \cap H_0^1(\Omega)$ to $L^2(\Omega)$ on convex domains, the above arguments show that the inf-sup condition

$$\sup_{w \in L^2(\Omega)\setminus\{0\}} \frac{\int_\Omega \gamma \mathcal{L} v w}{\|w\|_{L^2(\Omega)}} \geq (1-\sqrt{1-\epsilon})\|\Delta v\|_{L^2(\Omega)}, \quad \text{for all } v \in H^2(\Omega) \cap H_0^1(\Omega)$$

is satisfied. One can then appeal to the Babuška–Brezzi theorem to deduce the existence of strong solutions to (2.5). To our knowledge the use of this inf-sup condition for the numerical approximation has yet to be investigated.

Remark 2.41 (the case $p \neq 2$). It is possible to show (Maugeri *et al.* 2000, Theorem 1.2.3) that, if Ω is convex and A satisfies the Cordes condition, there are $1 < p_l < 2 < p_r < \infty$ such that if $p \in (p_l, p_r)$, $f \in L^p(\Omega)$ and $g = 0$, then problem (2.7) has a unique strong solution $u \in W^{2,p}(\Omega) \cap W_0^{1,p}(\Omega)$. We also have an *a priori* estimate in which the constant now depends on p.

Remark 2.42 (strong solutions under other conditions). It is possible to obtain the existence and uniqueness of strong solutions for problem (2.7) under other assumptions. Let us discuss two of them.

- Assuming that $A \in W^{1,d}(\Omega, \mathbb{S}^d)$ one can write the operator L in non-divergence form and extend the theory of weak solutions, described in

Section 2.3, to coefficients in this class. What is remarkable is that, in Example 2.31, for every $\varepsilon > 0$ the parameter λ can be chosen so that $A \in W^{1,d-\varepsilon}(\Omega, \mathbb{S}^d)$, thus showing that $A \in W^{1,d-\varepsilon}(\Omega, \mathbb{S}^d)$ is not sufficient for uniqueness. On the other hand, if $A \in W^{1,d+\varepsilon}(\Omega, \mathbb{S}^d)$, for some $\varepsilon > 0$, then $A \in C^{0,\alpha}(\bar{\Omega}, \mathbb{S}^d)$ and, consequently, the classical Schauder theory applies (see Theorem 2.17).

- In essence, the case of uniformly continuous coefficients boils down to realizing that, locally, their oscillation in the $L^\infty(\Omega)$-norm is small, and so they can be considered a constant. These ideas have been extended (see Maugeri *et al.* 2000, Chapter 2) to the case of a coefficient $A \in VMO(\Omega, \mathbb{S}^d)$, thus showing that this is a sufficient condition to obtain strong solutions. Since $W^{1,d}(\Omega)$ is a proper subset of $VMO(\Omega)$ this result truly extends the case of Sobolev coefficients detailed above.

2.5. Viscosity solutions

At this point we wish to introduce one final notion of solution, the one that will be suited to the study of fully nonlinear equations. This is that of a viscosity solution. The reader may recall that the notion of weak solutions, introduced in Section 2.3, was based on an integration-by-parts argument (2.2) and the positivity (2.3) of the resulting operators. While this proved sufficient for linear and quasi-linear operators in divergence form, different arguments are necessary for fully nonlinear operators such as those of Examples 2.12–2.14. The fundamental property that will be used to define solutions in this case, as in Corollary 2.2, will be a comparison principle.

2.5.1. Definition and first properties

Let us begin by motivating the definition following Katzourakis (2015). Let F be an elliptic operator in the sense of Definition 2.10 and let $u \in C^2(\Omega)$ be a classical solution to

$$F(x, u, Du, D^2u) = 0 \quad \text{in } \Omega. \tag{2.19}$$

Let $x_0 \in \Omega$ and assume that there is a smooth function $\varphi \in C^2(\Omega)$ that can *touch from above* the graph of u at x_0. More precisely, we assume that

$$u(x) \le \varphi(x) \quad \text{for all } x \in \Omega, \quad u(x_0) = \varphi(x_0).$$

These conditions imply that the function $u - \varphi$ has a local maximum at x_0 and, consequently,

$$D(u - \varphi)(x_0) = 0, \quad D^2(u - \varphi)(x_0) \le 0.$$

Since the operator F is assumed to be elliptic we obtain

$$\begin{aligned} 0 &= F(x_0, u(x_0), Du(x_0), D^2u(x_0)) = F(x_0, \varphi(x_0), D\varphi(x_0), D^2u(x_0)) \\ &\le F(x_0, \varphi(x_0), D\varphi(x_0), D^2\varphi(x_0)). \end{aligned}$$

Similar considerations will give us that if $\psi \in C^2(\Omega)$ *touches from below* the graph of u at x_0, we would obtain

$$F(x_0, \psi(x_0), D\psi(x_0), D^2\psi(x_0)) \leq 0.$$

Finally we note that it is possible to reach the same conclusions if we replace the equality in (2.19) with a corresponding inequality. These considerations motivate the following definition.

Definition 2.43 (viscosity solution). Let F be elliptic in the sense of Definition 2.10. We say that the function $u \in C(\Omega)$ is:

(a) a *viscosity subsolution* of (2.19) if, whenever $x_0 \in \Omega$, $\varphi \in C^2(\Omega)$ and $u - \varphi$ has a local maximum at x_0 we have that

$$F(x_0, \varphi(x_0), D\varphi(x_0), D^2\varphi(x_0)) \geq 0,$$

(b) a *viscosity supersolution* of (2.19) if, whenever $x_0 \in \Omega$, $\varphi \in C^2(\Omega)$ and $u - \varphi$ has a local minimum at x_0 we have that

$$F(x_0, \varphi(x_0), D\varphi(x_0), D^2\varphi(x_0)) \leq 0,$$

(c) a *viscosity solution* if it is a sub- and supersolution.

Remark 2.44 (viscosity solutions). Several remarks must immediately be made about Definition 2.43.

- While the motivation provided assumed that the function u is smooth, the definition only requires its continuity.
- By approximation and continuity of F, it is sufficient to verify the condition for quadratic polynomials $\varphi \in \mathbb{P}_2$; see Caffarelli and Cabré (1995, Proposition 2.4).
- If $u \in C^2(\Omega)$ is a classical solution then it is a viscosity solution. This follows from the ellipticity of F. Moreover, sufficiently smooth viscosity solutions are also classical; see Caffarelli and Cabré (1995, Lemma 2.5) and Katzourakis (2015, Theorem 2.11).
- This definition talks only about solutions to equation (2.19), not the boundary value problem (2.6). More details on this issue will be provided below; see Definition 2.48 and Section 2.5.3.
- The definition assumes that the candidate solution *can* be touched from above (below). At points where this is not possible there is nothing to verify, and the function automatically satisfies the equation at these points.
- For the divergence-form operators L and $\tilde{L}$ of Definition 2.3 and Example 2.4, respectively, it is known (Ishii 1995) that the concepts of

weak solution, in the sense of Definition 2.19, and viscosity solutions coincide.

- We will not provide a historical account of the origin and development of this definition. The interested reader can consult the classical reference by Crandall *et al.* (1992).

A remarkable property of viscosity solutions is their *stability*, which is detailed in the following two results. For a proof of the first one we refer to Caffarelli and Cabré (1995, Proposition 2.8), Katzourakis (2015, Theorem 3.2) or Crandall *et al.* (1992, Section 6). For the second one, we refer to Caffarelli and Cabré (1995, Proposition 2.7) or Katzourakis (2015, Theorem 3.12).

Theorem 2.45 (limits and viscosity solutions). Let $\{F_k\}_{k\in\mathbb{N}}$ be a sequence of uniformly elliptic operators in the sense of Definition 2.10, and let $\{u_k\}_{k\in\mathbb{N}} \subset C(\Omega)$, for each k, be viscosity subsolutions to the equations

$$F_k(x, u_k, Du_k, D^2u_k) = 0.$$

If, as $k \to \infty$, $F_k \to F$ uniformly on compact subsets of $\Omega \times \mathbb{R} \times \mathbb{R}^d \times \mathbb{S}^d$ and $u_k \to u$ uniformly in compact subsets of Ω, then u is a viscosity subsolution of

$$F(x, u, Du, D^2u) = 0.$$

Theorem 2.46 (suprema of subsolutions). Let $\mathcal{U} \subset C(\Omega)$ be a set of viscosity subsolutions of (2.19). For $x \in \Omega$ define

$$\bar{u}(x) = \sup\{u(x) : u \in \mathcal{U}\}.$$

Suppose that $\bar{u} \in C(\Omega)$ and is bounded. Then $\bar{u}$ is a viscosity subsolution of (2.19).

2.5.2. Existence and uniqueness

Let us now turn our attention to the existence and uniqueness of viscosity solutions to (2.6). To do so we must specify in which sense the boundary conditions are being understood. We begin by introducing the notion of semi-continuity.

Definition 2.47 (semi-continuity). We say the function $u \in LSC(\Omega)$ (is *lower semi-continuous*) if, for all $x \in \Omega$,

$$u(x) \leq \liminf_{y\to x} u(y).$$

On the other hand, we say $u \in USC(\Omega)$ (is *upper semi-continuous*) if $-u \in LSC(\Omega)$.

Note that, in Definition 2.43 and the discussion that followed, nothing would have changed if we had only required subsolutions and supersolutions to be upper and lower semi-continuous, respectively. With this definition at hand, we may define viscosity solutions to the Dirichlet problem.

Definition 2.48 (solution to the Dirichlet problem). Let F be elliptic in the sense of Definition 2.10 and $g \in C(\partial\Omega)$.

(a) The function $u_\star \in USC(\bar{\Omega})$ is a viscosity subsolution to (2.6) if it is a viscosity subsolution to the equation (2.19) and $u_\star(x) \leq g(x)$ for all $x \in \partial\Omega$.

(b) The function $u^\star \in LSC(\bar{\Omega})$ is a viscosity supersolution to (2.6) if it is a viscosity supersolution to (2.19) and $u^\star(x) \geq g(x)$ for all $x \in \partial\Omega$.

(c) The function $u \in C(\bar{\Omega})$ is a viscosity solution to (2.6) if it is a sub- and supersolution.

Note that this definition requires the boundary values to be attained in the *classical sense*. Different boundary conditions might require a different interpretation, and we will briefly comment on this below.

We now turn our attention to the existence of solutions and the so-called Perron's method. Simply put, this method provides existence under the assumption that the problem cannot have more than one solution. While uniqueness of classical solutions is immediate, as shown in Theorem 2.16, in this more general setting we need one additional condition.

Definition 2.49 (comparison). We say that the Dirichlet problem (2.6) satisfies a comparison principle if, whenever $u_\star \in USC(\Omega)$ and $u^\star \in LSC(\Omega)$ are sub- and supersolutions, respectively, we have

$$u_\star \leq u^\star \quad \text{in } \Omega.$$

Note that from Definition 2.49, it immediately follows that (2.6) cannot have more than one solution. Indeed, if u and v are two viscosity solutions, then, in particular, u is a subsolution and v a supersolution; consequently, $u \leq v$. An analogous reasoning yields the reverse inequality.

With these two conditions at hand, we proceed to show existence of solutions. For a proof, we refer the reader, for instance, to Crandall *et al.* (1992, Theorem 4.1) and Katzourakis (2015, Theorem 5.3).

Theorem 2.50 (Perron's method). Let F be elliptic in the sense of Definition 2.10 and $g \in C(\partial\Omega)$. Assume that the Dirichlet problem (2.6) satisfies a comparison principle in the sense of Definition 2.49. If there exist a subsolution $u_\star$ and a supersolution $u^\star$ to (2.6) that satisfy the boundary condition, then

$$u(x) = \sup\bigl\{v(x) : u_\star \leq v \leq u^\star \text{ and } v \text{ is a subsolution}\bigr\}$$

defines a viscosity solution to (2.6).

Note that, while Theorem 2.50 provides a somewhat explicit construction of the unique solution to (2.6), one still needs to verify the existence of sub- and supersolutions that satisfy the boundary condition in a *classical sense.* This must be done on a case-by-case basis, and we refer the reader to Crandall *et al.* (1992, Example 4.6) and Katzourakis (2015, Application 5.9) for two examples where these are constructed.

It remains to understand which operators satisfy the comparison principle of Definition 2.49. Loosely speaking, similar ideas to those presented in Theorem 2.16 should yield uniqueness of viscosity solutions. However, the arguments presented there cannot be applied directly since we are dealing with functions that are merely continuous, and additional structural conditions must be imposed. This is due to the subtle fact – which may have escaped the reader – that Definition 2.10 is *too general.* By this we mean that, for instance, first-order and parabolic equations fit into this definition. For this reason, many authors say that an operator is *degenerate elliptic* if it only satisfies the monotonicity condition with respect to the M variable. This is in contrast with uniform ellipticity, which precludes these two degenerate cases.

For the sake of illustration, let us concentrate our efforts on finding a comparison principle for uniformly elliptic equations. We begin by showing, following Kawohl and Kutev (2007, Example 1), that uniform ellipticity is not enough to ensure a comparison principle.

Example 2.51 (lack of comparison). Consider the Dirichlet problem

$$u'' + 18x(u')^4 = 0 \text{ in } (-1,1), \quad u(-1) = -b, \; u(1) = b,$$

with $b > 1$. Clearly, the equation is uniformly elliptic. It is easy to check that the functions

$$u_\star(x) = \begin{cases} \sqrt[3]{x} - 1 + b & x \in [0,1], \\ \sqrt[3]{x} + 1 - b & x \in [-1,0), \end{cases} \qquad u^\star(x) = \begin{cases} \sqrt[3]{x} - 1 + b & x \in (0,1], \\ \sqrt[3]{x} + 1 - b & x \in [-1,0], \end{cases}$$

which differ only at the origin, are viscosity sub- and supersolutions, respectively, and that they satisfy the boundary values. Note, however, that

$$\max_{x \in (-1,1)} \{u_\star(x) - u^\star(x)\} = u_\star(0) - u^\star(0) = 2b - 2 > 0.$$

While, to our knowledge, necessary and sufficient conditions for the existence of a comparison principle for a general elliptic operator are not known, there are several sufficient conditions. We collect these in the following result.

Theorem 2.52 (existence of comparison principle). If the Dirichlet problem (2.6) satisfies any of the structural conditions given below, then it satisfies a comparison principle in the sense of Definition 2.49.

(a) (Katzourakis 2015, Theorem 6.1) The dependence with respect to x is decoupled, that is, the equation reads

$$F(u, Du, D^2u) = f,$$

with $F \in C(\mathbb{R}, \mathbb{R}^d, \mathbb{S}^d)$, $f \in C(\bar{\Omega})$. The operator F is elliptic, and satisfies, for some $\gamma > 0$,

$$F(r, \mathbf{p}, M) \geq F(s, \mathbf{p}, M) + \gamma(s - r), \quad \text{for all } r \leq s.$$

(b) The operator F is elliptic, independent of the r and $\mathbf{p}$ variables, and satisfies, for some $\lambda > 0$,

$$F(x, M + tI) \geq F(x, M) + \lambda t, \quad \text{for all } t \geq 0.$$

(c) (Trudinger 1988) The operator F is uniformly elliptic, Lipschitz-continuous in $\mathbf{p}$ and the following continuity assumption holds:

$$|F(x, r, \mathbf{p}, M) - F(y, r, \mathbf{p}, M)| \leq \mu_2 |x - y|^{1/2} |M| + \omega(|x - y|),$$

for all $x, y \in \Omega$, r and $\mathbf{p}$ in a suitable ball and $\omega(a) \to 0$ as $a \downarrow 0$. Additionally, one must assume that sub- and supersolutions belong to $C^{0,1}(\Omega)$.

(d) (Koike 2004) The dependence with respect to the r variable is decoupled, that is, the equation reads

$$\nu u + F(x, Du, D^2u) = 0,$$

with either $\nu < 0$ and F elliptic, or $\nu \leq 0$, F uniformly elliptic and Lipschitz in the $\mathbf{p}$ variable, for $\mathbf{p} \in \mathbb{R}^d$.

(e) (Silvestre 2015) The operator F is independent of x and $\mathbf{p}$ and is strictly decreasing in r, that is, whenever $r > s$,

$$F(r, M) < F(s, M), \quad \text{for all } M \in \mathbb{S}^d.$$

Other conditions can be found in the literature.

2.5.3. Other boundary conditions

So far, for all notions of solutions, we have only discussed the Dirichlet problem (see the second equation in (2.6)). Moreover, for viscosity solutions we have assumed that the boundary conditions are attained in a classical sense. Let us here consider other types of boundary conditions as well as generalized notions for them. Consider

$$F(x, u, Du, D^2u) = 0 \ \text{ in } \Omega, \quad B(x, u, Du) = 0 \ \text{ on } \partial\Omega, \tag{2.20}$$

where, as before, the map F is elliptic, but its domain of definition on the x variable is now $\bar{\Omega}$. The function $B : \partial\Omega \times \mathbb{R} \times \mathbb{R}^d$ is assumed to be non-increasing in its second argument, that is,

$$r \geq s \Rightarrow B(x, r, \mathbf{p}) \leq B(x, s, \mathbf{p}), \quad \text{for all } x \in \partial\Omega, \ \mathbf{p} \in \mathbb{R}^d.$$

The Dirichlet problem obviously falls into this description with $B(x,r,\mathbf{p}) = g(x) - r$, but others are also admissible. For instance, let $\mathbf{n}(x)$ denote the outer normal to $\partial\Omega$ at x and let $\boldsymbol{\nu} : \partial\Omega \to \mathbb{R}^d$ be such that, for all $x \in \partial\Omega$, we have $\boldsymbol{\nu}(x) \cdot \mathbf{n}(x) > 0$. The boundary condition

$$B(x, \mathbf{p}) = \boldsymbol{\nu}(x) \cdot \mathbf{p} - g(x)$$

gives rise to the so-called *oblique derivative* problem; if $\boldsymbol{\nu} = \mathbf{n}$, this is the Neumann problem. A nonlinear example is the capillarity condition

$$B(x, r, \mathbf{p}) = \mathbf{n} \cdot \mathbf{p} - g(x,r)\sqrt{1 + |\mathbf{p}|^2}.$$

At the beginning of Section 2.5.1 the introduction of viscosity solution was motivated by the assumption that the function $u - \varphi$ had a local maximum (minimum) at $x_0 \in \Omega$. When dealing with boundary conditions, we must now allow for $x \in \partial\Omega$. At these points the relations that led to the definition of viscosity solution no longer hold, and a modification is necessary. It turns out that the correct notion is as follows.

Definition 2.53 (viscosity subsolution). With the functions F and B as above, we say that $u \in USC(\bar{\Omega})$ is a *viscosity subsolution* to (2.20) if it is a viscosity subsolution to (2.19) and, whenever there is a $\varphi \in C^2(\mathbb{R}^d)$ that touches the graph of u from above at $x_0 \in \partial\Omega$, then either

$$B(x_0, \varphi(x_0), D\varphi(x_0)) \geq 0 \quad \text{or} \quad F(x_0, \varphi(x_0), D\varphi(x_0), D^2\varphi(x_0)) \geq 0.$$

In an analogous manner we can consider supersolutions and, as before, a solution to (2.20) is a function $u \in C(\bar{\Omega})$ that is both a sub- and supersolution.

It is important to realize that boundary conditions in the viscosity sense, in general, are *not* equivalent to those in the classical sense. The reason for this, once more, is that Definition 2.10 is rather general and, for instance, allows us to consider first-order equations for which Dirichlet conditions cannot be imposed on the whole boundary. It is therefore natural to ask when a boundary condition in the viscosity sense is attained classically. Let us briefly elaborate on this issue for the Dirichlet problem (2.6). We begin with the definition of a barrier.

Definition 2.54 (barrier). We say that (2.6) has *barriers* at $x_0 \in \partial\Omega$ if there exist two continuous functions $\bar{u}$, $\underline{u}$ that are super- and subsolutions to (2.6), respectively, and that satisfy $\bar{u}(x_0) = \underline{u}(x_0) = g(x_0)$.

Proposition 2.55 (viscosity versus classical). Let u be a viscosity solution to (2.6) in the sense of Definition 2.53. If barriers exist at $x_0 \in \partial\Omega$, then $u(x_0) = g(x_0)$.

In other words, classical and viscosity conditions coincide at points where it is possible to construct a barrier. We conclude this discussion by providing a sufficient condition for the existence of barriers.

Proposition 2.56 (existence of barriers). Let Ω be such that it has a tangent ball from outside at every point of $\partial\Omega$. If F is uniformly elliptic, Lipschitz with respect to all its variables and, for every $x \in \bar{\Omega}$, we have $F(x, 0, \mathbf{0}, 0) = 0$, then barriers exist at every point $x_0 \in \partial\Omega$.

2.5.4. Regularity

To finalize the presentation on viscosity solutions, we elaborate on their regularity. This is important not only because these results will serve as a guide to establishing rates of convergence for numerical schemes, but also many of the ideas and techniques that we present here have a discrete analogue that will be detailed in subsequent sections.

We begin with a result by Nirenberg (1953) that shows that in two dimensions, essentially, all solutions to elliptic equations are locally $C^{2,\alpha}$.

Theorem 2.57 (regularity in two dimensions). Let $d = 2$. Assume that F is uniformly elliptic in the sense of Definition 2.10 and that it has bounded first derivatives with respect to all its arguments. If u is a solution to (2.19), then for every $\omega \Subset \Omega$ there are $C > 0$, $\alpha \in (0, 1)$ that depend only on the ellipticity constants of F, the bounds on its first derivatives and the distance between ω and $\partial\Omega$, for which

$$\|u\|_{C^{2,\alpha}(\omega)} \leq C\|u\|_{L^\infty(\Omega)}.$$

It is remarkable that this result was obtained long before the development of the theory of viscosity solutions.

To obtain global regularity or results in more dimensions we begin by introducing several notions of a more or less geometric nature. Recall that a function $u : \Omega \to \mathbb{R}$ is convex if

$$u(\alpha x + (1 - \alpha)y) \leq \alpha u(x) + (1 - \alpha)u(y), \quad \text{for all } x, y \in \Omega, \ \alpha \in [0, 1].$$

For a convex function we define its subdifferential as follows.

Definition 2.58 (subdifferential). Let $u \in C(\Omega)$. The subdifferential of u at the point $x \in \Omega$ is

$$\partial u(x) = \{\mathbf{p} \in \mathbb{R}^d : u(y) - u(x) \geq \mathbf{p} \cdot (y - x) \ \forall y \in \Omega\}.$$

It is well known (Ekeland and Témam 1999) that if u is convex then $\partial u(x) \neq \emptyset$, and that if u is differentiable at x then $\partial u(x) = \{Du(x)\}$. Given a function u, we can always construct the largest convex function lying below u. This gives rise to the *convex envelope*. In what follows we will only need this concept for the negative part of a function, so we define the convex envelope in this restricted setting.

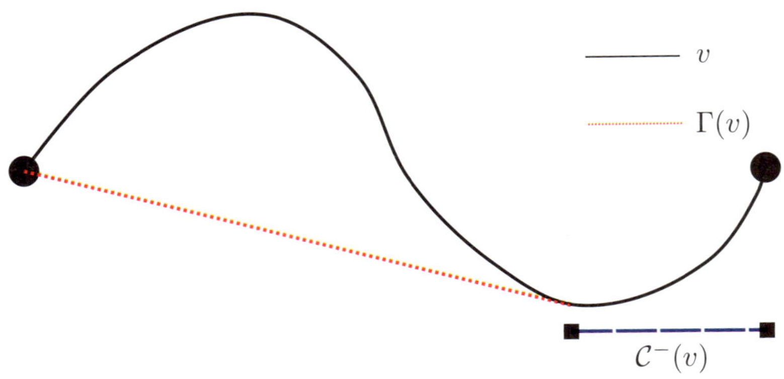

Figure 2.1. Convex envelope (dotted red) and contact set (dashed blue).

Definition 2.59 (convex envelope and contact set). Let B_r be a ball such that $\Omega \subset B_r$, and let $v \in C(\Omega)$ with $v \geq 0$ on $\partial\Omega$. Extend v^- by zero to $B_r \setminus \Omega$. The *convex envelope* of v is defined, for $x \in B_r$, by

$$\Gamma(v)(x) = \sup\{L(x) : L(z) \leq -v^-(z) \ \forall z \in B_r, \ L \in \mathbb{P}_1\}.$$

The points at which these two functions coincide are called *contact points*:

$$\mathcal{C}^-(v) = \{x \in B_r : v(x) = \Gamma(v)(x)\}.$$

An illustration of the convex envelope of a function and its contact set is given in Figure 2.1. From the figure it is intuitively clear that, for fixed values of v on the boundary, how deep the graph of v can go depends only on the values of v at $\mathcal{C}^-(v)$. The formalization of this observation is the so-called Alexandrov estimate.

Theorem 2.60 (Alexandrov estimate). Let $v \in C(\bar{B}_r)$ with $v \geq 0$ on ∂B_r. If $\Gamma(v) \in C^{1,1}(B_r)$, then

$$\sup_{B_r} v^- \leq Cr|\partial\Gamma(v)(\mathcal{C}^-(v))|^{1/d}.$$

In other words, there is a set $A \subset B_r$ that satisfies $|B_r \setminus A| = 0$ and for which we have

$$\sup_{B_r} v^- \leq Cr\left(\int_{A\cap\mathcal{C}^-(v)} \det D^2\Gamma(v)\right)^{1/d},$$

where the constant C depends only on d.

Proof. For completeness, we sketch the proof for $v \in C^2(B_r)$, since in this case $\partial\Gamma(v)$ is single-valued on $\mathcal{C}^-(v)$.

Let $M = \sup_{B_r} v^-/2r$ and assume that $B_M \subset \partial\Gamma(v)(\mathcal{C}^-(v))$. If that is the case,

$$M^d \leq C|\partial\Gamma(v)(\mathcal{C}^-(v))|,$$

for a constant that depends only on the dimension d. This shows the first estimate. On the other hand, a simple change of variables yields

$$|\partial\Gamma(v)(\mathcal{C}^-(v))| = \int_{\partial\Gamma(v)(\mathcal{C}^-(v))} = \int_{\mathcal{C}^-(v)} \det D^2\Gamma(v),$$

so that the second statement follows from the first one.

We now show the inclusion $B_M \subset \partial\Gamma(v)(\mathcal{C}^-(v))$. Let $z \in B_r$ be a point where $\sup_{B_r} v^-$ is attained. For $\mathbf{a} \in B_M$, define the affine function

$$L(x) = -\sup_{B_r} v^- + \mathbf{a} \cdot (x - z)$$

and note that $L(z) = -\sup_{B_r} v^-$ and, for all $x \in B_r$,

$$L(x) \leq -\sup_{B_r} v^- + |\mathbf{a}|\,|x - z| < -\sup_{B_r} v^- + 2Mr = 0.$$

Since $Dv(z) = 0$ there is a $x_1 \in B_r$ such that $v(x_1) < L(x_1) < 0$. In addition, we have that $v(x) \geq 0 > L(x)$ for $x \in \partial B_r$. This shows that, if $\bar{x} \in B_r$ is a point where $v - L$ attains its minimum, then $v(\bar{x}) < L(\bar{x}) \leq 0$ and $Dv(\bar{x}) = DL(\bar{x}) = \mathbf{a}$.

Define $\tilde{L}(x) = L(x) + v(\bar{x}) - L(\bar{x})$ and note that $v(\bar{x}) = \tilde{L}(\bar{x})$, $Dv(\bar{x}) = D\tilde{L}(\bar{x}) = \mathbf{a}$ and, for every $x \in B_r$ $v(x) \geq \tilde{L}(x)$. In other words, $\tilde{L}$ is a supporting hyperplane for v. This shows that $\bar{x} \in \mathcal{C}^-(v)$ and that $\mathbf{a} \in \partial\Gamma(v)(\bar{x})$, that is, $B_M \subset \partial\Gamma(v)(\mathcal{C}^-(v))$. □

Note that in Theorem 2.60 only the contact set is relevant. This is due to the fact that if for $x_0 \in B_r$ we have $\Gamma(v)(x_0) < v(x_0)$, then locally $\Gamma(v)$ is affine, and thus $D^2\Gamma(v) = 0$.

With this estimate at hand we proceed to obtain the fundamental *a priori* estimate for viscosity solutions, the so-called Alexandrov–Bakelman–Pucci estimate. We begin by providing some motivation for this result. To do so, assume that $u \in C^2(\bar{B}_r)$ with $u \geq 0$ on ∂B_r satisfies $\mathcal{L}u \leq f$ in B_r. In this setting we have that, for $x \in \mathcal{C}^-(u)$, $D^2u(x) \geq 0$, and consequently $A(x) : D^2u(x) \geq 0$ as well. This, in particular, implies that $f(x) \geq 0$. Let $D = \inf_{x\in\bar{B}_r} \det A(x) > 0$ and observe that

$$\det A(x)D^2u(x) = \det A(x) \det D^2u(x) \geq D \det D^2u(x) \geq 0.$$

Let $\sigma(A(x)D^2u(x)) = \{\mu_i(x)\}_{i=1}^d$. Then an application of the arithmetic–geometric inequality reveals that

$$\begin{aligned}
&\det A(x)D^2u(x) \\
&\quad = \prod_{i=1}^d \mu_i(x) = \left(\prod_{i=1}^d \mu_i(x)^{1/d}\right)^d \leq \left(\frac{1}{d}\sum_{i=1}^d \mu_i(x)\right)^d \\
&\quad = \left(\frac{1}{d}\operatorname{tr} A(x)D^2u(x)\right)^d = \left(\frac{1}{d}A(x) : D^2u(x)\right)^d \leq \left(\frac{1}{d}f(x)\right)^d,
\end{aligned}$$

where in the last step we used that $\mathcal{L}u \leq f$. Theorem 2.60 then yields that

$$\sup_{B_r} u^- \leq Cr\left(\int_{\mathcal{C}^-(u)} (f^+)^d\right)^{1/d},$$

for a constant that depends only on d and D.

While the considerations presented assumed that we were working with a linear equation, we essentially used that the matrix was uniformly positive definite and bounded, that is, the operator $\mathcal{L}$ is elliptic. A similar conclusion can be drawn from the fact that an operator is elliptic in the sense of Definition 2.10. We begin by observing (Caffarelli and Cabré 1995, Lemma 2.2) that F is uniformly elliptic if and only if

$$F(x, r, \mathbf{p}, M + N) - F(x, r, \mathbf{p}, M) \leq \Lambda|N^+| - \lambda|N^-|,$$

where $N = N^+ - N^-$ with $N^+, N^- \geq 0$ and $N^+N^- = 0$. Now, if u is a sufficiently smooth subsolution of (2.19), from the observation above we have

$$\begin{aligned} f(x) &:= -F(x, 0, \mathbf{0}, 0) \leq F(x, u(x), Du(x), D^2u(x)) - F(x, 0, \mathbf{0}, 0) \\ &\leq \Lambda \sum_{\lambda_i(u)>0} \lambda_i(u) + \lambda \sum_{\lambda_i(u)<0} \lambda_i(u), \end{aligned}$$

where $\sigma(D^2u(x)) = \{\lambda_i(u)\}_{i=1}^d$. Similarly, for a supersolution we have

$$f(x) \geq \Lambda \sum_{\lambda_i(u)<0} \lambda_i(u) + \lambda \sum_{\lambda_i(u)>0} \lambda_i(u).$$

This motivates the following definitions which, in a sense, describe the class of all possible viscosity solutions to uniformly elliptic equations.

Definition 2.61 (class $\mathcal{S}$). Let the operator F be uniformly elliptic in the sense of Definition 2.10 and let $f = -F(\cdot, 0, \mathbf{0}, 0)$. We say that $u \in \underline{\mathcal{S}}(\lambda, \Lambda, f)$ if $u \in C(\Omega)$ and the inequality

$$f(x) \leq \Lambda \sum_{\lambda_i(u)>0} \lambda_i(u) + \lambda \sum_{\lambda_i(u)<0} \lambda_i(u)$$

holds in the viscosity sense. Similarly, we say that $u \in \overline{\mathcal{S}}(\lambda, \Lambda, f)$ if $u \in C(\Omega)$ and

$$f(x) \geq \Lambda \sum_{\lambda_i(u)<0} \lambda_i(u) + \lambda \sum_{\lambda_i(u)>0} \lambda_i(u)$$

in the viscosity sense. Finally $\mathcal{S}(\lambda, \Lambda, f) = \underline{\mathcal{S}}(\lambda, \Lambda, f) \cap \overline{\mathcal{S}}(\lambda, \Lambda, f)$.

With this notation at hand we present the Alexandrov–Bakelman–Pucci (ABP) estimate.

Theorem 2.62 (ABP estimate). Let $u \in \overline{\mathcal{S}}(\lambda, \Lambda, f)$ in Ω with $u \geq 0$ on $\partial\Omega$ and assume that f is continuous and bounded in Ω. Then

$$\sup_{\Omega} u^- \leq Cr\left(\int_{\mathcal{C}^-(u)} (f^+)^d\right)^{1/d},$$

where the constant C depends only on d, λ and Λ, and r is such that $\Omega \subset B_{r/2}$, and we have extended u by zero outside Ω.

Note that, as in the Alexandrov estimate, only the contact set $\mathcal{C}^-(u)$ is relevant in this estimate. Note also that we obtain control of the L^∞-norm of u in terms of the L^d-norm of the data f. While Theorem 2.62 is a sort of stability estimate, it is also useful in establishing regularity of solutions. To do so, we begin with the Harnack inequality of Krylov and Safonov; see Safonov (1980). In what follows, we let Q_l denote a cube with sides parallel to the coordinate axes and of length l.

Theorem 2.63 (Harnack inequality). Let $u \in \mathcal{S}(\lambda, \Lambda, f)$ in Q_1 with $f \in C(Q_1) \cap L^\infty(Q_1)$. If $u \geq 0$ in Q_1, then

$$\sup_{Q_{1/2}} u \leq C\Big(\inf_{Q_1} u + \|f\|_{L^d(Q_1)}\Big),$$

where the constant C depends only on d, λ and Λ.

Since this will be useful later on, let us now show how one can obtain interior Hölder continuity of functions in $\mathcal{S}(\lambda, \Lambda, f)$ from a Harnack inequality. We begin with a technical result, commonly referred to as an iteration lemma; see Han and Lin (2011, Lemma 3.4) and Gilbarg and Trudinger (2001, Lemma 8.23).

Lemma 2.64 (iteration). Let $\varphi : (0, R] \to \mathbb{R}$ be non-decreasing. Assume that for some $A > 0$, $B \geq 0$ and $\alpha > \beta$ we have

$$\varphi(\rho) \leq A\left[\left(\frac{\rho}{r}\right)^\alpha + \varepsilon\right]\varphi(r) + Br^\beta$$

whenever $0 < \rho \leq r \leq R$. Then, for every $\gamma \in (\beta, \alpha)$ there exists ε_0 such that if $\varepsilon < \varepsilon_0$ then

$$\varphi(r) \leq C\left[\frac{\varphi(R_0)}{R_0} r^\gamma + Br^\beta\right], \quad \text{for all } r \in [0, R_0),$$

for some fixed constant C.

With this result at hand we obtain local Hölder continuity.

Theorem 2.65 (local Hölder regularity). Let $u \in \mathcal{S}(\lambda, \Lambda, f)$ in Q_1. Then there exists $\alpha \in (0, 1)$ for which $u \in C^\alpha(\bar{Q}_{1/2})$ and

$$\|u\|_{C^\alpha(\bar{Q}_{1/2})} \leq C\big(\|u\|_{L^\infty(Q_1)} + \|f\|_{L^d(Q_1)}\big),$$

where the constant C is independent of u and f.

Proof. The proof is rather standard, so we merely sketch it. Let $m_r = \inf_{Q_r} u$, $M_r = \sup_{Q_r} u$ and $\varpi_r = M_r - m_r$. Applying the Harnack inequality of Theorem 2.63 to the non-negative function $u - m_1$ yields

$$M_{1/2} - m_1 \leq C(m_{1/2} - m_1 + \|f\|_{L^d(Q_1)}).$$

Since the function $M_1 - u \geq 0$, we can once more apply the Harnack inequality to obtain

$$M_1 - m_{1/2} \leq C(M_1 - M_{1/2} + \|f\|_{L^d(Q_1)}).$$

Adding these two inequalities yields

$$\varpi_{1/2} \leq \mu\varpi_1 + 2\|f\|_{L^d(Q_1)},$$

where $\mu = (C-1)/(C+1) \in (0,1)$.

A similar argument for the functions $u_r(y) = u(ry)/r^2$ and $f_r(y) = f(ry)$ with $y \in Q_1$ reveals that

$$\varpi_r \leq \mu\varpi_{r/2} + 2r\|f\|_{L^d(Q_1)}.$$

An application of the iteration Lemma 2.64 with $\varphi(r) = \varpi_r$ immediately yields Hölder continuity and the estimate. $\square$

In a similar fashion, we can establish smoothness up to the boundary; see Caffarelli and Cabré (1995, Proposition 4.14).

Theorem 2.66 (global regularity). Let Ω be sufficiently smooth and $u \in \mathcal{S}(\lambda, \Lambda, f) \cap C(\bar{\Omega})$ with $f \in C(\Omega)$. Let $g = u_{|\partial\Omega}$ and ϱ be a modulus of continuity of g. Then there is a modulus of continuity ϱ^* of u in $\bar{\Omega}$ which depends only on $\lambda, \Lambda, \varrho, \|f\|_{L^d(\Omega)}$ and $\|g\|_{L^\infty(\Omega)}$.

We now focus on Hölder estimates for first and second derivatives. To simplify the presentation, in problem (2.6), the PDE takes the form

$$F(x, D^2u) = f \quad \text{in } \Omega. \tag{2.21}$$

To quantify the smoothness of F with respect to the x variable, we introduce the function

$$\beta(x) = \sup_{M \in \mathbb{S}^d} \frac{|F(M,x) - F(M,0)|}{|M| + 1}.$$

The local regularity is as follows.

Theorem 2.67 (local $C^{2,\alpha}$-regularity). Assume that F is uniformly elliptic in the sense of Definition 2.10, $\beta, f \in C^\alpha(B_1)$ and that there is a constant $\bar{\alpha} \in (0,1)$ such that for any $M \in \mathbb{S}^d$ with $F(0,M) = 0$ and $w_0 \in C(\partial B_1)$ there exists $w \in C^2(B_1) \cap C(\bar{B}_1) \cap C^{2,\bar{\alpha}}(B_{1/2})$ which satisfies

$$F(0, D^2w + M) = 0 \ \text{ in } B_1, \quad w = w_0 \ \text{ on } \partial B_1, \tag{2.22}$$

with an *a priori* estimate. If u is a viscosity solution of (2.21) then $u \in C^{2,\alpha}(\bar{B}_{1/2})$ for some $\alpha \in (0,1)$ with an *a priori* estimate.

To apply this theorem, one must verify that solutions to (2.22) have $C^{2,\bar{\alpha}}$-estimates. For an F that is convex, the Evans–Krylov theorem (Caffarelli and Silvestre 2010*a*) provides such an estimate.

Theorem 2.68 (Evans–Krylov). Let F be convex and depend only on M. If u is a viscosity solution of $F(D^2u) = 0$ in B_1, then

$$\|u\|_{C^{2,\bar{\alpha}}(\bar{B}_{1/2})} \leq C\big(\|u\|_{L^\infty(B_1)} + |F(0)|\big),$$

for some constants $\bar{\alpha} \in (0,1)$ and C that depend only on the dimension d and the ellipticity of F.

We mention that Theorem 2.67 can be applied to the Hamilton–Jacobi–Bellman operators of Example 2.12. With the aid of the so-called method of continuity, this allows us to show that solutions to the Dirichlet problem (2.6) for this class of operators are classical.

On the other hand, it is natural to ask if the convexity of F is essential for this result. To understand this, we begin by providing a $C^{1,\alpha}$-estimate without convexity assumptions.

Theorem 2.69 ($C^{1,\alpha}$-regularity). Let u be a viscosity solution of

$$F(D^2u) = 0 \quad \text{in } B_1.$$

Then

$$\|u\|_{C^{1,\bar{\alpha}}(\bar{B}_{1/2})} \leq C\big(\|u\|_{L^\infty(B_1)} + |F(0)|\big),$$

where the constants $\bar{\alpha} \in (0,1)$ and C depend only on the dimension and ellipticity of F.

A similar argument to Theorem 2.67 then allows us to conclude that, in this setting and under similar assumptions, solutions to (2.6) are locally $C^{1,\alpha}$ with an *a priori* estimate. However, there exists a series of counterexamples (Nadirashvili and Vlăduţ 2007, Nadirashvili and Vlăduţ 2008, Nadirashvili, Tkachev and Vlăduţ 2012) showing that, in general, the convexity assumption on F cannot be removed.

Example 2.70 (non-classical viscosity solution). Let $d = 5$. Define

$$P_5(x) = x_1^3 + \frac{3}{2}x_1(x_3^2 + x_4^2 - 2x_5^2 - 2x_2^2) + \frac{3\sqrt{3}}{2}(x_2x_3^2 - x_2x_4^2 + 2x_3x_4x_5),$$

and, for $\delta \in [0,1)$,

$$w(x) = P_5(x)/|x|^{1+\delta} \in C^{1,1-\delta}(\bar{B}_1) \setminus C^2(B_1).$$

There exists an Isaacs operator F that depends only on M and is Lipschitz, such that w is a viscosity solution of

$$F(D^2 w) = 0 \text{ in } B_1, \quad w = P_5 \text{ on } \partial B_1.$$

The existence of non-classical solutions in dimensions $3 \leq d < 5$ is an open problem.

We conclude by providing global regularity results in the general case; see Caffarelli and Souganidis (2008), Caffarelli and Silvestre (2010*b*) and Turanova (2015).

Theorem 2.71 (global regularity). Let Ω be sufficiently regular and let u be a viscosity solution to (2.6) with F of the form (2.21) being Lipschitz and uniformly elliptic. If $f \in C^{0,1}(\Omega)$ and, for some $\gamma \in (0,1]$, $g \in C^{1,\gamma}(\partial\Omega)$, then $u \in C^{1,\alpha}(\Omega) \cap C^{0,1}(\bar{\Omega})$ with

$$\|u\|_{C^{1,\alpha}(\Omega)} \leq C\big(\|f\|_{L^\infty(\Omega)} + \|g\|_{C^{1,\gamma}(\partial\Omega)}\big),$$

where $\alpha \in (0,1)$ and C depend only on d, λ, Λ and the smoothness of F.

3. Monotonicity in numerical methods

In this section we review some basic properties of numerical methods and state sufficient conditions to ensure that discrete approximations converge to the solutions of the underlying PDE. The underlying theme of this section is that as the notion of solution to the PDE becomes weaker, additional conditions of the numerical approximation are required to guarantee convergence. For example, for linear differential equations, the well-known Lax–Richtmyer equivalence theorem shows that any consistent scheme is convergent if and only if it is stable; these results extend to mildly nonlinear problems as well. However, in the fully nonlinear regime, consistency and stability are no longer sufficient in general. Rather, additional monotonicity conditions are required, that essentially mimic the comparison principles discussed in the previous section.

We note that while the content of this section deals with finite difference and finite element methods, the main ideas extend to other discretization techniques as well.

3.1. Stability, consistency and monotonicity implies convergence

As before, let Ω be an open subset of $\mathbb{R}^d$ with Lipschitz boundary $\partial\Omega$. We consider numerical approximations of elliptic problems of the most general form (2.20), which we recall for convenience:

$$F(x, u, Du, D^2 u) = 0 \text{ in } \Omega, \quad B(x, u, Du) = 0 \text{ on } \partial\Omega. \tag{3.1}$$

Here $F \in C(\Omega \times \mathbb{R} \times \mathbb{R}^d \times \mathbb{S}^d)$ is locally bounded and elliptic in the sense of Definition 2.10.

We further assume that $B \in C(\Omega \times \mathbb{R} \times \mathbb{R}^d)$ is non-increasing in its second argument. For the moment, we consider viscosity solutions that satisfy the boundary conditions only in a viscosity sense; see Definition 2.53. In this case, and to simplify the presentation, we define the operator

$$\mathsf{F}(x, r, \mathbf{p}, M) = \begin{cases} F(x, r, \mathbf{p}, M) & \text{if } x \in \Omega, \\ B(x, r, \mathbf{p}) & \text{if } x \in \partial\Omega, \end{cases}$$

so that (3.1) becomes

$$\mathsf{F}[u] := \mathsf{F}(x, u, Du, D^2 u) = 0 \quad \text{in } \bar{\Omega}. \tag{3.2}$$

Note that, since B is non-increasing in its second argument and F is elliptic, the operator F is elliptic in the sense of Definition 2.10. We further see that, according to Definition 2.53, $u \in C(\bar{\Omega})$ is a viscosity solution to (3.1) (equivalently (3.2)) if it is a viscosity solution to

$$F(x, u, Du, D^2 u) = 0 \quad \text{in } \Omega,$$

and

$$\begin{aligned} \max\{F(x, u, Du, D^2 u), B(x, u, Du)\} &\geq 0 \quad \text{on } \partial\Omega, \\ \min\{F(x, u, Du, D^2 u), B(x, u, Du)\} &\leq 0 \quad \text{on } \partial\Omega \end{aligned}$$

in the viscosity sense.

We now consider approximation schemes. Find $u_h \in X_h$ satisfying

$$\mathsf{F}_h[u_h](z) = 0 \quad \text{in } \bar{\Omega}_h, \tag{3.3}$$

where F_h is a locally bounded operator which we may think of as an approximation to F, and X_h is some finite-dimensional space. The operator is parametrized by $h > 0$, which we may view as a discretization parameter, or in some cases, a regularization parameter. The discrete domain $\overline{\Omega}_h$ is an approximation to $\overline{\Omega}$ with the property $\lim_{h\to 0^+} \bar{\Omega}_h = \bar{\Omega}$; namely, for all $z_0 \in \bar{\Omega}$, there exists a sequence $\{z_h\}_{h>0} \subset \bar{\Omega}_h$ such that $\lim_{h\to 0^+} z_h = z_0$.

We now address the well-posedness of (3.3), and the sufficient structure conditions on F_h to ensure that the discrete solutions to (3.3) (if they exist) converge. As a first step we state the fundamental notions of consistency and stability.

Definition 3.1 (consistency). The discrete problem (3.3) is said to be *consistent* with (3.2) if there exists an operator $I_h : C(\bar{\Omega}) \to X_h$ such that I_h converges uniformly to the identity operator as $h \to 0^+$, and for all sequences $\{z_h\}_{h>0}$ with $z_h \in \bar{\Omega}_h$ and $z_h \to z_0 \in \bar{\Omega}$ and $\phi \in C^2(\bar{\Omega})$,

$$\lim_{h\to 0^+} \mathsf{F}_h[I_h \phi](z_h) = \mathsf{F}[\phi](z_0).$$

Remark 3.2 (envelopes). If the operators are not continuous, then the notion of consistency is changed to

$$\limsup_{h\to 0^+} \mathsf{F}_h[I_h\phi](z_h) \le \mathsf{F}^*[\phi](z_0),$$
$$\liminf_{h\to 0^+} \mathsf{F}_h[I_h\phi](z_h) \ge \mathsf{F}_*[\phi](z_0),$$

where F^* (resp. F_*) denote the upper (resp. lower) semi-continuous envelope of F; see Definition 2.47.

Definition 3.3 (stability). We say that problem (3.3) is *stable* if, for all $h > 0$, there exists a solution $u_h \in X_h$ to (3.3), and moreover, if $w_h \in X_h$ satisfies $\mathsf{F}_h[w_h] = \epsilon_h$, then $\|u_h - w_h\|_{L^\infty(\bar{\Omega}_h)} \le C\|\epsilon_h\|_{L^\infty(\bar{\Omega}_h)}$, with $C > 0$ independent of h.

The following theorem states the well-known result that, for linear problems with classical solutions, consistent and stable schemes converge.

Theorem 3.4 (Lax–Richtmyer). Suppose that F_h is an affine operator, and that there exists a classical solution $u \in C^2(\bar{\Omega})$ satisfying (3.2). Suppose further that problem (3.3) is stable and that the operator in Definition 3.1 satisfies the stronger condition $\lim_{h\to 0^+} \|\mathsf{F}_h[I_h u]\|_{L^\infty(\bar{\Omega}_h)} = 0$. Then u_h converges locally uniformly to u.

Proof. By the given assumptions, there exists a linear operator G_h and a function l_h such that

$$\mathsf{F}_h[v_h](z) = \mathsf{G}_h[v_h](z) - \mathsf{l}_h(z) \quad z \in \bar{\Omega}_h.$$

The stability of the scheme shows that $\mathsf{G}_h[\cdot]$ is an isomorphism whose inverse is bounded independent of h. Since

$$\mathsf{G}_h[u_h - I_h u] = \mathsf{l}_h - \mathsf{G}_h[I_h u] = -\mathsf{F}_h[I_h u],$$

the consistency of the scheme implies that $\lim_{h\to 0^+} \|u_h - I_h u\|_{L^\infty(\bar{\Omega}_h)} = 0$, and thus, since I_h converges to the identity operator, u_h converges locally uniformly to u. □

While Theorem 3.4 is a useful result for a large class of problems, it is not applicable to fully nonlinear problems nor to weaker notions of solutions, in particular, viscosity solutions. The issue is that, if u is not a classical solution, then the approximation $I_h u$ may not be well-defined, and the consistency $\mathsf{F}_h[I_h u] \to 0$ used in the proof of Theorem 3.4 is no longer valid. Rather, to prove convergence to viscosity solutions, an additional structure condition is required. This requirement is summarized in the following definition.

Definition 3.5 (monotone operator). The discrete operator F_h is said to be monotone if, whenever $u_h - v_h$ has a global non-negative maximum at $z \in \bar{\Omega}_h$, we have

$$\mathsf{F}_h[u_h](z) \leq \mathsf{F}_h[v_h](z). \tag{3.4}$$

Remark 3.6 (monotonicity). The notion of monotonicity is essentially a discrete version of ellipticity. Indeed, following the proof of Theorem 2.16, if F is an elliptic operator in the sense of Definition 2.10, and if $u - v$, with $u, v \in C^2(\bar{\Omega})$, has a global non-negative maximum at $x \in \bar{\Omega}$, then $u(x) \geq v(x)$, $Du(x) = Dv(x)$ and $D^2u(x) \leq D^2v(x)$. Since F is non-increasing in its second argument, and non-decreasing in its fourth, we have

$$\mathsf{F}[u](x) = \mathsf{F}(x, u(x), Du(x), D^2u(x)) \leq \mathsf{F}(x, v(x), Dv(x), D^2v(x)) = \mathsf{F}[v](x).$$

Conversely, it can be shown that the operator F is elliptic if, whenever $u-v$ has a global non-negative maximum at $x \in \bar{\Omega}$, then $\mathsf{F}[u](x) \leq \mathsf{F}[v](x)$.

Consistency, stability and monotonicity are the three sufficient ingredients to guarantee convergence to viscosity solutions. The following result closely follows Barles and Souganidis (1991, Theorem 2.1).

Theorem 3.7 (Barles–Souganidis). Suppose that problem (3.3) is consistent, stable and monotone in the sense of Definitions 3.1, 3.3 and 3.5, respectively. Suppose further that F satisfies the comparison principle given in Definition 2.49. Then u_h converges locally uniformly to the unique continuous viscosity solution of (3.2).

Proof. Define $\bar{u} \in USC(\Omega)$ and $\underline{u} \in LSC(\Omega)$ by

$$\bar{u}(x) := \limsup_{\substack{y \to x \\ h \to 0^+}} u_h(y), \quad \underline{u}(x) := \liminf_{\substack{y \to x \\ h \to 0^+}} u_h(y), \quad x \in \bar{\Omega}. \tag{3.5}$$

Note that the stability of the scheme implies that both $\bar{u}$ and $\underline{u}$ are well-defined. The proof proceeds by showing that $\bar{u}$ and $\underline{u}$ are (viscosity) subsolutions and supersolutions to (3.2), respectively, and then appealing to the comparison principle.

To this end, suppose that $z_0 \in \Omega$ is a strict local maximum of $\underline{u} - \phi$ for some $\phi \in C^2(\Omega)$. Then standard arguments show that there exist sequences $\{h_n\}_{n=1}^{\infty}$ and $\{z_{h_n}\}_{n=1}^{\infty}$ such that

$$h_n \to 0, \quad z_{h_n} \to x_0, \quad \text{as } n \to \infty,$$

and $u_{h_n} - I_{h_n}\phi$ obtains a strict local maximum at z_{h_n}. Since $\mathsf{F}_{h_n}[u_{h_n}](z_{h_n}) = 0$, the monotonicity of the discrete operator implies that

$$\mathsf{F}_{h_n}[I_{h_n}\phi](z_{h_n}) \geq 0.$$

Passing to the limit, together with consistency of the scheme, yields

$$0 \leq \lim_{n \to \infty} \mathsf{F}_{h_n}[I_{h_n}\phi](z_{h_n}) = \mathsf{F}[\phi](z_0).$$

Similar arguments show that if $\bar{u} - \phi$ obtains a strict minimum at $z_0 \in \Omega$, then $0 \geq \mathsf{F}[\phi](z_0)$. Thus, $\bar{u}$ and $\underline{u}$ are subsolutions and supersolutions to (3.2), respectively. Since $\underline{u} \leq \bar{u}$, the comparison principle of F implies that $\underline{u} = \bar{u} = u$, and u is the viscosity solution to (3.2). $\square$

We once again mention that the problems and discretizations considered so far take into account the boundary conditions in a viscosity sense. While this set-up simplifies the proof of convergence, it may have practical limitations since the framework requires a consistent and monotone discretization of both the boundary conditions and the differential operator for all $x \in \partial\Omega$ and smooth functions ϕ. Also recall that, in general, viscosity boundary conditions are not equivalent to those imposed pointwise unless other conditions are assumed; see Proposition 2.55. Here we turn our attention to the elliptic boundary value problem (2.6), where the Dirichlet boundary condition is understood in the classical sense; see Definition 2.48.

To this end, we consider approximations of the form

$$F_h[u_h] = 0 \ \text{ in } \Omega_h^I, \quad u_h = g_h \ \text{ on } \Omega_h^B, \tag{3.6}$$

where $g_h = I_h g \in X_h$ is a discrete approximation to the Dirichlet data $g \in C(\bar{\Omega})$, and Ω_h^I and Ω_h^B are disjoint sets with $\Omega_h^I \to \Omega$ and $\Omega_h^B \to \partial\Omega$ as $h \to 0^+$. Note that with minor notational changes, the notions of consistency, stability, and monotonicity are applicable to the operator F_h. A natural question is whether the results of Theorem 3.7 carry over to the discrete problem (3.6). This issue is addressed in the next theorem.

Theorem 3.8 (convergence). Suppose that F_h is a consistent, stable and monotone operator. Suppose further that either

(i) $\underline{u}(x) \geq g(x)$ and $\bar{u}(x) \leq g(x)$ for all $x \in \partial\Omega$, where $\underline{u}, \bar{u}$ are given by (3.5), or

(ii) the sequence of solutions $\{u_h\}_{h>0}$ is equicontinuous.

Then u_h converges locally uniformly to the unique continuous viscosity solution of (3.1) with $B(x, u, Du) = g - u$.

Proof. The proof of case (i) follows directly from the arguments given in Theorem 3.7. The proof of case (ii) follows from the Arzelà–Ascoli theorem, and again appealing to the proof of Theorem 3.7. $\square$

3.2. Monotonicity in finite difference schemes

Here we discuss basic monotonicity results of finite difference schemes. The main message given in this section is that for any uniformly elliptic operator, one can construct a consistent and monotone finite difference scheme. The drawback, however, is that monotonicity requires a wide stencil, which may severely impact its practical use. Much of the material in this section is found in Kuo and Trudinger (1990, 1992), Kocan (1995), Motzkin and Wasow (1953) and Oberman (2006).

For simplicity we assume that the domain Ω is discretized on an equally spaced Cartesian grid and that each coordinate direction is discretized uniformly; in particular, by a possible change of coordinates, we assume that the grid is given by

$$\bar{\Omega}_h = \mathbb{Z}_h^d \cap \overline{\Omega}, \quad \text{with } \mathbb{Z}_h^d := \{he : e \in \mathbb{Z}^d\},$$

where $h > 0$ is the grid scale and $\mathbb{Z}^d$ is the set of d-tuples of integers. A finite subset $S \subset \mathbb{Z}^d \backslash \{0\}$ is called a *stencil*, and the space of *nodal functions*, denoted by X_h^{fd}, consists of real-valued functions with domain $\bar{\Omega}_h$. The canonical interpolant $I_h^{fd} : C^0(\bar{\Omega}) \to X_h^{fd}$ is the operator satisfying $I_h^{fd} v(z) = v(z)$ for all $z \in \bar{\Omega}_h$. We assume the existence of a positive integer m such that S is of the form

$$S = \{y : y \in \mathbb{Z}^d \backslash \{0\} : |y|_{\ell^\infty} \leq m\}. \tag{3.7}$$

The value m satisfying (3.7) is called the *stencil size* of S. The cardinality of S is $|S| := (2m+1)^d - 1$.

We consider finite difference schemes with stencil S acting on grid functions. These discrete operators are thus of the (implicit) form

$$F_h[v_h](z) = F_h(z, v_h(z), Tv_h(z)), \tag{3.8}$$

where $Tv_h(x) = \{v_h(x + hy) : y \in S\}$ is the set of translates of $v_h(x)$ with respect to the stencil. The method (3.8) is called a *one-step* scheme if $m = 1$, that is, the value $F_h[v_h](z)$ depends only on z, $v_h(z)$, and the values of v_h at neighbouring points of z. Otherwise, we call the scheme a *wide-stencil* scheme if $m \geq 2$.

To construct monotone schemes, we first reformulate this property so that it is easier to work with.

Definition 3.9 (non-negative operator). The operator F_h is of *non-negative type* (or simply non-negative) if

$$F_h(z, r, q + \tau) \geq F_h(z, r, q) \geq F_h(z, r + t, q + \tau), \tag{3.9}$$

for all $z \in \mathbb{R}^d$, $r, t \in \mathbb{R}$, and $q, \tau \in \mathbb{R}^{|S|}$ satisfying

$$0 \leq \tau_i \leq t \quad i = 1, 2, \ldots, |S|. \tag{3.10}$$

We see that if F_h is of non-negative type, then F_h is non-increasing in its second argument and non-decreasing in its third argument. If F_h is differentiable, then it is of non-negative type provided that

$$\frac{\partial F_h}{\partial q_i} \geq 0 \ (i = 1, 2, \ldots, |S|), \quad \frac{\partial F_h}{\partial r} + \sum_{i=1}^{|S|} \frac{\partial F_h}{\partial q_i} \leq 0.$$

Remark 3.10 (reformulation). Alternatively, as in Oberman (2006), one can consider finite difference schemes of the form

$$F_h[u](z) = G_h(z, u(z), u(z) - Tu(z)).$$

Let $\mathbf{1} \in \mathbb{R}^{|S|}$ denote the vector of ones. Then using the correspondence $F_h(z, r, q) = G_h(z, r, r\mathbf{1} - q)$, one sees that F_h is of non-negative type if and only if G_h is non-increasing in its second and third arguments.

Let us now show that non-negativity is simply a reformulation of monotonicity.

Lemma 3.11 (equivalence). A finite difference scheme of the form (3.8) is monotone if and only if it is of non-negative type.

Proof. Suppose that F_h is of non-negative type. Let u_h and v_h be two grid functions such that $u_h - v_h$ has a global non-negative maximum at some grid point z. Set $r = v_h(z)$, $t = u_h(z) - v_h(z) \geq 0$, $q_i = v_h(z + hy_i)$, and $\tau_i = \max\{0, u_h(z+hy_i) - v_h(z+hy_i)\}$, so that $t \geq \tau_i$. Noting that $q_i + \tau_i = v_h(z+hy_i) + \max\{0, u_h(z+hy_i) - v_h(z+hy_i)\} \geq u_h(z+hy_i)$, and F_h is non-decreasing in its third argument, we find that $F_h[u_h](z) \leq F_h(z, r+t, q+\tau)$. Therefore, by the second inequality in (3.9) we have

$$F_h[u_h](z) \leq F_h(z, r + t, q + \tau) \leq F_h(z, r, q) = F_h[v_h](z).$$

Thus, F_h is monotone.

Now suppose that F_h is monotone. Let $z \in \mathbb{R}^d$ be fixed, and let $r, t \in \mathbb{R}$ and $q, \tau \in \mathbb{R}^{|S|}$ satisfy (3.9). Then define the grid functions u_h, v_h (locally) as

$$v_h(z) = r, \quad v_h(z + hy_i) = q_i, \quad u_h(z) = r + t, \quad u_h(z + y_i) = q_i + \tau_i.$$

Then

$$u_h(z) - v_h(z) = t \geq \tau_i = u_h(z + hy_i) - v_h(z + hy_i),$$

that is, $u_h - v_h$ has a non-negative maximum at z. The monotonicity of F_h yields $F_h[u_h](z) \leq F_h[v_h](z)$; thus

$$F_h(z, r + t, q + \tau) \leq F_h(z, r, q). \tag{3.11}$$

On the other hand, with v_h as before, we consider the grid function w_h with

$$w_h(z) = r, \quad w_h(z + hy_i) = q_i + \tau_i.$$

Then $v_h - w_h$ has a global maximum at z and thus

$$F_h(z, r, q) = F_h[v_h](z) \leq F_h[w_h](z) = F_h(z, r, q + \tau). \tag{3.12}$$

We conclude from (3.11)–(3.12) that F_h is of non-negative type. □

Following the framework given in Kuo and Trudinger (1990, 1992) and Kocan (1995), we consider discrete operators constructed from the first- and second-order difference operators

$$\begin{aligned}
\delta^+_{y,h}u(z) &:= \frac{1}{h}(u(z+hy)-u(z)),\\
\delta^-_{y,h}u(z) &:= \frac{1}{h}(u(z)-u(z-hy)),\\
\delta_{y,h}u(z) &:= \frac{1}{2}(\delta^+_y+\delta^-_y)u(z) = \frac{1}{2h}(u(z+hy)-u(z-hy)),\\
\delta^2_{y,h}u(z) &:= \frac{1}{h^2}(u(z+hy)-2u(x)+u(z-hy)),
\end{aligned}$$

with $y \in S$. Taylor's theorem shows that the differences $\delta^{\pm}_{y,h}u(z)$ are first-order approximations to

$$\frac{\partial u(z)}{\partial y} := Du(z)\cdot y,$$

whereas $\delta_{y,h}u(z)$ and $\delta^2_{y,h}u(z)$, respectively, are second-order approximations to

$$\frac{\partial u(z)}{\partial y} \quad \text{and} \quad \frac{\partial^2 u(z)}{\partial y^2} := y\cdot D^2u(z)y;$$

by this we mean that

$$\begin{aligned}
\left|\frac{\partial^{\pm} u(z)}{\partial y} - \delta^{\pm}_{y,h}u(z)\right| &= O(h|y|^2),\\
\left|\frac{\partial u(z)}{\partial y} - \delta_{y,h}u(z)\right| &= O(h^2|y|^3),\\
\left|\frac{\partial^2 u(z)}{\partial y^2} - \delta^2_{y,h}u(z)\right| &= O(h^2|y|^4),
\end{aligned}$$

for sufficiently smooth u.

Let

$$\delta_h u_h(z) = \{\delta_{y,h}u_h(z) : y\in S\} \quad \text{and} \quad \delta^2_h u_h(z) = \{\delta^2_{y,h}u_h(z) : y \in S\}.$$

Then a consistent and monotone finite difference scheme can be constructed in the form

$$F_h[u_h](z) = \mathcal{F}_h(z, u_h(z), \delta_h u_h(z), \delta^2_h u_h(z)) \tag{3.13}$$

(Kuo and Trudinger 1990, Kuo and Trudinger 1992, Motzkin and Wasow 1953), where $\mathcal{F}_h : \Omega_h \times \mathbb{R} \times \mathbb{R}^{|S|} \times \mathbb{R}^{|S|} \to \mathbb{R}$. Let (z, r, q, s) denote points in the domain of $\mathcal{F}_h$, and assume that $\mathcal{F}_h$ is symmetric with respect to $\pm q_{\pm i}$ and $s_{\pm i}$. Then from Definition 3.9 and Lemma 3.11 we see that F_h of the form (3.13) is monotone provided that

$$\frac{h}{2}\left|\frac{\partial \mathcal{F}_h}{\partial q_i}\right| \le \frac{\partial \mathcal{F}_h}{\partial s_i} \quad i = 1, 2, \dots, |S|, \quad \text{and} \quad \frac{\partial \mathcal{F}_h}{\partial r} \le 0. \tag{3.14}$$

Below, we will require slightly stronger conditions on the operator $\mathcal{F}_h$.

Definition 3.12 (positive operator). An operator of the form (3.13) is of *positive type* (or simply positive) if (3.14) is satisfied and there exists $\lambda_{0,h} > 0$ and an orthogonal set of vectors $\{y_i\}_{i=1}^d \subset S$ such that

$$\lambda_{0,h} + \frac{h}{2}\left|\frac{\partial \mathcal{F}_h}{\partial q_i}\right| \le \frac{\partial \mathcal{F}_h}{\partial s_i}.$$

Remark 3.13 (discrete ellipticity). The discrete ellipticity constant $\lambda_{0,h}$ may depend on the stencil size m; see Theorem 3.36.

3.3. Finite difference stability estimates: Alexandrov estimates and Alexandrov–Bakelman–Pucci maximum principle

We now turn our attention to maximum principles of discrete operators, and correspondingly, stability estimates. As a starting point, we discuss monotone finite difference schemes for the linear non-divergence-form PDEs of Example 2.11

$$F[u] = \mathcal{L}u - f = A : D^2 u - f = 0. \tag{3.15}$$

While this setting may seem overly simplistic, as we shall see, the construction and theoretical results for the linear problem form all of the necessary tools to approximate viscosity solutions of nonlinear elliptic equations.

We assume that $f \in C(\bar{\Omega})$ and that the coefficient matrix A is bounded and uniformly symmetric positive definite. It is then reasonable to assume that F_h is linear and thus has the form

$$F_h[u_h](z) = \mathcal{L}_h u_h(z) - f(z) := \sum_{y \in S} a_y(z)\delta_{y,h}^2 u_h(z) - f(z) \tag{3.16}$$

for nodal functions (or coefficients) a_y. Applying Definition 3.9 to (3.16), we see that F_h is non-negative (and hence monotone) provided

$$a_y(z) \ge 0, \tag{3.17}$$

and of positive type if

$$a_{y_i}(z) \ge \lambda_{0,h} > 0 \quad i = 1, 2, \dots, d.$$

for some orthogonal basis $\{y_i\}_{i=1}^d \subset S$. These inequalities suggest that the negation of the ensuing system is an M-matrix, and hence solutions to the discrete problem satisfy certain maximum principles, analogous to the continuous setting. This issue is discussed in the next section.

3.3.1. Finite difference Alexandrov estimates
In this section we state and prove ABP maximum principles for grid functions. To get started, we first specify the fundamental notion of interior and boundary nodes used in this section.

Definition 3.14 (interior and boundary nodes). For a discrete operator F_h, we define the *set of interior nodes* Ω_h^I as the set of grid points $z \in \bar{\Omega}_h$ such that for any mesh function v_h, $F_h[v_h](z)$ depends only on the translates of v_h at points in $\bar{\Omega}_h$. The *set of boundary nodes* are given by $\Omega_h^B := \bar{\Omega}_h \backslash \Omega_h^I$.

Remark 3.15 (discrete domain). If F_h is a one-step method (*i.e.* $m = 1$), then $\Omega_h^B = \partial\Omega \cap \mathbb{Z}_h^d$ and $\Omega_h^I = \Omega \cap \mathbb{Z}_h^d$.

As a next step we introduce and discuss several basic properties of convexity and the subdifferential for discrete (nodal) functions.

Definition 3.16 (convex nodal function). We say that a nodal function $v_h \in X_h^{fd}$ is a *convex nodal function* if there is a supporting hyperplane of v_h at all interior nodes $z \in \Omega_h^I$.

Note that if v_h is the nodal interpolant of a convex function, then v_h is a convex nodal function.

Definition 3.17 (discrete convex envelope). Let $R > 0$ be sufficiently large so that $\bar{\Omega}$ (and hence $\bar{\Omega}_h$) is compactly contained in a ball B_R. For a nodal function (or continuous function) v_h with $v_h \geq 0$ on Ω_h^B, we extend v_h^- to $B_{R,h} \backslash \bar{\Omega}_h$ by zero, where $B_{R,h} = B_R \cap \mathbb{Z}_h^d$, and $v_h^- = -\max\{0, -v_h\}$ is the negative part of v_h. We define the *discrete convex envelope* of $-v_h^-$ as

$$\Gamma_h(v_h)(x) := \sup\{L(x) : L(z) \leq -v_h^-(z) \ \forall z \in B_{R,h}, \ L \in \mathbb{P}_1\} \qquad (3.18)$$

for all $x \in \bar{B}_R$.

Remark 3.18 (discrete convexity). There are subtle issues in the above definitions that require some elaboration.

- If $v_h \in X_h^{fd}$ is convex and $v_h \leq 0$, then we have
$$v_h(z) = \Gamma_h(v_h)(z) \quad \text{for all } z \in \Omega_h^I. \qquad (3.19)$$
Thus, $\Gamma_h(v_h)$ is a natural convex extension of v_h.
- Since for every $x \in \partial B_R$, there exists an affine function L with $L(z) \leq -v_h^-(z)$ for all $z \in \bar{\Omega}_h$ and $L(x) = 0$, we conclude that $\Gamma_h(v_h) = 0$ on ∂B_R.
- Definition 3.17 implies that $\Gamma_h(v_h)$ is a convex, piecewise linear function with respect to a simplicial triangulation. The vertices of the triangulation are a subset of the gridpoints $B_{R,h}$, and its configuration depends on v_h; see Examples 3.44–3.45.

- For $v_h \in X_h^{fd}$, let $\tilde{v}_h \in C(\bar{\Omega})$ denote the canonical multi-linear function. Then, since the inequality constraints in (3.18) are only posed on a discrete set of points, and since $\tilde{v}_h$ is not necessarily piecewise affine, we have $\Gamma(\tilde{v}_h) \neq \Gamma_h(v_h)$ in general (Carnicer and Dahmen 1994).

Next, we require the notion of a subdifferential acting on nodal functions. Recall from Definition 2.58 that the subdifferential requires function values at all points of the domain Ω, so this notion is not directly applicable to the discrete case. Instead, with a slight abuse of notation, we define its natural extension to nodal functions as follows:

$$\partial v_h(z) = \{\mathbf{p} \in \mathbb{R}^d : v_h(x) - v_h(z) \geq \mathbf{p} \cdot (x - z),\ \forall x \in \bar{\Omega}_h\}, \tag{3.20}$$

for all $z \in \bar{\Omega}_h$ and $v_h \in X_h^{fd}$.

Lemma 3.19 (discrete subdifferential). If $v_h \in X_h^{fd}$ is a convex nodal function, then $\partial v_h(z) = \partial \Gamma_h(v_h)(z)$ for all $z \in \Omega_h^I$.

Proof. Thanks to (3.19), if $\mathbf{p} \in \partial \Gamma_h(v_h)(z)$, that is,

$$\Gamma_h(v_h)(x) \geq \Gamma_h(v_h)(z) + \mathbf{p} \cdot (x - z), \quad \text{for all } x \in \Omega,$$

then $\mathbf{p} \in \partial v_h(z)$.

Conversely, let $\mathcal{T}_z$ be a local mesh induced by $\Gamma(u_h)(z)$, let $K \in \mathcal{T}_z$ be a d-dimensional simplex, and let $\{z_j\}_{j=1}^{d+1}$ be the vertices of K. If $\mathbf{p} \in \partial v_h(z)$, then we clearly have $v_h(x) \geq v_h(z) + \mathbf{p} \cdot (x - z)$ for all vertices $x \in \bar{\Omega}_h$. Again, thanks to (3.19), we have $\Gamma(v_h)(z_i) \geq \Gamma(v_h)(z) + \mathbf{p} \cdot (z_i - z)$ for all vertices. Since $\Gamma(v_h)$ is linear on element K, we have $\Gamma(v_h)(x) \geq \Gamma(v_h)(z) + \mathbf{p} \cdot (x - z)$ for any $x \in K$. This shows that $\mathbf{p} \in \partial \Gamma(u_h)(z)$ as well. □

Let us state two properties of subdifferentials. The proof of the first one follows directly from its definition.

Lemma 3.20 (monotonicity of subdifferential). Let $w_h, v_h \in X_h^{fd}$ be two convex nodal functions such that, for a fixed $z_* \in \bar{\Omega}_h$, $w_h(z_*) = v_h(z_*)$ and $w_h(z) \leq v_h(z)$ for all $z \neq z_*$. Then,

$$\partial w_h(z_*) \subset \partial v_h(z_*).$$

Lemma 3.21 (addition inequality). Let w_h and v_h be two convex nodal functions. Then

$$\partial w_h(z) + \partial v_h(z) \subset \partial(w_h + v_h)(z), \quad \text{for all } z \in \Omega_h^I,$$

where $\partial w_h(z) + \partial v_h(z)$ is the Minkowski sum:

$$\partial w_h(z) + \partial v_h(z) = \{\mathbf{p} + \mathbf{q} \in \mathbb{R}^d : \mathbf{p} \in \partial w_h(z), \mathbf{q} \in \partial v_h(z)\}.$$

Proof. We note that if $\mathbf{p} \in \partial w_h(z)$ and $\mathbf{q} \in \partial v_h(z)$, then

$$w_h(x) \geq w_h(z) + w \cdot (x - z) \quad \text{and} \quad v_h(x) \geq v_h(z) + v \cdot (x - z)$$

for all $x \in \bar{\Omega}_h$. Adding the two inequalities yields

$$w_h(x) + v_h(x) \geq w_h(z) + v_h(z) + (\mathbf{p} + \mathbf{q}) \cdot (x - z),$$

which implies that $(\mathbf{p} + \mathbf{q}) \in \partial(w_h + v_h)$. □

Given a convex nodal function u_h, computing its discrete subdifferential set is a non-trivial task. The following lemma shows that it involves computing the convex envelope of u_h.

Lemma 3.22 (characterization of subdifferential). Let u_h be a convex nodal function, and let $\mathcal{T}_h$ be the simplicial mesh induced by its convex envelope. The subdifferential of u_h at z is the convex hull of the piecewise gradient, that is,

$$\operatorname{conv}\{Du_h|_K, K \in \mathcal{T}_h,\ z \in \bar{K}\}.$$

Here, Du_h is the gradient of the piecewise linear polynomial induced by u_h and $\mathcal{T}_h$.

As final preparation for stating the finite difference version of the Alexandrov estimate, we define the nodal contact set.

Definition 3.23 (nodal contact set). Let v_h be either a nodal function or a continuous function with $v_h \geq 0$ on Ω_h^B. The (*lower*) *nodal contact set* of v_h is given by

$$\mathcal{C}_h^-(v_h) = \{z \in \Omega_h^I : \Gamma_h(v_h)(z) = v_h(z)\}.$$

Note that, for $x \in \mathcal{C}_h^-(v_h)$, we have

$$v_h(z) \geq \Gamma_h(v_h)(z) \geq v_h(x) + \mathbf{p} \cdot (z - x),$$

for all $z \in B_{R,h}$ and all $\mathbf{p} \in \partial\Gamma_h(v_h)(x)$.

We are now ready to state and prove the finite difference Alexandrov estimate. Recall that we assume Ω is compactly contained in a ball B_R of radius R, and that we set $B_{R,h} = B_R \cap \mathbb{Z}_h^d$.

Lemma 3.24 (finite difference Alexandrov estimate). Let $v_h \in X_h^{fd}$ with $v_h \geq 0$ on Ω_h^B. Then

$$\sup_{\bar{\Omega}_h} v_h^- \leq CR\Bigg(\sum_{z \in \mathcal{C}_h^-(v_h)} |\partial\Gamma_h(v_h)(z)|\Bigg)^{1/d}, \tag{3.21}$$

where the constant $C > 0$ depends only on d.

Proof. We follow the arguments given in Nochetto and Zhang (2017*a*, Proposition 5.1); see also Kuo and Trudinger (1990).

Let $z_* \in B_{R,h}$ satisfy $\sup_{B_{R,h}} v_h^- = v_h^-(z_*)$, and let L be a horizontal plane touching v_h from below at z_*. By Definition 3.17 we have

$$\Gamma_h(v_h)(z) \geq L(z) = L(z_*) = v_h(z_*), \quad \text{for all } z \in B_{R,h}.$$

Thus, $\sup_{B_{R,h}} \Gamma_h(v_h)^- \leq v_h^-(z_*)$. Since $\Gamma_h(v_h) \leq v_h$ on $B_{R,h}$ implies

$$\sup_{B_{R,h}} v_h^- \leq \sup_{B_{R,h}} \Gamma_h(v_h)^-,$$

we conclude that

$$\sup_{\bar{\Omega}_h} v_h^- = \sup_{B_{R,h}} v_h^- = \sup_{B_{R,h}} \Gamma_h(v_h)^- = \sup_{B_R} \Gamma_h(v_h)^-.$$

Therefore, to conclude the proof it suffices to show that

$$\max_{\bar{\Omega}_h} \Gamma_h(v_h)^- \leq CR \Bigg(\sum_{z \in \mathcal{C}_h^-(v_h)} |\partial \Gamma_h(v_h)(z)| \Bigg)^{1/d}.$$

This is done in three steps.

Step 1. Let $K(x)$ be the cone with vertex z_* satisfying

$$K(z_*) = -\sup_{B_R} \Gamma_h(v_h)^- =: -M \quad \text{and} \quad K(x) = 0 \text{ on } \partial B_R,$$

and assume that $M > 0$, for otherwise (3.21) is trivial. We note that for any vector $\mathbf{p} \in B_{M/(2R)}(0)$, the affine function $L(x) = -M + \mathbf{p} \cdot x - z_*$ is a supporting plane of $K(x)$ at point z_*, namely $L(x) \leq K(x)$ for all $x \in B_R$ and $L(z_*) = K(z_*)$. This implies that $\partial K(z_*) \supset B_{M/(2R)}(0)$, and therefore

$$|\partial K(z_*)| \geq C \left(\frac{M}{R} \right)^d.$$

Step 2. We claim that

$$\partial K(z_*) \subset \bigcup_{z \in \mathcal{C}_h^-(v_h)} \partial \Gamma_h(v_h)(z). \tag{3.22}$$

This is equivalent to showing that for any supporting plane L of K at z_*, there is a parallel supporting plane $\tilde{L}$ for $\Gamma_h(v_h)$ at some contact node $y \in \mathcal{C}_h^-(v_h)$.

Consider the (nodal) function $v_h - L$, and observe that $v_h \geq 0$ on Ω_h^B and $v_h(z_*) = K(z_*) = L(z_*)$, whence

$$v_h(z) - L(z) \geq K(z) - L(z) \geq 0 \text{ on } \Omega_h^B,$$
$$v_h(z_*) - L(z_*) = K(z_*) - L(z_*) = 0.$$

We infer that $v_h - L$ attains a non-positive minimum for some $x \in \Omega_h^I$. Hence, $\widetilde{L}(z) = L(z) + v_h(x) - L(x)$ satisfies $\tilde{L}(z) \leq v_h(z)$ for all $z \in B_{R,h}$ and

$\tilde{L}(x) = v_h(x)$. Applying Definition 3.17 we conclude that $\tilde{L} \le \Gamma_h(v_h) \le v_h$ and therefore $\Gamma_h(v_h)(x) = v_h(x)$; thus $x \in \mathcal{C}_h^-(v_h)$.

Step 3. Computing Lebesgue measures in (3.22) yields

$$C\left(\frac{M}{R}\right)^d \le |\partial K(z_*)| \le \sum_{z \in \mathcal{C}_h^-(v_h)} |\partial \Gamma_h(v_h)(z)|.$$

Finally, (3.44) follows from this last inequality and some simple algebraic manipulation. □

The following theorem states that positive finite difference operators satisfy a discrete Alexandrov–Bakelman–Pucci estimate (see Theorem 2.62 and Kuo and Trudinger 1990).

Theorem 3.25 (finite difference ABP estimate). Suppose that $\mathcal{L}_h$ is of positive type and of the form (3.16). Suppose that $u_h \in X_h^{fd}$ satisfies

$$\begin{cases} \mathcal{L}_h u_h \le f & \text{in } \Omega_h^I, \\ u_h = g_h & \text{on } \Omega_h^B \end{cases} \tag{3.23}$$

for some $g_h \in X_h^{fd}$. Then

$$\sup_{\bar{\Omega}_h} u_h^- \le \sup_{\Omega_h^B} g_h^- + C\frac{R}{\lambda_{0,h}}\left(\sum_{z \in \mathcal{C}_h^-(u_h)} h^d (f^+(z))^d\right)^{1/d},$$

where $\lambda_{0,h}$ is given in Definition 3.12, and the constant $C > 0$ depends only on d.

Proof. We follow the arguments given in Kuo and Trudinger (1990, Theorem 2.1).

Note that we can assume that $u_h \ge 0$ on Ω_h^B, by replacing u_h with $u_h + \max_{\Omega_h^B} g_h^-$. Moreover we can assume that $\max_{\bar{\Omega}_h} u_h^- > 0$, since otherwise the proof is trivial.

Let $z \in \mathcal{C}_h^-(u_h)$ and $y \in S$. As $\Gamma_h(u_h)$ is convex, we have $\delta_{y,h}^2 \Gamma_h(u_h)(z) \ge 0$. Thus, since the coefficients of F_h are positive, $\Gamma_h(u_h)(z) = u_h(z)$ and $\Gamma_h(u_h)(z \pm hy) \le u_h(z \pm hy)$, we have

$$0 \le a_y(z)\delta_{y,h}^2 \Gamma_h(u_h)(z) \le a_y(z)\delta_{y,h}^2 u_h(z).$$

Summing over $y \in S$ yields

$$0 \le a_y(z)\delta_{y,h}^2 \Gamma_h(u_h)(z) \le \sum_{y' \in S} a_{y'}(z)\delta_{y',h}^2 u_h(z) \le f(z) = f^+(z).$$

Take y to be the orthogonal set $\{y_i\}_{i=1}^d$ given in Definition 3.12. Expand the left-hand side of the above inequality to get

$$\delta^+_{y_i,h}\Gamma_h(u_h)(z) - \delta^-_{y_i,h}\Gamma_h(u_h)(z) = h\delta^2_{y_i,h}\Gamma_h(u_h)(z) \le \frac{h}{\lambda_{0,h}} f^+(z). \tag{3.24}$$

Now, let $\mathbf{p} \in \partial\Gamma_h(u_h)(z)$ so that $\Gamma_h(u_h)(z \pm hy_i) \ge \Gamma_h(u_h)(z) \pm h\mathbf{p}\cdot y_i$. By manipulating terms and applying inequality (3.24) we obtain

$$\delta^-_{y_i,h}\Gamma_h(u_h)(z) \le \mathbf{p}\cdot y_i \le \delta^+_{y_i,h}\Gamma_h(u_h)(z) \le \delta^-_{y_i,h}\Gamma_h(u_h)(z) + \frac{h}{\lambda_{0,h}} f^+(z).$$

Since $\{y_i/|y_i|\}_{i=1}^d$ is an orthonormal basis of $\mathbb{R}^d$, these two inequalities show that the Lebesgue measure of $\partial\Gamma_h(u_h)(z)$ is bounded by

$$|\partial\Gamma_h(u_h)(z)| \le \frac{h^d}{\lambda_{0,h}^d} |f^+(z)|^d,$$

and therefore

$$\sum_{z\in\mathcal{C}_h^-(u_h)} |\partial\Gamma_h(u_h)(z)| \le \sum_{z\in\mathcal{C}_h^-(u_h)} \frac{(hf^+(z))^d}{\lambda_{0,h}^d}. \tag{3.25}$$

Combining (3.25) and Lemma 3.24 yields the desired result. □

Remark 3.26 (extensions). The finite difference ABP estimate given in Theorem 3.25 has been extended to operators with lower-order terms and to general meshes in Kuo and Trudinger (1996, 2000).

Theorem 3.25 implies that if u_h solves

$$\begin{cases} \mathcal{L}_h u_h = f & \text{in } \Omega_h^I, \\ u_h = g_h & \text{on } \Omega_h^B, \end{cases} \tag{3.26}$$

then

$$\max_{\bar{\Omega}_h} |u_h| \le \max_{\Omega_h^B} |g_h| + \frac{CR}{\lambda_{0,h}} \left(\sum_{z\in\Omega_h^I} h^d |f(z)|^d \right)^{1/d}. \tag{3.27}$$

Since problem (3.26) is linear this estimate shows that there exists a unique solution to (3.26).

Similar to the continuous case (see Corollary 2.2), Theorem 3.25 implies a comparison principle.

Corollary 3.27 (discrete comparison). Suppose that $\mathcal{L}_h$ is of positive type and of the form (3.16). Let u_h and v_h be two nodal functions with $u_h \le v_h$ on Ω_h^B and $\mathcal{L}_h u_h \ge \mathcal{L}_h v_h$ in Ω_h^I. Then $u_h \le v_h$ in $\bar{\Omega}_h$.

Finally, since problems (3.16) and (3.26) are linear, the Lax–Richtmyer theorem immediately gives us error estimates.

Corollary 3.28 (rate of convergence). Let $I_h^{fd} : C(\bar{\Omega}) \to X_h^{fd}$ denote the canonical interpolant onto nodal functions. Let u be the solution to (3.15), and let $u_h \in X_h^{fd}$ be the unique solution to (3.26). Then

$$\|u_h - I_h^{fd}u\|_{L^\infty(\bar{\Omega}_h)} \le \|g_h - I_h^{fd}u\|_{L^\infty(\Omega_h^B)} + \frac{CR}{\lambda_{0,h}}\bigg(\sum_{z\in\Omega_h^I} h^d|\mathcal{L}_h I_h^{fd}u(z))|^d\bigg)^{1/d}.$$

Based on Corollary 3.28 and the consistency of the approximation scheme, one can derive error estimates with explicit dependence on the discretization parameter h. For example, if we can show that

$$\|g_h - I_h^{fd}u\|_{L^\infty(\Omega_h^B)} = O(h^k) \quad \text{and} \quad \|\mathcal{L}_h I_h^{fd}u\|_{L^\infty(\Omega_h^I)} = O(h^k)$$

for some positive integer $k \in \mathbb{N}$, then Corollary 3.28 shows that the error satisfies $\|u - I_h^{fd}u\|_{L^\infty(\bar{\Omega}_h)} \le Ch^k$. The value of k is determined by the consistency of the scheme, which typically follows from Taylor's theorem and the regularity of the exact solution. Unfortunately, the monotonicity of a scheme restricts the size of the order of convergence, as shown, for instance, by Oberman (2006) and Kossaczký, Ehrhardt and Günther (2016).

Theorem 3.29 (accuracy of monotone schemes). A monotone finite difference scheme of the form (3.8) is at most second-order accurate for second-order equations.

Finally, solutions to the discrete problem are Hölder-continuous (Kuo and Trudinger 1990, Corollary 4.6, Theorem 5.1).

Theorem 3.30 (Hölder continuity). Let $\mathcal{L}_h$ be of positive type, and let u_h satisfy $\mathcal{L}_h u_h = f$ in Ω_h^I. Assume that Ω satisfies a uniform exterior cone condition. Then there exist $\eta \in (0,1)$ and $C > 0$, independent of h, such that

$$|u_h(z) - u_h(y)| \le C|z-y|^\eta,$$

for all $z, y \in \bar{\Omega}_h$.

As in the continuous setting, the development of Hölder estimates depends on discrete Harnack inequalities.

3.4. Construction of monotone finite difference schemes

Theorem 3.25 shows that linear, positive finite difference schemes are indeed uniquely solvable with solutions uniformly bounded with respect to the data. We will also see that many of these results carry over to the fully nonlinear case, and thus, applying the Barles–Souganidis framework, such schemes converge to the viscosity solution of the nonlinear PDE. However, the theorem does not indicate how to construct such schemes. We now discuss this issue. First, we have the following classical results (Motzkin and Wasow 1953, Theorems 1 and 2).

Theorem 3.31 (impossibility). For a given (fixed) stencil width $m \in \mathbb{N}$, there exists an linear, elliptic operator $\mathcal{L}$ such that any linear and consistent finite difference scheme of the form (3.16) is not of positive type.

Theorem 3.32 (existence). Let $\mathcal{L}$ be a linear and uniformly elliptic operator satisfying (3.15). Then there exists, for sufficiently small h, a consistent finite difference scheme of the form (3.16) that is of positive type.

The main punchline of these theorems is that wide stencils are a necessary feature of consistent and positive-type finite difference discretizations, even for linear problems. The proof of Theorem 3.32, as presented in Motzkin and Wasow (1953, Theorem 2), is not constructive. On the other hand, the arguments given in Kuo and Trudinger (1990) and Kocan (1995) explicitly give an algorithm to construct consistent and positive schemes, and as a result, provide an estimate of the stencil width. We end this section by summarizing these results. As a first step, we state the following trivial observation.

Lemma 3.33 (positivity criterion). Suppose that $\mathcal{L}u = A : D^2u$, where the coefficient matrix is positive definite and has the form

$$A(x) = \sum_{\substack{y\in\mathbb{Z}^d\\ |y|_\infty\le M}} a_y(x) y \otimes y, \tag{3.28}$$

for some $M \in \mathbb{N}$ and with coefficients $a_y(x) \ge 0$ for all $x \in \Omega$. Assume further that there exists an orthogonal set $\{y_i\}_{i=1}^d \subset \mathbb{Z}^d$ with $|y_i|_\infty \le M$ such that $a_{y_i}(x) \ge c$ for some $c > 0$. Then the finite difference operator

$$\mathcal{L}_h u_h(z) = \sum_{\substack{y\in\mathbb{Z}^d\\ |y|_\infty\le M}} a_y(z)\delta^2_{y,h} u_h(z) \tag{3.29}$$

is of positive type and a consistent approximation to $\mathcal{L}$ with $\lambda_{0,h} = c$.

Of course, not all positive definite matrices are of the form (3.28). However, quite surprisingly, Lemma 3.33 provides the essential tools to construct consistent and positive finite difference schemes for linear (and nonlinear) elliptic equations. Let us consider an example.

Lemma 3.34 (diagonally dominant matrix). Suppose that $\mathcal{L}u = A : D^2u$ with

$$A(x) = \sum_{i,j=1}^d a_{i,j}(x) y_i \otimes y_j, \tag{3.30}$$

where $\{y_i\}_{i=1}^d \subset \mathbb{Z}^d$ is an orthogonal basis of $\mathbb{R}^d$, that is,

$$a_{i,j} = y_i \cdot Ay_j/(|y_i|\,|y_j|).$$

Suppose, in addition, that for some $c > 0$ we have

$$\sum_{\substack{i,j=1\\ j\neq i}}^{d} |a_{i,j}(x)| \leq a_{i,i}(x) - c \quad i = 1, 2, \ldots, \qquad x \in \Omega, \tag{3.31}$$

Then there exists a positive finite difference method $\mathcal{L}_h$ that is consistent with $\mathcal{L}$ and with $\lambda_{0,h} = c$.

Proof. By manipulating terms, we may write

$$\begin{aligned} A = &\sum_{i=1}^{d}\Bigg(a_{ii} - \sum_{\substack{j=1\\ j\neq i}}^{d} |a_{i,j}|\Bigg) y_i \otimes y_i \\ &+ \frac{1}{4}\sum_{\substack{i,j=1\\ i\neq j}}^{d} (|a_{i,j}| + a_{i,j})(y_i + y_j) \otimes (y_i + y_j) \\ &+ \frac{1}{4}\sum_{\substack{i,j=1\\ i\neq j}}^{d} (|a_{i,j}| - a_{i,j})(y_i - y_j) \otimes (y_i - y_j). \end{aligned}$$

Thus, A satisfies the conditions in Lemma 3.33 with $M \leq 2\max_i |y_i|_\infty$. The result now follows from Lemma 3.33. ☐

Remark 3.35 (stencil size). By taking $\{y_i\}_{i=1}^d \subset \mathbb{Z}^d$ as the standard basis of $\mathbb{R}^d$ in Lemma 3.34, we deduce that if A is strictly diagonally dominant, then there exists a consistent and positive one-step finite difference method.

Now, following Kocan (1995, Theorem 5), we can estimate the stencil size for a general positive definite matrix.

Theorem 3.36 (existence). Consider the elliptic operator $\mathcal{L}u = A : D^2u$, where A is uniformly positive definite in Ω with

$$\lambda I \leq A \leq \Lambda I$$

for positive constants $\lambda \leq \Lambda$. Then there exists a consistent and positive operator $\mathcal{L}_h$. The stencil size m satisfies $m \leq 2M_d(8d^3\mathcal{E})$, where $\mathcal{E} := \Lambda/\lambda$ and

$$M_d(s) = C\begin{cases} s & d = 2, \\ s^{5/2} & d = 3, \\ s^{2d-4} & d \geq 4. \end{cases}$$

Here the constant $C > 0$ depends only on the dimension d. Moreover, one can take the discrete ellipticity constant to be $\lambda_{0,h} = \lambda/(2dm^2)$.

Proof. Let $\{\lambda_i\}_{i=1}^d \subset [\lambda, \Lambda]$ denote the eigenvalues of A and let $\{\varphi_i\}_{i=1}^d$ be an orthonormal set of eigenvectors of A, labelled such that

$$A = \sum_{i=1}^d \lambda_i \varphi_i \otimes \varphi_i.$$

By Kocan (1995, Theorem 2), for $s > 0$ to be determined, there exists $y_i \in \mathbb{Z}^d$ such that

$$\left| \varphi_i - \frac{y_i}{|y_i|} \right|_\infty \le \frac{1}{s}, \quad \frac{M_d(s)}{2} \le |y_i|_\infty \le M_d(s). \tag{3.32}$$

Since $\{y_i \otimes y_j\}_{i,j=1}^d$ spans $\mathbb{S}^d$, we may write

$$\begin{aligned} A &= \sum_{i=1}^d \frac{\lambda_i}{|y_i|^2} y_i \otimes y_i + \sum_{i=1}^d \lambda_i \left(\varphi_i \otimes \varphi_i - \frac{y_i \otimes y_i}{|y_i|^2} \right) \\ &= \sum_{i=1}^d \frac{\lambda_i}{|y_i|^2} y_i \otimes y_i + \sum_{i,j=1}^d \frac{B_{i,j}}{|y_i|\,|y_j|} y_i \otimes y_j \end{aligned}$$

for some $B \in \mathbb{R}^{d \times d}$. Thus, A is of the form (3.30) with $a_{i,i} = (\lambda_i + B_{i,i})/|y_i|^2$ and $a_{i,j} = B_{i,j}/(|y_i|\,|y_j|)$.

Applying (3.32) and the inequalities $\lambda \le \lambda_i \le \Lambda$, we obtain

$$|B_{i,j}| \le \frac{2\Lambda d^{3/2}}{s},$$

and hence

$$|a_{i,i}| \ge \frac{1}{|y_i|^2} \left(\lambda - \frac{2\Lambda d^{3/2}}{s} \right) \ge \frac{\lambda}{|y_i|^2}, \quad |a_{i,j}| \le \frac{2d^{3/2}\Lambda}{s|y_i|\,|y_j|}.$$

Thus (3.31) will be satisfied if

$$\lambda \ge \max_i \left(\frac{2d^{3/2}\Lambda}{s} \sum_{j=1}^d \frac{|y_i|}{|y_j|} + c|y_i|^2 \right)$$

for some constant $c > 0$. Taking $c = \lambda/(2dm^2)$, and noting that $|y_i|/|y_j| \le 2d^{1/2}$, we conclude that if

$$s \ge 8d^3 \mathcal{E},$$

then (3.31) is satisfied. The desired result now follows from Lemma 3.34. $\square$

Finally we end this section with a result which shows that the uniform ellipticity of $\mathcal{L}$ condition in Theorem 3.36 cannot be relaxed; see Kocan (1995, Theorem 1).

Example 3.37 (uniform ellipticity is necessary). Suppose that $d = 2$ and $Lu = A : D^2u$, where

$$A = \begin{pmatrix} \alpha & \beta \\ \beta & \gamma \end{pmatrix},$$

$\alpha, \gamma > 0$ and $\alpha\gamma = \beta^2$ (so that $\det(A) = 0$). Then if $\sqrt{\alpha/\gamma}$ is irrational, there does not exist a consistent and non-negative scheme for the problem $\mathcal{L}u = f$.

3.5. Monotonicity in finite element methods

In this section we discuss the construction of monotone finite element methods for second-order elliptic problems. As in the previous section, we focus on the linear case, where the elliptic problem is given by (3.15), and extend these results to nonlinear problems in subsequent sections. We further simplify the presentation and analysis by assuming that the coefficient matrix is the identity matrix, $A = I$, and assume Dirichlet boundary conditions. Thus, we focus on the Poisson problem

$$\Delta u = f \quad \text{in } \Omega, \qquad u = g \quad \text{on } \partial\Omega. \tag{3.33}$$

Quite surprisingly, the results given here extend to fully nonlinear problems.

Let $\mathcal{T}_h$ be a simplicial, conforming and quasi-uniform triangulation of Ω (Ciarlet 2002). For simplicity, and to communicate the essential ideas, we shall ignore the approximation of Ω by the triangulation-induced polytope, and simply assume throughout the paper that $\bar{\Omega} = \cup_{T\in\mathcal{T}_h}\bar{T}$. Let X_h^l be the linear Lagrange finite element space, that is,

$$X_h^l = \{v_h \in C(\bar{\Omega}) : v_h|_T \in \mathbb{P}_1 \ \forall T \in \mathcal{T}_h\}. \tag{3.34}$$

We extend the definitions given in the previous sections to unstructured meshes by letting Ω_h^I and Ω_h^B denote the sets of vertices (or nodes) of the triangulation $\mathcal{T}_h$ that belong to Ω and $\partial\Omega$ respectively. Then any function $w_h \in X_h^l$ is uniquely determined by the values $w_h(z)$ for all $z \in \bar{\Omega}_h := \Omega_h^I \cup \Omega_h^B$.

To describe the finite element method and to facilitate further developments, we assume that the vertices are labelled such that $\bar{\Omega}_h = \{z_i\}_{i=1}^{N+M}$ for positive integers N, M, with $\Omega_h^I = \{z_i\}_{i=1}^N$ and $\Omega_h^B = \{z_i\}_{i=N+1}^M$. The fact that continuous piecewise linear polynomials are uniquely determined by their values at the vertices induces a basis of *hat functions* $\{\tilde{\phi}_i\} \subset X_h^l$, with the unique property $\tilde{\phi}_i(z_j) = \delta_{i,j}$. We define the *normalized hat functions* as $\phi_i = c_i^{-1}\tilde{\phi}_i$ with $c_i = \left(\int_\Omega \tilde{\phi}_i\right) > 0$, and note that

$$v_h = \sum_{i=1}^{N+M} c_i v_h(z_i)\phi_i, \quad \text{for all } v_h \in X_h^l,$$

and

$$v_h = \sum_{i=1}^{N} c_i v_h(z_i)\phi_i, \quad \text{for all } v_h \in X_{0,h}^l := X_h^l \cap H_0^1(\Omega). \tag{3.35}$$

We set $\omega_{z_i} = \text{supp}(\phi_i)$, which is the union of elements in $\mathcal{T}_h$ that have z_i as a vertex.

A finite element method for the Poisson problem simply restricts the variational formulation (2.9) (with $A = I$) onto the piecewise polynomial space X_h^l. Thus, we consider the following problem: find $u_h \in X_h^l$ with $u_h(z_i) = g(z_i)$ $(N+1 \le i \le N+M)$ and

$$-\int_\Omega Du_h \cdot Dv_h = \int_\Omega f v_h, \quad \text{for all } v_h \in X_{0,h}^l. \tag{3.36}$$

As in the continuous setting, the existence and uniqueness of u_h readily follows from the Lax–Milgram lemma. An application of Céa's lemma and interpolation results also show that the error satisfies $\|\nabla(u - u_h)\|_{L^2(\Omega)} = O(h)$ provided $u \in H^2(\Omega)$; we refer the reader to Ciarlet (2002), Brenner and Scott (2008) and Ern and Guermond (2004), for example, for proofs of these basic results.

To pose this problem in the operator framework of the previous sections, we first note that (3.36) is equivalent to the conditions

$$-\int_\Omega Du_h \cdot D\phi_i = \int_\Omega f\phi_i \quad i = 1, 2, \ldots, N. \tag{3.37}$$

Define $\mathcal{L}_h$ such that for all interior vertices $z_i \in \Omega_h^I$,

$$\mathcal{L}_h u_h(z_i) = \Delta_h u_h(z_i), \tag{3.38}$$

where the *finite element Laplacian* is defined by

$$\Delta_h u_h(z_i) := -\int_\Omega Du_h \cdot D\phi_i. \tag{3.39}$$

Set

$$f_h(z_i) = \int_\Omega f\phi_i.$$

We further define the piecewise linear function g_h on $\partial\Omega$ with the property

$$g_h(z_i) = g(z_i) \quad i = N+1, \ldots, N+M, \tag{3.40}$$

that is, $g_h = I_h^{fe} g$ is the nodal interpolant of g. With this notation, we see that the finite element method (3.36) is equivalent to problem (3.6) with $F_h = \mathcal{L}_h - f_h$.

Before discussing the monotonicity of the scheme (3.36), let us first point out that, unlike the finite difference scheme, one cannot take $I_h = I_h^{fe}$, the nodal interpolant, in Definition 3.1 to deduce the (operator) consistency of the finite element approximation. The next lemma exemplifies this point. For further details the reader is referred to Jensen and Smears (2013*b*) and Nochetto and Zhang (2017*a*).

Lemma 3.38 (finite element inconsistency). Let Δ_h be the finite element Laplacian, defined in (3.39), and denote by $I_h^{fe} : C(\bar{\Omega}) \to X_h^l$, the nodal interpolant onto the linear Lagrange finite element space. Then, in general, we have

$$\Delta_h(I_h^{fe}u)(z) \not\to \Delta u(z), \quad \text{for all } z \in \bar{\Omega}_h \ \text{ as } h \to 0$$

for all $u \in C^2(\Omega)$.

Proof. Let $d = 2$, and consider the triangulation $\mathcal{T}_h$ with four triangles and vertices $z_1 = (0,0)$, $z_2 = (h,0)$, $z_3 = (0,h)$, $z_4 = (-h,0)$ and $z_5 = (0,-h)$. Let u be a C^2-function that vanishes at the origin. Then a calculation shows that

$$\Delta_h I_h^{fe} u(z_1) = \frac{3}{2} h^{-2} \sum_{i=2}^{5} u(z_i).$$

Taking, for example, $u(x_1, x_2) = x_1^2$ then yields

$$\Delta_h I_h^{fe} u(z_1) = 3 \neq 2 = \Delta u(z_1), \quad \text{for all } h > 0.$$

In other words, Δ_h and I_h^{fe} are not a consistent approximation scheme. □

The inconsistency in Lemma 3.38 is caused by the wrong choice of interpolation operator. The so-called elliptic projection gives a correct one.

Definition 3.39 (elliptic projection). The elliptic projection

$$I_h^{ep} : H^1(\Omega) \cap C(\bar{\Omega}) \to X_h^l$$

is defined by

$$\Delta_h I_h^{ep} u(z_i) = -\int_\Omega Du \cdot D\phi_i \quad i = 1, 2, \ldots, N, \tag{3.41}$$

and $I_h^{ep} u = u$ on Ω_h^B.

The elliptic projection is (almost) quasi-optimal in the L^∞-norm (Schatz and Wahlbin 1982).

Proposition 3.40 (properties of I_h^{ep}). Let $I_h^{ep}u \in X_{0,h}^l$ be the elliptic projection of $u \in C(\bar{\Omega}) \cap H_0^1(\Omega)$ defined by (3.41). Then

$$\|u - I_h^{ep}u\|_{L^\infty(\Omega)} \leq C\,|\log h| \inf_{v_h \in X_{0,h}^l} \|u - v_h\|_{L^\infty(\Omega)}.$$

If $u \in W^{2,\infty}(\Omega)$, then

$$\|u - I_h^{ep}u\|_{L^\infty(\Omega)} \leq C\,|\log h| h^2 \|u\|_{W^{2,\infty}(\Omega)}.$$

More importantly, the finite element method is consistent when one uses the elliptic projection I_h^{fe}.

Lemma 3.41 (finite element consistency). Let $\{z_h\}_{h>0}$ with $z_h \in \bar{\Omega}_h^I$ and $z_h \to z_0 \in \Omega$. Then, for all $u \in C^2(\bar{\Omega})$,

$$\mathcal{L}_h I_h^{ep}u(z_h) = \Delta_h(I_h^{ep}u)(z_h) \to \Delta u(z_0), \quad \text{for all } z \in \bar{\Omega}_h,$$

as $h \to 0+$. Moreover, $I_h^{ep}u \to u$ on $\partial\Omega$.

Proof. The convergence $I_h^{ep}u \to u$ on $\partial\Omega$ follows from the definition of I_h^{ep} and standard interpolation theory.

Owing to the regularity of u, we have $\Delta u(z_h) \to \Delta u(z_0)$ as $h \to 0^+$. Now, let $\{\phi_h\}_{h>0} \subset X_{0,h}^l$ denote the normalized hat functions. Then, since $\|\phi_h\|_{L^1(\Omega)} = 1$, and $\phi_h \geq 0$, we have

$$\int_\Omega (\Delta u)\phi_h = \int_{\omega_{z_h}} (\Delta u)\phi_h \to \Delta u(z_0).$$

Therefore, by integration by parts,

$$\Delta_h I_h^{ep}u(z_h) = \int_\Omega (\Delta u)\phi_h \to \Delta u(z_0)$$

as $h \to 0+$, and consequently Δ_h is consistent. □

Lemma 3.42 (finite element monotonicity). Suppose that the bases satisfy

$$\int_\Omega D\phi_i \cdot D\phi_j \leq 0. \tag{3.42}$$

for $i, j = 1, 2, \ldots, N+M$ and $i \neq j$. Then $\mathcal{L}_h$, given by (3.38), is monotone.

Proof. Suppose that $v_h, w_h \in X_h^l$ and $w_h - v_h$ has a non-negative maximum at an interior vertex $z_i \in \Omega_h^B$. Without loss of generality we may assume that $w_h \leq v_h$ and $w_h(z_i) = v_h(z_i)$. Then we find that

$$\begin{aligned}\mathcal{L}_h w_h(z_i) - \mathcal{L}_h v_h(z_i) &= -\int_\Omega D(w_h - v_h) \cdot D\phi_i \\ &= -\sum_{j=1}^{N+M} c_j(w_h(z_j) - v_h(z_j)) \int_\Omega D\phi_j \cdot D\phi_i \leq 0.\end{aligned}$$

Thus, $\mathcal{L}_h w_h(z_i) \leq \mathcal{L}_h v_h(z_i)$, and therefore $\mathcal{L}_h$ is monotone. □

Lemma 3.42 indicates that the finite element method is monotone provided that a certain mesh condition is satisfied. Indeed, for an edge $E \subset \partial T$, we let θ_E^T denote the angle between the faces not containing E, and let κ_E^T be the $(d-2)$-dimensional simplex opposite to E. Then we have

$$\int_\Omega D\tilde{\phi}_i \cdot D\tilde{\phi}_j = -\frac{1}{d(d-1)} \sum_{T \supset E} |\kappa_E^T| \cot \theta_E^T$$

(Xu and Zikatanov 1999, Strang and Fix 1973), where E is the edge with vertices z_i and z_j. The condition (3.42) is satisfied if the mesh is weakly acute. For example, in two dimensions, this condition means that the sum of the angles opposite to any edge is less than or equal to π.

Corollary 3.43 (maximum principle). Let $u_h \in X_h^l$ solve (3.37). Suppose that (3.42) is satisfied and $f \geq 0$ and $g \leq 0$. Then $u_h \leq 0$.

Proof. Define the stiffness matrix $S \in \mathbb{R}^{(N+M)\times(N+M)}$ by

$$S_{i,j} = -\int_\Omega D\phi_i \cdot D\phi_j, \tag{3.43}$$

and note that condition (3.42) is equivalent to $S_{i,j} \geq 0$ for $i \neq j$. Moreover, since the hat functions form a partition of unity,

$$\sum_{j=1}^{N+M} c_j S_{i,j} = 0.$$

Now, suppose that u_h attains a strict positive maximum at an interior node z_i. We then find

$$\begin{aligned}0 = \mathcal{L}_h u_h(z_i) &= -\int_\Omega Du_h \cdot D\phi_i - \int_\Omega f\phi_i \\ &\leq -\int_\Omega Du_h \cdot D\phi_i \\ &= \sum_{j=1}^{N+M} c_j u_h(z_j) S_{i,j} = \sum_{j=1}^{N+M} c_j(u_h(z_j) - u_h(z_i)) S_{i,j} < 0,\end{aligned}$$

a contradiction. □

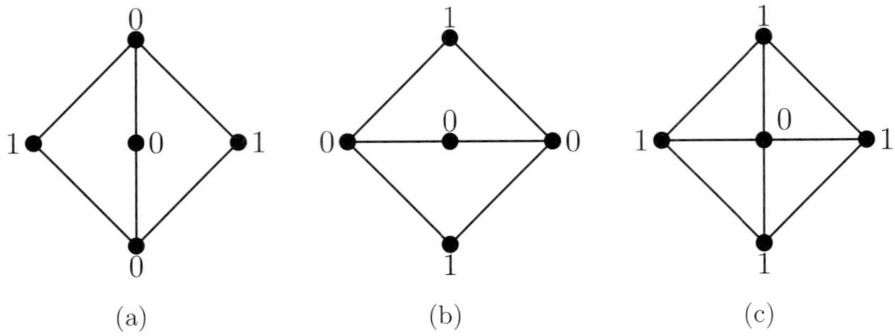

Figure 3.1. Meshes corresponding to convex envelopes $\Gamma(v_1) = |x_1|$ (a) and $\Gamma(v_2) = |x_2|$ (b), and $\Gamma(v_3) = |x_1| + |x_2|$ (c).

3.6. Finite element stability estimates: Alexandrov estimates and Alexandrov–Bakelman–Pucci maximum principle

As in the finite difference schemes discussed in the previous section, we develop some discrete Alexandrov estimates for finite element functions and analogous ABP maximum principles. Before stating and proving these results, it is useful to discuss some properties of the convex envelope of piecewise linear polynomials.

For $v_h \in X_h^l$ with $v_h \geq 0$ on $\partial\Omega$, let $\Gamma(v_h)$ and $\Gamma_h(v_h)$ denote the convex envelope and discrete convex envelope of v_h given in Definitions 2.59 and 3.17, respectively. Then, since v_h is piecewise affine, we find that $\Gamma_h(v_h) = \Gamma(v_h)$, and furthermore, $\Gamma_h(v_h)$ is also piecewise affine. However, perhaps unexpectedly, $\Gamma(v_h)$ is not necessarily piecewise linear subordinate to $\mathcal{T}_h$! The following examples illustrate this feature.

Example 3.44 (convex envelope). Consider a triangulation with vertices $z_1 = (1,0)$, $z_2 = (0,1)$, $z_3 = (-1,0)$, $z_4 = (0,-1)$ and $z_5 = (0,0)$. Consider the piecewise linear functions satisfying

$$v_1(z_1) = v_1(z_3) = 1, \quad v_2(z_2) = v_2(z_4) = 1,$$
$$v_3(z_1) = v_3(z_2) = v_3(z_3) = v_3(v_4) = 1,$$

and $v_j(z_i) = 0$ otherwise. The convex envelopes are $\Gamma(v_1) = |x_1|$, $\Gamma(v_2) = |x_2|$, and $\Gamma(v_3) = |x_1| + |x_2|$. The convex envelopes are subordinate to the meshes depicted in Figure 3.1.

As shown in the example above, since $\Gamma(v_h)$ is a piecewise linear function, it induces a mesh $\widetilde{\mathcal{T}}_h$ which depends on v_h. The following example shows that if v_h is the nodal interpolant of a function v, and if the Hessian D^2v is degenerate (or nearly degenerate), the induced mesh may be anisotropic.

Example 3.45 (anisotropy). Let $\Omega = \mathbb{R}^2$ and $\bar{\Omega}_h = \{(k,m)\}$. Let $v(x) = (x \cdot e)^2$, where $e = (1, A)$ for some integer $0 < A$ and $v_h(z) = v(z)$ for

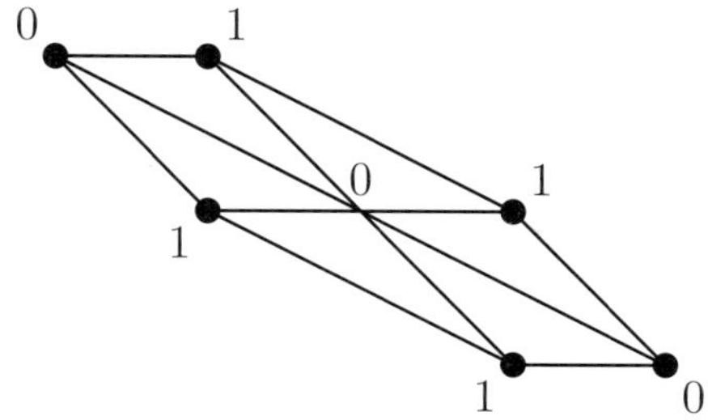

Figure 3.2. Mesh induced by the nodal interpolant of $v(x) = (x \cdot e)^2$ where $e = (1, 2)$. Its convex envelope equals $|x \cdot e|$ in the star of $(0, 0)$.

all $z \in \bar{\Omega}_h$. Then the convex envelope induces an anisotropic mesh depicted in Figure 3.2. The convex envelope in the star of the origin is $|x \cdot e|$.

Let us now state the Alexandrov estimate for finite element functions. Recall that Ω is compactly contained in a ball B_R, and the nodal contact set $\mathcal{C}_h^-(v_h)$ is given in Definition 3.23.

Lemma 3.46 (finite element Alexandrov estimate). For every $v_h \in X_h^l$ such that $v_h \geq 0$ on $\partial\Omega$, we have

$$\sup_{\bar{\Omega}} v_h^- \leq CR \Bigg(\sum_{z \in \mathcal{C}_h^-(v_h)} |\partial\Gamma(v_h)(z)| \Bigg)^{1/d}, \tag{3.44}$$

where the constant C depends only on the dimension d and the domain Ω.

Proof. The result directly follows from the proof of Lemma 3.24. Indeed it suffices to realize that, for every $v_h \in X_h^l$, $\sup_{\bar{\Omega}} v_h^- = \sup_{\bar{\Omega}_h} v_h^-$ and $\Gamma_h(v_h) = \Gamma(v_h)$; see Nochetto and Zhang (2017*a*, Proposition 5.1) for details. □

Let us point out that, with Lemma 3.46 in hand, it may be possible to extend the arguments given in Theorem 3.25 to develop ABP estimates for piecewise linear polynomials. Instead, following Nochetto and Zhang (2017*a*, Section 5), we outline a proof which is more geometric and is based on the characterization of the subdifferential of piecewise linear functions.

As a first step we define the local convex envelope and local subdifferential at a point $z \in \mathcal{C}_h^-(v_h)$ (see Definition 3.23).

Definition 3.47 (local convex envelope). For $v_h \in X_h^l$ and contact node $z \in \mathcal{C}_h^-(v_h)$, let ω_z denote the union of elements in $\mathcal{T}_h$ that have z as a vertex. We then define the *local convex envelope* by

$$\Gamma_z(v_h)(x) = \sup\{L(x) : L \leq v_h \ \text{ in } \omega_z, \ L \in \mathbb{P}_1, \ L(z) = v_h(z)\}$$

for all $x \in \omega_z$. Its *local subdifferential* is

$$\partial\Gamma_z(v_h)(z) = \{\mathbf{p} \in \mathbb{R}^d : \Gamma_z(v_h)(x) \geq \Gamma_z(v_h)(z) + \mathbf{p}\cdot(x-z),\ \forall x \in \omega_z\}. \tag{3.45}$$

Comparing (3.45) with Definition 2.59, we easily deduce that

$$\partial\Gamma(v_h)(z) \subset \partial\Gamma_z(v_h)(z), \quad \text{for all } z \in \mathcal{C}_h^-(v_h), \tag{3.46}$$

and therefore, by Lemma 3.46,

$$\sup_{\bar\Omega} v_h^- \leq CR\bigg(\sum_{z\in\mathcal{C}_h^-(v_h)} |\partial\Gamma_z(v_h)(z)|\bigg)^{1/d}. \tag{3.47}$$

Less obvious is the following result.

Proposition 3.48 (subordination). Suppose that $d = 2$ and, for $v_h \in X_h^l$ and contact point $z \in \mathcal{C}_h^-(v_h)$, let $\Gamma_z(v_h)$ be given by Definition 3.47. Then $\Gamma_z(v_h)$ is subordinate to $\mathcal{T}_h$.

We refer the reader to Nochetto and Zhang (2017*a*, Lemma 5.1) for a proof of this result. Here, instead, we will show that if $d \geq 3$ the assertion is no longer true. Set

$$z_0 = (0,0,-1), \quad z_1 = (-1,0,0), \quad z_2 = (0,1,0), \quad z_3 = (1,0,0),$$

and let T_1, T_2 be the convex hulls of z_0, z_1, z_2, z_3 and $z_0, z_1, -z_2, z_3$. Consider the piecewise linear function v_h with values $v_h(z_0) = -1$, $v_h(z_1) = v_h(z_3) = 0$ and $v_h(\pm z_2) = -1$. Then $\Gamma_{z_0}(v_h)(x) = |x_1| - 1$ is not affine on T_i for each $i = 1, 2$.

Next, to derive an ABP maximum principle, we state the relation between the subdifferential of a convex, piecewise linear polynomial with its finite element Laplacian. As a first step, we first integrate by parts in (3.38) to get the identity

$$\Delta_h v_h(z_i) = -\sum_{F\in\mathcal{F}_{z_i}} \int_F [[Dv_h]]\phi_i, \quad \text{for all } v_h \in X_h^l.$$

Here, $\mathcal{F}_{z_i}$ is the set of (interior) $(d-1)$-dimensional simplices that have z_i as a vertex, and, for a vector-valued function $\mathbf{w}$, the *jump of* $\mathbf{w}$ across the face F is given by

$$[[\mathbf{w}]]\big|_F := \begin{cases} \mathbf{n}_F^+\cdot\mathbf{w}^+\big|_F + \mathbf{n}_F^-\cdot\mathbf{w}^-\big|_F & \text{if } F = \partial K^+\cap\partial K^-, \\ \mathbf{n}_F^+\cdot\mathbf{w}^+\big|_F & \text{if } F = \partial K^+\cap\partial\Omega, \end{cases} \tag{3.48}$$

with $\mathbf{w}^\pm = \mathbf{w}|_{K^\pm}$, and $\mathbf{n}_F^\pm$ denoting the outward unit normal vectors of $K^\pm$

on F. We also define the jump of a scalar function v across F as

$$[[v]]\big|_F := \begin{cases} \mathbf{n}_F^+ v^+\big|_F + \mathbf{n}_F^- v^-\big|_F & \text{if } F = \partial K^+ \cap \partial K^-, \\ \mathbf{n}_F^+ v^+\big|_F & \text{if } F = \partial K^+ \cap \partial \Omega. \end{cases} \tag{3.49}$$

Now, since $[[Dv_h]]$ is constant on F, and

$$\int_F \phi_i = \frac{d+1}{d} \frac{|F|}{|\omega_{z_i}|},$$

we can obtain an expression on the finite element Laplacian with explicit dependence on the jumps:

$$\Delta_h v_h(z_i) = \frac{-(d+1)}{d} \sum_{F \in \mathcal{F}_{z_i}} \frac{|F|}{|\omega_{z_i}|} [[Dv_h]]\big|_F. \tag{3.50}$$

A relationship between the jumps of the gradients (and hence the discrete Laplacian) and the subdifferential of a convex, piecewise affine function is now given.

Proposition 3.49 (subdifferential versus jumps). Let γ be a piecewise affine convex function on a patch ω_z for some node $z \in \Omega_h^I$, and let $\mathcal{F}_z$ denote the set of $(d-1)$-dimensional simplices that touch z. Then, for any $F \in \mathcal{F}_z$, the jump $[[D\gamma]]|_F$ is non-positive and

$$|\partial\gamma(z)| \leq C\bigg(\sum_{F \in \mathcal{F}_z} -[[D\gamma]]\big|_F\bigg)^d. \tag{3.51}$$

We will not give a complete proof of Proposition 3.49, but rather give a rough idea of how such a result is obtained in two dimensions. Further details can be found in Nochetto and Zhang (2017*a*, Section 5.2).

Without loss of generality, assume that $z = 0$ and $\gamma(0) = 0$. We further denote by $\{z_j\}_{j=1}^m$ the set of nodes in ω_z. Now, since γ is a piecewise affine function, a vector $\mathbf{p} \in \partial\gamma(0)$ is characterized by the inequalities

$$\mathbf{p} \cdot z_j \leq \gamma(z_j) \quad 1 \leq j \leq m.$$

Therefore, we conclude that the subdifferential of $\gamma(0)$ is a convex polygon determined by the intersection of the half-spaces

$$S_j := \{\mathbf{p} \in \mathbb{R}^2 : \mathbf{p} \cdot z_j \leq \gamma(z_j)\}, \tag{3.52}$$

and that a vector $\mathbf{p}$ is in the interior of $\partial\gamma(0)$ if and only if

$$\mathbf{p} \cdot z_j < \gamma(z_j), \quad 1 \leq j \leq m,$$

and it is on the boundary of $\partial\gamma(0)$ if

$$\mathbf{p} \cdot z_j = \gamma(z_j)$$

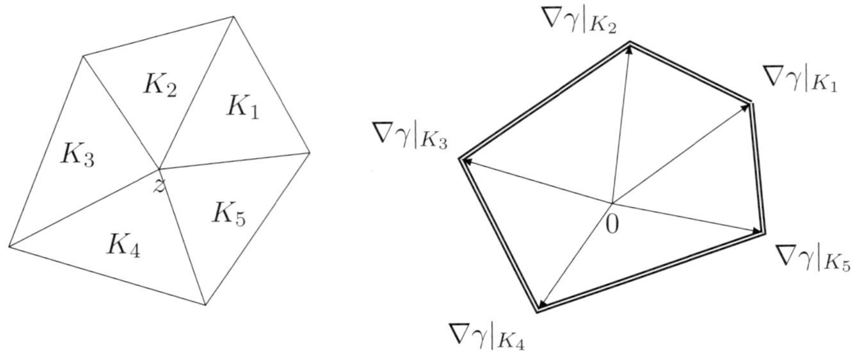

Figure 3.3. A pictorial description of the dual set of a patch.

for some j. This characterization of the boundary motivates the introduction of a $\partial\gamma(0)$-induced *dual mesh*, which we now explain.

Let T be an n-dimensional simplex in ω_z with $0 \le n \le 2$ such that $0 \in T$. We then define the $(2-n)$-dimensional dual set T^* as follows (see Figure 3.3).

- If $n = 0$, so that $T = \{0\}$, then we define T^* as the subdifferential $\partial\gamma(0)$.
- If $n = 2$, so that $T = K$ is an element of ω_z, then T^* is the vector $\partial\gamma\big|_K$.
- If $n = 1$, so that $T = F = \partial K^+ \cap \partial K^-$ is an (interior) edge in ω_z, then T^* is the line segment jointing the two vectors $\partial\gamma|_{K^\pm} = D\gamma_{K^\pm}$.

Note that T^* is a convex polytope contained in the $(2-n)$-dimensional plane

$$P_T = \{\mathbf{p} \in \mathbb{R}^2 : \mathbf{p} \cdot z = \gamma(z)\ \forall z \in T\},$$

and therefore, for arbitrary $\mathbf{p}_1, \mathbf{p}_2 \in P_T$, $(\mathbf{p}_1 - \mathbf{p}_2) \cdot z = 0$ for all $z \in T$, that is, P_T is orthogonal to T.

Now, the proceeding discussion implies that the boundary of $\partial\gamma(0)$ is given by

$$\bigcup \{F^* : \text{edges } F \text{ in } \omega_z \text{ with } 0 \in F\},$$

in other words, the boundary of $\partial\gamma$ is made of line segments that join $\partial\gamma(0)$ on neighbouring triangles; for further details see Nochetto and Zhang (2017a, Proposition 5.6). If $F = \partial K^+ \cap \partial K^-$, then it follows that the length of F^* is given by

$$|F^*| = \big|D\gamma|_{K^+} - D\gamma|_{K^-}\big| = -[[D\gamma]]|_F,$$

where we have used the fact that $D\gamma|_{K^+} - D\gamma_{K^-}$ is perpendicular to F

and the non-positivity of $[[D\gamma]]$. Putting everything together, we conclude that the boundary of $\partial\gamma(0)$ is bounded by $\sum_{F\in\mathcal{F}_z} -[[D\gamma]]\big|_F$. Thus, by the isoperimetric inequality,

$$|\partial\gamma(0)| \le C\biggl(\sum_{F\in\mathcal{F}_z} -[[D\gamma]]\big|_F\biggr)^2.$$

This last statement is (3.51) for $d = 2$.

Theorem 3.50 (finite element ABP estimate). Suppose that the simplicial mesh $\mathcal{T}_h$ satisfies (3.42) and that $u_h \in X_h^l$ satisfies

$$\begin{cases} \mathcal{L}_h u_h \le f_h & \text{in } \Omega_h^I, \\ u_h = g_h & \text{on } \Omega_h^B, \end{cases}$$

where $\mathcal{L}_h$ and g_h are given by (3.38) and (3.40), respectively, and X_h^l is the linear Lagrange space defined by (3.34). Then

$$\sup_{\bar{\Omega}} u_h^- \le \sup_{\Omega_h^B} g_h^- + CR\biggl(\sum_{z\in\mathcal{C}_h^-(u_h)} |\omega_z|(f_h^+(z))^d\biggr)^{1/d}. \tag{3.53}$$

Proof. We give the proof under the assumption that $\Gamma_z(u_h)$ is subordinate to $\mathcal{T}_h$ (which is the case in two dimensions). For the proof of the general case, we refer interested readers to Nochetto and Zhang (2017*a*).

As in the proof of Theorem 3.25, we may assume $u_h \ge 0$ on Ω_h^B and $\sup_{\bar{\Omega}} u_h^- > 0$.

Let $z \in \mathcal{C}_h^-(u_h)$, and note that $\Gamma_z(u_h)(x) \le u_h(x)$ for all $x \in \omega_z$ with equality at z. Then by (3.42) and Lemma 3.42, and since $\Gamma_z(u_h)$ is subordinate to $\mathcal{T}_h$, we find that

$$\Delta_h\Gamma_z(u_h)(z) \le \Delta_h u_h(z).$$

Moreover, by (3.50), we have

$$-\sum_{F\in\mathcal{F}_z} \frac{|F|}{|\omega_z|}[[D\Gamma_z(u_h)]]\big|_F = \frac{d}{d+1}\Delta_h\Gamma_z(u_h)(z) \le \frac{d}{d+1}\Delta_h u_h(z).$$

Applying Proposition 3.49, and using $|F| \approx |\omega_z|^{1-1/d}$, we conclude that

$$|\partial\Gamma_z(u_h)(z)| \le C\biggl(-\sum_{F\in\mathcal{F}_z} \frac{|F|}{|\omega_z|}[[D\Gamma_z]]\big|_F\biggr)^d |\omega_z| \le C\frac{|\omega_z|}{d+1}\Delta_h u_h(z).$$

Combining this last inequality with (3.47) yields the result. ∎

Remark 3.51 (contact set). The proof of Theorem 3.50 shows that if $\mathcal{L}_h u_h \le 0$ only on $\mathcal{C}_h^-(u_h)$, then (3.53) is still satisfied. This will be important in subsequent developments.

4. Finite element methods for elliptic problems in non-divergence form

In this section we summarize recent advances in finite element methods for elliptic problems in non-divergence form with non-smooth coefficients. For simplicity, and to illustrate the main ideas, we consider problems of the form (2.5) with no lower-order terms and with homogeneous Dirichlet boundary conditions. In this setting the problem reads

$$\mathcal{L}u = A : D^2 u = f \quad \text{in } \Omega, \qquad u = 0 \quad \text{on } \partial\Omega, \tag{4.1}$$

where A is a symmetric positive definite in $\bar{\Omega} \subset \mathbb{R}^d$ with either $A \in C(\bar{\Omega}, \mathbb{S}^d)$ or $A \in L^\infty(\Omega, \mathbb{S}^d)$. Further assumptions of the domain Ω, its boundary $\partial\Omega$, the coefficient matrix A, and the source function f will be made as they become necessary.

Recall from Section 2.1 that if the coefficient matrix is sufficiently smooth, then we can write the PDE in divergence form with A as the diffusion coefficient and $D\cdot A$ (taken column-wise) as the convective coefficient. In this setting, (weak) solutions are defined by an integration-by-parts argument (see Section 2.3), and as such, finite element methods are easily constructed. However, when A is not differentiable, the clear-cut methodology of Galerkin methods is no longer valid.

This section summarizes three classes of finite element methods for problem (4.1), each motivated by the different solution concepts presented in Sections 2.4–2.5. The first class considers problem (4.1) on convex domains with $A \in L^\infty(\Omega, \mathbb{S}^d)$ satisfying the Cordes condition. The second class of methods is motivated by the notion of strong solutions (see Section 2.4) under the assumption that $A \in C(\bar{\Omega}, \mathbb{S}^d)$. Finally, the third method is motivated by the notion of viscosity solutions (see Section 2.5), where comparison principles and monotonicity of the scheme are the central themes.

4.1. Discretization of non-divergence-form PDEs satisfying the Cordes condition

Here we discuss recent numerical methods for second-order elliptic PDEs in non-divergence form satisfying the Cordes condition (2.15). Recall from Section 2.4.1 that, if the domain is convex, this condition on the coefficient matrix ensures that the bilinear mapping

$$a(\cdot,\cdot) : (H^2(\Omega) \cap H_0^1(\Omega))^2 \ni (v, w) \to a(v, w) = \int_\Omega \gamma \mathcal{L}v \Delta w \tag{4.2}$$

is coercive, thus allowing one to define strong solutions via variational principles. Here, the function γ is given by (2.13). In this setting the existence of a strong solution $u \in H^2(\Omega) \cap H_0^1(\Omega)$ to (4.1) is deduced from the variational formulation

$$a(u,v) = \int_\Omega \gamma f \Delta v \, dx, \quad \text{for all } v \in H^2(\Omega) \cap H_0^1(\Omega) \tag{4.3}$$

and appealing to the Lax–Milgram lemma. Since the mapping $\Delta : H^2(\Omega) \cap H_0^1(\Omega) \to L^2(\Omega)$ is surjective on convex domains, one concludes that a function $u \in H^2(\Omega) \cap H_0^1(\Omega)$ satisfying (4.3) satisfies (4.1) almost everywhere, that is, $u \in H^2(\Omega) \cap H_0^1(\Omega)$ is a strong solution to (4.1). We refer the reader to Theorem 2.39 for details.

One immediately sees that the solution concept lends itself to a finite element approximation which would pose (4.3) over a finite-dimensional subspace of $H^2(\Omega) \cap H_0^1(\Omega)$ consisting of piecewise polynomials. To describe this procedure, we let $\mathcal{T}_h$ denote a simplicial, conforming and shape-regular triangulation of Ω parametrized by $h > 0$, and let $X_h^{cd} \subset H^2(\Omega) \cap H_0^1(\Omega)$ be a finite-dimensional subspace consisting of piecewise polynomials with respect to $\mathcal{T}_h$. A conforming finite element approximation to (4.3) seeks a function $u_h \in X_h^{cd}$ satisfying the discrete variational formulation

$$a(u_h, v_h) = \int_\Omega \gamma f \Delta v_h, \quad \text{for all } v_h \in X_h^{cd}. \tag{4.4}$$

Problem (4.4) represents a square linear system of equations. The coercivity of the bilinear form $a(\cdot,\cdot)$ over $H^2(\Omega) \cap H_0^1(\Omega)$ implies that this system is invertible, and thus there exists a unique solution $u_h \in X_h^{cd}$ to problem (4.4). Moreover, the continuity of $a(\cdot,\cdot)$ and Céa's lemma show that such approximations are quasi-optimal in the sense that

$$\|u - u_h\|_{H^2(\Omega)} \le \frac{C}{\alpha} \inf_{v_h \in X_h^{cd}} \|u - v_h\|_{H^2(\Omega)},$$

where $C > 0$ and $\alpha = 1 - \sqrt{1-\epsilon}$ are the continuity and coercivity constants of $a(\cdot,\cdot)$, respectively.

While method (4.4) is a stable and convergent numerical scheme for computing solutions to (4.3), it has some potential practical drawbacks. Piecewise polynomial subspaces of $H^2(\Omega)$ are difficult to construct and implement, and are not a practical option for solving second-order PDEs. These properties are further exacerbated in three dimensions, where, for example, polynomials of degree of at least nine are required to construct H^2-conforming finite element spaces on general simplicial partitions; see Lai and Schumaker (2007, Remark 1) and Ženíšek (1973). Nevertheless, this path is explored by Gallistl (2017), who also presents a mixed formulation.

4.1.1. C^0 finite element approximations

We now discuss finite element methods for problem (4.1) that use continuous basis functions, commonly used for second-order problems in divergence form. To this end, with $\mathcal{T}_h$ given in the previous section, we define the Lagrange finite element space

$$X_h^{cg} = \{v_h \in H_0^1(\Omega) : v_h\big|_T \in \mathbb{P}_k(T)\ \forall T \in \mathcal{T}_h\} \tag{4.5}$$

with $k \in \mathbb{N}$. Note that functions in X_h^{cg} are locally smooth, yet not globally $H^2(\Omega)$, and therefore second-order derivatives are only defined piecewise with respect to $\mathcal{T}_h$. To simplify the presentation, we shall write

$$\|v\|_{L^2(\mathcal{T}_h)} := \Bigg(\sum_{K\in\mathcal{T}_h} \|v\|_{L^2(K)}^2\Bigg)^{1/2}$$

for piecewise L^2-functions.

Since $X_h^{cg} \not\subset H^2(\Omega) \cap H_0^1(\Omega)$, the bilinear form $a(\cdot,\cdot)$ defined by (4.2) is not well-defined on $X_h^{cg} \times X_h^{cg}$. Moreover, a piecewise version of the bilinear form is not generally coercive on X_h^{cg}, since, for example,

$$\sum_{K\in\mathcal{T}_h} \int_K \gamma \mathcal{L} v_h \Delta v_h = 0$$

for all piecewise linear $v_h \in X_h^{cg}$. This stems from the fact that a piecewise version of the Miranda–Talenti estimate $\|D^2 v\|_{L^2(\Omega)} \le \|\Delta v\|_{L^2(\Omega)}$ is not satisfied on X_h^{cg}. As such, the coercivity proof of $a(\cdot,\cdot)$ found at the continuous level does not directly carry over to the discrete setting.

To overcome this we develop a discrete Miranda–Talenti estimate suitable for piecewise polynomials. To do so, we introduce some notation. Let $\mathcal{F}_h^I$ and $\mathcal{F}_h^B$ denote the set of interior and boundary edges/faces, respectively, and set $\mathcal{F}_h := \mathcal{F}_h^I \cup \mathcal{F}_h^B$ and $h_F := \operatorname{diam}(F)$ for $F \in \mathcal{F}_h$. We recall that the jump of a vector-valued function $\mathbf{w}$ is given by (3.48).

A discrete Miranda–Talenti estimate is based on the following result, whose proof can be found in Georgoulis, Houston and Virtanen (2011).

Lemma 4.1 (enrichment operator). Suppose that $d = 2$. Then there exists a finite-dimensional space $X_h^{cd} \subset H^2(\Omega) \cap H_0^1(\Omega)$ and an (enrichment) operator $E_h : X_h^{cg} \to X_h^{cd}$ satisfying

$$\|D^2(v_h - E_h v_h)\|_{L^2(\mathcal{T}_h)} \le C\Bigg(\sum_{F\in\mathcal{F}_h^I} h_F^{-1} \big\|[[Dv_h]]\big\|_{L^2(F)}^2\Bigg)^{1/2} \tag{4.6}$$

for all $v_h \in X_h^{cg}$. Here, the constant $C > 0$ depends on the polynomial degree k and the shape-regularity of $\mathcal{T}_h$, but is independent of h.

Theorem 4.2 (discrete Miranda–Talenti estimate). Suppose that $\Omega \subset \mathbb{R}^2$ is a convex polygon. Then, for all $v_h \in X_h^{cg}$,

$$\|D^2 v_h\|_{L^2(\mathcal{T}_h)} \leq \|\Delta v_h\|_{L^2(\mathcal{T}_h)} + C\Big(\sum_{F \in \mathcal{F}_h^I} h_F^{-1} \big\|[\![Dv_h]\!]\big\|_{L^2(F)}^2\Big)^{1/2}. \tag{4.7}$$

Proof. Fix $v_h \in X_h^{cg}$ and let $E_h : X_h^{cg} \to X_h^{cd}$ be the enrichment operator of Lemma 4.1. Since $E_h v_h \in H^2(\Omega) \cap H_0^1(\Omega)$ and Ω is convex, the Miranda–Talenti estimate

$$\|D^2 E_h v_h\|_{L^2(\Omega)} \leq \|\Delta E_h v_h\|_{L^2(\Omega)} = \|\Delta E_h v_h\|_{L^2(\mathcal{T}_h)}$$

is satisfied. Applying the triangle inequality, the inequality $\|\Delta E_h v_h\|_{L^2(\Omega)} \leq \sqrt{2}\|D^2 E_h v_h\|_{L^2(\Omega)}$, and estimate (4.6) yields

$$\begin{aligned}
\|D^2 v_h\|_{L^2(\mathcal{T}_h)} &\leq \|D^2 E_h v_h\|_{L^2(\Omega)} + \|D^2(v_h - E_h v_h)\|_{L^2(\mathcal{T}_h)} \\
&\leq \|\Delta E_h v_h\|_{L^2(\Omega)} + \|D^2(v_h - E_h v_h)\|_{L^2(\mathcal{T}_h)} \\
&\leq \|\Delta v_h\|_{L^2(\mathcal{T}_h)} + C\Big(\sum_{F \in \mathcal{F}_h^I} h_F^{-1} \big\|[\![Dv_h]\!]\big\|_{L^2(F)}^2\Big)^{1/2},
\end{aligned}$$

where in the last step we used that

$$\begin{aligned}
\|\Delta E_h v_h\|_{L^2(\Omega)} &\leq \|\Delta v_h\|_{L^2(\mathcal{T}_h)} + \|\Delta(E_h v_h - v_h)\|_{L^2(\mathcal{T}_h)} \\
&\leq \|\Delta v_h\|_{L^2(\mathcal{T}_h)} + \sqrt{2}\|D^2(E_h v_h - v_h)\|_{L^2(\mathcal{T}_h)}.
\end{aligned}$$

This concludes the proof. □

Motivated by Theorem 4.2, we define the bilinear form

$$a_h(v, w) := \sum_{K \in \mathcal{T}_h} \int_K \gamma \mathcal{L} v \Delta w + \sum_{F \in \mathcal{F}_h^I} \mu h_F^{-1} \int_F [\![Dv]\!]\,[\![Dw]\!], \tag{4.8}$$

where μ is a positive penalty parameter. We also define the discrete H^2-type norm

$$\|v\|_{H_h^2(\Omega)}^2 := \|\Delta v\|_{L^2(\mathcal{T}_h)}^2 + \sum_{F \in \mathcal{F}_h^I} h_F^{-1} \big\|[\![Dv]\!]\big\|_{L^2(F)}^2. \tag{4.9}$$

We consider the following finite element method: find $u_h \in X_h^{cg}$ satisfying

$$a_h(u_h, v_h) = \sum_{K \in \mathcal{T}_h} \int_K \gamma f \Delta v_h, \quad \text{for all } v_h \in X_h^{cg}. \tag{4.10}$$

Note that the additional penalization term does not affect the consistency of the scheme; that is, we obtain $a_h(u, v_h) = \sum_{K \in \mathcal{T}_h} \int_K \gamma f \Delta v_h$ for all $v_h \in X_h^{cg}$. The role of this term is to weakly enforce H^2-regularity and to ensure that the bilinear form $a_h(\cdot, \cdot)$ is coercive provided μ is sufficiently large.

Lemma 4.3 (coercivity). Suppose that $\Omega \subset \mathbb{R}^2$ is convex and that $A \in L^\infty(\Omega, \mathbb{S}^2)$ satisfies the Cordes condition (2.15) with parameter ϵ. There exists $\mu_* > 0$ depending on the shape-regularity of the mesh, polynomial degree k, and the parameter ϵ such that, for $\mu \geq \mu_*$,

$$\frac{\alpha}{2}\|v_h\|^2_{H^2_h(\Omega)} \leq a_h(v_h, v_h), \quad \text{for all } v_h \in X_h^{cg},$$

where $\alpha = 1 - \sqrt{1-\epsilon}$ is the coercivity constant of $a(\cdot,\cdot)$.

Proof. Adding a subtracting Δv_h and applying the Cordes condition yields

$$\begin{aligned}
a_h(v_h, v_h) &= \sum_{K\in\mathcal{T}_h} \int_K \gamma \mathcal{L} v_h \Delta v_h + \mu \sum_{F\in\mathcal{F}_h^I} h_F^{-1} \big\| [[Dv_h]] \big\|^2_{L^2(F)} \\
&= \|\Delta v_h\|^2_{L^2(\mathcal{T}_h)} + \sum_{K\in\mathcal{T}_h} \int_K (\gamma \mathcal{L} v_h - \Delta v_h)\Delta v_h \\
&\quad + \mu \sum_{F\in\mathcal{F}_h^I} h_F^{-1} \big\| [[Dv_h]] \big\|^2_{L^2(F)} \\
&\geq \|\Delta v_h\|^2_{L^2(\mathcal{T}_h)} - \sqrt{1-\epsilon}\|\Delta v_h\|_{L^2(\mathcal{T}_h)} \|D^2 v_h\|_{L^2(\mathcal{T}_h)} \\
&\quad + \mu \sum_{F\in\mathcal{F}_h^I} h_F^{-1} \big\| [[Dv_h]] \big\|^2_{L^2(F)}.
\end{aligned}$$

Applying the discrete Miranda–Talenti estimate and the Cauchy–Schwarz inequality, we obtain, for any $\tau > 0$,

$$\begin{aligned}
a_h(v_h, v_h) &\geq \left(\alpha - \frac{\tau}{2}\sqrt{1-\epsilon}\right) \|\Delta v_h\|^2_{L^2(\mathcal{T}_h)} \\
&\quad + \left(\mu - \frac{1}{2\tau}\sqrt{1-\epsilon}\right) \sum_{F\in\mathcal{F}_h^I} h_F^{-1} \big\| [[Dv_h]] \big\|^2_{L^2(F)}.
\end{aligned}$$

Taking $\tau = \alpha/\sqrt{1-\epsilon} = 1/\sqrt{1-\epsilon} - 1$ and $\mu_* = \alpha/2 + \sqrt{1-\epsilon}(1 - 1/\alpha)$ yields

$$\frac{\alpha}{2}\|v_h\|^2_{H^2_h(\Omega)} \leq a_h(v_h, v_h). \qquad \square$$

The coercivity stated in Lemma 4.3 shows that there exists a unique solution to the finite element method (4.10). Combined with the consistency of the scheme, we immediately obtain quasi-optimal error estimates in the discrete H^2-norm.

Theorem 4.4 (existence and error estimates). Suppose that $\Omega \subset \mathbb{R}^2$ is convex and that $A \in L^\infty(\Omega, \mathbb{S}^2)$ satisfies the Cordes condition (2.15) with parameter ϵ. Suppose that $\mu \geq \mu_*$, and let $u_h \in X_h^{cg}$ be the unique solution to (4.10) with X_h^{cg} given by (4.5). Then, if the solution to (4.1) satisfies

$u \in H^s(\Omega)$ with $2 \le s \le k+1$,

$$\alpha \|u - u_h\|_{H_h^2(\Omega)} \le C \inf_{v_h \in X_h^{cg}} \|u - v_h\|_{H_h^2(\Omega)} \le C h^{s-2} \|u\|_{H^s(\Omega)},$$

where $\alpha = 1 - \sqrt{1-\epsilon}$.

Remark 4.5 (linear case). Note that in the piecewise linear case ($k = 1$), Theorem 4.4 does not give a convergence result. In fact, it is easy to see that in this case the solution to (4.10) is the trivial one $u_h \equiv 0$.

Remark 4.6 (three dimensions). The results in Lemma 4.3 and Theorem 4.4 are restricted to the two-dimensional case due to Lemma 4.1. If there exists an enrichment operator satisfying (4.6) with $d = 3$, then these results carry over to the three-dimensional case.

4.1.2. Discontinuous Galerkin approximations

In this section we summarize the discretization developed and analysed in in Smears and Süli (2013), where a consistent discontinuous Galerkin (DG) method is constructed. Instead of developing a discrete Miranda–Talenti via penalization, the key idea of this approach is to add auxiliary terms in the bilinear form to bypass the Miranda–Talenti estimate found at the continuous level.

For $k \in \mathbb{N}$, define the piecewise polynomial space without continuity:

$$X_h^{dg} = \{v \in L^2(\Omega) : v|_T \in \mathbb{P}_k(T) \ \forall T \in \mathcal{T}_h\}. \tag{4.11}$$

We note that the method in Smears and Süli (2013) considers discretizations in an *hp*-framework where the polynomial degree is element-dependent. For simplicity, and to ease the presentation, we consider here only the h-version of the method, where the polynomial degree is globally fixed and we do not trace the dependence of the constants on the polynomial degree k.

To motivate the method, we again emphasize that a Miranda–Talenti estimate fails to hold for piecewise polynomials, and as such, the coercivity proof found at the continuous levels fails in the discrete setting. Indeed, mimicking the calculations in Lemma 2.38 element-wise over X_h^{dg} leads to

$$\sum_{K \in \mathcal{T}_h} \int_K \gamma \mathcal{L} v_h \Delta v_h = \sum_{K \in \mathcal{T}_h} \int_K \Delta v_h \Delta v_h + \sum_{K \in \mathcal{T}_h} \int_K (\gamma \mathcal{L} v_h - \Delta v_h) \Delta v_h. \tag{4.12}$$

Applying the Cordes condition and the Cauchy–Schwarz inequality yields the inequality

$$\sum_{K \in \mathcal{T}_h} \int_K \gamma \mathcal{L} v_h \Delta_h v \ge \|\Delta v_h\|_{L^2(\mathcal{T}_h)}^2 - \sqrt{1-\epsilon} \|D^2 v_h\|_{L^2(\mathcal{T}_h)} \|\Delta v_h\|_{L^2(\mathcal{T}_h)}.$$

Since the piecewise Hessian matrix of v_h cannot be controlled by its piecewise Laplacian (*e.g.* if v_h is piecewise harmonic), one concludes that the bilinear mapping $(v_h, w_h) \to \sum_{K\in\mathcal{T}_h} \int_K \gamma \mathcal{L} v_h \Delta w_h$ is not coercive over $X_h^{dg} \times X_h^{dg}$ in general.

The essential idea presented in Smears and Süli (2013) is to replace the bilinear form $(v_h, w_h) \to \sum_{K\in\mathcal{T}_h} \int_K \Delta v_h \Delta w_h$ implicit in the right-hand side of (4.12) with a consistent bilinear form that is coercive with a discrete H^2-type norm.

For a face $F \in \mathcal{F}_h$, let $\{t_i\}_{i=1}^{d-1}$ be an orthonormal coordinate system, and define the tangential gradient, tangential divergence, and tangential Laplacian, respectively, as

$$D_T v = \sum_{i=1}^{d-1} t_i \frac{\partial v}{\partial t_i}, \quad D_T \cdot \mathbf{w} = \sum_{i=1}^{d-1} \frac{\partial w_i}{\partial t_i}, \quad \Delta_T v = D_T \cdot D_T v.$$

We also define the average of a scalar or vector-valued function as

$$\{\!\{v\}\!\}|_F := \begin{cases} \frac{1}{2}(v_+ + v_-) & \text{if } F = \partial K^+ \cap \partial K^- \in \mathcal{F}_h^I, \\ v_+ & \text{if } F = \partial K^+ \cap \partial\Omega \in \mathcal{F}_h^B. \end{cases} \tag{4.13}$$

Then define the bilinear form

$$\begin{aligned} B_h(u_h, v_h) &= \sum_{K\in\mathcal{T}_h} \int_K \left(D^2 u_h : D^2 v_h + \Delta u_h \Delta v_h\right) \qquad (4.14) \\ &+ \sum_{F\in\mathcal{F}_h^I} \int_F \left(\{\!\{\Delta_T u_h\}\!\}[\![D v_h]\!] + \{\!\{\Delta_T v_h\}\!\}[\![D u_h]\!]\right) \\ &- \sum_{F\in\mathcal{F}_h} \int_F \left(D_T \{\!\{D u_h \cdot n\}\!\} \cdot [\![D_T v_h]\!]_T + D_T\{\!\{D v_h \cdot n\}\!\} \cdot [\![D_T u_h]\!]_T\right) \\ &+ \sum_{F\in\mathcal{F}_h^I} \mu h_F^{-1} \int_F [\![D u_h]\!][\![D v_h]\!] + \sum_{F\in\mathcal{F}_h} \mu h_F^{-3} \int_F [\![u_h]\!] \cdot [\![v_h]\!], \end{aligned}$$

where $[\![\mathbf{v}]\!]_T|_F := \mathbf{v}_+ - \mathbf{v}_-$ on $\mathcal{F}_h^I$ and $[\![\mathbf{v}]\!]_T|_F := \mathbf{v}_+$ on $\mathcal{F}_h^B$, and $\mu > 0$ is a penalization parameter.

For $\theta \in [0, 1]$, let us define the discrete DG-norm

$$\begin{aligned} \|v\|_{DG(\theta)}^2 &= (1-\theta)\|\Delta v\|_{L^2(\mathcal{T}_h)}^2 + \theta\|D^2 v\|_{L^2(\mathcal{T}_h)}^2 \qquad (4.15) \\ &\quad + c_*\left(\sum_{F\in\mathcal{F}_h^I} h_F^{-1}\|[\![Dv]\!]\|_{L^2(F)}^2 + \sum_{F\in\mathcal{F}_h} h_F^{-3}\|[\![v]\!]\|_{L^2(F)}^2\right). \end{aligned}$$

The seemingly abstruse bilinear form $B_h(\cdot, \cdot)$ is carefully defined to satisfy the following properties (Smears and Süli 2013, Lemmas 5 and 7).

Lemma 4.7 (properties of B_h). The bilinear form $B_h : X_h^{dg} \times X_h^{dg} \to \mathbb{R}$, defined in (4.14), satisfies the following properties.

- *Consistency.* If $u \in H^s(\Omega) \cap H_0^1(\Omega)$ for some $s > 5/2$, then
$$B_h(u, v_h) = 2 \sum_{K \in \mathcal{T}_h} \int_K \Delta u \Delta v_h$$
for all $v_h \in X_h^{dg}$.
- *Coercivity.* For any $\kappa > 1$, there exists a $\mu_* = C\kappa/(\kappa - 1)$ with $C > 0$ depending only on the shape-regularity of $\mathcal{T}_h$ and k such that for $\mu \geq \mu_*$,
$$2\|v_h\|_{DG(1/2)}^2 \leq \kappa B_h(v_h, v_h), \quad \text{for all } v_h \in X_h^{dg}, \tag{4.16}$$
for some constant $c_* > 0$ independent of the discretization parameter h and polynomial degree k.

The properties of $B_h(\cdot, \cdot)$ shown above allow us to define the following DG method: find $u_h \in X_h^{dg}$ such that, for all $v_h \in X_h^{dg}$,

$$\begin{aligned} a_h^{DG}(u_h, v_h) &:= \sum_{K \in \mathcal{T}_h} \int_K \gamma(\mathcal{L}u_h - \Delta u_h)\Delta v_h + \frac{1}{2} B_h(u_h, v_h) \\ &= \sum_{K \in \mathcal{T}_h} \int_K \gamma f \Delta v_h. \end{aligned} \tag{4.17}$$

Due to the consistency of $B_h(\cdot, \cdot)$ we see that the scheme is consistent provided the exact solution is sufficiently smooth. If u is the solution to (4.1) and satisfies $u \in H^s(\Omega)$ for some $s > 5/2$, then $a_h^{DG}(u, v_h) = \int_\Omega \gamma f \Delta_h v_h$ for all $v_h \in X_h^{dg}$. In addition, the coercivity of $B_h(\cdot, \cdot)$ implies the coercivity of $a_h^{DG}(\cdot, \cdot)$; see Smears and Süli (2013, Theorem 8).

Lemma 4.8 (coercivity). Suppose that $\Omega \subset \mathbb{R}^d$ is convex and that A satisfies the Cordes condition (2.15). Then there exists $\mu_* = O(\epsilon^{-1}) > 0$ such that

$$C\|v_h\|_{DG(1)}^2 \leq a_h^{DG}(v_h, v_h), \quad \text{for all } v_h \in X_h^{dg}.$$

Consequently, there exists a unique solution $u_h \in X_h^{dg}$ to (4.17).

Combined with consistency of $a_h^{DG}(\cdot, \cdot)$, Lemma 4.8 implies quasi-optimal error estimates in the discrete H^2-type norm (Smears and Süli 2013, Theorem 9).

Theorem 4.9 (existence and error estimates). Suppose the hypotheses of Lemma 4.8 hold. Further, suppose that the solution to (4.1) satisfies $u \in H^s(\Omega)$ for some $5/2 < s \leq k+1$. Then there exists an h-independent constant $C > 0$ such that

$$\|u - u_h\|_{DG(1)} \leq Ch^{s-2}\|u\|_{H^s(\Omega)}.$$

Remark 4.10 (regularity). The regularity assumption $u \in H^s(\Omega)$ with $s > 5/2$ ensures that $a_h^{DG}(u, v_h)$ is well-defined.

Remark 4.11 (extensions). Wang and Wang (2017) recently proposed and analysed a primal–dual discontinuous Galerkin method for second-order elliptic equations in non-divergence form.

4.2. Discrete finite element Calderón–Zygmund estimates

In this section we describe finite element discretizations to non-divergence-form elliptic operators based on the notion and theory of strong solutions. Recall from Definition 2.28 that a function $u \in W^{2,p}(\Omega)$ is a strong solution to (4.1) if the equation and boundary conditions hold almost everywhere in $\bar{\Omega}$. Such solutions exist provided the data is sufficiently regular and $A \in C(\bar{\Omega}, \mathbb{S}^d)$; see Theorem 2.29. This result is obtained by using the Calderón–Zygmund decomposition technique when $A = I$, and then extended to general $\mathcal{L}$ using the continuity of the coefficient matrix. In this section, we develop a discrete version of this theory to develop *a priori* estimates and convergence results of finite element solutions. Let us first present the derivation of the method.

Assume for the moment that the coefficient matrix A in (4.1) is sufficiently smooth. Then, as explained in Example 2.8, we can write problem (4.1) in divergence form:

$$D \cdot (ADu) - (D \cdot A) \cdot Du = f \quad \text{in } \Omega.$$

A standard finite element method for this problem (without stabilization) reads as follows: find $u_h \in X_h^{cg}$ such that

$$-\int_\Omega (ADu_h) \cdot Dv_h - \int_\Omega ((D \cdot A) \cdot Du_h)v_h = \int_\Omega fv_h, \quad \text{for all } v_h \in X_h^{cg}, \tag{4.18}$$

where $X_h^{cg} \subset H_0^1(\Omega)$ is the Lagrange finite element space of degree $k \geq 1$ defined by (3.34). It is well known that, for h sufficiently small, there exists a unique solution to (4.18).

If A is not sufficiently smooth and/or if the locations of the singularities are complex/unknown, then the classical finite element method (4.18) is not viable due to the differential operators acting on A. However, we easily circumvent this issue by using the integration-by-parts identity

$$\int_\Omega \boldsymbol{\tau} \cdot Dv_h = -\sum_{K \in \mathcal{T}_h} \int_K (D \cdot \boldsymbol{\tau})v_h + \sum_{F \in \mathcal{F}_h^I} \int_F [[\boldsymbol{\tau}]]v_h,$$

which holds for all piecewise smooth $\boldsymbol{\tau}$ and $v_h \in X_h^{cg}$. Taking $\boldsymbol{\tau} = ADu_h$ and applying the product rule yields

$$-\int_\Omega (ADu_h)\cdot Dv_h = \sum_{K\in\mathcal{T}_h}\int_K (A : D^2u_h)v_h + \int_\Omega ((D\cdot A)\cdot Du_h)v_h + \sum_{F\in\mathcal{F}_h^I}\int_F [[ADu_h]]v_h.$$

Substituting this identity into the (ill-posed) formulation leads to the finite element method

$$b_h(u_h, v_h) := \sum_{K\in\mathcal{T}_h}\int_K (A : D^2u_h)v_h - \sum_{F\in\mathcal{F}_h^I}\int_F [[ADu_h]]v_h = \int_\Omega fv_h \quad (4.19)$$

for all $v_h \in X_h^{cg}$. In contrast to (4.18), the formulation (4.19) is well-defined for non-differentiable A. Furthermore, by reversing the arguments, we see that (4.19) is equivalent to (4.18) if A is sufficiently smooth. In particular, when $A(x) \equiv \bar{A}$ is a constant SPD matrix, the method (4.19) reduces to the (well-posed) problem

$$b_{h,0}(u_h, v_h) := -\int_\Omega (\bar{A}Du_h)\cdot Dv_h = \int_\Omega fv_h, \quad \text{for all } v_h \in X_h^{cg}. \quad (4.20)$$

We point out that method (4.19) is consistent and meaningful in the piecewise linear case ($k = 1$).

While the derivation of the finite element method (4.19) is relatively simple, a stability and convergence analysis of the method is less obvious. The key difficulty is that integration by parts is not at our disposal, and it is unclear whether a clever choice of test function will render a coercivity or inf-sup condition. Rather, the stability analysis of (4.19) mimics the techniques found in the PDE theory, where Calderón–Zygmund estimates are the essential tools.

To describe the stability and convergence theory, we first define a discrete $W^{2,p}$-type norm:

$$\|v\|^p_{W_h^{2,p}(\Omega)} := \sum_{K\in\mathcal{T}_h}\|D^2v\|^p_{L^p(K)} + \sum_{F\in\mathcal{F}_h^I} h_F^{1-p}\|[[Dv]]\|^p_{L^p(F)}, \quad \text{for } 1 < p < \infty.$$

A discrete Calderón–Zygmund-type estimate with respect to this norm in the case of constant coefficients is now given.

Lemma 4.12 (discrete Calderón–Zygmund estimate). Let L be the elliptic, divergence-form operator (2.4), where the coefficient matrix A is constant and SPD. Suppose that the *a priori* estimate $C\|w\|_{W^{2,p}(\Omega)} \le \|Lw\|_{L^p(\Omega)}$ is satisfied for all $w \in W^{2,p}(\Omega)\cap W_0^{1,p}(\Omega)$. Let $b_{h,0}(\cdot,\cdot)$ be defined by (4.20). Then, for h sufficiently small,

$$C\|v_h\|_{W_h^{2,p}(\Omega)} \leq \sup_{w_h \in X_h^{cg}\setminus\{0\}} \frac{b_{h,0}(v_h, w_h)}{\|w_h\|_{L^{p'}(\Omega)}}, \quad \text{for all } v_h \in X_h^{cg},$$

where $1/p + 1/p' = 1$.

Proof. We prove the simpler case $p = 2$, and refer the reader to Feng, Hennings and Neilan (2017, Lemma 2.6) and Neilan (2013, Lemma 4.1) for general $p \in (1, \infty)$.

First, let $\mathcal{B}_h(v_h) \in X_h^{cg}$ be the unique solution to the problem

$$\int_\Omega \mathcal{B}_h(v_h) w_h \, dx = b_{h,0}(v_h, w_h), \quad \text{for all } w_h \in X_h^{cg},$$

and let $\varphi \in H_0^1(\Omega)$ be the unique (weak) solution to $L\varphi = -\mathcal{B}_h(v_h)$ in Ω. Then, for $w_h \in X_h^{cg}$, we find

$$b_{h,0}(v_h, w_h) = \int_\Omega \mathcal{B}_h(v_h) w_h = -\int_\Omega (AD\varphi) \cdot Dw_h = b_{h,0}(\varphi, w_h).$$

Thus, v_h is the elliptic projection of φ with respect to $b_{h,0}(\cdot,\cdot)$. Therefore, Céa's lemma and the hypothesis $\varphi \in H^2(\Omega)$ with

$$\|\varphi\|_{H^2(\Omega)} \leq C\|\mathcal{B}_h(v_h)\|_{L^2(\Omega)}$$

yield

$$\|\varphi - v_h\|_{H^1(\Omega)} \leq Ch\|\varphi\|_{H^2(\Omega)} \leq Ch\|\mathcal{B}_h(v_h)\|_{L^2(\Omega)}. \tag{4.21}$$

Next, for any $\varphi_h \in X_h^{cg}$, the triangle inequality and a scaling argument show that

$$\begin{aligned}
\|v_h\|_{H_h^2(\Omega)} &\leq \|v_h - \varphi_h\|_{H_h^2(\Omega)} + \|\varphi_h - \varphi\|_{H_h^2(\Omega)} + \|\varphi\|_{H_h^2(\Omega)} \\
&\leq Ch^{-1}(\|v_h - \varphi\|_{H^1(\Omega)} + \|\varphi - \varphi_h\|_{H^1(\Omega)}) + \|\varphi_h - \varphi\|_{H_h^2(\Omega)} + \|\varphi\|_{H^2(\Omega)}.
\end{aligned}$$

By taking φ_h to be the nodal interpolant of φ and applying (4.21) and the definition of $\mathcal{B}_h(v_h)$, we obtain

$$\begin{aligned}
\|v_h\|_{H_h^2(\Omega)} &\leq C\big[h^{-1}\|v_h - \varphi\|_{H^1(\Omega)} + \|\varphi\|_{H^2(\Omega)}\big] \leq C\|\mathcal{B}_h(v_h)\|_{L^2(\Omega)} \\
&= C \sup_{w_h \in X_h^{cg}\setminus\{0\}} \frac{\int_\Omega \mathcal{B}_h(v_h) w_h}{\|w_h\|_{L^2(\Omega)}} = C \sup_{w_h \in X_h^{cg}\setminus\{0\}} \frac{b_{h,0}(v_h, w_h)}{\|w_h\|_{L^2(\Omega)}}. \qquad \square
\end{aligned}$$

Remark 4.13 ($p = 2$). Note that when Ω is convex, the assumptions of Lemma 4.12 hold for $p = 2$.

A corollary of this result is a local stability estimate of the discrete adjoint problem.

Corollary 4.14 (local stability). For a domain $D \subset \Omega$, let ρ_D denote the radius of the largest ball inscribed in D, and let $X_h^{cg}(D)$ denote the set of functions in X_h^{cg} that vanish outside D. Suppose that the hypothesis of Lemma 4.12 are satisfied. Then, if h and ρ_D are sufficiently small,

$$C\|w_h\|_{L^{p'}(D)} \leq \sup_{v_h \in X_h^{cg}(D_h)\backslash\{0\}} \frac{b_h(v_h, w_h)}{\|v_h\|_{W_h^{2,p}(D_h)}}, \quad \text{for all } w_h \in X_h^{cg}(D).$$

with $D_h = \{x \in \Omega : \operatorname{dist}(x, D) \leq h\}$.

Proof. Again, we prove the case $p = p' = 2$ and refer the reader to Feng *et al.* (2017, Appendix B) for the general result.

Let $w_h \in X_h^{cg}(D)$, and let $\varphi_h \in X_h^{cg}$ satisfy

$$b_{h,0}(\varphi_h, v_h) = \int_\Omega w_h v_h, \quad \text{for all } v_h \in X_h^{cg},$$

where $b_{h,0}(\cdot,\cdot)$ is defined by (4.20) with

$$\bar{A} = \frac{1}{|D|}\int_D A.$$

Taking $v_h = w_h$ in the method yields

$$\|w_h\|_{L^2(D)}^2 = b_{h,0}(\varphi_h, w_h) = b_h(\varphi_h, w_h) + (b_{h,0}(\varphi_h, w_h) - b_h(\varphi_h, w_h)).$$

Using the (uniform) continuity of A, for any $\tau > 0$, we have

$$|b_{h,0}(\varphi_h, w_h) - b_h(\varphi_h, w_h)| \leq \tau\|\varphi_h\|_{H_h^2(\Omega)}\|w_h\|_{L^2(D)}$$

provided ρ_D is sufficiently small.

Applying the estimate

$$\|\varphi_h\|_{H_h^2(\Omega)} \leq C\|w_h\|_{L^2(\Omega)} = C\|w_h\|_{L^2(D)}$$

established in Theorem 4.12, we obtain

$$\begin{aligned}(1 - C\tau)\|w_h\|_{L^2(D)}^2 \leq b_h(\varphi_h, w_h) &= \left(\frac{b_h(\varphi_h, w_h)}{\|\varphi_h\|_{H_h^2(D_h)}}\right)\|\varphi\|_{H_h^2(\Omega)} \\ &\leq C\left(\sup_{v_h \in X_h^{cg}(D_h)\backslash\{0\}} \frac{b_h(v_h, w_h)}{\|v_h\|_{H_h^2(D_h)}}\right)\|w_h\|_{L^2(\Omega)}.\end{aligned}$$

Taking τ sufficiently small and manipulating terms in the last inequality yields the result. □

The local stability result for the discrete adjoint problem given in Corollary 4.14 leads to a stability estimate for method (4.19).

Theorem 4.15 (global stability). Suppose that $A \in C(\bar{\Omega}, \mathbb{S}^d)$, and that elliptic and divergence-form operators with constant coefficients inherit

$W^{2,p}$-regularity ($1 < p < \infty$). Then there exists $h_* > 0$ depending on the modulus of continuity of A and p such that, for $h \le h_*$,

$$C\|v_h\|_{W_h^{2,p}(\Omega)} \le \sup_{w_h \in X_h^{cg}\setminus\{0\}} \frac{b_h(v_h, w_h)}{\|w_h\|_{L^{p'}(\Omega)}}, \quad \text{for all } v_h \in X_h^{cg}. \tag{4.22}$$

Proof. We outline the main steps of the proof and refer to Feng *et al.* (2017) for details.

Combining Corollary 4.14 with cut-off functions techniques and a covering argument leads to the Gärding-type inequality

$$C\|w_h\|_{L^{p'}(\Omega)} \le \sup_{v_h \in X_h^{cg}\setminus\{0\}} \frac{b_h(v_h, w_h)}{\|v_h\|_{W_h^{2,p}(\Omega)}} + \|w_h\|_{W^{-1,p'}(\Omega)}, \quad \text{for all } w_h \in X_h^{cg}.$$

A standard duality argument then shows that, for h sufficiently small,

$$C\|w_h\|_{L^{p'}(\Omega)} \le \sup_{v_h \in X_h^{cg}\setminus\{0\}} \frac{b_h(v_h, w_h)}{\|v_h\|_{W_h^{2,p}(\Omega)}}, \quad \text{for all } w_h \in X_h^{cg}.$$

This estimate shows that, for fixed $v_h \in X_h^{cg}$, there exists a unique $w_h \in X_h^{cg}$ satisfying

$$\begin{aligned} b_h(z_h, w_h) &= \sum_{K \in \mathcal{T}_h} \int_K |D^2 v_h|^{p-2} D^2 v_h : D^2 z_h \\ &\quad + \sum_{F \in \mathcal{F}_h} h_F^{-1} \int_F |[[Dv_h]]|^{p-2} [[Dv_h]][[Dz_h]], \quad \text{for all } z_h \in X_h^{cg}. \end{aligned} \tag{4.23}$$

Applying the global stability estimate for the adjoint problem and Hölder's inequality, we obtain

$$C\|w_h\|_{L^{p'}(\Omega)} \le \sup_{z_h \in X_h^{cg}\setminus\{0\}} \frac{b_h(z_h, w_h)}{\|z_h\|_{W_h^{2,p}(\Omega)}} \le C\|v_h\|^{p-1}_{W_h^{2,p}(\Omega)}.$$

On the other hand, setting $z_h = v_h$ in (4.23) yields

$$\begin{aligned} \|v_h\|^p_{W_h^{2,p}(\Omega)} = b_h(v_h, w_h) &\le \left(\sup_{z_h \in X_h^{cg}\setminus\{0\}} \frac{b_h(v_h, z_h)}{\|z_h\|_{L^{p'}(\Omega)}} \right) \|w_h\|_{L^{p'}(\Omega)} \\ &\le C \left(\sup_{z_h \in X_h^{cg}\setminus\{0\}} \frac{b_h(v_h, z_h)}{\|z_h\|_{L^{p'}(\Omega)}} \right) \|v_h\|^{p-1}_{W_h^{2,p}(\Omega)}. \end{aligned}$$

Dividing by $\|v_h\|_{W_h^{2,p}(\Omega)}$, we obtain (4.22). □

Theorem 4.16 (existence and error estimates). Suppose that the hypotheses of Theorem 4.15 are satisfied. Then there exists a unique solution $u_h \in X_h^{cg}$ to (4.19). If the solution to (4.1) satisfies $u \in W^{s,p}(\Omega)$ for

$2 \le s \le k+1$, then

$$\|u - u_h\|_{W_h^{2,p}(\Omega)} \le Ch^{s-2}\|u\|_{W^{s,p}(\Omega)}.$$

Proof. The existence and uniqueness of a solution to (4.19) follows from the stability estimate (4.22). The error estimate follows from the stability and continuity of the bilinear form $b_h(\cdot,\cdot)$, the consistency of the scheme, and approximation properties of X_h^{cg} with respect to the discrete $W^{2,p}$-norm. □

Remark 4.17 (extensions). The ideas and analysis presented in this section has been extended to discontinuous Galerkin methods (Feng, Neilan and Schnake 2016) and mixed finite element methods (Lakkis and Pryer 2011, Neilan 2017).

4.3. Finite element method based on integro-differential approximation

In this section we consider a two-scale finite element discretization for problem (4.1) developed in Nochetto and Zhang (2017*a*), which is based on a regularized, integro-differential approximation proposed in Caffarelli and Silvestre (2010*b*). As in the previous sections we assume that the PDE operator $\mathcal{L}$ is uniformly elliptic, that is, there exist strictly positive constants λ, Λ satisfying $\lambda I \le A(x) \le \Lambda(x) I$, for all $x \in \bar{\Omega}$. We further make the simplifying assumption that $A \in C(\bar{\Omega}, \mathbb{S}^d)$, and make remarks when this regularity can be relaxed.

To explain and motivate the method, we first perform some algebraic manipulations, and rewrite the PDE as

$$A : D^2 u = \frac{\lambda}{2}\Delta u + A_\lambda^2 : D^2 u, \quad A_\lambda := \left(A - \frac{\lambda}{2} I\right)^{1/2}. \tag{4.24}$$

Let φ be a radially symmetric function with compact support in the unit ball satisfying $\int_{\mathbb{R}^d} |z|^2 \varphi(z) = d$. We then find that

$$\int_{\mathbb{R}^d} z_i z_j \varphi(z) = 0 \quad \text{for } i \ne j, \quad \text{and} \quad \int_{\mathbb{R}^d} z_i^2 \varphi(z) = 1.$$

Consequently, we have

$$\int_{\mathbb{R}^d} z \otimes z \varphi(z) = I,$$

and therefore

$$(A_\lambda(x))^2 : D^2 u(x) = A_\lambda(x)\left(\int_{\mathbb{R}^d} z \otimes z \varphi(z)\right) A_\lambda(x) : D^2 u(x).$$

For a regularization parameter $\epsilon > 0$, we make the change of variables

$y = \epsilon A_\lambda(x)z$ in the integral to obtain

$$(A_\lambda(x))^2 : D^2u(x) = \int_{\mathbb{R}^d} \frac{(y\otimes y) : D^2u(x)}{\epsilon^{d+2}\det(A_\lambda(x))}\varphi\Big(\frac{A_\lambda^{-1}(x)y}{\epsilon}\Big).$$

Set

$$Q = \Big(\Lambda - \frac{\lambda}{2}\Big)^{1/2}, \quad \Omega_\epsilon = \{x\in\Omega : \operatorname{dist}(x,\partial\Omega) > Q\epsilon\}, \quad \omega_\epsilon = \Omega\backslash\Omega_\epsilon, \tag{4.25}$$

and note that $\varphi(A_\lambda^{-1}(x)y/\epsilon)$ has support in the ball $B_{Q\epsilon}(0)$. For $x\in\Omega$, let $\theta = \theta(x) \in (0,1]$ be the largest number such that $x\pm\theta y\in\Omega$ for all $y\in B_{Q\epsilon}(0)$. Recall that the second difference operator is given by

$$\delta^2_{y,\theta}u(x) := \frac{u(x+\theta y) - 2u(x) + u(x-\theta y)}{\theta^2},$$

and note that $\delta^2_{\theta y,\theta}u(x) = (y\otimes y) : D^2u(x)$ if u is a quadratic polynomial, and that $\theta = 1$ for $x\in\Omega_\epsilon$.

Combining these calculations and identities, we are led to the approximation

$$(A_\lambda(x))^2 : D^2u(x) \approx \int_{\mathbb{R}^d} \frac{|y|^2\delta^2_{\theta y,\theta}u(x)}{\epsilon^{d+2}\det(A_\lambda(x))}\varphi\Big(\frac{A_\lambda^{-1}(x)y}{\epsilon}\Big) =: I_\epsilon u(x). \tag{4.26}$$

The approximation is quantified in the next lemma (Nochetto and Zhang 2017*a*, Lemma 2.1).

Lemma 4.18 (rate of convergence of integral transform). Let I_ϵ be the integral operator defined by (4.26), and let $U_{Q\epsilon}(x) := \bar{B}_{Q\epsilon}(x)\cap\bar{\Omega}$.

- If $u\in C^2(\bar{\Omega})$, then
$$I_\epsilon u(x) \to \Big(A(x) - \frac{\lambda}{2}I\Big) : D^2u(x) \quad \text{as } \epsilon\to 0^+,$$
for all $x\in\Omega$.
- If $u\in C^{2,\alpha}(U_{Q\epsilon}(x))$ for some $\alpha\in(0,1]$, then
$$\Big|I_\epsilon u(x) - \Big(A(x) - \frac{\lambda}{2}I\Big) : D^2u(x)\Big| \le C\|u\|_{C^{2,\alpha}(U_{Q\epsilon})}\theta^\alpha\epsilon^\alpha,$$
for all $x\in\Omega$.

We approximate equation (4.1) by the integro-differential equation

$$\mathcal{L}^\epsilon u^\epsilon := \frac{\lambda}{2}\Delta u^\epsilon + I_\epsilon u^\epsilon = f \quad \text{in } \Omega. \tag{4.27}$$

We refer the reader to Caffarelli and Silvestre (2010*b*) for details about the existence, uniqueness and regularity estimates of solution u^ϵ.

We now describe a convergent finite element scheme for the non-divergence-form problem (4.1) based on the regularized problem (4.27). To this end, we let $X_{0,h}^l = X_h^l \cap H_0^1(\Omega)$ be the linear, Lagrange finite element space with vanishing trace. Let $\phi_i \in X_{0,h}^l$ denote the normalized hat function with respect to the interior node $z_i \in \Omega_h^I$, and let Δ_h be the finite element Laplacian defined by (3.39). We consider the following finite element method: find $u_h \in X_{0,h}^l$ such that

$$\mathcal{L}_h^\epsilon u_h^\epsilon(z_i) := \frac{\lambda}{2}\Delta_h u_h^\epsilon(z_i) + I_\epsilon u_h^\epsilon(z_i) = f_i := \int_\Omega f\phi_i, \quad \text{for all } z_i \in \Omega_h^I. \tag{4.28}$$

Note that the formulation (4.28) is *not* obtained by testing (4.27) with ϕ_i (which would introduce the term $\int_\Omega I_\epsilon u_h^\epsilon \phi_i$). Instead, mass lumping is used to preserve the monotonicity of the scheme.

Lemma 4.19 (monotonicity). Suppose that $v_h, w_h \in X_{0,h}^l$ satisfy $v_h \le w_h$ with equality at $z \in \Omega_h^I$. Then $I_\epsilon v_h(z) \le I_\epsilon w_h(z)$. Consequently, if $\mathcal{T}_h$ satisfies (3.42), then $\mathcal{L}_h^\epsilon$ is monotone.

Proof. From the hypotheses and the definition of $\delta_{\theta y,\theta}^2$, we have $\delta_{\theta y,\theta}^2 v(z) \le \delta_{\theta y,\theta}^2 w(z)$, and therefore $I_\epsilon v_h(z_i) \le I_\epsilon w_h(z_i)$. The monotonicity of $\mathcal{L}_h^\epsilon$ then follows from Lemma 3.42. □

The monotonicity, along with the Alexandrov–Bakelman–Pucci estimate for the finite element Laplacian, yields the following maximum principle.

Theorem 4.20 (discrete ABP estimate for $\mathcal{L}_h^\epsilon$). Suppose that $\mathcal{T}_h$ satisfies (3.42). Then, for $v_h \in X_{0,h}^l$ satisfying

$$\mathcal{L}_h^\epsilon v_h(z_i) \le f_i, \quad \text{for all } z_i \in \Omega_h^I,$$

we obtain

$$\sup_\Omega v_h^- \le \frac{C}{\lambda}\bigg(\sum_{z_i \in \mathcal{C}_h^-(v_h)} |f_i^+|^d |\omega_{z_i}|\bigg)^{1/d},$$

where $\mathcal{C}_h^-(v_h)$ is the nodal contact set given in Definition 3.23.

Proof. Let $\Gamma_h(v_h) = \Gamma(v_h)$ be the convex envelope of v_h. Then, for a contact point $z_i \in \mathcal{C}_h^-(v_h)$,

$$0 \le I_\epsilon \Gamma(v_h)(z_i) \le I_\epsilon v_h(z_i),$$

where the first inequality follows from the convexity of $\Gamma(v_h)$ and the second follows from the monotonicity of I_ϵ in Lemma 4.19. Consequently,

$$\frac{\lambda}{2}\Delta_h v_h(z_i) \le \mathcal{L}_h^\epsilon v_h(z_i) \le f_i^+$$

since $f_i \geq 0$ for $z_i \in \mathcal{C}_h^-(v_h)$. The result now follows from this inequality and Theorem 3.50 (see Remark 3.51). $\square$

Since the method (4.28) is linear, a corollary of the ABP estimate is the existence and uniqueness of a solution u_h.

Corollary 4.21 (existence and uniqueness). Suppose that $\mathcal{T}_h$ satisfies (3.42). Then there exists a unique $u_h \in X_{0,h}^l$ satisfying (4.28).

We now turn our attention to error estimates of the finite element method (4.28) and derive a rate of convergence in the L^∞-norm. To do so we assume that the solution to (4.1) satisfies $u \in C^{2,\alpha}(\Omega)$. Recall (see Theorem 2.17) that this regularity is guaranteed provided that A is Hölder-continuous and $\partial\Omega$ is sufficiently smooth.

Now, since the method is linear and the problem is stable, such estimates reduce to the consistency of the method. However, as shown in Lemma 3.38, the finite element method is not consistent, in the sense of Definition 3.1, when supplemented with the canonical interpolant I_h^{fe}. Instead, we make use of the elliptic projection I_h^{ep} defined in (3.41) and its properties (see Lemma 3.41 and Proposition 3.40).

In conclusion, in order to derive error estimates, it suffices to derive upper bounds for the difference $u_h^\epsilon - I_h^{ep} u$. To this end, we apply the definition of the method (4.28) to obtain the *error equation*

$$\mathcal{L}_h^\epsilon[I_h^{ep} u - u_h^\epsilon](z_i) = \int_{\omega_{z_i}} (T_1^{(i)} + T_2^{(i)} + T_3^{(i)})\phi_i \tag{4.29}$$

with

$$\begin{aligned} T_1^{(i)} &= I_\epsilon[I_h^{ep} u](z_i) - I_\epsilon u(z_i), \\ T_2^{(i)} &= I_\epsilon u(z_i) - \left(A(z_i) - \frac{\lambda}{2} I\right) : D^2 u(z_i), \\ T_3^{(i)} &= \left(A(z_i) - \frac{\lambda}{2} I\right) : (D^2 u(z_i) - D^2 u(x)). \end{aligned}$$

Note that, with the finite element ABP estimate given in Theorem 4.20 and the approximation results of the elliptic projection stated in Proposition 3.40, upper-bound estimates of $T_j^{(i)}$ yield error estimates of $u - u_h$.

With the assumed regularity $u \in C^{2,\alpha}(\Omega)$, we immediately find that $T_3^{(i)}$ can be bounded by

$$|T_3^{(i)}| \leq C h^\alpha \|u\|_{C^{2,\alpha}(\Omega)}. \tag{4.30}$$

For $T_2^{(i)}$, we apply Lemma 4.18 to obtain

$$|T_2^{(i)}| \leq C\theta^\alpha \epsilon^\alpha \|u\|_{C^{2,\alpha}(\Omega)} \leq C\epsilon^\alpha \|u\|_{C^{2,\alpha}(\Omega)}. \tag{4.31}$$

Finally, we apply the approximation results of Proposition 3.40 and the definition of the second-order difference operator $\delta^2_{\theta y,\theta}$ to obtain

$$|\delta^2_{\theta y,\theta}(I_h^{ep}u(z_i) - u(z_i))| \le C\frac{h^2}{\theta^2}\,|\log h|\|u\|_{W^{2,\infty}(\Omega)}.$$

This leads to the estimate

$$|T_1^{(i)}| \le C\|u\|_{W^{2,\infty}(\Omega)}\bigg(\frac{h^2}{\epsilon^2}\,|\log h| + \frac{h^2}{\theta^2\epsilon^2}\,|\log h|\chi_{\omega_\epsilon}(z_i)\bigg), \tag{4.32}$$

where χ_{ω_ϵ} is the indicator function of ω_ϵ, which is defined in (4.25). Combining (4.30)–(4.32), we obtain

$$\begin{aligned} &|T_1^{(i)} + T_2^{(i)} + T_3^{(i)}| \\ &\quad\le C\bigg(h^\alpha + \epsilon^\alpha + \frac{h^2}{\epsilon^2}\,|\log h| + \frac{h^2}{\theta^2\epsilon^2}\,|\log h|\,|\chi_{\omega_\epsilon}(z_i)\bigg), \end{aligned} \tag{4.33}$$

where we have absorbed the factor $\|u\|_{C^{2,\alpha}(\Omega)}$ into the constant C.

Note that, owing to the last term on the right-hand side, estimate (4.33) reduces to order 1 in the boundary layer ω_ϵ. In order to derive meaningful estimates in this region, we introduce the barrier layer function

$$b(x) := \xi(\operatorname{dist}(x,\partial\Omega)), \text{ with } \xi(s) := \begin{cases} Q^{-2}(s-Q\epsilon)^2 - \epsilon^2 & \text{if } s \le Q\epsilon, \\ -\epsilon^2 & \text{if } s > Q_\epsilon, \end{cases} \tag{4.34}$$

The discrete boundary layer function is defined as $b_h := I_h^{fe}b$. The next result summarizes key properties of b_h; see Nochetto and Zhang (2017*a*, Lemma 6.1)

Lemma 4.22 (properties of barrier layer function). Let $b_h = I_h^{fe}b$, where b is given by (4.34). Then

$$\mathcal{L}_h^\epsilon b_h(z_i) \ge C\chi_{\omega_\epsilon(z_i)}, \quad |b_h(z_i)| \le C\epsilon^2.$$

Note that since every $z_i \in \omega_\epsilon$ is at most $O(h)$ from $\partial\Omega$, we have $\epsilon\theta \ge Ch$ on ω_ϵ. Consequently, the last term in (4.33) can be bounded by the discrete barrier function as follows:

$$\frac{h^2}{\theta^2\epsilon^2}\,|\log h|\chi_{\omega_\epsilon(z_i)}\|u\|_{C^{2,\alpha}(\Omega)} \le C\,|\log h|\chi_{\omega_\epsilon}(z_i) \le C\,|\log h|\mathcal{L}_h^\epsilon b_h(z_i).$$

Thus, combining this estimate with (4.33) and (4.29) leads to

$$\mathcal{L}_h^\epsilon\big[I_h^{ep}u - u_h^\epsilon - C\,|\log h|b_h\big](z_i) \le C\bigg(h^\alpha + \epsilon^\alpha + \frac{h^2}{\epsilon^2}\,|\log h|\bigg). \tag{4.35}$$

From this expression, we easily obtain estimates of $u - u_h$.

Theorem 4.23 (rate of convergence). Suppose that $\mathcal{T}_h$ satisfies (3.42), and that the solution to (4.1) satisfies $u \in C^{2,\alpha}(\Omega)$. Let $u_h \in X^l_{0,h}$ be the solution to (4.28) with $\epsilon = C(h^2 |\log h|)^{1/(2+\alpha)}$. Then

$$\|u - u_h\|_{L^\infty(\Omega)} \leq C(h^2 |\log h|)^{\alpha/(2+\alpha)}. \tag{4.36}$$

Proof. Applying the finite element ABP for $\mathcal{L}^\epsilon_h$ (see Theorem 4.20) to (4.35) yields

$$\sup_\Omega (I^{ep}_h u - u^\epsilon_h - C|\log h| b_h)^- \leq C\left(h^\alpha + \epsilon^\alpha + \frac{h^2}{\epsilon^2}|\log h|\right).$$

Therefore, since $|b_h| \leq \epsilon^2$, we get

$$\sup (I^{ep}_h u - u^\epsilon_h)^- \leq C\left(h^\alpha + \epsilon^\alpha + \left(\epsilon^2 + \frac{h^2}{\epsilon^2}\right)|\log h|\right).$$

Similar estimates are obtained for $\sup_\Omega (I^{ep}_h u - u^\epsilon_h)^+$, thus leading to

$$\|I^{ep}_h u - u^\epsilon_h\|_{L^\infty(\Omega)} \leq C\left(h^\alpha + \epsilon^\alpha + \left(\epsilon^2 + \frac{h^2}{\epsilon^2}\right)|\log h|\right).$$

If $\epsilon = C(h^2 |\log h|)^{1/(2+\alpha)}$, then

$$(h^2/\epsilon^2)|\log h| \leq C\epsilon^\alpha \leq C(h^2|\log h|)^{\alpha/(2+\alpha)},$$

and therefore

$$\|I^{ep}_h u - u^\epsilon_h\|_{L^\infty(\Omega)} \leq C(h^2|\log h|)^{\alpha/(2+\alpha)}.$$

Finally, applying Proposition 3.40 and the triangle inequality yields (4.36). The proof is complete. □

Theorem 4.23 shows that if $\alpha = 1$, then the error satisfies $\|u - u_h\|_{L^\infty(\Omega)} = O(h^{2/3-\tau})$ for arbitrary $\tau > 0$. If more regularity is assumed then an almost linear rate is obtained (Nochetto and Zhang 2017*a*, Corollary 6.8)

Theorem 4.24 (improved rate). Let h and ϵ satisfy $\epsilon = Ch^{2/(3+\alpha)}$. If the solution to (4.1) has the regularity $u \in C^{3,\alpha}(\Omega)$, and if $\mathcal{T}_h$ satisfies (3.42), then

$$\|u - u^\epsilon_h\|_{L^\infty(\Omega)} \leq Ch^{2(1+\alpha)/(3+\alpha)}|\log h|.$$

In the opposite direction, if we assume less regularity of the solution and data, then convergence is still obtained, although without rates (Nochetto and Zhang 2017*a*, Corollary 6.5).

Theorem 4.25 (convergence). Assume that $u \in C^2(\bar{\Omega})$, that the coefficient matrix satisfies $A \in VMO(\Omega, \mathbb{S}^d)$, and that the two scales ϵ and h satisfy $\epsilon = Ch|\log h|$. Let $u^\epsilon_h \in X^l_{0,h}$ satisfy (4.28) with $A(z_i)$ replaced by its average

$$\bar{A}(z_i) := \frac{1}{|\omega_{z_i}|} \int_{\omega_{z_i}} A(y).$$

If the mesh condition (3.42) is satisfied, then $\lim_{h \to 0^+} \|u - u_h^\epsilon\|_{L^\infty(\Omega)} = 0$.

5. Discretizations of convex second-order elliptic equations

Up to this point we have discussed general issues regarding stability and convergence of numerical methods for general fully nonlinear equations, as well as the construction of suitable schemes for linear problems in non-divergence form. For the rest of this overview we will merge the ideas presented in the previous sections. Since convergence of these schemes has already been discussed in Theorems 3.7 and 3.8, we will pay special attention to obtaining rates of convergence for them. In this section we will focus on *convex* equations. Moreover, as explained Section 1.3, there is no loss of generality in assuming that we are dealing with the Hamilton–Jacobi–Bellman equation, which we recall reads

$$\begin{cases} F(x, u, Du, D^2 u) := \sup_{\alpha \in \mathcal{A}} [\tilde{\mathcal{L}}^\alpha u(x) - f^\alpha(x)] = 0 & \text{in } \Omega, \\ u = g & \text{on } \partial\Omega, \end{cases} \tag{5.1}$$

where

$$\tilde{\mathcal{L}}^\alpha u(x) := A^\alpha(x) : D^2 u(x) + \mathbf{b}(x) \cdot Du(x) + c(x)u(x).$$

We assume that the linear elliptic operators are such that, for every $\alpha \in \mathcal{A}$,

$$A^\alpha \geq \lambda I, \quad \text{for all } \alpha \in \mathcal{A},$$

so that the operator F is uniformly elliptic.

The numerical schemes for problem (5.1) can be roughly classified as finite difference, finite element and semi-Lagrangian methods. In this section we will focus on the first two. Semi-Lagrangian schemes will be illustrated in a particular case, the Monge–Ampère equation, in Section 6.2.

5.1. Finite difference methods

The analysis of the convergence properties of finite difference schemes for (5.1) dates back to Krylov (1997, 2000), who considered the problem for $\Omega = \mathbb{R}^d$ and constant 'coefficients', that is, when the operators $\tilde{\mathcal{L}}^\alpha$ are x-independent. These results were later extended by Barles and Jakobsen (2002) to Lipschitz coefficients. Before embarking on the technical details of their results, let us give some intuition. Recall that, in general, a finite difference scheme is written in the form (3.13). Ideally, to approximate

u, the solution of (5.1), one would first construct a smooth function u_ε, parametrized by ε which, for a constant C independent of ε, satisfies

$$\|u - u_\varepsilon\|_{L^\infty(\Omega)} \le C\varepsilon^{\kappa_1}.$$

A strengthened notion of consistency (see Definition 3.1) would then imply that $F_h[I_h^{fd} u_\varepsilon] \approx h^{\kappa_2}$, where the hidden constants in this expression may depend on ε. By stability, Definition 3.3, and monotonicity, Definition 3.5, we obtain

$$\|I_h^{fd} u_\varepsilon - u_h\|_{L^\infty(\bar{\Omega}_h)} \le Ch^{\kappa_2}.$$

An application of the triangle inequality and relating the smoothing parameter ε to the discretization h would yield a rate of convergence.

Unfortunately, the construction of such a smooth approximation u_ε is not immediate in practice. The groundbreaking idea of Krylov was to 'shake the coefficients'. He introduced u_ε as the solution of

$$\inf_{|e|\le\varepsilon} F(x+e, u_\varepsilon, Du_\varepsilon, D^2 u_\varepsilon) = 0, \tag{5.2}$$

and, from u_ε, he was able to construct a smooth subsolution to (5.1) so that by comparison we can obtain an upper bound for $u_h - u$. To obtain a lower bound, the original work of Krylov invoked arguments that some authors have characterized as probabilistic. However, Krylov (2015, page 3) disagrees with this statement. Another line of reasoning was given by Barles and Jakobsen (2002) who, instead, proposed that the problem and scheme should play a symmetric role. In other words, they introduce u_h^ε which solves

$$\inf_{|e|\le\varepsilon} F_h[u_h^\varepsilon](z+e) = 0. \tag{5.3}$$

Under suitable assumptions they show that this family of operators possesses unique smooth solutions, where the smoothness is independent of ε. This then allows us to compare u and u_h^ε and obtain a rate of convergence. The assumptions that they rely on, however, must be checked for every particular instance.

Let us now proceed with the details. Recall that we are operating in the whole space $\mathbb{R}^d$ and that the mesh is given by $\bar{\Omega}_h = \mathbb{Z}_h^d$. We assume that all the coefficients of $\tilde{\mathcal{L}}^\alpha$ and f^α are Lipschitz-continuous uniformly in α and that $\sup_{\alpha\in\mathcal{A}} c^\alpha(x) \le c < 0$ for every $x \in \mathbb{R}^d$. This, as indicated in Theorem 2.52(b), implies that the operator has a comparison principle. The structural requirements on the discretization scheme are summarized below.

Assumption 5.1 (finite differences). The finite difference scheme

$$F_h = F_h(z, r, q)$$

satisfies the following.

(i) The scheme is consistent in a sense stronger than Definition 3.1, namely, for some $\kappa > 0$,

$$\|F_h[I_h^{fd}\phi] - F[\phi]\|_{L^\infty(\bar{\Omega}_h)} \le Ch^\kappa,$$

for all sufficiently smooth ϕ.

(ii) The scheme is of non-negative type: see Definition 3.9. In addition, there exists $\bar{c} > 0$ such that $F_h(z, r + t, q + \mathbf{1}t) \le F_h(z, r, q) - \bar{c}t$ for all $t \ge 0$.

(iii) The scheme is convex in the r and q variables.

(iv) The scheme is uniformly solvable and smooth under perturbations. In other words, for $h > 0$ small enough and $\varepsilon \in [0, 1]$, problem (5.3) has a unique solution u_h^ε and, moreover, for some $\delta > 0$ we have

$$\|u_h - u_h^\varepsilon\|_{L^\infty(\bar{\Omega}_h)} \le C\varepsilon^\delta.$$

Note that conditions (i)–(iii) of Assumption 5.1 are relatively easy to enforce. However, condition (iv) must be verified in each case. As a first step, we show that schemes satisfying the non-negativity condition satisfy a comparison principle.

Lemma 5.2 (discrete comparison principle). Let F_h satisfy Assumption 5.1(ii). Let $v_h, w_h \in X_h^{fd}$ be two bounded nodal functions that are sub- and supersolutions to the discrete problem $F_h[u_h] = 0$, respectively. Then $v_h \le w_h$.

Proof. We assume $r = \sup_{\bar{\Omega}_h}(v_h - w_h) > 0$ and derive a contradiction. For simplicity we assume there exists $z \in \bar{\Omega}_h$ such that $v_h(z) - w_h(z) = r$, for otherwise we can apply a standard limiting argument.

As in Section 3.2 we write $F_h[v_h](z) = F_h(z, v_h(z), Tv_h(z))$, where the set of translates of $v_h(z)$ is $Tv_h(z) = \{v_h(z + hy) : y \in S\}$, and S is the stencil. Note that $v_h(z) = w_h(z) + r$ and $v_h(z + hy) \le w_h(z + hy) + r$ for $y \in S$. Since F_h is increasing in its third argument (see Definition 3.9), we have

$$\begin{aligned} 0 &\le F_h[v_h](z) - F_h[w_h](z) \\ &= F_h(z, w_h(z) + r, Tv_h(z)) - F_h(z, w_h(z), Tw_h(z)) \\ &\le F_h(z, w_h(z) + r, Tw_h(z) + \mathbf{1}r) - F_h(z, w_h(z), Tw_h(z)). \end{aligned}$$

Applying Assumption 5.1(ii) then yields

$$0 \le (F_h(z, w_h(z), Tw_h(z)) - \bar{c}r) - F_h(z, w_h(z), Tw_h(z)) = -\bar{c}r,$$

which is a contradiction. □

Having shown the comparison principle, we now state the convergence rate of the scheme.

Theorem 5.3 (rate of convergence). Assume that the finite difference scheme satisfies Assumption 5.1. Then there are constants $C_1, C_2 > 0$ such that, for every $z \in \bar{\Omega}_h$,

$$I_h^{fd}u(z) - u_h(z) \le C_1 h^{\kappa_1}, \quad u_h(z) - I_h^{fd}u(z) \le C_2 h^{\kappa_2},$$

where the rates $\kappa_1, \kappa_2 > 0$ are not necessarily the same.

Proof. Let us sketch the proof of each of these bounds.

Proof of $I_h^{fd}u(z) - u_h(z) \le C_1 h^{\kappa_1}$. Let u_ε be the solution of (5.2). By definition, after the change of variables $y = x + e$, we realize that $u_\varepsilon(\cdot - e)$ is, for $|e| \le \varepsilon$, a subsolution to the equation, that is, it satisfies, in the viscosity sense,

$$F\big(y, u_\varepsilon(\cdot - e), Du_\varepsilon(\cdot - e), D^2 u_\varepsilon(\cdot - e)\big) \ge 0.$$

We regularize this function by the mollification $u^\varepsilon = u_\varepsilon \star \rho_\varepsilon$, where ρ_ε are the standard mollifiers. From the convexity of the operator F and Jensen's inequality, it follows that u^ε is also a subsolution. Since u^ε is now a smooth function, we can invoke consistency to obtain

$$F_h[I_h^{fd}u^\varepsilon](z) \ge F(z, u^\varepsilon(z), Du^\varepsilon(z), D^2u^\varepsilon(z)) - Ch^\kappa \ge -Ch^\kappa,$$

for some constant C that depends on the smoothness of u^ε which, in turn, scales like negative powers of ε, say $C \le \varepsilon^{-\delta_1}$. In conclusion, we have obtained that

$$F_h[I_h^{fd}u^\varepsilon](z) \ge -Ch^\kappa \varepsilon^{-\delta_1}.$$

Assumption 5.1(ii) shows that the function $I_h^{fd}(u^\varepsilon - Ch^\kappa\varepsilon^{-\delta_1})$ is, for a suitably chosen C, a subsolution of the scheme, that is, $F_h[I_h^{fd}(u^\varepsilon - Ch^\kappa\varepsilon^{-\delta_1})] \ge 0$. Therefore, by the comparison principle given in Lemma 5.2,

$$I_h^{fd}u^\varepsilon(z) - u_h(z) \le Ch^\kappa\varepsilon^{-\delta_1}.$$

In conclusion, using the continuity properties of equation (5.1) and properties of mollifiers, we obtain

$$\begin{aligned} I_h^{fd}u(z) - u_h(z) &= I_h^{fd}(u - u_\varepsilon)(z) + I_h^{fd}(u_\varepsilon - u^\varepsilon)(z) + I_h^{fd}(u^\varepsilon - u_h)(z) \\ &\le \frac{C_1}{2}(\varepsilon^{\delta_2} + h^\kappa\varepsilon^{-\delta_1}), \end{aligned}$$

where $\delta_2 > 0$ depends on the smoothness of u (see Theorem 2.67). Optimizing with respect to ε, we get the result.

Proof of $u_h(z)-I_h^{fd}u(z) \le C_2 h^{\kappa_2}$. We follow similar reasoning but this time, we interchange the roles that the equation and the scheme have played in the previous step. Indeed, by Assumption 5.1(iv) we know that there is a $u_h^\varepsilon \in X_h^{fd}$ that solves (5.3) and, again with the change of variables $\tilde{z} = z+e$, this function satisfies

$$F_h[u_h^\varepsilon](z) \ge 0,$$

so that it is a subsolution of the scheme. Now, convexity of F_h implies that $I_h^{fd}(u_h^\varepsilon \star \rho_\varepsilon)$ is also a subsolution of the scheme. Moreover, we have that $u_h^\varepsilon \star \rho_\varepsilon$ is a smooth function, and therefore consistency implies that for some $\delta_3 > 0$ we have

$$F[u_h^\varepsilon \star \rho_\varepsilon](z) \ge -Ch^\kappa \varepsilon^{-\delta_3}.$$

Monotonicity of the equation shows that, for a suitably chosen C, the function $u_h^\varepsilon \star \rho_\varepsilon - Ch^\kappa \varepsilon^{-\delta_3}$ is a subsolution which, by comparison, readily implies that

$$u_h^\varepsilon \star \rho_\varepsilon(z) - u(z) \le Ch^\kappa \varepsilon^{-\delta_3}.$$

Properties of convolutions together with Assumption 5.1(iv) then imply

$$\begin{aligned} u_h(z) - I_h^{fd}u(z) &= (u_h - u_h^\varepsilon)(z) + (u_h^\varepsilon - I_h^{fd}(u_h^\varepsilon \star \rho_\varepsilon))(z) \\ &\quad + I_h^{fd}(u_h^\varepsilon \star \rho_\varepsilon - u)(z) \le \frac{C_2}{3}(\varepsilon^\delta + \varepsilon^{\delta_2} + h^\kappa \varepsilon^{-\delta_3}). \end{aligned}$$

Optimization in ε once again yields the result. □

Let us now give an example of finite difference schemes for which the assumptions of Theorem 5.3 can be verified.

Example 5.4 (monotone finite differences). Let A^α be independent of x for every α and, possibly after a renormalization, verify

$$\sum_{i=1}^d \left[a_{i,i}^\alpha - \sum_{j \ne i} |a_{i,j}^\alpha| \right] \le 1.$$

More importantly, we assume that these matrices are diagonally dominant, that is, (3.31) holds. For simplicity, assume also that $\mathbf{b}^\alpha \equiv 0$ for all $\alpha \in \mathcal{A}$. As shown in Lemma 3.34, there exists a monotone finite difference $\mathcal{L}_h^\alpha + c^\alpha$ that is consistent with $\tilde{\mathcal{L}}^\alpha$. We then define

$$F_h[u_h](z) = \sup_{\alpha \in \mathcal{A}} \left[\mathcal{L}_h^\alpha u_h(z) + c^\alpha(z) u_h(z) - f^\alpha(z) \right] = 0.$$

In this case, Barles and Jakobsen (2002, Section 4) show that all the assumptions are verified and, moreover, that $\kappa_1 = \kappa_2 = 1/3$.

Remark 5.5 (examples and improvements). Let us comment on the various improvements and refinements of Theorem 5.3 as well as some extensions of Example 5.4.

- In Example 5.4 it is assumed that the leading coefficients A^α do not depend on the spatial variable x. Barles and Jakobsen (2005) studied the relation between (5.1) and a certain system of quasi-variational inequalities of compliance obstacle type (see (5.4) below). They used this system instead of Assumption 5.1(iv) to obtain an upper bound for $u_h - I_h^{fd}u$ for the case when A^α is Lipschitz-continuous in the space variable. Their results show that $\|I_h^{fd}u - u_h\|_{L^\infty(\bar{\Omega}_h)} \leq Ch^{1/5}$.
- Let $\{\mathbf{e}_i\}_{i=1}^d$ be an orthonormal basis of $\mathbb{R}^d$ and assume that the matrices have the form

$$A^\alpha(x) = \sum_{i=1}^{d} a_i^\alpha(x)\mathbf{e}_i \otimes \mathbf{e}_i$$

  for some a_i^α that is Lipschitz in x uniformly in α. Recall that, with this assumption, Lemma 3.33 guarantees the existence of a monotone finite difference scheme. In this setting Krylov (2005) shows that $\kappa_1 = \kappa_2 = 1/2$. Moreover, in the same setting but assuming that the coefficients are $C^{1,1}$, Krylov (2014) shows that $\kappa_1 = \kappa_2 = 2/3$.
- All the aforementioned results consider the case $\Omega = \mathbb{R}^d$, so that boundary conditions are not an issue: see Theorem 3.8. Dong and Krylov (2007) consider the boundary value problem (5.1) under the assumption that the boundary conditions are attained classically; see Definitions 2.48 and 2.53. Under the assumption that a *barrier* function can be constructed, that is, a smooth b such that $b > 0$ in Ω, $b = 0$ on $\partial\Omega$ and

$$F[b](x) \leq -1,$$

  they were able to extend the results presented here and show that the rate of convergence is $\|I_h^{fd}u - u_h\|_{L^\infty(\bar{\Omega}_h)} \leq Ch^{1/2}$.

Other results, which invoke the probabilistic interpretation of (5.1), can be found in the literature; see, for instance, Fleming and Soner (2006) and Menaldi (1989). The reader is also referred to the introduction of Krylov (2015) for a detailed account of the development of error estimates for (5.1).

5.2. Finite element methods

We now focus on the construction and analysis of finite element schemes for (5.1). We will divide the exposition in two cases. First we will discuss the discretization for a variant of the problem when the operators $\tilde{\mathcal{L}}^\alpha$ are replaced by operators in divergence form $\tilde{L}^\alpha$ with smooth coefficients. Then we will discuss the case of (5.1) where the coefficients for $\tilde{\mathcal{L}}^\alpha$ satisfy the Cordes condition of Definition 2.32, which is based on the discretization of non-divergence-form operators of Section 4.1. We must remark that it is possible to construct discrete schemes using the integro-differential approximation of Section 4.3. However, to avoid repetition, its discussion will be illustrated in Section 7.3 for a non-convex operator of Isaacs type (see Example 2.13). Setting $\#\mathcal{B} = 1$ there, we reduce the scheme and its analysis to the case we are concerned with here.

5.2.1. Discretization for divergence-form operators

Let us consider (5.1) but where the operators $\tilde{\mathcal{L}}^\alpha$ are replaced by divergence-form elliptic operators $\tilde{L}^\alpha$ with $C^2(\bar{\Omega})$ coefficients; see Definition 2.3. In addition, we assume that $\partial\Omega$ is sufficiently smooth, $g = 0$ and that, for every $\alpha \in \mathcal{A}$, we have $0 \leq f^\alpha \in L^\infty(\Omega)$. Finally, we assume that $\mathcal{A} = \{1, \dots, M\}$ for some $M \in \mathbb{N}$.

Remark 5.6 (smooth coefficients). Note that, if the coefficients of $\tilde{\mathcal{L}}^\alpha$ in (5.1) are sufficiently smooth, one can rewrite this operator in divergence form. Therefore, this reformulation is sufficiently general.

Recall that, by integration by parts, to every operator $\tilde{L}^\alpha$ we can associate the bilinear form $a^\alpha : H^1(\Omega) \times H^1(\Omega) \to \mathbb{R}$, defined by

$$a^\alpha(v, w) = \int_\Omega (A^\alpha Dv \cdot Dw + v^\alpha \mathbf{b}^\alpha \cdot Dw + \mathbf{c}^\alpha \cdot Dvw + d^\alpha vw).$$

To simplify the discussion, we will assume that these forms are coercive uniformly in α, that is, there is a constant λ_0 such that

$$\inf_{\alpha \in \mathcal{A}} a^\alpha(v, v) \geq \lambda_0 \|Dv\|^2_{L^2(\Omega)}, \quad \text{for all } v \in H^1_0(\Omega).$$

Given $k > 0$ and $w \in H^1_0(\Omega)$, define

$$\mathcal{K}(k, w) = \{v \in H^1_0(\Omega) : v \leq k + w\},$$

which is a closed and convex subset of $H^1_0(\Omega)$. For $k > 0$ we introduce a system of quasi-variational inequalities of compliance obstacle type as follows: find $u^\alpha_k \in \mathcal{K}(k, u^{\alpha+1})$ such that

$$a^\alpha(u^\alpha_k, u^\alpha_k - v) \leq (f^\alpha, u^\alpha_k - v), \quad \text{for all } v \in \mathcal{K}(k, u^{\alpha+1}), \tag{5.4}$$

with $u^{M+1}_k = u^1_k$. Standard results on quasi-variational inequalities, like those presented in Bensoussan and Lions (1984, Chapter 4), imply the existence and uniqueness of $\mathbf{u}_k := \{u^\alpha_k\}_{\alpha \in \mathcal{A}} \subset H^1_0(\Omega)$ that solves (5.4). In addition, adaptions of the results of Section 2.3 to the case of (quasi-) variational inequalities (*i.e.* by penalization and a limiting argument) allow us to conclude that $\mathbf{u}_k \subset W^{1,\infty}(\Omega) \cap W^{2,p}_{loc}(\Omega)$ $(p < \infty)$. The purpose of this system lies in the fact that, as $k \to 0$, the solutions to (5.4) converge to the solution to (5.1). For a proof of the following result see Evans and Friedman (1979, Theorem 7.2) and Bensoussan and Lions (1984, Section 4.6).

Theorem 5.7 (convergence as $k \to 0$). In the setting described above there exists a unique strong solution u to (5.1). Moreover, defining $\mathbf{u} = \{u\}_{\alpha \in \mathcal{A}}$, we have

$$\lim_{k \to 0} \|\mathbf{u} - \mathbf{u}_k\|_{\ell^\infty(\mathcal{A}, L^\infty(\Omega))} = 0.$$

It is remarkable that this result was shown before the development of the viscosity solutions of Section 2, and it was originally used to show the well-posedness of the Hamilton–Jacobi–Bellman equations. In addition, it can be used to propose finite element discretizations of (5.1) by instead discretizing (5.4). Then, if one is able to extend the standard $L^\infty(\Omega)$-norm estimates for variational inequalities (see Nochetto, Otárola and Salgado 2015, Theorem 2.9) and give a rate for the limit in Theorem 5.7, we obtain a convergent (with rates) finite element method. This programme has been carried out by Cortey-Dumont (1987), Boulbrachene and Haiour (2001) and Boulbrachene and Cortey-Dumont (2009), with some success.

With the notation of Section 3.5 we begin by defining, for $k > 0$ and $w_h \in X^l_{0,h}$, the set

$$\mathcal{K}_h(k, w_h) = \{v_h \in X^l_{0,h} : v_h \le k + w_h\},$$

which is a closed and convex subset of $X^l_{0,h}$. Moreover, we remark that it is sufficient to impose the inequality at the nodes $z \in \Omega^I_h$. We approximate the solution to (5.4) by the following set of discrete quasi-variational inequalities: find $u_h^{\alpha,k} \in \mathcal{K}_h(k, u_h^{\alpha+1,k})$ such that

$$a^\alpha(u_h^{\alpha,k}, u_h^{\alpha,k} - v_h) \le (f^\alpha, u_h^{\alpha,k} - v_h), \quad \text{for all } v_h \in \mathcal{K}_h(k, u_h^{\alpha+1,k}), \tag{5.5}$$

with $u_h^{M+1,k} = u_h^{1,k}$. Once again, it can be shown that this problem always has a unique solution $\mathbf{u}_{h,k} = \{u_h^{\alpha,k}\}_{\alpha \in \mathcal{A}} \subset X^l_{0,h}$.

To establish the $L^\infty(\Omega)$-norm convergence of the solutions to (5.5) to the solution to (5.4), we must assume that the ensuing stiffness matrices are M-matrices. Examining the proof of Lemma 3.42, we realize that for this to hold, it is sufficient to require that

$$a^\alpha(\phi_i, \phi_j) \le 0, \quad \text{for all } \alpha \in \mathcal{A},\ i \ne j, \tag{5.6}$$

which we assume below.

Remark 5.8 (lack of generality). The discussion of Section 3.5 shows that condition (5.6) is satisfied if the mesh is weakly acute *in the metric induced by A^α for all $\alpha \in \mathcal{A}$*. This is a severe restriction in practice, as it is not clear how to impose such a condition for one matrix, let alone for a family of them.

Although a rate of convergence for the limit in Theorem 5.7 does not seem possible, there is a rate for the approximation of (5.4) by (5.5).

Theorem 5.9 (rate of convergence). Assume that, for all $h > 0$, the family of triangulations $\mathcal{T}_h$ satisfies condition (5.6). Then we have

$$\|\mathbf{u}_k - \mathbf{u}_{h,k}\|_{\ell^\infty(\mathcal{A},L^\infty(\Omega))} \le Ch^2 \, |\log h|^3,$$

where the constant $C > 0$ depends on $M = \#\mathcal{A}$ and k.

Proof. We will follow the ideas of Boulbrachene and Haiour (2001) which, in turn, borrow from the iterative schemes used to prove existence of elliptic quasi-variational inequalities considered in Bensoussan and Lions (1984, Chapter 4).

Step 1. We introduce the following iterative scheme. Define $\hat{\mathbf{u}}^0 = \{\hat{u}^{\alpha,0}\}_{\alpha\in\mathcal{A}}$ as the solutions to the unconstrained problems, that is,

$$a^\alpha(\hat{u}^{\alpha,0}, v) = (f^\alpha, v), \quad \text{for all } v \in H_0^1(\Omega).$$

Assuming that, for $n \ge 0$, $\hat{\mathbf{u}}^n$ has been defined, we look for $\hat{u}^{\alpha,n+1} \in \mathcal{K}(k, \hat{u}^{\alpha+1,n})$ such that

$$a^\alpha(\hat{u}^{\alpha,n+1}, \hat{u}^{\alpha,n+1} - v) \le (f^\alpha, \hat{u}^{\alpha,n+1} - v), \quad \text{for all } v \in \mathcal{K}(k, \hat{u}^{\alpha+1,n}).$$

Using the positivity and order-preserving properties of the associated map, it is possible then to show that for $\lambda < \min\{1, k/\|\hat{\mathbf{u}}^0\|_{\ell^\infty(\mathcal{A},L^\infty(\Omega))}\}$ we have

$$\|\mathbf{u}_k - \hat{\mathbf{u}}^n\|_{\ell^\infty(\mathcal{A},L^\infty(\Omega))} \le (1-\lambda)^n \|\hat{\mathbf{u}}^0\|_{\ell^\infty(\mathcal{A},L^\infty(\Omega))}. \tag{5.7}$$

Step 2. Define $\tilde{\mathbf{u}}_h^0 = \{\tilde{u}_h^{\alpha,0}\}_{\alpha\in\mathcal{A}}$ as the finite element approximation of the unconstrained problems, that is,

$$a^\alpha(\tilde{u}_h^{\alpha,0}, v_h) = (f^\alpha, v_h), \quad \text{for all } v_h \in X_{0,h}^l.$$

Using (5.6), we can invoke standard finite element error estimates for linear problems to obtain

$$\|\hat{\mathbf{u}}^0 - \tilde{\mathbf{u}}_h^0\|_{\ell^\infty(\mathcal{A},L^\infty(\Omega))} \le Ch^2 \, |\log h|^2. \tag{5.8}$$

For each $n \ge 0$ we define $\tilde{u}_h^{\alpha,n+1} \in \mathcal{K}_h(k, \hat{u}^{\alpha+1,n})$ as the solution of

$$a^\alpha(\tilde{u}_h^{\alpha,n+1}, \tilde{u}_h^{\alpha,n+1} - v_h) \le (f^\alpha, \tilde{u}_h^{\alpha,n+1} - v_h), \quad \text{for all } v_h \in \mathcal{K}_h(k, I_h^{fe}\hat{u}^{\alpha+1,n}).$$

Note that this is simply the finite element approximation of $\hat{u}^{\alpha+1,n}$ as the solution to an obstacle problem. Once again using (5.6), we can invoke pointwise estimates for obstacle problems (see Baiocchi 1977, and Theorem 2.9 of Nochetto, Otárola and Salgado 2015) to conclude that

$$\|\hat{\mathbf{u}}^n - \tilde{\mathbf{u}}_h^n\|_{\ell^\infty(\mathcal{A},L^\infty(\Omega))} \le Ch^2 \, |\log h|^2, \tag{5.9}$$

where the constant is independent of n.

Step 3. Introduce a discrete iterative scheme analogous to that given in Step 1. In other words, set $\hat{\mathbf{u}}_h^0 = \tilde{\mathbf{u}}_h^0$ and, for $n \geq 0$, find $\hat{u}_h^{\alpha,n+1} \in \mathcal{K}_h(k, \hat{u}_h^{\alpha+1,n})$ as the solution of

$$a^\alpha(\hat{u}_h^{\alpha,n+1}, \hat{u}_h^{\alpha,n+1} - v_h) \leq (f^\alpha, \hat{u}_h^{\alpha,n+1} - v_h), \quad \text{for all } v \in \mathcal{K}_h(k, \hat{u}_h^{\alpha+1,n}).$$

Similar techniques to those that led to (5.7) yield

$$\|\mathbf{u}_{h,k} - \hat{\mathbf{u}}_h^n\|_{\ell^\infty(\mathcal{A},L^\infty(\Omega))} \leq (1-\lambda)^n \|\hat{\mathbf{u}}_h^0\|_{\ell^\infty(\mathcal{A},L^\infty(\Omega))}. \tag{5.10}$$

Step 4. By induction, it can be shown that

$$\|\hat{\mathbf{u}}^n - \hat{\mathbf{u}}_h^n\|_{\ell^\infty(\mathcal{A},L^\infty(\Omega))} \leq \sum_{k=0}^{n} \|\hat{\mathbf{u}}^k - \tilde{\mathbf{u}}_h^k\|_{\ell^\infty(\mathcal{A},L^\infty(\Omega))}.$$

With this at hand, the triangle inequality yields

$$\begin{aligned}\|\mathbf{u}_k - \mathbf{u}_{h,k}\|_{\ell^\infty(\mathcal{A},L^\infty(\Omega))} &\leq \|\mathbf{u}_k - \hat{\mathbf{u}}^n\|_{\ell^\infty(\mathcal{A},L^\infty(\Omega))} + \|\hat{\mathbf{u}}_h^n - \mathbf{u}_{h,k}\|_{\ell^\infty(\mathcal{A},L^\infty(\Omega))} \\ &\quad + \sum_{k=0}^{n} \|\hat{\mathbf{u}}^k - \tilde{\mathbf{u}}_h^k\|_{\ell^\infty(\mathcal{A},L^\infty(\Omega))},\end{aligned}$$

so that by using (5.7)–(5.10) we obtain

$$\|\mathbf{u}_k - \mathbf{u}_{h,k}\|_{\ell^\infty(\mathcal{A},L^\infty(\Omega))} \leq C\big[(1-\lambda)^n + nh^2\,|\log h|^2\big].$$

Now choose n so that $(1-\lambda)^n \approx h^2$ to obtain the result. □

Remark 5.10 ($k = 0$). An algorithm similar to that used in the proof of Theorem 5.9 is studied in Lions and Mercier (1980, Algorithme I) for $k = 0$. It is shown there that the ensuing iterates converge monotonically to $\mathbf{u}$, but no rate is given.

5.2.2. Discretization for HJB satisfying the Cordes condition

Let us now discuss the case of (5.1) with non-divergence-form operators without lower-order terms, that is, $\mathcal{L}^\alpha$. For simplicity, we set $g = 0$. More importantly, we will assume that the coefficient matrices A^α satisfy the Cordes condition of Definition 2.32. Furthermore, we will assume that the domain Ω is convex and, finally, that $\mathcal{A}$ is a compact metric space.

The stated assumptions imply, invoking Theorem 2.39, that each one of the operators $\mathcal{L}^\alpha$ is an isomorphism between $H^2(\Omega) \cap H_0^1(\Omega)$ and $L^2(\Omega)$. Consequently, we can apply the techniques of Section 4.1 to each one. Now, following the arguments of Smears and Süli (2014), let us show that these assumptions also imply that problem (5.1) is also well-posed and that its solution is strong.

To do so we must assume that the Cordes condition holds uniformly in $\mathcal{A}$ which, from (2.16), yields the existence of $\epsilon > 0$ for which

$$\sup_{\alpha\in\mathcal{A}} \|\gamma^\alpha \mathcal{L}^\alpha v - \Delta v\|_{L^2(\Omega)} \le \sqrt{1-\epsilon}\|D^2 v\|_{L^2(\Omega)}, \quad \text{for all } v \in H^2(\Omega). \tag{5.11}$$

This inequality motivates the definition of the (elliptic) nonlinear operator

$$F_\gamma[v] := \sup_{\alpha\in\mathcal{A}} \bigl[\gamma^\alpha(\mathcal{L}^\alpha v - f^\alpha)\bigr]. \tag{5.12}$$

The equivalence between problem $F_\gamma[u] = 0$ and (5.1) essentially follows from the continuity of the data and the positivity of γ^α. This result is summarized in the next lemma.

Lemma 5.11 (equivalence). Suppose that f^α and A^α are uniformly continuous for each $\alpha \in \mathcal{A}$, $\mathcal{A}$ is compact, and Ω is convex. The function $u \in H^2(\Omega) \cap H_0^1(\Omega)$ satisfies $F_\gamma[u] = 0$ a.e. Ω (*i.e.* u is a strong solution to this problem) if and only if it is a strong solution to (5.1).

Identity (5.11) and the algebraic identity $|\sup_{\alpha\in\mathcal{A}} x^\alpha - \sup_{\alpha\in\mathcal{A}} y^\alpha| \le \sup_{\alpha\in\mathcal{A}} |x^\alpha - y^\alpha|$ for bounded sequences $\{x^\alpha\}, \{y^\alpha\} \subset \mathbb{R}$ then yields the following result.

Lemma 5.12 (continuity). For all $u, v \in H^2(\Omega)$ we have

$$|F_\gamma[v] - F_\gamma[w] - \Delta(v-w)| \le \sqrt{1-\epsilon}|D^2(v-w)|.$$

Define

$$G(v,w) := \int_\Omega F_\gamma[v]\Delta w, \quad \text{for all } v, w \in H^2(\Omega) \cap H_0^1(\Omega).$$

The estimate of Lemma 5.12 leads to the following (strong) monotonicity property of G.

Theorem 5.13 (properties of G). Under the given assumptions, there is a positive constant C for which

$$G(v, v-w) - G(w, v-w) \ge C(1 - \sqrt{1-\epsilon})\|v-w\|_{H^2(\Omega)}^2 \tag{5.13}$$

for all $v, w \in H^2(\Omega) \cap H_0^1(\Omega)$. In addition, if $A^\alpha \in C(\bar{\Omega}, \mathbb{S}^d)$ and $f^\alpha \in C(\bar{\Omega})$, then

$$|G(v,u) - G(w,u)| \le C\|v-w\|_{H^2(\Omega)}\|u\|_{H^2(\Omega)}. \tag{5.14}$$

Proof. Applying Lemma 5.12 yields

$$\begin{aligned}
G(v, v-w) - G(w, v-w) &= \int_\Omega (F_\gamma[v] - F_\gamma[w])\Delta(v-w) \\
&= \|\Delta(v-w)\|_{L^2(\Omega)}^2 + \int_\Omega (F_\gamma[v] - F_\gamma[w] - \Delta(v-w))\Delta(v-w) \\
&\ge \|\Delta(v-w)\|_{L^2(\Omega)}^2 - \sqrt{1-\epsilon}\|D^2(v-w)\|_{L^2(\Omega)}\|\Delta(v-w)\|_{L^2(\Omega)}.
\end{aligned}$$

Since Ω is convex, we can apply the Miranda–Talenti estimate (2.17) to get

$$G(v, v-w) - G(w, v-w) \geq (1-\sqrt{1-\epsilon})\|\Delta(v-w)\|^2_{L^2(\Omega)}.$$

Inequality (5.13) now follows from the equivalence of $\|\Delta\cdot\|_{L^2(\Omega)}$ and $\|\cdot\|_{H^2(\Omega)}$ on convex domains.

Finally, the Lipschitz property (5.14) follows from the continuity of the data and the Cauchy–Schwarz inequality. □

Theorem 5.13 and Lemma 5.11, along with the Browder–Minty theorem and the fact that $\Delta : H^2(\Omega)\cap H^1_0(\Omega) \to L^2(\Omega)$ is surjective on convex domains, yield the following existence and uniqueness result for the HJB problem (5.1); see Smears and Süli (2014, Theorem 3) for details.

Theorem 5.14 (existence and uniqueness). Suppose that $\mathcal{A}$ is a compact metric space, $f^\alpha \in C(\bar{\Omega})$, and $A^\alpha \in C(\bar{\Omega}, \mathbb{S}^d)$ satisfies the Cordes condition for each $\alpha \in \mathcal{A}$. Then there exists a unique strong solution $u \in H^2(\Omega)\cap H^1_0(\Omega)$ to $F_\gamma[u] = 0$. Further, u is also the unique strong solution to (5.1).

Let us now discuss how the finite element methods for linear problems given in Section 4.1 are extended to the nonlinear problem (5.1). While any of the methods given in that section can be adopted for the nonlinear problem, here we focus on the DG approximation (Smears and Süli 2013, 2014, 2016).

Recall that X_h^{dg}, defined by (4.11), is the piecewise polynomial space of degree k with respect to a conforming and shape-regular, simplicial partition of Ω. As in Section 4.1, we will only consider the h-version of the method, where the polynomial degree is globally fixed, and we will not follow the dependence on the polynomial degree k of all the ensuing constants. We refer the reader to Smears and Süli (2014), who consider hp-DG methods.

The DG method for the linear problem (4.17) extends to the nonlinear problem (5.1) by essentially taking the supremum over $\mathcal{A}$. This yields the following scheme: find $u_h \in X_h^{dg}$ such that

$$G_h^{dg}(u_h, v_h) := \sum_{K\in\mathcal{T}_h}\int_K [F_\gamma[u_h] - \Delta u_h]\Delta v_h + \frac{1}{2}B_h(u_h, v_h) = 0, \tag{5.15}$$

where $B_h(\cdot,\cdot)$ is the bilinear form defined by (4.14) with penalty parameter $\mu > 0$. Recall from Section 4.1 that if $u \in H^s(\Omega)\cap H^1_0(\Omega)$ with $s > 5/2$, then

$$B_h(u, v_h) = 2\sum_{K\in\mathcal{T}_h}\int_K \Delta u \Delta v_h, \quad \text{for all } v_h \in X_h^{dg}.$$

Therefore (5.15) is a consistent method, that is, $G_h^{dg}(u, v_h) = 0$ for all $v_h \in X_h^{dg}$ provided u is sufficiently smooth. In addition, as in the linear case, the coercivity properties of B_h ensure that the structural properties found at the continuous level carry over. That is, we have the following result (see Smears and Süli 2014, Theorem 7).

Theorem 5.15 (properties of G_h^{dg}). Suppose that the hypotheses in Theorem 5.13 are satisfied. Then there exists $\mu_* = O(\epsilon^{-1})$ such that if $\mu \geq \mu_*$, then

$$C\|v_h - w_h\|_{DG(1)}^2 \leq G_h^{dg}(v_h, v_h - w_h) - G_h^{dg}(w_h, v_h - w_h),$$

for all $v_h, w_h \in X_h^{dg}$, where the discrete $H^2(\Omega)$-norm $\|\cdot\|_{DG(1)}$ is defined by (4.15). Moreover, there exists a constant $C > 0$ such that, for all $v_h, w_h, u_h \in X_h^{cg}$,

$$|G_h^{dg}(v_h, u_h) - G_h^{dg}(w_h, u_h)| \leq C\|v_h - w_h\|_{DG(1)}\|u_h\|_{DG(1)}.$$

Similar to the continuous setting, Theorem 5.15 yields existence and uniqueness for the DG method.

Corollary 5.16 (existence and uniqueness). If the hypotheses of Theorem 5.15 hold, then there exists a unique $u_h \in X_h^{dg}$ satisfying (5.15).

Finally, stability/monotonicity and the consistency of the method lead to the following error estimates. We refer to Smears and Süli (2014, Theorem 8) for a proof.

Theorem 5.17 (rates of convergence). Suppose that the hypotheses of Theorem 5.15 hold. Let $u_h \in X_h^{dg}$ be the solution to (5.15), and suppose that the solution to (5.1) has regularity $u \in H^s(\Omega)$ with $5/2 < s \leq k + 1$. Then the error satisfies

$$\|u - u_h\|_{DG(1)} \leq Ch^{s-2}\|u\|_{H^s(\Omega)}.$$

5.2.3. Finite element methods for parabolic isotropic HJB problems

In this section we depart from the elliptic framework and discuss the finite element method for *parabolic* (time-dependent) Hamilton–Jacobi–Bellman problems developed by Jensen and Smears (2013*b*). Assuming homogeneous Dirichlet boundary conditions, we consider numerical approximations of the problem:

$$\begin{cases} \dfrac{\partial u}{\partial t} - \sup\limits_{\alpha \in \mathcal{A}}[\tilde{\mathcal{L}}^\alpha u - f^\alpha] = 0 & \text{in } \Omega \times (0, T), \\ u = 0 & \text{on } \partial\Omega \times (0, T), \\ u = u_0 & \text{on } \bar{\Omega} \times \{0\}, \end{cases} \tag{5.16}$$

where $u_0 \in C(\bar{\Omega})$ is the initial time condition and T is the (finite) end-time. As before, we assume that $\mathcal{A}$ is compact and that the coefficients are uniformly continuous in $\bar{\Omega}$. In addition, we assume that the coefficients are time-independent, $f^\alpha \geq 0$, $c^\alpha \leq 0$, $u_0 \geq 0$, and, most importantly, the elliptic operator $\mathcal{L}^\alpha$ is isotropic, that is,

$$\tilde{\mathcal{L}}^\alpha u(x) = a^\alpha(x)\Delta u(x) + \mathbf{b}^\alpha(x) \cdot Du(x) + c^\alpha(x)u(x),$$

with $a^\alpha \geq 0$. Note that, even in the isotropic case, each elliptic operator is in non-divergence form.

Before discussing the finite element method let us first extend the definition of viscosity solutions to the parabolic setting (compare Definition 2.48).

Definition 5.18 (viscosity solution).

(a) A function $u_\star \in USC(\bar{\Omega} \times [0,T])$ is a viscosity subsolution to (5.16) if $u_\star \leq 0$ on $\partial\Omega \times (0,T)$, $u_\star \leq u_0$ on $\bar{\Omega} \times \{0\}$, and if, whenever

$$(x_0, t_0) \in \Omega \times (0,T), \quad \varphi \in C^2((0,T) \times \Omega)$$

and $u_\star - \varphi$ has a local maximum at (x_0, t_0), we have that

$$\frac{\partial \varphi}{\partial t}(x_0, t_0) - \sup_{\alpha \in \mathcal{A}}(\tilde{\mathcal{L}}^\alpha(x_0)\varphi(x_0, t_0) - f^\alpha(x_0)) \leq 0.$$

(b) A function $u^\star \in LSC(\bar{\Omega} \times [0,T])$ is a viscosity supersolution to (5.16) if $u^\star \geq 0$ on $\partial\Omega \times (0,T)$, $u^\star \geq u_0$ on $\bar{\Omega} \times \{0\}$, and if, whenever

$$(x_0, t_0) \in \Omega \times (0,T), \quad \varphi \in C^2((0,T) \times \Omega)$$

and $u_\star - \varphi$ has a local minimum at (x_0, t_0), we have that

$$\frac{\partial \varphi}{\partial t}(x_0, t_0) - \sup_{\alpha \in \mathcal{A}}(\tilde{\mathcal{L}}^\alpha(x_0)\varphi(x_0, t_0) - f^\alpha(x_0)) \geq 0.$$

(c) A function $u \in C(\bar{\Omega} \times [0,T])$ is a viscosity solution to (5.16) if it is both a sub- and supersolution.

The discretization of the non-divergence part of $\tilde{\mathcal{L}}^\alpha$ proposed by Jensen and Smears (2013*b*) is based on a 'freezing the coefficients' strategy. Let $\{\phi_i\} \subset X^l_{0,h}$ be the normalized hat functions defined in Section 3.5, where we recall that $X^l_{0,h}$ is the space of piecewise linear polynomials with vanishing trace (see (3.35)). Then, since ϕ_i is essentially a regularized Dirac delta distribution, if a point $x \in \Omega$ is close to a node $z_i \in \Omega^I_h$, then we have, at least formally,

$$a^\alpha(x)\Delta u(x) \approx -a^\alpha(z_i) \int_\Omega Du \cdot D\phi_i = a^\alpha(z_i)\Delta_h I^{ep}_h u(z_i),$$

where the finite element Laplacian Δ_h is given by (3.39), and I^{ep}_h is the elliptic projection. This heuristic approximation motivates the following semi-discrete finite element method: find $u_h : [0,T] \to X^l_{0,h}$ such that

$u_h(0) = I_h^{fe} u_0$ and

$$\frac{\partial u_h}{\partial t} - \sup_{\alpha\in\mathcal{A}}(\tilde{\mathcal{L}}_h^\alpha u_h - f_h^\alpha) = 0 \quad \text{in } \Omega_h^I \times (0,T), \tag{5.17}$$

where $I_h^{fe} : C^0(\bar{\Omega}) \to X_h^l$ is the nodal interpolant onto X_h^l, and the discrete operator and discrete source function are given respectively by

$$\begin{aligned}\tilde{\mathcal{L}}_h^\alpha u_h(z_i) &= a^\alpha(z_i)\Delta_h u_h(z_i) + \int_\Omega (\mathbf{b}^\alpha \cdot Du_h + c^\alpha u_h)\phi_i, \\ f_h^\alpha(z_i) &= \int_\Omega f^\alpha \phi_i.\end{aligned}$$

The fully discrete method proposed in Jensen and Smears (2013*b*) applies a one-step explicit–implicit ODE solver to a regularized version of (5.17). Let $\tau \in (0,1)$ be the (uniform) time step size and assume that $T/\tau =: M \in \mathbb{N}$. For a sequence of discrete functions $\{v_h^k\}_{k=0}^M \subset X_{0,h}^l$, we define the backward difference quotient as

$$d_\tau v_h^{k+1} := \frac{1}{\tau}(v_h^{k+1} - v_h^k) \in X_{0,h}^l.$$

For each $\alpha \in \mathcal{A}$, the discrete operator $\tilde{\mathcal{L}}_h^\alpha$ is approximately split into an explicit and implicit part,

$$\tilde{\mathcal{L}}_h^\alpha \approx \mathcal{E}_h^\alpha + \mathcal{I}_h^\alpha, \tag{5.18}$$

with

$$\begin{aligned}\mathcal{E}_h^\alpha u_h(z_i) &= a_e^\alpha(z_i)\Delta_h u_h(z_i) + \int_\Omega (\mathbf{b}_e^\alpha \cdot Du_h + c_e u_h)\phi_i, \\ \mathcal{I}_h^\alpha u_h(z_i) &= a_i^\alpha(z_i)\Delta_h u_h(z_i) + \int_\Omega (\mathbf{b}_i^\alpha \cdot Du_h + c_i u_h)\phi_i.\end{aligned}$$

We now consider the following fully discrete method: find the sequence $\{u_h^k\}_{k=0}^M \subset X_{0,h}^l$ with $u_h^0 = I_h^{fe} u_0$ and

$$d_\tau u_h^{k+1} - \sup_{\alpha\in\mathcal{A}}(\mathcal{E}_h^\alpha u_h^k + \mathcal{I}_h^\alpha u_h^{k+1} - f_h^\alpha) \quad \text{in } \Omega_h^I. \tag{5.19}$$

To show the well-posedness and to analyse method (5.19), we make several assumptions. First we quantify the approximation in (5.18) and make assumptions of the coefficients in the explicit and implicit operators.

Assumption 5.19 (coefficients). We assume that the coefficients satisfy

$$\begin{aligned}\lim_{h\to 0} \sup_{\alpha\in\mathcal{A}} \Big(&\sup_{z\in\Omega_h^I} \|a^\alpha - (a_e^\alpha(z) + a_i^\alpha(z))\|_{L^\infty(\omega_z)} \\ &+ \|\mathbf{b}^\alpha - (\mathbf{b}_e^\alpha + \mathbf{b}_i^\alpha)\|_{L^\infty(\Omega)} + \|c^\alpha - (c_e^\alpha + c_i^\alpha)\|_{L^\infty(\Omega)}\Big) = 0.\end{aligned}$$

In addition, when $\mathcal{A}$ is not finite, we also assume that the mappings $\alpha \mapsto (a_e^\alpha, \mathbf{b}_e^\alpha, c_i^\alpha)$ and $\alpha \mapsto (a_i^\alpha, \mathbf{b}_i^\alpha, c_i^\alpha)$ are continuous, $a_e, a_i \geq 0$, and $c_e, c_i \leq 0$.

Next we make assumptions regarding the monotonicity properties of the operators.

Assumption 5.20 (monotonicity). The splitting (5.18) has the following properties.

(a) The time-step explicit operators satisfy

$$\delta_{i,j} + \tau \mathcal{E}_h^\alpha(\tilde{\phi}_j)(z_i) \geq 0, \quad \text{for all } \alpha \in \mathcal{A}, \tag{5.20}$$

where $\{\tilde{\phi}_j\} \subset X_{0,h}^l$ are the (unnormalized) hat functions satisfying $\tilde{\phi}_j(z_i) = \delta_{i,j}$.

(b) The operator $\mathcal{I}_h^\alpha$ is monotone (in the sense of Definition 3.5) for each $\alpha \in \mathcal{A}$.

Condition (5.20) is essentially a time-step restriction and a monotonicity condition (if $\mathcal{E}_h^\alpha \not\equiv 0$ for all $\alpha \in \mathcal{A}$). For example, if $c_e^\alpha \equiv 0$, $\mathbf{b}_e^\alpha \equiv 0$, and $a_e^\alpha > 0$ for all $\alpha \in \mathcal{A}$, then the condition reads

$$\delta_{i,j} - \tau a_e^\alpha(z_i) \int_\Omega D\tilde{\phi}_j \cdot D\phi_i \geq 0.$$

Therefore, in this setting, (5.20) is satisfied if and only if (3.42) holds (so that Δ_h is monotone) and if

$$\tau \leq \left(a_e^\alpha(z_i) \int_\Omega D\phi_i \cdot D\tilde{\phi}_i \right)^{-1} = O(h^2).$$

The monotonicity of $\mathcal{I}_h^\alpha$ implies that the inverse of $(I - \tau\mathcal{I}_h^\alpha)$ is sign-preserving for any $\tau > 0$, that is, if $(I - \tau\mathcal{I}_h^\alpha)v_h \leq 0$ (≥ 0), then $v_h \leq 0$ (≥ 0). To see this, suppose that $v_h \in X_{0,h}^l$ satisfies $(I - \tau\mathcal{I}_h^\alpha)v_h \leq 0$ and v_h has a positive maximum at z_i. Then the monotonicity of $\mathcal{I}_h^\alpha$ implies $\mathcal{I}_h^\alpha v_h(z_i) \leq 0$ (see Definition 3.5), leading to $(I - \tau\mathcal{I}_h^\alpha)v_h(z_i) \geq v_h(z_i) > 0$, a contradiction. This argument also shows that $(I - \tau\mathcal{I}_h^\alpha)$ is invertible, and therefore the following problem is well-posed. For fixed $\alpha \in \mathcal{A}$, find $\{u_{h,\alpha}^k\}_{k=0}^M \subset X_{0,h}^l$ such that $u_{h,\alpha}^0 = I_h^{fe} u_0$ and $(k = 0, 1, \ldots, M-1)$

$$d_\tau u_{h,\alpha}^{k+1} - (\mathcal{E}_h^\alpha u_{h,\alpha}^k + \mathcal{I}_h^\alpha u_{h,\alpha}^{k+1} - f_h) \text{ in } \Omega_h^I, \quad u_{h,\alpha}^0 = I_h^{fe} u_0. \tag{5.21}$$

The construction of operators $\mathcal{E}_h^\alpha$, $\mathcal{I}_h^\alpha$ satisfying Assumptions 5.19–5.20 (on weakly acute triangulations) using the method of artificial diffusion can be found in Jensen and Smears (2013*b*, Section 8) and Jensen and Smears (2013*a*).

Theorem 5.21 (existence). Suppose that Assumptions 5.19 and 5.20 are satisfied. Then for each $k \in \{0, 1, \ldots, M-1\}$ there exists a unique solution $u_h^{k+1} \in X_{0,h}^l$ to (5.19). Moreover, if we let $\{u_{h,\alpha}^k\} \subset X_{0,h}^l$ denote the solution to (5.21), then $0 \le u_h^k \le u_{h,\alpha}^k$.

The proof of existence and uniqueness follows from the convergence results of Howard's method, which is discussed in the next section. The monotonicity and continuity properties of $\mathcal{E}_h^\alpha$, $\mathcal{I}_h^\alpha$ ensure that the hypotheses of Theorem 5.31 below are satisfied (see Jensen and Smears 2013*b*, Theorem 3.1).

The next result establishes the stability of the method. Its proof essentially follows from Assumption 5.20 and the non-positivity of c_e^α and c_i^α; see Jensen and Smears (2013*b*, Lemma 3.2 and Corollary 3.3) for details.

Lemma 5.22 (stability). Suppose that the assumptions of Theorem 5.21 hold, and let $\{u_{h,\alpha}^k\} \subset X_{0,h}^l$ denote the solution to (5.21) for a fixed $\alpha \in \mathcal{A}$. Then we have

$$\|\{u_{h,\alpha}^k\}_{k=0}^M\|_{\ell^\infty(\mathbb{N}\cap[0,M],L^\infty(\Omega))} \le \|I_h^{fe} u_0\|_{L^\infty(\Omega)} + T\|f^\alpha\|_{L^\infty(\Omega)}.$$

Therefore, by Theorem 5.21,

$$\|\{u_h^k\}_{k=0}^M\|_{\ell^\infty(\mathbb{N}\cap[0,M],L^\infty(\Omega))} \le \|I_h^{fe} u_0\|_{L^\infty(\Omega)} + T \inf_{\alpha\in\mathcal{A}} \|f^\alpha\|_{L^\infty(\Omega)}.$$

Next, to apply the Barles–Souganidis theory, we look at the consistency of the method. Essentially this result follows from Lemma 3.41 and the stability of the elliptic projection.

Lemma 5.23 (consistency). Let $I_h^{ep} : H^1(\Omega) \to X_h^l$ be the elliptic projection onto X_h^l. Let $\{(z_h, t_\tau)\}_{h>0,\tau>0}$ with $(z_h, t_\tau) \in \bar{\Omega}_h \times \mathbb{N} \cap [0, M]$ and $(z_h, t_\tau) \to (z_0, t_0) \in \bar{\Omega} \times [0, T]$ as $h, \tau \to 0^+$. Then, for any $\phi \in C^2(\bar{\Omega} \times [0, T])$,

$$\begin{aligned}\lim_{h,\tau\to 0^+} &\big(d_\tau I_h^{ep}\phi(t_\tau+\tau) - (\mathcal{E}_h^\alpha I_h^{ep}(\phi(t_\tau)) + \mathcal{I}_h^\alpha I_h^{ep}(\phi(t_\tau+\tau)) - f_h^\alpha)\big)(z_h) \\ &= \frac{\partial\phi}{\partial t}(z_0, t_0) - (\tilde{\mathcal{L}}^\alpha \phi(z_0, t_0) - f^\alpha(z_0)), \quad \text{for all } \alpha \in \mathcal{A},\end{aligned}$$

where $\phi(t) := \phi(\cdot, t)$, and the convergence is uniform with respect to $\alpha \in \mathcal{A}$.

Remark 5.24 (anisotropy). Note that the consistency result given in Lemma 5.23 does not extend to operators with anisotropic diffusion.

Finally to apply Theorem 3.8, the following assumption is made, which ensures that the limiting solution satisfies the boundary conditions in a classical sense.

Assumption 5.25 (limiting boundary values). Define $\bar{u}_\alpha \in USC(\bar{\Omega} \times [0,T])$:

$$\bar{u}_\alpha(x,t) := \limsup_{\substack{(z,k\tau)\to(x,t)\\ h,\tau\to 0^+}} u^k_{h,\alpha}(z),$$

where $u^k_{h,\alpha}$ is the solution to (5.21). Then

$$\inf_{\alpha\in\mathcal{A}} \bar{u}_\alpha(x,t) = 0, \quad \text{for all } (x,t) \in \partial\Omega \times [0,T].$$

Finally, combining Lemmas 5.23 and 5.22 with Theorem 3.8 yields the following convergence result; see Jensen and Smears (2013*b*, Theorem 6.2).

Theorem 5.26 (convergence). Suppose that Assumptions 5.19, 5.20 and 5.25 are satisfied. Suppose further that the HJB problem (5.16) satisfies a comparison principle (see Definition 2.49). Then

$$\lim_{h,\tau\to 0^+} \max_k \|u(\cdot,k\tau) - u^k_h\|_{L^\infty(\Omega)} = 0.$$

5.3. Solution of the discrete problems

To finalize the discussion concerning the approximation of convex equations, we must, at least briefly, discuss how to solve the system of nonlinear equations that arises after discretization, whether by finite differences or finite elements. We will present one of the most popular methods – known as policy iterations or Howard's algorithm – and discuss its convergence properties.

Once (5.1) is discretized by any of the methods described in the previous sections, we end up with the following system of nonlinear equations: find $\mathbf{x} \in \mathbb{R}^N$ that satisfies

$$\mathbf{F}(\mathbf{x}) = \sup_{\alpha\in\mathcal{A}} [\mathbf{K}^\alpha \mathbf{x} - \mathbf{f}^\alpha] = \mathbf{0}, \tag{5.22}$$

where the supremum is taken component-wise, N is the number of degrees of freedom in the discretization, $\{\mathbf{K}^\alpha\}_{\alpha\in\mathcal{A}}$ and $\{\mathbf{f}^\alpha\}_{\alpha\in\mathcal{A}}$ are discretizations of $\{\tilde{\mathcal{L}}^\alpha\}_{\alpha\in\mathcal{A}}$ and $\{f^\alpha\}_{\alpha\in\mathcal{A}}$, respectively, or as in (5.19), come from the implicit-explicit splitting of the operators (5.18).

In order to define Howard's algorithm we must introduce the following operations on $\mathcal{A}$, the matrices $\mathbf{K}^\alpha$ and vectors $\mathbf{f}^\alpha$. Given $\mathbf{y} \in \mathbb{R}^N$ and $i \in \{1,\ldots,N\}$ we define the element $\alpha(\mathbf{y},i) \in \mathcal{A}$ as the element that realizes the supremum in (5.22) when applied to $\mathbf{y}$ at component i, that is,

$$[\mathbf{K}^{\alpha(\mathbf{y},i)}\mathbf{y} - \mathbf{f}^{\alpha(\mathbf{y},i)}]_i = \sup_{\alpha\in\mathcal{A}} [\mathbf{K}^\alpha \mathbf{y} - \mathbf{f}^\alpha]_i.$$

We define $\boldsymbol{\alpha}(\mathbf{y}) \in \mathcal{A}^N$ with components $\boldsymbol{\alpha}(\mathbf{y})_i = \boldsymbol{\alpha}(\mathbf{y},i)$. Given $\boldsymbol{\alpha}(\mathbf{y})$ we

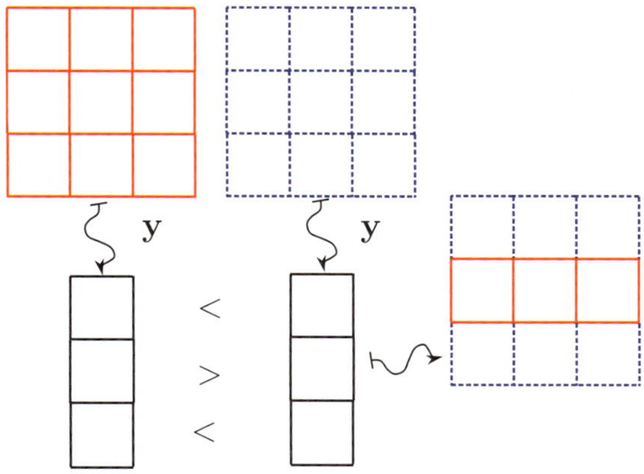

Figure 5.1. An illustration of the selection procedure defined in (5.23), for the simplified case of $\mathbf{f}^\alpha = \mathbf{0}$ for all $\alpha \in \mathcal{A}$ and $\#\mathcal{A} = 2$. We multiply each matrix (solid red and dashed blue in the figure) by the vector $\mathbf{y}$ and compare the components of the results. The matrix $\mathbf{K}^{\boldsymbol{\alpha}(\mathbf{y})}$ is constructed by choosing the row of the matrix that gives the largest result.

finally define the matrix $\mathbf{K}^{\boldsymbol{\alpha}(\mathbf{y})}$ and vector $\mathbf{f}^{\boldsymbol{\alpha}(\mathbf{y})}$ as follows:

$$\mathbf{K}^{\boldsymbol{\alpha}(\mathbf{y})}_{i,j} = \mathbf{K}^{\boldsymbol{\alpha}(\mathbf{y},i)}_{i,j}, \quad \mathbf{f}^{\boldsymbol{\alpha}(\mathbf{y})}_{i} = \mathbf{f}^{\boldsymbol{\alpha}(\mathbf{y},i)}_{i}. \tag{5.23}$$

An illustration of this selection procedure for $\#\mathcal{A} = 2$ is shown in Figure 5.1. With these operations at hand, Howard's algorithm, as presented in Bokanowski, Maroso and Zidani (2009), is described in Algorithm 5.1.

Note that at each step Algorithm 5.1 requires the solution to the linear system (5.24). Because of the way the matrices are constructed, it is necessary to ensure that they are invertible. This is guaranteed by the following condition.

Assumption 5.27 (monotonicity). For every $\mathbf{y} \in \mathbb{R}^N$ the matrix $\mathbf{K}^{\boldsymbol{\alpha}(\mathbf{y})}$ is monotone in the sense of Definition 3.5.

It is natural to wonder when such a condition is satisfied. As shown by Witte and Reisinger (2012), if, for every $\alpha \in \mathcal{A}$, the matrix $\mathbf{K}^\alpha$ is strictly diagonally dominant and monotone and its off-diagonal entries are non-positive, then this condition is satisfied. We comment finally that a monotone matrix is non-singular.

Having guaranteed that Howard's algorithm can proceed, we focus our attention on its convergence. While many references have studied it, we will follow Bokanowski *et al.* (2009) and Hintermüller, Ito and Kunisch (2002), and draw a connection between this method and active set strategies which, in turn, can be analysed as semismooth Newton methods. We begin with the definition of slant derivative.

Algorithm 5.1: Howard's algorithm.

Input : Set $\mathcal{A}$.
Matrices $\{\mathbf{K}^\alpha\}_{\alpha\in\mathcal{A}} \subset \mathbb{R}^{N\times N}$.
right-hand sides $\{\mathbf{f}^\alpha\}_{\alpha\in\mathcal{A}} \subset \mathbb{R}^N$.
Output: Vector $\mathbf{x} \in \mathbb{R}^N$, solution of (5.22).

Initialization: Choose $\mathbf{x}_{-1} \in \mathbb{R}^N$;
for $k \geq 0$ **do**
  Set $\boldsymbol{\alpha}_k = \boldsymbol{\alpha}(\mathbf{x}_{k-1})$;
  Compute $\mathbf{K}^{\boldsymbol{\alpha}_k} \in \mathbb{R}^{N\times N}$ and $\mathbf{f}^{\boldsymbol{\alpha}_k} \in \mathbb{R}^N$;
  Find: $\mathbf{x}_k \in \mathbb{R}^N$ that solves

$$\mathbf{K}^{\boldsymbol{\alpha}_k}\mathbf{x}_k = \mathbf{f}^{\boldsymbol{\alpha}_k} ; \tag{5.24}$$

  if $k \geq 1$ **and** $\mathbf{x}_k = \mathbf{x}_{k-1}$ **then**
    return $\mathbf{x}_k$;
  end
end

Definition 5.28 (slant derivative). Let X and Z be Banach spaces and let $D \subset X$ be open. The mapping $F : D \to Z$ is called *slantly differentiable* in the open subset $U \subset D$ if there is a family of mappings $G : U \to \mathfrak{L}(X, Z)$ such that, for every $x \in U$,

$$\lim_{h\to 0} \frac{1}{\|h\|_X} \|F(x+h) - F(x) - G(x+h)h\|_Z = 0.$$

We call G a *slanting function* for F.

To approximate $x^\star \in X$, the solution of $F(x) = 0$, one can then apply the semismooth Newton method which, starting from $x_0 \in D$, computes iterates via

$$x_{k+1} = x_k - G(x_k)^{-1}F(x_k). \tag{5.25}$$

The convergence of (5.25) is given in the following result, whose proof can be found in Hintermüller *et al.* (2002, Theorem 1.1).

Theorem 5.29 (convergence). Assume that F is slantly differentiable in a neighbourhood of $x^\star$. If $G(x)$ is non-singular for all $x \in U$ and $\{\|G(x)^{-1}\| : x \in U\}$ is bounded, then the semismooth Newton method (5.25) converges superlinearly provided x_0 is sufficiently close to $x^\star$.

The connection between Howard's algorithm and semismooth Newton methods is given by the following result; see Hintermüller *et al.* (2002, Lemma 3.1).

Lemma 5.30 (slant derivative of max). The mapping

$$\mathbf{M} : \mathbb{R}^N \ni \mathbf{y} \mapsto \max\{\mathbf{0}, \mathbf{y}\} \in \mathbb{R}^N$$

is slantly differentiable on $\mathbb{R}^N$ and a slanting function is

$$\mathbf{G_M}(\mathbf{y})_{i,j} = g(\mathbf{y}_j)\delta_{i,j}, \quad g(z) = \begin{cases} 0 & z \leq 0, \\ 1 & z > 0. \end{cases}$$

Let us explain how this result relates to Howard's algorithm in the (perhaps overly) simplistic case that $N = 1$, $\mathcal{A} = \{1, 2\}$ and $\mathbf{f}^1 = \mathbf{f}^2 = \mathbf{f}$. In this setting, we are looking for $\mathbf{x} \in \mathbb{R}$ that solves

$$\mathbf{0} = \mathbf{F}(\mathbf{x}) = \max\{\mathbf{K}^1\mathbf{x}, \mathbf{K}^2\mathbf{x}\} - \mathbf{f} = \mathbf{K}^1\mathbf{x} + \max\{\mathbf{0}, (\mathbf{K}^2 - \mathbf{K}^1)\mathbf{x}\} - \mathbf{f}.$$

By Lemma 5.30 a slanting function for $\mathbf{F}$ at $\mathbf{y}$ is

$$\mathbf{G}(\mathbf{y}) = \mathbf{K}^1 + \mathbf{G_M}((\mathbf{K}^2 - \mathbf{K}^1)\mathbf{y})(\mathbf{K}^2 - \mathbf{K}^1).$$

If $(\mathbf{K}^2 - \mathbf{K}^1)\mathbf{y} \geq 0$, then

$$\mathbf{G}(\mathbf{y}) = \mathbf{K}^1 + (\mathbf{K}^2 - \mathbf{K}^1) = \mathbf{K}^2,$$

and, similarly, if $(\mathbf{K}^2 - \mathbf{K}^1)\mathbf{y} \leq 0$ we get $\mathbf{G}(\mathbf{y}) = \mathbf{K}^1$. In other words, $\mathbf{G}(\mathbf{y})$ always coincides with the coefficient that gives the largest result. Having made this observation, we can now establish convergence for Howard's algorithm (Bokanowski *et al.* 2009).

Theorem 5.31 (convergence). If Assumption 5.27 is satisfied, then the sequence $\{\mathbf{x}_k\}_{k\geq 0}$, generated by Algorithm 5.1, satisfies $\mathbf{x}_k \geq \mathbf{x}_{k+1}$ for every $k \geq 0$ and converges to $\mathbf{x}$, the solution of (5.22). Moreover,

(1) If $\#\mathcal{A}$ is finite, the sequence converges in at most $(\#\mathcal{A})^N$ steps.

(2) If $\mathcal{A}$ is infinite, the convergence is asymptotically superlinear.

Proof. To draw a connection between Algorithm 5.1 and the semismooth Newton method (5.25), let us sketch the proof. Recall that we are looking for a solution of (5.22), which can be understood as looking for a zero of the map $\mathbf{F}$. Now, following Bokanowski *et al.* (2009, Theorem 3.8), let us show that for $\mathbf{y} \in \mathbb{R}^N$ the matrix $\mathbf{K}^{\boldsymbol{\alpha}(\mathbf{y})}$ is a slanting function for $\mathbf{F}$ at the point $\mathbf{y}$. For if that is the case, then (5.25) can be rewritten as

$$\mathbf{K}^{\boldsymbol{\alpha}(\mathbf{x}_k)}(\mathbf{x}_{k+1} - \mathbf{x}_k) = -\mathbf{F}(\mathbf{x}_k) = -\mathbf{K}^{\boldsymbol{\alpha}(\mathbf{x}_k)}\mathbf{x}_k + \mathbf{f}^{\boldsymbol{\alpha}(\mathbf{x}_k)},$$

which is clearly (5.24). By invoking Theorem 5.29 this will yield the superlinear convergence of $\{\mathbf{x}_k\}_{k\geq 0}$.

Now, to show the slant-differentiability of $\mathbf{F}$, let $\mathbf{y}, \mathbf{h} \in \mathbb{R}^N$ and note that

$$\begin{aligned} \mathbf{F}(\mathbf{y}) + \mathbf{K}^{\boldsymbol{\alpha}(\mathbf{y}+\mathbf{h})}\mathbf{h} &\geq \mathbf{K}^{\boldsymbol{\alpha}(\mathbf{y}+\mathbf{h})}\mathbf{y} - \mathbf{f}^{\boldsymbol{\alpha}(\mathbf{y}+\mathbf{h})} + \mathbf{K}^{\boldsymbol{\alpha}(\mathbf{y}+\mathbf{h})}\mathbf{h} = \mathbf{F}(\mathbf{y} + \mathbf{h}) \\ &\geq \mathbf{K}^{\boldsymbol{\alpha}(\mathbf{y})}(\mathbf{y} + \mathbf{h}) - \mathbf{f}^{\boldsymbol{\alpha}(\mathbf{y})} = \mathbf{F}(\mathbf{y}) + \mathbf{K}^{\boldsymbol{\alpha}(\mathbf{y})}\mathbf{h}. \end{aligned}$$

Consequently,

$$\begin{aligned} \mathbf{0} \geq \mathbf{F}(\mathbf{y}+\mathbf{h}) - \mathbf{F}(\mathbf{y}) - \mathbf{K}^{\boldsymbol{\alpha}(\mathbf{y}+\mathbf{h})}\mathbf{h} &\geq [\mathbf{K}^{\boldsymbol{\alpha}(\mathbf{y})} - \mathbf{K}^{\boldsymbol{\alpha}(\mathbf{y}+\mathbf{h})}]\mathbf{h} \\ &\geq -|\mathbf{K}^{\boldsymbol{\alpha}(\mathbf{y})} - \mathbf{K}^{\boldsymbol{\alpha}(\mathbf{y}+\mathbf{h})}|_\infty |\mathbf{h}|_\infty. \end{aligned}$$

Since Lemma 3.2 of Bokanowski *et al.* (2009) shows that $\lim_{|\mathbf{h}|_\infty \to 0} |\mathbf{K}^{\boldsymbol{\alpha}(\mathbf{y})} - \mathbf{K}^{\boldsymbol{\alpha}(\mathbf{y}+\mathbf{h})}|_\infty = 0$, the result follows.

When $\#\mathcal{A}$ is finite we remark that convergence in a finite number of steps follows from the monotonicity of the iterates. To show the monotonicity, we exploit the construction of the parameters $\boldsymbol{\alpha}_k$. Indeed, since

$$\mathbf{K}^{\boldsymbol{\alpha}_{k+1}}\mathbf{x}_k - \mathbf{f}^{\boldsymbol{\alpha}_{k+1}} = \mathbf{F}(\mathbf{x}_k) \geq \mathbf{K}^{\boldsymbol{\alpha}_k}\mathbf{x}_k - \mathbf{f}^{\boldsymbol{\alpha}_k} = \mathbf{0} = \mathbf{K}^{\boldsymbol{\alpha}_{k+1}}\mathbf{x}_{k+1} - \mathbf{f}^{\boldsymbol{\alpha}_{k+1}},$$

we conclude that $\mathbf{K}^{\boldsymbol{\alpha}_{k+1}}(\mathbf{x}_k - \mathbf{x}_{k+1}) \geq 0$. From Assumption 5.27 we have that $\mathbf{K}^{\boldsymbol{\alpha}_{k+1}}$ is monotone and therefore $\mathbf{x}_k \geq \mathbf{x}_{k+1}$. This concludes the proof. □

We conclude by commenting that in Algorithm 5.1 it is necessary to solve the linear system of equations (5.24) at every step, which, for large N, can be very time-consuming. We refer to Hoppe (1986), who describes a multilevel technique to solve this problem and shows its global convergence. Variations of this multigrid strategy have been explored by Alla, Falcone and Kalise (2015) and Han and Wan (2013). Numerical methods of penalty type have been studied by Witte and Reisinger (2012) and Lions and Mercier (1980, Algorithme III).

6. The Monge–Ampère equation

The focus of this section is the study of discretizations for a particular convex equation, the Monge–Ampère equation with Dirichlet boundary conditions:

$$\begin{cases} \det(D^2u) = f & \text{in } \Omega, \\ u = g & \text{on } \partial\Omega. \end{cases} \tag{6.1}$$

The Monge–Ampère equation finds relevance in several areas of mathematics and its applications, including differential geometry, calculus of variations, economics, meteorology and the optimal mass transportation problem (Gutiérrez 2001, Trudinger and Wang 2005, Villani 2003, Benamou and Brenier 1998, Benamou and Brenier 2000, Stojanovic 2005). As we have already mentioned (see also Remark 6.12 below), the Monge–Ampère problem fits into the framework of the previous section. However, due to its many applications, and correspondingly its many discretizations and convergence results, we find it appropriate to devote an entire section to a sample of detailed results. Before discussing these methods, let us give a brief synopsis of some discretizations for the Monge–Ampère problem that we do not cover in detail.

- *Monotone, wide-stencil finite difference schemes* (see Oberman 2008, Froese and Oberman 2011, Froese 2016 and Benamou *et al.* 2014). The basis of these methods is to use determinant identities (such as Hadamard's inequality) to construct consistent and monotone finite difference schemes for the Monge–Ampère problem. Such schemes conform to the Barles–Souganidis theory under certain assumptions and convergence to the viscosity solution is immediate. We refer the reader to Nochetto and Zhang (2017*b*), who have recently derived rates of convergence for these methods.

- *Regularization.* Feng and Neilan (2009, 2011) and Neilan (2010) first approximate the Monge–Ampère equation at the continuous level by a fourth-order quasi-linear PDE. In this framework, solutions are defined via variational principles, and therefore Galerkin methods are readily constructed. Error estimates between the finite element approximation and the regularized solution are derived, but a general convergence theory for viscosity solutions is an open problem.

- *Variational methods.* Classical Galerkin methodologies are used by Böhmer (2008), Brenner, Gudi, Neilan and Sung (2011), Brenner and Neilan (2012), Neilan (2013) and Awanou (2015*a*, 2015*b*, 2015*c*), for example, to construct finite element approximations for strong or classical solutions of the Monge–Ampère equation. Assuming that the solution to (6.1) is sufficiently regular, error estimates are typically derived in H^1- or H^2-type norms.

- *Optimal control.* Dean and Glowinski (2006*a*, 2006*b*) treat the Monge–Ampère equation as a constraint of a minimization problem, leading to a saddle-point reformulation. Computationally efficient algorithms are proposed, but convergence to the viscosity solution is not shown.

We refer to the review paper by Feng, Glowinski and Neilan (2013) for a summary of these discretization types.

6.1. *Solution concepts*

The Monge–Ampère equation has a very unique structure and, because of this, the solution to problem (6.1) can be understood not only with the concepts described in Section 2 but with some additional ones. To discuss these issues, we first make the (sometimes necessary) assumption that the domain Ω is convex. With this assumption, and with an abuse of notation, we (re)define the convex envelope of a function v with $v = 0$ on $\partial\Omega$ as

$$\Gamma(v)(x) := \sup\{L(x) : L(z) \leq v(z) \ \forall z \in \Omega, \ L \in \mathbb{P}_1\}.$$

As we previously described, in contrast to the problems discussed thus far, the Monge–Ampère operator $F(x, r, \mathbf{p}, M) := \det(M) - f(x)$ is *not* elliptic on $\Omega \times \mathbb{R} \times \mathbb{R}^d \times \mathbb{S}^d$ (see Definition 2.10 and Example 2.14). Indeed, we have

$$\det(M) \leq \det(N), \quad \text{for all } N \geq M$$

if and only if $M \geq 0$, and consequently F is only elliptic when restricted to the class of semi-positive definite matrices. In particular, if a classical solution u to (6.1) exists, then F is elliptic at u if and only if $D^2 u \geq 0$ in Ω, implying that u is a convex function. Further note that, besides ensuring ellipticity, the convexity assumption is required to have any hope of uniqueness, since, when d is even,

$$\det(D^2(-u)) = \det(D^2 u).$$

Let us now discuss the existence and uniqueness of different types of solutions of the Monge–Ampère problem (6.1) and the required conditions on the domain Ω and data, f and g, to guarantee that such solutions exist. First we have the following result regarding classical solutions (Wang 1996, Trudinger and Wang 2008).

Theorem 6.1 (classical solutions). Suppose that Ω is an open, bounded and strictly convex domain with boundary $\partial\Omega \in C^3$. Suppose that $g \in C^3(\bar{\Omega})$, $f \in C^\alpha(\Omega)$ for some $\alpha \in (0,1)$, and that $\min_{\bar{\Omega}} f > 0$. Then there exists a unique (classical) convex solution $u \in C^{2,\alpha}(\Omega)$ satisfying (6.1).

Remark 6.2 (optimality). The conditions on $\partial\Omega$, g and f stated in Theorem 6.1 are optimal (Trudinger and Wang 2008).

Next we turn our attention to viscosity solutions to (6.1). Since the operator F is only elliptic on the class of convex functions, we must modify the definition of viscosity solutions to reflect this fact.

Definition 6.3 (viscosity solution). A function $u \in C(\bar{\Omega})$ is called a viscosity subsolution (resp. supersolution) of (6.1) on the set of convex functions if u is convex and if, for all convex $\varphi \in C^2(\Omega)$ such that $u - \varphi$ has a local maximum (resp. minimum) at $x_0 \in \Omega$, we have $\det(D^2\varphi(x_0)) \geq f(x_0)$ (resp. $\det(D^2\varphi(x_0)) \leq f(x_0)$). A function $u \in C(\bar{\Omega})$ is a viscosity solution to (6.1) on the class of convex functions if it is simultaneously a viscosity subsolution and supersolution on the set of convex functions.

Remark 6.4 (f is non-negative). Note that the definition of viscosity solution implicitly requires the source function f to be continuous and non-negative.

Before stating existence and uniqueness results for viscosity solutions, let us give an example of a non-classical viscosity solution, following Froese and Oberman (2011). Note that this example shows that $f \in C^\alpha(\Omega)$ and smooth Ω is not sufficient to guarantee the existence of a classical solution.

Example 6.5 (viscosity solution). Let $\Omega = B_2(0) \subset \mathbb{R}^2$, and consider the function

$$u(x) = \begin{cases} 0 & \text{in } |x| \leq 1, \\ \frac{1}{2}(|x|-1)^2 & \text{in } 1 \leq |x| \leq 2. \end{cases} \tag{6.2}$$

Then u is a viscosity solution of (6.1) with $g = 1/2$ and

$$f(x) = \begin{cases} 0 & \text{in } |x| \leq 1, \\ 1 - |x|^{-1} & \text{in } 1 \leq |x| \leq 2. \end{cases}$$

Theorem 6.6 (existence). Assume that Ω is convex. Assume further that $g \in C(\partial\Omega)$, $f \in C(\bar{\Omega})$ and $f \geq 0$. Then there exists a unique viscosity solution $u \in C(\bar{\Omega})$ to (6.1) in the class of convex functions.

Apart from the viscosity solution, another type of weak solution of the Monge–Ampère equation is based on geometric arguments, which we now describe. To motivate it, suppose for the moment that f is uniformly positive in Ω and that $u \in C^2(\bar{\Omega})$ is a strictly convex function satisfying (6.1) pointwise. Let $\partial u(x)$ denote the subdifferential of u at the point $x \in \Omega$ given by Definition 2.58, that is, $\partial u(x)$ is the set of slopes of the supporting hyperplanes at x. From the convexity and smoothness assumptions we infer that $\partial u(x) = \{Du(x)\}$ and that the subdifferential, viewed as a map from Ω to $\mathbb{R}^d$, is injective. Consequently, a change of variables reveals that

$$\int_D f = \int_D \det(D^2u) = \int_{\partial u(D)} = |\partial u(D)| \quad \text{for all Borel sets } D \subset \Omega,$$

where we recall that $\partial u(D) = \cup_{x \in D} \partial u(x)$ and $|\partial u(D)|$ is the d-dimensional Lebesgue measure of $\partial u(D)$. Since ∂u is well-defined for non-smooth convex functions, the above identity allows us to widen the class of admissible solutions.

Definition 6.7 (Alexandrov solution). A convex function $u \in C(\bar{\Omega})$ is an *Alexandrov solution* to (6.1) if $u|_{\partial\Omega} = g$ and

$$|\partial u(D)| = \int_D f \tag{6.3}$$

for all Borel sets $D \subset \Omega$.

Note that the continuity of the source term f is no longer required for (6.3) to be well-defined. The following example illustrates this feature.

Example 6.8 (Alexandrov solution). Let $\Omega = B_1(0) \subset \mathbb{R}^2$. Then the function

$$u(x) = |x| - 1$$

is an Alexandrov solution of the Monge–Ampère equation

$$\det(D^2u)(x) = \pi\delta_{(x=0)},$$

where $\delta_{(x=0)}$ is the Dirac measure at origin. However, u is not a viscosity solution because the right-hand side is not a (continuous) function.

The existence and uniqueness of Alexandrov solutions is summarized in the next theorem: see Gutiérrez (2001, Theorem 1.6.2).

Theorem 6.9 (existence of Alexandrov solutions). Let $\Omega \subset \mathbb{R}^d$ be an open, bounded and strictly convex domain. Suppose that the data satisfies $g \in C(\partial\Omega)$ and $\int_\Omega f < \infty$. Then there exists a unique Alexandrov solution $u \in C(\bar{\Omega})$ to (6.1) in the class of convex functions.

The next result states situations where viscosity solutions and Alexandrov solutions coincide (Gutiérrez 2001, Propositions 1.3.4 and 1.7.1). The result also shows that the notion of Alexandrov solution is strictly weaker than that of a viscosity solution.

Proposition 6.10 (equivalence). Suppose that $u \in C(\bar{\Omega})$ is an Alexandrov solution to (6.1). Then if f is continuous, u is a viscosity solution to (6.1). Conversely, if u is a viscosity solution to (6.1) and if $f \in C(\bar{\Omega})$ with $f > 0$ in $\bar{\Omega}$, then u is an Alexandrov solution.

The rest of our presentation will deal separately with numerical methods for (6.1) depending on the type of solution we aim to approximate: viscosity or Alexandrov solutions. Before we discuss them, it is useful to review a property of the determinant. For convenience of notation we define the set of symmetric non-negative definite matrices as

$$\mathbb{S}^d_+ := \{A \in \mathbb{S}^d : A \geq 0\}.$$

Proposition 6.11 (determinant inequality). The following inequality holds:

$$\det(A)^{1/d}\det(B)^{1/d} \leq \frac{1}{d}\operatorname{tr} AB = \frac{1}{d}(A : B), \quad \text{for all } A, B \in \mathbb{S}^d_+$$

with equality holding if and only if $B^{1/2}AB^{1/2} = cI$ for some scalar $c \geq 0$.

Proof. Let $C = B^{1/2}AB^{1/2}$, and note that $C \in \mathbb{S}^d_+$. The arithmetic–geometric inequality yields

$$\det(C)^{1/d} \leq \frac{1}{d}\operatorname{tr} C,$$

with equality holding if and only if $B^{1/2}AB^{1/2} = C = cI$. Since $A, B \in \mathbb{S}^d_+$, we find that $c \geq 0$. The inequality follows from the algebraic identities $\det C = \det(AB) = \det A \det B$ and $\operatorname{tr} C = \operatorname{tr} AB = A : B$. □

Remark 6.12 (Monge–Ampère is concave). The determinant inequality of Proposition 6.11 implies that for any positive definite matrix A,

$$\det(A)^{1/d} = \inf_{B\in\mathbb{S}^d_+} \left\{ \frac{1}{d} \operatorname{tr} AB : \det B = 1 \right\}.$$

Therefore, for any symmetric positive definite A and $B \in \mathbb{S}^d_+$, and for $0 < t \leq 1$, we have

$$\begin{aligned}\det(tA + (1-t)B)^{1/d} &= \inf_{K\in\mathbb{S}^d_+} \left\{ \frac{1}{d} \operatorname{tr}((tA + (1-t)B)K) : \det K = 1 \right\} \\ &\geq t \inf_{K\in\mathbb{S}^d_+} \left\{ \frac{1}{d} \operatorname{tr}(AK) : \det K = 1 \right\} \\ &\quad + (1-t) \inf_{K\in\mathbb{S}^d_+} \left\{ \frac{1}{d} \operatorname{tr}(BK) : \det K = 1 \right\} \\ &= t \det(A)^{1/d} + (1-t)\det(B)^{1/d}.\end{aligned}$$

The inequality

$$\det(tA + (1-t)B)^{1/d} \geq t \det(A)^{1/d} + (1-t)\det(B)^{1/d}$$

is also satisfied if $t = 0$, and if $\det(A) = \det(B) = 0$. Thus the function $A \to \det A^{1/d}$ is concave over $\mathbb{S}^d_+$.

6.2. *Approximation of viscosity solutions: Hamilton–Jacobi–Bellman reformulation*

In this section we summarize the results of Feng and Jensen (2017), where a numerical method based on an HJB reformation of the Monge–Ampère problem is developed. Before stating the method we first note that Remark 6.12 implicitly gives such an HJB reformation, that is, if u satisfies the first equation in (6.1), then formally

$$f(x)^{1/d} = (\det(D^2u)(x))^{1/d} = \frac{1}{d} \inf_{\substack{B\in\mathbb{S}^d_+ \\ \det(B)=1}} D^2u(x) : B.$$

Note that the constraint in the control set is nonlinear, and furthermore, the control set is not compact. These two features make the PDE and numerical constructions less obvious. Instead, the method proposed in Feng and Jensen (2017) is based on the following result (Krylov 1987).

Proposition 6.13 (HJB reformulation of MA). Define

$$\mathbb{S}^d_1 = \{A \in \mathbb{S}^d_+ : \operatorname{tr} A = 1\},$$

and let $f \in \mathbb{R}$ with $f \geq 0$. Then a matrix $A \in \mathbb{S}^d$ satisfies

$$H(A,f) := \sup_{B\in\mathbb{S}_1^d} \left(-\frac{1}{d} B : A + f^{1/d} \det(B)^{1/d} \right) = 0 \tag{6.4}$$

if and only if $\det(A) = f$ and $A \in \mathbb{S}_+^d$. Moreover, there exists a maximizer $B_* \in \mathbb{S}_1^d$ that commutes with A.

Proof. If $A \in \mathbb{S}_+^d$ and $\det(A) = f$, then Proposition 6.11 implies

$$-\frac{1}{d} B : A + f^{1/d} \det(B)^{1/d} \leq (f^{1/d} - \det(A)^{1/d}) \det(B)^{1/d} = 0$$

and equality holds if and only if $B^{1/2} A B^{1/2} = cI$ for some $c \geq 0$. If $A > 0$ then we take $B = A^{-1}/(\operatorname{tr} A^{-1})$. Otherwise, if $\det(A) = 0$, we use the eigendecomposition of a matrix to construct $B \in \mathbb{S}_+^d$ that commutes with A, $\operatorname{tr} B = 1$, and $B^{1/2} A B^{1/2} = 0$. In either case we conclude that A satisfies (6.4).

Now our goal is to show that (6.4) implies that $\det(A) = f$ and $A \in \mathbb{S}_+^d$. We prove the statement in two cases.

Case I. If $A > 0$, then take $B = cA^{-1}$ with $c = 1/(\operatorname{tr} A^{-1})$ in (6.4). Using the identities

$$\frac{1}{d} B : A = c \quad \text{and} \quad \det(B) = c^d \det(A)^{-1},$$

we find that

$$c(f^{1/d} - \det(A)^{1/d}) \det(A)^{-1/d} = \frac{-1}{d} B : A + f^{1/d} \det(B)^{1/d} \leq 0$$

and therefore $\det(A) \geq f$. On the other hand Proposition 6.11 implies that

$$0 = \sup_{B\in\mathbb{S}_1^d} \left(-\frac{1}{d} B : A + f^{1/d} \det(B)^{1/d} \right) \leq \sup_{B\in\mathbb{S}_1^d} (f^{1/d} - \det(A)^{1/d}) \det(B)^{1/d}.$$

Now if $\det(A) > f$, then we must have $\det(B) = 0$ for the matrix B to attain the supremum. However, since $A > 0$, Proposition 6.11 implies that the inequality above is strict if $B \neq cA^{-1}$. This leads to a contradiction. Therefore $\det(A) = f$. This proves the first case.

Case II. If A is not strictly positive definite, then without loss of generality, we assume that A is of the diagonal form

$$A = \lambda_1 \mathbf{v}_1 \otimes \mathbf{v}_1 + \cdots + \lambda_d \mathbf{v}_d \otimes \mathbf{v}_d \quad \lambda_1 \leq \cdots \leq \lambda_d,$$

where $\{\mathbf{v}_i\}_{i=1}^d$ are the eigenvectors of A with unit length, and $\lambda_1 \leq 0$.

Consider the matrix

$$B = (1-\epsilon)\mathbf{v}_1 \otimes \mathbf{v}_1 + \frac{\epsilon}{d-1} \sum_{i=2}^{d} \mathbf{v}_i \otimes \mathbf{v}_i \in \mathbb{S}_1^d,$$

for some parameter $\epsilon \in (0,1)$. We then have

$$
\begin{aligned}
-\frac{1}{d}B : A + f^{1/d}\det(B)^{1/d} &= -\frac{1}{d}\bigg((1-\epsilon)\lambda_1 + \frac{\epsilon}{d-1}\sum_{i=2}^{d}\lambda_i\bigg) \\
&\quad + (1-\epsilon)^{1/d}\frac{\epsilon^{(d-1)/d}}{(d-1)^{(d-1)/d}}f^{1/d} \\
&\geq \frac{-1}{d}((1-\epsilon)\lambda_1 + \epsilon\lambda_d) + \left[\frac{(1-\epsilon)\epsilon^{(d-1)}}{(d-1)^{(d-1)}}f\right]^{1/d}.
\end{aligned}
$$

Note that if $f > 0$ and for sufficiently small ϵ, the dominating term is

$$
\left[\frac{(1-\epsilon)\epsilon^{(d-1)}}{(d-1)^{(d-1)}}f\right]^{1/d}.
$$

Thus, if $f > 0$, then we deduce

$$
-\frac{1}{d}B : A + f^{1/d}\det(B)^{1/d} > 0,
$$

which contradicts (6.4). If $f = 0$, then

$$
-\frac{1}{d}B : A + f^{1/d}\det(B)^{1/d} = -\frac{1}{d}\bigg((1-\epsilon)\lambda_1 + \frac{\epsilon}{d-1}\sum_{i=2}^{d}\lambda_i\bigg).
$$

Equation (6.4) holds for any $\epsilon > 0$ if and only if $\lambda_1 = 0$. Thus, $A \geq 0$ and $\det(A) = f = 0$. This completes the proof. □

Lemma 6.13 leads to the following result, showing that viscosity solutions to the Monge–Ampère equation are viscosity solutions of an HJB problem. We refer the reader to Feng and Jensen (2017, Theorem 3.3) for a proof.

Theorem 6.14 (equivalence of viscosity solutions). Let $f \in C(\Omega)$ be a non-negative function, and let u be a viscosity subsolution (resp. supersolution) of the Monge–Ampère problem (6.1) on the set of convex functions. Then u is a viscosity subsolution (resp. supersolution) to the HJB problem

$$
\begin{cases} H(D^2u, f) = 0 & \text{in } \Omega, \\ u = g & \text{on } \partial\Omega. \end{cases} \tag{6.5}
$$

Remark 6.15 (convexity of solution). We emphasize that the convexity of the solution to (6.5) is not assumed *a priori*; it occurs implicitly from the structure of the HJB problem.

Remark 6.16 ($\mathbb{S}_1^d$ is not Cordes). It is worth mentioning that matrices in $\mathbb{S}_1^d$ do not necessarily satisfy the Cordes condition if $d \geq 3$ (see Definition 2.32). To see this, consider $B \in \mathbb{S}_1^3$ with eigenvalues $\lambda_1 = 1 - \tau$,

$\lambda_2 = \lambda_3 = \tau/2$ for some $\tau \in (0, 1/3]$. We then find

$$\frac{|B|^2}{(\operatorname{tr} B)^2} = \lambda_1^2 + \lambda_2^2 + \lambda_3^2 = \frac{3}{2}\tau^2 - 2\tau + 1 \geq \frac{1}{2},$$

and thus B does not satisfy the Cordes condition.

6.2.1. Discretization of the HJB-reformulation

To describe a discretization method for problem (6.5), we first note that if $B \in \mathbb{S}_1^d$, then $B = YDY^\intercal$, where $D \in \mathbb{S}_1^d$ is a diagonal matrix and $Y = [\mathbf{y}_1, \cdots, \mathbf{y}_d] \in \mathbb{SO}^d$, where $\mathbb{SO}^d$ is the special orthogonal group. Thus, letting $\mathbb{D}_1^d$ denote the space of diagonal non-negative definite matrices with unit trace, we find that

$$H(A, f) = \sup_{\substack{Y \in \mathbb{SO}^d \\ D \in \mathbb{D}_1^d}} \left(-\frac{1}{d} YDY^\intercal : A + f^{1/d} \det(D)^{1/d} \right).$$

Let $\mathbf{y}_i \in \mathbb{R}^d$ denote the ith column of $Y \in \mathbb{SO}^d$ and let $\lambda_i = D_{i,i}$. Then the solution to the Monge–Ampèreequation satisfies

$$\sup_{\substack{Y \in \mathbb{SO}^d \\ D \in \mathbb{D}_1^d}} \left(-\frac{1}{d} \sum_{i=1}^{d} \lambda_i \mathbf{y}_i \otimes \mathbf{y}_i : D^2 u(x) + f(x)^{1/d} \left(\prod_{i=1}^{d} \lambda_i \right)^{1/d} \right) = 0$$

in the viscosity sense.

Let $\mathcal{T}_h$ be a quasi-uniform and shape-regular mesh of Ω and let X_h^l be the continuous piecewise linear polynomials over $\mathcal{T}_h$ defined by (3.34). We let Ω_h^I denote the set of interior vertices of $\mathcal{T}_h$, and let Ω_h^B be the set of boundary vertices. Let $v_h \in X_h^l$. Recall that, for a discretization parameter $k > 0$ and vector y, the second-order difference operator is defined by

$$\delta_{y,k}^2 v_h(z) = \frac{1}{k^2}(v_h(z + ky) - 2v_h(z) + v_h(z - ky))$$

provided that z is sufficiently far from the boundary, namely, $z \pm ky \in \bar{\Omega}$. Otherwise we set

$$\delta_{y,k}^2 v_h(z) = \frac{2}{k^+ + k^-} \left(\frac{v_h(z + k^+ y) - v_h(z)}{k^+} + \frac{v_h(z - k^- y) - v_h(z)}{k^-} \right),$$

where $k^\pm$ is the only element in $(0, k]$ such that $z \pm k^\pm y \in \partial\Omega$.

With these definitions at hand we introduce, on X_h^l, the following approximation of the HJB reformulation:

$$H_h^k[u_h, f](z) = \sup_{\substack{Y \in \mathbb{SO}^d \\ D \in \mathbb{D}_1^d}} \left(\frac{-1}{d} \sum_{i=1}^{d} \lambda_i \delta_{y,k}^2 u_h(z) + f(z)^{1/d} \det(D)^{1/d} \right),$$

where $z \in \Omega_h^I$.

The numerical method based on the HJB reformulation seeks a piecewise linear function $u_h \in X_h^l$ such that

$$\begin{cases} H_h^k[u_h, f] = 0 & \text{in } \Omega_h^I, \\ u_h = g & \text{on } \Omega_h^B. \end{cases} \tag{6.6}$$

Note that the boundary conditions uniquely determine u_h on $\partial\Omega$, and therefore the problem is well-defined. The scheme (6.6) is a semi-Lagrangian method because the points $z \pm ky$, which are used to evaluate $\delta_{y,h}^2 u_h(z)$, may not belong to $\bar{\Omega}_h$. Thus, additional effort is needed to evaluate $u_h(z \pm ky)$.

It is straightforward to check that $\delta_{y,k}^2 u_h(z)$ is monotone for any direction y, implying that operator $H_h^k[u_h, f]$ is monotone. The monotonicity also leads to stability of the method (Feng and Jensen 2017, Lemmas 6.2 and 6.4).

Lemma 6.17 (stability). Problem (6.6) is stable in the sense that there exists a unique solution $u_h \in X_h^l$ to (6.6) and an h-independent constant $C > 0$ such that $\|u_h\|_{L^\infty(\Omega)} \le C$. Moreover, if we set

$$\bar{u}(x) := \limsup_{\substack{y \to x \\ h \to 0^+}} u_h(y), \quad \underline{u}(x) := \liminf_{\substack{y \to x \\ h \to 0^+}} u_h(y), \quad x \in \bar{\Omega},$$

then $\bar{u}(x) = \underline{u}(x) = g(x)$ for all $x \in \partial\Omega$ provided that Ω is strictly convex.

To show the consistency of the method, we recall that $I_h^{fe} : C(\bar{\Omega}) \to X_h^l$ denotes the nodal interpolant. We then have the following result.

Lemma 6.18 (consistency). Let $\phi \in C^{2,\alpha}(\Omega)$. Then there is a constant C such that for every $z \in \Omega_h^I$ we have

$$\left| H(D^2\phi(z), f(z)) - H_h^k[I_h^{fe}\phi, f](z) \right| \le C\left(k^\alpha + \frac{h^2}{k_{\min}^2}\right),$$

where $k_{\min} = \min\{k^+, k^-\}$. Consequently, the method is consistent if $k \to 0$ and $h/k_{\min} \to 0$.

Proof. It suffices to show that

$$|\delta_{y,k}^2 I_h^{fe}\phi(z) - (y \otimes y) : D^2\phi(z)| \le C\left(k^\alpha + \frac{h^2}{k_{\min}^2}\right).$$

First, by Taylor's theorem, we have

$$|\delta_{y,k}^2 \phi(z) - (y \otimes y) : D^2\phi(z)| \le C k^\alpha \|\phi\|_{C^{2,\alpha}(\Omega)}. \tag{6.7}$$

Recalling that

$$\|\phi - I_h^{fe}\phi\|_{L^\infty(\Omega)} \le C h^2 \|\phi\|_{W^{2,\infty}(\Omega)}$$

(Ciarlet 2002), we have

$$|\delta_{y,k}^2 \phi(z) - \delta_{y,k}^2 I_h^{fe} \phi(z)| \le C \frac{h^2}{k_{\min}^2} \|\phi\|_{W^{2,\infty}(\Omega)}. \tag{6.8}$$

The desired result now follows from (6.7)–(6.8) and the triangle inequality. □

Remark 6.19 (discrete controls). To implement the method (6.6), it remains to specify a discrete set $\mathbb{S}_{1,h}^d \subset \mathbb{S}_1^d$ of symmetric positive definite matrices with unit trace. To ensure consistency of the method, we require the discrete set $\mathbb{S}_{1,h}^d$ to be dense as $h \to 0$, that is, for any $B \in \mathbb{S}_1^d$, there exists $B_h \in \mathbb{S}_{1,h}^d$ such that $B_h \to B$ as $h \to 0$.

Since the method is monotone and consistent, the convergence of the numerical solution now follows from the Barles–Souganidis theory; see Theorem 3.8 and Feng and Jensen (2017, Theorem 6.5).

Theorem 6.20 (convergence). Let $\Omega \subset \mathbb{R}^d$ be a strictly convex domain. Assume that $f \in C(\Omega)$ with $f \ge 0$ and $g \in C(\partial\Omega)$. Then as $h \to 0$, $h/k_{\min} \to 0$, the solutions $u_h \in X_h^l$ of (6.6) converge uniformly to the unique viscosity solution on the set of convex functions of the Monge–Ampère problem (6.1).

We conclude the discussion on semi-Lagrangian schemes by commenting that rates of convergence for a general semi-Lagrangian scheme for (5.1) have been obtained by Debrabant and Jakobsen (2013, Corollary 7.3).

6.3. Approximation of Alexandrov solutions

Now we discuss numerical methods based on the Alexandrov solution concept presented in Definition 6.7. Essentially, numerical methods of this class are finite-dimensional analogues of (6.3).

Let $\{\omega_i\}_{i=1}^N$ be an open, disjoint partition of the domain, that is, $\omega_i \cap \omega_j = \emptyset$ for $i \ne j$ and $\bar{\Omega} = \cup_{i=1}^N \bar{\omega}_i$. Let $\Omega_h^I := \{z_i\}_{i=1}^N$ be a collection of points with the property that $z_i \in \omega_j$ if and only if $i = j$, and let $\Omega_h^B := \{z_i\}_{i=N+1}^{M+N}$ be a set of distinct points on $\partial\Omega$. As before, we let $\bar{\Omega}_h = \Omega_h^I \cup \Omega_h^B$ and we will call its elements nodes or grid points.

Recall that for a nodal function $v_h \in X_h^{fd}$, its subdifferential at a grid point $z \in \bar{\Omega}_h$ is given by (3.20). Now, a natural generalization of (6.3), and the discrete problem proposed in Oliker and Prussner (1988) reads as follows: find a convex nodal function $u_h \in X_h^{fd}$ satisfying

$$\begin{cases} |\partial u_h(z_i)| = \displaystyle\int_{\omega_i} f & \text{for all } z_i \in \Omega_h^I, \\ u_h(z_i) = g(z_i) & \text{for all } z_i \in \Omega_h^B. \end{cases} \tag{6.9}$$

Note that since the partition $\{\omega_i\}_{i=1}^N$ is non-overlapping, for all Borel sets $D \subset \Omega$ we have

$$|\partial u_h(D)| = \sum_{z_i \in D} f_i, \quad f_i = \int_{\omega_i} f.$$

Thus, the scheme is obtained by replacing f in (6.3) with a family of Dirac measures supported at the nodes and by replacing g with its nodal interpolant on the boundary.

One special case of this method is when the interior nodal set is a lattice, that is, for some basis $\{\tilde{\mathbf{e}}_j\}_{j=1}^d$ of $\mathbb{R}^d$ we have

$$\Omega_h^I = \left\{ z = h \sum_{j=1}^d z^j \tilde{\mathbf{e}}_j : z^j \in \mathbb{Z} \right\} \cap \Omega.$$

We remark that this property applies to interior nodes only. For the boundary nodes, we only require that their spacing is of order h, namely, $\partial\Omega \subset \cup_{z \in \Omega_h^B} B_{h/2}(z)$. In this case, the partition $\{\omega_i\}$ of the domain can be taken as parallelotopes

$$\omega_i = \left\{ z_i + \sum_{j=1}^d h^j \tilde{\mathbf{e}}_j : h^j \in \mathbb{R}, \ |h^j| \leq \frac{h}{2} \right\} \cap \Omega. \tag{6.10}$$

The length of the coordinate vectors $\{\tilde{\mathbf{e}}_j\}_{j=1}^d$ is such that the parallelotope ω_i is inside the ball $B_h(z_i)$ centred at z_i and of radius h. Note that, by construction, $\omega_j = z_j - z_i + \omega_i$; consequently, the radius of the largest ball inscribed in ω_i does not depend on i and we denote it by ρ. We define the shape-regularity of the nodal set as

$$\sigma = h\rho^{-1}. \tag{6.11}$$

Remark 6.21 (meshless nodal function). It is worth mentioning again that the solution u_h is only defined at the nodes $\bar{\Omega}_h$. Its convex envelope induces a triangulation of the domain Ω and a piecewise linear function. However, this triangulation is not known *a priori*. See Remark 3.18 and Examples 3.44–3.45.

In view of Example 3.45, if the solution of the Monge–Ampère equation is nearly degenerate, wide stencils are needed to compute the subdifferential. Thus, method (6.9) may be regarded as a wide-stencil finite difference scheme when the solution is nearly degenerate. On the other hand, if the solution to (6.1) is strictly convex, that is, for $0 < \lambda \leq \Lambda$ we have $\lambda I \leq D^2 u \leq \Lambda I$, then the subdifferential of u_h at node z depends only on the values of u_h at the adjacent nodes of z. To make this last statement rigorous, we state a definition.

Definition 6.22 (adjacent set). Let $u_h \in X_h^{fd}$ and $z \in \Omega_h^I$. The *adjacent set* A_z of u_h at z is the collection of nodes $z_i \in \bar{\Omega}_h$ such that there exists a supporting hyperplane ℓ of u_h at z and $\ell(z_i) = u_h(z_i)$.

Note that A_z is the set of nodes of a star of z which is induced by the discrete convex envelope of u_h; see Figure 3.2. In particular, we have that the subdifferential $\partial u_h(z)$ is determined by the values $u(z_i)$ for $z_i \in A_z$.

Let us now estimate the size of A_z.

Lemma 6.23 (size of $\boldsymbol{A_z}$). Let $v \in C^2(\bar{\Omega})$ be a strictly convex function with $\lambda I \leq D^2 v(x) \leq \Lambda I$ for some constants $0 < \lambda \leq \Lambda$. Assume that $E := \{\tilde{\mathbf{e}}_j\}_{j=1}^d$ is a basis of $\mathbb{R}^d$ such that Ω_h^I is a lattice spanned by E with shape-regularity constant σ defined in (6.11). Let the nodal function $v_h \in X_h^{fd}$ be defined by $v_h = I_h^{fd} v$. Then, if A_z is the adjacent set of v_h at $z \in \Omega_h^I$,

$$A_z \subset B_{Rh}(z),$$

where

$$R \geq \bar{R} := \frac{\Lambda}{\lambda}\sigma^2 \left| \sum_{j=1}^d \tilde{\mathbf{e}}_j \right|^2. \tag{6.12}$$

Proof. Without loss of generality, we may assume that $z = 0$, $v(z) = 0$ and $Dv(z) = \mathbf{0}$. Let $\hat{z} \in \Omega_h^I \cap \partial \operatorname{conv}(A_z)$ and let ω be the parallelotope defined as in (6.10) with centre $z = 0$. By convexity of ω, there is a $c \in (0,1)$ such that $c\hat{z} \in \partial\omega$. Thus, we can express $\hat{z}$ as a multiple of a convex combination of $\{\zeta_j\}_{j=1}^{2^d}$, the vertices of ω. In other words, for $R = 1/c$ we have

$$\hat{z} = R \sum_{j=1}^{2^d} k_j \zeta_j, \quad k_j \geq 0, \quad \sum_{j=1}^{2^d} k_j = 1.$$

This representation shows that $|\hat{z}| \leq Rh$; thus, to obtain the result, it remains to estimate R.

Since $c\hat{z} \in \partial\omega$ we have, using (6.11), that $|c\hat{z}| \geq \rho = h\sigma^{-1}$ which can be rewritten as $|\hat{z}| \geq Rh\sigma^{-1}$. Using that $D^2 v \geq \lambda I$ we estimate

$$v_h(\hat{z}) = v(\hat{z}) \geq \frac{1}{2}\lambda R^2 \sigma^{-2} h^2.$$

Let us now obtain an upper bound for $v_h(\hat{z})$. To do so, let us introduce $\hat{\omega}$ as the (unique) smallest parallelotope with vertices $\{\bar{z}_m\}_{m=1}^{2^d} \subset \Omega_h^I$ such that $z \in \{\bar{z}_m\}_{m=1}^{2^d}$ is a vertex and $c\hat{z} \in \hat{\omega}$. This parallelotope can be thought of as belonging to the *dual mesh*. We now invoke Caratheodory's theorem (Dacorogna 2008, Theorem 2.13) to conclude that there is a subset of $\{\bar{z}_m\}_{m=1}^{2^d}$, of cardinality $d+1$, for which $c\hat{z}$ can be expressed as a convex combination of these vertices. In other words, up to a permutation in

$\{1,\dots,2^d\}$, we have

$$\hat{z} = R\sum_{m=1}^{d+1} \alpha_m \bar{z}_m, \quad \alpha_m \geq 0, \quad \sum_{m=1}^{d+1} \alpha_m = 1. \tag{6.13}$$

We now invoke that $\hat{z} \in A_z$. This implies that there is an affine function ℓ that verifies

$$\ell(z) = v_h(z) = 0, \quad \ell(\hat{z}) = v_h(\hat{z}), \quad \ell(\bar{z}_m) \leq v_h(\bar{z}_m), \ m = 1,\dots,d+1.$$

Using representation (6.13) of $\hat{z}$ and that $D^2 v \leq \Lambda I$ we then obtain

$$v_h(\hat{z}) = R\sum_{m=1}^{d+1} \alpha_m \ell(\bar{z}_m) \leq \frac{1}{2}\Lambda R \sum_{m=1}^{d+1} \alpha_m |\bar{z}_m|^2.$$

It remains to observe now that

$$\bar{z}_m = z + \sum_{j=1}^{d} \epsilon_j h \tilde{\mathbf{e}}_j$$

with $\epsilon_j \in \{-1, 0, 1\}$, and therefore

$$|\bar{z}_m| \leq h\left|\sum_{j=1}^{d} \tilde{\mathbf{e}}_j\right|.$$

A combination of the obtained upper and lower bounds for $v_h(\hat{z})$ yields

$$\frac{1}{2}\lambda R^2 h^2 \sigma^{-2} \leq \frac{1}{2}\Lambda R h^2 \left|\sum_{j=1}^{d} \tilde{\mathbf{e}}_j\right|^2,$$

from which (6.12) follows. □

Remark 6.24 (Cartesian lattice). In the setting of Lemma 6.23, if E is the canonical basis of $\mathbb{R}^d$, then Mirebeau (2015) and Benamou, Collino and Mirebeau (2016) have improved estimate (6.12) to $\bar{R} = (\Lambda/\lambda)\sigma^2$.

If Ω_h^I is a lattice, then we are able to show consistency of the method (6.9) in the following sense.

Lemma 6.25 (consistency). Let $E = \{\tilde{\mathbf{e}}_j\}_{j=1}^d$ be a basis of $\mathbb{R}^d$ and Ω_h^I be a lattice spanned by E. Let p be a strictly convex quadratic polynomial with $\lambda I \leq D^2 p \leq \Lambda I$. If $z \in \Omega_h^I$ is such that $\operatorname{dist}(z, \partial\Omega) \geq \bar{R}h$, with $\bar{R}$ as in (6.12), then we have

$$|\partial I_h^{fd} p(z)| = \det(D^2 p)|\omega_z|, \tag{6.14}$$

where ω_z is the parallelotope defined by (6.10).

Proof. We divide the proof into two steps.

Step 1. We first show that (6.14) holds when the domain is $\Omega = \mathbb{R}^d$. Without loss of generality, we assume that $z = 0$ and $p(x) = \frac{1}{2}x \cdot Mx$ for some $M \geq \lambda I$. For a vector $\mathbf{q} \in \mathbb{R}^d$ we define the norm $|\mathbf{q}|_M^2 := \mathbf{q} \cdot M\mathbf{q}$, and define the set

$$
\begin{aligned}
V &:= \{\mathbf{q} \in \mathbb{R}^d : |\mathbf{q}|_M \leq |\mathbf{q} - z_j|_M,\ \forall z_j \in \bar{\Omega}_h\} \\
&= \left\{\mathbf{q} \in \mathbb{R}^d : \mathbf{q} \cdot Mz_j \leq \frac{1}{2}z_j \cdot Mz_j,\ \forall z_j \in \bar{\Omega}_h\right\}.
\end{aligned}
$$

It can be shown (see Lemma 2.3 of Mirebeau 2015 for details) that translations of V tile $\mathbb{R}^d$ and that $|V| = |\omega_z|$. Moreover, by a simple algebraic manipulation,

$$
V = \{M^{-1}\mathbf{q} \in \mathbb{R}^d : \mathbf{q} \cdot z_j \leq I_h^{fd} p(z_j),\ \forall z_j \in \bar{\Omega}_h\}.
$$

Thus, $V = M^{-1}[\partial I_h^{fd} p(0)]$, that is, it is the image of subdifferential $\partial I_h^{fd} p(0)$ under the linear map M^{-1}. Taking measures on both sides yields

$$
|\omega_z| = |V| = \det(M)^{-1} |\partial I_h^{fd} p(0)|.
$$

The proof of Step 1 is now completed by rearranging terms.

Step 2. We now consider a bounded domain and show that (6.14) holds for nodes sufficiently far away from the boundary, that is, $\operatorname{dist}(z, \partial\Omega) \geq \bar{R}h$. To do so, we observe that the subdifferential $\partial I_h^{fd} p(z)$ is determined only by the function values in the adjacent set A_z. Since, as shown in Lemma 6.23, $A_z \subset B_{\bar{R}h}(z)$, we deduce that if the node z is bounded away from the boundary with $\operatorname{dist}(z, \partial\Omega) \geq \bar{R}h$, then $A_z \subset \Omega$. Thus, (6.14) holds.

This concludes the proof. □

6.3.1. A truncated version

When Ω_h^I is a Cartesian lattice, scheme (6.9) is is closely related to the finite difference method of Mirebeau (2015) and Benamou *et al.* (2016) which we now describe. For simplicity, suppose that $d = 2$ and that the interior grid points are given by

$$
\Omega_h^I = \Omega \cap \mathbb{Z}_h^2.
$$

Let $S \subset \mathbb{Z}^2 \backslash \{0\}$ denote a stencil. For any $y \in S$ and $z \in \Omega_h^I$ sufficiently far from $\partial\Omega$, we recall that the second-order difference operator in the direction y is given by

$$
\delta_{y,h}^2 v(z) = \frac{v(z + hy) - 2v(z) + v(z - hy)}{h^2}.
$$

When $z \in \Omega_h^I$ is close to $\partial\Omega$, the point $z \pm hy$ may not belong to $\bar{\Omega}_h$. In such cases, we define

$$\delta^2_{y,h}v(z) := \frac{2}{h^+ + h^-}\left(\frac{v(z+h^+y)-v(z)}{h^+} + \frac{v(z-h^-y)-v(z)}{h^-}\right),$$

where $h^{\pm}$ is the only element in $[0,h]$ such that $z \pm h^{\pm}y \in \partial\Omega$. This construction implicitly defines the set of boundary points Ω_h^B.

We define the set of *superbases* of S as

$$Y_h := \{(y_0, y_1, y_2) \in S^3 : |\det(y_0, y_1, y_2)| = 1,\ y_0 + y_1 + y_2 = \mathbf{0}\}.$$

Note that for $z \in \Omega_h^I$ and $\mathbf{y} = (y_0, y_1, y_2) \in Y_h$, the convex hull $\mathcal{H}_{z,\mathbf{y}} = \operatorname{conv}\{z \pm hy_i\}_{i=0}^3$ is a hexagon. Given a nodal function v_h, superbasis $\mathbf{y} = (y_0, y_1, y_2) \in Y_h$ and a point $z \in \Omega_h^I$, we let $\Gamma_{z,\mathbf{y}}(v_h)$ denote the maximal convex map bounded above by v_h at the points z and $\{z \pm hy_i\}_{i=0}^3$. As before, $\Gamma_{z,\mathbf{y}}(v_h)$, restricted to $\mathcal{H}_{z,\mathbf{y}}$, is a piecewise linear function with respect to some triangulation of $\mathcal{H}_{z,\mathbf{y}}$, which depends on the values of v_h on the extreme points of $\mathcal{H}_{z,\mathbf{y}}$ and z.

We define the discrete Monge–Ampère operator

$$\gamma(\Delta^+_{y_0}v_h(z), \Delta^+_{y_1}v_h(z), \Delta^+_{y_2}v_h(z)),$$

with $\Delta^+_y v(z) = \max\{\delta^2_{y,h}v_h(z), 0\}$ and

$$\gamma(\delta_0, \delta_1, \delta_2) := \begin{cases} \delta_{i+1}\delta_{i+2} & \text{if } \delta_i \geq \delta_{i+1} + \delta_{i+2},\ i = 0, \dots, 2 \mod 3, \\ \gamma_1(\delta_0, \delta_1, \delta_2) & \text{otherwise,} \end{cases}$$

with

$$\gamma_1(\delta_0, \delta_1, \delta_2) := \frac{1}{2}(\delta_0\delta_1 + \delta_1\delta_2 + \delta_0\delta_2) - \frac{1}{4}(\delta_0^2 + \delta_1^2 + \delta_2^2).$$

As shown by Benamou *et al.* (2016, Remark 1.10), from the definition of subdifferential and some geometric arguments it follows that

$$\gamma(\Delta^+_{y_0}v_h(z), \Delta^+_{y_1}v_h(z), \Delta^+_{y_2}v_h(z)) = h^2|\partial\Gamma_{z,\mathbf{y}}(v_h)(z)|.$$

The scheme proposed in Benamou *et al.* (2016) reads as follows: find the nodal function $u_h \in X_h^{fd}$ such that

$$\begin{cases} \min\limits_{\mathbf{y}\in Y_h} \gamma(\Delta^+_{y_0}u_h(z), \Delta^+_{y_1}u_h(z), \Delta^+_{y_2}u_h(z)) = f(z) & \text{for all } z \in \Omega_h^I, \\ u_h(z) = g(z) & \text{for all } z \in \Omega_h^B. \end{cases} \tag{6.15}$$

Lemma 6.26 (consistency). Let M be a positive definite matrix and $p(x) = \frac{1}{2}x \cdot Mx$ be a convex quadratic polynomial. Then

$$\min_{\mathbf{y}\in Y_h} |\partial\Gamma_{z,\mathbf{y}}(I_h^{fd}p)(z)| = \det(M)$$

if and only if there is an M-obtuse superbasis (y_0, y_1, y_2), that is,

$$y_i \cdot My_j \leq 0, \quad \text{for all } i \neq j.$$

Moreover, if $B_R \subset \operatorname{conv} S$ with $R^2 = 2|M|\,|M^{-1}|$, then such an M-obtuse basis exists in S.

We refer to Benamou *et al.* (2016, Propositions 1.12 and 2.2) for a proof. Note that the previous result shows that if the matrix M is anisotropic, that is, $|M|\,|M^{-1}|$ is large, then a wide stencil is needed to ensure the existence of an M-obtuse superbasis.

Remark 6.27 (three dimensions). In three space dimensions, to the best of our knowledge, there is no explicit formula to compute $|\partial\Gamma_{z,\mathbf{y}}(v_h)(z)|$. As shown in Mirebeau (2015), if $p(x) = \frac{1}{2}x \cdot Mx$ and the stencil S is such that $B_R \subset \operatorname{conv} S$, for some R that depends on $|M|\,|M^{-1}|$, then we have that

$$\det(M) = h^3 |\partial\Gamma_{z,\mathbf{y}}(I_h^{fd} p)(z)|.$$

This result is consistent with Lemma 6.25.

6.3.2. Stability of (6.9)

While the convergence of monotone and consistent schemes, like (6.9), can be obtained using the framework described in Section 3.1, few results are known on the rate of convergence of such approximations. Here, following Nochetto and Zhang (2017*b*), we discuss some recent results on the L^∞-rate of convergence of scheme (6.9).

In addition to the discrete Alexandrov estimates of Lemmas 3.24 and 3.46, the derivation of these error estimates involves a discrete barrier argument as in Section 4.3 and the Brunn–Minkowski inequality, which we now state.

Let D and E be two non-empty compact subsets of $\mathbb{R}^d$. We define their (Minkowski) sum as

$$D + E := \{\mathbf{v} + \mathbf{w} \in \mathbb{R}^d : \mathbf{v} \in D,\, \mathbf{w} \in E\}.$$

Proposition 6.28 (Brunn–Minkowski). Let A and B be two non-empty compact subsets of $\mathbb{R}^d$. Then the following inequality holds:

$$|A + B|^{1/d} \geq |A|^{1/d} + |B|^{1/d}$$

where $|\cdot|$ denotes the Lebesgue measure on $\mathbb{R}^d$.

Remark 6.29 (concavity). The Brunn–Minkowski inequality implies that the function $D \to |D|^{1/d}$ is concave, in the sense that for $0 \leq t \leq 1$,

$$|tA + (1-t)B|^{1/d} \geq t|A|^{1/d} + (1-t)|B|^{1/d}.$$

The discrete Alexandrov estimate Lemma 3.24 shows that the L^∞-norm of nodal or piecewise linear function v_h is controlled by the measure of the subdifferential $|\partial v_h|$. Now suppose u_h and w_h are two nodal functions. The following stability estimate shows that the difference $v_h - w_h$ measured in the L^∞-norm is controlled by the difference of the measure of their subdifferentials. This can be recast as a stability estimate for scheme (6.9).

Proposition 6.30 (stability). Let v_h and w_h be two nodal functions with $v_h \geq w_h$ on Ω_h^B. Then

$$\sup_{\bar{\Omega}_h}(v_h - w_h)^- \leq C\biggl(\sum_{z\in\mathcal{C}_h^-(v_h-w_h)} \bigl(|\partial v_h(z)|^{1/d} - |\partial w_h(z)|^{1/d}\bigr)^d\biggr)^{1/d}.$$

Proof. Consider the discrete convex envelope of the difference $v_h - w_h$, which we denote by $\Gamma_h(v_h - w_h)$ and its lower nodal contact set

$$\mathcal{C}_h^-(v_h - w_h) = \{z \in \Omega_h^I : \Gamma_h(v_h - w_h)(z) = (u_h - w_h)(z)\}.$$

By Lemma 3.24 (finite difference Alexandrov estimate), we have

$$\sup_{\bar{\Omega}_h}(v_h - w_h)^- \leq C\biggl(\sum_{z\in\mathcal{C}_h(v_h-w_h)^-} |\partial\Gamma_h(v_h - w_h)(z)|\biggr)^{1/d}. \tag{6.16}$$

Thus, we only need to estimate $|\partial\Gamma_h(v_h - w_h)(z)|$ for all z in the contact set, which we do as follows. We first note that Lemma 3.21 implies that

$$\partial w_h(z) + \partial\Gamma_h(v_h - w_h)(z) \subset \partial v_h(z). \tag{6.17}$$

From this, and the Brunn–Minkowski inequality (Proposition 6.28), we obtain

$$\begin{aligned} &|\partial w_h(z)|^{1/d} + |\partial\Gamma_h(v_h - w_h)(z)|^{1/d} \\ &\qquad \leq |\partial w_h(z) + \partial\Gamma_h(v_h - w_h)(z)|^{1/d} \leq |\partial v_h(z)|^{1/d}, \end{aligned} \tag{6.18}$$

which clearly implies that

$$|\partial\Gamma_h(v_h - w_h)(z)| \leq \bigl(|\partial v_h(z)|^{1/d} - |\partial w_h(z)|^{1/d}\bigr)^d.$$

This is the desired estimate for $|\partial\Gamma_h(u_h - w_h)(z)|$. Inserting it into (6.16) yields the claimed result. □

A direct consequence of this stability result is a maximum principle for nodal functions, which we state below.

Corollary 6.31 (maximum principle). Let v_h and w_h be two nodal functions associated with $\bar{\Omega}_h$. If $v_h \geq w_h$ on Ω_h^B and $|\partial v_h(z)| \leq |\partial w_h(z)|$ for all $z \in \Omega_h^I$, then

$$w_h(z) \leq v_h(z), \quad \text{for all } z \in \bar{\Omega}_h.$$

Proof. By (6.17) and (6.18), we have $|\partial w_h(z)| \le |\partial v_h(z)|$ for any $z \in \mathcal{C}_h^-(v_h - w_h)$. Since $|\partial v_h(z)| \le |\partial w_h(z)|$ by assumption, we have $|\partial v_h(z)| = |\partial w_h(z)|$ at contact points. Thus, by Proposition 6.30, we get

$$\sup_{\bar{\Omega}_h}(v_h - w_h)^- = 0.$$

Consequently $v_h - w_h \ge 0$, which proves the result. □

6.3.3. Error estimates for (6.9)

Let us now derive rates of convergence in the L^∞-norm for method (6.9). To do so, we will use all the tools we have developed in previous sections, namely the discrete Alexandrov estimate of Proposition 3.24, the Brunn–Minkowski inequality of Proposition 6.28 and the stability result of Proposition 6.30.

Owing to Proposition 6.30 we only need to study the consistency error. Since the method is consistent for convex quadratic polynomials at nodes bounded away from the boundary $\partial\Omega$ (see Lemma 6.25), if we expect that u can be well approximated by quadratic polynomials, then the consistency error will also be small. Let us make this intuition rigorous.

Lemma 6.32 (interior consistency). Let $E = \{\tilde{\mathbf{e}}_j\}_{j=1}^d$ be a basis of $\mathbb{R}^d$ and let Ω_h^I be a lattice spanned by E. Given u, strictly convex, let $z \in \Omega_h^I$ with $\operatorname{dist}(z, \partial\Omega) \ge \bar{R}h$, where $\bar{R}$ is defined in (6.12), and set $\bar{B} = B_{\bar{R}h}(z)$. If $u \in C^{2,\alpha}(\bar{B})$, then we have

$$\left| |\partial I_h^{fd} u(z)| - \int_{\omega_z} \det(D^2 u) \right| \le Ch^\alpha |\omega_z|,$$

where the constant C depends on $|u|_{C^{2,\alpha}(\bar{B})}$, and ω_z is defined in (6.10).

Proof. Let us show that

$$|\partial I_h^{fd} u(z)| \le \int_{\omega_z} \det(D^2 u) + Ch^\alpha |\omega_z|.$$

The other inequality can be obtained in a similar fashion.

Since $u \in C^{2,\alpha}(\bar{B})$, there is a convex quadratic polynomial $p \in \mathbb{P}_2$ that satisfies $p(z) = u(z)$, $Dp(z) = Du(z)$, $D^2 p = D^2 u(z)$ and, moreover,

$$u(x) \le p(x) + |u|_{C^{2,\alpha}(\bar{B})} h^{2+\alpha}, \quad \text{for all } x \in \bar{B}.$$

Define $q(x) = p(x) + h^\alpha |u|_{C^{2,\alpha}(\bar{B})} |x - z|^2$ and note that, by construction, $q(z) = u(z)$ and for all nodes $z_j \in \bar{B} \cap \Omega_h^I$ we have $u(z_j) \le q(z_j)$. Since q is convex, its nodal interpolant $I_h^{fd} q$ is also convex (see Definition 3.16). Thus, we can apply Lemma 3.20 (monotonicity), to get $|\partial I_h^{fd} u(z)| \le |\partial I_h^{fd} q(z)|$. From these considerations we see that it is sufficient to show that

$$|\partial I_h^{fd} q(z)| \le \int_{\omega_z} \det(D^2 u) + Ch^\alpha |\omega_z|.$$

Since $\lambda + Ch^\alpha \le D^2 q \le \Lambda + Ch^\alpha$ and

$$\frac{\Lambda + Ch^\alpha}{\lambda + Ch^\alpha} \le \frac{\Lambda}{\lambda},$$

we invoke the consistency result of Lemma 6.25 and the regularity $u \in C^{2,\alpha}(\bar{B})$ to obtain

$$\begin{aligned} |\partial I_h^{fd} q(z)| &= \det(D^2 q)|\omega_z| \le (\det(D^2 p) + Ch^\alpha)|\omega_z| \\ &\le \int_{\omega_z} \det(D^2 u) + Ch^\alpha |\omega_z|. \end{aligned}$$

This concludes the proof. □

The previous result establishes a consistency error at nodes bounded away from the boundary. For nodes close to the boundary, we have the following estimate.

Lemma 6.33 (boundary consistency). Let $E = \{\tilde{\mathbf{e}}_j\}_{j=1}^d$ be a basis of $\mathbb{R}^d$ and let Ω_h^I be a lattice spanned by E. Given u, strictly convex, let $z \in \Omega_h^I$ satisfy $\mathrm{dist}(z, \partial\Omega) \le \bar{R}h$, where $\bar{R}$ is defined in (6.12) and set $\bar{B} = B_{\bar{R}h}(z) \cap \Omega$. If $u \in C^{1,1}(\bar{B})$, then

$$\left| |\partial I_h^{fd} u(z)| - \int_{\omega_z} \det D^2 u \right| \le C|\omega_z|, \tag{6.19}$$

where the constant C depends only on $|u|_{C^{1,1}(\bar{B})}$.

Proof. As in Lemma 6.32, it suffices to show that

$$|\partial I_h^{fd} u(z)| \le \int_{\omega_z} \det D^2 u + C|\omega_z|.$$

Since $\lambda I \le D^2 u \le \Lambda I$, Lemma 6.23 yields $A_z \subset B_{\bar{R}h}(z) \cap \Omega$. Recall now that $\Gamma(I_h^{fd} u)$ is piecewise linear with respect to a triangulation that has as nodes $\bar{\Omega}_h$. The $C^{1,1}$-regularity assumption of u implies that, if $K \subset \omega_z \subset \bar{B}$ is an element of this triangulation, we have

$$D\Gamma(I_h^{fd} u)|_K = Du(z) + \mathbf{v}_K \quad |\mathbf{v}_K| \le Ch|u|_{C^{1,1}(\bar{B})}.$$

Thus, by Lemma 3.22 (characterization of subdifferential), we deduce that the piecewise gradient $D\Gamma(I_h^{fd} u)|_K$ is contained in a ball centred at $Du(z)$ and with radius $Ch|u|_{C^{1,1}(\bar{B})}$. Thus, we have

$$|\partial I_h^{fd} u(z)| \le \int_{\omega_z} \det D^2 u + C|u|^d_{C^{1,1}(\bar{B})}|\omega_z|.$$

This completes the proof. □

To control the L^∞-error caused by the $O(1)$ error near the boundary, we construct a discrete barrier function below. We refer to Nochetto and Zhang (2017*b*) for a proof.

Lemma 6.34 (discrete barrier). Let Ω be uniformly convex and let Ω_h^I be a translation-invariant nodal set in Ω. Given a constant $M > 0$, for each node $z \in \Omega_h^I$ with $\operatorname{dist}(z, \partial\Omega) \leq \bar{R}h$, there exists a convex nodal function $p_z \in X_h^{fd}$ such that $|\partial p_z(z_i)| \geq M|\omega_z|$ for all $z_i \in \bar{\Omega}_h$, $p_z(z_i) \leq 0$ on $z_i \in \Omega_h^B$ and

$$|p_z(z)| \leq CRM^{1/d}h,$$

for sufficiently small h.

Now we are ready to prove the L^∞-error estimate.

Theorem 6.35 (rate of convergence I). Assume that Ω is uniformly convex, and let u be the strictly convex (Alexandrov) solution of Monge–Ampère equation (6.1) with $f \geq \lambda^d > 0$. Suppose that the nodes Ω_h^I are translation-invariant, and let $u_h \in X_h^{fd}$ be the solution of (6.9). If $\lambda I \leq D^2 u \leq \Lambda I$ and $u \in C^{2,\alpha}(\Omega)$, then

$$\|u - \Gamma(u_h)\|_{L^\infty(\Omega)} \leq Ch^\alpha,$$

where the constant

$$C = C(d, \Omega, \lambda, \Lambda)(|u|_{C^{2,\alpha}(\bar{\Omega})} + |u|_{C^{1,1}(\bar{\Omega})}).$$

Proof. We begin by constructing a piecewise linear approximation of u. Recall that $\Gamma(I_h^{fd}u)$, the convex envelope of the nodal function $I_h^{fd}u$, is a piecewise linear function over a mesh that has $\bar{\Omega}_h$ as nodes. Thus, classic interpolation theory yields

$$\|\Gamma(I_h^{fd}u) - u\|_{L^\infty(\Omega)} \leq Ch^2|u|_{C^{1,1}(\bar{\Omega})}.$$

Therefore, we only need to prove that

$$\sup_{\bar{\Omega}_h}(I_h^{fd}u - u_h)^- \leq Ch^\alpha. \tag{6.20}$$

A similar inequality, which controls the positive part of $I_h^{fd}u - u_h$, can be derived in an analogous fashion.

Step 1. We first show that for all $z \in \bar{\Omega}_h$ such that $\operatorname{dist}(z, \partial\Omega) \leq \bar{R}h$,

$$(I_h^{fd}u - u_h)(z) \geq -Ch|u|_{C^{1,1}(\bar{\Omega})}. \tag{6.21}$$

Let p_z be the discrete barrier defined in Lemma 6.34 with free parameter M and consider the function $u_h + p_z$. Since Lemma 3.21 (addition inequality) implies

$$\partial u_h(z_i) + \partial p_z(z_i) \subset \partial(u_h + p_z)(z_i),$$

by Proposition 6.28 (Brunn–Minkowski inequality), we obtain

$$|\partial(u_h + p_z)(z_i)| \geq \big(|\partial u_h(z_i)|^{1/d} + |\partial p_z(z_i)|^{1/d}\big)^d.$$

Therefore, by Lemmas 6.33 and 6.34, we have

$$\begin{aligned} |\partial(u_h + p_z)(z_i)| &\geq \left(\left(\int_{\omega_{z_i}} \det(D^2 u)\right)^{1/d} + (M|\omega_z|)^{1/d}\right)^d \\ &\geq |\partial I_h^{fd} u(z_i)|, \quad \text{for all } z_i \in \Omega_h^I \end{aligned}$$

provided that $M = C|u|^d_{C^{1,1}(\bar{B})}$ Since $p_z \leq 0$ on Ω_h^B, we have $u_h + p_z \leq I_h^{fd} u$ on Ω_h^B. Moreover, because $|\partial(u_h + p_z)(z_i)| \geq |\partial I_h^{fd} u(z_i)|$ for all $z_i \in \Omega_h^I$, we have, by the maximum principle of Corollary 6.31,

$$u_h(z_i) + p_z(z_i) \leq I_h^{fd} u(z_i), \quad \text{for all } z_i \in \bar{\Omega}_h.$$

The estimate on the discrete barrier function, given in Lemma 6.34, yields

$$u_h(z) - Ch|u|_{C^{1,1}(\bar{\Omega})} \leq u_h(z) + p_z(z) \leq I_h^{fd} u(z), \tag{6.22}$$

thus proving (6.21).

Step 2. For all nodes z with $\operatorname{dist}(z, \partial\Omega) \geq \bar{R}h$, thanks to the consistency of the method, Lemma 6.32, we have

$$\big||\partial I_h^{fd} u(z)| - |\partial u_h(z)|\big| \leq Ch^\alpha |u|_{C^{2,\alpha}(\bar{\Omega})} |\omega_z|. \tag{6.23}$$

We apply the stability result of Proposition 6.30 on a smaller domain,

$$\bar{\Omega}_{*,h} = \{z \in \Omega_h^I : \operatorname{dist}(z, \partial\Omega) \geq \bar{R}h\}, \tag{6.24}$$

and on the nodal functions $I_h^{fd} u$ and $u_h - Ch|u|_{C^{1,1}(\bar{\Omega})}$, where C is the constant in (6.21) so that $I_h^{fd} u \geq u_h - Ch|u|_{C^{1,1}(\bar{\Omega})}$ on the boundary nodes of $\bar{\Omega}_{*,h}$. Letting $\mathcal{C}_{h,*}^-$ denote the contact set of $I_h^{fd} u - (u_h - Ch|u|_{C^{1,1}(\bar{\Omega})})$ with respect to $\bar{\Omega}_{h,*}$, we get

$$\begin{aligned} &\sup_{\bar{\Omega}_{*,h}} \big(I_h^{fd} u - (u_h - Ch|u|_{C^{1,1}(\bar{\Omega})})\big)^- \\ &\qquad \leq C\left(\sum_{z \in \mathcal{C}_{h,*}^-} \big(|\partial I_h^{fd} u(z)|^{1/d} - |\partial u_h(z)|^{1/d}\big)^d\right)^{1/d}. \end{aligned} \tag{6.25}$$

Note now that $t \to t^{1/d}$ is a concave function and therefore

$$(t + \delta)^{1/d} \leq t^{1/d} + \frac{1}{d} t^{(1-d)/d} \delta.$$

Thus, by setting $t = |\partial u_h(z)|$, $\delta = |\partial I_h^{fd} u(z)| - |\partial u_h(z)|$ and applying (6.23), we find

$$\begin{aligned} |\partial I_h^{fd} u(z)|^{1/d} - |\partial u_h(z)|^{1/d} &\leq \frac{1}{d}|\partial u_h(z)|^{(1-d)/d}(|\partial u_h(z)| - |\partial I_h^{fd} u(z)|) \\ &\leq Ch^\alpha |u|_{C^{2,\alpha}(\bar\Omega)} |\omega_z| \left(\int_{\omega_z} f\right)^{(1-d)/d} \\ &\leq Ch^\alpha |u|_{C^{2,\alpha}(\bar\Omega)} |\omega_z|^{1/d} \end{aligned}$$

because $f \geq \lambda^d > 0$. Inserting this estimate into (6.25) yields

$$\begin{aligned} \sup_{\bar\Omega_{h,*}} (I_h^{fd} u - (u_h - Ch|u|_{C^{1,1}(\bar\Omega)}))^- &\leq Ch^\alpha |u|_{C^{2,\alpha}(\bar\Omega)} \Bigg(\sum_{z \in \mathcal{C}_{h,*}^-} |\omega_z| \Bigg)^{1/d} \\ &\leq Ch^\alpha |u|_{C^{2,\alpha}(\bar\Omega)} |\Omega|^{1/d}, \end{aligned}$$

or equivalently

$$\sup_{\bar\Omega_{h,*}} (I_h^{fd} u - u_h)^- \leq Ch|u|_{C^{1,1}(\bar\Omega)} + Ch^\alpha |u|_{C^{2,\alpha}(\bar\Omega)}.$$

This inequality, together with (6.22), proves the lower bound for $I_h^{fd} u - u_h$. The upper bound can be proved in a similar fashion. □

Note that Proposition 6.30 controls the L^∞-error by the L^d-norm of the consistency error. Thus, if large consistency errors occur only in regions with small measure, we may still derive a rate of convergence. This observation may be used for solutions that are not $C^2(\Omega)$-regular.

To state the result, we first introduce the Minkowski–Bouligand dimension. Let U be a subset of Ω. Let $\{\omega_i\}_{z_i \in \Omega_h^I}$ be a translation-invariant partition covering Ω where ω_i is as in (6.10). Define $m = m(h)$ to be the number of elements of $\{\omega_{z_i}\}$ required to cover U. We define the (Minkowski–Bouligand) dimension of U as

$$\dim U = -\lim_{h \to 0} \frac{\log m(h)}{\log h}.$$

For example, it is easy to check that ∂B_1, the discontinuity set of $D^2 u$ in Example 6.5, is of dimension one. Note that in this example the solution $u \in C^{1,1}(\bar\Omega) \setminus C^2(\bar\Omega)$.

The next result addresses the rate of convergence for piecewise smooth solutions such that the discontinuity set of $D^2 u$ is of low dimension.

Theorem 6.36 (rate of convergence II). Let $u \in C^{1,1}(\bar\Omega)$ be strictly convex with $\lambda I \leq D^2 u \leq \Lambda I$, and solve (6.1). Assume that $D^2 u$ is piecewise Hölder-continuous (with exponent $\alpha > 0$) and its discontinuity set U has dimension $n < d$. Let u_h be the solution of (6.9), over a lattice Ω_h^I of translation-invariant nodes. Then

$$\| u - \Gamma(u_h) \|_{L^\infty(\Omega)} \le Ch^\alpha |u|_{C^{2,\alpha}(\bar\Omega\setminus U)} + Ch^{(d-n)/d}|u|_{C^{1,1}(\bar\Omega)}.$$

Proof. Following the estimate of Theorem 6.35, we first note that for $z \in \bar\Omega_h$ with $\mathrm{dist}(z, \partial\Omega) \le \bar R h$,

$$|(u_h - I_h^{fd} u)(z)| \le Ch. \tag{6.26}$$

Since $D^2 u$ is Hölder-continuous except on the set U and the aspect ratio Λ/λ of $D^2 u$ is bounded, by Lemmas 6.32 and 6.33 we have that

$$|\partial I_h^{fd} u(z)| \le \begin{cases} \displaystyle\int_{\omega_z} \det D^2 u + Ch^\alpha |\omega_z|\, |u|_{C^{2,\alpha}(\bar\Omega\setminus U)} & z \in \bar\Omega_{h,*} \setminus U_{\bar R h}, \\ \displaystyle\int_{\omega_z} \det D^2 u + C|\omega_z|\, |u|_{C^{1,1}(\bar\Omega)} & z \in U_{\bar R h}, \end{cases}$$

where $\bar\Omega_{h,*}$ is given by (6.24) and

$$U_{\bar R h} = \{z \in \bar\Omega_{h,*} : \mathrm{dist}(z, U) \le \bar R h\}.$$

Now, the stability estimate of Proposition 6.30 yields

$$\begin{aligned} &\sup_{\bar\Omega_{h,*}} \big(I_h^{fd} u - (u_h - Ch|u|_{C^{1,1}(\bar\Omega)})\big)^- \qquad (6.27) \\ &\quad \le C\bigg(\sum_{z \in \mathcal{C}_{h,*}^-} \big(|\partial I_h^{fd} u(z)|^{1/d} - |\partial u_h(z)|^{1/d}\big)^d \bigg)^{1/d}, \end{aligned}$$

where $C > 0$ is the constant in (6.26).

For $z \in \bar\Omega_{h,*} \setminus U_{\bar R h}$, we apply the same arguments as in the proof of Theorem 6.35, to get

$$|\partial I_h^{fd} u(z)|^{1/d} - |\partial u_h(z)|^{1/d} \le Ch^\alpha |u|_{C^{2,\alpha}(\bar\Omega\setminus U)} |\omega_z|^{1/d}.$$

For $z \in U_{\bar R h}$ similar arguments (essentially take $\alpha = 0$) yield

$$|\partial I_h^{fd} u(z)|^{1/d} - |\partial u_h(z)|^{1/d} \le C|u|_{C^{1,1}(\bar\Omega)} |\omega_z|^{1/d}.$$

Inserting these estimates into (6.27), we deduce that

$$\begin{aligned} \sup_{\bar\Omega_{h,*}} (I_h^{fd} u - u_h)^- &\le C\bigg(h^{\alpha d} |u|^d_{C^{2,\alpha}(\bar\Omega\setminus U)} |\Omega| + |u|^d_{C^{1,1}(\bar\Omega)} \sum_{z \in U_{\bar R h}} |\omega_z| \bigg)^{1/d} \\ &\le Ch^\alpha |u|_{C^{2,\alpha}(\bar\Omega\setminus U)} + C|u|_{C^{1,1}(\bar\Omega)} \bigg(\sum_{z \in U_{\bar R h}} |\omega_z| \bigg)^{1/d}. \end{aligned}$$

Since $\dim U = n < d$, we have

$$\sum_{z \in U_{\bar R h}} |\omega_z| = N|\omega_z| \le CNh^d$$

with $N \leq Ch^{-n}$. Thus, we conclude that

$$\sup_{\bar{\Omega}_{h,*}} (I_h^{fd} u - u_h)^- \leq C(h^\alpha |u|_{C^{2,\alpha}(\bar{\Omega}\setminus U)} + h^{(d-n)/d} |u|_{C^{1,1}(\bar{\Omega})}).$$

This completes the proof. □

Remark 6.37 (extensions). The developments of this section have found the following extensions.

(1) Nochetto *et al.* (2017) have recently applied the error analysis developed for method (6.9) to the wide-stencil schemes by Oberman (2008) and Froese and Oberman (2011).

(2) The error estimate stated in Theorems 6.35 and 6.36 applies only to structured nodes (lattices). It remains an open problem how to extend the analysis to unstructured nodes and to the degenerate case, that is, when $f = 0$ in some region.

(3) The error analysis of monotone schemes for the Monge–Ampère equation in other norms, such as the H^1-norm, remains an open problem.

7. Discretizations of non-convex second-order elliptic equations

This section is a continuation of the developments of Section 5, where we apply the discretizations and results for uniformly elliptic linear PDEs to fully nonlinear problems. However, in contrast to Section 5, we shall not assume convexity (or concavity) of the differential operator, but rather, only assume it is uniformly elliptic. As explained in Section 1.3, it suffices to consider numerical approximations of the Isaacs equation given in Example 2.13. For simplicity and to communicate the essential points in the discussion, we shall assume that the nonlinear problem does not have lower-order terms, and in addition has homogeneous Dirichlet boundary conditions. Thus we consider numerical approximations for the problem

$$\begin{cases} F(x, D^2u) := \inf_{\beta\in\mathcal{B}} \sup_{\alpha\in\mathcal{A}} [\mathcal{L}^{\alpha,\beta} u(x) - f^{\alpha,\beta}(x)] = 0 & \text{in } \Omega, \\ u = 0 & \text{on } \partial\Omega, \end{cases} \tag{7.1}$$

with $\mathcal{L}^{\alpha,\beta} u(x) = A^{\alpha,\beta}(x) : D^2 u(x)$, and the coefficient matrices satisfy

$$\lambda I \leq A^{\alpha,\beta} \leq \Lambda I, \quad \text{for all } \alpha \in \mathcal{A},\ \beta \in \mathcal{B},$$

so that F is uniformly elliptic. We assume that for each $x \in \Omega$, the mapping $M \mapsto F(x, M)$ is continuous and locally Lipschitz-continuous on $\mathbb{S}^d$, and that $f^{\alpha,\beta} \in C(\bar{\Omega})$ for each $\alpha \in \mathcal{A}$ and $\beta \in \mathcal{B}$. Further assumptions may be made in subsequent developments.

As in the convex case, we can roughly classify numerical schemes as finite difference, finite element and semi-Lagrangian methods. The construction and analysis of finite difference schemes as detailed in Section 7.1, in principle, follow the convex case of Section 5 and the Barles–Souganidis theory as presented in Theorems 3.7 and 3.8, but are clouded by numerous technicalities that have for many years prevented researchers from obtaining rates of convergence. In fact – as described in the introduction of Kuo and Trudinger (1992) – this was considered an important unsolved open problem for several years.

The heart of the issue can be captured by examining the proof of Theorem 5.3. An important step there is the construction of a smooth subsolution to the equation (scheme) which, in the convex case, can be obtained by mollification of a subsolution to a perturbed equation. Convexity of the operator allows us to claim that this is a subsolution to the original problem and can thus be used to carry out the programme detailed at the beginning of Section 5.1. However, without convexity, it is not clear how to construct a smooth approximation to the solution of (7.1), which can be used to invoke the consistency of the scheme. This is particularly important in the non-convex case since, as shown in Example 2.70, one cannot assume smoothness of the solution to (7.1).

For many years, all of the available results were rather specialized. For instance, Jakobsen (2004) considers a one-dimensional problem and clearly shows that the arguments do not extend to more dimensions. A particular case of an Isaacs equation – an obstacle problem for an HJB equation – is discussed in Jakobsen (2006), where this special structure is exploited.

The derivation of rates of convergence for general schemes remained an unsolved problem until Caffarelli and Souganidis (2008) showed how to obtain a rate of convergence, within the Barles–Souganidis framework, for approximations of (7.1) in the case that, for all $\alpha \in \mathcal{A}$ and $\beta \in \mathcal{B}$, the matrices $A^{\alpha,\beta}$ do not depend on x. We detail these results in Section 7.2, where we also comment on extensions and variations to these estimates. Simply put, the estimates assert that there exists an algebraic rate of convergence. An explicit rate, however, is not available at the moment.

Quite recently, Salgado and Zhang (2016) have extended the results of the finite element method in Section 4.3 to the case of (7.1), and obtained an algebraic rate of convergence which, as in the finite difference case, is not explicit. These developments are detailed in Section 7.3.

We also comment that, as of the time of writing, no rates of convergence are available for semi-Lagrangian schemes. The only known result is convergence, as obtained by Debrabant and Jakobsen (2013).

We conclude our discussion on non-convex equations in Section 7.4 by describing how to solve the nonlinear system of equations that results after discretization, by any of the schemes discussed before.

7.1. Finite difference methods

Here, following the framework given in Section 3.2 and in Kuo and Trudinger (1992), we construct finite difference approximations to the fully nonlinear problem (7.1), and study the stability and convergence of these discretizations. We consider the following problem: find $u_h \in X_h^{fd}$ satisfying

$$\begin{cases} F_h[u_h] = 0 & \text{in } \Omega_h^I, \\ u_h = 0 & \text{on } \Omega_h^B, \end{cases} \tag{7.2}$$

where the interior and boundary nodes are as in Definition 3.14. We assume that F_h is consistent and that the scheme is of the form $F_h[u_h](z) = \mathcal{F}_h(z, \delta_h^2 u_h(z))$ with $\delta_h^2 u_h(z) = \{\delta_{y,h}^2 u_h(z) : y \in S\}$. We further assume that $\mathcal{F}_h$ is of positive type in the sense of Definition 3.12, so that F_h is monotone, and that

$$\frac{\partial \mathcal{F}_h}{\partial s_y} \leq \Lambda_0,$$

for some $\Lambda_0 > 0$. For example, schemes that satisfy these properties, and the ones that we have in mind, are

$$F_h[u_h] = \inf_{\beta \in \mathcal{B}} \sup_{\alpha \in \mathcal{A}} (\mathcal{L}_h^{\alpha,\beta} u_h - f^{\alpha,\beta}),$$

where each (linear) discrete operator is given by

$$\mathcal{L}_h^{\alpha,\beta} u_h(z) = \sum_{y \in S} a_y^{\alpha,\beta}(z) \delta_{y,h}^2 u_h(z),$$

and is of positive type and consistent with $\mathcal{L}^{\alpha,\beta}$; Section 3.4 describes how to construct linear operators with these properties.

Before discussing the solvability of the finite difference scheme (7.2), let us show first that solutions to the scheme are uniformly bounded. Let v_h, w_h be two grid functions, and consider the linearization

$$\begin{aligned} F_h[v_h](z) - F_h[w_h](z) &= \mathcal{L}_h(v_h - w_h)(z) \\ &:= \sum_{y \in S} a_y(z) \delta_{y,h}^2 (v_h - w_h)(z), \end{aligned} \tag{7.3}$$

with

$$a_y(z) = \int_0^1 \frac{\partial \mathcal{F}_h}{\partial s_y}(z, \delta_{h,y}^2 q_t(z)) \,\mathrm{d}t, \quad q_t = t v_h + (1-t) w_h.$$

In particular, if we set $v_h = u_h$, $w_h = 0$, and $f = -F(x, 0)$, then the solution to (7.2) satisfies

$$\mathcal{L}_{h,u_h} u_h = f \quad \text{in } \Omega_h^I, \tag{7.4}$$

where the coefficients in the operator $\mathcal{L}_{h,u_h}$ depend on u_h. Since $\mathcal{L}_{h,u_h}$ is a positive operator, Theorem 3.25 yields the follow stability result.

Theorem 7.1 (uniqueness and stability). In this setting, solutions to (7.2) are unique, and satisfy

$$\|u_h\|_{L^\infty(\bar{\Omega}_h)} \leq C\bigg(\sum_{z\in\Omega_h^I} h^d|f(z)|^d\bigg)^{1/d} \leq C.$$

In addition to uniqueness and stability, these *a priori* estimates also imply the existence of solutions.

Theorem 7.2 (existence). Under these conditions, problem (7.2) has a unique solution.

Proof. Consider the map $Q_h : X_h^{fd} \to X_h^{fd}$ satisfying

$$\begin{cases} \mathcal{L}_{h,v_h} Q_h(v_h) = f & \text{in } \Omega_h^I, \\ v_h = 0 & \text{on } \Omega_h^B. \end{cases}$$

Theorem 3.25 ensures that Q_h is well-defined and that $\|Q_h(v_h)\|_{L^\infty(\bar{\Omega}_h)} \leq C$. Brouwer's fixed-point theorem shows that Q_h has a fixed point $u_h \in X_h^{fd}$, which is a solution to (7.4), and hence (7.2). □

Remark 7.3 (other proofs). Other (constructive) proofs of existence of solutions can be found in Kuo and Trudinger (1992, Section 4) and in Section 7.4.

Finally we apply Theorem 3.30 to deduce that solutions to (7.2) are Hölder-continuous.

Proposition 7.4 (discrete Hölder continuity). Suppose that $u_h \in X_h^{fd}$ solves (7.2). Then there exists $\eta \in (0,1)$ and $C > 0$ depending on the data, such that, for $z_1, z_2 \in \bar{\Omega}_h$,

$$|u_h(z_1) - u_h(z_2)| \leq C|z_1 - z_2|^\eta.$$

Proposition 7.4 implies that the sequence of solutions $\{u_h\}_{h>0}$ is equicontinuous. Thus, applying Theorem 3.8 we obtain convergence to the viscosity solution.

Theorem 7.5 (convergence). The solutions u_h to (7.2) converge locally uniformly to the viscosity solution of (7.1).

7.2. Rates of convergence for finite difference schemes with constant coefficients

Here we summarize the results regarding rates of convergence of finite difference schemes for fully nonlinear, non-convex PDEs, as originally shown by Caffarelli and Souganidis (2008). As a starting point, we focus on the case where the operator in (7.1) is such that, for every $\alpha \in \mathcal{A}$ and $\beta \in \mathcal{B}$, the matrices $A^{\alpha,\beta}$ are independent of x, and, moreover, we have $f^{\alpha,\beta} = f \in C^{0,1}(\Omega)$. Under these conditions, problem (7.1) satisfies a comparison principle (see Theorem 2.52), and there exists a unique viscosity solution with regularity $u \in C^{1,s}(\bar{\Omega})$ for some universal constant $s > 0$; see Theorem 2.71.

In order to derive rates, we shall, with an abuse of notation, extend the domain of F_h to continuous functions and to all

$$z \in \Omega^I := \{x \in \Omega : \operatorname{dist}\{x, \partial\Omega\} \geq mh\} \supset \Omega_h^I,$$

where m is the stencil size. We also assume, in this section, the following (strengthened) consistency criterion:

$$|F(z, D^2\phi(z)) - F_h[\phi](z)| \leq Ch\|D^3\phi\|_{L^\infty(\Omega)}, \tag{7.5}$$

for all smooth ϕ and $z \in \Omega_h^I$.

Now, as in the convex case, the key idea to obtain rates of convergence is to construct a smooth function u_ϵ that is a subsolution to (7.1) and then apply the consistency and monotonicity of the finite difference operator to get a one-sided bound of the error; see the proof of Theorem 5.3 for details. However, unlike the convex case, it is not immediate how to construct a sufficiently smooth subsolution to carry out this programme. For example, the standard mollification $u^\epsilon = u_\epsilon \star \rho_\epsilon$ used in Theorem 5.3 is no longer a subsolution due to the lack of convexity of F (and F_h).

Instead, we employ the so-called *sup- (inf-) convolutions* of the viscosity solution u.

Definition 7.6 (sup-convolution). Let $u \in C(\bar{\Omega})$ and $\tau > 0$, The *sup- (inf-) convolution* $u_\tau^\pm$ of u is

$$u_\tau^+(x) = \sup_{y \in \bar{\Omega}} \left[u(y) - \frac{1}{h^\tau}|x-y|^2 \right],$$

$$u_\tau^-(x) = \inf_{y \in \bar{\Omega}} \left[u(y) + \frac{1}{h^\tau}|x-y|^2 \right].$$

Remark 7.7 (alternative definition). Note that the definition we give here is tied to a mesh size h. We do so because this is the scale that suits our needs. In general the literature defines the sup-convolution of a function by

$$u_\varrho^+ = \sup_{y \in \bar{\Omega}} \left[u(y) - \frac{1}{2\varrho}|x-y|^2 \right], \quad \varrho > 0.$$

The change of variables $\varrho = \frac{1}{2}h^\tau$ shows the relation between these two. Similar reasoning can be used for the inf-convolution.

We show two examples of sup-convolution of functions to motivate the introduction of this useful concept.

Example 7.8 (sup-convolution I). Let $u = |x|^2$ be defined in $\mathbb{R}^d$ and consider its sup-convolution

$$u_\tau^+(x) = \sup_{y\in\mathbb{R}^d}\left[|y|^2 - \frac{1}{h^\tau}|x-y|^2\right].$$

Let $y_* = y_*(x)$ be a point where the supremum is attained. Then a simple calculation shows that

$$2y_* + 2h^{-\tau}(x - y_*) = 0,$$

which implies that $y_* = (1-h^\tau)^{-1}x$. Inserting y_* into the definition of $u_\tau^+(x)$, we obtain

$$u_\tau^+(x) = \left(\frac{1}{1-h^\tau}\right)^2|x|^2 - h^{-\tau}\left(\frac{h^\tau}{1-h^\tau}\right)^2|x|^2 = \frac{|x|^2}{(1-h^\tau)}.$$

See Figure 7.1 and note that $D^2u_\tau^+ > D^2u$.

Next, we consider a function that is less smooth.

Example 7.9 (sup-convolution II). Let $u = -|x|$ be defined in $\mathbb{R}^d$. The maximum is attained at $y_* = y_*(x)$ if and only if

$$\mathbf{0} \in -\partial|y_*| + 2h^{-\tau}(x - y_*)$$

or, equivalently,

$$x \in y_* + \frac{h^\tau}{2}\partial|y_*|,$$

where $\partial|y_*|$ denotes the subdifferential of $|\cdot|$ at y_*. We note that if $|x| \le h^\tau/2$, then $y_* = \mathbf{0}$ because $\partial|\mathbf{0}| = B_1(0)$. Otherwise, we have

$$x = y_* + \frac{h^\tau y_*}{2|y_*|}$$

because $\partial|y_*| = \{y_*/|y_*|\}$ for $|y_*| > 0$. Therefore, we conclude that

$$u_\tau^+(x) = \begin{cases} -h^{-\tau}|x|^2 & |x| \le \dfrac{h^\tau}{2}, \\ -|x| + \dfrac{h^\tau}{4} & \text{otherwise.} \end{cases}$$

Note that u_τ^+ is $C^{1,1}$, while u is only Lipschitz. Moreover, near the singularity $x = 0$, u_τ^+ behaves like a paraboloid with $D^2u_\tau^+ = -2h^{-\tau}I$; see Figure 7.2.

Example 7.9 shows that, intuitively speaking, the sup-convolution u_τ^+ 'opens up' the kinks of u.

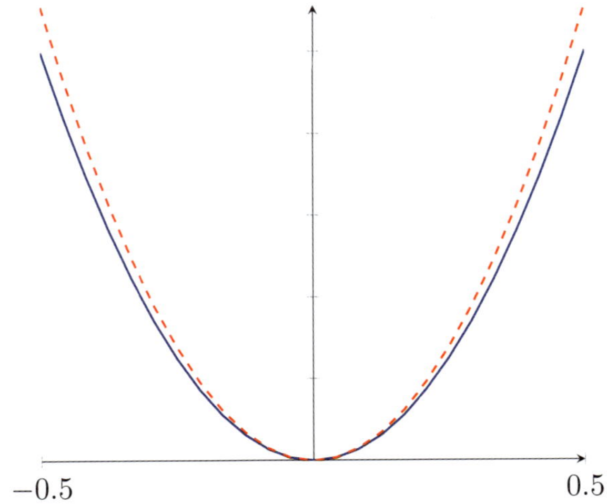

Figure 7.1. The graph of the function $u(x) = x^2$ (solid blue) and its sup-convolution (dashed red) with parameters $h = 0.1$ and $\tau = 1$.

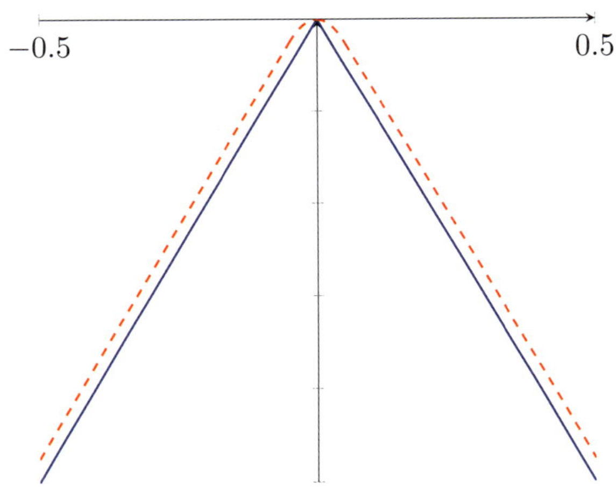

Figure 7.2. The graph of the function $u(x) = -|x|$ (solid blue) and its sup-convolution (dashed red) with parameters $h = 0.1$ and $\tau = 1$.

Figures 7.1 and 7.2 illustrate some of the general properties of sup- and inf-convolutions. To state them concisely, we begin with a definition.

Definition 7.10 (opening t). We say that $p \in \mathbb{P}_2$ is a paraboloid of *opening* t if, for some $\ell \in \mathbb{P}_1$, we have

$$p(x) = \ell(x) \pm \frac{t}{2}|x|^2.$$

The main properties of sup- and inf-convolutions are as follows. To simplify the presentation, we let $\bar{\partial} u(x)$ denote the superdifferential of u at the point x, that is,

$$\mathbf{v} \in \bar{\partial} u(x) \Leftrightarrow -\mathbf{v} \in \partial(-u)(x).$$

Proposition 7.11 (properties of u_τ^+). Let $u \in C^{0,1}(\bar{\Omega})$ and, for $x \in \Omega$, let $y_* = y_*(x)$ denote the point where the supremum in the definition of u_τ^+ is attained. The following statements hold:

(i) $x = y_* + \frac{1}{2}h^\tau \mathbf{v}$ for some vector $\mathbf{v} \in \bar{\partial} u(y_*)$ and, therefore, $|x - y_*| \leq Ch^\tau$,

(ii) $\|u - u_\tau^+\|_{L^\infty(\Omega)} \leq Ch^\tau$,

(iii) there exists a paraboloid of opening $2h^{-\tau}$, that touches u_τ^+ (resp. u_τ^-) from below (resp. above) at x,

(iv) if u is the viscosity solution to (7.1), then $|x_1 - x_2| \leq C|y_*(x_1) - y_*(x_2)|$.

Proof. Let us prove each statement separately.

(i) For any fixed x, if $u(y) - h^{-\tau}|x-y|^2$ attains its maximum at y_*, then

$$\mathbf{0} \in \bar{\partial} u(y_*) - 2h^{-\tau}(x - y_*).$$

Thus, we have $x = y_* + \frac{1}{2}h^\tau \mathbf{v}$ for some $\mathbf{v} \in \bar{\partial} u(y_*)$.

(ii) By definition $u_\tau^+ \geq u$. Moreover, property (i) implies that

$$0 \leq u_\tau^+(x) - u(x) = u(y_*) - u(x) - h^{-\tau}\left|\frac{1}{2}h^\tau \mathbf{v}\right|^2 \leq |u(x) - u(y_*)| + \frac{1}{4}h^\tau|\mathbf{v}|^2$$

for some $\mathbf{v} \in \bar{\partial} u(y_*)$. Since u is Lipschitz-continuous, we conclude that, for every $x \in \Omega$,

$$|u(x) - u_\tau^+(x)| \leq Ch^\tau.$$

(iii) Since, for any $x \in \Omega$, $y_* = y_*(x)$ is the point where the supremum is attained, let us define the paraboloid

$$p(z) = u(y_*) - \frac{1}{h^\tau}|z - y_*|^2,$$

and note that $p(x) = u_\tau^+(x)$. Moreover,

$$p(z) \leq \sup_{y\in\bar{\Omega}}\left[u(y) - \frac{1}{h^\tau}|z-y|^2\right] =: u_\tau^+(z), \quad \text{for all } z \in \bar{\Omega}.$$

Thus, the paraboloid p touches the graph of u_τ^+ at point x from below.

(iv) Since, by definition,

$$u_\tau^+(x) = u(y_*(x)) - \frac{1}{h^\tau}|x - y_*(x)|^2,$$

upon defining the paraboloid

$$p(z) = u_\tau^+(x) + \frac{1}{h^\tau}|x-z|^2$$

we have that

$$p(y_*(x)) = u(y_*(x)), \quad p(z) \geq u(z), \quad \text{for all } z \in \bar{\Omega}.$$

In other words, at $y_*(x)$, the function u is touched from below by a paraboloid with opening $2h^{-\tau}$. On the other hand, since u solves (7.1), the Harnack inequality of Theorem 2.63 implies that, at $y_*(x)$, the function u can be touched from above by a paraboloid of opening $Ch^{-\tau}$ with C depending only on d, λ and Λ. Invoking Caffarelli and Cabré (1995, Proposition 1.2) we conclude that for every $x_1, x_2 \in \bar{\Omega}$ the function u is differentiable at $y_*(x_i)$, $i = 1, 2$ and, moreover,

$$|Du(y_*(x_1)) - Du(y_*(x_2))| \leq Ch^{-\tau}|y_*(x_1) - y_*(x_2)|.$$

We now invoke (i) to obtain that

$$|x_1 - x_2| = \left|y_*(x_1) - y_*(x_2) + \frac{1}{2}h^\tau Du(y_*(x_1)) - \frac{1}{2}h^\tau Du(y_*(x_2))\right|.$$

The previous two inequalities yield the claim.

This completes the proof. □

Next we establish a lower bound for $F_h[u_\tau^+](\cdot)$. We define

$$\Omega_h^{I,\tau} = \{z \in \Omega_h^I : \operatorname{dist}(z, \partial\Omega) \geq Ch^\tau\},$$

where the constant $C > 0$ depends on the stencil size m and is chosen so that $F_h[u](x)$ is well-defined for x satisfying $|x-z| \leq Ch^\tau$ for some $z \in \Omega_h^{I,\tau}$; see Figure 7.3.

Proposition 7.12 (bound on $F_h[u_\tau^+](\cdot)$). Assume that $u \in C^{0,1}(\bar{\Omega})$ and that $z \in \Omega_h^{I,\tau}$. Let $y_* = y_*(z) \in \Omega$ be chosen so that

$$u_\tau^+(z) = u(y_*) - \frac{1}{h^\tau}|z - y_*|^2. \tag{7.6}$$

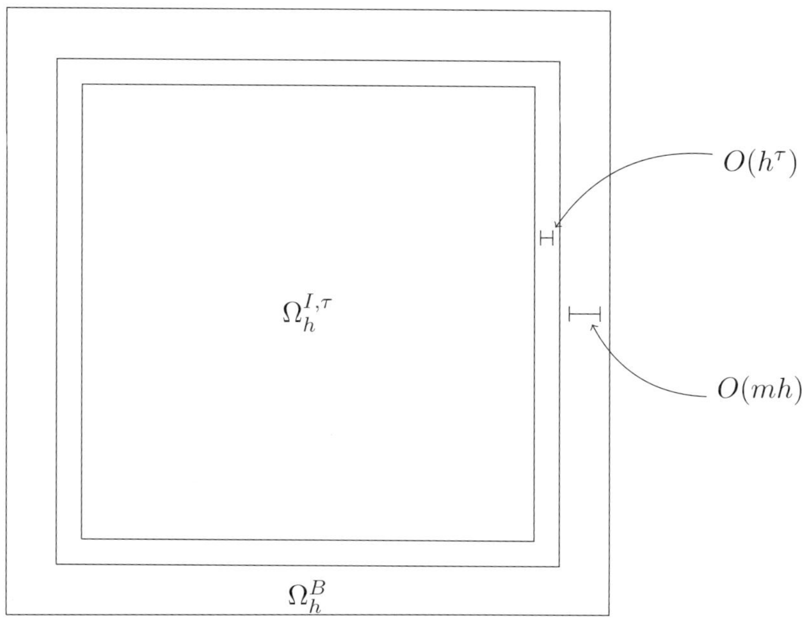

Figure 7.3. A pictorial description of the set $\Omega_h^{I,\tau}$. The values $F_h[u](x)$ are well-defined for x within a distance of Ch^τ of this set.

We then have

$$F_h[u_\tau^+](z) \geq F_h[u](y_*) + (f(y_*) - f(z)).$$

Moreover, for all $z \in \Omega_h^I$, we have

$$F_h[u_\tau^+](z) \geq -Ch^{-\tau}. \tag{7.7}$$

Proof. First note that, by Proposition 7.11(i) we have that $|z - y_*| \leq Ch^\tau$, and therefore $F_h[u](y_*)$ is well-defined. Now, owing to the monotonicity of F_h, to show the first estimate it suffices to show that

$$\delta_{y,h}^2 u_\tau^+(z) \geq \delta_{y,h}^2 u(y_*) \quad y \in S.$$

Define the function

$$v(x) = u(x - z + y_*) - \frac{1}{h^\tau}|z - y_*|^2,$$

and note that

$$v(z) = u(y_*) - \frac{1}{h^\tau}|z - y_*|^2 = u_\tau^+(z).$$

The change of variables $\zeta = x - z + y_*$ reveals that

$$v(x) = u(\zeta) - \frac{1}{h^\tau}|\zeta - x|^2 \leq u_\tau^+(x),$$

and therefore we conclude that, for any $y \in S$, $\delta^2_{y,h} v(z) \leq \delta^2_{y,h} u^+_\tau(z)$. This, together with the fact that, for any $y \in S$, $\delta^2_{y,h} v(z) = \delta^2_{y,h} u(y_*)$, yields

$$\delta^2_{y,h} u(y_*) = \delta^2_{y,h} v(z) \leq \delta^2_{y,h} u^+_\tau(z).$$

Since f is independent of α and β, this proves the estimate $F_h[u^+_\tau](z) + f(z) \geq F_h[u](y_*) + f(y_*)$.

To prove (7.7), let

$$w(x) = u(y_*) - \frac{1}{h^\tau}|x - y_*|^2.$$

Clearly $w(z) = u^+_\tau(z)$ and $w(x) \leq u^+_\tau(x)$. Therefore, monotonicity and the identity

$$\delta^2_{y,h} w(z) = -\frac{2}{h^\tau}|y|^2$$

yields

$$F_h[u^+_\tau](z) \geq F_h[w](z) \geq -Ch^{-\tau},$$

which concludes the proof. □

Our goal now is to compare u^+_τ with the numerical solution u_h through $F_h[u_h]$ and $F_h[u^+_\tau]$ and a comparison principle for the discrete operator F_h. From Proposition 7.2 it follows that $F_h[u^+_\tau]$ at z depends on the consistency error of $F_h[u]$ at $y_* := y_*(z)$. The consistency of F_h, as stated in (7.5), implies that

$$F_h[p](y_*) = F(y_*, D^2 p), \quad \text{for all } p \in \mathbb{P}_2.$$

Consequently, the consistency error $F_h[u](y_*)$ depends on the quality of the approximation of the solution u by a quadratic polynomial at y_*. This, in turn, depends on the regularity of u.

The next result asserts that in our setting, outside sets of arbitrarily small measure, viscosity solutions to (7.1) have pointwise second-order Taylor expansions with an error that is controlled by the size of the singular set. For a proof, the reader is referred to Caffarelli and Souganidis (2008, Theorem A).

Theorem 7.13 (quadratic approximation). Assume that F has constant coefficients, is uniformly elliptic with ellipticity constants $\lambda, \Lambda > 0$, and that $f \in C^{0,1}(\bar{\Omega})$. Let u be a Lipschitz-continuous solution of $F(x, D^2u) = 0$ in $B_{2r}(x_0)$. There exist positive constants σ, t_0 and C that depend on λ, Λ and d such that, for all $t > t_0$, it is possible to find an open set $A_t \subset \Omega$ such that if $x \in A_t$, there exists a quadratic polynomial $p_x \in \mathbb{P}_2$ of opening less than t that satisfies

$$F(x, D^2 p_x(x)) = 0.$$

Moreover,

$$|u(y) - u(x) - p_x(y-x)| \le Ct|x-y|^3, \quad \text{for all } y \in B_{2r}(x_0), \tag{7.8}$$

and

$$|\Omega \backslash A_t| \le Ct^{-\sigma}.$$

This theorem essentially says that, for every t, except in a singular set of measure $t^{-\sigma}$, which we denote by $\mathcal{S}_t$, the function u has a second-order Taylor expansion. We define the regular set $\mathcal{R}_t := \bar{\Omega} \backslash \mathcal{S}_t$.

Let us now explain how Theorem 7.13 can be used to obtain a rate of convergence. By consistency, (7.5), we have that, for every $p \in \mathbb{P}_2$,

$$F_h[p](z) = F(z, D^2 p(z)), \quad \text{for all } z \in \Omega.$$

Therefore, if $t > 0$, $z \in \Omega$ and $B_h(z) \cap \mathcal{R}_t \neq \emptyset$, we choose $y \in B_h(z) \cap \mathcal{R}_t$ and set $p = p_y$, the second-order expansion of u at y from Theorem 7.13. We then estimate the consistency error as follows:

$$|F_h[u](z)| \le |F_h[u](z) - F_h[p_y](z)| + |F_h[p_y](z) - F_h[p_y](y)| + |F_h[p_y](y)|.$$

From estimate (7.8) and the fact that $F_h[p_y](z) - F_h[p_y](y) = f(y) - f(z)$ and $F_h[p_y](y) = 0$, it follows that

$$|F_h[u](z)| \le Cht + |f(z) - f(y)| \le C_1 ht + C_2 h$$

where $t > 1$, C_1 depends on the constant in (7.8) and stencil size m, and C_2 depends on the Lipschitz constant of f. Thus, the consistency error is of the order $O(th)$ on Ω_h^I except for a set that has measure $O(t^{-\sigma})$. To be more precise, we introduce the discrete regular and singular sets

$$\begin{aligned} \mathcal{R}_{t,h} &:= \{z \in \Omega_h^{I,\tau} : B_h(y_*(z)) \cap \mathcal{R}_t \neq \emptyset\}, \\ \mathcal{S}_{t,h} &:= \{z \in \Omega_h^{I,\tau} : z \notin \mathcal{R}_{t,h}\}. \end{aligned} \tag{7.9}$$

We now estimate the size of the discrete singular set.

Lemma 7.14 (cardinality of $\mathcal{S}_{t,h}$). Let $\mathcal{S}_{t,h}$ be defined by (7.9). Then

$$\#\mathcal{S}_{t,h} \le C(h^{-d} t^{-\sigma} + h^{1-d}).$$

Proof. Note that

$$Ch^d(\#\mathcal{S}_{t,h}) \le \sum_{z \in \mathcal{S}_{t,h}} |B_h(y_*(z))| \le C \left| \bigcup_{z \in \mathcal{S}_{t,h}} B_h(z) \right| \le C(t^{-\sigma} + h).$$

Indeed, the first inequality is obvious. The second one follows from Proposition 7.11(iv), which shows that, for $z_1, z_2 \in \bar{\Omega}_h$ with $z_1 \neq z_2$, we have $|y_*(z_1) - y_*(z_2)| \ge Ch$, so the overlap of the balls can be controlled independently of h. Finally, the last inequality follows from Theorem 7.13. Rearranging terms in this last inequality yields the result. □

We now estimate the consistency error of $F_h[u_\tau^+]$ and have the following corollary of Proposition 7.2.

Corollary 7.15 (consistency). For every $z \in \Omega_h^{I,\tau}$ we have

$$F_h[u_\tau^+](z) \geq -C \begin{cases} (ht + h^\tau) & z \in \mathcal{R}_{t,h}, \\ h^{-\tau} & z \in \mathcal{S}_{t,h}. \end{cases}$$

Proof. For $z \in \mathcal{R}_{t,h}$, since $|z - y_*| \leq Ch^\tau$ and f is Lipschitz-continuous, we have by Proposition 7.2

$$F_h[u_\tau^+](z) \geq F_h[u](y_*) - Ch^\tau \geq -C(ht + h^\tau).$$

The estimate on the singular set $\mathcal{S}_{t,h}$ was already established in (7.7). □

The previous result controls the consistency error in the interior of the domain. To obtain a rate of convergence for the scheme, we also need to control this error near the boundary. This is accomplished with the help of the following discrete barrier function.

Lemma 7.16 (discrete barrier). Let the domain Ω satisfy an exterior ball condition. If h is sufficiently small, then for any point $z \in \Omega_h^I$ with $\operatorname{dist}(z, \partial\Omega) \leq \delta$, there exists a discrete function $b_{z,h} \in X_h^{fd}$ such that

$$F_h[b_{z,h}](\zeta) \leq 0, \quad \text{for all } \zeta \in \Omega_h^I, \qquad b_{z,h}(\zeta) \geq 0, \quad \text{for all } \zeta \in \Omega_h^B.$$

Moreover, we have

$$|b_{z,h}(z)| \leq C\delta,$$

where C depends only on λ, Λ, d and Ω.

Proof. Let $z_* \in \partial\Omega$ such that $|z - z_*| = \operatorname{dist}(z, \partial\Omega)$. Since Ω satisfies an exterior ball condition, there exists a open ball $B_r(z')$ centred at some point z' such that

$$z_* \in \partial B_r(z') \quad \text{and} \quad B_r(z') \cap \Omega = \emptyset.$$

We construct a barrier function

$$b_z(x) = B(r^{-q} - |x - z'|^{-q}),$$

where B and q are some positive constant to be determined later. We observe that

$$\begin{aligned} Db_z(x) =& qB|x - z'|^{-q-2}(x - z'), \\ D^2 b_z(x) =& qB|x - z'|^{-q-2}\left(I - (2+q)\frac{(x - z')}{|x - z'|} \otimes \frac{(x - z')}{|x - z'|}\right). \end{aligned}$$

Note that $D^2 b_z(x)$ is a diagonal perturbation of a rank-one matrix that has eigenvalues 0 (with multiplicity $d - 1$) and $-q(2+q)B|x - z'|^{-q-2}$

(with multiplicity 1). Therefore $D^2 b_z$ has eigenvalues $qB|x-z'|^{-q-2}$ (with multiplicity $d-1$) and $-q(1+q)B/|x-z'|^{q+2}$ (with multiplicity 1). In particular the smallest eigenvalue of $D^2 b_z(x)$ is $-q(1+q)B/|x-z'|^{q+2}$ and the largest is $qB|x-z'|^{-q-2}$.

Since $\lambda I \leq A^{\alpha,\beta} \leq \Lambda I$ we have

$$\inf_{\beta\in\mathcal{B}} \sup_{\alpha\in\mathcal{A}} \big[A^{\alpha,\beta} : D^2 b_z(x)\big] \leq \frac{Bq}{|x-z'|^{q+2}}[(d-1)\Lambda - (1+q)\lambda].$$

Taking

$$q = 2(d-1)\frac{\Lambda}{\lambda} - 1,$$

we obtain that

$$\begin{aligned}\inf_{\beta\in\mathcal{B}} \sup_{\alpha\in\mathcal{A}} \big[A^{\alpha,\beta} : D^2 b_z(x)\big] &\leq \frac{Bq}{|x-z'|^{q+2}}(1-d)\Lambda \\ &\leq \frac{Bq}{|r+\operatorname{diam}(\Omega)|^{q+2}}(1-d)\Lambda.\end{aligned}$$

Since the right-hand side of the last inequality is negative, we conclude that, for B sufficiently large,

$$\inf_{\beta\in\mathcal{B}} \sup_{\alpha\in\mathcal{A}} \big[A^{\alpha,\beta} : D^2 b_z(x)\big] \leq -2\|f\|_{L^\infty(\Omega)}.$$

The discrete barrier function is then defined as $b_{h,z} = I_h^{fd} b_z$. It is easy to check that

$$b_{h,z}(\zeta) \geq 0, \quad \text{for all } \zeta \in \Omega_h^B, \quad |b_{h,z}(z)| \leq C\delta,$$

so that it remains to verify that $F_h[b_h](\zeta) \leq 0$ for all $\zeta \in \Omega_h^I$.

Since b_z is a smooth function, the consistency condition (7.5) implies

$$F_h[b_{h,z}](\zeta) - F(\zeta, D^2 b_z)(\zeta) = O(h).$$

Finally, because $F(\zeta, D^2 b_z)(\zeta) \leq -\|f\|_{L^\infty(\Omega)}$ we deduce that, for h sufficiently small,

$$F_h[b_{h,z}](\zeta) \leq 0, \quad \text{for all } \zeta \in \Omega_h^I.$$

This completes the proof. □

Remark 7.17 (bound on u_h). Corollary 7.15 implies that for any $z \in \Omega_h^I$ with $\operatorname{dist}(z, \partial\Omega) \leq \delta$, we have

$$|u_h(z)| \leq C\delta.$$

Indeed, since $b_{z,h} \geq u_h = 0$ on Ω_h^B and $F_h[b_{z,h}](z) \leq F_h[u_h](z)$ for all $z \in \Omega_h^I$, invoking a comparison principle for F_h, we obtain

$$u_h(z) \leq b_{z,h}(z) \leq C\delta.$$

Similarly we can show that $u_h(z) \geq -C\delta$.

Combining Corollary 7.15 and Lemma 7.14, we obtain a rate of convergence for the finite difference scheme (7.2).

Theorem 7.18 (rate of convergence). Let the domain Ω satisfy an exterior ball condition and $f \in C^{0,1}(\bar{\Omega})$. Let $u_h \in X_h^{fd}$ be the solution to (7.2) and $u \in C^{0,1}(\bar{\Omega})$ the viscosity solution to (7.1). Then we obtain

$$\|I_h^{fd} u - u_h\|_{L^\infty(\bar{\Omega}_h)} \le C h^{\sigma/(2d+\sigma)},$$

where σ is the exponent in Theorem 7.13.

Proof. By Proposition 7.11(ii) we have that $\|u - u_\tau^+\|_{L^\infty(\Omega)} \le Ch^\tau$. Thus it is sufficient to bound $v_h = u_h - I_h^{fd} u_\tau^+$. We shall show that $\sup_{\bar{\Omega}_h} v_h^- \le Ch^{\sigma/(2d+\sigma)}$; the proof of the complementary estimate is similar. We divide the proof into two steps.

Step 1 (*boundary estimate*). We first show that, for any $z \in \bar{\Omega}_h \backslash \Omega_h^{I,\tau}$,

$$|v_h(z)| \le Ch^\tau$$

for some constant $C > 0$. By the definition of v_h, we have

$$\begin{aligned} |v_h(z)| &\le |u_h(z) - u(z)| + |u(z) - I_h^{fd} u_\tau^+(z)| \\ &\le |u_h(z)| + |u(z)| + |u(z) - I_h^{fd} u_\tau^+(z)|. \end{aligned}$$

Since u is Lipschitz-continuous and $u = 0$ on $\partial\Omega$, we have that $|u(z)| \le Ch^\tau$. In addition, owing to Remark 7.17 and Proposition 7.11(ii), we conclude that

$$|v_h(z)| \le Ch^\tau.$$

Step 2 (*interior estimate*). Now

$$-F_h[u_\tau^+](z) = F_h[u_h](z) - F_h[u_\tau^+](z) \ge \mathcal{L}_h(u_h - I_h^{fd} u_\tau^+)(z) = \mathcal{L}_h v_h(z),$$

where $\mathcal{L}_h$ is the linear(ized) operator (7.3) with coefficients that depend on u_τ^+ and u_h. Thus, setting $t = h^{\tau-1}$, we have by Corollary 7.15

$$\mathcal{L}_h v_h(z) \le C \begin{cases} h^\tau & z \in \mathcal{R}_{h^{\tau-1},h}, \\ h^{-\tau} & z \in \mathcal{S}_{h^{\tau-1},h}, \end{cases} \tag{7.10}$$

and, by Lemma 7.14,

$$\#\mathcal{S}_{h^{\tau-1},h} \le C(h^{-d+\sigma(1-\tau)} + h^{1-d}) \le Ch^{-d+\sigma(1-\tau)}.$$

Applying the ABP estimate given in Theorem 3.25 to (7.10) (with Ω_h^I replaced with $\Omega_h^{I,\tau}$) yields

$$\begin{aligned}\sup_{\Omega_h^{I,\tau}} v_h^- &\le C\bigg(\sum_{z\in\mathcal{R}_{h^{\tau-1},h}} h^{d+d\tau} + \sum_{z\in\mathcal{S}_{h^{\tau-1},h}} h^{d-d\tau}\bigg)^{1/d}\\ &\le C\big(h^{d\tau} + h^{-d\tau+\sigma(1-\tau)}\big)^{1/d}.\end{aligned}$$

Thus, setting τ so that $d\tau = -d\tau + \sigma(1-\tau)$, that is, $\tau = \sigma/(2d+\sigma)$, we obtain

$$\sup_{\Omega_h^{I,\tau}} v_h^- \le C h^{\sigma/(2d+\sigma)},$$

and therefore

$$\sup_{\Omega_h^{I,\tau}} (u_h - I_h^{fd} u_\tau^+)^- \le C h^{\sigma/(2d+\sigma)}.$$

Gathering the obtained bounds implies the result. □

Remark 7.19 (extensions). The works by Krylov (2015) and Turanova (2015) have obtained rates of convergence for finite difference schemes for non-convex PDEs with variable coefficients and low-order terms.

7.3. Finite element methods

Now, following Salgado and Zhang (2016), let us focus on the development of a finite element scheme for (7.1) based on the integro-differential approximation presented in Section 4.3. The idea is that after choosing an $\epsilon > 0$, we replace each of the operators $\mathcal{L}^{\alpha,\beta}$ with their integro-differential approximations $\mathcal{L}_\epsilon^{\alpha,\beta}$, as defined in (4.27). In so doing we obtain – according to Caffarelli and Silvestre (2010*b*) – a smooth approximation u^ϵ of u, the solution to (7.1). Moreover, the difference $u - u^\epsilon$ can be controlled in terms of ϵ. We can now proceed to approximate u^ϵ by simply taking the inf-sup over discrete problems of the form (4.28), thus obtaining a discrete function $u_h^\epsilon \in X_{0,h}^l$. Using the regularity of u^ϵ, we can control the difference $u^\epsilon - u_h^\epsilon$ in terms of the mesh size and, possibly, negative powers of ϵ. Optimizing with respect to ϵ yields a rate of convergence.

Let us now proceed with the details. To set ideas we will assume that for every $\alpha \in \mathcal{A}$ and $\beta \in \mathcal{B}$ we have $f^{\alpha,\beta} = f \in C^{0,1}(\bar\Omega)$ and that the matrices $A^{\alpha,\beta}$ are constant. The approximate problem is then

$$\begin{cases} \dfrac{\lambda}{2}\Delta u^\epsilon + \inf\limits_{\beta\in\mathcal{B}} \sup\limits_{\alpha\in\mathcal{A}} I_\epsilon^{\alpha,\beta} u^\epsilon = f & \text{in } \Omega_\epsilon, \\ u = 0 & \text{on } \omega_\epsilon, \end{cases} \tag{7.11}$$

where, in analogy to (4.26), the integral operators are defined by

$$I_\epsilon^{\alpha,\beta} w(x) = \frac{1}{\epsilon^{d+2} \det A_\lambda^{\alpha,\beta}} \int_{\mathbb{R}^d} |y|^2 \delta^2_{\theta y,\theta} w(x) \varphi\Big(\frac{1}{\epsilon}(A_\lambda^{\alpha,\beta})^{-1} y\Big),$$

with

$$A_\lambda^{\alpha,\beta} = \Big(A^{\alpha,\beta} - \frac{\lambda}{2} I\Big)^{1/2},$$

and the domains Ω_ϵ and ω_ϵ are given by (4.25).

Let us now state the existence, uniqueness and smoothness properties of u^ϵ and its rate of convergence to u. For a proof see Theorem 3.5, Theorem 4.8 and Theorem 6.1 of Caffarelli and Silvestre (2010*b*).

Proposition 7.20 (properties of u^ϵ). Problem (7.11) has a unique classical solution u^ϵ. Moreover, there exists $s \in (0,1)$ that depends on λ, Λ and d but not on ϵ such that, for every $\omega \Subset \Omega$, we have

$$\|u^\epsilon\|_{C^{1,s}(\omega)} + \|u^\epsilon\|_{C^{0,1}(\bar{\Omega})} \leq C\big(\|u^\epsilon\|_{L^\infty(\Omega)} + \|f\|_{L^\infty(\Omega)}\big),$$

where the constant $C > 0$ depends on the distance between ω and $\partial\Omega$. Additionally, there exists a $\gamma > 0$ such that

$$\|u - u^\epsilon\|_{L^\infty(\Omega)} \leq C\epsilon^\gamma \|f\|_{C^{0,1}(\Omega)},$$

where the constant $C > 0$ depends only on λ, Λ, d and Ω.

With the properties of u^ϵ at hand we introduce a finite element scheme to approximate it. Namely, given a quasi-uniform triangulation for which (3.42) holds, we seek $u_h^\epsilon \in X_{0,h}^l$ such that (compare with (4.28))

$$F_h^\epsilon[u_h^\epsilon](z_i) := \frac{\lambda}{2}\Delta_h u_h^\epsilon(z_i) + \inf_{\beta\in\mathcal{B}} \sup_{\alpha\in\mathcal{A}} I_\epsilon^{\alpha,\beta} u_h^\epsilon(z_i) - f_i = 0, \quad \text{for all } z_i \in \Omega_h^I, \tag{7.12}$$

where, as in (4.28), $f_i = \int_\Omega f\phi_i$ and ϕ_i are the normalized hat functions.

Let us now address the existence, uniqueness and approximation properties of u_h^ϵ. We begin by noting that, since each one of the operators

$$w_h \mapsto \frac{\lambda}{2}\Delta_h w_h + I_\epsilon^{\alpha,\beta} w_h$$

is monotone (see Lemma 4.19), the same holds for F_h^ϵ. This is the content of the next result.

Corollary 7.21 (monotonicity). Assume that $\mathcal{T}_h$ is such that Δ_h is monotone. Then the operator F_h^ϵ, defined in (7.12) is monotone.

Remark 7.22 (flexibility). Note that, in Corollary 7.21, all that is necessary is a comparison principle *for the finite element Laplacian* Δ_h. A sufficient condition for this is given in Lemma 3.42. This is in stark contrast with the methods of Section 5.2, where either a restrictive mesh condition is needed (see Remark 5.8) or the coefficients are assumed to be isotropic as in (5.16).

In addition to monotonicity we must also consider the consistency of the scheme which, owing to the consistency of Δ_h (see Lemma 3.41), reduces to the study of the consistency of the inf-sup of integral transforms when applied to Lipschitz functions (recall that $u^\epsilon \in C^{0,1}(\bar{\Omega})$ uniformly in ϵ). To measure this we define, for $z_i \in \Omega_h^I$,

$$\mathcal{R}_{h,\epsilon}[w](z_i) = \int_\Omega \left[\inf_{\beta \in \mathcal{B}} \sup_{\alpha \in \mathcal{A}} I_\epsilon^{\alpha,\beta} I_h^{ep} w(z_i) - \inf_{\beta \in \mathcal{B}} \sup_{\alpha \in \mathcal{A}} I_\epsilon^{\alpha,\beta} w(x) \right] \phi_i(x).$$

The consistency of the scheme is then encoded in the following result, whose proof mainly follows Lemma 4.18 and Theorem 4.23 but must take into account the reduced regularity of the functions. This is the reason for the reduced rate.

Lemma 7.23 (consistency). Let $w \in C^{0,1}(\bar{\Omega})$. Then for every $z \in \Omega_h^I$ we have

$$|\mathcal{R}_{h,\epsilon}[w](z)| \leq C \frac{h}{\epsilon^2} |\log h| \|w\|_{C^{0,1}(\bar{\Omega})},$$

where the constant $C > 0$ is independent of z, h, ϵ and w.

Monotonicity and consistency allow us to conclude existence and uniqueness of solutions. The proof of the following result is either a corollary of Proposition 7.27 below or follows from a discrete version of Perron's method.

Theorem 7.24 (existence and uniqueness). Let the family of meshes be such that Δ_h is monotone. Then, for every $h > 0$ and $\epsilon > 0$, the scheme (7.12) has a unique solution.

To conclude, we study the rates of convergence. Since Proposition 7.20 provides a rate for $\|u - u^\epsilon\|_{L^\infty(\Omega)}$ and Proposition 3.40 provides one for $\|u^\epsilon - I_h^{ep} u^\epsilon\|_{L^\infty(\Omega)}$, it remains to compare the discrete solution u_h^ϵ to the Galerkin projection $I_h^{ep} u^\epsilon$. Upon denoting $e_h = I_h^{ep} u^\epsilon - u_h^\epsilon \in X_{0,h}^l$, and after some manipulations, it turns out that the error e_h satisfies, for every $z \in \Omega_h^I$,

$$\frac{\lambda}{2} \Delta_h e_h(z) + \inf_{\beta \in \mathcal{B}} \sup_{\alpha \in \mathcal{A}} I_\epsilon^{\alpha,\beta} I_h^{ep} u^\epsilon(z) - \inf_{\beta \in \mathcal{B}} \sup_{\alpha \in \mathcal{A}} I_\epsilon^{\alpha,\beta} u_h^\epsilon(z) = \mathcal{R}_{h,\epsilon}[u^\epsilon](z). \quad (7.13)$$

With this equation at hand we are ready to establish a rate of convergence.

Theorem 7.25 (rate of convergence). Let $u \in C^{1,s}(\Omega) \cap C^{0,1}(\bar{\Omega})$ be the solution of problem (7.1) and let $u_h^\epsilon \in X_{0,h}^l$ be the solution to scheme

(7.12) with $\epsilon \geq Ch^{1/2} |\log h|$. There exists $\gamma > 0$ such that

$$\|u - u_h^\epsilon\|_{L^\infty(\Omega)} \leq C\left(\epsilon^\gamma + \frac{h}{\epsilon^2} |\log h|\right) \|f\|_{C^{0,1}(\bar{\Omega})},$$

where the constant $C > 0$ depends only on λ, Λ, Ω and d.

Proof. As already discussed, the rates of convergence reduce to estimating the difference e_h. To do so we employ the error equation (7.13) and note that, if z belongs to $\mathcal{C}_h^-(e_h)$, the contact set of e_h, then we must have

$$\inf_{\beta\in\mathcal{B}} \sup_{\alpha\in\mathcal{A}} I_\epsilon^{\alpha,\beta} I_h^{ep} u^\epsilon(z) \geq \inf_{\beta\in\mathcal{B}} \sup_{\alpha\in\mathcal{A}} I_h^{\alpha,\beta} u_h^\epsilon(z).$$

In other words, for $z \in \mathcal{C}_h^-(e_h)$, the error e_h satisfies the inequality

$$\frac{\lambda}{2}\Delta_h e_h(z) \leq \mathcal{R}_{h,\epsilon}[u^\epsilon](z).$$

The discrete ABP estimate of Theorem 4.20 then implies that

$$\sup_\Omega e_h^- \leq C\left(\sum_{z\in\mathcal{C}_h^-(e_h)} |\omega_z| \, |\mathcal{R}_{h,\epsilon}[u^\epsilon](z)|^d\right)^{1/d}.$$

We must now invoke the consistency estimates of Lemma 7.23 to obtain an upper bound for e_h^-. A similar argument will yield the reverse bound. The theorem is thus proved. ☐

Remark 7.26 (algebraic rate). If the actual value of $\gamma > 0$ is known, then one can choose $\epsilon^{\gamma+2} = h|\log h|$ to obtain

$$\|u - u_h^\epsilon\|_{L^\infty(\Omega)} \leq Ch^{\gamma/(\gamma+2)} |\log h|^{\gamma/(\gamma+2)},$$

which in the best scenario, $\gamma = 1$, would yield $O(h^{1/3} |\log h|^{1/3})$.

7.4. Solution of the discrete problems

Having discussed discretization schemes for Isaacs equations (7.1) and detailed their convergence properties, to finalize our discussion we will focus on how to solve the ensuing nonlinear systems of equations. We will see – as should by now be clear to the reader – that the approaches are extensions of the convex case, which we discussed in Section 5.3, but the results are much more modest.

After discretization, in complete analogy to (5.22), we must solve the following problem: find $\mathbf{x} \in \mathbb{R}^N$, such that

$$\mathbf{F}(\mathbf{x}) = \inf_{\beta\in\mathcal{B}} \sup_{\alpha\in\mathcal{A}} \left[\mathbf{K}^{\alpha,\beta}\mathbf{x} - \mathbf{f}^{\alpha,\beta}\right] = \mathbf{0}, \tag{7.14}$$

where the inf-sup is computed component-wise, $\{\mathbf{K}^{\alpha,\beta} : \alpha \in \mathcal{A}, \beta \in \mathcal{B}\}$ and

$\{\mathbf{f}^{\alpha,\beta} : \alpha \in \mathcal{A}, \beta \in \mathcal{B}\}$ are, respectively, discretizations of $\{\mathcal{L}^{\alpha,\beta} : \alpha \in \mathcal{A}, \beta \in \mathcal{B}\}$ and $\{f^{\alpha,\beta} : \alpha \in \mathcal{A}, \beta \in \mathcal{B}\}$. Once again, the dimension N equals the number of degrees of freedom in the discretization.

One would be tempted, following Algorithm 5.1, to define, for $\mathbf{y} \in \mathbb{R}^N$, the indices $\boldsymbol{\alpha}(\mathbf{y})$ and $\boldsymbol{\beta}(\mathbf{y})$ by conditions similar to those that led to (5.23) and use them in each iteration of a possible extension of Howard's algorithm. However, this will not produce a convergent method; see, for instance, Bokanowski *et al.* (2009, Remark 5.8). The reason for this, simply put, is that the map $\mathbf{y} \mapsto \min\{\mathbf{g}, \max\{\mathbf{y}, \mathbf{h}\}\}$ is not slantly differentiable and, moreover, since the map $\mathbf{F}$, defined in (7.14), is neither convex nor concave, any generalized notion of derivative for this function will not possess any monotonicity properties.

For the reasons outlined above, we now describe, following Bokanowski *et al.* (2009, Section 5), a convergent scheme that cannot be cast as a Newton-type method but can be understood as a two-level Howard's algorithm. We begin by defining, for $\mathbf{y} \in \mathbb{R}^N$ and $i \in \{1, \dots, N\}$, the element $\beta(\mathbf{y}, i) \in \mathcal{B}$ by the condition

$$\mathbf{F}(\mathbf{y})_i = \sup_{\alpha \in \mathcal{A}} \big[\mathbf{K}^{\alpha,\beta(\mathbf{y},i)}\mathbf{y} - \mathbf{f}^{\alpha,\beta(\mathbf{y},i)}\big]_i. \tag{7.15}$$

With this at hand, we define $\boldsymbol{\beta}(\mathbf{y}) \in \mathcal{B}^N$ by $\boldsymbol{\beta}(\mathbf{y})_i = \boldsymbol{\beta}(\mathbf{y}, i)$, and the mapping $\mathbf{F}^{\boldsymbol{\beta}(\mathbf{y})} : \mathbb{R}^N \to \mathbb{R}^N$ is such that

$$\mathbf{F}^{\boldsymbol{\beta}(\mathbf{y})}(\mathbf{w})_i = \sup_{\alpha \in \mathcal{A}} \big[\mathbf{K}^{\alpha,\beta(\mathbf{y},i)}\mathbf{w} - \mathbf{f}^{\alpha,\beta(\mathbf{y},i)}\big]_i, \quad i = 1, \dots, N. \tag{7.16}$$

The generalization of Howard's method is described in Algorithm 7.1.

Note that, owing to the definition of $\mathbf{F}^{\boldsymbol{\beta}(\mathbf{y})}$ given in (7.15)–(7.16), every step of Algorithm 7.1 requires the solution of a discrete Hamilton–Jacobi–Bellman equation (7.17). This can be done by applying Algorithm 5.1 and justifies calling this method a *two-level* one. The convergence properties of Algorithm 7.1 are described below.

Proposition 7.27 (convergence of two-level Howard). Let $\mathcal{A}$ and $\mathcal{B}$ be finite sets. Assume that, for every $\boldsymbol{\beta} \in \mathcal{B}^N$, the map $\mathbf{F}^{\boldsymbol{\beta}}$ is monotone in the sense of Definition 3.5. Then the sequence of iterates, given by Algorithm 7.1 satisfies $\mathbf{x}_k \geq \mathbf{x}_{k+1}$ and converges, in a finite number of steps, to $\mathbf{x}$, the solution of (7.14).

Proof. Using equations (7.15)–(7.16), line 3 of Algorithm 7.1 and identity (7.17), we observe that the iterates $\{\boldsymbol{\beta}_k\}_{k\in\mathbb{N}}$ and $\{\mathbf{x}_k\}_{k\in\mathbb{N}}$ satisfy

$$\mathbf{F}^{\boldsymbol{\beta}_{k+1}}(\mathbf{x}_k) = \mathbf{F}^{\boldsymbol{\beta}(\mathbf{x}_k)}(\mathbf{x}_k) = \mathbf{F}(\mathbf{x}_k) \leq \mathbf{F}^{\boldsymbol{\beta}_k}(\mathbf{x}_k) = \mathbf{0} = \mathbf{F}^{\boldsymbol{\beta}_{k+1}}(\mathbf{x}_{k+1}),$$

so that, from the monotonicity of $\mathbf{F}^{\boldsymbol{\beta}_{k+1}}$, we conclude that $\mathbf{x}_k \geq \mathbf{x}_{k+1}$.

Algorithm 7.1: Two-level Howard's algorithm for non-convex problems.

Input : Sets $\mathcal{A}$ and $\mathcal{B}$.
Matrices $\{\mathbf{K}^{\alpha,\beta} : \alpha \in \mathcal{A}, \beta \in \mathcal{B}\} \subset \mathbb{R}^{N \times N}$.
right-hand sides $\{\mathbf{f}^{\alpha,\beta} : \alpha \in \mathcal{A}, \beta \in \mathcal{B}\} \subset \mathbb{R}^{N}$.
Output: Vector $\mathbf{x} \in \mathbb{R}^N$, solution of (7.14).

Initialization: Choose $\mathbf{x}_{-1} \in \mathbb{R}^N$;
for $k \geq 0$ **do**
  Set $\boldsymbol{\beta}_k = \boldsymbol{\beta}(\mathbf{x}_{k-1})$;
  Find: $\mathbf{x}_k \in \mathbb{R}^N$ that solves

$$\mathbf{F}^{\boldsymbol{\beta}_k}(\mathbf{x}_k) = \mathbf{0} \tag{7.17}$$

  ;
  if $\mathbf{F}(\mathbf{x}_k) = \mathbf{0}$ **then**
    return $\mathbf{x}_k$;
  end
end

As in Theorem 5.31, the fact that $\#\mathcal{B}$ is finite together with the monotonicity of iterates imply that Algorithm 7.1 converges in a finite number of steps. □

We conclude the discussion on Algorithm 7.1 by commenting that the case when $\mathcal{A}$ and $\mathcal{B}$ are compact spaces and when equation (7.17) is only solved approximately are also discussed by Bokanowski *et al.* (2009).

In addition to Algorithm 7.1, we present the Richardson-type iterative scheme of Kuo and Trudinger (1992, Section 4.2). Starting from $\mathbf{x}_0 \in \mathbb{R}^N$, the iterates are computed via

$$\mathbf{x}_{k+1} = \mathbf{G}(\mathbf{x}_k) := \mathbf{x}_k - \frac{1}{\Lambda_N}\mathbf{F}(\mathbf{x}_k), \tag{7.18}$$

where $\Lambda_N > 0$ is a sufficiently large constant that depends on N. It is shown there that, for $\Lambda_N \geq CN^{2/d}$, the mapping $\mathbf{G}$ is a contraction, that is, there is a constant $C > 0$ such that

$$|\mathbf{G}(\mathbf{v}) - \mathbf{G}(\mathbf{w})| \leq \left(1 - \frac{N^{-2/d}}{C}\right)|\mathbf{v} - \mathbf{w}|,$$

so that (7.18) is convergent. Some strategies that combine Algorithm 7.1 and (7.18) are suggested in González and Tidball (1992).

8. Outlook

Overall we know too much about linear PDE and too little about nonlinear PDE.

L. C. Evans

In this paper we have summarized some of the recent trends and advances in the discretizations and convergence analysis for strongly nonlinear PDEs, with an emphasis on convex and non-convex fully nonlinear equations. While these two classes of equations and discretization types have fundamentally different structure conditions, common themes permeate the analysis; for example, consistency, monotonicity and comparison principles, Alexandrov–Bakelman–Pucci estimates, wide stencils and smooth approximations. It is also evident from our discussion that, due to the pointwise definition of viscosity solutions and, correspondingly, the monotonicity criterion given in the Barles–Souganidis framework, there is a stark contrast between finite difference schemes and finite element methods for such problems; only within the last 10 years have *any* significant advances been made in the convergence analysis of finite element-type methods.

Despite the recent flurry of results for numerical fully nonlinear PDEs, there still remain fundamental open problems in the field. Probably the most pressing, at least on the finite element front, is an alternative framework that bypasses or relaxes the monotonicity requirement found in the Barles–Souganidis theory. Other completely open problems include, but are not limited to:

(i) derivation of *sharp* rates of convergence for non-convex fully nonlinear equations,

(ii) rates of convergence for finite difference schemes on *unstructured* grids,

(iii) rates of convergence of finite element methods for convex and non-convex PDEs in other norms, for example Sobolev norms such as H^1,

(iv) *a posteriori* error estimation and adaptive methods for fully nonlinear problems,

(v) rates of convergence for non-convex degenerate problems.

In light of these questions we feel that the quote from the preface of Evans (2010) given above nicely summarizes the current state of numerical PDEs. We hope that this overview will motivate current and future researchers to work on the numerical approximation of nonlinear problems.

Acknowledgements

The authors are grateful that they were given the opportunity to write this review.

The first two authors were supported in part by the US National Science Foundation grants DMS-1417980 (Neilan) and DMS-1418784 (Salgado), and the Alfred Sloan Foundation (Neilan). This work was initiated at the conference 'Nonlinear PDEs, Numerical Analysis, and Applications' at the University of Pittsburgh, which was supported by NSF grant DMS-1541585, an IMA conference grant, and the Mathematical Research Center, University of Pittsburgh.

REFERENCES[1]

A. Alla, M. Falcone and D. Kalise (2015), 'An efficient policy iteration algorithm for dynamic programming equations', *SIAM J. Sci. Comput.* **37**, A181–A200.

G. Awanou (2015*a*), 'Pseudo transient continuation and time marching methods for Monge–Ampère type equations', *Adv. Comput. Math.* **41**, 907–935.

G. Awanou (2015*b*), 'Quadratic mixed finite element approximations of the Monge–Ampère equation in 2D', *Calcolo* **52**, 503–518.

G. Awanou (2015*c*), 'Standard finite elements for the numerical resolution of the elliptic Monge–Ampère equations: Classical solutions', *IMA J. Numer. Anal.* **35**, 1150–1166.

C. Baiocchi (1977), Estimations d'erreur dans L^∞ pour les inéquations à obstacle. In *Mathematical Aspects of Finite Element Methods: Proc. Conf., Consiglio Naz. delle Ricerche (C.N.R.), Rome, 1975*, Vol. 606 of *Lecture Notes in Mathematics*, Springer, pp. 27–34.

G. Barles and E. Jakobsen (2005), 'Error bounds for monotone approximation schemes for Hamilton–Jacobi–Bellman equations', *SIAM J. Numer. Anal.* **43**, 540–558.

G. Barles and E. Jakobsen (2002), 'On the convergence rate of approximation schemes for Hamilton–Jacobi–Bellman equations', *M2AN Math. Model. Numer. Anal.* **36**, 33–54.

G. Barles and P. Souganidis (1991), 'Convergence of approximation schemes for fully nonlinear second order equations', *Asymptotic Anal.* **4**, 271–283.

R. Bellman (2010), *Dynamic Programming*, Princeton Landmarks in Mathematics, Princeton University Press. Reprint of the 1957 edition.

J.-D. Benamou and Y. Brenier (1998), 'Weak existence for the semigeostrophic equations formulated as a coupled Monge–Ampère/transport problem', *SIAM J. Appl. Math.* **58**, 1450–1461.

J.-D. Benamou and Y. Brenier (2000), 'A computational fluid mechanics solution to the Monge–Kantorovich mass transfer problem', *Numer. Math.* **84**, 375–393.

[1] The URLs cited in this work were correct at the time of going to press, but the publisher and the authors make no undertaking that the citations remain live or are accurate or appropriate.

J.-D. Benamou, F. Collino and J.-M. Mirebeau (2016), 'Monotone and consistent discretization of the Monge–Ampère operator', *Math. Comp.* **85**(302), 2743–2775.

J.-D. Benamou, B. Froese and A. Oberman (2014), 'Numerical solution of the optimal transportation problem using the Monge–Ampère equation', *J. Comput. Phys.* **260**, 107–126.

A. Bensoussan and J.-L. Lions (1984), *Impulse Control and Quasivariational Inequalities* (translated from the French by J. M. Cole), Gauthier-Villars.

K. Böhmer (2008), 'On finite element methods for fully nonlinear elliptic equations of second order', *SIAM J. Numer. Anal.* **46**, 1212–1249.

O. Bokanowski, S. Maroso and H. Zidani (2009), 'Some convergence results for Howard's algorithm', *SIAM J. Numer. Anal.* **47**, 3001–3026.

M. Boulbrachene and P. Cortey-Dumont (2009), 'Optimal L^∞-error estimate of a finite element method for Hamilton–Jacobi–Bellman equations', *Numer. Funct. Anal. Optim.* **30**, 421–435.

M. Boulbrachene and M. Haiour (2001), 'The finite element approximation of Hamilton–Jacobi–Bellman equations', *Comput. Math. Appl.* **41**, 993–1007.

S. Brenner and M. Neilan (2012), 'Finite element approximations of the three dimensional Monge–Ampère equation', *ESAIM Math. Model. Numer. Anal.* **46**, 979–1001.

S. Brenner and L. Scott (2008), *The Mathematical Theory of Finite Element Methods*, third edition, Vol. 15 of *Texts in Applied Mathematics*, Springer.

S. Brenner, T. Gudi, M. Neilan and L.-Y. Sung (2011), 'C^0 penalty methods for the fully nonlinear Monge–Ampère equation', *Math. Comp.* **80**(276), 1979–1995.

L. Caffarelli and X. Cabré (1995), *Fully Nonlinear Elliptic Equations*, Vol. 43 of *American Mathematical Society Colloquium Publications*, AMS.

L. Caffarelli and L. Silvestre (2010*a*), 'On the Evans–Krylov theorem', *Proc. Amer. Math. Soc.* **138**, 263–265.

L. Caffarelli and L. Silvestre (2010*b*), Smooth approximations of solutions to non-convex fully nonlinear elliptic equations. In *Nonlinear Partial Differential Equations and Related Topics*, Vol. 229 of *American Mathematical Society Translations, Series 2*, AMS, pp. 67–85.

L. Caffarelli and P. Souganidis (2008), 'A rate of convergence for monotone finite difference approximations to fully nonlinear, uniformly elliptic PDEs', *Comm. Pure Appl. Math.* **61**, 1–17.

A. Calderón and A. Zygmund (1952), 'On the existence of certain singular integrals', *Acta Math.* **88**, 85–139.

J. Carnicer and W. Dahmen (1994), 'Characterization of local strict convexity preserving interpolation methods by C^1 functions', *J. Approx. Theory* **77**, 2–30.

P. Ciarlet (2002), *The Finite Element Method for Elliptic Problems*, Vol. 40 of *Classics in Applied Mathematics*, SIAM. Reprint of the 1978 original.

P. Cortey-Dumont (1987), 'Sur l'analyse numérique des équations de Hamilton–Jacobi–Bellman', *Math. Methods Appl. Sci.* **9**, 198–209.

M. Crandall and P.-L. Lions (1983), 'Viscosity solutions of Hamilton–Jacobi equations', *Trans. Amer. Math. Soc.* **277**, 1–42.

M. Crandall, H. Ishii and P.-L. Lions (1992), 'User's guide to viscosity solutions of second order partial differential equations', *Bull. Amer. Math. Soc.* (N.S.) **27**, 1–67.

B. Dacorogna (2008), *Direct Methods in the Calculus of Variations*, second edition, Vol. 78 of *Applied Mathematical Sciences*, Springer.

E. Dean and R. Glowinski (2006*a*), 'An augmented Lagrangian approach to the numerical solution of the Dirichlet problem for the elliptic Monge–Ampère equation in two dimensions', *Electron. Trans. Numer. Anal.* **22**, 71–96.

E. Dean and R. Glowinski (2006*b*), 'Numerical methods for fully nonlinear elliptic equations of the Monge–Ampère type', *Comput. Methods Appl. Mech. Engrg* **195**, 1344–1386.

K. Debrabant and E. Jakobsen (2013), 'Semi-Lagrangian schemes for linear and fully non-linear diffusion equations', *Math. Comp.* **82**(283), 1433–1462.

H. Dong and N. Krylov (2007), 'The rate of convergence of finite-difference approximations for parabolic Bellman equations with Lipschitz coefficients in cylindrical domains', *Appl. Math. Optim.* **56**, 37–66.

I. Ekeland and R. Témam (1999), *Convex Analysis and Variational Problems*, English edition, Vol. 28 of *Classics in Applied Mathematics*, SIAM.

A. Ern and J.-L. Guermond (2004), *Theory and Practice of Finite Elements*, Vol. 159 of *Applied Mathematical Sciences*, Springer.

L. Evans (1980), 'On solving certain nonlinear partial differential equations by accretive operator methods', *Israel J. Math.* **36**, 225–247.

L. Evans (2010), *Partial Differential Equations*, second edition, Vol. 19 of *Graduate Studies in Mathematics*, AMS.

L. Evans and A. Friedman (1979), 'Optimal stochastic switching and the Dirichlet problem for the Bellman equation', *Trans. Amer. Math. Soc.* **253**, 365–389.

X. Feng and M. Jensen (2017), 'Convergent semi-Lagrangian methods for the Monge–Ampère equation on unstructured grids', *SIAM J. Numer. Anal.*, to appear.

X. Feng and T. Lewis (2014), 'Local discontinuous Galerkin methods for one-dimensional second order fully nonlinear elliptic and parabolic equations', *J. Sci. Comput.* **59**, 129–157.

X. Feng and M. Neilan (2009), 'Mixed finite element methods for the fully nonlinear Monge–Ampère equation based on the vanishing moment method', *SIAM J. Numer. Anal.* **47**, 1226–1250.

X. Feng and M. Neilan (2011), 'Analysis of Galerkin methods for the fully nonlinear Monge–Ampère equation', *J. Sci. Comput.* **47**, 303–327.

X. Feng, R. Glowinski and M. Neilan (2013), 'Recent developments in numerical methods for fully nonlinear second order partial differential equations', *SIAM Rev.* **55**, 205–267.

X. Feng, L. Hennings and M. Neilan (2017), 'C^0 discontinuous Galerkin finite element methods for second order linear elliptic partial differential equations in non-divergence form', *Math. Comp.*, to appear.

X. Feng, M. Neilan and S. Schnake (2016), Interior penalty discontinuous Galerkin methods for second order linear non-divergence form elliptic PDEs. arXiv:1605.04364

W. Fleming and H. Soner (2006), *Controlled Markov Processes and Viscosity Solutions*, second edition, Vol. 25 of *Stochastic Modelling and Applied Probability*, Springer.

W. Fleming and P. Souganidis (1989), 'On the existence of value functions of two-player, zero-sum stochastic differential games', *Indiana Univ. Math. J.* **38**, 293–314.

B. Froese (2016), Convergent approximation of surfaces of prescribed Gaussian curvature with weak Dirichlet conditions. arXiv:1601.06315

B. Froese and A. Oberman (2011), 'Convergent finite difference solvers for viscosity solutions of the elliptic Monge–Ampère equation in dimensions two and higher', *SIAM J. Numer. Anal.* **49**, 1692–1714.

D. Gallistl (2017), 'Variational formulation and numerical analysis of linear elliptic equations in nondivergence form with Cordes coefficients', *SIAM J. Numer. Anal.*, to appear.

E. Georgoulis, P. Houston and J. Virtanen (2011), 'An *a posteriori* error indicator for discontinuous Galerkin approximations of fourth-order elliptic problems', *IMA J. Numer. Anal.* **31**, 281–298.

S. Gerschgorin (1930), 'Fehlerabschätzung für das Differenzenverfahren zur Lösung partieller Differentialgleichungen', *Z. Angew. Math. Mech.* **10**, 373–382.

D. Gilbarg and N. Trudinger (2001), *Elliptic Partial Differential Equations of Second Order*, Classics in Mathematics, Springer. Reprint of the 1998 edition.

R. González and M. Tidball (1992), Fast solution of general nonlinear fixed point problems. In *System Modelling and Optimization (Zurich, 1991)*, Vol. 180 of *Lecture Notes in Control and Information Sciences*, Springer, pp. 35–44.

P. Grisvard (2011), *Elliptic Problems in Nonsmooth Domains*, Vol. 69 of *Classics in Applied Mathematics*, SIAM. Reprint of the 1985 original.

C. Gutiérrez (2001), *The Monge–Ampère Equation*, Vol. 44 of *Progress in Nonlinear Differential Equations and their Applications*, Birkhäuser.

D. Han and J. Wan (2013), 'Multigrid methods for second order Hamilton–Jacobi–Bellman and Hamilton–Jacobi–Bellman–Isaacs equations', *SIAM J. Sci. Comput.* **35**, S323–S344.

Q. Han and F. Lin (2011), *Elliptic Partial Differential Equations*, second edition, Vol. 1 of *Courant Lecture Notes in Mathematics*, AMS.

M. Hintermüller, K. Ito and K. Kunisch (2002), 'The primal–dual active set strategy as a semismooth Newton method', *SIAM J. Optim.* **13**, 865–888.

H. Holden and N. Risebro (2015), *Front Tracking for Hyperbolic Conservation Laws*, second edition, Vol. 152 of *Applied Mathematical Sciences*, Springer.

R. Hoppe (1986), 'Multigrid methods for Hamilton–Jacobi–Bellman equations', *Numer. Math.* **49**, 239–254.

A. Iserles (2009), *A First Course in the Numerical Analysis of Differential Equations*, second edition, Cambridge Texts in Applied Mathematics, Cambridge University Press.

H. Ishii (1995), 'On the equivalence of two notions of weak solutions, viscosity solutions and distribution solutions', *Funkcial. Ekvac.* **38**, 101–120.

E. Jakobsen (2004), 'On error bounds for approximation schemes for non-convex degenerate elliptic equations', *BIT* **44**, 269–285.

E. Jakobsen (2006), 'On error bounds for monotone approximation schemes for multi-dimensional Isaacs equations', *Asymptot. Anal.* **49**, 249–273.

M. Jensen and I. Smears (2013*a*), Finite element methods with artificial diffusion for Hamilton–Jacobi–Bellman equations. In *Numerical Mathematics and Advanced Applications 2011* (A. Cangiani *et al.*, eds), Springer, pp. 267–274.

M. Jensen and I. Smears (2013*b*), 'On the convergence of finite element methods for Hamilton–Jacobi–Bellman equations', *SIAM J. Numer. Anal.* **51**, 137–162.

M. Katsoulakis (1995), 'A representation formula and regularizing properties for viscosity solutions of second-order fully nonlinear degenerate parabolic equations', *Nonlinear Anal.* **24**, 147–158.

N. Katzourakis (2015), *An Introduction to Viscosity Solutions for Fully Nonlinear PDE with Applications to Calculus of Variations in L^∞*, Springer Briefs in Mathematics, Springer.

B. Kawohl and N. Kutev (2007), 'Comparison principle for viscosity solutions of fully nonlinear, degenerate elliptic equations', *Comm. Partial Diff. Equations* **32**, 1209–1224.

M. Kocan (1995), 'Approximation of viscosity solutions of elliptic partial differential equations on minimal grids', *Numer. Math.* **72**, 73–92.

S. Koike (2004), *A Beginner's Guide to the Theory of Viscosity Solutions*, Vol. 13 of *MSJ Memoirs*, Mathematical Society of Japan.

S. Koike and A. Święch (2013), 'Representation formulas for solutions of Isaacs integro-PDE', *Indiana Univ. Math. J.* **62**, 1473–1502.

I. Kossaczký, M. Ehrhardt and M. Günther (2016), 'On the non-existence of higher order monotone approximation schemes for HJB equations', *Appl. Math. Lett.* **52**, 53–57.

N. Krylov (1987), *Nonlinear Elliptic and Parabolic Equations of the Second Order* (translated from the Russian by P. L. Buzytsky), Vol. 7 of *Mathematics and its Applications (Soviet Series)*, Reidel.

N. Krylov (1996), *Lectures on Elliptic and Parabolic Equations in Hölder Spaces*, Vol. 12 of *Graduate Studies in Mathematics*, AMS.

N. Krylov (1997), 'On the rate of convergence of finite-difference approximations for Bellman's equations', *Algebra i Analiz* **9**, 245–256.

N. Krylov (2000), 'On the rate of convergence of finite-difference approximations for Bellman's equations with variable coefficients', *Probab. Theory Rel. Fields* **117**, 1–16.

N. Krylov (2005), 'The rate of convergence of finite-difference approximations for Bellman equations with Lipschitz coefficients', *Appl. Math. Optim.* **52**, 365–399.

N. Krylov (2008), *Lectures on Elliptic and Parabolic Equations in Sobolev Spaces*, Vol. 96 of *Graduate Studies in Mathematics*, AMS.

N. Krylov (2014), 'On the rate of convergence of difference approximations for uniformly nondegenerate elliptic Bellman's equations', *Appl. Math. Optim.* **69**, 431–458.

N. Krylov (2015), 'On the rate of convergence of finite-difference approximations for elliptic Isaacs equations in smooth domains', *Comm. Partial Diff. Equations* **40**, 1393–1407.

H.-J. Kuo and N. Trudinger (1990), 'Linear elliptic difference inequalities with random coefficients', *Math. Comp.* **55**(191), 37–53.
H.-J. Kuo and N. Trudinger (1992), 'Discrete methods for fully nonlinear elliptic equations', *SIAM J. Numer. Anal.* **29**, 123–135.
H.-J. Kuo and N. Trudinger (1996), 'Positive difference operators on general meshes', *Duke Math. J.* **83**, 415–433.
H.-J. Kuo and N. Trudinger (2000), 'A note on the discrete Aleksandrov–Bakelman maximum principle', *Taiwanese J. Math.* **4**, 55–64.
M.-J. Lai and L. Schumaker (2007), 'Trivariate C^r polynomial macroelements', *Constr. Approx.* **26**, 11–28.
O. Lakkis and T. Pryer (2011), 'A finite element method for second order nonvariational elliptic problems', *SIAM J. Sci. Comput.* **33**, 786–801.
P.-L. Lions (1982), *Generalized Solutions of Hamilton–Jacobi Equations*, Vol. 69 of *Research Notes in Mathematics*, Pitman.
P.-L. Lions (1983), 'Optimal control of diffusion processes and Hamilton–Jacobi–Bellman equations II: Viscosity solutions and uniqueness', *Comm. Partial Diff. Equations* **8**, 1229–1276.
P.-L. Lions and B. Mercier (1980), 'Approximation numérique des équations de Hamilton–Jacobi–Bellman', *RAIRO Anal. Numér.* **14**, 369–393.
A. Maugeri, D. Palagachev and L. Softova (2000), *Elliptic and Parabolic Equations with Discontinuous Coefficients*, Vol. 109 of *Mathematical Research*, Wiley–VCH.
J.-L. Menaldi (1989), 'Some estimates for finite difference approximations', *SIAM J. Control Optim.* **27**, 579–607.
G. Mingione (2006), 'Regularity of minima: An invitation to the dark side of the calculus of variations', *Appl. Math.* **51**, 355–426.
J.-M. Mirebeau (2015), 'Discretization of the 3D Monge–Ampere operator, between wide stencils and power diagrams', *ESAIM Math. Model. Numer. Anal.* **49**, 1511–1523.
T. Motzkin and W. Wasow (1953), 'On the approximation of linear elliptic differential equations by difference equations with positive coefficients', *J. Math. Phys.* **31**, 253–259.
N. Nadirashvili and S. Vlăduţ (2007), 'Nonclassical solutions of fully nonlinear elliptic equations', *Geom. Funct. Anal.* **17**, 1283–1296.
N. Nadirashvili and S. Vlăduţ (2008), 'Singular viscosity solutions to fully nonlinear elliptic equations', *J. Math. Pures Appl.* (9) **89**, 107–113.
N. Nadirashvili, V. Tkachev and S. Vlăduţ (2012), 'A non-classical solution to a Hessian equation from Cartan isoparametric cubic', *Adv. Math.* **231**, 1589–1597.
M. Neilan (2010), 'A nonconforming Morley finite element method for the fully nonlinear Monge–Ampère equation', *Numer. Math.* **115**, 371–394.
M. Neilan (2013), 'Quadratic finite element approximations of the Monge–Ampère equation', *J. Sci. Comput.* **54**, 200–226.
M. Neilan (2017), 'Convergence analysis of a finite element method for second order non-variational elliptic problems', *J. Numer. Math.*, to appear.
L. Nirenberg (1953), 'On nonlinear elliptic partial differential equations and Hölder continuity', *Comm. Pure Appl. Math.* **6**, 103–156; addendum, 395.

R. Nochetto and W. Zhang (2017*a*), 'Discrete ABP estimate and convergence rates for linear elliptic equations in non-divergence form', *Found. Comput. Math.*, to appear.

R. Nochetto and W. Zhang (2017*b*), Pointwise rates of convergence for the Oliker–Prussner's method for the Monge–Ampère equation. arXiv:1611.02786

R. Nochetto, D. Ntogakas and W. Zhang (2017), Two-scale method for the Monge–Ampère equation: Convergence rates. Preprint.

R. Nochetto, E. Otárola and A. J. Salgado (2015), 'Convergence rates for the classical, thin and fractional elliptic obstacle problems', *Philos. Trans. A* **373**(2050), 20140449.

A. Oberman (2006), 'Convergent difference schemes for degenerate elliptic and parabolic equations: Hamilton–Jacobi equations and free boundary problems', *SIAM J. Numer. Anal.* **44**, 879–895.

A. Oberman (2008), 'Wide stencil finite difference schemes for the elliptic Monge–Ampère equation and functions of the eigenvalues of the Hessian', *Discrete Contin. Dyn. Syst. Ser. B* **10**, 221–238.

B. Øksendal (2003), *Stochastic Differential Equations: An Introduction with Applications*, sixth edition, Universitext, Springer.

V. Oliker and L. Prussner (1988), 'On the numerical solution of the equation $(\partial^2 z/\partial x^2)(\partial^2 z/\partial y^2) - ((\partial^2 z/\partial x \partial y))^2 = f$ and its discretizations, I', *Numer. Math.* **54**, 271–293.

M. Safonov (1980), 'Harnack's inequality for elliptic equations and Hölder property of their solutions', *Zap. Nauchn. Sem. Leningrad. Otdel. Mat. Inst. Steklov. (LOMI)* **96**, 272–287, 312.

M. Safonov (1987), 'Unimprovability of estimates of Hölder constants for solutions of linear elliptic equations with measurable coefficients', *Mat. Sb.* (N.S.) **132**(174), 275–288.

A. J. Salgado and W. Zhang (2016), Finite element approximation of the Isaacs equation. arXiv:1512.09091v1

A. Schatz and L. Wahlbin (1982), 'On the quasi-optimality in L_∞ of the $\dot{H}^1$-projection into finite element spaces', *Math. Comp.* **38**(157), 1–22.

L. Silvestre (2015), Viscosity solutions of elliptic equations. http://math.uchicago.edu/~luis/preprints/viscosity-solutions.pdf

I. Smears and E. Süli (2013), 'Discontinuous Galerkin finite element approximation of nondivergence form elliptic equations with Cordès coefficients', *SIAM J. Numer. Anal.* **51**, 2088–2106.

I. Smears and E. Süli (2014), 'Discontinuous Galerkin finite element approximation of Hamilton–Jacobi–Bellman equations with Cordes coefficients', *SIAM J. Numer. Anal.* **52**, 993–1016.

I. Smears and E. Süli (2016), 'Discontinuous Galerkin finite element methods for time-dependent Hamilton–Jacobi–Bellman equations with Cordes coefficients', *Numer. Math.* **133**, 141–176.

S. Stojanovic (2005), 'Risk premium and fair option prices under stochastic volatility: The HARA solution', *CR Math. Acad. Sci. Paris* **340**, 551–556.

G. Strang and G. Fix (1973), *An Analysis of the Finite Element Method*, Prentice-Hall Series in Automatic Computation, Prentice-Hall.

L. Tolstoy (1998), *Anna Karenina*, Project Gutenberg, EBook #1399.

N. Trudinger (1988), 'Comparison principles and pointwise estimates for viscosity solutions of nonlinear elliptic equations', *Rev. Mat. Iberoamericana* **4**, 453–468.

N. Trudinger and X.-J. Wang (2005), 'The affine Plateau problem', *J. Amer. Math. Soc.* **18**, 253–289.

N. Trudinger and X.-J. Wang (2008), 'Boundary regularity for the Monge–Ampère and affine maximal surface equations', *Ann. of Math.* (2) **167**, 993–1028.

O. Turanova (2015), 'Error estimates for approximations of nonhomogeneous nonlinear uniformly elliptic equations', *Calc. Var. Partial Diff. Equations* **54**, 2939–2983.

C. Villani (2003), *Topics in Optimal Transportation*, Vol. 58 of *Graduate Studies in Mathematics*, AMS.

C. Wang and J. Wang (2017), 'A primal–dual weak Galerkin finite element method for second order elliptic equations in non-divergence form', *Math. Comp.*, to appear.

X.-J. Wang (1996), 'Regularity for Monge–Ampère equation near the boundary', *Analysis* **16**, 101–107.

J. Witte and C. Reisinger (2012), 'Penalty methods for the solution of discrete HJB equations: Continuous control and obstacle problems', *SIAM J. Numer. Anal.* **50**, 595–625.

J. Xu and L. Zikatanov (1999), 'A monotone finite element scheme for convection-diffusion equations', *Math. Comp.* **68**(228), 1429–1446.

A. Ženíšek (1973), Hermite interpolation on simplexes in the finite element method. In *Proc. Equadiff III: Third Czechoslovak Conf. Differential Equations and their Appl., Brno, 1972*, Purkyně University, Brno, pp. 271–277. Vol. 1 of *Folia Fac. Sci. Natur. Univ. Purkynianae Brunensis, Ser. Monograph.*

Acta Numerica (2017), pp. 305–364
doi:10.1017/S096249291700006X

A survey of structure from motion

Onur Özyeşil
INTECH Investment Management LLC,
One Palmer Square, Suite 441,
Princeton, NJ 08542, USA
E-mail: oozyesil@intechjanus.com

Vladislav Voroninski
Helm.ai, Menlo Park,
CA 94025, USA
E-mail: vlad@helm.ai

Ronen Basri
Department of Computer Science and Applied Mathematics,
Weizmann Institute of Science,
Rehovot, 76100, Israel
E-mail: ronen.basri@weizmann.ac.il

Amit Singer
Department of Mathematics and PACM,
Princeton University,
Princeton, NJ 08544-1000, USA
E-mail: amits@math.princeton.edu

The structure from motion (SfM) problem in computer vision is to recover the three-dimensional (3D) structure of a stationary scene from a set of projective measurements, represented as a collection of two-dimensional (2D) images, via estimation of motion of the cameras corresponding to these images. In essence, SfM involves the three main stages of (i) extracting features in images (*e.g.* points of interest, lines, *etc.*) and matching these features between images, (ii) camera motion estimation (*e.g.* using relative pairwise camera positions estimated from the extracted features), and (iii) recovery of the 3D structure using the estimated motion and features (*e.g.* by minimizing the so-called *reprojection error*). This survey mainly focuses on relatively recent developments in the literature pertaining to stages (ii) and (iii). More specifically, after touching upon the early factorization-based techniques for motion and structure estimation, we provide a detailed account of some of the recent camera *location* estimation methods in the literature, followed by discussion of notable techniques for 3D structure recovery. We also cover the basics of the *simultaneous localization and mapping* (SLAM) problem, which can be viewed as a specific case of the SfM problem. Further, our survey includes a review of the fundamentals of feature extraction and matching (*i.e.* stage (i) above), various recent methods for handling ambiguities in 3D scenes, SfM techniques involving relatively uncommon camera models and image features, and popular sources of data and SfM software.

CONTENTS

1. Introduction

Recovering the three-dimensional structure of a scene from images is a fundamental goal of computer vision. A particularly effective approach to three-dimensional (3D) reconstruction involves the use of many images of a stationary scene. This problem, commonly referred to as *multiview structure from motion* (multiview SfM) (depicted in Figure 1.1), is the subject of a large body of research in computer vision, starting with the seminal paper of Longuet-Higgins (1981). A comprehensive and in-depth summary of this enormous body of work can be found in Hartley and Zisserman (2000).

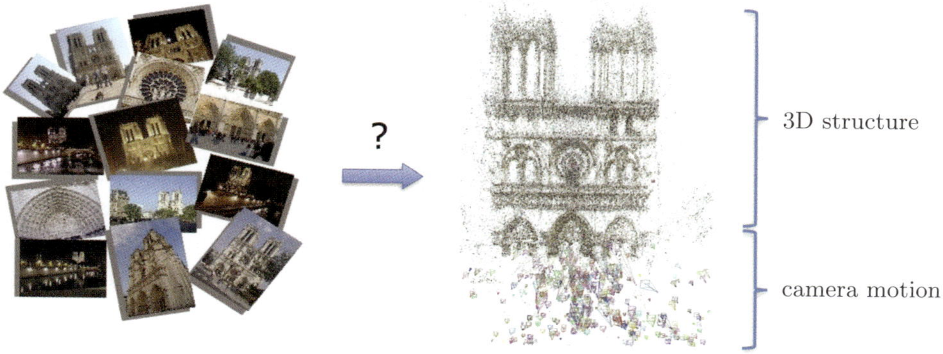

Figure 1.1. The structure from motion (SfM) problem.

Modern methods usually solve the multiview SfM problem via *bundle adjustment* techniques, which aim to optimize a cost function known as the *total reprojection error*; see Section 4 for a more detailed and general discussion of bundle adjustment. With this cost function, given n images of a stationary scene, the objective is to simultaneously determine the structure (3D coordinates of scene points) and the calibration parameters of each of the n cameras that minimize the discrepancy between image measurements and their *predictive model*. For instance, in the specific case of pairwise point correspondences between images, we let $\{\mathbf{P}_j\}_{j=1}^{p} \subseteq \mathbb{R}^3$ denote the unknown positions of scene points, and let $\{C_i\}_{i=1}^{n}$ be the 3×4 camera matrices, where the entries of each C_i encode the location of camera i, its orientation and internal calibration parameters. Also, let $(x_{ij}, y_{ij}) \in \mathbb{R}^2$ denote the measured projection of point $\mathbf{P}_j$ onto the image plane of camera C_i. Then we can define a (relatively simple) bundle adjustment problem based on minimizing the sum of squares cost function

$$\underset{\{\mathbf{P}_j\},\{C_i\}}{\text{minimize}} \sum_{i\sim j} \left(x_{ij} - \frac{C_{i1}^T \mathbf{P}_j}{C_{i3}^T \mathbf{P}_j} \right)^2 + \left(y_{ij} - \frac{C_{i2}^T \mathbf{P}_j}{C_{i3}^T \mathbf{P}_j} \right)^2, \tag{1.1}$$

where $C_{ik} \in \mathbb{R}^4$ denotes the kth row of C_i $(1 \leq k \leq 3)$, $i \sim j$ means that the jth scene point is visible by the ith camera, and we abuse notation and identify the representation $[\mathbf{P}_j, 1] \in \mathbb{R}^4$ of $\mathbf{P}_j$ in homogeneous coordinates by $\mathbf{P}_j$. Unfortunately, as a result of the special structure of the camera matrices and the cost function in (1.1), the bundle adjustment problem (1.1) is not convex, and its (naïve) optimization in realistic settings typically converges to an often undesired local minimum. It is therefore critical to develop methods that can initialize bundle adjustment, that is, provide initial camera motion (and intrinsic calibration) and 3D structure estimates, as closely as possible to the true solution.

Snavely, Seitz and Szeliski (2006) presented a sequential pipeline for SfM, demonstrating that it can produce accurate reconstructions in practical scenarios where hundreds or even thousands of independently captured photographs are provided, and sparked huge interest in the development of efficient SfM techniques for large unordered image sets (see Section 4). The suggested pipeline begins by detecting keypoints in each image. It then uses the SIFT descriptor (Lowe 2004) to compare those keypoints across images and to produce a set of potential matches. Random sampling and consensus (RANSAC: Bolles and Fischler 1981) is then used to robustly estimate essential matrices between pairs of images (for computation of relative motion of camera pairs) and to discard outlier matches. Then, starting with a pair of images for which the largest number of inlier matches were found and then greedily adding one image at a time, bundle adjustment is solved repeatedly. Although this sequential pipeline is computationally challenging, it successfully deals with large collections of images, in many cases producing highly accurate reconstructions. However, it is based on greedy steps that may not result in an optimal solution. Clearly, global approaches, which consider all images simultaneously (at least for initial camera motion estimation), may potentially yield improved solutions. Indeed, a number of recent methods attempt to estimate camera locations and orientations globally. In this survey, for the initial camera motion estimation part of SfM, we focus on methods for camera location estimation in Section 3, and refer the reader to other surveys that review the problems of camera orientation and calibration recovery (Tron, Zhou and Daniilidis 2016).

The outline of this survey is as follows. In Section 2 we briefly discuss some of the early works in the literature that introduced the basic SfM framework and the relatively simple factorization methods. Section 3 covers some of the recent work on camera location estimation from pairwise keypoint matches, such as the simpler least-squares methods based on the (homogeneous) epipolar constraints, relatively stable convex methods based on a linearized representation (see (3.8)) of pairwise direction measurements and heuristic methods for outlier rejection in pairwise directions. In Section 4 we provide some of the key ideas and methods in the literature for 3D structure estimation, with an emphasis on contemporary methods aimed at processing large unordered sets of images. In Section 5 we discuss details of influential works on the *simultaneous localization and mapping* (SLAM) problem, which has recently experienced a dramatic increase in popularity. Section 6 considers some of the remaining crucial topics in the vast SfM literature, including methods for feature point extraction and matching, works on handling instabilities induced by symmetries and ambiguities in the images, methods for relatively uncommon types of cameras (*e.g.* omnidirectional cameras), techniques using alternative basic measurements (*e.g.* based on lines instead of feature points), and some of the important software packages and popular datasets. We conclude in Section 7 with a discussion of the current state of the art and possible future directions for the SfM problem.

Rather unfortunately, as is the case for any survey of a topic for which there is an extensive body of literature, we are unable to cover every important work in the field. In essence, we try to focus on relatively recent SfM techniques. For earlier works in the literature, we refer the reader to other sources, for example Triggs, McLauchlan, Hartley and Fitzgibbon (2000), Hartley and Zisserman (2000) and Oliensis (2000).

2. Early works

This section covers some of the relatively early works on SfM, which introduced the basic problem formalism and the factorization-based methods. We first consider the seminal work by Longuet-Higgins (1981), who introduced the first linear method based on point correspondences, later named the *eight-point algorithm*, to solve the SfM problem for a pair of cameras. Specifically, for a pair of cameras, Longuet-Higgins (1981) aimed to estimate the relative camera motion, that is, relative rotation and translation, and the 3D coordinates of the scene points captured by these cameras. Here we represent the ith camera by $(R_i, \mathbf{t}_i)$, where $R_i \in \mathrm{SO}(3)$ is the camera's orientation matrix and $\mathbf{t}_i \in \mathbb{R}^3$ is its position. Considering that we can always fix the (intrinsic) coordinate system of one of the two cameras to be the global coordinate system (*i.e.* we can set $R_i = I$ and $\mathbf{t}_i = \mathbf{0}$, since an absolute coordinate system cannot be determined from point correspondences), solving for the relative motion for a camera pair, and for the scene points based on the computed motion, then corresponds to actually solving the SfM problem for the pair. The problem setup[1] involves a simple pinhole camera model, in which a scene point $\mathbf{P} \in \mathbb{R}^3$ is represented in the ith image plane by $\mathbf{p}_i \in \mathbb{R}^3$, as shown in Figure 2.1; see also Özyeşil *et al.* (2015, Section 4) and Hartley and Zisserman (2000), for example. Here, we obtain $\mathbf{p}_i$ by first representing $\mathbf{P}$, in the ith camera's coordinate system, as $\mathbf{P}_i = R_i^T(\mathbf{P} - \mathbf{t}_i) = (\mathbf{P}_i^x, \mathbf{P}_i^y, \mathbf{P}_i^z)^T$, and then projecting it onto the ith image plane via $\mathbf{p}_i = (f_i/P_i^z)\mathbf{P}_i$, where $R_i \in \mathrm{SO}(3)$ denotes the orientation matrix, $\mathbf{t}_i \in \mathbb{R}^3$ denotes the location of the focal point, and $f_i \in \mathbb{R}^+$ is the

[1] Our notation and type of projective measurement (or equivalently the global coordinate system) are different from those of Longuet-Higgins (1981), even though the two choices are equivalent, that is, any result obtained by using one can be represented in terms of the other. Our choices (more or less) reflect the common terminology in the literature: see *e.g.* Özyeşil, Singer and Basri (2015), Özyeşil and Singer (2015) and Arie-Nachimson, Kovalsky, Kemelmacher-Shlizerman, Singer and Basri (2012). We also represent the epipolar constraints in their general form for a pair (i, j), to make use of them in the following sections.

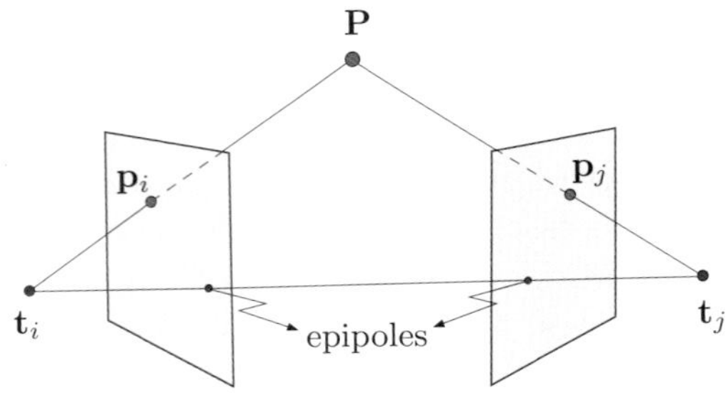

Figure 2.1. The pinhole camera model.

focal length of the ith camera. For a pair of cameras i and j as in Figure 2.1, we can restate the coplanarity of the points $\mathbf{P}, \mathbf{t}_i$ and $\mathbf{t}_j$, that is, $[(\mathbf{P}-\mathbf{t}_i) \times (\mathbf{P}-\mathbf{t}_j)]^T(\mathbf{t}_i - \mathbf{t}_j) = 0$, in terms of the observable corresponding point measurements $\mathbf{p}_i, \mathbf{p}_j$ (see Figure 3.1 below for a depiction of a set of corresponding points between a pair of images) and the camera parameters as

$$\mathbf{p}_i^T([R_i^T(\mathbf{t}_j - \mathbf{t}_i)]_\times R_i^T R_j)\mathbf{p}_j =: \mathbf{p}_i^T E_{ij} \mathbf{p}_j = 0, \tag{2.1}$$

where $[\mathbf{t}]_\times$ is the skew-symmetric matrix corresponding to the cross-product with $\mathbf{t}$, and $E_{ij} = [R_i^T(\mathbf{t}_j - \mathbf{t}_i)]_\times R_i^T R_j$ is the *essential matrix* for the cameras i and j. Also, let $T_{ij} := [R_i^T(\mathbf{t}_j - \mathbf{t}_i)]_\times$ and $R_{ij} := R_i^T R_j$, yielding $E_{ij} = T_{ij}R_{ij}$. Longuet-Higgins (1981) made the crucial observation that, among the various equivalent restatements of the coplanarity of $\mathbf{P}, \mathbf{t}_i$ and $\mathbf{t}_j$, the useful property of (2.1), known as the *epipolar constraint*, is that it provides a basis for the estimation of the structured essential matrices in terms of the observable corresponding points. In other words, by fixing the undetermined scale for the entries of E_{ij} (*e.g.* the Frobenius norm $\|E_{ij}\|_F = 1$, or as in Longuet-Higgins (1981) $\|\mathbf{t}_j - \mathbf{t}_i\| = 1$, which is equivalent to $\|E_{ij}\|_F = \sqrt{2}$), we can solve for $E_{ij} \in \mathbb{R}^{3\times 3}$ from eight linearly independent epipolar constraints (2.1) corresponding to eight 3D points.[2] Longuet-Higgins (1981) did not provide any algorithm for this purpose, nor did he consider the effects of uncertainties (*i.e.* noise) in the corresponding point measurements; however, the usual approach in the literature is simply to minimize the sum of squared errors in the epipolar constraints subject to the scale constraint, which is equivalent to finding the singular vector corresponding to the smallest singular value of the resulting data matrix. After estimating E_{ij} (up to a sign to be determined later), Longuet-Higgins (1981) solved for T_{ij} (up to another sign, again to be determined later) by

[2] In fact, due to the special structure of E_{ij}, it is well known (see *e.g.* Hartley and Zisserman 2000) that only five corresponding 3D points are sufficient to estimate E_{ij}. However, this requires us to solve a non-linear system of equations.

eliminating R_{ij} using $E_{ij}E_{ij}^T = T_{ij}T_{ij}^T$. He then solved for R_{ij}, based on its orthogonality, and for the scene points using simple algebraic equations (see Longuet-Higgins 1981 for details). Lastly, the undetermined signs of E_{ij} and T_{ij} are fixed by requiring the scene points to lie in front of both cameras. If the scene points are behind both cameras, the sign of T_{ij} is altered, and the relative rotation R_{ij} and the scene points are recalculated. If the scene points fall behind one of the two cameras, the sign of E_{ij} is altered, and T_{ij}, R_{ij} and the scene points are recalculated. Further, Longuet-Higgins (1981) briefly discussed some of the degenerate cases of eight 3D points for which the eight-point estimation cannot be used, for example the cases of all points corresponding to the vertices of a cube, at least seven points lying on a plane, six of the eight points corresponding to the vertices of a regular hexagon, or at least four points lying on a line. We note that although Longuet-Higgins (1981) argued for the accuracy of the eight-point algorithm, it was in fact observed to be sensitive to noise, which resulted in the construction of alternative methods, such as a normalized version of the eight-point algorithm (Hartley 1997), for relative pairwise motion estimation. For further reading on alternative methods to obtain epipolar geometries between images, see for example Zhang (1998) and Hartley and Zisserman (2000).

Another early work in the literature is that of Tomasi and Kanade (1992), who introduced the *factorization method* for SfM. To model the SfM problem for objects that are relatively distant compared to their sizes, Tomasi and Kanade (1992) assumed an *orthographic* camera model, in which the 3D points are measured via parallel projections onto the image plane, consequently ignoring the camera translation along the optical axis. Let the location of the jth 3D point on the jth camera plane be given by $\mathbf{q}_{ij} = (x_{ij}, y_{ij}) \in \mathbb{R}^2$. In the noiseless case, these orthographic measurements of m 3D points by n cameras are represented in terms of a *measurement matrix* $W \in \mathbb{R}^{2n\times m}$ satisfying

$$W = \begin{bmatrix} x_{11} & \cdots & x_{1m} \\ \vdots & \ddots & \vdots \\ x_{n1} & \cdots & x_{nm} \\ y_{11} & \cdots & y_{1m} \\ \vdots & \ddots & \vdots \\ y_{n1} & \cdots & y_{nm} \end{bmatrix}. \tag{2.2}$$

The crucial observation made by Tomasi and Kanade (1992) is that if the origin of the global reference system is chosen to be at the centre of the 3D structure points (implying that each row of W sums to zero), then W can be decomposed into

$$W = RS, \quad \text{for } R = \begin{bmatrix} \mathbf{u}_1^T \\ \vdots \\ \mathbf{u}_n^T \\ \mathbf{v}_1^T \\ \vdots \\ \mathbf{v}_n^T \end{bmatrix} \in \mathbb{R}^{2n \times 3} \quad \text{and} \quad S = [\mathbf{P}_1 \ldots \mathbf{P}_m] \in \mathbb{R}^{3 \times m}, \tag{2.3}$$

where $\mathbf{u}_k$ and $\mathbf{v}_k$ represent the orientation of the kth camera (since, in the orthographic model, the third direction given by $\mathbf{u}_k \times \mathbf{v}_k$ is parallel to the viewing direction of the cameras) and $\mathbf{P}_k$ is the kth 3D point. As a result, W has rank 3. In the noisy case, which may result in a higher rank for W, Tomasi and Kanade (1992) considered a valid measurement matrix to be given by the best rank-3 approximation to the noisy W in the Frobenius norm sense, which is given by the singular value decomposition (SVD). Note that replacing R and S with RQ and $Q^{-1}S$, for an arbitrary invertible matrix $Q \in \mathbb{R}^{3\times 3}$, results in the same measurement matrix W. As a result, in order to compute R and S from W, Tomasi and Kanade (1992) proposed first computing a decomposition of W using SVD, and then solving for Q using the orthonormality of the n pairs $(\mathbf{u}_k, \mathbf{v}_k)$. That is, for $W = U_W \Sigma_W V_W^T$ representing the SVD of W, they first let $\hat{R} = U_W \Sigma^{1/2}$ and $\hat{S} = \Sigma^{1/2} V_W^T$, and then computed the solution $R = \hat{R}Q$ and $S = Q^{-1}\hat{S}$, where Q is given to be a solution to the non-linear system of $3n$ equations

$$\hat{\mathbf{u}}_k^T Q Q^T \hat{\mathbf{u}}_k = \hat{\mathbf{v}}_k^T Q Q^T \hat{\mathbf{v}}_k = 1, \quad \hat{\mathbf{u}}_k^T Q Q^T \hat{\mathbf{v}}_k = 0, \quad k = 1, \ldots, n, \tag{2.4}$$

where $\hat{\mathbf{u}}_k$ and $\hat{\mathbf{v}}_k$ represent the kth and $(k+n)$th rows of $\hat{R}$ (see (2.3)). They also extended the factorization method to the case of *occlusions* (*i.e.* to the case when some scene points are not visible by all of the cameras) and provided empirical evidence demonstrating the stability of their factorization approach on various experimental scenarios.

Sturm and Triggs introduced an extension of the factorization methodology for the multiview SfM with perspective cameras. They considered the basic image projection equations

$$\lambda_{ij} \mathbf{p}_{ij} = C_i \mathbf{P}_j, \quad i = 1, \ldots, n, \quad j = 1, \ldots, m, \tag{2.5}$$

where, for the jth 3D point represented (by abuse of notation) in the homogeneous coordinates by $\mathbf{P}_j \in \mathbb{R}^4$ and C_i denoting the ith camera matrix, $\mathbf{p}_{ij} \in \mathbb{R}^3$ is the projection of the jth point onto the ith camera plane (in the homogeneous coordinates) and λ_{ij} are the undetermined scales, called *projective depths*. Similarly to Tomasi and Kanade (1992), Sturm and Triggs (1996) collected these projective measurements into a rank-4 measurement matrix $Y \in \mathbb{R}^{3n \times m}$ given by

$$Y = \begin{bmatrix} \lambda_{11}\mathbf{p}_{11} & \dots & \lambda_{1m}\mathbf{p}_{1m} \\ \vdots & \ddots & \vdots \\ \lambda_{n1}\mathbf{p}_{n1} & \dots & \lambda_{nm}\mathbf{p}_{nm} \end{bmatrix} = \begin{bmatrix} C_1 \\ \vdots \\ C_n \end{bmatrix} \begin{bmatrix} \mathbf{P}_1 & \dots \mathbf{P}_m \end{bmatrix}. \tag{2.6}$$

The crucial point here is that we do not have direct access to the projective depths λ_{ij}, and if λ_{ij} were to be accurately estimated, which constitutes the main contribution of Sturm and Triggs (1996), a factorization technique similar to that of Tomasi and Kanade (1992) could easily be applied. In order to achieve this goal, Sturm and Triggs (1996) considered the linear system of equations for pairs of cameras (i, j) given by

$$F_{ij}\mathbf{p}_{jk}\lambda_{jk} = (\mathbf{e}_{ij} - \mathbf{p}_{ik})\lambda_{ik}, \tag{2.7}$$

where the *fundamental matrix* $F_{ij} \in \mathbb{R}^{3\times 3}$ satisfies $E_{ij} = K_i^T F_{ij} K_j$, for K_i denoting the ith camera calibration matrix, E_{ij} denotes the essential matrix for the pair (i, j) (see (2.1)), and $\mathbf{e}_{ij}$ is the epipole on the ith image plane (see Figure 2.1). Considering the homogeneous linear equations (2.7) for the minimal number of $n - 1$ camera pairs, Sturm and Triggs (1996) estimated the projective depths up to global scale. After substitution of the recovered λ_{ij} in (2.6), they proceeded similarly to Tomasi and Kanade (1992) and factorized Y into its camera and structure components using SVD. However, this time there is no need to compute an undetermined multiplier Q since the camera and structure parts of Y do not have particular forms. Lastly, Sturm and Triggs (1996) concluded with empirical results to evaluate the performance of their algorithm. A closely related approach, with an additional section for the extension of the algorithm in Sturm and Triggs (1996) to the case of lines, instead of points, as feature measurements, is given by Triggs (1996).

An excellent account of factorization-based methods, for various camera models and alternative feature measurements, can also be found in the survey by Kanade and Morris (1998).

3. Camera location estimation

In this section we discuss the existing literature focusing primarily on the camera location estimation part of the SfM problem. We consider the camera location estimation methods based on corresponding point estimates between pairs of images. In the majority of the methods we discuss, the corresponding point estimates are used to make relative motion measurements between pairs of images, which can be decomposed into pairwise rotational and translational measurements. The generic approach for these methods is to estimate the camera orientations separately based on the pairwise rotational measurements, and to use these camera orientation estimates together with the pairwise translational measurements in order to solve for the camera locations. On the other hand, we also discuss some of the methods in the literature that, to some degree, deviate from the generic recipe mentioned above, and aim to jointly estimate camera orientations and locations, use local structure estimates for location estimation, or consider triplets of cameras.

In order to illustrate the generic approach mentioned above, consider the simple pinhole camera model introduced in Section 2 and depicted in Figure 2.1. As discussed in Section 2, the essential matrix estimates, computed using the epipolar constraints (2.1) in the presence of sufficiently many corresponding scene points (see Figure 3.1 for a real example of pairwise point correspondences) and knowledge of the intrinsic camera parameters, can be uniquely factorized into relative rotational and translational parts, that is, into estimates of $R_i^T R_j$ and $[R_i^T(\mathbf{t}_j - \mathbf{t}_i)]_\times$. In general, the relative rotational parts are used for the estimation of camera orientations R_i; see Tron *et al.* (2016) for a detailed account of the existing methods for camera orientation estimation. In order to estimate the camera locations, the estimated orientations can be used directly with the translational parts to obtain estimates of the *pairwise directions* $\gamma_{ij} = (\mathbf{t}_i - \mathbf{t}_j)/\|\mathbf{t}_i - \mathbf{t}_j\|$. An alternative approach is to go back to the epipolar constraints (2.1) to rewrite them as a set of linear homogeneous equations in the camera locations $\mathbf{t}_i$ given by

$$(\mathbf{t}_i - \mathbf{t}_j)^T \mathbf{y}_{ij}^k = 0, \quad k = 1, \ldots, m_{ij}, \tag{3.1}$$

where $\mathbf{y}_{ij}^k$ are functions of R_i, R_j and the kth corresponding points $\mathbf{p}_i^k, \mathbf{p}_j^k$, and m_{ij} denotes the number of corresponding points for the cameras i and j. These equations can then be used directly for camera location estimation, or to obtain pairwise direction estimates.

A crucial point to note about the estimates of the pairwise directions γ_{ij} is that they lack any *scale information* pertaining to the distance $\|\mathbf{t}_i - \mathbf{t}_j\|$ between camera pairs. The difficulties[3] arising from this homogeneity can be regarded as the most important contributor to the diversity of the camera location estimation methods in the literature.

A relatively direct approach for location estimation from pairwise directions is based on the homogeneous system of equations

$$(I - \gamma_{ij}\gamma_{ij}^T)(\mathbf{t}_i - \mathbf{t}_j) = \mathbf{0}, \tag{3.2}$$

[3] Note that the pairwise relative motion measurements are also invariant to global orientation and location shifts, that is, any global rotational shift of the form $R_i \to RR_i$ and any global translation of the form $\mathbf{t}_i \to \mathbf{t}_i + \mathbf{t}$ for $R \in \mathrm{SO}(3)$ and $\mathbf{t} \in \mathbb{R}^3$ produce equivalent solutions. However, contrary to the case of invariance to scale transformations of the form $\mathbf{t}_i \to c\mathbf{t}_i$ for $c \in \mathbb{R}^+$, these invariances are not observed to result in any difficulties in motion estimation.

Figure 3.1. Two images from the Notre Dame Cathedral set in Snavely *et al.* (2006), with corresponding feature points, extracted using the scale-invariant feature transform (SIFT: Lowe 2004).

which encapsulates the fact that the difference vectors $\mathbf{t}_i - \mathbf{t}_j$ are parallel to the pairwise directions γ_{ij}. Brand, Antone and Teller (2004) estimated the locations by minimizing the *sum of squared errors* in the system obtained by replacing γ_{ij} in (3.2) with their estimates[4] and using constraints to fix the global scale and translation, given respectively by $\sum_i \|\mathbf{t}_i\|^2 = 1$ and $\sum_i \mathbf{t}_i = \mathbf{0}$, to prevent the trivial solution of $\mathbf{t}_i \equiv \mathbf{t}$ for some $\mathbf{t} \in \mathbb{R}^3$, which is summarized as

$$\begin{aligned} &\underset{\{\mathbf{t}_i\}}{\text{minimize}} \quad \sum_{i\sim j} (\mathbf{t}_i - \mathbf{t}_j)^T (I - \gamma_{ij}\gamma_{ij}^T)(\mathbf{t}_i - \mathbf{t}_j) \\ &\text{subject to} \quad \sum_i \|\mathbf{t}_i\|^2 = 1; \quad \sum_i \mathbf{t}_i = \mathbf{0}. \end{aligned} \tag{3.3}$$

Note that, considering the n locations stacked into a vector $\mathbf{t} \in \mathbb{R}^{3n}$, whose ith 3×1 block is equal to $\mathbf{t}_i$, we can rewrite the cost function of (3.3) as $\mathbf{t}^T L \mathbf{t}$, for some appropriate $L \in \mathbb{R}^{3n\times 3n}$, and the constraints as $\|\mathbf{t}\|^2 = 1$ and $V\mathbf{t} = \mathbf{0}$ (for $V = I_3 \otimes \mathbf{1}_n^T$). Then the solution to the least-squares problem (3.3) is given in terms of the (normalized) eigenvector corresponding to the fourth smallest eigenvalue of L, whose null space includes the three rows of V. Brand *et al.* (2004) also (heuristically) identified various classes of problems, which they referred to as 'problem pathologies', resulting in relatively unstable solutions. Among the listed 'pathologies', an important

[4] Brand *et al.* (2004) actually assumed a slightly more general setup, where the directional measurements are not necessarily unit length, that is, they may have some scale-related part (the source of which was unspecified). They also proposed a *maximum covariance spectral solution*, maximizing the alignment of each $\mathbf{t}_i - \mathbf{t}_j$ with γ_{ij}, instead of penalizing their deviation. We do not discuss this method due to its inferior performance, as reported by Brand *et al.* (2004).

one is the 'underconstrained' class of pathologies which possess additional degrees of freedom resulting in more than one set of location estimates (not related to each other via a global translation and/or scale transformation) that may be significantly different. As sources of these underconstrained pathologies, Brand *et al.* (2004) identified two (rather extreme) conditions, namely disconnectedness of the measurement graph[5] and the collinearity of all the directions at a single node, and also provide heuristic solutions to resolve the extra degrees of freedom; these two conditions are in fact only specific examples of a rich set of algebraic conditions defining underconstrained problems, as was fully characterized later by Özyeşil *et al.* (2015) and Özyeşil and Singer (2015). A similar location estimation method based on the system of equations (3.2) is given in Govindu (2001). In this method, instead of directly minimizing the sum of squared errors in the system (3.2), an iteratively reweighted least-squares solver, where in the $(k+1)$th iteration each equation in (3.2) is weighted with $1/\|\mathbf{t}_i^k - \mathbf{t}_j^k\|$, where $\mathbf{t}_i^k$ denotes the solution of the kth iteration, is used to democratize the contribution of each pair of cameras in the total cost function. Another approach, similar in spirit to these least-squares methods, is introduced in Arie-Nachimson *et al.* (2012). This time, instead of the system of equations (3.2), a least-squares method is used to minimize the sum of squared errors in the epipolar constraints (3.1).

Even though the least-squares methods of Brand *et al.* (2004), Govindu (2001) and Arie-Nachimson *et al.* (2012) are computationally very efficient and, for relatively small datasets, produce acceptable location estimates, the quality of their estimates degrades significantly for large, sparse, unordered and noisy datasets, that is, for most of the real datasets studied in the literature. Typically, they tend to produce spurious location estimates that cluster tightly around a small number of points: see Figure 3.2 for an example. This undesired tendency can be intuitively explained by considering the least-squares optimization problem (3.3), for which a clustering part of the solution results in a very small contribution to the cost function, and a few locations, usually corresponding to nodes in the measurement graph having few edges, are separated from the cluster to satisfy the constraints. In order to resolve this difficulty, several different methods have been proposed in the literature. A relatively straightforward starting point, introduced by Özyeşil *et al.* (2015), is to require the locations to satisfy a maximal proximity condition by incorporating 'repulsion constraints' of the form $\|\mathbf{t}_i - \mathbf{t}_j\| \geq c$, for some fixed $c > 0$ (taken to be $c = 1$ in Özyeşil *et al.* 2015), into the least-squares formulation of Brand *et al.* (2004). These non-convex constraints convert the least-squares formulation into the notoriously

[5] The measurement graph is composed of nodes for each camera i and has an edge between nodes i and j if there is a direction estimate for this pair.

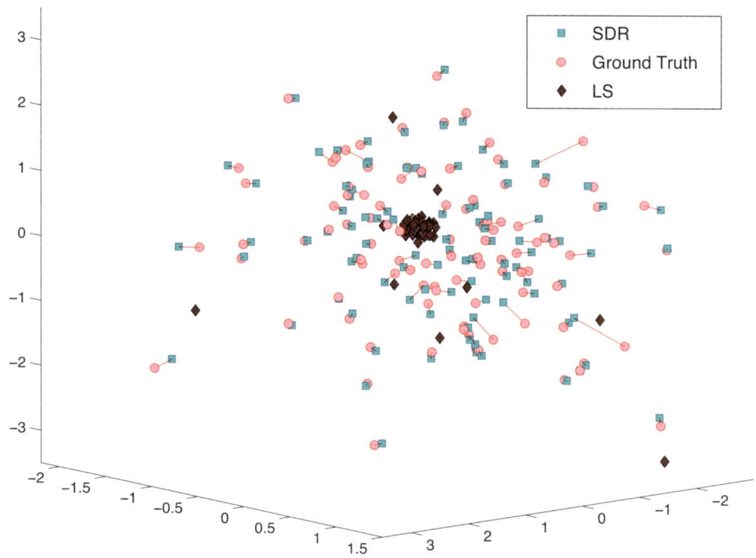

Figure 3.2. Estimates for a noisy, synthetic example of the camera location estimation problem for 100 cameras, demonstrating the clustering estimates of the least-squares (LS) method of Brand *et al.* (2004) versus estimates of the relatively stable semidefinite relaxation (SDR) solver of Özyeşil *et al.* (2015). The segments represent the error incurred by the SDR solution compared to the ground truth.

difficult problem

$$\begin{aligned} \underset{\{\mathbf{t}_i\}_{i\in V_t}\subseteq\mathbb{R}^d}{\text{minimize}} \quad & \sum_{i\sim j} \mathrm{Tr}((\mathbf{t}_i-\mathbf{t}_j)(\mathbf{t}_i-\mathbf{t}_j)^T(I-\gamma_{ij}\gamma_{ij}^T)) \\ \text{subject to} \quad & \|\mathbf{t}_i-\mathbf{t}_j\|_2^2 \geq 1, \quad \text{for all } (i,j)\in E_t, \\ & \sum_i \mathbf{t}_i = \mathbf{0}. \end{aligned} \tag{3.4}$$

To approximate this computationally difficult problem, Özyeşil *et al.* (2015) first introduce the rank-1, positive semidefinite matrix variable $T\in\mathbb{R}^{3n\times 3n}$ (where n denotes the number of cameras), the ijth 3×3 block T_{ij} of which is given by $T_{ij}=\mathbf{t}_i\mathbf{t}_j^T$. Using T, the cost function and the non-convex constraints of (3.4) are given, respectively, by linear functions

$$\sum_{i\sim j}\mathrm{Tr}(\mathcal{L}_{ij}(T)(I-\gamma_{ij}\gamma_{ij}^T))$$

and $\mathrm{Tr}(\mathcal{L}_{ij}(T))\geq 1$ of T, where the operator $\mathcal{L}_{ij}(\cdot)$ satisfies

$$\mathcal{L}_{ij}(T)=T_{ii}+T_{jj}-T_{ij}-T_{ji}=(\mathbf{t}_i-\mathbf{t}_j)(\mathbf{t}_i-\mathbf{t}_j)^T. \tag{3.5}$$

As a result of the linearity in T of the cost function and the constraints, the non-convexity of (3.4) is fully represented by the non-convex rank-1 constraint on T (note that positive semidefiniteness is a convex constraint). As a result, Özyeşil *et al.* (2015) remove this final non-convex constraint to obtain the *semidefinite relaxation* formulation given by

$$\begin{aligned} \underset{T}{\text{minimize}} \quad & \sum_{i\sim j} \operatorname{Tr}(\mathcal{L}_{ij}(T)(I - \gamma_{ij}\gamma_{ij}^T)) \\ \text{subject to} \quad & \operatorname{Tr}(\mathcal{L}_{ij}(T)) \geq 1, \quad \text{for all } i \sim j, \\ & \operatorname{Tr}(VT) = 0, \\ & T \succeq 0. \end{aligned} \tag{3.6}$$

The approximate solution $\mathbf{t}_i$ in (3.4) is obtained to be the ith 3×1 block of the leading eigenvector of the solution of the SDR (3.6). The tightness[6] of this relaxation up to relatively high levels of noise is demonstrated empirically by Özyeşil *et al.* (2015), who also prove a *stability of recovery* result, which quantifies the amount of distortion in the location estimates to be of the order of the noise present in the pairwise directions, under fairly general assumptions for the noise model. To improve the quality of their location estimates, Özyeşil *et al.* (2015) propose a 'robust pairwise direction estimation' method based on the system (3.1), where the $\mathbf{y}_{ij}^k$ are viewed as noisy samples from the two-dimensional subspace orthogonal to γ_{ij} and efficient robust subspace recovery techniques are used to estimate each pairwise direction γ_{ij}. Further, and perhaps more importantly, Özyeşil *et al.* (2015) provide a complete characterization of *well-posed* location recovery from pairwise directions[7] by demonstrating its equivalence to the existing results in the field of *parallel rigidity theory.* For instance, while camera orientation estimation only requires the connectivity of the measurement graph, connectivity alone is insufficient to make a camera location estimation well-posed; see Figure 3.3 for an example. The complete characterization of well-posed location recovery problems (for arbitrary dimensions $d \geq 2$), via (generically) parallel rigid measurement graphs, is summarized in Theorem 3.1.

[6] Tightness here refers to obtaining the solution of the non-convex problem as the solution of the relaxed convex problem, that is, the two globally optimal solutions are the same and hence there is no *relaxation gap.*

[7] To be more precise, in Özyeşil *et al.* (2015), the information assumed to be available is actually a collection of potentially incomplete and noisy versions of the *pairwise lines* (or unsigned directions), which can be represented by the projection matrices $(\mathbf{t}_i - \mathbf{t}_j)(\mathbf{t}_i - \mathbf{t}_j)^T / \|\mathbf{t}_i - \mathbf{t}_j\|^2$. However, the well-posed examples of this case turn out to be the same as in the case of pairwise directions.

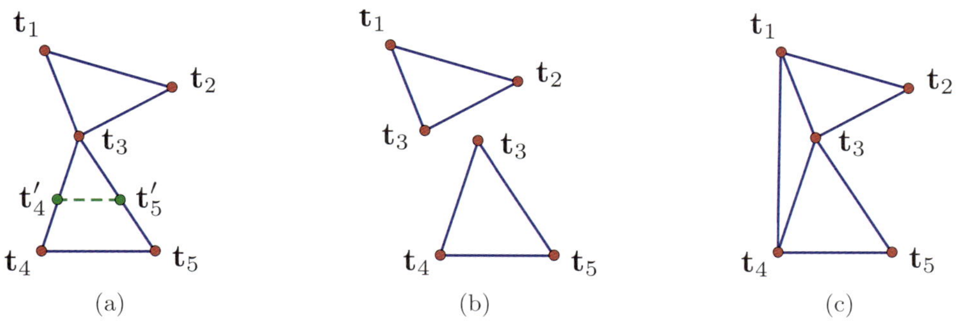

Figure 3.3. (a) A noiseless example of the camera location estimation problem for five locations on a connected graph (we assume noiseless direction measurements between location pairs having an edge), which is *not* parallel rigid (in $\mathbb{R}^2$ and $\mathbb{R}^3$). Non-uniqueness is demonstrated by two non-congruent location solutions $\{\mathbf{t}_1, \mathbf{t}_2, \mathbf{t}_3, \mathbf{t}_4, \mathbf{t}_5\}$ and $\{\mathbf{t}_1, \mathbf{t}_2, \mathbf{t}_3, \mathbf{t}'_4, \mathbf{t}'_5\}$, each of which can be obtained from the other by an *independent rescaling* of the solution for one of its maximally parallel rigid components. (b) Maximally parallel rigid components of the formation in (a). (c) A parallel rigid formation (in $\mathbb{R}^2$ and $\mathbb{R}^3$) obtained from the formation in (a) by adding the extra edge $(1, 4)$ linking its maximally parallel rigid components.

Theorem 3.1. A graph $G = (V, E)$ is generically parallel rigid in $\mathbb{R}^d$ if and only if it contains a non-empty set of edges $E' \subseteq E$, with $(d-1)|E'| = d|V| - (d+1)$, such that for all subsets E'' of E' we have

$$(d-1)|E''| \leq d|V(E'')| - (d+1), \tag{3.7}$$

where $V(E'')$ denotes the vertex set of the edges in E''.

Özyeşil *et al.* (2015) also provide an efficient randomized algorithm for testing well-posedness. Further, they introduce an efficient *alternating direction augmented Lagrangian method* (ADM) to solve the SDR (3.6), and a distributed algorithm to apply the SDR method to large sets of images.

An alternative two-step procedure, studied by Wilson and Snavely (2014), is composed of the detection of outliers among the pairwise direction measurements, via a procedure called 1DSfM, followed by a (non-convex) location estimation method; see Figure 3.4 for a simplified illustration of 1DSfM. The main idea in the outlier detection step is to project all of the pairwise directions onto a one-dimensional (random) subspace and try to solve the resulting ordering problem[8] as consistently as possible, from which

[8] This ordering problem, namely the *minimum feedback arc set problem*, is known to be an NP-complete problem, so Wilson and Snavely (2014) use a heuristic/approximate method observed to perform well in the literature.

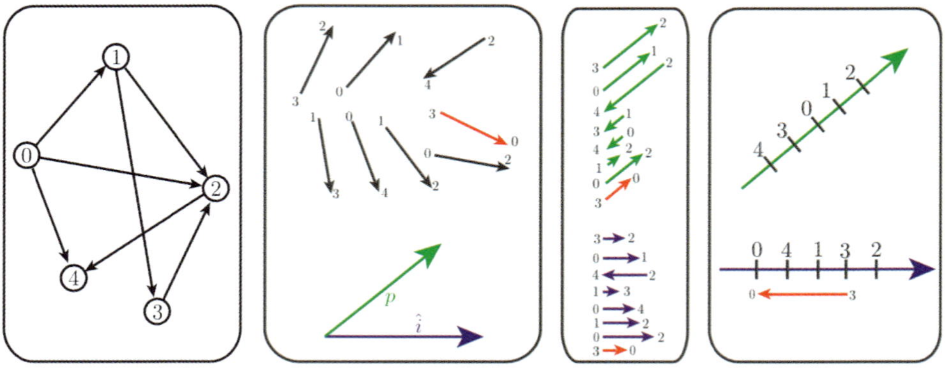

(a) (b) (c) (d)

Figure 3.4. A simplified illustration (courtesy of Wilson and Snavely 2014) of outlier detection via 1DSfM. (a) Ground truth camera locations with (potentially noisy) pairwise directions available for camera pairs connected with arrows. (b) Available pairwise directions, with an outlier edge shown in red, and two directions, $\hat{\imath}$ and p, for one-dimensional (1D) projections. (c) Projections of pairwise directions onto $\hat{\imath}$ and p, resulting in two variants of the minimum feedback arc set problem. (d) Solutions to the 1D problems in (c), depicting the violation (satisfaction) of the 1D constraints by the outlier edge $(3,0)$ in the 1D problem induced by $\hat{\imath}$ (p), since outlier edges may be consistent with only some of the solutions to the 1D subproblems.

potential outliers can be identified as having larger inconsistencies. The detection procedure 'solves' several of these randomized one-dimensional problems and identifies the outliers as those directions having an inconsistency score larger than a certain value. After removal of outlier direction measurements, the locations are estimated by an unconstrained maximization, using the Levenberg–Marquardt algorithm, of the non-convex cost function $\sum_{ij} \hat{\gamma}_{ij}^T(\mathbf{t}_i - \mathbf{t}_j)/\|\mathbf{t}_i - \mathbf{t}_j\|$ that measures the alignment of the pairwise directions $(\mathbf{t}_i - \mathbf{t}_j)/\|\mathbf{t}_i - \mathbf{t}_j\|$ with their 'cleaned' estimates $\hat{\gamma}_{ij}$. Wilson and Snavely (2014) also provide extensive experimental results for large datasets, demonstrating the accuracy and efficiency of their method.

Instead of the systems of equations (3.1) or (3.2), on which the various least-squares methods are based, an alternative approach aiming to prevent the clustering phenomena is to consider the system of equations

$$\mathbf{t}_i - \mathbf{t}_j - \alpha_{ij}\gamma_{ij} = \mathbf{0}, \tag{3.8}$$

where $\alpha_{ij} := \|\mathbf{t}_i - \mathbf{t}_j\|$. Note that if we ignore this defining (non-convex) relation between the α_{ij} and $\mathbf{t}_i$, (3.8) turns out to be a linear system in α_{ij} and $\mathbf{t}_i$ for given γ_{ij}. This system forms the basis of the location estimation method used by Tron and Vidal (2014), where the cost function is the sum of squares of the errors in (3.8) and the additional constraints $\alpha_{ij} \geq 1$ are used to prevent the trivially collapsing solution $\mathbf{t}_i \equiv \mathbf{0}, \alpha_{ij} \equiv 0$. That is, for location estimation, Tron and Vidal (2014) solve

$$\begin{aligned} &\underset{\{\mathbf{t}_i\},\{\alpha_{ij}\}}{\text{minimize}} && \sum_{i \sim j} \|\mathbf{t}_i - \mathbf{t}_j - \alpha_{ij}\gamma_{ij}\|^2 \\ &\text{subject to} && \sum_i \mathbf{t}_i = \mathbf{0}; \quad \alpha_{ij} \geq 1. \end{aligned} \tag{3.9}$$

In fact, the main theme in Tron and Vidal (2014) is to provide a distributed algorithm for camera motion estimation, that is, joint estimation of orientations and locations. In this respect, they provide a *consensus*-based three-step algorithm, and local convergence and partial stability results (for the rotational part). The quadratic location estimation method (3.9) is used to obtain an initial set of locations (*i.e.* it constitutes the second step of their distributed algorithm), and it is then fused with the rotational part to obtain a joint motion estimation method. In terms of the camera location estimation problem with given pairwise directions, the crucial contribution of Tron and Vidal (2014) is using (3.8) as an alternative to (3.1) or (3.2), together with the relaxed repulsion constraints $\alpha_{ij} \geq 1$. As reported by Özyeşil and Singer (2015), for example, this framework provides location estimates with acceptable quality, which do not tend to cluster for large and noisy datasets, while preserving computational efficiency. A closely related optimization method based on (3.8) and the constraints $\alpha_{ij} \geq 1$ has been introduced by Moulon, Monasse and Marlet (2013). The cost function of this method is given by the *maximum absolute deviation* in the system (3.8).[9] Moulon *et al.* (2013) also introduce robust rotation and pairwise direction estimation methods. We note, however, that such maximal deviation penalty methods (*i.e.* ℓ_∞-norm minimization methods) are highly sensitive to outliers in the pairwise direction estimates.

3.1. Robust convex methods

The existence of a large proportion of outliers among pairwise direction estimates, for example stemming from systematically mismatched features between images as a result of symmetries or ambiguities in the images (see also Section 6.2), manifests itself as large errors in the estimated locations. For some datasets this difficulty may be tamed by preprocessing the pairwise directions for rejection of outliers (as in Wilson and Snavely 2014),

[9] The method of Moulon *et al.* (2013) actually assumes the possibility of multiple pairwise direction measurements for each camera pair i and j, computed from each triangle containing the pair.

or by making use of consistencies in locally constructed 3D structures. However, generally applicable efficient algorithms, which are resilient to a large number of outliers and which exhibit provable convergence to globally optimal solutions and admit theoretical guarantees of robustness, are of high value. In this section we will consider two recent algorithms that have some of these desired properties.

The first location estimation algorithm, based on the system (3.8) and closely related to Tron and Vidal (2014) and Moulon *et al.* (2013), is given in Özyeşil and Singer (2015), where this time, instead of using a cost function given as the sum of squared errors (as in Tron and Vidal 2014) or the maximum error (as in Moulon *et al.* 2013) in the system of equations (3.8), the method minimizes the sum of *unsquared* errors in (3.8) (hence the name *least unsquared deviations* (LUD) method) by employing the same relaxed repulsion constraints $\alpha_{ij} \geq 1$. In other words, the $\mathbf{t}_i$ are estimated by solving

$$\begin{aligned} &\underset{\{\mathbf{t}_i\},\{\alpha_{ij}\}}{\text{minimize}} && \sum_{i\sim j} \|\mathbf{t}_i - \mathbf{t}_j - \alpha_{ij}\gamma_{ij}\| \\ &\text{subject to} && \sum_i \mathbf{t}_i = \mathbf{0}; \quad \alpha_{ij} \geq 1. \end{aligned} \tag{3.10}$$

The motivation for this formulation (mainly inspired by compressed sensing methods designed to recover corrupted signals from a minimal set of linear measurements), is to maintain robustness to outliers in the measurements of the pairwise directions γ_{ij}. As a convex *second-order cone program* (SOCP), the LUD solver (3.10) is a computationally efficient location estimator with guaranteed convergence to the globally optimal solution. More importantly, the rather interesting phenomenon of exact location recovery[10] by the LUD method in the presence of sufficiently many exact direction measurements (*i.e.* sufficiently few outlier direction measurements), the number of which varies with the level of sparsity in the measurement graph and the number of cameras, was first empirically demonstrated by Özyeşil and Singer (2015) (see Figure 3.5). An efficient iteratively weighted least-squares (IRLS) solver for the LUD problem (3.10) is provided in Özyeşil and Singer (2015).

Moreover, Özyeşil and Singer (2015) introduce a (non-convex) direction estimation algorithm based on the system (3.1) (similar to that of Özyeşil *et al.* 2015 but more efficient). The pairwise directions are estimated in a two-step procedure that first produces estimates of unsigned directions $\gamma_{ij}^0 = b_{ij}\gamma_{ij}$, and then computes the initially undetermined signs $b_{ij} \in \{-1, +1\}$

[10] Here, 'exact location recovery' corresponds to the estimation of locations up to a global scale and translation (*i.e.* up to a gauge freedom of the form $\mathbf{t}_i \to c\mathbf{t}_i + \mathbf{t}$ for arbitrary $c > 0$ and $\mathbf{t} \in \mathbb{R}^3$), whose ground truth values are known, with a total error smaller than a fixed value, which may be chosen to represent machine precision errors.

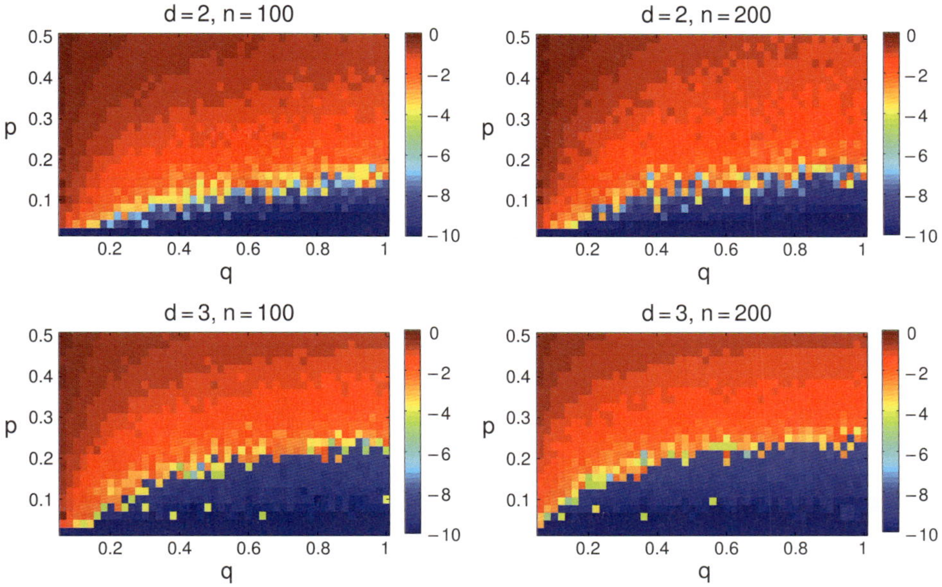

Figure 3.5. Recovery error of the LUD solver (3.10), demonstrating exact recovery regimes in the presence of outliers for dimensions $d = 2, 3$ and number of cameras $n = 100, 200$. Measurement graphs, with an edge between cameras having pairwise directions, are sampled from the Erdős–Rényi ensemble. Camera locations are i.i.d. Gaussian samples, independent of the measurement graph. The colour intensity of each pixel represents the logarithmic recovery error $\log_{10}(\text{NRMSE})$ (see equation (14) in Özyeşil and Singer (2015) for the definition of the normalized root mean squared error, NRMSE), depending on the probability q of the existence of an edge for each pair (x-axis), and the probability p of each existing edge being an outlier (y-axis). NRMSE values are averaged over 10 trials.

via the chirality requirement on the 3D points. In order to maintain robustness to outliers among estimates $\hat{\mathbf{y}}_{ij}^k$ of $\mathbf{y}_{ij}^k$ in (3.1) for the estimation of unsigned directions γ_{ij}^0, Özyeşil and Singer (2015) consider the following (non-convex) problem:

$$\begin{aligned} \underset{\gamma_{ij}^0}{\text{minimize}} \quad & \sum_{k=1}^{m_{ij}} |(\gamma_{ij}^0)^T \hat{\mathbf{y}}_{ij}^k| \\ \text{subject to} \quad & \|\gamma_{ij}^0\| = 1. \end{aligned} \tag{3.11}$$

To obtain the estimate $\hat{\gamma}_{ij}^0$ they use a (heuristic) IRLS method, which is not guaranteed to converge to global optima (since the optimization problem (3.11) is not convex). However, they provide empirical evidence for the

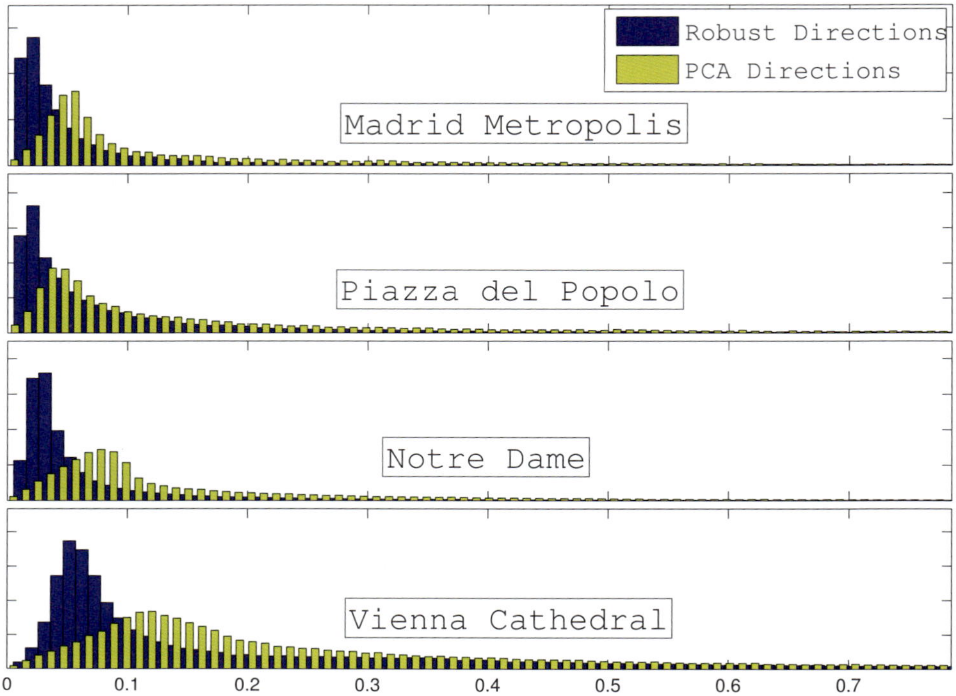

Figure 3.6. Histogram plots of the errors in direction estimates computed by the robust method (3.11) and the PCA method, which replaces the cost function in (3.11) with the sum of squares version $\sum_{k=1}^{m_{ij}}((\gamma_{ij}^0)^T\hat{\mathbf{y}}_{ij}^k)^2$, for some of the datasets studied in Özyeşil and Singer (2015). The errors represent the angles between the estimated directions and the corresponding ground truth directions, computed from a sequential SfM method based on Snavely *et al.* (2006), and provided in Wilson and Snavely (2014). The errors take values in $[0, \pi]$, but the histograms are restricted to $[0, \pi/4]$ to emphasize the difference in quality for the estimated directions.

robustness of the solution to (3.11), while preserving computational efficiency (see Figure 3.6). Finally, Özyeşil and Singer (2015) demonstrate the high quality and efficiency of the overall method on large datasets from Wilson and Snavely (2014).

The excellent empirical performance of the LUD method has recently inspired the development of a new method for location estimation from pairwise directions called *ShapeFit* (Hand, Lee and Voroninski 2017). ShapeFit is based on a convex program that minimizes the sum of the *unsquared* errors in the system (3.2) subject to the constraint $\sum_{ij}(\mathbf{t}_i - \mathbf{t}_j)^T\gamma_{ij} = 1$, which fixes the scale and sign of the location estimates, as well as encouraging correlation with the true relative directions. Specifically, the ShapeFit program consists of the following:

$$\begin{aligned} \underset{\{\mathbf{t}_i\}}{\text{minimize}} \quad & \sum_{i\sim j} \|(I - \gamma_{ij}\gamma_{ij}^T)(\mathbf{t}_i - \mathbf{t}_j)\|_2 \\ \text{subject to} \quad & \sum_i \mathbf{t}_i = \mathbf{0}, \quad \sum_{i\sim j} (\mathbf{t}_i - \mathbf{t}_j)^T \gamma_{ij} = 1. \end{aligned} \tag{3.12}$$

Similar to the LUD method of Özyeşil and Singer (2015), the motivation behind the cost function of ShapeFit is to improve robustness to outliers among the pairwise direction measurements. Also, ShapeFit is amenable to theoretical analysis, and the results of Hand *et al.* (2017) provide the first rigorous results guaranteeing exact location recovery from corrupted relative direction observations. More concretely, Theorem 1 in Hand *et al.* (2017) considers a high-dimensional version of the location recovery problem (*i.e.* $\mathbf{t}_i \in \mathbb{R}^d$ for sufficiently large d) and proves exact location recovery with high probability for i.i.d. Gaussian locations in the presence of a measurement graph sampled from an Erdős–Rényi ensemble, for which at most a constant fraction of direction measurements for any particular location is arbitrarily corrupted. For the same setup, with the exception of requiring the fraction of corrupted directions measurements for any particular location to be at most a constant, *up to poly-logarithmic factors*, Theorem 2 of Hand *et al.* (2017) proves exact location recovery for the physically relevant case $d = 3$. The physically relevant theorem is as follows.

Theorem 3.2. There exists $n_0 \in \mathbb{N}$ and $c \in \mathbb{R}$ such that the following holds for all $n \geq n_0$. Let the measurement graph $G([n], E)$ be drawn from the Erdős–Rényi ensemble $G(n, p)$ for some $p = \Omega(n^{-1/5} \log^{3/5} n)$. Let $\mathbf{t}_1^{(0)}, \ldots, \mathbf{t}_n^{(0)} \in \mathbb{R}^3$ denote the ground truth location, where $\mathbf{t}_i^{(0)} \sim \mathcal{N}(0, I_{3\times 3})$ are i.i.d. independent from G. There exists $\gamma = \Omega(p^5 / \log^3 n)$ and an event of probability at least $1 - 1/n^4$ on which the following holds.

For arbitrary subgraphs $E_b \subset E$ satisfying that, for any vertex i, the number of edges emanating from i which are corrupted is bounded by γn, and for arbitrary pairwise direction corruptions $\mathbf{v}_{ij} \in \mathbb{S}^2$ for $(i, j) \in E_b$, the convex program (3.12) has a unique minimizer equal to

$$\left\{\alpha\left(\mathbf{t}_i^{(0)} - \bar{\mathbf{t}}^{(0)}\right)\right\}_{i\in[n]}$$

for some positive α and for

$$\bar{\mathbf{t}}^{(0)} = \frac{1}{n} \sum_{i\in[n]} \mathbf{t}^{(0)}.$$

Both of these theorems follow from the much more general Theorem 3 in Hand *et al.* (2017), which establishes that ShapeFit succeeds in recovering locations from corrupted relative directions under broad deterministic assumptions on the measurement graph G and the geometry of locations. Note that the setting for ShapeFit theorems is that of robust statistics, in that the corruptions are assumed to be adversarial, and thus for a fixed set of locations and observations the theorem above establishes that ShapeFit succeeds uniformly in the data corruptions, provided there is a bound on the number of corrupted edges emanating from any particular location node.

The proof strategy for ShapeFit theorems is a direct analysis of the optimality conditions of the convex objective function, which relies on the combinatorial propagation of various local geometric properties. For instance, one such property is that if a collection of triangles share the same base and the locations opposite the base are sufficiently 'well-distributed', then an infinitesimal rotation of the base induces infinitesimal rotations in edges of many of the triangles. It is then ensured that for each corrupted edge (k, l), one can find sufficiently many triangles in the observation graph with two good edges and base (k, l), with locations at the opposing vertices being 'well-distributed'. However, the full proof is more nuanced, and needs the constraints of the ShapeFit program. The other important local property is that for a tetrahedron in $\mathbb{R}^3$ with well-distributed vertices, any discordant parallel deviations on two of its disjoint edges induce enough infinitesimal rotational motion on some other edge of the tetrahedron. Combinatorial propagation is then handled separately in two regimes of the relative balance of parallel deviations on the good and corrupted subgraphs.

Hand *et al.* (2017) also provide empirical evidence supporting the main theoretical results about ShapeFit. These results demonstrate that the ShapeFit method can tolerate a larger fraction of pure outliers as compared to the LUD estimator for exact location recovery from randomly generated data, whereas in a mixed noise setting the LUD solver has a smoother change in its recovery performance (as corruption probability varies) and lower (higher) recovery errors in the high (low) corruption probability regime. More importantly, Goldstein, Hand, Lee, Voroninski and Soatto (2016) test ShapeFit empirically on a dozen standard 'photo tourism' datasets. They introduce a novel numerical framework for solving programs like ShapeFit and LUD, leading to an ADMM-based approach called ShapeKick, which solves location recovery problems orders of magnitude faster than competing methods. Goldstein *et al.* (2016) provide a comparative study, with respect to competing methods such as LUD and 1DSfM, demonstrating the accuracy and high efficiency of ShapeFit and ShapeKick.

3.2. Other methods

One of the methods deviating from the generic recipe considered above is studied in Govindu (2004). In this method, camera orientations and locations are jointly estimated using an *iterative averaging method on the*

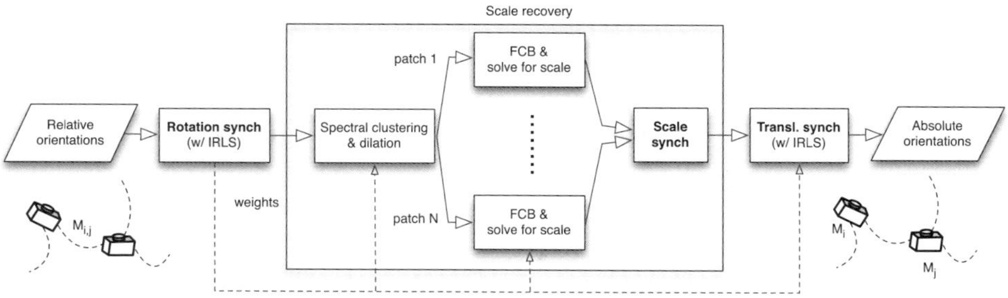

Figure 3.7. Depiction of the group synchronization pipeline (GSP) in Arrigoni *et al.* (2016), including the group synchronization stages in bold. Figure courtesy of Arrigoni *et al.* (2016).

Lie algebra of the group of Euclidean motions. Although Govindu's paper lacks theoretical and strong empirical stability guarantees, its approach is an efficient alternative for joint motion estimation in the low-noise regime. Another recent method slightly deviating from the generic recipe has been introduced by Arrigoni, Fusiello and Rossi (2016). Called the *group synchronization pipeline* (GSP), this method is based on a three-step estimation, via (robust) synchronization, of the camera orientations in SO(3), baseline scales[11] (*i.e.* α_{ij} in (3.8)) in $\mathbb{R}^+$, and the camera locations in $\mathbb{R}^3$; see Figure 3.7 for a depiction of the GSP. The term *synchronization* here refers to a class of methods aiming to (globally) estimate a set of mathematical group elements $\{g_i\}$ from a potentially noisy and incomplete set of observations of their pairwise ratios[12] $g_i^{-1}g_j$. The term was originally coined by Singer (2011), who studied the problem of estimating planar rotations from measurements of their pairwise ratios. As an example of a synchronization problem, in order to estimate the camera orientations $R_i \in \mathrm{SO}(3)$ (represented as 3×3 orthonormal matrices with determinant one) from measurements $\hat{R}_{ij}$ of their ratios $R_{ij} = R_i^T R_j$, we can consider the problem

$$\begin{aligned} \underset{\{R_i\}}{\text{minimize}} \quad & \sum_{i\sim j} \|R_i - R_j\hat{R}_{ij}^T\|_F^2 \\ \text{subject to} \quad & \{R_i\} \subseteq \mathrm{SO}(3), \end{aligned} \tag{3.13}$$

which is a difficult non-convex problem due to the set of constraints. However, the problem (3.13) (and its variations using robust cost functions

[11] Note that Arrigoni *et al.* (2016) perform the synchronization of the baseline scales via a distributed two-step method in order to mitigate error propagation, where each step uses spectral methods to produce its estimates.

[12] Such an estimate can be obtained only up to a global multiplication gg_i of the group elements by a group element g, since the ratios $g_i^{-1}g_j$ are invariant to such transformations.

instead of the sum of squared deviations in (3.13)) has been shown in the literature to admit high-quality approximations via convex relaxations and spectral methods; see for example Arie-Nachimson *et al.* (2012), Özyeşil *et al.* (2015), Chatterjee and Govindu (2013), Wang and Singer (2013) and Cucuringu, Singer and Cowburn (2012). The estimates of Arrigoni *et al.*'s first two steps are fused into the available data in the last step to produce the final location estimates via an iteratively reweighted least-squares (IRLS) solver for the resulting linear system of equations. The high quality of the overall approach in Arrigoni *et al.* (2016) is empirically presented on large datasets. Instead of relying solely on relations between pairs of cameras, Jiang, Cui and Tan (2013) introduce a linear method based on minimizing the error in a set of geometrical relations among triplets of cameras (for details see Jiang *et al.* 2013, Section 4.1), which requires knowledge of *ratios* of baseline lengths computed from locally reconstructed scene points. Furthermore, to improve the quality of the location estimates in the presence of outliers among pairwise measurements, Jiang *et al.* (2013) employ various heuristic outlier rejection techniques. We note that although this linear method has some desirable properties, such as being applicable to collinear motion, computational efficiency, *etc.*, it is rather sensitive to outlier measurements, and hence the outlier rejection techniques proposed by Jiang *et al.* are of critical importance to maintaining stability. Another alternative location estimation approach, introduced by Sinha, Steedly and Szeliski (2010), which aims to tame the instability resulting from unknown scales, uses two-view reconstructions for camera pairs. The main idea is to obtain relative scales and translation between the two-view reconstructions sharing sufficiently many 3D points, and to use this information in the camera location and scale estimation. The pipeline in Sinha *et al.* (2010) also produces an initial 3D structure estimate, which, together with its motion estimate, can be used to initialize reprojection error minimization algorithms for final structural refinement.

4. Structure estimation

This section considers some of the key contributions in the literature mainly concerning the 3D structure estimation part of the SfM problem. These works have a relatively wide spectrum, including incremental and global methods, joint structure and motion estimation methods, and methods specifically targeting very large-scale SfM. Typically, the number of 3D structure points in SfM is much greater than the number of corresponding cameras: for an unordered collection of about 10^3 cameras, for example, it is reasonable to expect of the order of 10^6 structure points. Therefore, the fundamental challenge is the *efficient* processing of the given data in a consistent way, while maintaining robustness to noise in the data.

The primary algorithms for joint refinement of the 3D structure, the camera motion and possibly also the intrinsic camera parameters (also called the camera calibration parameters) are known as *bundle adjustment methods*. As indicated in Triggs *et al.*'s excellent survey of bundle adjustment (see Triggs *et al.* 2000, Figure 9), their origin can be traced back to the works of Gauss and Legendre on least-squares estimation in the late eighteenth and early nineteenth centuries. However, in this section we will focus on more recent developments, and more generally on structure estimation methods that mainly aim to facilitate the processing of large unordered sets of images.

Triggs *et al.* (2000) discussed several different aspects of bundle adjustment, including the basic projection model and problem parametrization, error modelling and the choice of cost function, the main types of numerical optimization algorithms, the sparse problem structure as induced by the network of variables (and how to exploit it to improve reconstruction accuracy), gauge invariance, and quality control to detect outliers and characterize the overall accuracy of the estimated parameters. They presented the projection model and the problem parametrization in a generalized, rather abstract framework to incorporate various scene models (*e.g.* relatively simple isolated 3D features made up of points, lines or planes, or more complicated models involving dynamics, photometry or complex objects linked by constraints) and camera models (*e.g.* perspective, affine or orthographic projections, or less frequently encountered models such as rational polynomial cameras). To define a *feature prediction error* (also sometimes referred to as the *total reprojection error*), Triggs *et al.* (2000) considered a simple scene model comprising static, individual (abstract) 3D features $\mathbf{X}_j$ imaged by a set of cameras corresponding to the position and intrinsic calibration parameters $\mathbf{C}_i$. In this framework we are given noisy measurements $\bar{\mathbf{x}}_{ij}$ of the image features $\mathbf{x}_{ij}$ representing the true image of features $\mathbf{X}_j$ in image i, that are assumed to satisfy a *predictive model* given by $\mathbf{x}_{ij} = \mathbf{x}(\mathbf{C}_i, \mathbf{X}_j)$. The feature prediction errors for the measurements $\bar{\mathbf{x}}_{ij}$ are then defined by

$$\Delta\mathbf{x}_{ij}(\mathbf{C}_i, \mathbf{X}_j) := \bar{\mathbf{x}}_{ij} - \mathbf{x}(\mathbf{C}_i, \mathbf{X}_j). \tag{4.1}$$

To reconstruct the scene, a measure of these prediction errors is minimized, where bundle adjustment corresponds to the *refinement* part of this optimization, and requires an initial set of estimates for the camera and structure parameters (*e.g.* computed using a simpler technique). In summary, as noted by Triggs *et al.* (2000, p. 302), bundle adjustment 'is essentially a matter of optimizing a complicated nonlinear cost function (the total prediction error) over a large nonlinear parameter space (the scene and camera parameters)', the accuracy of which is highly dependent on the initial estimates of camera and structure parameters (explaining, for example, the wide variety of initial motion estimation methods), and on the choice of the parametrization of the parameter space; for a more detailed discussion see Triggs *et al.* (2000, Section 2.2). Moreover, Triggs *et al.* studied the choice of the cost function defining the measure of the errors in (4.1), reaching the main conclusion that robust, statistically based error metrics allowing the presence of outliers in the measurements are appropriate choices. They discussed three main categories of algorithms for the numerical optimization of the chosen cost function in detail, namely the second-order Newton-style methods, first-order methods (demonstrating linear asymptotic convergence) and the sequential methods incorporating a series of observations one by one (as opposed to solving for the whole system globally). For second-order methods, which have fast asymptotic convergence rates but relatively high costs per iteration, they explored several different techniques for improving efficiency, mainly exploiting the sparsity patterns of the Hessian of the cost function. They also discussed various first-order methods, including the simple gradient descent, linear and non-linear Gauss–Seidel methods and Krylov subspace methods. For the third category of sequential algorithms, they considered low-rank update techniques for updating the Hessian and various filtering techniques. Triggs *et al.* (2000) also provided methods for handling theoretical and algorithmic difficulties that arise from several gauge symmetries involved in the problem, and (essentially) concluded with various analytical and heuristic techniques for quality control, that are used to maintain algorithmic stability and robustness to outlier measurements, and to characterize the overall accuracy of the estimates.

A relatively recent work on large-scale bundle adjustment (for tens of thousands of images) is that of Agarwal, Snavely, Seitz and Szeliski (2010). The main challenge identified by Agarwal *et al.* (2010) is the low sparsity levels (*i.e.* relatively high density) present in the connectivity graphs[13] of *unstructured* sets of images, such as community photo collections, as opposed to the higher sparsity of the *structured* image sets (*e.g.* street-level datasets such as images captured from a moving truck); for an example see Figure 4.1. The decrease in sparsity drastically increases the computational cost of the sparse factorization methods employed in classical bundle adjustment methods. To address this difficulty, the inexact (truncated) Newton-type algorithm introduced by Agarwal *et al.* (2010) makes use of conjugate gradients for the Newton step computation combined with various preconditioners. In more detail, they consider a non-linear least-squares[14]

[13] Here, 'connectivity graph' refers to the graph, the nodes of which denote the cameras corresponding to the images, and the edges of which exist between camera pairs having sufficiently many corresponding feature points in their corresponding images.

[14] As noted by Agarwal *et al.* (2010), the simple non-linear least-squares setup encapsulates cost functions more general than the ℓ_2-norm. For example, robust cost functions

(a) structured: 6375 photos (b) unstructured: 4585 photos

Figure 4.1. Sparsity patterns of the adjacency matrices of the connectivity graphs for two sets of images: (a) a structured set of photos (taken from a moving truck), and (b) an unstructured collection of community photos corresponding to a search for the term 'Dubrovnik' in Flickr. Figure courtesy of Agarwal *et al.* (2010).

bundle adjustment problem stated (abstractly) as

$$\underset{\mathbf{x}}{\text{minimize}} \quad \frac{1}{2}\|\mathbf{F}(\mathbf{x})\|^2, \tag{4.2}$$

where $\mathbf{x} \in \mathbb{R}^p$, $\mathbf{F} : \mathbb{R}^p \to \mathbb{R}^q$. Let $J(\mathbf{x}) \in \mathbb{R}^{q \times p}$ denote the Jacobian of $\mathbf{F}$ at $\mathbf{x}$ and let

$$\mathbf{g}(\mathbf{x}) := \nabla \frac{1}{2}\|\mathbf{F}(\mathbf{x})\|^2 = J(\mathbf{x})^T \mathbf{F}(\mathbf{x}).$$

Agarwal *et al.* (2010) consider a Levenberg–Marquardt (LM) approach to solving (4.2), which, to compute the update $\mathbf{x} \to \mathbf{x} + \Delta\mathbf{x}$ at each iteration (while ensuring convergence by controlling the step size $\|\Delta\mathbf{x}\|$), solves the regularized least-squares problem

$$\underset{\Delta\mathbf{x}}{\text{minimize}} \quad \frac{1}{2}\left(\|J(\mathbf{x})\Delta\mathbf{x} + \mathbf{F}(\mathbf{x})\|^2 + \mu\|D(\mathbf{x})\Delta\mathbf{x}\|^2\right), \tag{4.3}$$

where $D(\mathbf{x})$ is a non-negative diagonal matrix, for example, given by

$$D(\mathbf{x})_{ii} = \sqrt{\sum_j J(\mathbf{x})_{ji}^2},$$

and $\mu > 0$ determines the regularization strength. Note that if the regularized Hessian $H_\mu(\mathbf{x}) := J(\mathbf{x})^T J(\mathbf{x}) + \mu D(\mathbf{x})^T D(\mathbf{x})$ is positive definite, we can solve (4.3) via the *normal equations*

$$H_\mu(\mathbf{x})\Delta\mathbf{x} = -\mathbf{g}(\mathbf{x}). \tag{4.4}$$

such as the Huber norm can be used by casting the problem as reweighted non-linear least-squares.

Here, the crucial point is that $H_\mu(\mathbf{x})$ usually turns out to have a special block structure induced by the sparsity pattern of the bundle adjustment problem, which simplifies the solution to (4.4). For instance, assuming that the components of $\mathbf{F}(\cdot)$ only depend on a single camera or a single 3D feature, which is usually the case, then we get

$$H_\mu = \begin{bmatrix} B & E \\ E^T & C \end{bmatrix}, \tag{4.5}$$

where B and C are block diagonal (with blocks of size < 10, and hence can be efficiently inverted), and E is block sparse. Therefore, by also considering the appropriate block representations $\Delta\mathbf{x} = [\Delta\mathbf{y}, \Delta\mathbf{z}]$ and $-g = [v, w]$, and solving for $\Delta\mathbf{z}$ by $\Delta\mathbf{z} = C^{-1}(w - E^T\Delta\mathbf{y})$, we obtain

$$\big[B - EC^{-1}E^T\big]\Delta\mathbf{y} = v - EC^{-1}w. \tag{4.6}$$

Here the Schur complement S of C in H_μ given by $S = B - EC^{-1}E^T$, and known as the *reduced camera matrix*, has a block sparse structure with a non-zero ijth block if and only if the ith and the jth images have at least one common feature point. As noted by Agarwal *et al.* (2010), computationally speaking, the reduction of (4.4) to (4.6), via the *Schur complement trick* given above, allows us to obtain a solution to (4.4) using classical Cholesky factorization methods (that do not exploit the sparsity pattern in S) in $O(n^2)$ space and $O(n^3)$ time complexity, where n denotes the number of cameras. These computational costs become prohibitive for large datasets, and hence an alternative is to use decomposition algorithms that exploit the sparsity in S. However, as the datasets grow and the level and simplicity of sparsity in S decreases (as argued by Agarwal *et al.* 2010 for the case of community photo collections), even the construction and storage of S becomes computationally challenging. As a result, instead of trying to solve (4.3) exactly, for example by using the Schur complement trick discussed above, they propose using an inexact Newton method based on an approximate iterative linear system solver, such as conjugate gradient (CG), possessing a termination rule given by

$$\|H_\mu(\mathbf{x})\Delta\mathbf{x} + \mathbf{g}(\mathbf{x})\| \le \eta_k \|\mathbf{g}(\mathbf{x})\|, \tag{4.7}$$

where $0 < \eta_k \le \eta_0 < 1$, k denoting the LM iteration number, is used as a forcing sequence to guarantee convergence. Further, Agarwal *et al.* (2010) consider several *preconditioners* M for which the matrix $M^{-1}H_\mu(\mathbf{x})$ of the equivalent linear system $M^{-1}H_\mu(\mathbf{x})\Delta\mathbf{x} = -M^{-1}\mathbf{g}(\mathbf{x})$ has reduced condition number, which therefore improves the CG convergence rate. Agarwal *et al.* (2010) conclude their paper by extensively comparing the experimental stability and efficiency of their preconditioned CG method with other preconditioners, and with the classical methods factoring the Schur complement, obtaining significant improvements in accuracy and efficiency, in both time and memory, using their proposed algorithm.

For large, unordered and hence computationally challenging datasets, Crandall, Owens, Snavely and Huttenlocher (2011) have introduced an alternative global approach, whose formulation is based on first computing a coarse initial solution via a hybrid *discrete-continuous optimization*, using discrete belief propagation (BP) within a Markov random field (MRF) formulation of pairwise camera and camera-point constraints to estimate camera parameters and a continuous LM improvement, and then refining the resulting solution with classical bundle adjustment. This method also employs various alternative sources of data such as noisy geotags and vanishing point estimates extracted from images. More concretely, the method first solves for camera orientations by minimizing a robust cost function of the error in the (classical) pairwise rotational consistency equations (*i.e.* the equations $R_{ij} = R_i^T R_j$, for which we have access to noisy measurements of some of the relative rotations R_{ij}), using a mixture of BP (on a discretization of the 3D sphere) and LM refinement. Having solved for the orientations R_i, the joint estimation of the camera locations and structure points is based on the minimization of (a robust cost function of) the total *angles of deviation* between pairwise translations $\mathbf{t}_i - \mathbf{t}_j$ and pairwise directions γ_{ij} (see *e.g.* (3.2)) and between 3D point-to-camera displacements (given by $\mathbf{P}_k - \mathbf{t}_i$ between the kth 3D point $\mathbf{P}_k$ and camera i) and 'ray directions' (given in the noiseless case by $\mathbf{r}_{ik} = R_i^T K_i^{-1} \mathbf{p}_{ik}$, for the known intrinsic matrix K_i of camera i, and the projective measurement $\mathbf{p}_{ik}$ of $\mathbf{P}_k$ by camera i). This minimization is again performed with a hybrid BP and LM method. Even though, in principle, such non-convex methods are highly sensitive to initial points, and hence susceptible to local minima, Crandall *et al.* (2011) have been able to improve its sensitivity and obtain, as empirically demonstrated, high-quality reconstructions by also making use of external, additional sources of information like geotags and vanishing points. Further, the parallelizable structure of the methods in Crandall *et al.* (2011) allows the realization of significant improvements in computational efficiency.

A relatively popular main approach to handling challenging datasets is to employ *sequential processing* of the available images, that is, to incorporate the images one by one, or in small groups, to obtain a solution for the whole set based on a 'core solution' computed from a small subset of the available data. A systematic method of this sequential type was introduced by Snavely, Seitz and Szeliski (2008*b*), whose main idea was to identify a small subset of the available images, called the *skeletal set*, by first estimating the accuracy of pairwise reconstructions of overlapping images and then extracting a subset whose reconstruction accuracy approximates that of the total set. The remaining images are then incorporated by position estimation, and a final bundle adjustment can be used to improve accuracy. The computation of the skeletal set is rather complicated,[15] but the main idea is to identify a (directed) measure of proximity between cameras, identified as nodes of the *image graph*, given by the (translational) uncertainty in the pairwise reconstruction (obtained by fixing the location and orientation of one of the cameras in the pair and then repeating it for the other one) as quantified by the trace of the *covariance* matrix given by the translational part of the inverse of the Hessian of the reduced camera system (4.6) (*i.e.* the Schur complement $S = B - EC^{-1}E^T$ for the camera pair); see Figure 4.2 for a simplified depiction. The skeletal set problem is to identify the smallest subset of cameras, subject to the constraint that the distance (length of the shortest path, as defined by the total uncertainty of the path) between any pair of cameras in the (directed) subgraph induced by this subset is at most t times longer than their distance in the original digraph (induced by the set of all cameras). The factor t is called the *stretch factor*, and controls the amount of additional uncertainty introduced between camera pairs by removing cameras from the available dataset. Also, the skeletal set is required to satisfy the constraint of being uniquely realizable, which is modelled and controlled via a structure that they call the *pair graph*. Unfortunately, this problem (even if we ignore the additional realizability constraint), known as *computing a minimum t-spanner* for the digraph of camera pairs, is NP-complete for general graphs. Hence, Snavely *et al.* (2008*b*, Section 4) employ a two-step heuristic approximation to compute the skeletal set. The empirical results provided by Snavely *et al.* (2008*b*) demonstrate a significant increase in overall SfM efficiency, and the accuracy of the reconstructed structure as compared to applying existing SfM techniques to the full set of images.

Another incremental method, similar in spirit to Snavely *et al.* (2008*b*), has been studied by Havlena, Torii and Pajdla (2010). Their main idea for improving the efficiency of the overall SfM pipeline is to identify a small subset of input images by computing an approximate *minimally connected dominating set*, for which each of the remaining images has significant overlaps with at least one of the images in this set, and to employ *task prioritization* to promote easier image matching. Moreover, instead of computing actual image matchings, which is computationally prohibitive for thousands of cameras if performed on each pair of images, they employ fast image indexing techniques based on large image vocabularies to measure visual overlaps between images. To identify image similarities, they use a

[15] For brevity, we skip the finer details of the computation of the skeletal set in Snavely *et al.* (2008*b*), such as removal of outlier pairwise reconstructions, removal of (duplicate) images with similar content, or identification of *infeasible paths* between triplets of images using higher-level connectivity structures called the *pair graph* and its augmentation with the *image graph*.

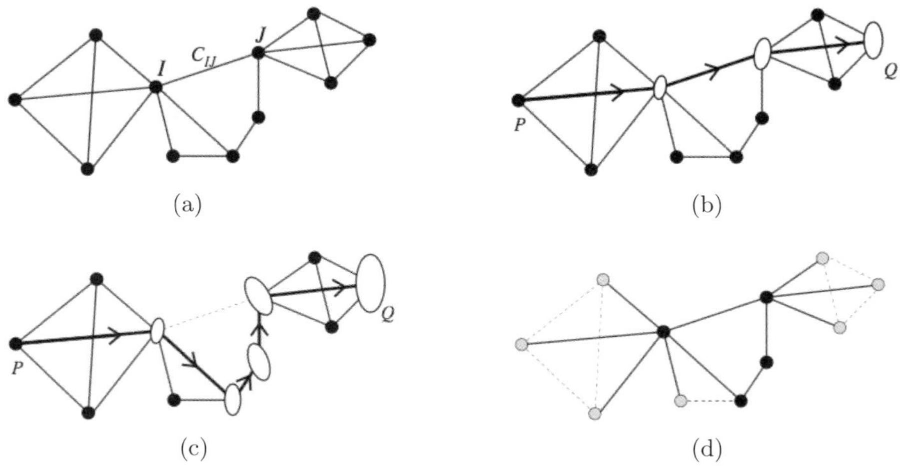

Figure 4.2. A toy example of the construction of the skeletal graph in Snavely *et al.* (2008*b*). (a) Nodes of the image graph represent images and its edges represent pairwise reconstructions (in fact the image graph is directed, and each connected pair has two directed edges; only one is shown for simplicity). Each (directed) edge (I, J) is endowed with a covariance matrix C_{IJ} representing the uncertainty in image J relative to I. (b) Relative uncertainty between the nodes P and Q is quantified by the trace of the covariance matrices (ellipses) chained up along the shortest path (arrows) between the nodes. (c) Removing an edge lengthens the shortest path, resulting in larger estimated covariance (larger ellipses). (d) The solid edges are those of the skeletal graph, while the dotted edges have been removed. The black (interior) nodes form the skeletal set, and are to be reconstructed first, while the grey (leaf) nodes are added to it using position estimation. The skeletal graph is constructed by trying to minimize the number of interior nodes, while bounding the increase in relative uncertainty between all pairs of nodes in the original graph in (a). Figure courtesy of Snavely *et al.* (2008*b*).

'bag-of-words approach' (Havlena *et al.* 2010, Section 2.1). The computation of a minimally connected dominating set for the induced (unweighted, undirected) graph, defined as a minimum-sized connected subgraph, of which each vertex of the original graph is either a member or a neighbour, is in general NP-hard. Hence, they use a polynomial-time approximation algorithm from the literature for this task (Havlena *et al.* 2010, Algorithm 1). For the 3D structure, they make use of the *atomic 3D models from image triplets* approach of Havlena, Torii, Knopp and Pajdla (2009), and grow the structure by using a prioritization queue, which orders the different tasks of either creating a new atomic 3D model or incorporating one image into a given 3D model (Havlena *et al.* 2010, Section 3). They conclude with various experimental results, including image sets captured by omnidirectional cameras, demonstrating the efficiency and accuracy of their approach.

(a) (b) (c)

Figure 4.3. Photo tourism system of Snavely *et al.* (2006): (a) An unstructured collection of images, *e.g.* downloaded from the Internet, as input to the system. (b) SfM step composed of camera motion estimation and 3D structure reconstruction. (c) Content exploration, such as 3D scene visualization, object-based photo browsing, geolocation from images, object annotation, *etc.* Figure courtesy of Snavely *et al.* (2006).

A notable work of an incremental nature is the *photo tourism* system of Snavely *et al.* (2006), which was also discussed in Section 1. Although the SfM part of their paper is primarily based on a rather classical, generic type of sequential incorporation of images, one by one, using the sparse bundle adjustment algorithm of Lourakis and Argyros (2009), it provides a very rich set of tools for content exploration, such as 3D scene visualization, object-based photo browsing, geolocation from images, or object annotation; see Figure 4.3 for a simplified illustration of the photo tourism system. For sequential SfM, Snavely *et al.* identify an initial pair of cameras sharing a large set of feature points, which also have a known baseline to improve the accuracy of the reconstructed initial 3D points. The selection criterion for a new camera is to have a maximal number of feature points that correspond to the already reconstructed 3D points. This iterative procedure is repeated until no remaining camera observes any reconstructed 3D point; for additional details, such as detection of outlier tracks, geo-registration, and specific algorithms used, see Snavely *et al.* (2006, Section 4). Even though the overall computational cost of the proposed SfM pipeline can be fairly high for relatively large datasets,[16] it provides an accurate alternative to SfM, and more importantly, the rich set of tools it provides for content exploration strikingly demonstrates the wide range of applicability and usefulness of SfM.

Yet another remarkable SfM approach for processing large (specifically *city-scale*) unordered datasets has been introduced by Agarwal, Snavely, Simon, Seitz and Szeliski (2009). The key paradigm they explore is the

[16] For example, it is reported to take *two weeks* to process and match the 2635 images in the Notre Dame dataset and register 597 of them.

parallel processing of the various stages in SfM, and hence its main contributions are the identification of challenges and introduction of solutions in the parallelization of the SfM pipeline. In its multi-stage image matching method, where each stage consists of a proposal step (of possible image pairs sharing common features) and a verification step, they employ several different techniques, such as vocabulary tree-based whole-image similarity and query expansion, while simultaneously controlling the computational efficiency of the distributed framework; for details see Agarwal *et al.* (2009, Sections 2.2, 2.3). For the SfM part, they use the skeletal set method of Snavely *et al.* (2008*b*) to reduce the computational cost significantly, and they compute the initial camera motion and 3D structure for the skeletal set via the sequential SfM method of Snavely *et al.* (2006). To improve upon the computed motion and structure, they perform bundle adjustment, where, depending on the size of the problem, they choose between a preconditioned CG method (as in Agarwal *et al.* 2010) for large datasets, and an exact step LM algorithm for smaller datasets. Finally, Agarwal *et al.* (2009) demonstrate the accuracy and efficiency of their overall approach via extensive experimental results for various city-scale datasets.

Interestingly, after camera rotations have been estimated, the rest of the SfM pipeline may be reformulated as a bipartite location recovery problem. Namely, camera locations and 3D structure points form the two sets of nodes of the bipartite graph, and directions between cameras and scene points obtained from correspondences between pairs of views form the edges. This observation, coupled with fast algorithms for ShapeFit and LUD, has potentially substantial implications for real-time robotics applications, as discussed in Section 5. Hand, Lee and Voroninski (2015), in follow-up work to Hand *et al.* (2017), have proved a set of corruption-robust exact recovery theorems for ShapeFit in the bipartite setting. As obtained after the correspondence estimation step of the SfM pipeline, Hand *et al.* (2015) assume the availability of measurements (partially corrupted with adversarial noise) of camera-to-point directions $(\mathbf{t}_i - \mathbf{p}_j)/\|\mathbf{t}_i - \mathbf{p}_j\|$, where $\mathbf{t}_i \in \mathbb{R}^d$ is the ith location and $\mathbf{p}_j \in \mathbb{R}^d$ the jth structure point. Using the convex ShapeFit algorithm introduced in Hand *et al.* (2017) for the simultaneous estimation of locations and structure points, they prove a (high-dimensional) exact recovery result for a set of i.i.d. Gaussian locations and structure points with high probability if the underlying bipartite measurement graph is sampled from an Erdős–Rényi ensemble, the problem dimension d is sufficiently large, and at most a constant fraction of direction measurements involving any particular location or structure point are corrupted; for a precise statement of this main result see Hand *et al.* (2015, Theorem 1). Although this result is in the high-dimensional regime, it is the first *simultaneous* exact location and structure recovery result for partially corrupted camera-to-point direction measurements. The proofs in Hand *et al.* (2017) crucially rely on the existence of sufficiently many triangles in random graphs, while any bipartite graph cannot contain a triangle. The results in Hand *et al.* (2015) establish the necessary properties for exact recovery in high dimensions from corrupted relative-direction observations, similar to Hand *et al.* (2017, Theorem 1), by relying instead on the existence of randomly distributed four-cycles in random bipartite graphs. Hand *et al.* (2015) also provide empirical evidence on simulated data supporting their theoretical findings, and Goldstein *et al.* (2016) evaluate bipartite ShapeFit on real datasets, providing evidence of robust recovery in the bipartite setting.

5. Simultaneous localization and mapping (SLAM)

Simultaneous localization and mapping (SLAM) consists in estimating a depth map of the 3D environment, as well as localizing the agent which is moving throughout this environment. SLAM is a fundamental problem in robotics, providing crucial information for autonomous navigation of cars, drones and consumer robots. Pre-built maps of the environment can be used to provide much useful information for navigation, provided that localization within those maps can be performed accurately. The correct assertion that a vehicle or robotic entity is in a previously observed location is called a loop closure. SLAM is effectively a special case of the SfM problem, in that it is specific to robotics and augmented/virtual reality applications, yet retains many of the same difficulties.

A multitude of combinations of sensing techniques have been used for SLAM, including LIDAR, inertial sensors, GPS and vision. LIDAR is a form of active sensing, which consists of measuring laser time of flight to infer depth in real time. While LIDAR to a large extent trivializes the depth estimation problem and also eases the task of loop closure, it is an expensive sensor with low resolution, and suffers performance degradation in sub-optimal weather conditions such as rain and fog. Nonetheless, LIDAR is the centrepiece of many of today's approaches to autonomous navigation, by allowing pre-mapping of urban environments and accurate loop closures in such maps, which provides information about the static part of the environment in real time. Localization can also be aided by differential GPS, which has very high accuracy but is also expensive. The holy grail of SLAM research is a purely vision-based approach, since vision sensors are cheap and ubiquitous, and provide very rich and high-resolution information about the environment. However, due to the aperture problem and various other ambiguities of the image formation process, using vision alone for SLAM is a difficult task and remains at the forefront of robotics research, in particular due to its high computational complexity. In this section we will cover state-of-the-art vision-based and visual-inertial SLAM approaches as well as some of the particular difficulties faced in robotics applications.

5.1. New approaches to visual SLAM and their limitations

As outlined above, the SfM pipeline (and similarly a SLAM pipeline that does not rely on pre-built maps) begins by (i) establishing feature correspondences and then (ii) estimating camera orientations from relative orientations. Once camera orientations have been estimated, the task is to recover camera locations and 3D structure points, which can be done in several ways. The standard pipeline involves first solving for camera locations from relative directions between cameras, obtained from epipolar geometry and orientation estimates. Several algorithms, as described in Section 3, can be used for this step. Once camera locations have been recovered, 3D structure may be obtained via triangulation. Often, the resulting camera position and 3D structure estimates are not as accurate as desired, and there is a global step which estimates these quantities simultaneously given suitable initial data, which is called bundle adjustment (see Section 4). Standard approaches to bundle adjustment are computationally prohibitive for real-time applications. Recent progress in algorithms and numerical methods allow for fast solutions of the location recovery problem. In particular, Goldstein *et al.* (2016) have developed a numerical framework based on ADMM and tailored to SfM, which allows solution of the LUD and ShapeFit programs at 10–100 times the speed of competing approaches. This capability offers a new path towards real-time SLAM by reformulating the SfM pipeline: that is, after obtaining camera orientations, the point correspondences give relative directions between camera and 3D structure point locations. Thus, we may frame estimating camera locations and 3D structure points as a location recovery problem on a bipartite graph. ShapeFit and LUD can be applied to such a bipartite formulation, and ShapeFit has been proved to recover locations and 3D structure in this regime exactly, under suitable assumptions. Empirical results demonstrate the effectiveness of ShapeFit and LUD in the bipartite regime. Bundle adjustment can potentially be avoided if the output of this pipeline is accurate enough, which would enable usage in real-time robotics applications. The entire process depends on the accuracy of the estimated camera orientations, and in practice, accelerometers are relatively accurate at providing such estimates.

The main difficulty in vision-based SfM systems is the abundance of outliers in correspondence estimates, which are due to the highly ambiguous nature of local photometric information due to illumination changes, occlusions, specularities, repetitive structures and objects that move independently of the scene. In offline applications such as photo tourism (Snavely *et al.* 2006), it is possible to lower the outlier rates in correspondences to still high but acceptable levels (say, 40%) by using RANSAC (Bolles and Fischler 1981), which is a probabilistic technique that randomly samples a set of observations, fits a model, and validates that model by how many of the other observations it approximately fits. In the case of SfM, RANSAC exploits the eight-point algorithm, which, given a set of eight accurately estimated correspondences between a pair of views, estimates the relative orientation and direction between camera locations. Having obtained a candidate relative orientation and translation between camera views, it is possible to validate correspondences. While effective, RANSAC is ultimately a combinatorial procedure, and thus it is only tractable to run to completion in offline applications such as photo tourism. In real-time settings, trade-offs have to be made with the use of RANSAC, at times leaving the fraction of outliers in correspondences as high as 80 or 90 %. These very high outlier ratios make SfM intractable via methods such as ShapeFit and LUD, and extra information needs to be taken into account to enable efficient SfM in such regimes.

5.2. Visual-inertial navigation systems

To alleviate the issue outlined above, it is common practice to combine visual information with input from inertial sensors (accelerometers and gyrometers), to provide estimates of 3D orientation and position of the sensor platform, also called visual-inertial navigation (VINS). Such an approach is prevalent in applications such as augmented and virtual reality. It is common to model the visual inertial navigation problem as the evolution of a dynamical system, the states of which signify 3D positions of the sensor platform as an element of the Euclidean group of motions SE(3), which produces sensor measurements as outputs, up to some uncertainty. It is a natural first step to analyse the observability of such a system, under assumptions on model parameters such as calibration coefficients, the sensor biases, and a generative 'noisy' input which drives the evolution of the system. While most models treat sensor bias as noise independent of other states, this assumption is frequently violated in practice. Hernandez, Tsotsos and Soatto (2015) show that in the absence of such an independence assumption, the resulting model is not observable. The problem is then recast as one of sensitivity analysis, with results of Hernandez *et al.* (2015) deriving an explicit characterization of the set of trajectories indistinguishable under the observed measurements, as a function of the properties of the sensor and sufficient excitation conditions (the properties of motion of the sensor platform), which can be used for analysis and validation purposes.

Tsotsos, Chiuso and Soatto (2015) consider the VINS task while specifically focusing on the handling of outliers. This task falls under the umbrella of robust statistics, and most VINS systems employ some form of outlier/inlier testing. Tsotsos *et al.* (2015) derive the optimal outlier discriminant, showing that it is intractable, motivating state augmentation with a delay-line in the model and examining various analytical and empirical approximations. Compared to decoupled systems, that is, those in which position estimates are computed from visual and inertial measurements individually and then fused, the approach in Tsotsos *et al.* (2015) is guaranteed to stay within a bounded set of the state trajectories.

5.3. Application-specific difficulties

Structure from motion is a well-developed subfield of computer vision, and has led to a number of successful commercial applications. However, the distribution of camera locations strongly affects the performance of SfM and SLAM algorithms, and is dictated by the application at hand. For instance, photo tourism has been successfully commercialized and is one of the easier cases of SfM, because camera locations tend to be well-distributed in space, thus providing sufficiently diverse views of the scene which can be used to infer depth information. On the other hand, the application of autonomous driving requires handling scenarios in which the camera moves in essentially a straight line for long periods of time. There are many issues with this special case.

- (i) The most informative features (the ones that undergo the most motion) are in the periphery and quickly move out of the field of view, which explains the popularity of omnidirectional cameras for such applications.
- (ii) There exist many local minima in the least-squares reprojection error, which is the basis for many algorithms, each corresponding to various ambiguities such as Necker-reversal, plane-translation and bas-relief.
- (iii) Considering the worst case of moving exactly in parallel with the optical axis, it is evident that depth at the centre pixel of the image is unrecoverable, and this phenomenon induces many local minima around the true direction of translation.

Semi-global approaches aim to overcome these issues, but they are too computationally intensive for real-time applications. One may ponder whether these ambiguities are inherent to the problem or are a by-product of the least-squares reprojection functional. Vedaldi, Guidi and Soatto (2007) show that the answer is at least partly the latter, by demonstrating that enforcing a bound on the depth of the scene makes the reproduction error continuous. However, local minima are still present, as evidenced by empirical results.

5.4. Direct versus feature-based methods

One challenge of applications, such as robotics and augmented/virtual reality, is transitioning between differently scaled scenes, for instance from indoors to outdoors. Sensors which provide scale measurements, such as stereo or LIDAR, have a limited scale range in which they perform well. For instance, depth that far exceeds the baseline of a stereo rig is difficult to infer via stereo without sufficiently high image resolution, which in turn impacts real-time computational tractability. Meanwhile, monocular SLAM is immune to this issue since it is inherently scale-invariant. At the same time, due to the fact that absolute scale cannot be inferred in monocular SLAM, scale estimates are subject to drift over time.

SfM approaches fall into two categories: feature-based and direct methods. Feature-based approaches, as described above, decouple the problem into two steps: (i) features are extracted from images, and (ii) camera position and 3D scene structure are computed based on (i) only. Naturally, the efficacy of such approaches depends crucially on the admissible set of features, which in many approaches is limited to a sparse set of keypoints in each image. In contrast, direct methods for visual odometry attempt to utilize all the photometric information in images by optimizing for camera position and 3D structure directly. While direct methods have an information-theoretic advantage over feature-based ones, they are typically more computationally intensive. A method recently proposed by Engel, Schöps and Cremers (2014), called large-scale direct monocular SLAM (LSD-SLAM), is capable of building consistent large-scale maps of a 3D scene in real time, by estimating semi-dense depth maps via filtering and direct image alignment. The map of the scene is represented as a collection of keyframes with 3D similarity transformations as constraints between positions at keyframes. This representation allows scale changes to be taken into account naturally, and corrects accumulated drift.

For alternative methods and further discussion of SLAM, we refer the reader to Aulinas, Petillot, Salvi and Lladó (2008), Fuentes-Pacheco, Ruiz-Ascencio and Rendón-Mancha (2015), Younes, Asmar and Shammas (2016), and the references therein.

6. Other topics

This section briefly discusses some of the additional topics in the SfM literature, not considered in previous sections. These topics include methods for feature extraction and matching, alternative core measurement types and camera models, methods aiming to handle symmetries and ambiguities in images, and widely utilized sources of data and SfM software.

6.1. Feature description and matching

Although the processes of detection, description and matching of image features are usually considered to precede the SfM steps, here we touch upon some of the most frequently used methods for these tasks.[17] One of these methods is *scale-invariant feature transform* (SIFT) (Lowe 1999, 2004), which has been extensively used as a building block for many of the existing corresponding point-based SfM methods; see, for example, Snavely *et al.* (2006), Wu (2011), Crandall *et al.* (2011), Özyeşil *et al.* (2015), Özyeşil and Singer (2015) and Arie-Nachimson *et al.* (2012). In essence, SIFT transforms an input image into a collection of *local features* (*e.g.* represented as vectors in some Euclidean space), which are invariant to scaling and rotation of the underlying image, and partially invariant to 3D camera viewpoint and to changes in illumination. The detection and description stages of SIFT are mainly composed of four steps. In the first step, potential keypoints, which are invariant to scale and rotation, are determined via detection of extrema in a *scale-space representation* of the image, using difference of Gaussians (DoG). The scale-space representation of an image, depicted in Figure 6.1, is a parametrized collection of smoothed images, where the single parameter corresponds to the size of a kernel used to filter out fine structures in the image. SIFT computes scale-space representations, using DoG kernels, repeatedly for down-sampled copies of the original image, and then identifies locally extremal points on these images by simply comparing the pixel values with the nearest neighbours in the current and adjacent scales. After identifying these candidate points, that is, the extrema in the scale-space representation, the next step is to compute (approximately) location, scale and the ratio of principal curvatures (in order to identify and reject the candidates having low contrast or that are poorly localized on an edge: for details see Lowe 2004, Section 4). The third step aims to achieve invariance to rotation in the descriptor by assigning an orientation (to be phased out in the representation of the descriptor) to each keypoint based on local image gradient directions. At this stage, SIFT has assigned a scale, an orientation and a location to each remaining keypoint in the image, and the next step involves the computation of a descriptor for (the local regions around) each keypoint, while maintaining invariance to changes in illumination and 3D view-point, as much as possible. By operating on the image having a scale closest to that of a keypoint to fix a level of Gaussian blur, SIFT first samples gradient magnitudes and orientations around the keypoint location, and rotates the coordinates of the descriptor and the gradient orientations relative to the keypoint orientation (computed in the previous step) in order to achieve rotational invariance; see Figure 6.2 for a depiction. The

[17] The literature on the (classical) problem of *detection* of local feature points (sometimes also referred to as feature extraction) is a vast accumulation of works of roughly the last five decades, including techniques for detection of edges, corners, blobs and regions, segmentation-based methods and machine-learning-based methods, which we do not intend to discuss here in detail. For an exposition of some of these methods, see for example Tuytelaars and Mikolajczyk (2008).

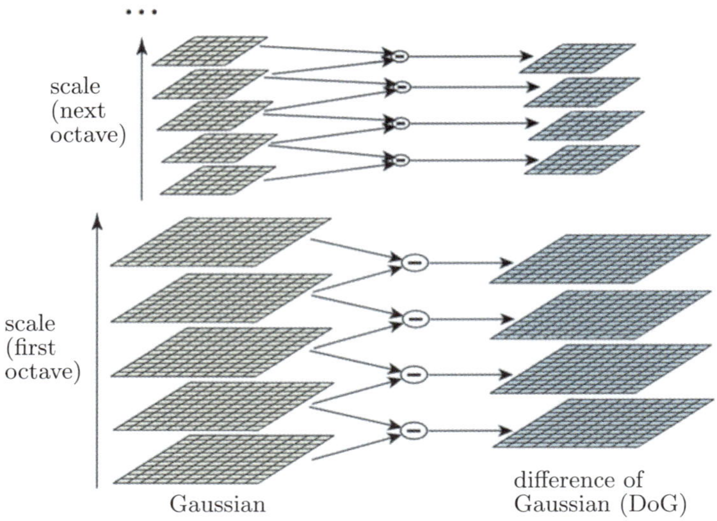

Figure 6.1. The scale-space representation of an image in Lowe (2004) using difference of Gaussians (DoG). For each octave of the scale space, the original image is first convolved with Gaussians to obtain the set of scale space images (on the left), and then adjacent Gaussian images are subtracted to produce the DoG images (on the right). After each octave, the Gaussian image is down-sampled by a factor of 2, and the process is repeated. Figure courtesy of Lowe (2004).

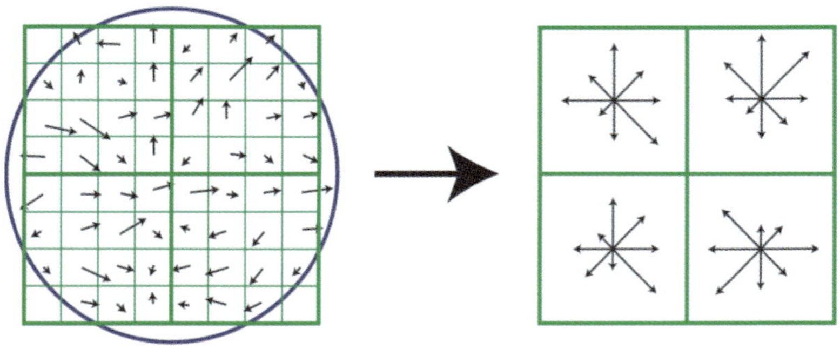

(a) image gradients (b) keypoint descriptor

Figure 6.2. To obtain a keypoint descriptor, first the gradient (magnitude and orientation) at each image sample point in a region around the keypoint location is computed (a) and then scaled by a Gaussian window (represented by the overlaid circle). These samples are then collected into orientation histograms representing the contents over 4×4 subregions (b), where the length of each arrow is given by the sum of the gradient magnitudes quantized to that direction within the region. For simplicity, the figure shows a 2×2 descriptor array computed from an 8×8 set of samples, whereas 4×4 descriptors computed from 16×16 sample arrays are used in practice. Figure courtesy of Lowe (2004).

descriptor is given as a collection of orientation histograms (having 8 bin values corresponding to a weighted sum of magnitudes of gradients, whose directions are close to one of 8 quantized directions) of 4×4 subregions around the keypoint location. The resulting 128-dimensional descriptor is then normalized to have unit norm to attain invariance to homogeneous changes in illumination. To also improve invariance to non-linear changes in illumination, the entries of this unit norm feature vector are thresholded (to compensate for large relative changes in gradient magnitudes), and are normalized to have unit norm once more. In the last stage of matching the extracted feature vectors (*i.e.* descriptors) between images, the Euclidean distance between feature vectors is used to compute nearest neighbours. Matched points are then identified to be those closer to their nearest neighbours than a fixed threshold.

Several works in the literature build upon the SIFT method. One such method, called PCA-SIFT (Ke and Sukthankar 2004), is different from SIFT only in its computation of descriptors from local neighbourhoods around keypoints. Similar to SIFT, for a given keypoint with assigned scale and orientation, PCA-SIFT extracts a 41×41 patch around the keypoint (in the image corresponding to the assigned scale) and rotates the patch using the assigned orientation to attain rotational invariance. We then compute a $2 \times 39 \times 39 = 3042$-dimensional vector of concatenated vertical and horizontal gradients for the 41×41 patch centred at the keypoint. As in SIFT, this vector is normalized to have unit length, to improve invariance to changes in illumination. A low-dimensional representation of this unit norm vector is the desired descriptor. To compute this low-dimensional representation, PCA-SIFT initially computes (offline) an *eigenspace* spanned by the top n eigenvectors (say, $n = 20$) of the covariance matrix of a large number (taken to be 21 000 in Ke and Sukthankar 2004) of 3042-dimensional vectors corresponding to patches extracted from a diverse collection of images. The descriptor for each keypoint is then computed to be the orthogonal projection of the 3042-dimensional vector corresponding to its local patch onto the n-dimensional precomputed eigenspace. The computed descriptors are matched between images via the same method in SIFT. Ke and Sukthankar (2004) also demonstrate empirically that PCA-SIFT is resilient to various deformations and noise in the images; see also Mikolajczyk and Schmid (2005) for a comparative study.

An alternative variation of SIFT, called the *gradient location and orientation histogram* (GLOH), was introduced by Mikolajczyk and Schmid (2005).[18] In essence, GLOH uses a similar approach to PCA-SIFT to compute descriptors. The difference is mainly in its representation of the

[18] The main focus of Mikolajczyk and Schmid (2005) is actually the performance evaluation of various descriptors, based on local regions, with respect to various factors

locations of the gradients in the neighbourhood of the keypoints. Instead of dividing the local neighbourhood (in the image having the assigned scale) of the keypoint using a rectangular grid, GLOH uses *logarithmic polar coordinates* with three bins in the radial component and eight bins in the angular component (without dividing the central bin into angular components), resulting in a total of 17 locations. Further, using a quantization of the gradient orientations into 16 bins, GLOH computes a histogram of $16 \times 17 = 272$ bins. Similar to PCA-SIFT, this histogram is represented in a 128-dimensional eigenspace computed from a collection of 47 000 image patches (via PCA of the corresponding covariance matrix). Mikolajczyk and Schmid (2005) also provide extensive empirical results demonstrating the performance of GLOH relative to nine different methods including SIFT and PCA-SIFT.

For details and performance evaluation of some of the other approaches in the literature (including the computationally efficient *speeded up robust features* (SURF) method of Bay, Tuytelaars and Van Gool 2006) for detection and description of local features, we refer the reader to Gauglitz, Höllerer and Turk (2011).

6.2. Symmetries and ambiguities in images

A fundamental difficulty for SfM methods based on *local features* in images is the presence of distinct objects in the 3D structure that look similar. These similarities may arise from (relatively local) symmetries present in the 3D structure (*e.g.* four similar sides of a clock tower, spherically symmetric dome of a cathedral, *etc.*) or repeated structures in the scene (*e.g.* multiple buildings with similar windows). Such *ambiguous* structures may cause unrelated features in different images to be incorrectly matched, and hence result in major errors in structure recovery, for example partial structure estimates incorrectly superimposed on top of each other, *phantom* structures floating in the scene, or broken structural integrity due to clustering. As a result, for scenes with ambiguous structures, special care must be taken to disentangle incorrect correspondences of local features. In this section we briefly discuss some of the recent progress in the literature aiming to handle such ambiguities.

A disambiguation method based on a global integration of geometric relations between images in a Bayesian framework was introduced by Zach, Klopschitz and Pollefeys (2010). In order to detect and remove inconsistently matched features, the main idea employed by Zach *et al.* (2010) is to utilize consistency relations induced by (invertible) pairwise geometric transformations (*e.g.* homographies, relative camera rotations, similarity

influencing their performance. Hence, for details of some of the other methods for feature extraction, we refer the reader to their paper.

transformations between local 3D reconstructions) along cycles of related images, that is, images presumably sharing measurements of similar parts of the scene. More concretely, Zach *et al.* (2010) consider the underlying graph of related images (having edges endowed with the pairwise transformations) and propose detecting inconsistent edges by using the fact that (in the noiseless case) chaining all pairwise transformations along any cycle must result in the identity transformation. In order to identify inconsistent edges from cycle consistencies, they use a Bayesian inference approach, where the inconsistency of each edge is represented using a latent binary variable. The objective is then to maximize the joint posterior probability of these latent variables given inconsistencies of each cycle[19] they belong to, that is, given some measure of the deviations, from the identity transformation, of the transformations corresponding to these variables chained along cycles. In order to obtain an approximate (more precisely, a locally optimal) solution to this difficult problem, they use *loopy belief propagation*. Also, to obtain quality certificates, they propose approximating the optimization of the log-likelihood function corresponding to the posterior probability using a (standard) convex relaxation approach (for details see Zach *et al.* 2010, Section 3). Finally, they provide experimental evidence of the improvements gained by using their overall disambiguation approach.

Relying solely on pairwise geometric relations for disambiguation may be insufficient for scenes involving large duplicate structures, which result in large, self-consistent sets of incorrectly matched feature pairs. A relatively recent method studied by Roberts, Sinha, Szeliski, Steedly and Szeliski (2011), which fuses an *expectation maximization* (EM) algorithm to estimate camera positions and a sampling technique (similar to RANSAC: Bolles and Fischler 1981) to identify erroneous matches, aims to disambiguate such cases. In designing their method, the basic argument of Roberts *et al.* (2011) is that merely obtaining a geometrically consistent set of edges (as in Zach *et al.* 2010) may be insufficient for disambiguation, since such an approach inherently assumes the *statistical independence of erroneous matches*. For camera motion estimation, they use a Gaussian mixture model (involving hidden binary variables to represent correct/erroneous matches as in Zach *et al.* 2010) to infer the erroneous matches that are *geometrically inconsistent* with the majority of the others. The inference employs an iterative EM algorithm applied first to relative camera rotations, to obtain camera orientations in SO(3), and then to (a spanning tree of) triplet 3D reconstructions, to obtain a camera motion estimate (for details see Roberts *et al.* 2011,

[19] Since, in principle, the underlying graph can have an exponentially large number of cycles, Zach *et al.* (2010) restrict the number of inspected cycles to the set of all triangles together with cycles of maximum length six induced by a set of cycle bases (specifically, spanning tree bases); for further details see Zach *et al.* (2010, Section 3).

Section 3). Since such an approach applied to the whole pairwise measurements with an underlying assumption of independence is argued to be ineffective for disambiguation, they use a specifically designed *random sampling* of spanning trees of the pairwise measurement graphs, which constitutes the main contribution of the paper. To increase the probability of sampling desired spanning trees (*i.e.* minimal sets of pairwise measurements with possibly higher accuracy), they use an efficient sampling method from the literature to generate random spanning trees according to a specific distribution of spanning trees. The distribution is based on a set of edge weights, where the probability of each tree is proportional to the product of its edge weights. To determine the edge weights, Roberts *et al.* (2011) consider two sources of information from the images.

(i) The ratio of the number of matched features between an image pair to the number of all features (in one of the images in the pair) matched to any other image (adjusted by the distance of the missing correspondences to the matched features between the image pair).

(ii) In the presence of *time stamps* for the cameras, the ratio between the time difference of the matched image pair and the smallest time difference of any matched image pair including one of the images in the pair.

For each sampled minimal hypothesis, Roberts *et al.* (2011) compute a set of matches consistent with the hypothesis and update the estimated camera motion (using the EM approach discussed above, while fixing the edges of the sampled spanning tree to be correct edges), and select the hypothesis with the highest score to obtain the final camera motion estimate; for the numerical details, see Roberts *et al.* (2011, Section 5). A final bundle adjustment is applied, only to the matched pairs inferred as correct, to obtain the refined camera motion and the 3D reconstruction. Experimental results provided in Roberts *et al.* (2011), using relatively small sets of highly similar images, demonstrate improvements obtained via the disambiguation approach, and also suggest an increase in computational time required to obtain the final results. For example, disambiguation steps for the CUP dataset of 64 images and the BLDG dataset of 76 images take 44 and 90 minutes, respectively.

Another disambiguation method based on the optimization of a measure of missing image projections of potential 3D structures is introduced in Jiang, Tan and Cheong (2012). To quantify the quality of a 3D reconstruction, Jiang *et al.* (2012) employ the fact that a 3D point has similar appearances in images in which it is captured. A SIFT descriptor for a 3D point is defined in Jiang *et al.* (2012) to be the SIFT descriptor of the corresponding feature point in its *foremost parallel image* (determined by using an approximate surface normal computed for the 3D point). Using this definition, one can hope to identify the visibility of a 3D point in an image by matching its SIFT descriptor (as defined above) to the SIFT descriptor of its projection on the image. Hence, the quality of a 3D reconstruction can be quantified by measuring the average probability $P_{\text{miss}}(p, i)$ (averaged over images and 3D points) for the points p that are missing in the images i, that is, the average probability that the SIFT descriptors of its 3D points do not match with those of their image projections (see Jiang *et al.* 2012, Section 3 for the parametric assumptions they make in order to compute $P_{\text{miss}}(p, i)$). To minimize the average $P_{\text{miss}}(p, i)$, they first compute pairwise 3D reconstructions and relative camera positions. The optimization is performed on spanning trees of the match graph (induced by edges corresponding to camera pairs with a computed relative position), where they employ various heuristics in each iteration to improve the efficiency of its search (for details see Jiang *et al.* 2012, Section 5). They conclude with experimental results demonstrating the accuracy, efficiency and limitations of their disambiguation approach.

Most of the disambiguation methods in the literature are designed to operate on relatively small sets of images created to exemplify relatively difficult SfM disambiguation problems. However, the level of difficulty and the efficiency requirements of large unordered image collections can be significantly different. A recent method, based purely on a graph-theoretic approach, designed to disambiguate such large datasets is studied in Wilson and Snavely (2013). The main concept in Wilson and Snavely (2013) is the bipartite *visibility graph* $G = (I, T, E)$, where nodes are the images I and the tracks T (*i.e.* sets of matched features across several images), and an edge $(i, t) \in E$ exists if the track t includes a feature in image i. The main idea in Wilson and Snavely (2013) is to identify and remove *bad tracks* in T (*i.e.* tracks which correspond to more than one 3D point, possibly due to ambiguities in the scene), which tend to form *bridges* between unrelated parts of the scene (and hence between the corresponding images and tracks of these scenes in G). To quantify the quality of a track t (*i.e.* its dissimilarity to a bridge-type node), they use the *bipartite local clustering coefficient* of t defined to be the ratio of the number of closed 4-paths centred at t to the number of all 4-paths centred at t. Also, in order to eliminate the sensitivity of bipartite local clustering coefficients to the variations in the sizes of local clusters in the visibility graph (*e.g.* resulting from largely varying popularities of different parts of a tourist scene), they compute a covering subgraph G' of the visibility graph G with improved uniformity; see Wilson and Snavely (2013, Section 5) for technical details and specific parameters. The bad tracks of G' are determined to be those having bipartite local clustering coefficients less than a threshold, and are removed from G' (the same procedure is also applied to G). The remaining (good) tracks and their corresponding images are used in an incremental SfM pipeline from

Agarwal *et al.* (2009) to obtain an initial (skeletal) reconstruction, and then remaining tracks and images in G are used to enrich the reconstruction. Although this purely graph-theoretic approach may be insufficient for difficult SfM disambiguation problems, Wilson and Snavely (2013) provide extensive empirical results (for large Internet photo collections) demonstrating the accuracy and the high computational efficiency of their disambiguation technique.

For alternative approaches aiming to resolve the difficulties in SfM for symmetric and ambiguous scenes, we refer the reader to other sources, for example Cohen, Zach, Sinha and Pollefeys (2012), Snavely, Seitz and Szeliski (2008*a*), Schaffalitzky and Zisserman (2002), Martinec and Pajdla (2007) and Pachauri, Kondor, Sargur and Singh (2014).

6.3. Alternative features and camera models

Our discussion in the previous sections has mainly focused on SfM techniques based on *point features* extracted from images formed by *perspective projection*. In this section we touch upon some of the SfM methods in the literature designed for alternative features and camera models.

A relatively popular alternative to using correspondences of point features for SfM is to use corresponding *lines* across multiple images. A solution to line-based SfM is composed of a camera motion estimate and 3D line reconstructions, which may be more suitable, as compared to 3D point reconstructions, for applications such as modelling of urban scenes or augmented reality. A crucial point for line-based SfM methods is the mathematical representation of the four-dimensional set of 3D lines and a corresponding model for their measurements in images, since the usefulness of an existing representation may depend on the particular problem considered. Relatively early factorization methods for line-based SfM (see *e.g.* Martinec and Pajdla 2003, Quan and Kanade 1997) are usually constrained in their accuracy and applicability to scenes with occlusions. A direct reprojection error minimization approach (susceptible to local minima) is proposed in Taylor and Kriegman (1995), where a line l is represented as $l = (\mathbf{d}, \mathbf{v})$ using a *closest point* $\mathbf{d}$ *to the origin* and a unit norm *direction* $\mathbf{v}$.[20] Two different representations of lines are used in Bartoli and Sturm (2005) for the two problems of triangulation of matched features and bundle adjustment. For triangulation, Plücker coordinates are used to represent lines, mainly due to the resulting linearity of the reprojected lines as functions of the 3D lines. For a pair of 3D points $\mathbf{x} = [\bar{\mathbf{x}}\ x_0]$ and $\mathbf{y} = [\bar{\mathbf{y}}\ y_0]$ in homogeneous coordinates, the six-dimensional vector $[\mathbf{a}\ \mathbf{b}]$ representing the Plücker

[20] Note that this representation is a two-to-one correspondence, between the four-dimensional manifold of parameters $(\mathbf{d}, \mathbf{v})$ satisfying $\|\mathbf{v}\| = 1, \mathbf{v}^T\mathbf{d} = 0$ and the set of 3D lines, since $(\mathbf{d}, \mathbf{v})$ and $(\mathbf{d}, -\mathbf{v})$ correspond to the same line.

coordinates of the line through the points $\mathbf{x}$ and $\mathbf{y}$ satisfies $\mathbf{a} = \bar{\mathbf{x}} \times \bar{\mathbf{y}}$, $\mathbf{b} = x_0\bar{\mathbf{y}} - y_0\bar{\mathbf{x}}$ and $\mathbf{a}^T\mathbf{b} = 0$. A linear algorithm and two quasi-linear algorithms (combined with a final step of rounding to the Plücker coordinates) for triangulation of line features are introduced in Bartoli and Sturm (2005). For bundle adjustment, Bartoli and Sturm (2005) employ the *orthonormal representation* of 3D lines $l = (U, W)$, for $U \in \mathrm{SO}(3)$ and $W \in \mathrm{SO}(2)$; see Bartoli and Sturm (2005, Section 5.3) for details of the computation of U, W. This is mainly due to the completeness and minimal updatability (*i.e.* only four parameters are required to update the representation) of the orthonormal representation, which also lacks gauge freedoms and internal constraints, making it a numerically stable choice for the non-linear, unconstrained bundle adjustment step. Bartoli and Sturm (2005) also provide experimental results to evaluate the performance of their overall approach. An alternative method specifically designed for line-based reconstruction of 3D urban scenes has been introduced by Schindler, Krishnamurthy and Dellaert (2006). Their main idea is to classify the lines in the images into different orientation classes (*e.g.* horizontal lines, vertical lines) using an expectation-maximization approach to estimate 3D vanishing point assignments (representing vanishing directions, at least three of which are assumed to be mutually orthogonal) for every pixel in the images. They argue that such a classification improves the accuracy of matching of lines between images (since only lines in separate classes are not allowed to be matched), and also increases the efficiency and accuracy of the 3D structure reconstruction by reducing the degrees of freedom of 3D lines during optimization. To represent a 3D line, they consider an embedding of the four-dimensional manifold of lines in the 11-dimensional space $\mathrm{SO}(3) \times \mathbb{R}^2$, where the $\mathrm{SO}(3)$ and $\mathbb{R}^2$ parts essentially represent plane of orthogonality of the line and the point of intersection of the line and the plane of orthogonality, respectively. In order to overcome the singularities induced by this over-parametrization during non-linear optimization for 3D structure recovery, Schindler *et al.* (2006) use local parametrizations around each point in $\mathrm{SO}(3) \times \mathbb{R}^2$ (encoded by the exponential map of the Lie group $\mathrm{SO}(3)$) with varying degrees of freedom for each class of lines: for example, horizontal lines have three degrees of freedom whereas vertical lines have only two degrees of freedom. They also give experimental results providing the 3D line reconstructions (for relatively small image sets). For additional reading on urban scene reconstruction, we refer the reader to Musialski *et al.* (2013), Pollefeys *et al.* (2008) and the references therein.

There exist several works in the SfM literature based on relatively uncommon camera models that aim to solve the SfM problem with specific applications in mind or improve the accuracy of existing 3D recovery techniques by using different measurement apparatus (*i.e.* different cameras). One such study is that of Chang and Hebert (2000), who consider the SfM

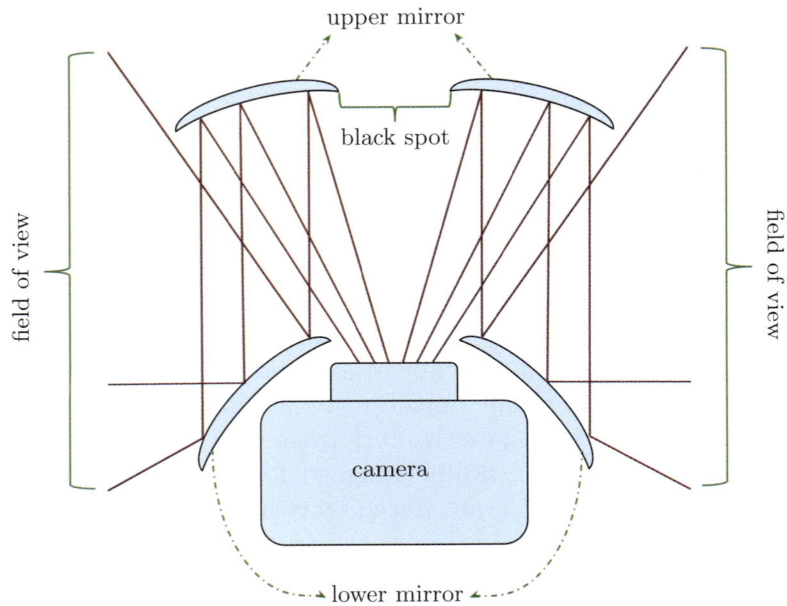

Figure 6.3. (Simplified) diagram of a 2D slice of an omnidirectional camera with two mirrors (upper and lower).

problem for *omnidirectional cameras*, that is, cameras with large visual coverage, possibly having a 360° field of view in the horizontal plane, or one that approximately covers the entire visual sphere: see Figure 6.3 for a simplified diagram. Chang and Hebert (2000) derived the epipolar constraints for catadioptric omnidirectional cameras. To estimate the pairwise camera motions they employed a two-step minimization of the error in these constraints, composed of an initial (classical) estimation of the relative motion parameters from the essential matrix estimates, together with direct minimization of the epipolar errors with respect to the relative motion parameters. The solution of the first step is taken as initial point. They also provided a brief analysis to quantify the errors in the estimated motion parameters induced by measurement uncertainties, along with experimental results demonstrating (occasional) improvements in accuracy obtained by using omnidirectional cameras, instead of conventional ones. A more systematic and detailed study of SfM from variants of omnidirectional cameras (*i.e.* fish-eye lenses and catadioptric cameras with conical mirrors) has been provided by Micusik and Pajdla (2006), covering two-view geometry estimation, camera calibration and 3D reconstruction problems. Micusik and Pajdla (2006) introduce different parametric measurement models for para-catadioptric cameras and fish-eye lenses, and solve the corresponding camera auto-calibration problems by minimizing an angular error measure via *polynomial eigenvalue* computation. For other works on SfM in the omnidirectional setup, see Gluckman and Nayar (1998), Shakernia, Vidal and Sastry (2003), Kannala and Brandt (2006) and Aliaga (2001), for example, and the references therein. An alternative approach in the literature is to use *generic camera models*, representing cameras as sets of projection rays, to incorporate various kinds of cameras in a unified framework, for example cameras with radial distortion, non-central catadioptric cameras, pinhole cameras or stereo rig cameras. Interestingly, such a generic SfM framework allows simultaneous processing of images captured by different camera types. The main approach to motion estimation in the generic setting is to assign projection rays to points of interest in images, by using camera calibration information, and to minimize the error in the *generalized epipolar constraints* (also known as the *Pless equations*) written in terms of the corresponding projection rays (*e.g.* represented using Plücker coordinates). Further, 3D structure points can be estimated using geometric relations between the projection rays and the 3D points of interest; see Ramalingam, Lodha and Sturm (2006, Sections 3 and 4), for example. For experimental results and technical details of these generic SfM methods, we refer the reader to Ramalingam *et al.* (2006), Mouragnon, Lhuillier, Dhome, Dekeyser and Sayd (2009), Kannala and Brandt (2006) and the references therein.

6.4. Popular software packages and data sources

For most of the existing SfM methods, the most computationally involved part is the bundle adjustment step. As a result, efficient and accurate implementations for bundle adjustment are crucial for high-quality SfM solutions. One of the most popular software packages for bundle adjustment is the *Bundler* package of Snavely *et al.* (2006). Essentially, Bundler is based on an incremental application of a modified version of the *sparse bundle adjustment* (SBA) package of Lourakis and Argyros (2009), and, being an incremental method, it incorporates a few images at a time to produce a camera motion and a 3D structure estimate from a set of images, image features and image matches provided as input. In general, for large, unordered, weakly connected sets of images, incremental methods may suffer from accumulation of estimation errors, which result in large deformations in the structure estimate. Further, several intermediate applications of bundle adjustment may hinder the overall computational efficiency. Nevertheless, the Bundler package has been empirically shown to produce accurate solutions that are usually considered as reference solutions for experimental comparisons (see *e.g.* Özyeşil and Singer 2015, Wilson and Snavely 2014). As the name suggests, the efficiency of the crucial SBA method of Lourakis and Argyros (2009) at the core of Bundler essentially arises from exploitation of the sparsity of the Jacobian in the normal equations (4.4) of the Levenberg–Marquardt iterations. We note that, as a standalone package independent of Bundler, SBA is an efficient and stable alternative for final refinements of initial motion and structure estimates, that is, for a final bundle adjustment step.

Wu, Agarwal, Curless and Seitz (2011) have introduced a highly efficient multicore CPU and GPU implementation of the preconditioned conjugate gradient (CG) solver for bundle adjustment (similar to Agarwal *et al.* 2010; see Section 4), called *multicore bundle adjustment*. The multicore bundle adjustment methodology is reported to maintain high accuracy (*i.e.* comparable to alternatives) while requiring less memory usage and providing runtime improvements of up to ten times for the CPU, and up to 30 times for the GPU implementations. This improvement in efficiency is solely based on the optimized usage of parallel computational resources (*i.e.* multicore CPUs and GPUs). An incremental SfM package, called *VisualSfM*, based on the multicore bundle adjustment algorithm of Wu *et al.* (2011), was later provided by Wu (2011). Endowed with a very useful graphical user interface (GUI) that allows explicit visualization of several different stages of SfM, VisualSfM can be considered to be an efficient and relatively stable incremental SfM solver (albeit possessing the problems inherent in the incremental approach). Further, by the integration of the GPU implementation of SIFT given in Wu (2007), VisualSfM can be used as an end-to-end SfM engine taking as input a set of images and producing as output a motion and a 3D structure estimate (however, such a *black-box* usage may result in degradation of accuracy in the final estimates).

Estimation of a 3D structure in SfM usually refers to the computation of a *point cloud* (also known as the *sparse structure*) corresponding to points from the stationary scene. However, we might also need a *dense structure*, for visualization or 3D printing, for example, corresponding to a relatively continuous representation of the corresponding scene, in the form of a dense set of rectangular patches covering the surfaces visible in the input images; see Figure 6.4 for an example. This problem is known as *multiview stereopsis*. An accurate algorithm for this purpose, namely *patch-based multiview stereo software* (PMVS), which is based on a three-step procedure of matching, expansion and filtering of patches of keypoints,[21] is given in Furukawa and Ponce (2010*a*, 2010*b*). When provided with accurate initial camera motion and calibration estimates, PMVS produces high-quality, well-connected dense 3D reconstructions. However, dense reconstruction can be

[21] Here we do not provide any detailed description of this particular algorithm for the optional procedure of dense 3D reconstruction: see Furukawa and Ponce (2010*a*, 2010*b*) for details.

(a) (b)

Figure 6.4. Three-dimensional structures for the Montreal Notre-Dame dataset from Wilson and Snavely (2014). (a) Sparse 3D structure obtained by using the method of Özyeşil and Singer (2015) for camera motion estimation and the PBA package of Wu *et al.* (2011) for final refinement. (b) A snapshot of the dense 3D reconstruction obtained by using the PMVS package of Furukawa and Ponce (2010*a*, 2010*b*) with the motion estimate of Özyeşil and Singer (2015).

computationally very expensive for large sets of images. To tame this computational inefficiency, Furukawa, Curless, Seitz and Szeliski (2010*a*, 2010*b*) propose a highly scalable (to tens of thousands of images) distributed approach based on the decomposition of the images into overlapping sets to be processed in parallel, that is, to produce local dense 3D structures using PMVS, for example. The emerging dense structures are then merged to produce the final dense 3D estimate. The empirical results of Furukawa, Curless, Seitz and Szeliski (2010*b*) demonstrate the high accuracy and improved efficiency of this distributed approach. We also note that the software packages PMVS and CMVS are integrated into the VisualSfM package.

A popular open-source software library consisting of several hundreds of computer vision algorithms, including those for the SfM problem, is *OpenCV* (Open Source Computer Vision Library).[22] The SfM module of the OpenCV library provides a rich selection of algorithms for various stages of the problem ranging from fundamental and essential matrix estimation, pairwise camera pose estimation to triangulation and 3D structure reconstruction. Being a well documented and supported library, OpenCV is particularly valuable for relatively well-known algorithms frequently used by existing SfM solvers.

[22] The homepage of the OpenCV library is http://opencv.org.

An alternative open-source library providing a complete SfM pipeline, from feature detection and matching to 3D structure reconstruction, is the openMVG library by Moulon *et al.* The well documented and supported openMVG library provides efficient implementations of multiple algorithms for each stage of the SfM problem, and also includes accurate solvers for classical methods (*e.g.* the eight- or five-point methods to estimate essential matrices) used in various stages of the problem.

Another extensive software library providing a complete SfM pipeline, also including the dense 3D reconstruction stage, is the *COLMAP* library of Schönberger and Frahm (2016) and Schönberger, Zheng, Frahm and Pollefeys (2016).[23] The COLMAP library includes well documented implementations of several methods in the literature for each stage of the SfM problem. COLMAP also supports various camera models and allows a rich selection of parameters for various stages. Additionally, COLMAP includes a graphical user interface that allows easy access to most of its functionality and a real-time visualization of the reconstruction process.

A recent end-to-end library, called the *Theia Vision Library*, which incorporates efficient, reliable and supported implementations of various alternative methods for each stage of SfM, is available in Sweeney (2016). Theia provides implementations of various methods and useful routines for image handling, feature detection, description and matching, relative pairwise position estimation, camera modelling, global camera orientation and location estimation, initial structure estimation, incremental SfM, motion and structure refinement using bundle adjustment, and gauge symmetry computation for performance evaluation. Theia also provides extensive documentation for all of its components and is open to contributions for extensions and improvements.

Evaluation of the accuracy and the efficiency of many of the existing methods for SfM are solely based on empirical observations. Consequently, easily accessible datasets (*e.g.* sets of raw images, extracted and matched features, or intrinsic camera calibrations) possessing specific characteristics, ranging from benchmark image sets with ground truth motion parameters to large unordered Internet photo collections, are of great value for comparative performance evaluations. Moreover, and perhaps more importantly, collection and preprocessing (*e.g.* feature extraction and matching) of such datasets can be fairly time-consuming and expensive. Hence, the unavailability of such datasets may significantly hinder the development and testing of new methods for SfM. A relatively early work providing several benchmark datasets (now considered to have small sizes) including high-resolution images, intrinsic camera parameters, ground truth camera motion and resulting dense 3D constructions, is that of Strecha, Hansen, Van Gool, Fua

[23] The COLMAP library is accessible at https://colmap.github.io.

and Thoennessen (2008).[24] The availability of ground truth intrinsic and extrinsic camera parameters in these datasets makes them useful for evaluating camera motion recovery accuracy (see *e.g.* Arie-Nachimson *et al.* 2012, Özyeşil *et al.* 2015) with *high precision.* As discussed in Section 4, more recently there has been growing interest in SfM using large and unordered images. One highly popular source of data is the *BigSFM* platform,[25] which includes large SfM examples with original images, extracted and matched features, relative camera positions and intrinsic camera parameters, and is based on the work of Agarwal *et al.* (2009, 2010), Snavely *et al.* (2006), Wilson and Snavely (2013, 2014) and others; see also the links in the data section of the BigSFM web page. As a widely used source of data, the BigSFM platform provides a relatively standardized basis for comparative performance evaluation of various methods in the literature, targeting efficient feature extraction and matching, global camera motion estimation, incremental and global bundle adjustment, handling symmetries and ambiguities, geo-positioning, and so on.

For an extensive account of other open-source software packages and data resources, we refer the reader to the *awesome 3D reconstruction list*[26] by Pierre Moulon.

7. Conclusion

Our survey has covered various methods in the SfM literature pertaining to camera location estimation, 3D structure and motion refinement, simultaneous localization and mapping (SLAM), feature extraction and matching, symmetric and ambiguous scenes, and alternative features and camera models. As also mentioned in the previous sections, for additional topics and methods in the literature, we refer the reader to other works, for example Hartley and Zisserman (2000), Triggs *et al.* (2000), Oliensis (2000), Zhang (1998), Tron *et al.* (2016), Kanade and Morris (1998), Musialski *et al.* (2013), Mikolajczyk and Schmid (2005) and Gauglitz *et al.* (2011). In this section we conclude by briefly mentioning potential areas of progress in the SfM problem.

For most of the existing methods in the SfM literature, one of the most computationally intensive steps (usually assumed to have been accomplished by an existing method) is the feature extraction and matching. Even though there exist sufficiently accurate methods for this task, the computational cost of these methods can be prohibitive for specific applications, and some

[24] Datasets are available at http://cvlabwww.epfl.ch/data/multiview/.
[25] Accessible at http://www.cs.cornell.edu/projects/bigsfm/.
[26] See https://github.com/openMVG/awesome_3DReconstruction_list/.

of the relatively efficient techniques turn out to induce significant degradation in accuracy. In other words, the existing methods seem to have peaked in improving the computational efficiency without suffering losses in accuracy. In this respect, more efficient feature extraction and matching methods, which have improved invariance properties and enable fusion of local and global image characteristics (*e.g.* to maintain higher resilience to ambiguities in images), are of high value.

As discussed in Section 6.2, existing SfM methods for ambiguous scenes can be classified into the two main categories of relatively computationally inefficient methods designed for high proportions of mismatched features (usually observed for small, experimental image sets) and relatively efficient techniques for large image sets (*e.g.* community photo collections) that tend to fail at disambiguation of scenes with large ambiguous structures. Hence, we believe that there is potential for alternative methods with acceptable levels of computational efficiency that have improved robustness to higher numbers of mismatched features.

Considering the recent boost of interest in 3D scene perception and reconstruction (*e.g.* for augmented and virtual reality applications or self-driving cars), we believe that specific SfM algorithms targeting efficient solutions of relatively tightly constrained problems (*e.g.* accurate and fast depth estimation with known camera motion) will attract more attention in the future. Consequently, the design and analysis of efficient SfM techniques, with predictable stability in accuracy and provable robustness, are an important potential area of development.

Finally, even though there exist various theoretical works in the literature that study fundamental problems in SfM and/or provide rigorous analysis of stability and robustness of specific methods (see *e.g.* the early works of Chiuso, Brockett and Soatto 2000, Ma, Košecká and Sastry 2001, Soatto 1997 and the relatively recent works of Özyeşil *et al.* 2015, Hand *et al.* 2017, Hand *et al.* 2015), we believe that the SfM community would still benefit greatly from rigorous results on fundamental problems, for example, what is the theoretically maximal amount of mismatched features or level of noise in images that can be tolerated for stable structure recovery, and can this be achieved efficiently? Further, there is a need for theoretical analysis of stability, robustness and the computational efficiency of existing and new methods.

Acknowledgements

The authors wish to thank Noah Snavely, David G. Lowe and Federica Arrigoni for their permission to include Figures 3.4, 4.1, 4.2, 4.3, Figures 6.1, 6.2 and Figure 3.7, respectively; see the captions of these figures for the corresponding publications of these authors.

A.S. was partially supported by Award Number R01GM090200 from the NIGMS, FA9550-12-1-0317 from AFOSR, the Simons Foundation Investigator Award and Simons Collaboration on Algorithms and Geometry, and the Moore Foundation Data-Driven Discovery Investigator Award. R.B. was supported in part by the Israel Science Foundation grant 1265/14 and by the Minerva Foundation with funding from the Federal German Ministry for Education and Research.

REFERENCES[27]

S. Agarwal, N. Snavely, S. Seitz and R. Szeliski (2010), Bundle adjustment in the large. In *ECCV 2010: 11th European Conference on Computer Vision*, part II, Vol. 6312 of Lecture Notes in Computer Science, Springer, pp. 29–42.

S. Agarwal, N. Snavely, I. Simon, S. Seitz and R. Szeliski (2009), Building Rome in a day. In *ICCV 2009: 12th IEEE International Conference on Computer Vision*.

D. Aliaga (2001), Accurate catadioptric calibration for real-time pose estimation in room-size environments. In *ICCV 2001: 8th IEEE International Conference on Computer Vision*, pp. 127–134.

M. Arie-Nachimson, S. Kovalsky, I. Kemelmacher-Shlizerman, A. Singer and R. Basri (2012), Global motion estimation from point matches. In *3DimPVT 2012: 2nd IEEE International Conference on 3D Imaging, Modeling, Processing, Visualization and Transmission*, pp. 81–88.

F. Arrigoni, A. Fusiello and B. Rossi (2016), Camera motion from group synchronization. In *3DV 2016: 4th IEEE International Conference on 3D Vision*, pp. 546–555.

J. Aulinas, Y. Petillot, J. Salvi and X. Lladó (2008), The SLAM problem: A survey. In *2008 Conference on Artificial Intelligence Research and Development; 11th International Conference of the Catalan Association for Artificial Intelligence*, IOS Press, pp. 363–371.

A. Bartoli and P. Sturm (2005), 'Structure-from-motion using lines: Representation, triangulation, and bundle adjustment', *Comput. Vision Image Underst.* **100**, 416–441.

H. Bay, T. Tuytelaars and L. Van Gool (2006), SURF: Speeded up robust features. In *ECCV 2006: 9th European Conference on Computer Vision*, Vol. 3951 of Lecture Notes in Computer Science, Springer, pp. 404–417.

R. Bolles and M. Fischler (1981), A RANSAC-based approach to model fitting and its application to finding cylinders in range data. In *IJCAI '81: 7th International Joint Conference on Artificial intelligence*, part 2, pp. 637–643.

M. Brand, M. Antone and S. Teller (2004), Spectral solution of large-scale extrinsic camera calibration as a graph embedding problem. In *ECCV 2004: 8th European Conference on Computer Vision*, Vol. 3022 of Lecture Notes in Computer Science, Springer, pp. 262–273.

[27] The URLs cited in this work were correct at the time of going to press, but the publisher and the authors make no undertaking that the citations remain live or are accurate or appropriate.

P. Chang and M. Hebert (2000), Omni-directional structure from motion. In *2000 IEEE Workshop on Omnidirectional Vision*, pp. 127–133.

A. Chatterjee and V. Govindu (2013), Efficient and robust large-scale rotation averaging. In *ICCV 2013: IEEE International Conference on Computer Vision*, pp. 521–528.

A. Chiuso, R. Brockett and S. Soatto (2000), 'Optimal structure from motion: Local ambiguities and global estimates', *Int. J. Comput. Vision* **39**, 195–228.

A. Cohen, C. Zach, S. Sinha and M. Pollefeys (2012), Discovering and exploiting 3D symmetries in structure from motion. In *CVPR 2012: IEEE Conference on Computer Vision and Pattern Recognition*, pp. 1514–1521.

D. Crandall, A. Owens, N. Snavely and D. Huttenlocher (2011), Discrete-continuous optimization for large-scale structure from motion. In *CVPR 2011: IEEE Conference on Computer Vision and Pattern Recognition*, pp. 3001–3008.

M. Cucuringu, A. Singer and D. Cowburn (2012), 'Eigenvector synchronization, graph rigidity and the molecule problem', *Inf. Inference* **1**, 27–67.

J. Engel, T. Schöps and D. Cremers (2014), LSD-SLAM: Large-scale direct monocular SLAM. In *ECCV 2014: 13th European Conference on Computer Vision*, Vol. 8690 of Lecture Notes in Computer Science, Springer, pp. 834–849.

J. Fuentes-Pacheco, J. Ruiz-Ascencio and J. Rendón-Mancha (2015), 'Visual simultaneous localization and mapping: A survey', *Artificial Intelligence Review* **43**, 55–81.

Y. Furukawa and J. Ponce (2010*a*), 'Accurate, dense, and robust multiview stereopsis', *IEEE Trans. Pattern Anal. Mach. Intel.* **32**, 1362–1376.

Y. Furukawa and J. Ponce (2010*b*), PMVS: Patch-based multi-view stereo software. http://www.di.ens.fr/pmvs/

Y. Furukawa, B. Curless, S. Seitz and R. Szeliski (2010*a*), CMVS: Clustering views for multi-view stereo. http://www.di.ens.fr/cmvs/

Y. Furukawa, B. Curless, S. Seitz and R. Szeliski (2010*b*), Towards Internet-scale multi-view stereo. In *CVPR 2010: IEEE Conference on Computer Vision and Pattern Recognition*, pp. 1434–1441.

S. Gauglitz, T. Höllerer and M. Turk (2011), 'Evaluation of interest point detectors and feature descriptors for visual tracking', *Int. J. Comput. Vision* **94**, 335–360.

J. Gluckman and S. Nayar (1998), Ego-motion and omnidirectional cameras. In *Sixth International Conference on Computer Vision*, IEEE cat. no. 98CH36271, pp. 999–1005.

T. Goldstein, P. Hand, C. Lee, V. Voroninski and S. Soatto (2016), ShapeFit and ShapeKick for robust, scalable structure from motion. In *ECCV 2016: 14th European Conference on Computer Vision*, Vol. 9911 of Lecture Notes in Computer Science, Springer, pp. 289–304.

V. Govindu (2001), Combining two-view constraints for motion estimation. In *CVPR 2001: IEEE Computer Society Conference on Computer Vision and Pattern Recognition*, part 2, pp. II-218–II-225.

V. Govindu (2004), Lie-algebraic averaging for globally consistent motion estimation. In *CVPR 2004: IEEE Computer Society Conference on Computer Vision and Pattern Recognition*, part 1, pp. I-684–I-691.

P. Hand, C. Lee and V. Voroninski (2015), Exact simultaneous recovery of locations and structure from known orientations and corrupted point correspondences. arXiv:1509.05064

P. Hand, C. Lee and V. Voroninski (2017), 'ShapeFit: Exact location recovery from corrupted pairwise directions', *Comm. Pure Appl. Math.*, to appear.

R. Hartley (1997), 'In defense of the eight-point algorithm', *IEEE Trans. Pattern Anal. Mach. Intel.* **19**, 580–593.

R. Hartley and A. Zisserman (2000), *Multiple View Geometry in Computer Vision*, Cambridge University Press.

M. Havlena, A. Torii and T. Pajdla (2010), Efficient structure from motion by graph optimization. In *ECCV 2010: 11th European Conference on Computer Vision*, Vol. 6312 of Lecture Notes in Computer Science, Springer, pp. 100–113.

M. Havlena, A. Torii, J. Knopp and T. Pajdla (2009), Randomized structure from motion based on atomic 3D models from camera triplets. In *CVPR 2009: IEEE Conference on Computer Vision and Pattern Recognition*, pp. 2874–2881.

J. Hernandez, K. Tsotsos and S. Soatto (2015), Observability, identifiability and sensitivity of vision-aided inertial navigation. In *ICRA 2015: IEEE International Conference on Robotics and Automation*, pp. 2319–2325.

N. Jiang, Z. Cui and P. Tan (2013), A global linear method for camera pose registration. In *ICCV 2013: IEEE International Conference on Computer Vision*, pp. 481–488.

N. Jiang, P. Tan and L. Cheong (2012), Seeing double without confusion: Structure-from-motion in highly ambiguous scenes. In *CVPR 2012: IEEE Conference on Computer Vision and Pattern Recognition*, pp. 1458–1465.

T. Kanade and D. Morris (1998), 'Factorization methods for structure from motion', *Philos. Trans. Royal Soc. London* **356**, 1153–1173.

J. Kannala and S. Brandt (2006), 'A generic camera model and calibration method for conventional, wide-angle, and fish-eye lenses', *IEEE Trans. Pattern Anal. Mach. Intel.* **28**, 1335–1340.

Y. Ke and R. Sukthankar (2004), PCA-SIFT: A more distinctive representation for local image descriptors. In *CVPR 2004: IEEE Computer Society Conference on Computer Vision and Pattern Recognition*, part 2, pp. 506–513.

H. Longuet-Higgins (1981), 'A computer algorithm for reconstructing a scene from two projections', *Nature* **293**, 133–135.

M. Lourakis and A. Argyros (2009), 'SBA: A software package for generic sparse bundle adjustment', *ACM Trans. Math. Softw.* **36**, 2:1–2:30.

D. Lowe (1999), Object recognition from local scale-invariant features. In *ICCV 1999: 7th IEEE International Conference on Computer Vision*, part 2, pp. 1150–1157.

D. Lowe (2004), 'Distinctive image features from scale-invariant keypoints', *Int. J. Comput. Vision* **60**, 91–110.

Y. Ma, J. Košecká and S. Sastry (2001), 'Optimization criteria and geometric algorithms for motion and structure estimation', *Int. J. Comput. Vision* **44**, 219–249.

D. Martinec and T. Pajdla (2003), Line reconstruction from many perspective images by factorization. In *CVPR 2003: IEEE Computer Society Conference on Computer Vision and Pattern Recognition*, part 1, pp. 497–502.

D. Martinec and T. Pajdla (2007), Robust rotation and translation estimation in multiview reconstruction. In *CVPR 2007: IEEE Conference on Computer Vision and Pattern Recognition*, pp. 1–8.

B. Micusik and T. Pajdla (2006), 'Structure from motion with wide circular field of view cameras', *IEEE Trans. Pattern Anal. Mach. Intel.* **28**, 1135–1149.

K. Mikolajczyk and C. Schmid (2005), 'A performance evaluation of local descriptors', *IEEE Trans. Pattern Anal. Mach. Intel.* **27**, 1615–1630.

P. Moulon, P. Monasse and R. Marlet (2013), Global fusion of relative motions for robust, accurate and scalable structure from motion. In *ICCV 2013: IEEE International Conference on Computer Vision*, pp. 3248–3255.

P. Moulon, P. Monasse, R. Marlet *et al.*, OpenMVG: An open multiple view geometry library. https://github.com/openMVG/openMVG

E. Mouragnon, M. Lhuillier, M. Dhome, F. Dekeyser and P. Sayd (2009), 'Generic and real-time structure from motion using local bundle adjustment', *Image and Vision Computing* **27**, 1178–1193.

P. Musialski, P. Wonka, D. Aliaga, M. Wimmer, L. Van Gool and W. Purgathofer (2013), 'A survey of urban reconstruction', *Computer Graphics Forum* **32**, 146–177.

J. Oliensis (2000), 'A critique of structure-from-motion algorithms', *Comput. Vision Image Underst.* **80**, 172–214.

O. Özyeşil and A. Singer (2015), Robust camera location estimation by convex programming. In *CVPR 2015: IEEE Conference on Computer Vision and Pattern Recognition*, pp. 2674–2683.

O. Özyeşil, A. Singer and R. Basri (2015), 'Stable camera motion estimation using convex programming', *SIAM J. Imaging Sci.* **8**, 1220–1262.

D. Pachauri, R. Kondor, G. Sargur and V. Singh (2014), Permutation diffusion maps (PDM) with application to the image association problem in computer vision. In *Advances in Neural Information Processing Systems 27* (Z. Ghahramani *et al.*, eds), Curran Associates, pp. 541–549.

M. Pollefeys, D. Nistér, J.-M. Frahm, A. Akbarzadeh, P. Mordohai, B. Clipp, C. Engels, D. Gallup, S.-J. Kim, P. Merrell, C. Salmi, S. Sinha, B. Talton, L. Wang, Q. Yang, H. Stewénius, R. Yang, G. Welch and H. Towles (2008), 'Detailed real-time urban 3D reconstruction from video', *Int. J. Comput. Vision* **78**, 143–167.

L. Quan and T. Kanade (1997), 'Affine structure from line correspondences with uncalibrated affine cameras', *IEEE Trans. Pattern Anal. Mach. Intel.* **19**, 834–845.

S. Ramalingam, S. Lodha and P. Sturm (2006), 'A generic structure-from-motion framework', *Comput. Vision Image Underst.* **103**, 218–228.

R. Roberts, S. Sinha, R. Szeliski, D. Steedly and R. Szeliski (2011), Structure from motion for scenes with large duplicate structures. In *CVPR 2011: IEEE Conference on Computer Vision and Pattern Recognition*, pp. 3137–3144.

F. Schaffalitzky and A. Zisserman (2002), Multi-view matching for unordered image sets, or 'How do I organize my holiday snaps?' In *ECCV 2002: 7th European*

Conference on Computer Vision, part 1, Vol. 2350 of Lecture Notes in Computer Science, Springer, pp. 414–431.

G. Schindler, P. Krishnamurthy and F. Dellaert (2006), Line-based structure from motion for urban environments. In *Third International Symposium on 3D Data Processing, Visualization, and Transmission*, IEEE, pp. 846–853.

J. Schönberger and J.-M. Frahm (2016), Structure-from-motion revisited. In *CVPR 2016: IEEE Conference on Computer Vision and Pattern Recognition*, pp. 4104–4113.

J. Schönberger, E. Zheng, J.-M. Frahm and M. Pollefeys (2016), Pixelwise view selection for unstructured multi-view stereo. In *ECCV 2016: 14th European Conference on Computer Vision*, part III, Springer, pp. 501–518.

O. Shakernia, R. Vidal and S. Sastry (2003), Omnidirectional egomotion estimation from back-projection flow. In *CVPRW 2003: Computer Vision and Pattern Recognition Workshop*, part 7, pp. 82–82.

A. Singer (2011), 'Angular synchronization by eigenvectors and semidefinite programming', *Appl. Comput. Harmon. Anal.* **30**, 20–36.

S. Sinha, D. Steedly and R. Szeliski (2010), A multi-stage linear approach to structure from motion. In *ECCV 2010 Workshops: Trends and Topics in Computer Vision*, part II, Vol. 6554 of Lecture Notes in Computer Science, Springer, pp. 267-281.

N. Snavely, S. Seitz and R. Szeliski (2006), 'Photo tourism: Exploring photo collections in 3D', In *ACM Trans. Graph.* **25**, 835–846.

N. Snavely, S. Seitz and R. Szeliski (2008*a*), 'Modeling the world from internet photo collections', *Int. J. Comput. Vision* **80**, 189–210.

N. Snavely, S. Seitz and R. Szeliski (2008*b*), Skeletal graphs for efficient structure from motion. In *CVPR 2008: IEEE Conference on Computer Vision and Pattern Recognition*, pp. 1–8.

S. Soatto (1997), '3-D structure from visual motion: Modeling, representation and observability', *Automatica* **33**, 1287–1312.

C. Strecha, W. Hansen, L. Van Gool, P. Fua and U. Thoennessen (2008), On benchmarking camera calibration and multi-view stereo for high resolution imagery. In *CVPR 2008: IEEE Conference on Computer Vision and Pattern Recognition*, pp. 1–8.

P. Sturm and B. Triggs (1996), A factorization based algorithm for multi-image projective structure and motion. In *ECCV '96: 4th European Conference on Computer Vision*, part II, Vol. 1065 of Lecture Notes in Computer Science, Springer, pp. 709–720.

C. Sweeney (2016), Theia multiview geometry library: Tutorial and reference. http://theia-sfm.org

C. Taylor and D. Kriegman (1995), 'Structure and motion from line segments in multiple images', *IEEE Trans. Pattern Anal. Mach. Intel.* **17**, 1021–1032.

C. Tomasi and T. Kanade (1992), 'Shape and motion from image streams under orthography: A factorization method', *Int. J. Comput. Vision* **9**, 137–154.

B. Triggs (1996), Factorization methods for projective structure and motion. In *CVPR '96: IEEE Computer Society Conference on Computer Vision and Pattern Recognition*, pp. 845–851.

B. Triggs, P. McLauchlan, R. Hartley and A. Fitzgibbon (2000), Bundle adjustment: A modern synthesis. In *Vision Algorithms: Theory and Practice*, Vol. 1883 of Lecture Notes in Computer Science, Springer, pp. 298–375.

R. Tron and R. Vidal (2014), 'Distributed 3-D localization of camera sensor networks from 2-D image measurements', *IEEE Trans. Automatic Control* **59**, 3325–3340.

R. Tron, X. Zhou and K. Daniilidis (2016), A survey on rotation optimization in structure from motion. In *IEEE Conference on Computer Vision and Pattern Recognition (CVPR) Workshops*, pp. 77–85.

K. Tsotsos, A. Chiuso and S. Soatto (2015), Robust inference for visual-inertial sensor fusion. In *ICRA 2015: IEEE International Conference on Robotics and Automation*, pp. 5203–5210.

T. Tuytelaars and K. Mikolajczyk (2008), 'Local invariant feature detectors: A survey', *Found. Trends Comput. Graphics Vision* **3**, 177–280.

A. Vedaldi, G. Guidi and S. Soatto (2007), Moving forward in structure from motion. In *CVPR 2007: IEEE Conference on Computer Vision and Pattern Recognition*, pp. 1–7.

L. Wang and A. Singer (2013), 'Exact and stable recovery of rotations for robust synchronization', *Inf. Inference* **2**, 145–193.

K. Wilson and N. Snavely (2013), Network principles for SfM: Disambiguating repeated structures with local context. In *ICCV 2013: IEEE International Conference on Computer Vision*, pp. 513–520.

K. Wilson and N. Snavely (2014), Robust global translations with 1DSfM. In *ECCV 2014: 13th European Conference on Computer Vision*, part III, Vol. 8691 of Lecture Notes in Computer Science, Springer, pp. 61–75.

C. Wu (2007), SiftGPU: A GPU implementation of scale invariant feature transform (SIFT). http://cs.unc.edu/~ccwu/siftgpu/

C. Wu (2011), VisualSFM: A visual structure from motion system. http://ccwu.me/vsfm/

C. Wu, S. Agarwal, B. Curless and S. Seitz (2011), Multicore bundle adjustment. In *CVPR 2011: IEEE Conference on Computer Vision and Pattern Recognition*, pp. 3057–3064.

G. Younes, D. Asmar and E. Shammas (2016), A survey on non-filter-based monocular visual SLAM systems. arXiv.org:1607.00470

C. Zach, M. Klopschitz and M. Pollefeys (2010), Disambiguating visual relations using loop constraints. In *CVPR 2010: IEEE Conference on Computer Vision and Pattern Recognition*, pp. 1426–1433.

Z. Zhang (1998), 'Determining the epipolar geometry and its uncertainty: A review', *Int. J. Comput. Vision* **27**, 161–195.

Acta Numerica (2017), pp. 365–590
doi:10.1017/S0962492917000046

The cardiovascular system: Mathematical modelling, numerical algorithms and clinical applications

A. Quarteroni[1,2], A. Manzoni[1] and C. Vergara[2]

[1] *Chair of Modelling and Scientific Computing (CMCS),*
Institute of Mathematics,
Ecole Polytechnique Fédérale de Lausanne,
Station 8, CH-1015 Lausanne, Switzerland
E-mail: alfio.quarteroni@epfl.ch, andrea.manzoni@epfl.ch

[2] *MOX, Dipartimento di Matematica, Politecnico di Milano,*
Piazza Leonardo da Vinci 32, 20133, Milan, Italy
E-mail: christian.vergara@polimi.it

Mathematical and numerical modelling of the cardiovascular system is a research topic that has attracted remarkable interest from the mathematical community because of its intrinsic mathematical difficulty and the increasing impact of cardiovascular diseases worldwide. In this review article we will address the two principal components of the cardiovascular system: arterial circulation and heart function. We will systematically describe all aspects of the problem, ranging from data imaging acquisition, stating the basic physical principles, analysing the associated mathematical models that comprise PDE and ODE systems, proposing sound and efficient numerical methods for their approximation, and simulating both benchmark problems and clinically inspired problems. Mathematical modelling itself imposes tremendous challenges, due to the amazing complexity of the cardiocirculatory system, the multiscale nature of the physiological processes involved, and the need to devise computational methods that are stable, reliable and efficient. Critical issues involve filtering the data, identifying the parameters of mathematical models, devising optimal treatments and accounting for uncertainties. For this reason, we will devote the last part of the paper to control and inverse problems, including parameter estimation, uncertainty quantification and the development of reduced-order models that are of paramount importance when solving problems with high complexity, which would otherwise be out of reach.

CONTENTS

1. Introduction

This is a review paper on the mathematical and numerical modelling of the cardiovascular system, a research topic that has attracted remarkable interest from both the mathematical and bioengineering communities over the past 25 years. The driving motivation for this interest is the increasing impact of cardiovascular diseases in our lives. According to Mozaffarian *et al.* (2015), cardiovascular diseases are the major cause of death worldwide, leading to more than 17.3 million deaths per year, a number that is expected to grow to more than 23.6 million by 2030. In Europe this now corresponds to nearly half of all deaths (47%).

In this paper we focus on the two principal components of the cardiovascular system: *arterial circulation* and *heart function*, with its electrical and mechanical activities. Geometric complexity, lack of data to feed the mathematical models, and the multiphysics and multiscale nature of the processes at hand present major challenges when trying to reproduce both function and malfunction.

Owing to its composite nature, the cardiovascular system is first modelled by means of stand-alone *core components*, each describing a single functionality, for example arterial fluid dynamics, the electrical activity of the heart, and the fluid dynamics in the left ventricle. Each core model needs careful mathematical analysis and efficient numerical approximation, often via specifically devised methods. The next step is integration of the core models into global, coupled integrated models suitable for describing a meaningful and coherent part of the cardiovascular system – or even the entire system. This step requires the introduction of suitable coupling conditions and novel numerical strategies for a stable, robust and computationally effective solution of the global problem.

Clinical data play a decisive role in models of the cardiovascular system, and at the same time they present a formidable challenge. Clinical radiological images (such as *computer tomography* and *magnetic resonance imaging*) are necessary to construct the computational domains. The procedure of geometric reconstruction is difficult and, especially for the heart, requires advanced mathematical and numerical tools. Standard radiological images can sometimes be useless: some cardiovascular components may be smaller than the spatial resolution of the imaging device (this is the case for the Purkinje network, for example); in other cases the elevated brightness gap between fluid and vessel wall makes the detection of the latter very hard. Boundary data are also difficult to obtain. When the computational domain results from an artificial truncation, specific physical quantities (*e.g.* fluid velocity or pressure) should be provided at those locations of the arterial tree corresponding to the artificial boundaries. However, this would require invasive measurements that cannot be easily carried out. Finally, the huge inter- and intra-patient data variability and uncertainty are further sources of concern regarding model calibration and validation.

In spite of all these difficulties, a wealth of models has already been successfully used to address both physiological and pathological instances. The aim is, on one hand, a better understanding of the physical and quantitative processes governing the cardiovascular system, and on the other hand the opening of new frontiers in therapeutic planning and the design of implantable devices (*e.g.* medical stents and cardiac defibrillators).

The literature on the mathematical and numerical modelling of the cardiovascular system is huge, as readers will see by browsing our references, a tiny subset of the total. In the forthcoming sections we will try to provide a perspective on the main contributions to this field. Here, among the several books, monographs and review papers published so far, we mention Formaggia, Quarteroni and Veneziani (2009*a*), Taylor and Figueroa (2009) and Quarteroni, Veneziani and Vergara (2016*c*) for the circulatory system, and Peskin (2002), Smith, Nickerson, Crampin and Hunter (2004), Colli

Franzone, Pavarino and Scacchi (2014), Quarteroni (2015) and Quarteroni, Lassila, Rossi and Ruiz-Baier (2017) for the heart.

This review paper consists of three main parts: in Part 1 we model the *arterial circulation* (Sections 2, 3 and 4), in Part 2 we model the *heart function* (Sections 5, 6 and 7), and in Part 3 we treat *inverse problems* and include *uncertainty* (Sections 8, 9, 10 and 11). Both Parts 1 and 2 consist of an introductory section on physiology (Sections 2 and 5), a section describing the available data and their use (Sections 3 and 6), and a final section on mathematical and numerical modelling (Sections 4 and 7). In Part 3 we begin by emphasizing the need to move beyond a single (forward) simulation in some applications (Section 8). This represents the common denominator of three topics recently applied to cardiovascular mathematics: *control and optimization* (Section 9), *parameter estimation* (Section 10) and *uncertainty quantification* (Section 11).

When appropriate (in particular in Sections 4, 7, 9, 10 and 11), we report some numerical results to highlight the effectiveness of the numerical strategies presented here. Unless otherwise specified, all our numerical results have been obtained using the finite element library LifeV; see www.lifev.org for more details.

PART ONE
Arterial circulation

2. Basic facts on quantitative physiology

The cardiovascular system is a closed circuit that carries oxygenated blood to all the tissues and organs of the body. Functionally, it can be regarded as made up of three compartments: the *heart*, the *systemic* and *pulmonary circulations*, and the *microvasculature*. In this section we will recall the most important features of the physiology of the systemic circulation characterizing the mathematical models that will be introduced later on. We will also highlight the main peculiarities of the pulmonary circulation. Heart physiology will be addressed in Section 5.

The systemic circulation is made up of the arteries, which carry oxygenated blood ejected by the left heart to the living tissues, and the veins, which allow non-oxygenated blood to return to the right heart. The exchange of oxygen between blood and the body tissues occurs in the microvasculature, which in fact separates the systemic arterial tree from the venous systems. In the pulmonary circulation, non-oxygenated blood ejected by the right heart flows in the pulmonary arteries towards the lungs where it becomes oxygenated and goes back to the left heart through the pulmonary veins.

Blood is composed of *plasma* (about 55% of its total volume), which consists of water (about 92% of plasma volume), proteins and ions. The remainder corresponds to the blood cells, of which 97% of the volume is made up of *erythrocytes* (red blood cells), which carry the oxygen in oxygenated blood. The other cells are *leucocytes* (white blood cells) and *platelets*. The diameter of blood cells is approximately 10^{-3} cm, whereas that of the smallest arteries and veins is about 10^{-1} cm. This is why blood in the systemic and pulmonary circulations is often considered to be Newtonian, that is, characterized by a linear relationship between internal forces and velocity gradients (Perktold and Hilbert 1986, Formaggia *et al.* 2009*a*). However, in the smallest arteries, such as coronary arteries (the arteries perfusing the heart and the corresponding veins: see Figure 2.1(c)), or in the presence of vessel narrowing (stenosis), non-Newtonian blood rheology is more appropriate: see *e.g.* Chen, Lu and Wang (2006) and references therein.

Thanks to heart contraction, blood flow is pulsatile, and blood is pumped into the two circulations by means of discrete pulses with a pressure usually varying during a heartbeat in the ranges 70–130 mmHg and 20–30 mmHg for the systemic and pulmonary networks, respectively (1 mmHg $\simeq$ 133.3 Pa = 1333 g cm^{-1} s^{-2}).

In the systemic circulation, blood first enters the *aorta* (the largest artery with diameter equal to about 2.5 cm in adults: see Figure 2.1(a)) and then flows through a network of hundreds of branching arteries of decreasing size, reaching all the regions of the body. Dimensions and numbers of veins are comparable with those of arteries. The waveform of the flow rate as a function of time is characterized by different peak values when moving downstream towards the smallest arteries. In particular, the flow rate peak value is about 200 cm^3 s^{-1} in the aorta, 80 cm^3 s^{-1} in the abdominal aorta, 15 cm^3 s^{-1} in the carotid arteries (the arteries supplying blood to the brain: see Figure 2.1(b)), and 1 cm^3 s^{-1} in the coronary arteries (corresponding to a maximum blood velocity of about 150 cm s^{-1} in the aorta, 100 cm s^{-1} in the abdominal aorta, 80 cm s^{-1} in the carotid arteries and 40 cm s^{-1} in the coronary arteries). Further, the shape of the waveforms changes while moving downstream: see Figure 2.2(a). In particular, in the ascending aorta, after the systolic peak the flow rate decelerates assuming null or even negative values, whereas in the abdominal aorta and in carotid arteries it is more spread out and always positive. In any case, we can distinguish the *systolic phase* – the interval of acceleration and deceleration of blood flow – and the *diastolic phase* – the interval of almost constant or negative flow.[1] A different situation occurs in coronary arteries, where the peak flow

[1] The above definition of systole and diastole is formulated from the perspective of the arteries. An almost equivalent definition could be given from the perspective of the heart: see Section 5.

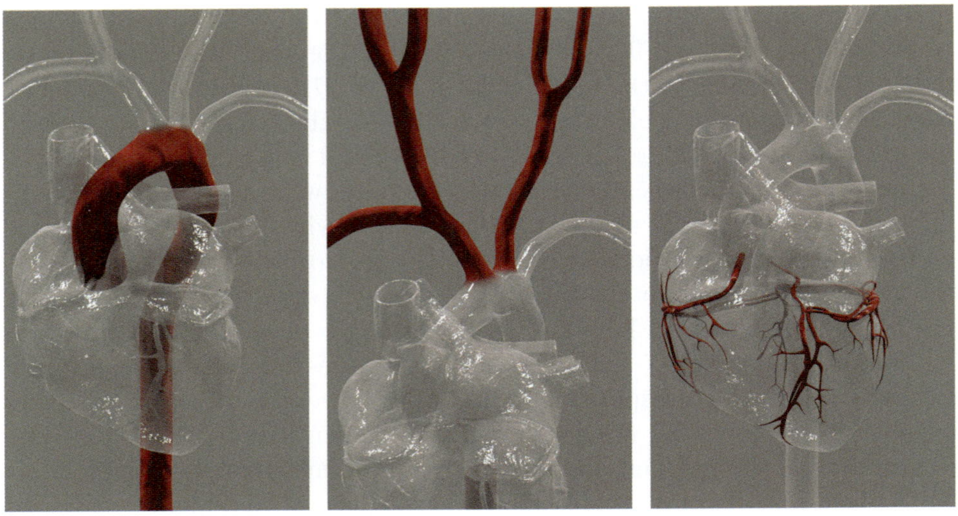

(a) (b) (c)

Figure 2.1. The aorta (a), the carotid arteries (b) and (a subset of) the coronary arteries (c).

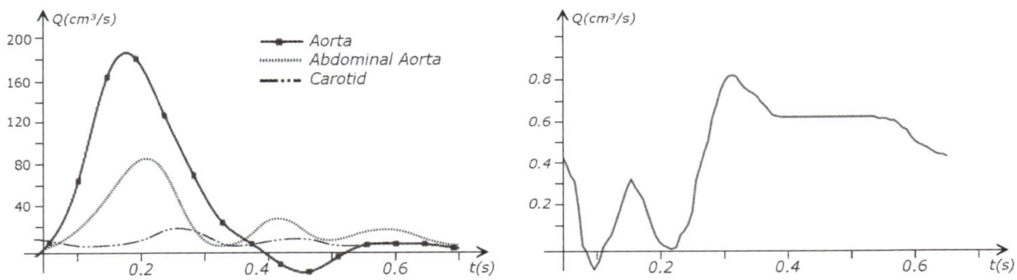

(a) (b)

Figure 2.2. Typical flow rate waveforms in the ascending aorta, abdominal aorta and carotid arteries (a), and in the coronary arteries (b).

rate is reached during diastole: see Figure 2.2(b). The coronary arteries are not directly fed by the heart; indeed, blood in the proximal part of the aorta (the sinuses of Valsalva from which the coronary arteries originate) during diastole is allowed to enter the coronary arteries thanks to the elastic response of the aorta (see below for more details).

In the pulmonary circulation blood first enters the *pulmonary artery* (diameter equal to about 3.0 cm in adults) and then flows into another network of branching arteries of decreasing size reaching the lungs. The waveforms and peak intensities are similar to those of the systemic arteries.

The different characteristics of blood flow in the arteries of the systemic circulation result in different values of the *Reynolds number* $Re = (\rho_f \, D \, U)/\mu$ (where ρ_f is the blood density, D and U are the characteristic vessel dimension and blood velocity, respectively, and μ is the fluid viscosity), a dimensionless number which quantifies the importance of inertial forces over the viscous forces. In particular, $Re \simeq 4000$ in the aorta and $Re \simeq 400$ in coronary arteries, with intermediate values when moving downstream along the aorta. Thus, blood covers a range of Reynolds numbers where both the inertial and the viscous components of the flow are relevant. Although in the aorta Re is higher than the critical value of 2000 above which the flow would no longer be laminar in a straight pipe, the pulsatile nature of blood flow does not allow development of full transition to turbulence. It is debatable whether transition to turbulence effects occur in the aorta. Some authors speculate that the helicoidal velocity pattern in the aorta, induced by the torsion of the heart's contraction, inhibits any transition to turbulence, thus supporting the thesis that in healthy conditions turbulence is never observed in the cardiovascular system (Morbiducci *et al.* 2009). This is not necessarily the case for some pathological conditions, such as carotid stenosis, yielding a narrowing of the vessel lumen and increased complexity of the geometry together with higher Reynolds numbers: see *e.g.* Ahmed and Giddens (1984), Lee *et al.* (2008), Kefayati, Holdsworth and Poepping (2014) and Lancellotti *et al.* (2015). The *Womersley number* $W = \sqrt{(2 \, A \, f)/\mu}$ (where A and f are the characteristic cross-section vessel area and time frequency of the flow rate signal, respectively) is a dimensionless number quantifying the pulsatility of flow. We find decreasing values in the systemic circulation moving downstream: $W \simeq 10$ in the aorta, $W \simeq 3$ in the carotid arteries. Similar values of Re and W are found in the pulmonary arteries.

In the veins of the systemic circulation, we find values of the flow rate, Reynolds and Womersley numbers comparable to the arteries, the only difference being that the blood flow waveform is more spread out than for the corresponding arteries. Another major difference is given by blood pressure values. In the arteries the range of pressure is almost the same, independent of the location in the tree (70–130 mmHg), whereas in the veins it reduces, assuming a mean value of about 10 mmHg. This is due to the high resistance experienced by blood flow in the microvasculature. The latter is composed of thousands of *arterioles* and *venules* and billions of *capillaries*. The blood velocity and vessel dimensions are here greatly reduced (about 10^{-1} cm s^{-1} in the former and 10^{-2} cm in the latter). This means that Re is very small in comparison with the systemic circulation, so viscous forces completely dominate the inertial forces. As a result, the highest resistance to flow is found in the microvasculature, thus provoking a big decrease in the blood pressure. Since the typical dimension of capillaries is comparable to that of erythrocytes, a multiphase model seems appropriate for their mathematical description (Enden and Popel 1992). Finally, we observe that most veins are supplied with valves that prevent backflow of blood, and venous flow is highly sensitive to muscle contraction and respiratory effects.

As observed, blood pressure assumes the same range of values along the entire systemic arterial tree, 70–130 mmHg. More precisely, negligible dissipation is experienced by the pressure signal in large and medium sized vessels before reaching the small vessels and microvasculature. Of course, at a given instant the pressure is not constant in space along the tree. Indeed, a time shift characterizes the pressure waveforms at different locations which generate gradient pressures between proximal and distal regions facilitating blood movement. These spatial gradients are due to the propagating nature of the pressure, which is in fact a wave travelling along the arterial network. The wave speed ranges from about 500 cm s^{-1} in the aorta to 1200 cm s^{-1} in the coronary arteries. The presence of bifurcations or high-resistance regions, such as the microvasculature, produces wave reflections that propagate back towards the heart.

The propagation of a pressure wave along the vascular tree is due to vessel *compliance*, that is, the ability of the vessel to distend under the forces exerted by blood pressure. Vessel wall displacements are quite large, reaching up to 10% of the lumen diameter. This is possible thanks to the structure of the vessel walls: their total thickness is about 10% of the lumen diameter and they are composed of three layers: the *intima*, the *media* and the *adventitia*. The inner part of the intima is the *endothelium* (facing the blood), whereas the remaining part is made up of connective tissue. The media and the adventitia play a major role in characterizing the mechanical response of the vessel wall. Their main structural components are *elastin* and *collagen*. The media is also formed of *smooth muscle cells* which provide tone to the vessel wall. Elastin forms complex networks that are very distensible, providing the elasticity of the vessel wall at small strain. In contrast, collagen forms stiff fibres oriented in a helical form providing tensile strength at large strain. Thus, the artery vessel wall is characterized by highly non-linear elastic properties. The quantity of elastin and collagen decreases going downstream along the arterial network, whereas the quantity of smooth muscle cells increases. This allows the proximal arteries (to the heart), in particular the aorta, to be very extensible and, thanks to the high peripheral resistances due to the elevated tone of the distal arteries and to the microvasculature, to store about 50% of the blood entering during systole. This blood reserve is then discharged during diastole owing to the vessel wall elastic response (the *windkessel effect*). This is responsible for the smoothing of the blood flow waveform discussed above, going downstream along the arterial network, which guarantees nearly continuous peripheral blood flow and thus an almost continuous exchange of oxygen with the tissues. Further, pulmonary artery walls are extensible (with muscular tone increasing downstream), even though their thickness is only about 1% of the lumen diameter.

As already observed, there is mutual exchange of energy between blood and extensible vessel walls: the latter accumulate elastic potential energy under the forces exerted by the blood pressure, which is then transferred to the blood as kinetic energy. From the mechanical point of view, this gives rise to a *fluid–structure interaction* problem. This process occurs at short time scales, proportional to the duration of a heartbeat (~ 1 s). Other interaction mechanisms may take place at larger time scales yielding wall modifications of vessel properties. This occurs in the case of several arterial diseases, such as *atherosclerosis* and *aneurysm* formation. In the first case, an increased permeability of vessel wall to lipoprotein provokes a cascade of events at the cellular level which leads to the accumulation of fatty material in the intima, just below the endothelium, and then to plaque formation in the media. Preferential sites of atherosclerotic plaque formation are the carotid arteries and the coronary arteries. The main complications are partial occlusion of the lumen with consequent (cerebral or cardiac) ischaemia, or even total occlusion resulting in (cerebral or cardiac) infarction. An aneurysm consists in the dilatation of the vessel wall with formation of a (possibly huge) bulge, mainly in the aorta and cerebral arteries, due to a loss of elastin and to the consequent remodelling of collagen, resulting in a weakening of the arterial wall; 80–90% of ruptured abdominal aortic aneurysms and 45% of ruptured cerebral aneurysms result in death. The role of blood fluid dynamics has been recognized as crucial for the development of both of these diseases (Glagov, Zarins, Giddens and Ku 1988, Bagci *et al.* 2008). In particular, *wall shear stresses*, that is, the viscous/friction forces exerted by the blood on the endothelium, despite being 100 times smaller in magnitude than pressure, regulate the permeability of the wall to lipoprotein and the loss of elastin, thus playing an important role in atherosclerosis and aneurysm development. For both these arterial diseases, this supplementary interaction between fluid and structure occurs at time scales of several years.

More on the physiology of the systemic and pulmonary circulations and microvasculature in view of mathematical modelling is available in Nichols and O'Rourke (2005), Quarteroni, Tuveri and Veneziani (2000*c*) and Formaggia *et al.* (2009*a*), for example.

3. All about data

The ultimate ambition of mathematical models in medicine is to provide quantitative results to enhance the understanding of biophysical processes and hence to support clinicians in their diagnostic and therapeutic procedures. To this end, we must consider data that are *patient-specific*, to use the bioengineering jargon – that is, related to real patients. Obtaining and processing patient-specific data is a major issue which deserves a specific review paper in its own right. Here, we provide a brief overview of the most common techniques for acquisition and analysis of 'clinical' data. This data *preprocessing* is essential prior to the set-up of a numerical simulation.

In this section we address the case of data related to the arterial (or venous) circulation, whereas in Section 6 we will discuss cardiac data. In arteries we have two interacting processes: the blood flow in the *vessel lumen* (the region occupied by the blood, which is referred to as the *fluid domain*) and the displacement of the *vessel wall* (referred to as *structure*). We need *geometric*, *boundary* and *biological* data, which are discussed below.

3.1. Geometric vascular data

Geometric data are necessary to build the geometry of the computational domains wherein the differential problems are numerically solved. At the end of the geometric preprocessing step, we obtain the fluid computational domain for the blood fluid dynamics problem, and the structure computational domain for the vessel wall displacement problem.

The processing of geometric data for blood flow simulations is a major task since vessels exhibit high morphological variability due, for example, to the evident vessel tortuosity and the presence of several bifurcations. Moreover, in unhealthy cases, this variability is further emphasized, because of the possible presence of calcifications, stenoses, aneurysms or even prostheses (such as stents).

Geometric preprocessing consists of the following steps, which are usually performed in sequence (Antiga *et al.* 2008, Antiga, Peiró and Steinman 2009): acquisition of clinical images, image enhancement, image segmentation, and generation of the computational mesh. These items are addressed below.

3.1.1. Acquisition of clinical images

Angiography is an imaging technique successfully used to 'identify' the vessel lumen. It exploits the property that a liquid inside the vessel appears brighter than the vessel wall and the surrounding tissue. Angiographies are usually acquired as two-dimensional (2D) images, corresponding to different slices of the domain of interest, but three-dimensional (3D) acquisitions of volumes are also possible.

One of the most common techniques for obtaining an angiography is *X-ray imaging*, based on the projection of X-ray beams through the body onto suitable screens, and on the contrast produced in the 2D image by the different absorption properties of body structures. To highlight the vessel lumen, a radio-opaque dye is inserted into the bloodstream through the arterial system. To reconstruct tortuous geometries, a *rotational angiography* (RA) is performed, where X-ray sources and detectors are rapidly rotated around the patient, allowing one to acquire many projections within a few seconds. The excellent spatial resolution of projection angiography (about 0.2 mm, 0.4 mm for RA) makes this technique the gold standard for most vascular imaging applications. Another X-ray angiography technique, widely used for blood flow simulation, is based on *computed tomography* (CT) technology, where multiple X-ray sources and detectors are rotated rapidly around the patient, allowing one to acquire 3D images with excellent spatial resolution (less than 1 mm in *computed tomography angiography*, CTA). Unlike projection angiography, another advantage of CTA is the possibility of using intravenous rather than arterial injections. Recently, temporally resolved CTA (4D-CTA) has become feasible. This allows one to obtain several (15–20) 3D images during a heartbeat.

Difficulties may arise in the presence of metal artifacts due to metallic prostheses such as pacemakers, resulting in streaks on the images obscuring anatomical details; see Robertson, Yuan, Wang and Vannier (1997) and Faggiano, Lorenzi and Quarteroni (2014) for possible mathematical treatments.

Another widely used technique to obtain angiographies is *magnetic resonance imaging* (MRI), based on the different decay rates exhibited by body structures on exposure to radio frequency (RF) energy. This is called *magnetic resonance angiography* (MRA). The generated contrast in the images can be tuned by selecting different RF stimuli. This allows MRA to be suitably tuned to detect soft tissues. Another advantage of MRA is that angiography can be generated without using exogenous agents. However, usually an intravenous injection of a paramagnetic contrast agent is used to improve the blood signal and reduce the acquisition time (*contrast-enhanced* (CE)-MRA).

Finally, we mention *ultrasound* (US) imaging, based on the reflections of high-frequency sound waves (a few MHz) transmitted into the body. Ultrasound is the least expensive and invasive of the techniques discussed here, and allows real-time acquisition of 2D images. In contrast, its spatial resolution is the poorest. Recently it has even been possible to acquire 3D images (3D US) by reconstructing a 3D volume from 2D slices.

On the other hand, only a few techniques currently allow us to obtain images of vessel walls. Among these we cite *black blood* (BB)-MRA, by which the vessel wall and the surrounding tissue can also be viewed, and *intravascular ultrasound* (IVUS), which is however very invasive since the transducer is placed directly into the artery (typically a coronary artery) via a catheter.

No matter which technique is used, from a mathematical standpoint we can assume that at the end of the acquisition step we obtain a vector $\boldsymbol{I}^{clin}$, whose jth component, I_j^{clin}, corresponds to the intensity of the image at the point $\boldsymbol{x}_j$ in grey-scale representing the contrast generated by the imaging technique. The collection of the points $\boldsymbol{x}_j$, $j = 1, \dots, N^{clin}$, forms the lattice $\mathcal{L}^{clin}$, where N^{clin} is the total number of acquisition points (pixels or voxels) where the image contrast has been evaluated. Here and below, a lattice is a simple collection of points determined by the point coordinates. It can be useful to associate a corresponding image intensity (scalar) function with the image intensity vector $\boldsymbol{I}^{clin}$, which is typically obtained by interpolation, and will be denoted by $I^{clin}(\boldsymbol{x})$.

3.1.2. Image enhancement

Medical images are often affected by noise and artifacts that may interfere with the quality of the final results of the preprocessing step. Thus, prior to the reconstruction of the 3D geometry, an imaging enhancement is usually performed.

One popular enhancement technique is *resampling*, consisting in suitably changing the resolution of the images in one or more directions. In practice, an interpolation of image intensity values $\boldsymbol{I}^{clin}$ onto a more refined lattice is performed. The most commonly used methods are constant interpolation, first-order composite Lagrangian interpolation, B-spline (Unser 1999), and windowed sinc interpolation.

The noise in the medical images may be due to thermal effects in the signal processing electronics or to other undesired sources. Reduction of noise could be obtained by means of a *smoothing filter*, which does not require any prior information about the nature of the noise and has a regularizing effect on the image. This technique is the most commonly used both for CT and MRI images. A very popular filter is the *Gaussian* filter, consisting in performing a discrete convolution with a Gaussian kernel over the lattice $\mathcal{L}^{clin}$ of the image intensity $\boldsymbol{I}^{clin}$. Unfortunately, together with the noise, smoothing could also filter significant high-frequency image contents. Moreover, since the image is separated from the background by sharp boundaries, characterized by high-frequency content, the smoothing filter could blur and move the boundaries. To prevent this, *anisotropic diffusion filtering* has been introduced (Perona and Malik 1990): the heat equation is solved for a new image intensity function, with diffusion coefficient decreasing for increasing values of the gradient magnitude of intensity. By so doing, the filtering is not performed at the boundaries where the gradient is large.

Another technique, called *multiscale vessel enhancement* (Frangi *et al.* 1999), exploits the specific tubular shape of vascular geometries, and therefore assumes that the minimum modulus eigenvalue of the Hessian matrix of the image intensity function I^{clin} is small, while the other two are large and of equal sign.

At the end of this substep we obtain a new image intensity vector $\boldsymbol{I}^{en}$ whose jth component, I_j^{en}, denotes the intensity of the enhanced image in grey-scale at the point $\boldsymbol{x}_j$, $j = 1, \ldots, N^{en}$, in the lattice $\mathcal{L}^{en}$ (and correspondingly an associated enhanced image intensity function $I^{en}(\boldsymbol{x})$ via interpolation). Here, N^{en} is the total number of points in the enhanced image intensity vector. Usually, $N^{en} > N^{clin}$.

3.1.3. Image segmentation

Image segmentation is the cornerstone of the preprocessing step. It consists in the construction of the shape of a vascular district from the image obtained after the enhancement substep. In particular, the segmentation allows one to detect those points of the lattice $\mathcal{L}^{en}$ which – presumably – belong to the boundary of the vessel lumen. The precise definition of the boundary of the lumen is a challenging task which generally requires considerable experience on the part of the user.

The first technique we describe is *thresholding*, consisting in selecting a threshold k to identify the points $\boldsymbol{x}_j \in \mathcal{L}^{en}$ such that $I_j^{en} > k$. This is motivated by the assumption that k separates different anatomical structures, in our case the vessel lumen (characterized by intensity values larger than k) and the background, obtained by the collection of points for which $I_j^{en} \leq k$. The value of k is determined either manually or via a suitable algorithm. In the latter case, one commonly used strategy is *full width at half maximum* (FWHM), consisting in setting the threshold halfway between the peak intensity within the lumen and the intensity of the background. For the segmentation of special structures, such as calcifications or stents, higher-bound thresholds are used (Boldak, Rolland and Toumoulin 2003).

A more sophisticated class of segmentation methods than thresholding is given by *front propagation methods*, where the propagation of a suitable wavefront is tracked. The speed of the wave is small in regions where $\boldsymbol{I}^{en}$ changes rapidly and high for other regions, so the wavefront slows down when approaching the boundary. The most popular front propagation method is the *fast marching method*, which provides an efficient solution to the eikonal problem

$$\|\nabla T(\boldsymbol{x})\| = \frac{1}{V(I^{en}(\boldsymbol{x}))}, \quad \boldsymbol{x} \in \mathcal{D}^{en},$$

where $\mathcal{D}^{en} \subset \mathbb{R}^3$ is a region that contains all $\boldsymbol{x}_j \in \mathcal{L}^{en}$, and where suitable boundary conditions are prescribed on a selected boundary where the propagation starts (Zhu and Tian 2003). In the above equation, V is the speed of the wavefront and $T(\boldsymbol{x})$ is the first arrival time at point $\boldsymbol{x}$. In fact, T consists of isocontours, denoting a collection of surfaces describing the shape of the waveform. The vessel boundary is then represented by the points $\boldsymbol{x}_j \in \mathcal{L}^{en}$ such that $T(\boldsymbol{x}_j) = T^b$ (up to a given tolerance), where T^b is a suitable value selected by the user.

Another class of segmentation methods is that of *deformable models*, where a suitable energy is minimized, allowing the deformation of the body (in our case the boundary of the vessel lumen) to reach a final state with smallest energy, accounting for external terms derived from the image and internal terms constraining the boundary to be regular. The most widely used class of deformable models is the *level set method*, where a deformable surface is represented implicitly as the zero level of a higher-dimensional embedding function (Sethian 1999). Deformable models, for example those based on cylindrically parametrized surface meshes, incorporate anatomical knowledge of the vessel shape (Frangi *et al.* 1999, Yim *et al.* 2001).

For the segmentation of the vessel wall, Steinman *et al.* (2001), starting from BB-MRA images, segmented the vessel wall outer boundary using the same deformable model as used for the vessel lumen segmentation. Usually, BB-MRA or other images detecting the vessel wall are not routinely acquired in clinical practice. In this case, a reasonable approach to obtaining the vessel wall is to 'extrude' the reconstructed boundary lumen along the outward unit vector by using a suitable function specifying the vessel wall thickness in the different regions of the district of interest.

In those cases where the image intensity vectors $\boldsymbol{I}^{clin}$ and $\boldsymbol{I}^{en}$ refer to 2D slices, application of the above segmentation strategies leads to identification of several vessel boundaries (contours), one for each slice, which now need to be connected to obtain the 3D boundary surface. This operation is called *surface reconstruction*. A simple procedure is to connect successive contours with straight lines defining surface triangle edges. This strategy is not suitable in the presence of changes of shape such as in bifurcations. Better surface reconstruction is provided by least-squares fitting of polynomial surfaces to the contour set (Wang, Dutton and Taylor 1999). This strategy is suited to managing bifurcations whose branches are fitted separately with a successive extension into the parent vessel. A variant of this approach has been proposed in Geiger (1993), where contours are first filled with triangles which are then connected to the triangles of the adjacent contours by means of tetrahedra. The final lumen surface is then represented by the boundary of this tetrahedral mesh (formed by triangles). We also mention *shape-based interpolation* where, for each contour, a characteristic function with positive (resp. negative) values for points located inside (resp. outside) the contour is constructed. The final lumen boundary surface is then represented by the zero level set of the interpolation of all these characteristic functions (Raya and Udupa 1990). Finally, we briefly describe interpolation by means of *radial basis functions* (RBFs), which provide a flexible way of interpolating data in multi-dimensional spaces, even for unstructured data where interpolation nodes are scattered and/or do not form a regular grid, and for which it is often impossible to apply polynomial or spline interpolation (Carr, Fright and Beatson 1997, Fornefett, Rohr and Stiehl 2001). The coefficients in the linear combination with respect to the RBF basis are determined by solving a suitable linear system, which is invertible under very mild conditions (Peiró *et al.* 2007).

A special mention must go to *centreline* reconstruction. The centreline is a one-dimensional curve centred inside the vessel lumen. Many segmentation tools use the centreline as the starting point, making the assumption that the shape of the section around each centreline location is known (O'Donnell, Jolly and Gupta 1998). Centreline reconstruction allows complete reconstruction of the computational domain when using *one-dimensional modelling* of blood flow: see Section 4.5.1.

In any case, at the end of the segmentation step we obtain the lattice $\mathcal{L}^{surf}$ which collects the points $\boldsymbol{x}_j$, $j = 1, \ldots, N^{surf}$, classified as belonging to the lumen vessel surface or to the outer wall, where N^{surf} denotes the total number of points of the surface lattice.

3.1.4. Building the computational mesh

Once the final boundary lattice $\mathcal{L}^{surf}$ (made up of points on the lumen boundary) is made available, we are ready to build the volumetric mesh $\mathcal{T}^{vol}$ in the lumen. This mesh usually consists of unstructured tetrahedra, because of their flexibility in filling volumes of complex shape.

Unstructured volumetric meshes are constructed starting from an analytical expression, say $\mathcal{S}(\boldsymbol{x})$, representing the surface associated with the boundary lattice $\mathcal{L}^{surf}$. This expression can derive from an explicit representation, for instance a bivariate parametric function built as a collection of adjacent polygons. The latter are typically triangles, generated by Lagrangian shape functions, or patches, generated by high-degree polynomials such as NURBS (Sun, Starly, Nam and Darling 2005). Alternatively, the surface is represented implicitly as the isosurface of an embedding function. Note that some of the segmentation strategies described above, such as deformable models and those used for the surface reconstruction, directly provide an analytical expression $\mathcal{S}(\boldsymbol{x})$ of the lumen boundary surface.

For the construction of unstructured volumetric meshes $\mathcal{T}^{vol}$, we mention two possible approaches. In the first, a boundary surface mesh $\mathcal{T}^{surf}$ is generated. To this end, we start from a lattice $\widetilde{\mathcal{L}}^{surf}$ (in principle different to $\mathcal{L}^{surf}$) composed of points of $\mathcal{S}$. Then, the *Voronoi diagram* for $\widetilde{\mathcal{L}}^{surf}$ is constructed. This is a partition of $\mathcal{S}$ into non-overlapping regions, each one containing exactly one point (node) of $\widetilde{\mathcal{L}}^{surf}$ and composed of all the points of $\mathcal{S}$ that are closer to that node than to any other node. Starting from the Voronoi diagram, it is possible to generate a *Delaunay mesh* $\mathcal{T}^{surf}$: see Thompson, Soni and Weatherill (1999). We emphasize that the vertices of the mesh $\mathcal{T}^{surf}$ do not necessarily coincide with the points of the lattice $\widetilde{\mathcal{L}}^{surf}$.

Popular algorithms for generating a Delaunay mesh have been proposed by Watson (1981) and Weatherill and Hassan (1994). Once a surface mesh $\mathcal{T}^{surf}$ is made available, the volumetric mesh $\mathcal{T}^{vol}$ is generated. The latter could be obtained by *advancing front* methods, where, starting from the triangles of the surface mesh, a front composed of internal nodes is generated. These new nodes allow us to identify tetrahedra, whose validity is verified by checking that they do not intersect the front (Bentley and Friedman 1979).

The second approach relies on directly generating the volumetric mesh $\mathcal{T}^{vol}$, for example by means of Delaunay 3D mesh generation, where a starting volumetric lattice $\widetilde{\mathcal{L}}^{vol}$ is obtained by locating the nodes in the volume $\mathcal{V}(\boldsymbol{x})$ contained in $\mathcal{S}(\boldsymbol{x})$. One of the main problems related to this approach is that boundary meshing is often difficult, since the related surface triangulation could not be of Delaunay type. An alternative approach is given by *octree mesh generation*, where $\mathcal{V}(\boldsymbol{x})$ is embedded in a box and successive subdivisions are performed until the smallest cells permit accurate description of the boundary. Despite being faster, this strategy generates meshes with poor quality near the boundary.

When a volumetric mesh $\mathcal{T}^{vol}$ is obtained, a further step (*mesh optimization*) could be introduced prior to the generation of the final mesh, so as to improve its quality. This prevents mesh distortion, for example the presence of very small angles, which could reduce the convergence of algorithms for the solution of the PDE of interest and thus their accuracy. Mesh optimization is incorporated in the strategies described above; it leads to an *optimal mesh*, providing the best accuracy for a given number of nodes.

A mesh is deemed valid for blood flow simulations if it allows recovery of outputs of physical interest. In arteries, the mesh should be fine enough to capture wall shear stress (WSS) (Celik *et al.* 2008) and, to this end, the construction of a boundary layer mesh is essential, even at low Reynolds numbers (Bevan *et al.* 2010). Here *WSS* expresses the magnitude of tangential viscous forces exerted by the fluid on the lumen boundary Σ^t, defined by

$$WSS = \mu \sqrt{\sum_{j=1}^{2} \big((\nabla \boldsymbol{v}\, \boldsymbol{n}) \cdot \boldsymbol{\tau}^{(j)}\big)^2} \quad \text{on } \Sigma^t,$$

where $\boldsymbol{v}$ is the fluid velocity, $\boldsymbol{n}$ is the outward unit vector, and $\boldsymbol{\tau}^{(j)}$, $j = 1, 2$, represent the tangential unit vectors. Note that *WSS* is a scalar function of $\boldsymbol{x} \in \Sigma^t$ and $t > 0$. In componentwise notation,

$$WSS = \mu \left(\sum_{j=1}^{2} \left(\sum_{i,k=1}^{3} \left(\frac{\partial v_i}{\partial x_k} n_k \right) \tau_i^{(j)} \right)^2 \right)^{1/2} \quad \text{on } \Sigma^t.$$

For the structure domain, hexahedral meshing is preferable so as to prevent the locking phenomenon, whereas tetrahedral meshes are used when conforming meshes at the boundary lumen interface are needed, in view of fluid–structure interaction problems (see Section 4.3). Usually, three or four layers of elements are enough to obtain an accurate result (Younis *et al.* 2004).

For recent reviews on geometric reconstruction for blood flow simulation, see Antiga *et al.* (2009), Sazonov *et al.* (2011) and Lesage, Angelini, Bloch and Funka-Lea (2009).

3.2. Boundary vascular data

The differential equations we will treat in the following sections need appropriate boundary conditions. For our problems (incompressible Navier–Stokes equations for the fluid and finite elasticity for the structure), we anticipate the kind of boundary conditions that should ideally be prescribed, namely

$$\boldsymbol{v} = \boldsymbol{g}_f \quad \text{on } \Gamma_f^{D,t}, \qquad -p\boldsymbol{n} + \mu(\nabla \boldsymbol{v} + (\nabla \boldsymbol{v})^T)\boldsymbol{n} = \boldsymbol{h}_f \quad \text{on } \Gamma_f^{N,t}$$

for the fluid problem and

$$\boldsymbol{d} = \boldsymbol{g}_s \quad \text{on } \Gamma_s^{D,t}, \qquad \boldsymbol{T}_s\boldsymbol{n} = \boldsymbol{h}_s \quad \text{on } \Gamma_s^{N,t}$$

for the structure problem. In the above equations, the Dirichlet and Neumann boundaries, $\Gamma_j^{D,t}$ and $\Gamma_j^{N,t}$, respectively, are such that $\Gamma_j^{D,t} \cap \Gamma_j^{N,t} = \emptyset$, $\Gamma_j^{D,t} \cup \Gamma_j^{N,t} = \partial\Omega_j^t$, $j = f, s$, where Ω_f^t and Ω_s^t are the fluid and structure domains at time t: see Figure 3.1(a). Moreover, p denotes the fluid pressure, $-p\boldsymbol{n} + \mu(\nabla \boldsymbol{v} + (\nabla \boldsymbol{v})^T)\boldsymbol{n}$ is the fluid normal Cauchy stress, $\boldsymbol{d}$ is the structure displacement, $\boldsymbol{T}_s$ is the Cauchy stress tensor of the wall material, and $\boldsymbol{g}_f$, $\boldsymbol{g}_s$, $\boldsymbol{h}_f$, $\boldsymbol{h}_s$ are data: see Section 4. We use the superscript ^{t} to indicate time-dependence.

As we will see below, the boundary of the computational domain (for either the fluid or the structure) will be composed of two parts, namely the *physical boundary* and the *artificial boundary*. On the physical boundary, suitable conditions are often suggested by physical principles. For example, for the fluid problem, *no-slip* Dirichlet conditions should be prescribed at the lumen boundary, since it is assumed that the fluid particles perfectly adhere to the vessel wall. This leads to a homogeneous Dirichlet condition ($\boldsymbol{v} = \boldsymbol{0}$) in the case of rigid walls, and to a kinematic interface condition ($\boldsymbol{v} = \dot{\boldsymbol{d}}$) for fluid–structure interaction problems (see Section 4.3). As for the structure problem, at the internal physical boundary (*i.e.* at the lumen boundary) the fluid pressure is often prescribed. This leads to a Neumann boundary condition ($\boldsymbol{T}_s\boldsymbol{n} = -P\boldsymbol{n}$, where P is a measurement of the fluid pressure) for a pure structure problem, and to a dynamic interface

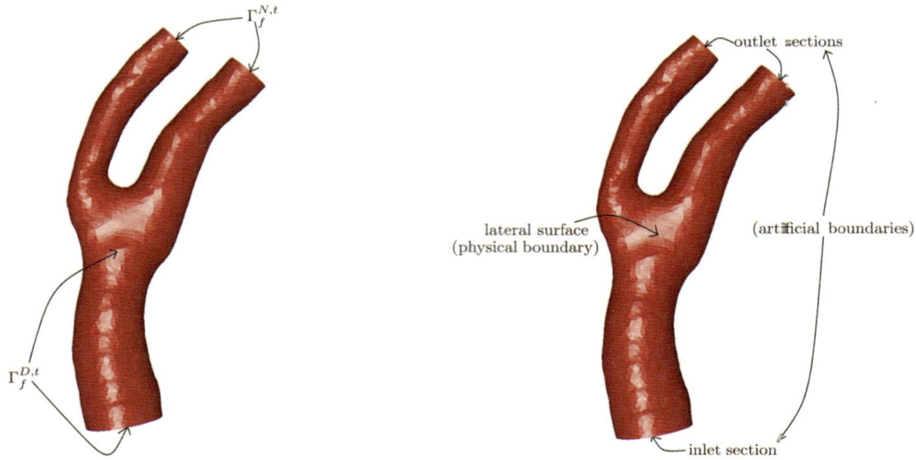

(a) (b)

Figure 3.1. Possible choices of the Dirichlet and Neumann boundaries (a) and physical and artificial boundaries (b) for a carotid domain in the fluid stand-alone problem (reconstructed from MRA images).

condition ($\boldsymbol{T}_s\boldsymbol{n} = -p\boldsymbol{n} + \mu(\nabla\boldsymbol{v} + (\nabla\boldsymbol{v})^T)\boldsymbol{n}$) for fluid–structure interaction. On the outer wall boundary Γ_{ext}, the interaction with the surrounding tissue should be considered. This is often modelled by means of a Robin boundary condition of the form

$$\alpha_{ST}\boldsymbol{d} + \boldsymbol{T}_s\boldsymbol{n} = P_{ext}\boldsymbol{n} \quad \text{on } \Gamma_{ext}, \tag{3.1}$$

which assimilates the surrounding tissue to a sequence of elastic springs with rigidity α_{ST} and where P_{ext} is the external pressure (Moireau *et al.* 2012).

In contrast, the artificial sections are those introduced by the truncation of the computational domains: see Figure 3.1(b). Truncation is done in order to focus on a specific domain of interest. Ideally, the boundary conditions to be used on artificial sections should derive from clinical measurements.

The technique mainly used to obtain boundary data on artificial boundaries is ultrasound. This is because of its non-invasiveness and the fact that it is used daily in clinical practice. If the ultrasound beam is swept through a plane or sector (unlike in geometric acquisitions where it is kept fixed), it is possible to measure the blood velocity at a single point of a cross-section Γ^t in the direction of the ultrasound beam by exploiting the Doppler effect. The velocity is then converted into a flow rate measure Q across Γ^t for each time (this is the principle of the so-called *echo-colour-Doppler* technique: Savader, Lund and Osterman 1997) and can then be used to prescribe a *flow rate condition*

$$\rho_f \int_{\Gamma^t} \boldsymbol{v} \cdot \boldsymbol{n} \, \mathrm{d}\gamma = Q. \tag{3.2}$$

Condition (3.2) is *defective* since it is not enough to ensure well-posedness of the fluid problem. The treatment of defective conditions will be addressed in Section 4.4. Another technique allowing measurement of cross-sectional flow rates is based on the acquisition of thermal images (Mahmoud *et al.* 2013). Another quantity easily measurable by means of ultrasound is the lumen area $\int_{\Gamma^t} \mathrm{d}\gamma$. This information could be used to prescribe a defective condition for the vessel wall.

More sophisticated techniques could acquire velocity measurements in several points on the same cross-section, thus in principle leading to a Dirichlet boundary condition (possibly after suitable regularization). This is the case for *phase contrast* (PC)-MRA, where the blood velocities are encoded into images taken at several instants during a heartbeat (Morbiducci *et al.* 2009). The spatial resolution of modern PC-MRA is of the order of 1–2 mm on each cross-section and 4 mm along the longitudinal axis. The required mesh size is often less than 1 mm, however, so interpolation is needed to obtain a usable Dirichlet condition. The temporal resolution is about 0.03 s.

If the lumen artificial cross-section is orthogonal to the axial direction, then the viscous terms in the fluid normal Cauchy stress are very small (Heywood, Rannacher and Turek 1996). In this case, a measurement of the pressure $P(t)$ could be used to prescribe a Neumann boundary condition. This could be obtained for the arterial system non-invasively by means of a *sphygmomanometer*, which usually measures the pressure at the level of the radial aorta (remember that the pressure peak could be considered constant along the arterial tree, at least until the capillary net: see Section 2). To have continuous monitoring of the pressure (e.g. during hospital recovery) or to take a measurement in the venous system (where the pressure reduces), a catheter with a transducer could be placed in the district of interest. In any case, the average pressure over the cross-section is measured. This leads at each time to the following defective boundary condition for the normal component of the normal Cauchy stress of the fluid:

$$\frac{1}{|\Gamma^t|} \int_{\Gamma^t} \big(p\boldsymbol{n} - \mu(\nabla \boldsymbol{v} + (\nabla \boldsymbol{v})^T)\boldsymbol{n}\big) \cdot \boldsymbol{n} \, \mathrm{d}\gamma = P. \tag{3.3}$$

Unfortunately, no measurement is available at the artificial sections of the structure, so 'empirical' and 'practical' choices are made (see Section 4.4).

Finally, we observe that measurements acquired at several instants during a heartbeat could be used in principle for physical boundaries as well. For example, from 4D-CTA the boundary lumen displacement could be used as the Dirichlet condition for the structure problem, whereas from PC-MRA the blood velocity at the boundary lumen could be used to prescribe a Dirichlet condition for the fluid problem. Since, at the physical boundaries, physical principles are used to prescribe boundary conditions, these 'extra' data could be used for validation of the numerical results or in a *parameter estimation* fashion, as described in Section 10. PC-MRA also allows us to acquire internal measurements of blood velocity. This could also be used for validation or parameter estimation.

3.3. Biological vascular data

Finally, we need to know the values of physical parameters appearing in the differential equations. For the fluid, two parameters characterize the problem, namely blood *density* and blood *viscosity*. Although density is easily measurable in principle, no patient-specific acquisition is made in general because its value never significantly departs from an average value of 1.06 g cm^{-3}. In contrast, the range of variability of viscosity is larger. Indeed, its value depends on the shear rate (non-Newtonian behaviour) and on the physical state of the patient. When the assumption of Newtonian fluid (holding for medium and large healthy vessels) is made, typical values of the viscosity range in the interval 0.03–0.04 P (= 1 g cm^{-1} s^{-1}). Again, no patient-specific viscosity measurements are usually made and, except for pathological situations, a value in the previous range is selected.

The parameters characterizing the vessel wall depend (also in number) on the constitutive law used to represent its behaviour. More typically, they are the density, the *elasticity* (or compliance) and the degree of *incompressibility*. For linear elasticity models, the latter two are quantified by *Young's* modulus and the *Poisson modulus*, respectively. The density value is very similar to that of blood, the range most commonly used reaching values up to 1.2 g cm^{-3}. The compliance of the vessel could be qualitatively seen as the ratio between the volume and pressure variations (Nichols and O'Rourke 2005). Patient-specific measurements could be obtained from simultaneous (invasive) measurement of pressure and cross-sectional area at different locations, by measuring the rate of propagation of flow waves (Boese, Bock, Schoenberg and Schad 2000), or by *elastography*, where the elasticity properties are analysed by images before and after a deformation (Oberai, Gokhale and Feijóo 2003). The range of variability of vessel compliance is quite wide (200–800 kPa in normal conditions), with great dependence on the location and on the possible presence of aneurysms or calcifications. As for the Poisson modulus, only *ex vivo* measurements are possible. An acceptable value used by the community is 0.49, meaning that vessels could be considered as quasi-incompressible. Finally, let us mention that the coefficient α_{ST} in (3.1) can be regarded as representative of the

Young's modulus of the surrounding tissue. As such, it could thus be measured, though this is a difficult endeavour. Estimates have been provided in Liu *et al.* (2007), for example.

When not available from measurements, patient-specific values of biological data could alternatively be obtained by means of parameter estimation mathematical techniques. This will be the topic of Section 10.

4. Modelling blood flow

4.1. The fluid problem

In large and medium sized arteries – those more typically affected by vascular diseases – blood can be modelled by means of the Navier–Stokes (NS) equations for incompressible homogeneous Newtonian fluids (Perktold, Thurner and Kenner 1994, Taylor, Hughes and Zarins 1996, Taylor, Hughes and Zarins 1998, Formaggia *et al.* 2009*a*). For non-Newtonian rheological models necessary to describe some specific flow processes, such as clotting or sickle cell diseases, or more generally flow in capillaries, we refer to Robertson, Sequeira and Owens (2009) and Fasano, Santos and Sequeira (2012), for example.

For the mathematical formulation of the problem, we write the fluid equations with respect to an Eulerian frame of reference, and we let $\Omega_f^t = \Omega_f(t) \subset \mathbb{R}^3$ denote the time-varying arterial lumen, at time $t > 0$ (see Figure 4.1(a)).

Then, at each time $t > 0$, we look for fluid velocity $\boldsymbol{v}$ and fluid pressure p such that

$$\rho_f\left(\frac{\partial \boldsymbol{v}}{\partial t} + (\boldsymbol{v}\cdot\nabla)\boldsymbol{v}\right) - \nabla\cdot\boldsymbol{T}_f(\boldsymbol{v},p) = \boldsymbol{0} \quad \text{in } \Omega_f^t, \tag{4.1a}$$

$$\nabla\cdot\boldsymbol{v} = 0 \quad \text{in } \Omega_f^t. \tag{4.1b}$$

Note that volumetric forces acting in the fluid domain (*e.g.* due to gravity) have been set to zero since they are quite often negligible. Moreover,

$$\boldsymbol{T}_f(\boldsymbol{v},p) = -p\boldsymbol{I} + \mu(\nabla\boldsymbol{v} + (\nabla\boldsymbol{v})^T) \tag{4.2}$$

is the fluid Cauchy stress tensor, where μ is the blood viscosity. As we consider only Newtonian rheology here, μ is assumed to be constant.

Finally, problem (4.1) is completed by the initial condition

$$\boldsymbol{v}|_{t=0} = \boldsymbol{v}_0 \quad \text{in } \Omega_f,$$

where $\Omega_f = \Omega_f^0$, and boundary conditions. The latter typically prescribe no-slip conditions on the physical boundary Σ^t,

$$\boldsymbol{v} = \boldsymbol{\phi} \quad \text{on } \Sigma^t, \tag{4.3}$$

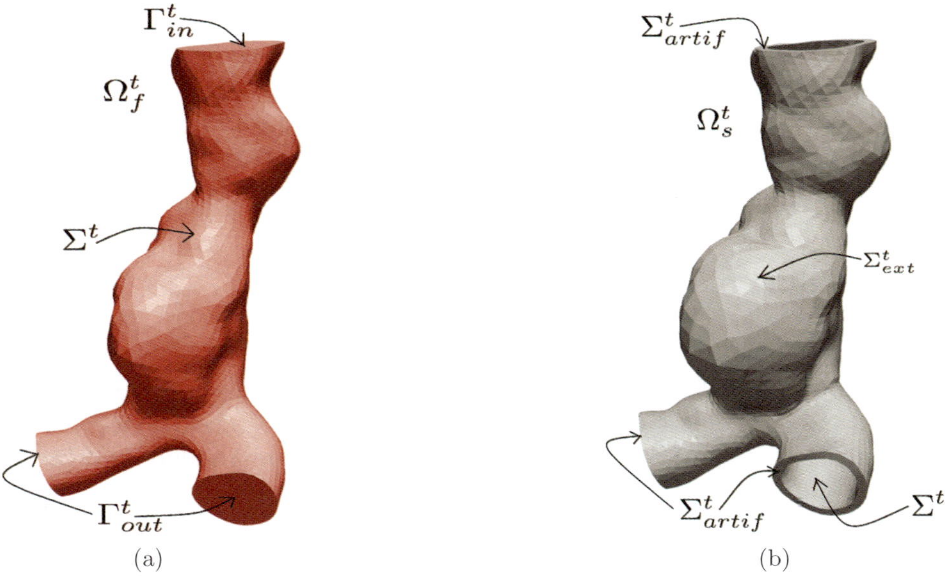

Figure 4.1. Representation of the fluid domain (a) and structure domain (b). The fluid domain illustrated is that of an abdominal aorta in the presence of an aneurysm, reconstructed from CTA images. The structure domain was obtained via extrusion of the fluid domain.

the upstream velocity on the proximal boundaries, say Γ^t_{in},

$$\boldsymbol{v} = \boldsymbol{v}_{up} \quad \text{on } \Gamma^t_{in}, \tag{4.4}$$

and traction conditions on the distal boundaries, say Γ^t_{out},

$$\boldsymbol{T}_f\,\boldsymbol{n} = \boldsymbol{h}_f \quad \text{on } \Gamma^t_{out}. \tag{4.5}$$

Here, $\boldsymbol{v}_0$, $\boldsymbol{v}_{up}$, $\boldsymbol{\phi}$ and $\boldsymbol{h}_f$ are suitable functions with the required regularity (Quarteroni and Valli 1994). Note that the lumen boundary displacement $\boldsymbol{\phi}$ at this level is a known function of space and time. For rigid boundaries, we have $\boldsymbol{\phi} = \boldsymbol{0}$.

When patient-specific measurements are available, other conditions might be prescribed. However, measurements seldom provide a complete dataset to be used in the computation: see our discussion in Section 3.2. This prompts the issue of solvability of Navier–Stokes equations, which we address in Section 4.4.

For each $t > 0$ (a.e. $t > 0$), the weak form of (4.1) together with the boundary conditions (4.3, 4.4, 4.5) reads as follows: find $\boldsymbol{v} = \boldsymbol{v}(t) \in [H^1(\Omega^t_f)]^3$, $\boldsymbol{v} = \boldsymbol{v}_{up}$ on Γ^t_{in}, $\boldsymbol{v} = \boldsymbol{\phi}$ on Σ^t, $\boldsymbol{v} = \boldsymbol{v}_0$ for $t = 0$ in Ω_f, and

$p = p(t) \in L^2(\Omega_f^t)$ such that

$$\rho_f \int_{\Omega_f^t} \frac{\partial \boldsymbol{v}}{\partial t} \cdot \boldsymbol{w} \,\mathrm{d}\omega + \mathcal{A}_f^t(\boldsymbol{v}, \boldsymbol{v}, \boldsymbol{w}) + \mathcal{B}^t(p, \boldsymbol{w}) = \int_{\Gamma_{out}^t} \boldsymbol{h}_f \cdot \boldsymbol{n} \,\mathrm{d}\gamma, \tag{4.6a}$$

$$\mathcal{B}^t(q, \boldsymbol{v}) = 0, \tag{4.6b}$$

for all

$$\boldsymbol{w} \in \boldsymbol{V}^t = \{[H^1(\Omega_f^t)]^3 : \boldsymbol{w} = \boldsymbol{0} \text{ on } \partial\Omega \setminus \Gamma_{out}^t\}$$

and $q \in L^2(\Omega_f^t)$, and where we have set

$$\mathcal{A}_f^t(\boldsymbol{z}, \boldsymbol{v}, \boldsymbol{w}) = \rho_f \int_{\Omega_f^t} (\boldsymbol{z} \cdot \nabla)\boldsymbol{v} \cdot \boldsymbol{w} \,\mathrm{d}\omega + \mu \int_{\Omega_f^t} (\nabla \boldsymbol{v} + (\nabla \boldsymbol{v})^T) : \nabla \boldsymbol{w} \,\mathrm{d}\omega$$

and

$$\mathcal{B}^t(q, \boldsymbol{w}) = -\int_{\Omega_f^t} q \nabla \cdot \boldsymbol{w} \,\mathrm{d}\omega.$$

The existence of a global-in-time weak solution of the above problem was proved by Leray (1934) for the case $\Omega_f = \mathbb{R}^3$ and by Hopf (1951) for the case of bounded domain. The uniqueness has been proved only for the two-dimensional case (Lions and Prodi 1959); for the three-dimensional case, only local-in-time uniqueness results are available (Prodi 1962).

4.2. Mechanical wall models

The problem that models the deformation of vessel walls is given by the elastodynamics equation, which is usually written in a reference domain $\Omega_s = \Omega_s(0) \subset \mathbb{R}^3$ using a Lagrangian framework. For any $t > 0$, the material domain $\Omega_s^t = \Omega_s(t)$ (depicted in Figure 4.1(b)) is the image of Ω_s by a proper Lagrangian map $\mathcal{L} : \Omega_s \to \Omega_s^t$. We use the notation $\widehat{g} = g \circ \mathcal{L}$ to denote the function $\widehat{g}$ induced on Ω_s by the function g defined in the current solid configuration Ω_s^t.

For simplicity, we assume the arterial wall obeys a (possibly non-linear) finite elastic law relating stress to strain in the arterial tissue; for more complex behaviours of arterial walls see Holzapfel, Gasser and Ogden (2000) and Holzapfel and Ogden (2010), for example. In more realistic settings, stress is a function of the strain but also of the past loading history (Fung, Fronek and Patitucci 1979).

The problem we consider is as follows: find, at each time $t > 0$, the structure displacement $\boldsymbol{d}$ such that

$$\rho_s \frac{\partial^2 \widehat{\boldsymbol{d}}}{\partial t^2} - \nabla \cdot \widehat{\boldsymbol{T}}_s(\widehat{\boldsymbol{d}}) = \boldsymbol{0} \quad \text{in } \Omega_s, \tag{4.7}$$

where ρ_s is the structure density. Note that volumetric forces acting in the solid domain (*e.g.* due to muscle forces) have been set to zero since they are quite often negligible.

The above equation is stated in terms of the first Piola–Kirchhoff tensor $\widehat{\boldsymbol{T}}_s(\widehat{\boldsymbol{d}})$, which is related to the Cauchy tensor $\boldsymbol{T}_s(\boldsymbol{d})$ via the relation $\widehat{\boldsymbol{T}}_s = J\boldsymbol{T}_s\boldsymbol{F}^{-T}$. Here, $\boldsymbol{F} = \nabla\boldsymbol{x}$ is the deformation tensor, where the gradient is taken with respect to the reference space coordinates and $\boldsymbol{x}$ denotes point coordinates in the current configuration. Correspondingly, $J = \det(\boldsymbol{F})$ denotes the change of volume between the reference and the current configurations; note that $\boldsymbol{F}$ (and thus J) depends on the current configuration Ω_s^t.

For a *hyperelastic* material, the first Piola–Kirchhoff stress tensor is obtained by differentiating a suitable strain energy density function Θ:

$$\widehat{\boldsymbol{T}}_s = \frac{\partial\Theta}{\partial\boldsymbol{F}}. \tag{4.8}$$

Several non-linear elastic energy functions have been proposed for arteries. For the *Saint Venant–Kirchhoff* material,

$$\begin{aligned}\Theta(\boldsymbol{C}) = {} & \frac{E\nu}{2(1+\nu)(1-2\nu)}\left(\operatorname{tr}\left(\frac{1}{2}(\boldsymbol{C}^T - \boldsymbol{I})\right)\right)^2 \\ & + \frac{E}{2(1+\nu)}\operatorname{tr}\left(\left(\frac{1}{2}(\boldsymbol{C}^T - \boldsymbol{I})\right)^2\right),\end{aligned} \tag{4.9}$$

where $\boldsymbol{C} = \boldsymbol{F}^T\boldsymbol{F}$, E is *Young's modulus* and ν is the *Poisson modulus* of the vessel wall. More complex and accurate functions widely used for arteries are based on separating the isotropic elastic behaviour due to the *elastin* and the anisotropic behaviour accounting for the stiffening increment at large displacements due to the *collagen*:

$$\Theta = \Theta^{iso} + \Theta^{aniso}. \tag{4.10}$$

A common choice for the isotropic part is the *neo-Hookean* law

$$\Theta^{iso}(\boldsymbol{C}) = \frac{G}{2}(\operatorname{tr}(\boldsymbol{C}) - 3), \tag{4.11}$$

whereas for the anisotropic part, an exponential law is often considered,

$$\Theta^{aniso}(\boldsymbol{C}) = \frac{k_1}{2k_2}\left(\mathrm{e}^{(k_2(\boldsymbol{a}\cdot(\boldsymbol{C}\boldsymbol{a})-1)^2)} - 1\right) \tag{4.12}$$

where $\boldsymbol{a}$ is the unit vector identifying the preferred direction of the collagen fibres, G is the *shear modulus*, and k_1, k_2 are material parameters, where k_2 characterizes the stiffness of the material for large displacements (Fung 1993, Holzapfel *et al.* 2000, Raghavan and Vorp 2000). More complete laws also account for the symmetrical helical arrangement of the collagen fibres, introducing a second predominant direction of fibres (Holzapfel and Ogden 2010). When not available from medical images, the fibre directions are computed by means of suitable algorithms under the assumption that their orientation is mainly governed by the principal stresses: see *e.g.* Hariton, de Botton, Gasser and Holzapfel (2006) and Kuhl and Holzapfel (2006). For distal arteries of muscular type, viscoelastic and pseudoelastic terms are also considered (Holzapfel and Gasser 2001).

Sometimes the arterial tissue is modelled as being incompressible by imposing the incompressibility constraint $J = 1$ on the strain energy function via Lagrange multipliers (Holzapfel 2000), that is,

$$\Theta^{inc} = \Theta + p_s(J-1). \tag{4.13}$$

Here p_s is the hydrostatic pressure related to the vessel wall displacement, which plays the role of the Lagrange multiplier for the incompressibility constraint. Correspondingly, the Cauchy stress tensor is augmented as follows:

$$\boldsymbol{T}_s^{inc}(\boldsymbol{d}, p_s) = \boldsymbol{T}_s(\boldsymbol{d}) + p_s\boldsymbol{I},$$

where $\boldsymbol{T}_s$ is the component arising from the energy Θ. However, experimental studies have shown that the arterial tissue in fact exhibits *nearly incompressible* behaviour (Carew, Vaishnav and Patel 1968). This means that the strain energy function could be decomposed into two terms

$$\Theta(\boldsymbol{C}) = \Theta_{vol}(J) + \Theta_{isoc}(\bar{\boldsymbol{C}}),$$

where $\bar{\boldsymbol{C}} = J^{-2/3}\boldsymbol{C}$, $\det\bar{\boldsymbol{C}} = 1$. The *isochoric* part Θ_{isoc} is given by the general function (4.10) provided that $\boldsymbol{C}$ is replaced by $\bar{\boldsymbol{C}}$ and characterizes the mechanical response of the material to incompressible deformations. The *volumetric* part is usually given by

$$\Theta_{vol}(J) = \frac{\kappa}{2}(J-1)^2,$$

where κ (*bulk modulus*) is a penalty parameter to enforce the incompressibility constraint $J = 1$ (Li and Robertson 2013). In this case, the related Cauchy stress tensor is the same as in the compressible case. For a review of arterial vessel wall models we refer the reader to Holzapfel and Ogden (2010).

Problem (4.7) has to be completed by the initial conditions

$$\boldsymbol{d}|_{t=0} = \boldsymbol{d}_0, \quad \left.\frac{\partial \boldsymbol{d}}{\partial t}\right|_{t=0} = \boldsymbol{d}_1 \quad \text{in } \Omega_s,$$

and boundary conditions. The latter typically prescribe on the artificial sections Γ^t_{artif} either

$$\boldsymbol{d} = \boldsymbol{0} \quad \text{on } \Gamma^t_{artif} \tag{4.14}$$

(fixed boundary) or $\boldsymbol{d}\cdot\boldsymbol{n} = 0$ together with $(\boldsymbol{T}_s\boldsymbol{n})\cdot\boldsymbol{\tau}^{(j)} = \boldsymbol{0}$, $j = 1, 2$, where $\boldsymbol{\tau}^{(j)}$ are the unit tangential directions (displacement allowed in the tangential direction), whereas at the internal physical boundary Σ^t they prescribe the solid traction

$$\boldsymbol{T}_s\boldsymbol{n} = \boldsymbol{h}_s \quad \text{on } \Sigma^t. \tag{4.15}$$

Here $\boldsymbol{d}_0$, $\boldsymbol{d}_1$ and $\boldsymbol{h}_s$ are suitable given functions. When considering the fluid–structure (FS) coupling, $\boldsymbol{h}_s$ is of course provided by the normal Cauchy stress from the fluid side: see Section 4.3. To account for the effect of the tissues surrounding the artery, the algebraic law (3.1) is often prescribed at the external lateral surface Γ_{ext} of Ω_s, to mimic the elastic behaviour of this tissue (Moireau *et al.* 2012).

For each time $t > 0$, the weak form of (4.7) together with the boundary conditions (3.1), (4.14) and (4.15), in the case for fixed boundaries, reads as follows: find $\widehat{\boldsymbol{d}} = \widehat{\boldsymbol{d}}(t) \in \boldsymbol{D}$, $\widehat{\boldsymbol{d}} = \boldsymbol{0}$ on Γ_{artif}, $\widehat{\boldsymbol{d}} = \boldsymbol{d}_0$ and $\partial\widehat{\boldsymbol{d}}/\partial t = \boldsymbol{d}_1$ for $t = 0$ in Ω_s, such that

$$\begin{aligned}\rho_s \int_{\Omega_s} \frac{\partial^2 \widehat{\boldsymbol{d}}}{\partial t^2} \cdot \widehat{\boldsymbol{e}}\, \mathrm{d}\omega + \int_{\Omega_s} \widehat{\boldsymbol{T}}_s(\widehat{\boldsymbol{d}}) : \nabla \widehat{\boldsymbol{e}}\, \mathrm{d}\omega + \int_{\Gamma_{ext}} \alpha_{ST} \widehat{\boldsymbol{d}} \cdot \widehat{\boldsymbol{e}}\, \mathrm{d}\sigma \\ = \int_{\Gamma_{ext}} P_{ext} \widehat{\boldsymbol{n}} \cdot \widehat{\boldsymbol{e}}\, \mathrm{d}\sigma + \int_{\Sigma} \widehat{\boldsymbol{h}}_s \cdot \widehat{\boldsymbol{e}}\, \mathrm{d}\sigma,\end{aligned} \tag{4.16}$$

for all $\widehat{\boldsymbol{e}} \in \boldsymbol{D} = \boldsymbol{D}^0$, where

$$\boldsymbol{D}^t = \{\boldsymbol{e} \in [H^1(\Omega_s^t)]^3 : \boldsymbol{e} = \boldsymbol{0} \text{ on } \Gamma_{artif}^t\}.$$

The existence of strong (steady) solutions of the above problems could be proved using the theory developed in Ball (1976). For example, this is the case for the Saint Venant–Kirchhoff constitutive law given in (4.9) (Ciarlet and Necas 1985). The existence and uniqueness of weak solutions are guaranteed by the coercivity and convexity of the energy Θ (Ciarlet 1988, Dacorogna 2000). However, a convex strain energy function is not generally able to describe instabilities such as buckling (Ball 1977). Thus, to avoid the use of convex functions, the property of *polyconvexity* was introduced (Ball 1976). Its fulfilment guarantees physically admissible solutions (Dacorogna 2000). Both the neo-Hookean law (4.11) and the exponential law (4.12) satisfy the polyconvexity property (Balzani, Neff, Schroder and Holzapfel 2006).

4.2.1. Modelling the structure as a 2D membrane

In some circumstances, because of the thinness of the vessel wall, a non-linear shell model has been proposed: see *e.g.* Leuprecht *et al.* (2002) and Zhao *et al.* (2000). In this case, the structure problem is described by two-dimensional equations defined with respect to the middle surface, consisting of the computation of the deformation of this surface.

A simpler equation may be obtained if the structure is modelled as a 2D *membrane* whose position in space at any time exactly coincides with internal boundary Σ^t, yielding the so-called *generalized string model*

$$\rho_s H_s \frac{\partial^2 \widehat{d}_r}{\partial t^2} - \nabla \cdot (\boldsymbol{P} \nabla \widehat{d}_r) + \chi H_s \widehat{d}_r = \widehat{f}_s \quad \text{in } \Sigma \tag{4.17}$$

(Quarteroni *et al.* 2000*c*). Here Σ denotes the reference membrane configuration, d_r is the radial displacement and H_s is the structure thickness. The tensor $\boldsymbol{P}$ accounts for *shear deformations* and, possibly, for *pre-stress*,

$$\chi = \frac{E}{1-\nu^2}(4\rho_1^2 - 2(1-\nu)\rho_2),$$

where $\rho_1(\boldsymbol{x})$ and $\rho_2(\boldsymbol{x})$ are the *mean* and *Gaussian curvatures* of Σ, respectively (Nobile and Vergara 2008), and f_s is the forcing term, given by a measurement of the fluid pressure. Equation (4.17) is derived from *Hooke's law* for linear elasticity under the assumptions of small thickness, *plane stresses*, and negligible *elastic bending terms* (Zienkiewicz and Taylor 2005). To account for the effect of the surrounding tissue, the term χ in (4.17) needs to be augmented by the elastic coefficient of the tissue α_{ST} (Formaggia, Quarteroni and Vergara 2013).

A further simplification arises when Σ denotes the lateral surface of a cylinder. By discarding any dependence on the circumferential coordinate, model (4.17) reduces to

$$\rho_s H_s \frac{\partial^2 \widehat{d}_r}{\partial t^2} - kGH_s \frac{\partial^2 \widehat{d}_r}{\partial z^2} + \frac{EH_s}{(1-\nu^2)R_0^2} H_s \widehat{d}_r = \widehat{f}_s \quad \text{in } \Sigma, \tag{4.18}$$

where k is the *Timoshenko correction factor*, G is the *shear modulus*, R_0 is the initial cylinder radius and z is the axial coordinate. Often, in the latter case, a *viscoelastic term* of the form $\gamma_v(\partial^3 d_r / \partial^2 z \partial t)$ is also added, with γ_v denoting a suitable viscoelastic parameter (Quarteroni *et al.* 2000*c*).

4.3. The coupled fluid–structure interaction problem

Blood flow in the vessel lumen and deformation of the vessel wall are intimately connected via a fluid–structure interaction (FSI). In particular, fluid and structure interact via the fluid–structure (FS) interface Σ^t which coincides with the physical fluid boundary and the internal vessel wall boundary introduced in the previous sections. The coupled problem is obtained by combining (4.1) and (4.7) at each $t > 0$, as follows:

$$\rho_f \left(\frac{\partial \boldsymbol{v}}{\partial t} + (\boldsymbol{v} \cdot \nabla)\boldsymbol{v} \right) - \nabla \cdot \boldsymbol{T}_f(\boldsymbol{v}, p) = \boldsymbol{0} \quad \text{in } \Omega_f^t, \tag{4.19a}$$

$$\nabla \cdot \boldsymbol{v} = 0 \quad \text{in } \Omega_f^t, \tag{4.19b}$$

$$\boldsymbol{v} = \frac{\partial \boldsymbol{d}}{\partial t} \quad \text{at } \Sigma^t, \tag{4.19c}$$

$$\boldsymbol{T}_s(\boldsymbol{d})\,\boldsymbol{n} = \boldsymbol{T}_f(\boldsymbol{v},p)\,\boldsymbol{n} \qquad \text{at } \Sigma^t, \tag{4.19d}$$

$$\rho_s \frac{\partial^2 \widehat{\boldsymbol{d}}}{\partial t^2} - \nabla \cdot \widehat{\boldsymbol{T}}_s(\widehat{\boldsymbol{d}}) = \boldsymbol{0} \qquad \text{in } \Omega_s, \tag{4.19e}$$

$$\boldsymbol{d}_f = \boldsymbol{d} \qquad \text{at } \Sigma^t, \tag{4.19f}$$

together with the boundary conditions (3.1), (4.4), (4.5), (4.14), and where $\boldsymbol{d}_f$ is the displacement of the fluid domain Ω_f^t at the FS interface and we have used the convention that $\boldsymbol{n}$ is the structure outward unit normal. The matching conditions (4.19c, 4.19d) enforced at the FS interface express the *continuity of velocities* (kinematic condition) and the *continuity of normal stresses* (dynamic condition), respectively, whereas condition (4.19f) guarantees the geometry adherence between the fluid and structure domains (geometric condition). The well-posedness analysis of the coupled problem (4.19) (supplemented by the relevant boundary conditions) has been carried out under several regularity assumptions. For a comprehensive description of this topic we refer to Beirão da Veiga (2004), Grandmont (1998), Bodnár, Galdi and Nečasová (2014) and Maday (2009), for example.

For each time $t > 0$, the weak formulation of the FSI problem (4.19) together with its boundary conditions (for simplicity we set $\boldsymbol{v}_{up} = \boldsymbol{0}$ in (4.4)) reads as follows: find

$$\begin{aligned}(\boldsymbol{v}(t), \widehat{\boldsymbol{d}}(t)) \in \boldsymbol{W}^t = \{&(\boldsymbol{w}, \widehat{\boldsymbol{e}}) \in [H^1(\Omega_f^t)]^3 \times [H^1(\Omega_s)]^3 : \\ &(\boldsymbol{w}, \widehat{\boldsymbol{e}}) = (\boldsymbol{0}, \boldsymbol{0}) \text{ on } \Gamma_{in}^t \times \Gamma_{artif} \text{ and } \boldsymbol{w} = \boldsymbol{e} \text{ on } \Sigma^t\},\end{aligned}$$

$\boldsymbol{v} = \boldsymbol{v}_0$ for $t = 0$ in Ω_f, $\widehat{\boldsymbol{d}} = \boldsymbol{d}_0$ and $\partial\widehat{\boldsymbol{d}}/\partial t = \boldsymbol{d}_1$ for $t = 0$ in Ω_s, and $p(t) \in L^2(\Omega_f^t)$, such that

$$\begin{aligned}&\rho_f \int_{\Omega_f^t} \frac{\partial \boldsymbol{v}}{\partial t} \cdot \boldsymbol{w}\,\mathrm{d}\omega + \mathcal{A}_f^t(\boldsymbol{v}, \boldsymbol{v}, \boldsymbol{w}) + \mathcal{B}^t(p, \boldsymbol{w}) + \rho_s \int_{\Omega_s} \frac{\partial^2 \widehat{\boldsymbol{d}}}{\partial t^2} \cdot \widehat{\boldsymbol{e}}\,\mathrm{d}\omega \\ &\qquad + \int_{\Omega_s} \widehat{\boldsymbol{T}}_s(\widehat{\boldsymbol{d}}) : \nabla \widehat{\boldsymbol{e}}\,\mathrm{d}\omega + \int_{\Gamma_{ext}} \alpha_{ST} \widehat{\boldsymbol{d}} \cdot \widehat{\boldsymbol{e}}\,\mathrm{d}\sigma \\ &\qquad\qquad = \int_{\Gamma_{out}^t} \boldsymbol{h}_f \cdot \boldsymbol{n}\,\mathrm{d}\gamma + \int_{\Gamma_{ext}} P_{ext} \widehat{\boldsymbol{n}} \cdot \widehat{\boldsymbol{e}}\,\mathrm{d}\sigma, \\ &\mathcal{B}^t(q, \boldsymbol{v}) = 0, \\ &\boldsymbol{d}_f = \boldsymbol{d} \quad \text{at } \Sigma^t,\end{aligned} \tag{4.20}$$

for all $(\boldsymbol{w}, \widehat{\boldsymbol{e}}) \in \boldsymbol{W}^t$ and $q \in L^2(\Omega_f^t)$. Note that in the above weak formulation, the two terms arising after integration by parts and involving the normal Cauchy stresses $\boldsymbol{T}_f\boldsymbol{n}$ and $\boldsymbol{T}_s\boldsymbol{n}$ at the interface Σ^t cancel out, thanks to (4.19d) and to the special choice of the test functions in $\boldsymbol{W}^t$.

After introducing the total energy for the FSI problem, that is,

$$\mathcal{E}_{3D}(t) = \frac{\rho_f}{2}\int_{\Omega_f^t} |\boldsymbol{v}|^2 \,\mathrm{d}\omega + \int_{\Omega_s} \frac{\rho_s}{2}|\dot{\widehat{\boldsymbol{d}}}|^2 \,\mathrm{d}\omega + \int_{\Omega_s} \Theta(\widehat{\boldsymbol{d}}) \,\mathrm{d}\omega + \int_{\Gamma_{ext}} \alpha_{ST}|\widehat{\boldsymbol{d}}|^2 \,\mathrm{d}\gamma, \tag{4.21}$$

the following conservation property holds true for the case of homogeneous boundary conditions:

$$\frac{\mathrm{d}}{\mathrm{d}t}\mathcal{E}_{3D}(t) + \frac{\mu}{2}\int_{\Omega_f^t} (\nabla \boldsymbol{v} + (\nabla \boldsymbol{v})^t)^2 \,\mathrm{d}\omega = 0$$

(Formaggia, Gerbeau, Nobile and Quarteroni 2001, Formaggia *et al.* 2013).

When the membrane model (4.17) is used instead of (4.19e), the matching conditions (4.19c, 4.19d) are replaced by

$$\begin{aligned} \boldsymbol{v}\cdot\boldsymbol{n} &= \frac{\partial d_r}{\partial t} && \text{at } \Sigma^t, \\ \boldsymbol{T}_f(\boldsymbol{v},p)\boldsymbol{n}\cdot\boldsymbol{n} &= -f_s && \text{at } \Sigma^t, \end{aligned}$$

where d_r is the membrane displacement written in the current configuration and f_s is the forcing term of the membrane equation (see (4.17)), acting only at the FS interface; in this case it also plays the role of structure stress exerted at the FS interface. Owing to (4.17) itself, we can rewrite the above interface conditions as follows:

$$\boldsymbol{v}\cdot\boldsymbol{n} = \frac{\partial d_r}{\partial t} \qquad \text{at } \Sigma^t, \tag{4.23a}$$

$$\boldsymbol{T}_f(\boldsymbol{v},p)\boldsymbol{n}\cdot\boldsymbol{n} = -\left(\rho_s H_s \frac{\partial^2 d_r}{\partial t^2} - \nabla\cdot(\boldsymbol{P}\nabla d_r) + \chi H_s d_r\right) \qquad \text{at } \Sigma^t. \tag{4.23b}$$

Since the coupling only occurs in the radial direction, we have to complete the conditions at Σ^t for the fluid problem in the tangential directions by prescribing further equations on the fluid variables, for example homogeneous Dirichlet or Neumann conditions (Nobile 2001).

Figueroa *et al.* (2006) propose an effective formulation to solve the FSI problem with a membrane structure, while Colciago, Deparis and Quarteroni (2014) discuss the accuracy of the FSI-membrane problem in comparison to a full 3D/3D simulation. In particular, for Hooke's law, the wall shear stresses computed with these two FSI models are in good agreement for a distal arterial tract such as a femoropopliteal bypass. In contrast, when larger displacements are considered such as in the ascending aorta, the discrepancies between the two FSI models increase.

4.4. The boundary issue

According to the mathematical theory of the incompressible Navier–Stokes equations, three scalar conditions need to be prescribed at each point of the boundary. This, however, is seldom realistic in clinical practice. For instance, PC-MRA provides velocity data, but this technique is not routinely used and *ad hoc* studies are required (Morbiducci *et al.* 2009). Alternatively, the flow rate $Q = Q(t)$ can be obtained across a boundary cross-section Γ^t, by suitable postprocessing of data retrieved by echo-colour-Doppler or by thermal images: see Section 3.2. This yields the flow rate condition (3.2). In other situations, at both the inlet and outlet cross-sections, pressure measurements $P = P(t)$ may be considered as representative of an average estimate, leading to condition (3.3).

From a mathematical perspective, (3.2) and (3.3) are *defective* conditions as they prescribe only one scalar function over the entire section Γ^t (Formaggia, Gerbeau, Nobile and Quarteroni 2002). Several strategies have been proposed so far to supplement (3.2) or (3.3) with other conditions that allow us to 'close' the system. For clarity, we can classify them according to three different strategies, which are reported below.

4.4.1. Conjecturing velocity and pressure profiles

A common trick to effectively prescribe the flow rate condition (3.2) consists in prescribing a velocity profile

$$\boldsymbol{v}(t) = \boldsymbol{g}(t) \quad \text{on } \Gamma^t, \tag{4.24}$$

where $\boldsymbol{g} = \boldsymbol{g}(t, \boldsymbol{x})$ is chosen in such a way as to satisfy (3.2), that is,

$$\rho_f \int_{\Gamma^t} \boldsymbol{g}(t) \cdot \boldsymbol{n} \, \mathrm{d}\gamma = Q(t). \tag{4.25}$$

The flow rate condition (3.2) is therefore replaced with the standard (vectorial) Dirichlet condition (4.24). A classical choice for $\boldsymbol{g}$ is a parabolic profile (*e.g.* for flow simulations in the carotid arteries: Campbell *et al.* 2012), a constant profile (often used for the ascending aorta: Moireau *et al.* 2012), or that obtained from the Womersley solution (He, Ku and Moore 1993). Both the parabolic and Womersley profiles require a circular section to be prescribed, while non-circular sections require an appropriate morphing (He *et al.* 1993).

In spite of its straightforward implementation, this choice has a major impact on the solution, particularly in the neighbourhood of the section Γ^t and for elevated values of the Reynolds number (Veneziani and Vergara 2007). To reduce the sensitivity to the arbitrary choice of the profile, the computational domain can be artificially elongated by operating what is called a *flow extension* (Moore, Steinman and Ethier 1997).

A similar approach could be applied to the mean normal Cauchy stress condition (3.3) as well. In the case at hand, we can postulate that the pressure on Γ^t is constant and that the viscous stress can be neglected, that is, we can prescribe

$$p\boldsymbol{n} - \mu(\nabla \boldsymbol{v} + (\nabla \boldsymbol{v})^T)\,\boldsymbol{n} = P\boldsymbol{n} \quad \text{on } \Gamma^t. \tag{4.26}$$

Note that the above condition in particular satisfies the defective condition (3.3). Condition (4.26) is generally acceptable because the pressure changes in arteries mainly occur along the axial direction and the viscous stresses are negligible on orthogonal cross-sections.

Since $P\boldsymbol{n}$ plays the role of boundary normal Cauchy stress when implemented in the framework of finite element approximations, no further action is required beyond assembling the matrix for Neumann conditions. For this reason, the treatment has been called the '*do-nothing*' approach (Heywood *et al.* 1996). As pointed out by Veneziani (1998*a*, 1998*b*), this procedure is in fact not completely 'innocent'. The do-nothing approach corresponds to the following weak formulation (for simplicity we assume homogeneous Dirichlet conditions, $\boldsymbol{v}_{up} = \boldsymbol{0}$): for each $t > 0$, find $\boldsymbol{v} \in \widetilde{\boldsymbol{V}}^t$, $\boldsymbol{v} = \boldsymbol{v}_0$ for $t = 0$ in Ω_f, and $p \in L^2(\Omega_f^t)$ such that

$$\begin{aligned}
&\rho_f \int_{\Omega_f^t} \frac{\partial \boldsymbol{v}}{\partial t} \cdot \boldsymbol{w}\,\mathrm{d}\omega + \mathcal{A}_f^t(\boldsymbol{v}, \boldsymbol{v}, \boldsymbol{w}) + \mathcal{B}^t(p, \boldsymbol{w}) \\
&\qquad = \int_{\Gamma_{out}^t} \boldsymbol{h}_f \cdot \boldsymbol{n}\,\mathrm{d}\gamma - P \int_{\Gamma^t} \boldsymbol{w} \cdot \boldsymbol{n}\,\mathrm{d}\gamma, \\
&\mathcal{B}^t(q, \boldsymbol{v}) = 0,
\end{aligned}$$

for all

$$\boldsymbol{w} \in \widetilde{\boldsymbol{V}}^t = \{[H^1(\Omega_f^t)]^3 : \boldsymbol{w} = \boldsymbol{0} \quad \text{on } \partial\Omega \setminus (\Gamma_{out}^t \cup \Gamma^t)\}$$

and $q \in L^2(\Omega_f^t)$.

A do-nothing formulation for the flow rate conditions is possible too: see Heywood *et al.* (1996) and Veneziani (1998*b*).

Note that as an alternative to (3.3), other defective conditions involving the fluid pressure could be considered as well. This is the case for *mean pressure conditions* (Heywood *et al.* 1996), or conditions involving the total pressure (Formaggia *et al.* 2013), defined by $p_{tot} = p + (\rho_f/2)|\boldsymbol{v}|^2$. For a comprehensive review of these conditions, we refer the interested reader to Quarteroni *et al.* (2016*c*).

4.4.2. Augmented formulation

An alternative approach is to regard the flow rate boundary condition (3.2) as a constraint for the solution of the fluid problem and then enforce it via a Lagrange multiplier approach. As a scalar constraint, we need a scalar multiplier $\lambda = \lambda(t)$ at each time, resulting in the following weak formulation (we again consider the case of homogeneous Dirichlet conditions): for each $t > 0$, find $\boldsymbol{v} \in \widetilde{\boldsymbol{V}}^t$, $\boldsymbol{v} = \boldsymbol{v}_0$ for $t = 0$ in Ω_f, $p \in L^2(\Omega_f^t)$, and $\lambda \in \mathbb{R}$ such that

$$\begin{aligned}
&\rho_f \int_{\Omega_f^t} \frac{\partial \boldsymbol{v}}{\partial t} \cdot \boldsymbol{w}\, \mathrm{d}\omega + \mathcal{A}_f^t(\boldsymbol{v}, \boldsymbol{v}, \boldsymbol{w}) + \mathcal{B}^t(p, \boldsymbol{w}) + \mathcal{C}^t(\lambda, \boldsymbol{w}) = \int_{\Gamma_{out}^t} \boldsymbol{h}_f \cdot \boldsymbol{w}\, \mathrm{d}\gamma,\\
&\mathcal{B}^t(q, \boldsymbol{v}) = 0,\\
&\mathcal{C}^t(\psi, \boldsymbol{v}) = \psi \frac{Q}{\rho_f},
\end{aligned} \tag{4.27}$$

for all $\boldsymbol{w} \in \widetilde{\boldsymbol{V}}^t$, $q \in L^2(\Omega_f^t)$, and $\psi \in \mathbb{R}$, and where we have set

$$\mathcal{C}^t(\psi, \boldsymbol{w}) = \psi \int_{\Gamma^t} \boldsymbol{w} \cdot \boldsymbol{n}\, \mathrm{d}\gamma.$$

See Formaggia *et al.* (2002) and Veneziani and Vergara (2005), who also analyse the well-posedness of this problem.

Besides prescribing the flow rate condition (3.2), the above augmented formulation enforces at each time a constant-in-space normal Cauchy stress on Γ^t aligned with its normal direction, which precisely coincides with the Lagrange multiplier λ, that is,

$$-p\boldsymbol{n} + \mu(\nabla \boldsymbol{u} + (\nabla \boldsymbol{v})^T)\, \boldsymbol{n} = \lambda \boldsymbol{n} \quad \text{on } \Gamma^t.$$

This method is particularly suitable when the artificial cross-section is orthogonal to the longitudinal axis, so that vector $\boldsymbol{n}$ is truly aligned along the axial direction.

Since the velocity spatial profile is not prescribed *a priori*, this technique has been used to improve the parabolic-based law implemented in the Doppler technology for the estimation of the flow rate starting from the peak velocity (Ponzini, Vergara, Redaelli and Veneziani 2006, Vergara *et al.* 2010, Ponzini *et al.* 2010).

Extension of the augmented formulation to the case of compliant walls is addressed in Formaggia, Veneziani and Vergara (2009*b*) and to the quasi-Newtonian case in Ervin and Lee (2007).

An augmented formulation has been proposed by Formaggia *et al.* (2002) to prescribe condition (3.3) as well. However, as noticed by Formaggia *et al.* (2002), in this case it yields at each time the condition

$$\boldsymbol{v} = \lambda \boldsymbol{n} \quad \text{at } \Gamma^t,$$

where $\lambda = \lambda(t)$ is again the constant-in-space Lagrange multiplier. This is a non-homogeneous Dirichlet boundary condition for the fluid velocity, which is generally incompatible with the no-slip condition $\boldsymbol{v} = \boldsymbol{\phi}$ prescribed at the physical boundary Σ^t. For this reason, the approach is no longer used.

4.4.3. A control-based approach

A different strategy for the fulfilment of condition (3.2) is based on the minimization of the mismatch functional

$$J(\boldsymbol{v}) = \frac{1}{2}\left(\int_{\Gamma^t} \boldsymbol{v}\cdot\boldsymbol{n}\,\mathrm{d}\gamma - Q\right)^2, \tag{4.28}$$

constrained by the fact that $\boldsymbol{v}$ satisfies the incompressible Navier–Stokes equations (Formaggia, Veneziani and Vergara 2008). This PDE-constrained optimization – which can be regarded as the dual of the above augmented strategy – yields a system of optimality conditions to be fulfilled (also referred to as the Karush–Kuhn–Tucker (KKT) system): see Section 9.1.2 for further details. In particular, Formaggia *et al.* (2008) used the normal component of the normal Cauchy stress on Γ^t as the control variable for the minimization of the mismatch functional. This approach was considered by Formaggia *et al.* (2009*b*) for the compliant case, whereas Lee (2011), Galvin and Lee (2013) and Galvin, Lee and Rebholz (2012) address the non-Newtonian, quasi-Newtonian and viscoelastic cases.

The same approach can also be used to fulfil the defective condition (3.3) provided that a suitable functional to be minimized is introduced (Formaggia *et al.* 2008). This allows us to prescribe (3.3) on a section oblique with respect to the longitudinal axis too. In this case the control variable is the complete normal Cauchy stress vector, that is, the direction of the normal Cauchy stress is also *a priori* unknown.

Boundary data may also be lacking for the cross-section of the vessel wall. In this case we end up with defective BC issues for the vessel wall: see *e.g.* Quarteroni *et al.* (2016*c*).

4.5. Geometric reduced models and the geometric multiscale approach

4.5.1. The 1D and 0D models

Numerical modelling of the entire cardiovascular system by means of 3D models is currently not affordable because of the complexity of the computational domain, which is composed of thousands of arteries and veins and billions of arterioles, capillaries and venules (Nichols and O'Rourke 2005). In many applications, reduced-dimensional models are used instead, either as stand-alone models or coupled with the 3D ones.

The first one-dimensional (1D) model was introduced almost 250 years ago by Euler (1775). Subsequently this approach was brought into the engineering environment by Barnard, Hunt, Timlake and Varley (1966), Hughes (1974) and Hughes and Lubliner (1973). These models allow the description of blood flow in a compliant vessel where the only space coordinate is that of the vessel axis.

One-dimensional models may be derived from 3D models by making simplifying assumptions on the behaviour of the flow, the structure, and their

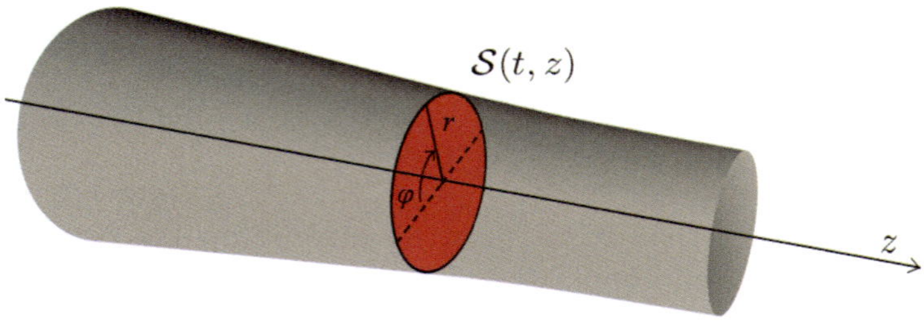

Figure 4.2. Fluid domain for the derivation of the 1D model.

interaction (Quarteroni and Formaggia 2004, Peiró and Veneziani 2009). The starting fluid domain is a truncated cone: see Figure 4.2. Referring to cylindrical coordinates (r, φ, z), we make the following simplifying assumptions: (i) the axis of the cylinder is fixed; (ii) for any z, the cross-section $\mathcal{S}(t, z)$ is a circle with radius $R(t, z)$; (iii) the solution of both fluid and structure problems does not depend on φ; (iv) the pressure is constant over each section $\mathcal{S}(t, z)$; (v) the axial fluid velocity v_z dominates the other velocity components; (vi) only radial displacements are allowed, so the structure deformation takes the form $\boldsymbol{d} = d\boldsymbol{e}_r$, where $\boldsymbol{e}_r$ is the unit vector in the radial direction, and $d(t, z) = R(t, z) - R_0(z)$, where $R_0(z)$ is the reference radius at the equilibrium; (vii) the fluid is assumed to obey Poiseuille's law, so that the viscous effects are modelled by a linear term proportional to the flow rate; (viii) the vessel structure is modelled as a membrane with constant thickness.

We introduce the following quantities:

$$
\begin{aligned}
&A(t, z) = |\mathcal{S}(t, z)| = \pi R(t, z)^2 && \text{lumen section area,}\\
&\bar{v}(t, z) = A^{-1} \int_{\mathcal{S}(t,z)} v_z(t, z) \, \mathrm{d}S && \text{mean velocity,}\\
&s(r/R) \quad \text{such that } v_z(t, r, z) = \bar{v}(t, z) s(r/R(t, z)) && \text{velocity profile,}\\
&Q(t, z) = \rho_f \int_{\mathcal{S}(t,z)} v_z \, \mathrm{d}S = \rho_f A(t, z) \bar{v}(t, z) && \text{flow rate,}\\
&P(t, z) = A^{-1} \int_{\mathcal{S}(t,z)} p(t, z) \, \mathrm{d}S && \text{mean pressure.}
\end{aligned}
$$

As for the structure and its interaction with the fluid, we need to introduce a membrane law, which in fact prescribes a relation between the pressure and the lumen area (which is determined by d_r) of the form

$$P(t, z) = P_{ext} + \psi(A(t, z), A_0(z), \boldsymbol{\beta}(z)). \tag{4.29}$$

where ψ is a given function satisfying $\partial\psi/\partial A > 0$, $\psi(A_0) = 0$. Here $\boldsymbol{\beta}$ is a vector of parameters describing the mechanical properties of the membrane.

By integrating over the sections $\mathcal{S}$ the momentum fluid equation (4.19a) in the z-direction and the mass conservation law (4.19b), we obtain the system

$$\frac{\partial \boldsymbol{U}}{\partial t} + \mathrm{H}(\boldsymbol{U})\frac{\partial \boldsymbol{U}}{\partial z} + \boldsymbol{B}(\boldsymbol{U}) = \boldsymbol{0} \quad z \in (0, L),\, t > 0, \tag{4.30}$$

where $\boldsymbol{U} = [A\ Q]^T$ is the vector of the unknowns,

$$\alpha = \frac{\int_S u_z^2}{A\bar{v}^2} = \frac{1}{A}\int_0^1 s^2(y)\,\mathrm{d}y$$

is the *Coriolis coefficient*, $K_r = -2\pi\mu s'(1)$ is the *friction parameter*,

$$c_1 = \sqrt{\frac{A}{\rho_f}\frac{\partial\psi}{\partial A}},$$

while

$$\mathrm{H}(\boldsymbol{U}) = \begin{bmatrix} 0 & 1 \\ c_1^2 - \alpha\left(\dfrac{Q}{A}\right)^2 & 2\alpha\dfrac{Q}{A} \end{bmatrix}, \tag{4.31a}$$

$$\boldsymbol{B}(\boldsymbol{U}) = \begin{bmatrix} 0 \\ K_r\frac{Q}{A} + \dfrac{A}{\rho_f}\dfrac{\partial\psi}{\partial A_0}\dfrac{\partial A_0}{\partial z} + \dfrac{A}{\rho_f}\dfrac{\partial\psi}{\partial\boldsymbol{\beta}}\dfrac{\partial\boldsymbol{\beta}}{\partial z} \end{bmatrix} \tag{4.31b}$$

represent the flux matrix and the dissipation vector term, respectively. A complete derivation of the model can be found in Pedley (1980), Hughes (1974) and Peiró and Veneziani (2009), for example. Classical choices of the velocity profile s are flat ($\alpha = 1$) and parabolic ($\alpha = 4/3$).

The term $\partial A_0/\partial z$ in $\boldsymbol{B}$ is typically non-positive, accounting for vessel 'tapering', that is, reduction of the area of the lumen when proceeding from proximal to distal arteries. The term $\partial\boldsymbol{\beta}/\partial z$ originates from possibly different mechanical properties along the vessel, to describe, for example, the presence of atherosclerotic plaques or vascular prostheses.

If $A > 0$, system (4.30) has two real distinct eigenvalues,

$$\lambda_{1,2} = \alpha\bar{v} \pm \sqrt{c_1^2 + \bar{v}^2\alpha(\alpha - 1)} \tag{4.32}$$

(see *e.g.* Quarteroni and Formaggia 2004), so it is *strictly hyperbolic* (see *e.g.* LeVeque 1992). Under physiological conditions, $c_1 \gg \alpha\bar{v}$, yielding $\lambda_1 > 0$ and $\lambda_2 < 0$, we thus have two waves travelling in opposite directions, associated with corresponding characteristic variables. An explicit expression of these variables as a function of the physical variables could, in general, be derived, that is,

$$W_i = \zeta_i(A, Q), \quad i = 1, 2. \tag{4.33}$$

A simple membrane law (4.29) can be obtained by the algebraic relation

$$\psi(A, A_0, \eta) = \eta \frac{\sqrt{A} - \sqrt{A_0}}{A_0}, \quad \text{with } \eta = \frac{\sqrt{\pi} H_s E}{1 - \nu^2} \tag{4.34}$$

(Formaggia, Lamponi and Quarteroni 2003, Formaggia *et al.* 2013), where ν is the Poisson modulus of the membrane, E is its Young's modulus, and H_s is its thickness, yielding

$$c_1 = \sqrt{\frac{\eta\sqrt{A}}{2\rho_f A_0}}.$$

This simple law, stating that the membrane radial displacement d_r is linearly proportional to the fluid pressure, has been considered successfully in many applications: see *e.g.* Steele *et al.* (2003), Matthys *et al.* (2007) and Grinberg *et al.* (2010). Other laws have been proposed to account for additional features of arterial walls, such as viscoelasticity, wall-inertia and longitudinal pre-stress (Quarteroni *et al.* 2000*c*, Formaggia *et al.* 2003).

Remark 4.1. One-dimensional models do not allow us to describe secondary flows, such as vortices or recirculation regions. However, they provide average quantities about the axial component of the velocity, the radial vessel wall displacements and the pressure of a complex network at computational costs that are orders of magnitude lower than those of the corresponding three-dimensional FSI models (Kufahl and Clark 1985, Hillen, Hoogstraten and Post 1986, Blanco *et al.* 2015, Boileau *et al.* 2015, Malossi *et al.* 2013).

The accuracy of the solution provided by 1D models is addressed by Avolio (1980), Steele *et al.* (2003) and Reymond *et al.* (2009), who successfully compare the numerical results of different networks with clinical measurements, and by Matthys *et al.* (2007) and Muller and Toro (2013), who perform a comparison with *in vitro* measurements for a complete network of the cardiovascular system.

A further geometric reduction is represented by the so-called *lumped parameter models*, which are zero-dimensional (0D) models obtained by integrating the 1D problem over the axial direction. These are typically used to describe the peripheral part of the arterial and venous tree, such as the capillaries and the arterioles.

In this case only dependence on time is allowed, and nominal values of the unknowns are used as representative of the entire compartment. To this end, we introduce the *average* flow rate and pressure in the district at hand, defined by

$$\widehat{Q}(t) = \frac{1}{l}\int_{z_p}^{z_d} Q(t,z)\,\mathrm{d}z = \frac{\rho_f}{l}\int_{z_p}^{z_d}\int_{\mathcal{S}(t,z)} v_z(t,z)\,\mathrm{d}S\,\mathrm{d}z,$$
$$\widehat{P}(t) = \frac{1}{l}\int_{z_p}^{z_d} P(t,z)\,\mathrm{d}z = \frac{1}{V}\int_{z_p}^{z_d}\int_{\mathcal{S}(t,z)} p(t,z)\,\mathrm{d}S\,\mathrm{d}z$$

respectively, where z_p and z_d are the proximal and longitudinal abscissas of the segment, respectively, l is its length and V its volume. The convective term is dropped since in the peripheral sites the velocity is small.

If we take the longitudinal average of the momentum equation given by the first of (4.30) and combine it with (4.34), we obtain the ordinary differential equation (ODE)

$$\frac{\rho_f l}{A_0}\frac{\mathrm{d}\widehat{Q}}{\mathrm{d}t} + \frac{\rho_f K_R l}{A_0^2}\widehat{Q} + \widehat{P}_d - \widehat{P}_p = 0, \tag{4.35}$$

where $\widehat{P}_d$ and $\widehat{P}_p$ are the distal and proximal pressure, respectively. When taking the longitudinal average of the mass conservation law given by the second of (4.30) and using (4.34), we obtain

$$\frac{\sqrt{A_0}l}{\eta}\frac{\mathrm{d}\widehat{P}}{\mathrm{d}t} + \widehat{Q}_d - \widehat{Q}_p = 0, \tag{4.36}$$

where $\widehat{Q}_d$ and $\widehat{Q}_p$ are the distal and proximal flow rate, respectively. See Peiró and Veneziani (2009).

The two ODEs (4.35, 4.36) can be regarded as the starting point for a 0D description of a compartment model of an arterial tract. In fact, the term $L(\mathrm{d}\widehat{Q}/\mathrm{d}t)$, with $L = \rho_f l/A_0$, corresponds to the blood acceleration, $R\widehat{Q}$, with $R = \rho_f K_R l/A_0^2$, stems from the blood resistance due to the viscosity, while $C(\mathrm{d}\widehat{P}/\mathrm{d}t)$, with $C = \sqrt{A_0}l/\eta$, is due to the compliance of the vessel wall. Usually an electrical analogy is used to easily interpret 0D models. In particular, the flow rate plays the role of the current whereas the pressure is the potential. Accordingly, the acceleration term is represented by an inductance, the viscosity term by a resistance, and the compliance term by a capacitance.

To close the system (4.35, 4.36) (featuring four unknowns) we also need to include the boundary conditions originally prescribed on the 1D model. For instance, we can assume a Dirichlet condition at the inlet and a Neumann condition at the outlet. Thus, we may localize the unknown pressure $\widehat{P}$ at the proximal section ($\widehat{P} \approx \widehat{P}_p$), assuming that the distal pressure $\widehat{P}_d$ is given. Similarly we assume that the flow rate $\widehat{Q}$ is approximated by $\widehat{Q}_d$ and

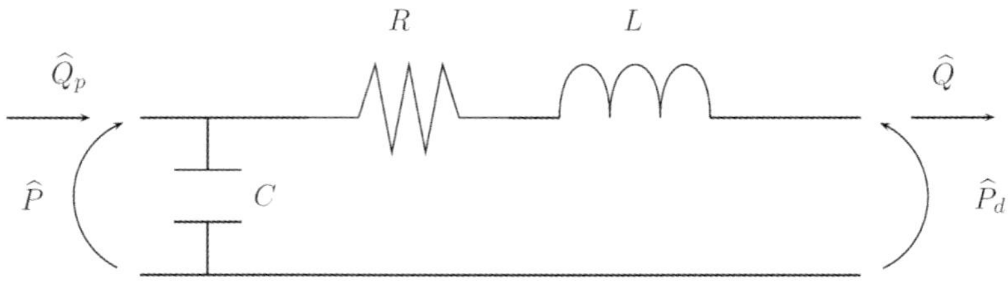

Figure 4.3. Example of lumped parameter scheme for an arterial tract.

that the proximal flow rate $\widehat{Q}_p$ is given. Then, from (4.35, 4.36), we obtain

$$\begin{aligned} \widehat{P} - L\frac{\mathrm{d}\widehat{Q}}{\mathrm{d}t} - R\widehat{Q} &= \widehat{P}_d, \\ C\frac{\mathrm{d}\widehat{P}}{\mathrm{d}t} + \widehat{Q} &= \widehat{Q}_p, \end{aligned} \tag{4.37}$$

corresponding to the electrical circuit drawn in Figure 4.3. Other sequences corresponding to different boundary conditions and then to different state variables are also possible (see *e.g.* Quarteroni *et al.* 2016*c*). Even though these schemes are equivalent in terms of functionality, they play a different role when coupled with higher-dimensional models (see *e.g.* Quarteroni *et al.* 2016*c*).

For a description of more complex vascular districts, we may combine several 0D elementary tracts, by gluing them via classical continuity arguments. However, the lumped parameter models have mainly been used to provide suitable boundary conditions at the distal artificial sections of 3D and 1D models. In this case, one simple compartment is enough to describe the entire arterial system downstream of the region of interest. Examples are provided by the *windkessel model* (Westerhof, Lankhaar and Westerhof 2009), featuring an average resistance and capacitance, the 3-*element windkessel model* (Westerhof *et al.* 2009), where a second resistance is added before the windkessel compartment, and the 4-*element windkessel model* (Stergiopulos, Westerhof, Meister and Westerhof 1996, Stergiopulos, Westerhof and Westerhof 1999), where an inductance element is added to the 3-element windkessel model. A 0D model given simply by a resistance is used to provide absorbing boundary conditions at the outlets of the fluid domain in FSI simulations: see *e.g.* Nobile and Vergara (2008). Instead, more sophisticated approaches account for the propagative dynamics associated with the peripheral circulation, such as the *structured tree* model (Olufsen *et al.* 2000), which assumes an asymmetric self-similar structure for the peripheral network.

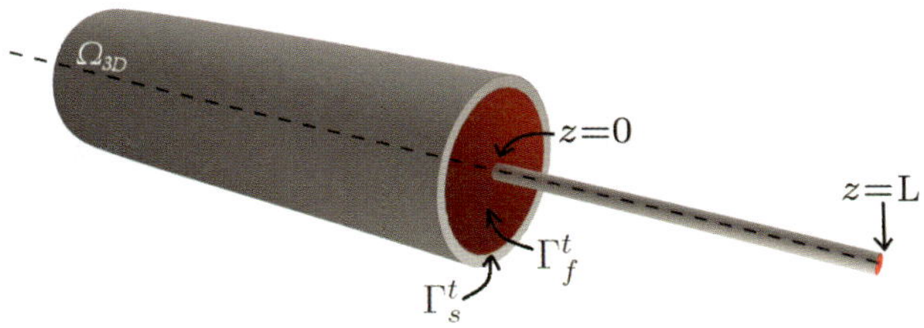

Figure 4.4. Schematic representation of the reference 3D/1D coupled model.

4.5.2. Geometric multiscale coupling

The geometric multiscale approach, first introduced in Quarteroni and Veneziani (1997), consists in the coupling between the 3D, 1D and 0D models. The idea is to use higher-dimensional models in those regions where a very detailed description is required, and lower-dimensional models in the remaining part of the region of interest. This allows us to describe much of the circulatory system.

As discussed earlier, 0D models are typically used to provide boundary conditions for 3D and 1D models. For this reason, the coupling between 3D or 1D models with an extended 0D network has rarely been considered in applications. Instead, the 3D/1D coupling has received a great deal of attention. For this reason, we will detail only the latter case, while referring to Quarteroni, Ragni and Veneziani (2001), Quarteroni and Veneziani (2003), Vignon-Clementel, Figueroa, Jansen and Taylor (2006), Kim *et al.* (2009), Migliavacca *et al.* (2006) and Haggerty *et al.* (2013) for the 3D/0D coupling, and Formaggia, Nobile, Quarteroni and Veneziani (1999) and Fernández, Milisic and Quarteroni (2005) for the 1D/0D coupling.

As shown in Figure 4.4, we consider a 3D/FSI problem (4.19) in a 3D cylindrical domain together with initial conditions and boundary conditions at the proximal boundaries and at the external structure.

At the distal boundaries, the 3D problem is coupled with the 1D model (4.30) written in the domain $z \in [0, L]$, together with initial conditions and a boundary condition at the distal boundary. We let $\Gamma^t = \Gamma^t_f \cup \Gamma^t_s$ be the coupling interface from the 3D side, which corresponds to the point $z = 0$ from the 1D side (see Figure 4.4).

A major mathematical issue is how to couple 3D and 1D models at the common interface. Several strategies can be pursued, yielding many (alternative) sets of interface conditions. For a rigorous derivation of the 3D/1D problem and a detailed discussion of the interface conditions, we refer the interested reader to the recent review article by Quarteroni *et al.* (2016*c*).

Here we will follow the guiding principle described in Formaggia *et al.* (2013), where suitable interface conditions are derived from a global energy estimate. In particular, we introduce, together with the 3D energy (4.21), the 1D energy

$$\mathcal{E}_{1D}(t) = \frac{\rho_f}{2}\int_0^L A\,\bar{v}^2\,\mathrm{d}x + \int_0^L \chi(A)\,\mathrm{d}x$$

(Formaggia *et al.* 2001), where $\chi(A) = \int_{A_0}^{A}\psi(\tau)\,\mathrm{d}\tau$ and ψ is the vessel law (see (4.29)). Note that the stand-alone 1D problem satisfies bounds for this energy functional as proved in Formaggia *et al.* (2001).

Let $P_{tot} = \psi(A) + (\rho_f/2)\bar{v}^2$ be the total pressure for the 1D model, and let $p_{tot} = p + (\rho_f/2)|\boldsymbol{v}|^2$ be that of the 3D model. For the interface coupling conditions holding at Γ^t, let us assume that the following inequality holds:

$$\int_{\Gamma_f^t} \boldsymbol{T}_f(\boldsymbol{v}, p_{tot})\boldsymbol{n}\cdot\boldsymbol{v}\,\mathrm{d}\gamma + \int_{\Gamma_s^t} \boldsymbol{T}_s(\boldsymbol{d})\boldsymbol{n}\cdot\dot{\boldsymbol{d}}\,\mathrm{d}\gamma + Q|_{z=0}\,P_{tot}|_{z=0} \leq 0. \tag{4.38}$$

Then, for all $t > 0$, the coupled 3D/1D problem (4.19, 4.30) with homogeneous boundary conditions satisfies the energy decay property

$$\frac{\mathrm{d}}{\mathrm{d}t}(\mathcal{E}^{3D}(t) + \mathcal{E}^{1D}(t)) \leq 0.$$

For the proof see Formaggia *et al.* (2013).

The above result provides an indication of how to find suitable interface conditions for the 3D/1D coupled problem. In particular, for inequality (4.38) to be fulfilled it is sufficient that the interface conditions

$$\rho_f \int_{\Gamma_f^t} \boldsymbol{v}\cdot\boldsymbol{n}\,\mathrm{d}\gamma = Q|_{z=0}, \tag{4.39a}$$

$$(\boldsymbol{T}_f(\boldsymbol{v}, p_{tot})\boldsymbol{n})|_{\Gamma_f^t} = -P_{tot}|_{z=0}\boldsymbol{n} \tag{4.39b}$$

hold for the fluid, together with

$$\boldsymbol{T}_s(\boldsymbol{d})\boldsymbol{n} = \boldsymbol{0} \quad \text{on } \Gamma_s^t \tag{4.40}$$

for the structure (Formaggia, Moura and Nobile 2007, Formaggia *et al.* 2013). Similarly, inequality (4.38) holds if relation (4.40) is replaced by

$$\boldsymbol{d}\cdot\boldsymbol{n} = \boldsymbol{0} \quad \text{on } \Gamma_s^t, \tag{4.41a}$$

$$(\boldsymbol{T}_s(\boldsymbol{d})\boldsymbol{n})\times\boldsymbol{n} = \boldsymbol{0} \quad \text{on } \Gamma_s^t. \tag{4.41b}$$

The interface conditions (4.39) prescribe the continuity of the flow rate (kinematic condition) and a dynamic condition involving the total pressures. Note that (4.40) and (4.41) are in fact independent of the 1D model, resulting in boundary conditions for the 3D structure problems only. This allows a discontinuity to manifest in the displacement between the 3D and 1D models.

Usually, dynamic interface conditions involving the pressure (instead of the total pressure) are considered in place of (4.39b), such as

$$\frac{1}{|\Gamma_f^t|}\int_{\Gamma_f^t} \boldsymbol{T}_f(\boldsymbol{v},p)\boldsymbol{n}\,\mathrm{d}\gamma = -\psi(A|_{z=0})\boldsymbol{n}. \tag{4.42}$$

However, the above condition does not satisfy the compatibility condition (4.38). More precisely, in this case we have

$$\frac{\mathrm{d}}{\mathrm{d}t}(\mathcal{E}^{3D}(t)+\mathcal{E}^{1D}(t)) = \frac{\rho_f}{2}\left(\frac{(Q(t)|_{z=0})^3}{(A(t)|_{z=0})^2} - \int_{\Gamma_f^t} |\boldsymbol{v}(t)|^2\boldsymbol{v}(t)\cdot\boldsymbol{n}\,\mathrm{d}\gamma\right).$$

Even though the right-hand side is not necessarily (always) negative, numerical evidence indicates that condition (4.42) leads to stable results for haemodynamic applications: see Malossi *et al.* (2013). This interface condition is indeed the most commonly used among dynamic ones.

4.6. Numerical strategies

In this section we comment on the numerical solution of the problems presented in the previous subsections. Because of the complexity of the problem, we do not seek to be exhaustive. Instead, we review some of most suitable methods for haemodynamics. In particular, for space discretization, we will focus on Galerkin-type methods, such as finite elements, spectral elements and discontinuous Galerkin methods.

We will use the following notation. Let Δt and h be the time and space discretization parameters. In our examples Δt is assumed to be fixed, but adaptive strategies could be considered as well: see *e.g.* Veneziani and Villa (2013). Correspondingly, the discrete times are $t^n = n\Delta t$. As usual, h is instead defined as a representative value of the mesh size, for example $h = \min\{h_K : K \in \mathcal{K}\}$, where $\mathcal{K}$ is the set of all tetrahedra in the mesh and h_K is the radius of the sphere inscribed in $K \in \mathcal{K}$. Given the functions $w(t)$ and $z(\boldsymbol{x})$, we let w^n denote the approximation of $w(t^n)$ and let $z_h(\boldsymbol{x})$ be the Galerkin approximation of $z(\boldsymbol{x})$.

4.6.1. Numerical methods for the fluid problem

We start by reviewing some numerical methods for the fluid problem (4.1) together with its initial and boundary conditions.

As for the time discretization, we usually consider implicit methods with a semi-implicit treatment of the convective term and (for a moving domain) of the fluid domain. The problem is solved at discrete time t^{n+1} in the domain Ω_f^* and with convective term $\rho_f(\boldsymbol{v}^*\cdot\nabla)\boldsymbol{v}^{n+1}$, where Ω_f^* and $\boldsymbol{v}^*$ are suitable extrapolations of Ω_f^{n+1} and $\boldsymbol{v}^{n+1}$ of the same order as the time discretization. This choice introduces a CFL-like (Courant–Friedrichs–Lewy) restriction on the time step to preserve absolute stability (Quarteroni and Valli 1994).

However, this condition is very mild in haemodynamic applications, since, for accuracy purposes, the pulsatility of the blood signal and the quick dynamics around systole can only be accommodated by choosing a small Δt. Usually, a second-order approximation is considered a good choice in haemodynamics; in this respect, the second-order backward difference formula (BDF2) and Crank–Nicolson are the most widely used methods (Quarteroni, Sacco and Saleri 2000*a*).

The first class of methods we present is based on a decomposition of the semi-discrete problem at the spatial continuous level (*differential splitting* or *projection* methods). The basic idea underlying these methods is to split the computation of velocity and pressure, with a final step aiming to recover the incompressibility constraint. In what follows we detail the Chorin–Teman method (Chorin 1968, Temam 1969), originally proposed for homogeneous Dirichlet conditions and fixed domain, which is the progenitor of these methods. We only detail the case of the backward Euler discretization.

Chorin–Temam method
For $n \geq 0$, at time t^{n+1}:

(1) Solve in Ω_f the advection-reaction-diffusion problem with homogeneous Dirichlet condition for the intermediate unknown velocity $\widetilde{\boldsymbol{v}}^{n+1}$:

$$\rho_f \frac{\widetilde{\boldsymbol{v}}^{n+1} - \boldsymbol{v}^n}{\Delta t} - \mu(\nabla \widetilde{\boldsymbol{v}}^{n+1} + (\nabla \widetilde{\boldsymbol{v}}^{n+1})^T) + \rho_f (\boldsymbol{v}^n \cdot \nabla)\widetilde{\boldsymbol{v}}^{n+1} = \mathbf{0}.$$

(2) Solve in Ω_f the pressure problem with homogeneous Neumann conditions:

$$\triangle p^{n+1} = \frac{\rho_f}{\Delta t} \nabla \cdot \widetilde{\boldsymbol{v}}^{n+1}.$$

(3) Correct the velocity:

$$\boldsymbol{v}^{n+1} = \widetilde{\boldsymbol{v}}^{n+1} - \frac{\Delta t}{\rho_f} \nabla p^{n+1}.$$

This splitting method is based on the *Ladyzhenskaya theorem* (Girault and Raviart 1986), which states that a vector function belonging to $[L^2(\Omega_f)]^3$ can always be decomposed as the sum of a solenoidal part and a gradient term. In fact, the correction step corresponds to projecting the intermediate velocity onto

$$\boldsymbol{H} = \{\boldsymbol{w} \in [L^2(\Omega_f)]^3 : \nabla \cdot \boldsymbol{w} = 0, \, \boldsymbol{w} \cdot \boldsymbol{n}|_{\partial \Omega_f} = 0\}.$$

Thus, it is possible to show that $\boldsymbol{v}^{n+1}$ and p^{n+1} are in fact solutions of the original semi-discrete problem. The Chorin–Temam method is very effective since it overcomes the saddle-point nature of the problem and solves two standard uncoupled elliptic problems. However, it suffers from inaccuracies at the boundary. In particular, the tangential velocity cannot be controlled (see the definition of $\boldsymbol{H}$) and spurious pressure values appear as a consequence of the artificial Neumann condition for the pressure problem (Rannacher 1992). This has an effect on the accuracy of the semi-discrete solution. In particular, the following error estimate holds true (Rannacher 1992):

$$\|\boldsymbol{v}(t^n,\boldsymbol{x}) - \boldsymbol{v}^n(\boldsymbol{x})\|_{[H^1]^3} + \|p(t^n,\boldsymbol{x}) - p^n(\boldsymbol{x})\|_{L^2} \lesssim \sqrt{\Delta t}.$$

The use of higher-order time approximations leads to the same accuracy.

An improvement of the above method is given by the *rotational incremental* variant of the Chorin–Temam scheme (Timmermans, Minev and van de Vosse 1996). Below we detail the case of BDF2 and second-order extrapolation of the convective term, since the first-order approximation does not lead to any improvement.

Rotational–incremental Chorin–Temam method
For $n \geq 0$, at time t^{n+1}:

(1) Solve in Ω_f the advection-reaction-diffusion problem with homogeneous Dirichlet condition in the intermediate unknown velocity $\widetilde{\boldsymbol{v}}^{n+1}$:

$$\begin{aligned}\rho_f \frac{3\widetilde{\boldsymbol{v}}^{n+1} - 4\boldsymbol{v}^n + \boldsymbol{v}^{n-1}}{2\Delta t} &- \mu(\nabla \widetilde{\boldsymbol{v}}^{n+1} + (\nabla \widetilde{\boldsymbol{v}}^{n+1})^T) \\ &+ \rho_f((2\boldsymbol{v}^n - \boldsymbol{v}^{n-1})\cdot\nabla)\widetilde{\boldsymbol{v}}^{n+1} + \nabla p^n = \mathbf{0}.\end{aligned}$$

(2) Solve the pressure problem:

$$\begin{aligned}\triangle p^{n+1} &= \triangle p^n + \left(\frac{3\rho_f}{2\Delta t} - \mu\right)\nabla\cdot\widetilde{\boldsymbol{v}}^{n+1}, && \boldsymbol{x}\in\Omega_f,\\ \frac{\partial p^{n+1}}{\partial \boldsymbol{n}} &= \mu(\nabla\times\nabla\times\widetilde{\boldsymbol{v}}^{n+1})\cdot\boldsymbol{n}, && \boldsymbol{x}\in\partial\Omega_f.\end{aligned}$$

(3) Correct the velocity:

$$\boldsymbol{v}^{n+1} = \widetilde{\boldsymbol{v}}^{n+1} - \frac{2}{3}\frac{\Delta t}{\rho_f}\nabla(p^{n+1} - p^n + \mu\nabla\cdot\widetilde{\boldsymbol{v}}^{n+1}).$$

Unlike the classical Chorin–Temam scheme, in the above method the boundary conditions for the pressure problem are consistent, and no numerical boundary layer for the pressure is observed. This is confirmed by the improved error estimate

$$\|\boldsymbol{v}(t^n,\boldsymbol{x}) - \boldsymbol{v}^n(\boldsymbol{x})\|_{[H^1]^3} + \|p(t^n,\boldsymbol{x}) - p^n(\boldsymbol{x})\|_{L^2} \lesssim \Delta t^{3/2}$$

(Guermond and Shen 2006).

The above two methods belong to the general class of *pressure-correction* methods; see also Codina and Badia (2006), Guermond, Minev and Shen (2006) and Guermond and Quartapelle (1998), for example. A different class is obtained by switching the role of velocity and pressure in the splitting, that is, the viscous term is now ignored or treated explicitly in the first step and the velocity is then corrected accordingly (*velocity-correction* schemes: Orszag, Israeli and Deville 1986, Karniadakis, Israeli and Orszag 1991). These schemes are characterized by the same non-optimal error estimates as the pressure-correction schemes due to artificial Neumann conditions for the pressure problem. Again, an improvement could be obtained by considering a rotational-incremental variant (Guermond and Shen 2003).

In haemodynamics it is often the case that Neumann boundary conditions are prescribed at some artificial section. The extension of the differential splitting methods to this case is addressed in Guermond, Minev and Shen (2005): on the Neumann boundary we have an artificial Dirichlet condition for the pressure, which again diminishes the optimal rate of convergence with respect to Δt.

Given the next methods we are going to review, it is convenient to introduce the algebraic problem arising from the application of a Galerkin-like method to the semi-discrete-in-time problem. First of all we notice that the solvability of the discretized-in-space problem is guaranteed by a suitable compatible choice of the approximation spaces for the velocity and the pressure in order to satisfy the discrete *inf-sup stability condition* (Quarteroni and Valli 1994). As is well known, an example of finite elements for a tetrahedral mesh is provided by piecewise polynomials of order 2 for the velocity approximation and of order 1 for the pressure approximation. This choice guarantees the existence and uniqueness of the solution to the linearized fully discrete problem and is often used to provide a stable solution in haemodynamics. In this case, we have the optimal error estimate

$$\|\boldsymbol{v}^n(\boldsymbol{x}) - \boldsymbol{v}_h^n(\boldsymbol{x})\|_{[H^1]^3} + \|p^n(\boldsymbol{x}) - p_h^n(\boldsymbol{x})\|_{L^2} \lesssim h^2,$$

provided that $\boldsymbol{v}^n$ and p^n are sufficiently regular. (For other stable choices see Quarteroni and Valli 1994 and Boffi, Brezzi and Fortin 2013.) Alternatively, suitable stabilization terms could be added to the problem, circumventing the inf-sup condition and allowing the use of polynomials of equal order. In this case, additional terms are added to the mass conservation equation and, if needed, to the momentum conservation equation. Usually, these techniques also stabilize convected-dominated problems arising when the Reynolds number is high, for example in the aorta or in stenotic carotid arteries. One technique is *streamline upwind/Petrov–Galerkin* (SUPG) (Muller *et al.* 2005). A generalization of SUPG is the *variational multiscale* (VMS) method (Hughes 1995, Hughes, Mazzei and Jansen 2000), which is based on decomposition of the unknown into two terms, one accounting for the large scales and the other for the small scales. The same decomposition is used for the test functions, so that a system of two coupled problems is obtained. The VMS method is also useful since it allows modelling of transition to turbulence effects, which may occur in some pathological conditions such as stenoses (Akkerman *et al.* 2008). For an application to haemodynamics, see Forti and Dede' (2015).

Here we introduce the algebraic problem related to the fully discretized linearized problem. For the sake of exposition, we limit ourselves to the cases without stabilization terms. For the more general case, we refer the interested reader to Elman, Silvester and Wathen (2005) and Benzi, Golub and Liesen (2005), for example. At each time step we have

$$\begin{bmatrix} A & B^T \\ B & 0 \end{bmatrix} \begin{bmatrix} \boldsymbol{V} \\ \boldsymbol{P} \end{bmatrix} = \begin{bmatrix} \boldsymbol{F}_f \\ \boldsymbol{0} \end{bmatrix},$$

where $\boldsymbol{V}$ and $\boldsymbol{P}$ are the vectors collecting the velocity and pressure unknowns, $A = \rho_f(\alpha/\Delta t)M_f + \rho_f N(\boldsymbol{V}^*) + \mu K$ (M_f is the mass matrix, N is the matrix related to the linearized convective term, K is the stiffness matrix), $\boldsymbol{F}_f$ accounts for non-homogeneous Dirichlet and Neumann conditions and the terms coming from time discretization, α depends on the time discretization scheme, and where we have omitted the current temporal index ${}^{n+1}$.

The above non-symmetric linear system can be solved by means of a Krylov method, for example the GMRES method. Suitable preconditioners are required. A classical choice is given by block preconditioners, which again split the solution of the velocity and of the pressure:

$$P = \begin{bmatrix} P_A & B^T \\ 0 & -P_\Sigma \end{bmatrix}.$$

If $P_A = A$ and $P_\Sigma = \Sigma = BA^{-1}B^T$ (the *Schur complement*), then the solution is achieved in three GMRES iterations (Elman *et al.* 2005). In fact, this choice is equivalent to formally solving the momentum equation for the velocity and substituting its expression in the mass equation. However, in practice this preconditioner is not efficient, since the linear system involving the Schur complement is too onerous, Σ being a full matrix whose explicit construction requires knowledge of A^{-1}. Efficient preconditioners can be obtained by approximating Σ (and, if needed, A). For low Reynolds numbers (less than 10, say), an effective choice is given by $P_\Sigma = \mu^{-1}M_P$, where M_P is the pressure mass matrix (or even its diagonal: Elman and Silvester 1996). Therefore this is a good choice in haemodynamics for small vessels. For increasing Reynolds numbers, the convergence properties of this preconditioner deteriorates since it does not account for the convective term. A better choice for medium and large vessels is given by $P_\Sigma = A_P F_P^{-1} M_P$, where A_P is the pressure stiffness matrix, and $F_P = \mu A_P + \rho_f N_P(\boldsymbol{V}^*)$, where N_P is the matrix related to the convective term defined on the pressure space (*pressure convection-diffusion* preconditioner: Elman *et al.* 2005, Benzi *et al.* 2005). For the solution of the velocity problem, suitable preconditioners for the advection-reaction-diffusion problem could be introduced. Alternatively, fast solutions such as V-cycle multigrid can be considered (Turek 1999).

Another class of preconditioners is obtained by an inexact block LU factorization of the fluid matrix. The starting point is the exact factorization

$$\begin{bmatrix} A & B^T \\ B & 0 \end{bmatrix} = \begin{bmatrix} A & 0 \\ B & -\Sigma \end{bmatrix} \begin{bmatrix} I & A^{-1}B^T \\ 0 & I \end{bmatrix}.$$

Again, different preconditioners are obtained by suitable approximations $\widehat{A}_1$ and $\widehat{A}_2$ of A and $\widehat{\Sigma}$ of Σ, leading to

$$P = \begin{bmatrix} \widehat{A}_1 & 0 \\ B & -\widehat{\Sigma} \end{bmatrix} \begin{bmatrix} I & \widehat{A}_2^{-1}B^T \\ 0 & I \end{bmatrix}.$$

A convenient choice is $\widehat{A}_1 = \widehat{A}_2 = D_A$ and $\widehat{\Sigma} = BD_A^{-1}B^T$, where D_A is the diagonal of A (*SIMPLE* preconditioner: Patankar and Spalding 1972, Li and Vuik 2004). Note that in this case $\widehat{\Sigma}$ is sparse and could be explicitly assembled. This is an effective choice when the fluid matrix is diagonally dominant, that is, when small values of Δt are used. Another choice is the *Yosida* preconditioner, where

$$\widehat{\Sigma} = \frac{\Delta t}{\rho_f \alpha} B M_f^{-1} B^T, \; \widehat{A}_1 = A \quad \text{and} \quad \widehat{A}_2 = \frac{\rho_f \alpha}{\Delta t} M_f$$

(Veneziani 2003). Again, the efficiency deteriorates for increasing Δt. The Yosida preconditioner was originally introduced as a solver in Veneziani (1998*b*); see also Quarteroni, Saleri and Veneziani (1999, 2000*b*). This led to a splitting of the velocity and of the pressure computation, which could be seen as the algebraic counterpart of the Chorin–Temam method (*algebraic pressure-correction* methods). In particular, we have the following steps:

$$\begin{aligned} & A\widetilde{\boldsymbol{V}} = \boldsymbol{F}_f && \text{computation of the velocity,} \\ & \Delta t B M_f^{-1} B^T \boldsymbol{P} = B\widetilde{\boldsymbol{V}} && \text{computation of the pressure,} \\ & \boldsymbol{V} = \widetilde{\boldsymbol{V}} - \frac{\Delta t}{\rho_f \alpha} M_f^{-1} B^T \boldsymbol{P} && \text{correction of the velocity.} \end{aligned}$$

Again, an incremental version of the algebraic pressure-correction methods could also be considered (Quarteroni, Saleri and Veneziani 2000*b*). An extension to spectral methods is provided in Gervasio, Saleri and Veneziani (2006).

For a recent comparison of the performance of different preconditioners (all described above) used for haemodynamic applications, see Deparis, Grandperrin and Quarteroni (2014).

In healthy conditions, blood flow is mainly laminar. Transitional flow may develop in some pathological instances, or with the assistance of devices. In these circumstances, suitable mesh refinement is often employed, possibly accompanied by the use of turbulence models. We mention the case of stenotic carotid arteries, for which Stroud, Berger and Saloner (2002) and Grinberg, Yakhot and Karniadakis (2009) use Reynolds-averaged Navier–Stokes (RANS) models, Lee *et al.* (2008), Fischer *et al.* (2007) and Cheung *et al.* (2010) use *direct numerical simulation* (DNS), and Rayz, Berger and Saloner (2007) and Lancellotti *et al.* (2015) use *large eddy simulation* (LES). Bazilevs *et al.* (2009) use the VMS formulation to describe transitional effects in the ascending aorta under the influence of the left ventricular assist device (LVAD).

In Figure 4.5 we show some examples of numerical results obtained in four real geometries reconstructed from radiological images (see the caption for details). These results highlight the complex pattern of blood flow induced by the geometry and by the heart pulsatility. To highlight the transitional effects in stenotic carotid arteries, Figure 4.5(c) plots the Q criterion, defined by

$$Q = -\frac{1}{2}\sum_{i,j} S_{ij}^2 - \Omega_{ij}^2,$$

where $\boldsymbol{S} = \nabla \boldsymbol{u} + (\nabla \boldsymbol{u})^T$ and $\boldsymbol{\Omega} = \nabla \boldsymbol{u} - (\nabla \boldsymbol{u})^T$ (Lee *et al.* 2008). Positive values of Q indicate locations where rotations dominate strain and shear.

4.6.2. Numerical methods for the vessel wall problem

In this section we review some of the most commonly used numerical approaches for the solution of problem (4.7), endowed with its initial and boundary conditions.

For the time discretization, a popular family of schemes is that of *Newmark*, which is characterized by two parameters θ and ζ (Newmark 1959). The special combination $\theta = 1/2$ and $\zeta = 1/4$ yields the following semi-discrete form of (4.16) (for simplicity we set $P_{ext} = 0$ and $\boldsymbol{h}_s = \boldsymbol{0}$):

$$\begin{aligned}\rho_s \int_{\Omega_s} \frac{4\widehat{\boldsymbol{d}}^{n+1} - 4\widehat{\boldsymbol{d}}^{n}}{\Delta t^2} \cdot \widehat{\boldsymbol{e}}\,\mathrm{d}\omega + \int_{\Omega_s} \widehat{\boldsymbol{T}}_s(\widehat{\boldsymbol{d}}^{n+1}) : \nabla \widehat{\boldsymbol{e}}\,\mathrm{d}\omega + \int_{\Gamma_{ext}} \alpha_{ST} \widehat{\boldsymbol{d}}^{n+1} \cdot \widehat{\boldsymbol{e}}\,\mathrm{d}\sigma \\ = \rho_s \int_{\Omega_s} \left(\frac{4\widehat{\boldsymbol{w}}^n}{\Delta t} + \widehat{\boldsymbol{a}}^n\right) \cdot \widehat{\boldsymbol{e}}\,\mathrm{d}\omega,\end{aligned} \tag{4.43}$$

$$\widehat{\boldsymbol{w}}^{n+1} = \frac{2}{\Delta t}(\widehat{\boldsymbol{d}}^{n+1} - \widehat{\boldsymbol{d}}^{n}) - \widehat{\boldsymbol{w}}^n, \quad \widehat{\boldsymbol{a}}^{n+1} = \frac{4}{\Delta t^2}(\widehat{\boldsymbol{d}}^{n+1} - \widehat{\boldsymbol{d}}^{n}) - \frac{4}{\Delta t}\widehat{\boldsymbol{w}}^n - \widehat{\boldsymbol{a}}^n,$$

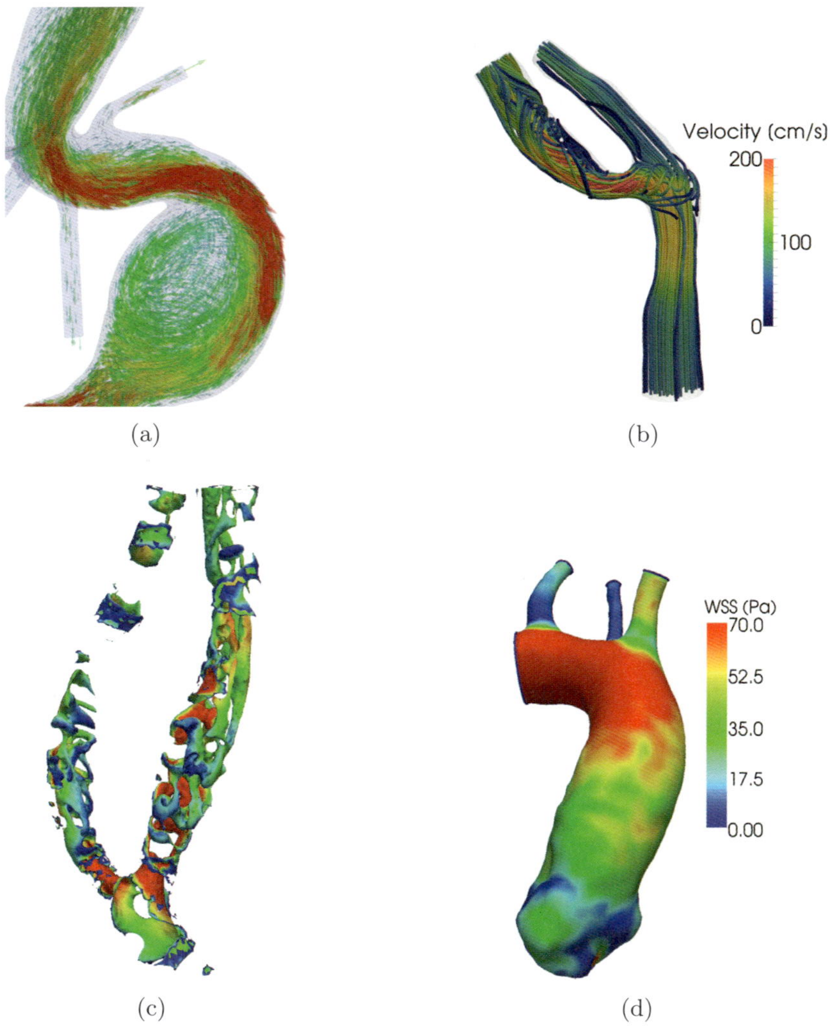

Figure 4.5. (a) Velocity vectors in the aneurysm of an abdominal aorta (CT images from the Vascular Surgery and Radiology Divisions at Fondazione IRCSS Cà Granda, Ospedale Maggiore Policlinico, Milan, Italy). (b) Velocity streamlines in a stenotic carotid artery (MRI images from the Vascular Surgery and Radiology Divisions at Ospedale Maggiore Policlinico, Milan). (c) Coherent vortical structures by Q criterion in a stenotic carotid artery (we show only the regions with $Q > 50\,000$ shaded by the velocity magnitude; CT images from the Vascular Surgery and Radiology Divisions at Ospedale Maggiore Policlinico, Milan). (d) Wall shear stress in an ascending aorta (MRI images from the Cardio Surgery and Radiology Divisions at Ospedale Borgo Trento, Verona, Italy). Numerical results were obtained using the finite element library LifeV, P2/P1 finite elements, the backward Euler scheme for the time discretization with a semi-implicit treatment of the non-linear term, and the Yosida preconditioner. For the stenotic carotid arteries an LES model has been used.

where $\widehat{\boldsymbol{w}}^{n+1}$ and $\widehat{\boldsymbol{a}}^{n+1}$ denote approximations of vessel wall velocity and acceleration, respectively. This method is unconditionally absolutely stable and second-order accurate with respect to Δt. An extension of Newmark schemes is provided by the *generalized-alpha* method (Chung and Hulbert 1993); see *e.g.* Isaksen *et al.* (2008) for an application to haemodynamics.

Space discretization is typically based on finite elements. Whatever the implicit temporal scheme chosen, a system of non-linear algebraic equations is obtained after space and time discretization, that is,

$$\frac{\rho_s \beta}{\Delta t^2} M_s \boldsymbol{D} + \boldsymbol{\Gamma}(\boldsymbol{D}) + \alpha_{ST} M_s^{ext} \boldsymbol{D} = \boldsymbol{G}_s,$$

where β depends on the time discretization (*e.g.* $\beta = 4$ for the Newmark method (4.43)), $\boldsymbol{D}$ is the vector collecting the vessel wall displacement unknowns, M_s is the mass matrix, M_s^{ext} is the boundary mass matrix related to Σ_{ext}, $\boldsymbol{\Gamma}$ is the non-linear operator defined by

$$\Gamma_i = \int_{\Omega_s} \widehat{\boldsymbol{T}}_s(\widehat{\boldsymbol{d}}) : \nabla \widehat{\boldsymbol{e}}_i \, \mathrm{d}\omega,$$

where $\widehat{\boldsymbol{e}}_i$ is the ith basis function, and $\boldsymbol{G}_s$ is the vector related to the right-hand side of the time-discretized equation. Notice that we have omitted the temporal index, which is understood. The above system is linearized by means of the Newton method, obtaining at each time step a sequence of linear systems of the form

$$\left(\frac{\rho_s \beta}{\Delta t^2} M_s + T(\boldsymbol{D}_{(k-1)}) + \alpha_{ST} M_s^{ext}\right) \delta \boldsymbol{D}_{(k)}$$
$$= \boldsymbol{G}_s - \frac{\rho_s \beta}{\Delta t^2} M_s \boldsymbol{D}_{(k-1)} - \boldsymbol{\Gamma}(\boldsymbol{D}_{(k-1)}) - \alpha_{ST} M_s^{ext} \boldsymbol{D}_{(k-1)},$$

where $k \geq 1$ is the Newton iteration index, to be solved until convergence occurs. Here $\delta \boldsymbol{D}_{(k)} = (\boldsymbol{D}_{(k)} - \boldsymbol{D}_{(k-1)})$, T is the matrix related to the linearization of the first Piola–Kirchhoff tensor, that is,

$$(T)_{ij} = \int_{\Omega_s} (D_F \widehat{\boldsymbol{T}}_s(\widehat{\boldsymbol{d}}_{(k-1)}) : \nabla \widehat{\boldsymbol{e}}_j) : \nabla \widehat{\boldsymbol{e}}_i,$$

where D_F is the Gâteaux derivative with respect to $\boldsymbol{F}$.

For the solution of the above linear system, domain decomposition (DD) methods are often used as efficient preconditioners for iterative Krylov methods. Since the matrix T is symmetric, the conjugate gradient method is usually considered for iterations. Among DD preconditioners, *finite element tearing and interconnect* (FETI) methods (Farhat and Roux 1991) are very often used in structural mechanics. In particular, *all floating* FETI methods have been considered for vessel wall problems, for example in Augustin,

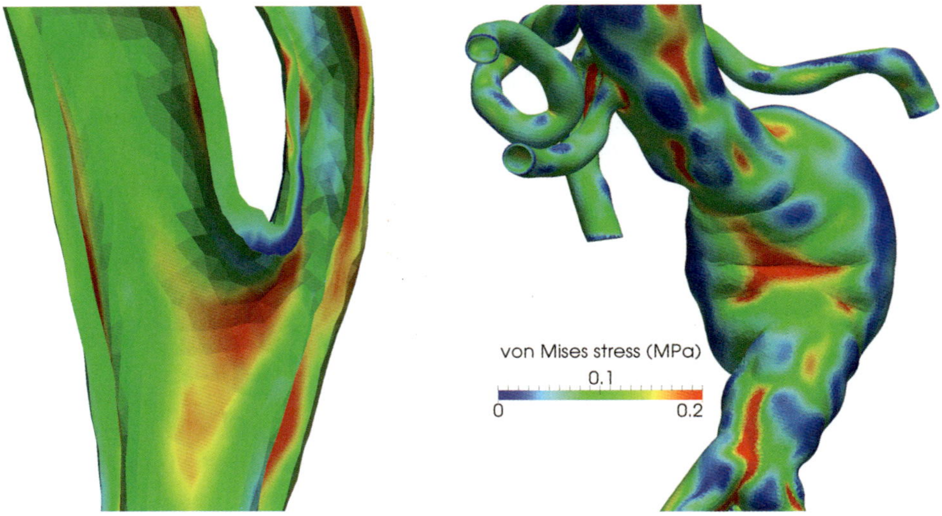

(a) (b)

Figure 4.6. (a) Von Mises internal stresses in a carotid artery (MRI images from the Vascular Surgery and Radiology Divisions at Ospedale Maggiore Policlinico, Milan, Italy). (b) Von Mises stresses in an abdominal aortic aneurysm (mesh from www.vascularmodel.com/sandbox/doku.php?id=start). Numerical results were obtained using LifeV (carotid artery) and the finite element library redbKIT v2.1 (github.com/redbKIT/redbKIT/releases) (AAA), P2 finite elements, a Newmark unconditionally stable scheme for the time discretization and an exponential vessel wall law.

Holzapfel and Steinbach (2014). As in classical FETI methods, Lagrange multipliers are introduced to glue the solution together at the subdomain interfaces. In addition, Lagrange multipliers are also used to prescribe Dirichlet boundary conditions. This simplifies the implementation of the FETI method since all the subdomains are treated in the same way. A variant successfully used for arterial vessel walls is the so-called *dual–primal FETI* method: see *e.g.* Balzani *et al.* (2010). Finally, we mention yet another class of DD methods considered for this problem, a two-level overlapping Schwarz method with an energy minimization coarse space: see Dohrmann and Widlund (2009).

In Figure 4.6 we illustrate some examples of numerical results obtained in real geometries reconstructed from radiological images (see the caption for details). These results highlight the anisotropic internal stresses characterizing vascular vessel walls.

4.6.3. Numerical methods for the fluid–structure interaction problem

The numerical solution of the coupled FSI problem (4.19) requires us to manage three sources of non-linearities, namely,

(i) the fluid domain is unknown (*geometric non-linearity*),

(ii) the fluid subproblem is non-linear (*fluid constitutive non-linearity*),

(iii) the vessel displacement subproblem is non-linear (*structure constitutive non-linearity*),

together with two different kinds of coupling,

(iv) the displacement of the fluid domain at the FS interface needs to match the displacement of the vessel wall (*geometric adherence*: see (4.19f)),

(v) the fluid and vessel displacement subproblems are coupled by means of the kinematic and dynamic conditions (4.19c, 4.19d) (*physical coupling*).

Arbitrary-Lagrangian formulation. As for points (i) and (iv), a classical numerical strategy relies on extending the FS interface displacement $\boldsymbol{d}_f = \boldsymbol{d}|_{\Sigma^t}$ to the whole fluid domain, thus associating a displacement with its internal points as well. This is obtained by solving an extra problem for the fluid mesh displacement, usually a harmonic extension of the FS interface datum $\boldsymbol{d}_f$, with homogeneous Dirichlet conditions at $\partial\Omega_f^t \setminus \Sigma^t$. The fluid domain displacement is then used to move the points of the fluid mesh accordingly, obtaining the new computational fluid domain. With this aim, the Navier–Stokes equations are reformulated on a frame of reference that moves with the fluid mesh. This is neither a Lagrangian description (where the frame moves with the fluid particles) nor an Eulerian description (where a fixed frame would be used). For this reason, it is called the *arbitrary Lagrangian–Eulerian* (ALE) approach (Hirt, Amsden and Cook 1974, Donea 1982). To write the Navier–Stokes equations in an ALE configuration, we use the Reynolds transport formula to express the ALE material time derivative $\delta/\delta t$ of a function $\boldsymbol{z}$ with respect to the Eulerian derivative, that is,

$$\frac{\delta \boldsymbol{z}}{\delta t} = \frac{\partial \boldsymbol{z}}{\partial t} + (\boldsymbol{v}_f \cdot \nabla)\boldsymbol{z},$$

where $\boldsymbol{v}_f = \dot{\boldsymbol{d}}_f$ is the velocity of the points of the fluid mesh. Thus, the FSI problem together with its initial and boundary conditions becomes

$$\rho_f \left(\frac{\delta \boldsymbol{v}}{\delta t} + ((\boldsymbol{v} - \boldsymbol{v}_f) \cdot \nabla)\boldsymbol{v} \right) - \nabla \cdot \boldsymbol{T}_f(\boldsymbol{v}, p) = \boldsymbol{0} \quad \text{in } \Omega_f^t, \tag{4.44a}$$

$$\nabla \cdot \boldsymbol{v} = 0 \quad \text{in } \Omega_f^t, \tag{4.44b}$$

$$\boldsymbol{v} = \frac{\partial \boldsymbol{d}}{\partial t} \quad \text{on } \Sigma^t, \tag{4.44c}$$

$$\boldsymbol{T}_s(\boldsymbol{d})\boldsymbol{n} = \boldsymbol{T}_f(\boldsymbol{v}, p)\boldsymbol{n} \quad \text{on } \Sigma^t, \tag{4.44d}$$

$$\rho_s \frac{\partial^2 \widehat{\boldsymbol{d}}}{\partial t^2} - \nabla \cdot \widehat{\boldsymbol{T}}_s(\widehat{\boldsymbol{d}}) = \boldsymbol{0} \qquad \text{in } \Omega_s, \tag{4.44e}$$

$$\widehat{\boldsymbol{d}}_f = \widehat{\boldsymbol{d}} \qquad \text{on } \Sigma, \tag{4.44f}$$

$$- \triangle \widehat{\boldsymbol{d}}_f = \boldsymbol{0} \qquad \text{in } \Omega_f. \tag{4.44g}$$

This choice is particularly suitable when finite elements are considered for the space discretization. Indeed, the terms involving spatial derivatives are as usual expressed with respect to the current configuration, whereas the (material) time derivative term is written in the reference configuration. In particular, given the nodal basis functions $\boldsymbol{\phi}_j$, we have

$$\frac{\delta(\boldsymbol{v}_h(t,\boldsymbol{x}))}{\delta t} = \frac{\delta}{\delta t}\Bigl(\sum_j v_h^j(t)\boldsymbol{\phi}_j(\boldsymbol{x}(t))\Bigr) = \sum_j \frac{\mathrm{d}v_h^j(t)}{\mathrm{d}t}\boldsymbol{\phi}_j(\boldsymbol{x}(t)),$$

since the time variations of the basis functions with respect to the reference domain vanish. This makes the computation of the fluid velocity on the nodes of the fluid mesh easy: see *e.g.* Nobile (2001).

For time discretization of (4.44), a common choice is to discretize the fluid and the vessel wall problems with two schemes of equal order (of order p, say), for example BDF2 or Crank–Nicolson for the fluid and the Newmark scheme (4.43) for the vessel problem.

Treatment of geometric coupling, geometric adherence and fluid non-linearity. As for issue (ii) above (constitutive fluid non-linearity), following the approach commonly used for a stand-alone fluid problem, the convective field is usually treated explicitly, by using a suitable extrapolation $\boldsymbol{v}^* - \boldsymbol{v}_f^*$ of order p from previous time steps. This choice introduces a CFL-like condition for Δt to preserve absolute stability, which however is always satisfied for the values of Δt usually considered in haemodynamics. Another common choice in haemodynamics is the explicit treatment of geometric coupling and adherence (issues (i) and (iv) above). In particular, the fluid problem coupled with the structure problem are solved in a domain Ω_f^* obtained by a suitable extrapolation from previous time steps of order p, and the fluid geometry problem is then solved in sequence (Swim and Seshaiyer 2006, Fernández, Gerbeau and Grandmont 2007, Badia, Quaini and Quarteroni 2008*c*, Nobile and Vergara 2008). Again, a limitation on Δt is required to ensure absolute stability. In particular, Fernández *et al.* (2007) prove for a model problem that stability is guaranteed under a CFL-like condition. Numerical evidence in real scenarios indicates that this is also a mild condition in the haemodynamics regime: see *e.g.* Moireau *et al.* (2012) and Nobile, Pozzoli and Vergara (2014). Based on the considerations above, the following temporal discretization of the FSI problem (4.44) can be considered.

Explicit scheme for geometric coupling and adherence
For $n \geq 1$, at time step t^n:

(1) Solve the FSI problem:

$$\frac{\rho_f \alpha}{\Delta t}\boldsymbol{v}^n + \rho_f((\boldsymbol{v}^* - \boldsymbol{v}_f^*)\cdot\nabla)\boldsymbol{v}^n - \nabla\cdot\boldsymbol{T}_f(\boldsymbol{v}^n, p^n) = \boldsymbol{g}_f^n \quad \text{in } \Omega_f^*, \quad (4.45a)$$

$$\nabla\cdot\boldsymbol{v}^n = 0 \quad \text{in } \Omega_f^*, \quad (4.45b)$$

$$\boldsymbol{v}^n = \frac{\alpha}{\Delta t}\boldsymbol{d}^n + \boldsymbol{g}_{fs}^n \quad \text{on } \Sigma^*, \quad (4.45c)$$

$$\boldsymbol{T}_s(\boldsymbol{d}^n)\boldsymbol{n}^* = \boldsymbol{T}_f(\boldsymbol{v}^n, p^n)\boldsymbol{n}^* \quad \text{on } \Sigma^*, \quad (4.45d)$$

$$\frac{\rho_s \beta}{\Delta t^2}\widehat{\boldsymbol{d}}^n - \nabla\cdot\widehat{\boldsymbol{T}}_s(\widehat{\boldsymbol{d}}^n) = \widehat{\boldsymbol{g}}_s^n \quad \text{in } \Omega_s, \quad (4.45e)$$

where α still depends on the time discretization scheme, and

$$\boldsymbol{g}_f^n = \boldsymbol{g}_f^n(\boldsymbol{v}^{n-1}, \boldsymbol{v}^{n-2}, \ldots),$$
$$\boldsymbol{g}_s^n = \boldsymbol{g}_s^n(\boldsymbol{d}^{n-1}, \boldsymbol{d}^{n-2}, \ldots),$$
$$\boldsymbol{g}_{fs}^n = \boldsymbol{g}_{fs}^n(\boldsymbol{d}^{n-1}, \boldsymbol{d}^{n-2}, \ldots)$$

account for the terms at previous time steps coming from time discretizations of order p of the corresponding equations (4.44a), (4.44e) and (4.44c), respectively.

(2) Then, solve the fluid geometry problem,

$$-\triangle\widehat{\boldsymbol{d}}_f^n = \boldsymbol{0} \quad \text{in } \Omega_f, \quad (4.46a)$$

$$\widehat{\boldsymbol{d}}_f^n = \widehat{\boldsymbol{d}}^n \quad \text{on } \Sigma, \quad (4.46b)$$

and build Ω_f^* accordingly.

In the above substeps, the FSI problem (4.45) is still coupled by means of the physical coupling given by the interface conditions (4.45c, 4.45d): see issue (v) above. For the solution of this problem, both *partitioned* and *monolithic* procedures have been successfully considered so far in haemodynamics. In partitioned schemes, the fluid and vessel wall subproblems are solved separately, one or more times per time step. Each of the two problems is equipped with a suitable boundary condition at the FS interface Σ^* derived by splitting the physical interface conditions (4.45c, 4.45d).

Partitioned algorithms of explicit type. From the computational point of view, an attractive class of partitioned schemes is that of the *loosely coupled algorithms* where the two subproblems are solved only once per time step. The classic loosely coupled algorithm, widely used in aerodynamics (Piperno and Farhat 2001), is the *explicit Dirichlet–Neumann* scheme, where condition (4.45c) is prescribed explicitly as a Dirichlet condition in the fluid subproblem, whereas condition (4.45d) is prescribed as a Neumann condition on the vessel wall subproblem, leading to the following.

Explicit Dirichlet–Neumann scheme
For $n \geq 1$, at time step t^n:

(1) Solve the fluid Oseen problem with a Dirichlet condition at the FS interface:

$$\frac{\rho_f \alpha}{\Delta t}\boldsymbol{v}^n + \rho_f((\boldsymbol{v}^* - \boldsymbol{v}_f^*)\cdot\nabla)\boldsymbol{v}^n - \nabla\cdot\boldsymbol{T}_f(\boldsymbol{v}^n, p^n) = \boldsymbol{g}_f^n \quad \text{in } \Omega_f^*, \quad (4.47a)$$

$$\nabla\cdot\boldsymbol{v}^n = 0 \quad \text{in } \Omega_f^*, \quad (4.47b)$$

$$\boldsymbol{v}^n = \frac{\alpha}{\Delta t}\boldsymbol{d}^{n-1} + \boldsymbol{g}_{fs}^{n-1} \quad \text{on } \Sigma^*. \quad (4.47c)$$

(2) Then, solve the non-linear vessel wall problem with a Neumann condition at the FS interface:

$$\frac{\rho_s \beta}{\Delta t^2}\widehat{\boldsymbol{d}}^n - \nabla\cdot\widehat{\boldsymbol{T}}_s(\widehat{\boldsymbol{d}}^n) = \widehat{\boldsymbol{g}}_s^n \quad \text{in } \Omega_s, \quad (4.48a)$$

$$\widehat{\boldsymbol{T}}_s(\widehat{\boldsymbol{d}}^n)\widehat{\boldsymbol{n}} = \widehat{\boldsymbol{T}}_f(\widehat{\boldsymbol{v}}^n, \widehat{p}^n)\widehat{\boldsymbol{n}} \quad \text{on } \Sigma. \quad (4.48b)$$

Note that the time-discretized kinematic condition (4.47c) differs from (4.45c) since now we are considering an explicit Dirichlet condition for the fluid subproblem, so the right-hand side is computed at the previous time step.

A 'parallel' version of this scheme is obtained by replacing $\widehat{\boldsymbol{T}}_f(\widehat{\boldsymbol{v}}^n, \widehat{p}^n)\widehat{\boldsymbol{n}}$ with $\widehat{\boldsymbol{T}}_f(\widehat{\boldsymbol{v}}^{n-1}, \widehat{p}^{n-1})\widehat{\boldsymbol{n}}$ in (4.48b). Notice that in the monolithic FSI problem (4.45), the dynamic continuity condition (4.45d) is written in the current configuration Σ^*, whereas for the structure subproblem alone (4.48) is written in the reference configuration Σ. Accordingly, in what follows the structure interface quantities will be written in the current configuration in monolithic FSI problems and in the reference configuration when the structure problem is uncoupled in view of a partitioned scheme.

Unfortunately, the explicit Dirichlet–Neumann scheme can be unconditionally absolutely unstable. In particular, Causin, Gerbeau and Nobile (2005) prove that this happens if the fluid and structure densities are comparable, which is precisely the case for haemodynamics. This is due to the so-called high *added mass effect*. See also Forster, Wall and Ramm (2007) and Nobile and Vergara (2012).

Stable loosely coupled algorithms have been introduced recently. To this end, in the FSI problem (4.45) the interface conditions (4.45c, 4.45d) are replaced by two linear independent combinations

$$\sigma_f \boldsymbol{v}^n + \boldsymbol{T}_f(\boldsymbol{v}^n, p^n)\boldsymbol{n}^* = \sigma_f \left(\frac{\alpha}{\Delta t} \boldsymbol{d}^n + \boldsymbol{g}_{fs}^n \right) + \boldsymbol{T}_s(\boldsymbol{d}^n)\boldsymbol{n}^* \quad \text{on } \Sigma^*, \tag{4.49a}$$

$$\sigma_s \left(\frac{\alpha}{\Delta t} \boldsymbol{d}^n + \boldsymbol{g}_{fs}^n \right) + \boldsymbol{T}_s(\boldsymbol{d}^n)\boldsymbol{n}^* = \sigma_s \boldsymbol{v}^n + \boldsymbol{T}_f(\boldsymbol{v}^n, p^n)\boldsymbol{n}^* \quad \text{on } \Sigma^*, \tag{4.49b}$$

where $\sigma_f \neq \sigma_s$ are, in general, two functions of space and time. This naturally leads to the following.

Explicit Robin–Robin scheme
For $n \geq 1$, at time step t^n:

(1) Solve the Oseen problem with a Robin condition at the FS interface:

$$\frac{\rho_f \alpha}{\Delta t} \boldsymbol{v}^n + \rho_f((\boldsymbol{v}^* - \boldsymbol{v}_f^*) \cdot \nabla)\boldsymbol{v}^n - \nabla \cdot \boldsymbol{T}_f(\boldsymbol{v}^n, p^n) = \boldsymbol{g}_f^n \quad \text{in } \Omega_f^*, \tag{4.50a}$$

$$\nabla \cdot \boldsymbol{v}^n = 0 \quad \text{in } \Omega_f^*, \tag{4.50b}$$

$$\sigma_f \boldsymbol{v}^n + \boldsymbol{T}_f(\boldsymbol{v}^n, p^n)\boldsymbol{n}^* = \sigma_f \left(\frac{\alpha}{\Delta t} \boldsymbol{d}^{n-1} + \boldsymbol{g}_{fs}^{n-1} \right) + \boldsymbol{T}_s(\boldsymbol{d}^{n-1})\boldsymbol{n}^* \quad \text{on } \Sigma^*. \tag{4.50c}$$

(2) Then, solve the non-linear vessel wall problem with a Robin condition at the FS interface:

$$\frac{\rho_s \beta}{\Delta t^2} \widehat{\boldsymbol{d}}^n - \nabla \cdot \widehat{\boldsymbol{T}}_s(\widehat{\boldsymbol{d}}^n) = \widehat{\boldsymbol{g}}_s^n \quad \text{in } \Omega_s, \tag{4.51a}$$

$$\frac{\sigma_s \alpha}{\Delta t} \widehat{\boldsymbol{d}}^n + \widehat{\boldsymbol{T}}_s(\widehat{\boldsymbol{d}}^n)\widehat{\boldsymbol{n}} = \sigma_s \widehat{\boldsymbol{v}}^n + \widehat{\boldsymbol{T}}_f(\widehat{\boldsymbol{v}}^n, \widehat{p}^n)\widehat{\boldsymbol{n}} - \sigma_s \widehat{\boldsymbol{g}}_{fs}^n \quad \text{on } \Sigma. \tag{4.51b}$$

Burman and Fernández (2009) propose a discontinuous Galerkin (DG)-like mortaring of the interface coupling conditions (4.45c, 4.45d). The corresponding block Gauss–Seidel explicit algorithm could be reinterpreted as an explicit Dirichlet–Robin scheme, where in (4.50c, 4.51b) we have $\sigma_f = +\infty$, $\sigma_s = -\gamma\mu/h$, where γ is the DG penalty parameter. Burman and Fernández show that the DG interface penalty and the viscous dissipation are not able to control the pressure fluctuations at the FS interface appearing in the discrete energy estimate. For this reason, they propose adding to the fluid problem a consistent stabilization term penalizing the pressure fluctuations, which is proved to be absolutely stable under a CFL-like condition. Banks, Henshaw and Schwendeman (2014) have introduced a stable explicit Robin–Robin scheme, setting the parameters in the Robin interface conditions (4.50c, 4.51b) after analysing the incoming and outgoing characteristic variables of the vessel wall problem. In particular, the above parameters are defined in terms of the outgoing characteristic variable for the fluid subproblem and in terms of the incoming characteristic variable for the vessel wall subproblem. For linear elasticity, the following values are obtained: $\sigma_{f,\text{norm}} = \sqrt{\rho_s(\mu_1 + 2\mu_2)}$ and $\sigma_{f,\text{tang}} = \sqrt{\rho_s \mu_2}$ in the normal and tangential direction, respectively (μ_1 and μ_2 are the Lamé constants), and $\sigma_s = -\sigma_f$. This choice allows the travelling information contained in the characteristic variables to provide a tighter coupling of the fluid and structure problems than that enforced by (4.45c, 4.45d).

Partitioned algorithms of semi-implicit type. Fernández *et al.* (2007) prove that to achieve stability without stabilization terms for the Dirichlet–Neumann scheme, an implicit coupling between the fluid pressure and the vessel wall displacement is required. In particular, a projection scheme is used where the fluid ALE-advection-diffusion step is solved explicitly, whereas the fluid pressure and vessel wall subproblems are coupled and solved in an iterative framework until convergence occurs (see also Astorino and Grandmont 2009 for a convergence analysis).

Semi-implicit pressure-vessel wall coupled scheme
For $n \geq 1$, at time step t^n:

(1) Solve the ALE-advection-diffusion problem with a Dirichlet condition at the FS interface:

$$\frac{\rho_f \alpha}{\Delta t}\widetilde{\boldsymbol{v}}^n + \rho_f((\boldsymbol{v}^* - \boldsymbol{v}_f^*) \cdot \nabla)\widetilde{\boldsymbol{v}}^n - \nabla \cdot \mu(\nabla \widetilde{\boldsymbol{v}}^n + (\nabla \widetilde{\boldsymbol{v}}^n)^T) = \boldsymbol{g}_f^n \quad \text{in } \Omega_f^*, \tag{4.52a}$$

$$\widetilde{\boldsymbol{v}}^n = \frac{\alpha}{\Delta t}\boldsymbol{d}^{n-1} + \boldsymbol{g}_{fs}^{n-1} \quad \text{on } \Sigma^*. \tag{4.52b}$$

(2) Then, solve the coupled pressure-vessel wall problem. To this end, introduce the following iterations on index $k \geq 1$.

(a) Solve the pressure problem with a Neumann condition at the FS interface:

$$\Delta p_{(k)}^n = \frac{\rho_f \alpha}{\Delta t}\nabla \cdot \widetilde{\boldsymbol{v}}_{(k)}^n \quad \text{in } \Omega_f^*,$$

$$\frac{\partial p_{(k)}^n}{\partial \boldsymbol{n}^*} = \frac{\alpha}{\Delta t}\boldsymbol{d}_{(k-1)}^n + \boldsymbol{g}_{fs}^n \quad \text{on } \Sigma^*.$$

(b) Then, solve the non-linear vessel wall problem with a Neumann condition at the FS interface:

$$\frac{\rho_s \beta}{\Delta t^2}\widehat{\boldsymbol{d}}_{(k)}^n - \nabla \cdot \widehat{\boldsymbol{T}}_s(\widehat{\boldsymbol{d}}_{(k)}^n) = \widehat{\boldsymbol{g}}_s^n \quad \text{in } \Omega_s,$$

$$\widehat{\boldsymbol{T}}_s(\widehat{\boldsymbol{d}}_{(k)}^n)\widehat{\boldsymbol{n}} = \widehat{\boldsymbol{T}}_f(\widehat{\widetilde{\boldsymbol{v}}}^n, \widehat{p}_{(k)}^n)\widehat{\boldsymbol{n}} \quad \text{on } \Sigma.$$

Astorino, Chouly and Fernández (2009) apply the DG mortaring approach to this projection scheme, leading to a Robin–Robin-like scheme. An algebraic version of the projection scheme proposed by Fernández *et al.* (2007) is introduced in Badia *et al.* (2008*c*).

Partitioned algorithms of implicit type. More generally, a fully implicit treatment of (4.45c, 4.45d) (or (4.49)) by means of partitioned algorithms is often considered. In this case, the fluid and vessel wall subproblems are solved iteratively until the whole interface conditions are satisfied within a prescribed tolerance. A general scheme is given by the following.

Implicit Robin–Robin scheme
For $n \geq 1$, $k \geq 1$, at time step t^n/iteration k:

(1) Solve the Oseen problem with a Robin condition at the FS interface:

$$\frac{\rho_f \alpha}{\Delta t}\boldsymbol{v}^n_{(k)} + \rho_f((\boldsymbol{v}^* - \boldsymbol{v}^*_f)\cdot\nabla)\boldsymbol{v}^n_{(k)} - \nabla\cdot\boldsymbol{T}_f(\boldsymbol{v}^n_{(k)}, p^n_{(k)}) = \boldsymbol{g}^n_f \quad \text{in } \Omega^*_f, \tag{4.55a}$$

$$\nabla\cdot\boldsymbol{v}^n_{(k)} = 0 \quad \text{in } \Omega^*_f, \tag{4.55b}$$

$$\sigma_f\boldsymbol{v}^n_{(k)} + \boldsymbol{T}_f(\boldsymbol{v}^n_{(k)}, p^n_{(k)})\boldsymbol{n}^* = \sigma_f\left(\frac{\alpha}{\Delta t}\boldsymbol{d}^n_{(k-1)} + \boldsymbol{g}^n_{fs}\right) + \boldsymbol{T}_s(\boldsymbol{d}^n_{(k-1)})\boldsymbol{n}^* \quad \text{on } \Sigma^*. \tag{4.55c}$$

(2) Then, solve the non-linear vessel wall problem with a Robin condition at the FS interface:

$$\frac{\rho_s \beta}{\Delta t^2}\widehat{\boldsymbol{d}}^n_{(k)} - \nabla\cdot\widehat{\boldsymbol{T}}_s(\widehat{\boldsymbol{d}}^n_{(k)}) = \widehat{\boldsymbol{g}}^n_s \quad \text{in } \Omega_s, \tag{4.56a}$$

$$\frac{\sigma_s \alpha}{\Delta t}\widehat{\boldsymbol{d}}^n_{(k)} + \widehat{\boldsymbol{T}}_s(\widehat{\boldsymbol{d}}^n_{(k)})\widehat{\boldsymbol{n}} = \sigma_s\widehat{\boldsymbol{v}}^n_{(k)} + \widehat{\boldsymbol{T}}_f(\widehat{\boldsymbol{v}}^n_{(k)}, \widehat{p}^n_{(k)})\widehat{\boldsymbol{n}} - \sigma_s\widehat{\boldsymbol{g}}^n_{fs} \quad \text{on } \Sigma. \tag{4.56b}$$

As proved in Causin *et al.* (2005), a small relaxation parameter is needed to achieve convergence in the implicit Dirichlet–Neumann scheme (corresponding to setting $\sigma_f = +\infty$, $\sigma_s = 0$ in (4.55c, 4.56b)). In practice, an Aitken relaxation procedure is often used to dynamically estimate an efficient relaxation parameter (Deparis 2004, Kuttler and Wall 2008). A better situation is obtained by properly selecting the parameters in the Robin interface conditions (4.55c, 4.56b). In particular, the choice

$$\sigma_f = \frac{\beta\rho_s H_s}{\alpha\Delta t} + \frac{E H_s \Delta t}{(1-\nu^2)R^2}, \quad \sigma_s = 0$$

(where, as usual, E and ν are Young's modulus and the Poisson ratio for the vessel material at small deformations, and H_s and R are the representative thickness and radius of the vessel) yields fast convergence without any relaxation (*Robin–Neumann* scheme: Badia, Nobile and Vergara 2008*a*, Nobile, Pozzoli and Vergara 2013). Gerardo-Giorda, Nobile and Vergara (2010) and Gigante and Vergara (2015) have characterized the optimal value of σ_s, leading to further improvement in the convergence history. Yu, Baek and Karniadakis (2013) have derived a Dirichlet–Robin scheme by means of a *generalized fictitious* method, where the coefficients of the fluid pressure and vessel wall acceleration are changed to account for the added mass effect. This again allows one to obtain good convergence properties for haemodynamic parameters without any relaxation. Another class of implicit methods with good convergence properties for high added mass effect is based on adding a suitable *interface artificial compressibility* (IAC) consistent term to the fluid problem, proportional to the jump of pressure between two successive iterations (Degroote *et al.* 2010). Degroote (2011) showed that for a finite volume approximation, the IAC method based on Dirichlet–Neumann iterations is equivalent to a Robin–Neumann scheme for a suitable choice of the parameter σ_f.

Partitioned algorithms for the FSI problem with a membrane structure. Recently, several papers have analysed algorithms for FSI problems featuring a reduced membrane model for the vessel wall. In this case, the FSI problem is given by the fluid problem (4.19a), the kinematic continuity condition (4.23a) and the membrane equation (4.23b), which in this case also plays the role of dynamic continuity condition. Moreover, homogeneous Dirichlet or Neumann conditions in the tangential direction need to be prescribed for the fluid problem at the interface Σ. By considering an implicit time discretization of (4.23b) and an explicit treatment of the geometry coupling, we obtain at each time step the following linearized FSI problem:

$$\frac{\rho_f \alpha}{\Delta t}\boldsymbol{v}^n + \rho_f((\boldsymbol{v}^* - \boldsymbol{v}_f^*)\cdot\nabla)\boldsymbol{v}^n - \nabla\cdot\boldsymbol{T}_f(\boldsymbol{v}^n, p^n) = \boldsymbol{g}_f^n \quad \text{in } \Omega_f^*, \tag{4.57a}$$

$$\nabla\cdot\boldsymbol{v}^n = 0 \quad \text{in } \Omega_f^*, \tag{4.57b}$$

$$\boldsymbol{v}^n\cdot\boldsymbol{n}^* = \frac{\alpha}{\Delta t}{d_r}^n + g_{fs}^n \quad \text{on } \Sigma^*, \tag{4.57c}$$

$$\boldsymbol{v}^n - (\boldsymbol{v}^n\cdot\boldsymbol{n}^*)\boldsymbol{n}^* = \boldsymbol{0} \quad \text{on } \Sigma^*, \tag{4.57d}$$

$$\frac{\rho_s H_s \beta}{\Delta t^2} d_r^n - \nabla\cdot(\boldsymbol{P}\nabla d_r^n) + \chi H_s d_r^n = -\boldsymbol{T}_f(\boldsymbol{v}^n, p^n)\boldsymbol{n}^*\cdot\boldsymbol{n}^* + g_s^n \quad \text{on } \Sigma^*, \tag{4.57e}$$

where, as usual, $\boldsymbol{g}_f^n$, g_s^n and g_{fs}^n account for the terms at previous time steps coming from time discretization.

The explicit Dirichlet–Neumann scheme applied to the previous monolithic problem reads as follows.

Explicit Dirichlet–Neumann scheme for the FSI problem with membrane structure

Given the quantities at previous time steps, at time step t^n:

(1) Solve the Oseen problem with a Dirichlet condition at the FS interface:

$$\begin{aligned} &\frac{\rho_f \alpha}{\Delta t}\boldsymbol{v}^n + \rho_f((\boldsymbol{v}^* - \boldsymbol{v}_f^*)\cdot\nabla)\boldsymbol{v}^n - \nabla\cdot\boldsymbol{T}_f(\boldsymbol{v}^n, p^n) = \boldsymbol{g}_f^n && \text{in } \Omega_f^*, \\ &\nabla\cdot\boldsymbol{v}^n = 0 && \text{in } \Omega_f^*, \\ &\boldsymbol{v}^n\cdot\boldsymbol{n}^* = \frac{\alpha}{\Delta t} d_r^{n-1} + g_{fs}^{n-1} && \text{on } \Sigma^*, \\ &\boldsymbol{v}^n - (\boldsymbol{v}^n\cdot\boldsymbol{n}^*)\boldsymbol{n}^* = \boldsymbol{0} && \text{on } \Sigma^*. \end{aligned}$$

(2) Then, solve the membrane problem:

$$\frac{\rho_s H_s \beta}{\Delta t^2}\widehat{d_r^n} - \nabla\cdot(\boldsymbol{P}\nabla\widehat{d_r^n}) + \chi H_s \widehat{d_r^n} = -\widehat{\boldsymbol{T}}_f(\widehat{\boldsymbol{v}}^n, \widehat{p}^n)\widehat{\boldsymbol{n}}\cdot\widehat{\boldsymbol{n}} + \widehat{g}_s^n \quad \text{on } \Sigma.$$

As in the case of scheme (4.47, 4.48), unfortunately this scheme is unconditionally absolutely unstable in the haemodynamic regime (Causin *et al.* 2005).

Different algorithms are obtained by linearly combining the interface conditions (4.57c) and (4.57e) and by substituting the new condition in (4.57c). In this case, we have to solve a coupled problem consisting of the following.

Robin–Neumann coupling for the FSI problem with membrane structure

Given the quantities at previous time steps, at time step t^n:

(1) Solve the Oseen problem with a Robin condition at the FS interface:

$$\frac{\rho_f \alpha}{\Delta t}\boldsymbol{v}^n + \rho_f((\boldsymbol{v}^* - \boldsymbol{v}_f^*)\cdot\nabla)\boldsymbol{v}^n - \nabla\cdot\boldsymbol{T}_f(\boldsymbol{v}^n, p^n) = \boldsymbol{g}_f^n \quad \text{in } \Omega_f^*, \tag{4.59a}$$

$$\nabla\cdot\boldsymbol{v}^n = 0 \quad \text{in } \Omega_f^*, \tag{4.59b}$$

$$\begin{aligned} &(\sigma_f\boldsymbol{v}^n + \boldsymbol{T}_f(\boldsymbol{v}^n, p^n)\boldsymbol{n}^*)\cdot\boldsymbol{n}^*\sigma_f\left(\frac{\alpha}{\Delta t}d_r^n + g_{fs}^n\right) \\ &\qquad - \left(\frac{\rho_s H_s\beta}{\Delta t^2}d_r^n - \nabla\cdot(\boldsymbol{P}\nabla d_r^n) + \chi H_s d_r^n\right) \quad \text{on } \Sigma^*, \end{aligned} \tag{4.59c}$$

$$\boldsymbol{v}^n - (\boldsymbol{v}^n\cdot\boldsymbol{n}^*)\boldsymbol{n}^* = \boldsymbol{0} \quad \text{on } \Sigma^*. \tag{4.59d}$$

(2) Solve the membrane problem:

$$\frac{\rho_s H_s \beta}{\Delta t^2}\widehat{d_r^n} - \nabla\cdot(\boldsymbol{P}\nabla\widehat{d_r^n}) + \chi H_s \widehat{d_r^n} = -\widehat{\boldsymbol{T}}_f(\widehat{\boldsymbol{v}}^n, \widehat{p}^n)\widehat{\boldsymbol{n}}\cdot\widehat{\boldsymbol{n}} + \widehat{g}_s^n \quad \text{on } \Sigma. \tag{4.60}$$

The above problem could be solved either monolithically or by means of a block Gauss–Seidel method that in fact introduces sub-iterations splitting the solution of (4.59) and (4.60). When $\boldsymbol{P} = \mathbf{0}$, the special choice

$$\sigma_f = \widetilde{\sigma}_f = \frac{\rho_s H_s \beta}{\alpha \Delta t} + \frac{\chi H_s \Delta t}{\alpha}, \tag{4.61}$$

introduced in (4.59c), yields

$$(\widetilde{\sigma}_f \boldsymbol{v}^n + \boldsymbol{T}_f(\boldsymbol{v}^n, p^n)\boldsymbol{n}^*) \cdot \boldsymbol{n}^* = \widetilde{\sigma}_f g_{fs}^n \quad \text{on } \Sigma^*. \tag{4.62}$$

At this stage, this is a Robin condition for the fluid problem without any explicit dependence on d_r^n. Thus, the monolithic problem given by (4.59, 4.60, 4.61) is equivalent to the stand-alone fluid problem (4.59a, 4.59b, 4.59d, 4.62): see Nobile and Vergara (2008). The solution of this fluid problem can then be used to feed the right-hand side of (4.60) and to get the structure displacement $\widehat{d_r^n}$. In this way, the fluid and structure problems are in fact decoupled, even if the coupling conditions are treated implicitly. This provides a smart and efficient way to solve the monolithic problem (4.57) exactly, at the expense of a single fluid problem solve (note that the membrane problem (4.60) is solved very cheaply).

Starting from this result, Guidoboni, Glowinski, Cavallini and Canic (2009) have proposed a stable Robin–Neumann scheme based on an operator splitting, for a general membrane law ($\boldsymbol{P} \neq \mathbf{0}$). The inertial vessel wall term is treated implicitly as in the previous case, leading to a Robin boundary condition for the fluid with $\sigma_f = \rho_s H_s \beta / \alpha \Delta t$, whereas the elastic and algebraic contributions are treated explicitly. Fernández (2013) proposes an incremental version of this scheme, where the elastic and algebraic parts of the membrane law are included in the Robin condition for the fluid problem by means of a suitable extrapolation from previous time steps. Finally, we mention Colciago *et al.* (2014), who treat the whole membrane law implicitly, leading to a generalized Robin condition, which however requires an *ad hoc* implementation.

Partitioned algorithms based on the Schur complement. Here we discuss partitioned schemes arising from an interface equation written for the FSI problem and introduced by Deparis, Discacciati, Fourestey and Quarteroni (2006). For the sake of exposition, we introduce the algebraic counterpart of (4.45) related to a finite element discretization for the case of linear elasticity for the vessel wall problem:

$$\left[\begin{array}{cc:cc} C^{ff} & C^{f\Sigma} & 0 & 0 \\ 0 & M^{\Sigma} & -M^{\Sigma} & 0 \\ \hdashline C^{\Sigma f} & C^{\Sigma\Sigma} & G^{\Sigma\Sigma} & G^{\Sigma s} \\ 0 & 0 & G^{s\Sigma} & G^{ss} \end{array}\right] \left[\begin{array}{c} \boldsymbol{V}^f \\ \boldsymbol{V}^{\Sigma} \\ \hdashline \boldsymbol{U}^{\Sigma} \\ \boldsymbol{D}^s \end{array}\right] = \left[\begin{array}{c} \boldsymbol{b}^f \\ \mathbf{0} \\ \hdashline \boldsymbol{b}^{\Sigma} \\ \boldsymbol{b}^s \end{array}\right], \tag{4.63}$$

where the superscript f refers to internal fluid variables, s to internal variables for the vessel wall, Σ to the FS interface variables; C (with superscripts) refers to fluid matrices, G to vessel wall matrices, M^Σ is the FS interface mass matrix; $\boldsymbol{V}$ includes the fluid and pressure variables, $\boldsymbol{D}^s$ the vessel wall displacement internal variables, and $\boldsymbol{U}^\Sigma$ the vessel wall velocity interface variables; the vectors $\boldsymbol{b}$ denote the right-hand sides. The first row corresponds to the momentum and mass conservation for the fluid, the second and third rows to the interface kinematic and dynamic conditions, and the last row to the vessel wall problem. By eliminating $\boldsymbol{V}^f$, $\boldsymbol{V}^\Sigma$ and $\boldsymbol{D}^s$ from (4.63), we obtain the interface equation

$$(\widetilde{C}^\Sigma + \widetilde{G}^\Sigma)\boldsymbol{U}^\Sigma = \widetilde{\mathbf{b}}^\Sigma, \tag{4.64}$$

where

$$\widetilde{C}^\Sigma = C^{\Sigma\Sigma} - C^{\Sigma f}(C^{ff})^{-1}C^{f\Sigma} \quad \text{and} \quad \widetilde{G}^\Sigma = G^{\Sigma\Sigma} - G^{\Sigma s}(G^{ss})^{-1}G^{s\Sigma}$$

are the fluid and vessel wall Schur complement matrices and

$$\widetilde{\mathbf{b}}^\Sigma = \mathbf{b}^\Sigma - C^{\Sigma f}(C^{ff})^{-1}\mathbf{b}^f - G^{\Sigma s}(G^{ss})^{-1}\mathbf{b}^s$$

is the corresponding right-hand side.

The Robin–Robin method (4.55, 4.56) could be obtained by applying the Richardson method to the interface equation (4.64) preconditioned by the matrix

$$P = \frac{1}{\sigma_f + \sigma_s}(\widetilde{C}^\Sigma + \sigma_f M^\Sigma)(M^\Sigma)^{-1}(\widetilde{G}^\Sigma + \sigma_s M^\Sigma). \tag{4.65}$$

This leads to a new family of partitioned schemes obtained by applying other Krylov methods to (4.64) with the same preconditioner. For example, GMRES preconditioned by the Dirichlet–Neumann preconditioner $P = \widetilde{G}^\Sigma$ is considered in Badia, Quaini and Quarteroni (2008*b*), whereas GMRES preconditioned with the RR preconditioner (4.65) is introduced in Badia, Nobile and Vergara (2009). The first of these two schemes highlighted better convergence properties with respect to the classical Dirichlet–Neumann method, whereas the second is more robust with respect to the choice of the interface parameters σ_f, σ_s than the classical Robin–Robin method.

In some works, a non-linear interface equation is written directly for the non-linear problem (4.44) and a Newton method is then applied to this equation (Fernández and Moubachir 2005, Degroote and Vierendeels 2011).

Remark 4.2. In the partitioned schemes introduced above, the vessel wall subproblem appearing at each iteration is still non-linear, that is, we have not discussed how to tackle point (iii) in the list of issues reported at the beginning of the subsection. The simplest way to treat vessel wall non-linearity – very appropriate when a non-linear structural solver is available – is to introduce inner Newton sub-iterations at each Robin–Robin iteration. Alternatively, Kuttler *et al.* (2010) have proposed a scheme based on integrating the management of the five sources of non-linearities/coupling (geometric coupling and adherence, fluid and structure constitutive non-linearities, physical coupling) in a single inexact-Newton loop. Other possible combinations will be discussed later on as specific monolithic procedures. For a detailed overview with numerical comparisons we refer to Nobile *et al.* (2014).

Monolithic solvers. As an alternative to partitioned schemes, monolithic procedures have been successfully considered in haemodynamics for the solution of the FSI problem (4.45). These methods consist in applying an exact or inexact Newton method to the whole non-linear FSI problem. Referring to the notation introduced above, by considering a finite element space discretization and the inexact Newton method, we obtain from (4.45)

$$\widetilde{J}(\boldsymbol{x}_{(k-1)})\delta\boldsymbol{x}_{(k)} = \boldsymbol{G} - \boldsymbol{A}(\boldsymbol{x}_{(k-1)}) \tag{4.66}$$

(for simplicity we omit the temporal index n), where $\widetilde{J}$ is the Jacobian matrix or a suitable approximation of it, $\boldsymbol{x} = [\boldsymbol{V}^f \; \boldsymbol{V}^\Sigma \; \boldsymbol{U}^\Sigma \; \boldsymbol{D}^s]^T$, and $\boldsymbol{A}(\boldsymbol{x}) = \boldsymbol{G}$ is the non-linear system related to (4.45). The exact Jacobian matrix is given by

$$J(\boldsymbol{x}_{(k-1)}) = \left[\begin{array}{cc:cc} C^{ff} & C^{f\Sigma} & 0 & 0 \\ 0 & M^{\Sigma} & -M^{\Sigma} & 0 \\ \hdashline C^{\Sigma f} & C^{\Sigma\Sigma} & J_G^{\Sigma\Sigma}(\boldsymbol{U}^{\Sigma}_{(k-1)}) & J_G^{\Sigma s}(\boldsymbol{D}^{s}_{(k-1)}) \\ 0 & 0 & J_G^{s\Sigma}(\boldsymbol{U}^{\Sigma}_{(k-1)}) & J_G^{ss}(\boldsymbol{D}^{s}_{(k-1)}) \end{array}\right], \tag{4.67}$$

where the submatrices J_G are the exact Jacobians of $\boldsymbol{A}$ related to the structure variables. For the solution of the linear systems (4.66) with the exact Jacobian, classical strategies used so far for haemodynamics include GMRES preconditioned by a one-level additive Schwarz method (Barker and Cai 2010*a*) and global algebraic multigrid (Gee, Kuttler and Wall 2011). Barker and Cai (2010*b*) use a two-level Newton method in combination with a two-level hybrid Schwarz preconditioner, where the solution on a coarse grid is used to provide a good initial guess to the Newton method. Alternatively, inexact Newton methods have been considered, for example by means of block approximations $\widetilde{J}$ of J splitting the fluid velocity, pressure and vessel wall unknowns (Heil 2004, Crosetto, Deparis, Fourestey and Quarteroni 2011).

A particular class of inexact Newton methods is obtained by neglecting the term $-M_\Sigma$ appearing in the upper-right block of the exact Jacobian (4.67). This yields in fact a class of partitioned schemes of Dirichlet–

Neumann type where issues (iii) and (v) (vessel wall non-linearity and physical coupling) are treated within the same iterations. An example is given by the preconditioner proposed in Crosetto *et al.* (2011), where the fluid and structure blocks in (4.67) are approximated by the corresponding algebraic additive Schwarz preconditioners. This method is strongly scalable for haemodynamic applications. Recently, Deparis, Forti, Grandperrin and Quarteroni (2016) have introduced a variant of the above preconditioner. This new preconditioner, named FaCSI, is based on operating a static condensation of the fluid interface variables and using a SIMPLE preconditioner for the fluid block.

For the sake of exposition, we have discussed numerical strategies for the FSI problem based on finite elements for the space discretization and finite differences for the time discretization. Other strategies considered so far in haemodynamics are space-time finite elements (see *e.g.* Tezduyar *et al.* 2007, Bazilevs, Takizawa and Tezduyar 2012) and the methods based on isogeometric analysis (see Bazilevs, Calo, Zhang and Hughes 2006, Bazilevs *et al.* 2009).

In Figure 4.7 we illustrate some examples of numerical results obtained in real geometries reconstructed from radiological images (see the caption for details).

4.6.4. Numerical methods for defective boundary problems

For the numerical solution of the augmented formulation (4.27), one could rely either on a monolithic strategy, where the full augmented matrix is constructed and solved, or on splitting techniques. For the latter, Formaggia *et al.* (2002) and Veneziani and Vergara (2005) propose writing the Schur complement equation with respect to the Lagrange multiplier of the linearized and discretized (in time and space) augmented formulation. This is a linear system whose dimension is equal to the number of flow rate conditions, say $m \geq 1$. By using the GMRES method to solve this system iteratively, the exact solution is reached after exactly m iterations (in exact arithmetic). At each iteration, the solution of a standard fluid problem with Neumann conditions is needed (*exact splitting technique*). The solution of a further standard fluid problem is required to compute the initial residual in the GMRES algorithm. This approach is quite expensive, even for the case $m = 1$, which requires the solution of two fluid problems per time step. However, it preserves *modularity*; indeed, it can be implemented using available standard fluid solvers in a *black box* mode. This is an interesting property when applications to cases of real interest are addressed: see Viscardi *et al.* (2010), Vergara, Viscardi, Antiga and Luciani (2012), Piccinelli *et al.* (2013) and Guerciotti *et al.* (2015, 2017).

To reduce the computational time required by the exact splitting approach, Veneziani and Vergara (2007) propose a different (inexact) splitting

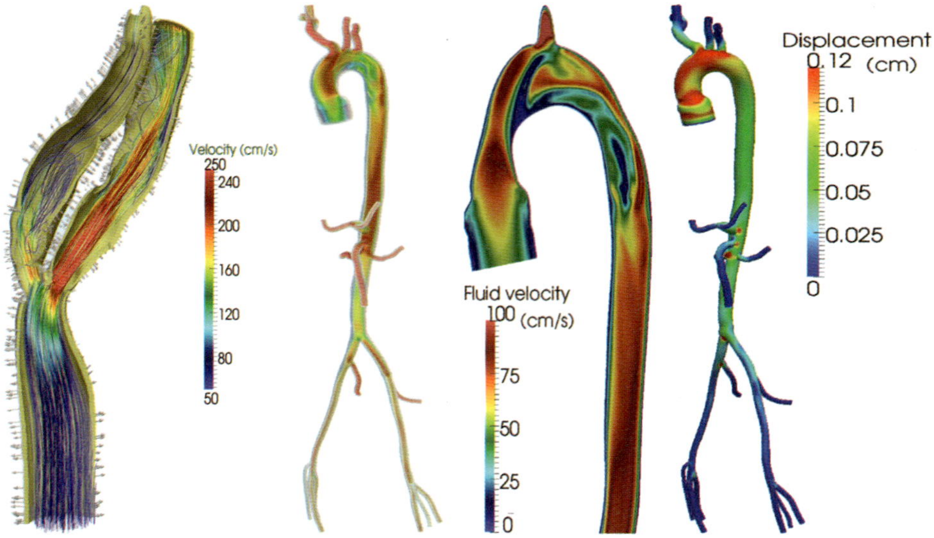

(a) (b) (c) (d)

Figure 4.7. (a) Blood velocity streamlines and vessel wall displacement vectors in a stenotic carotid artery (MRI images from the Vascular Surgery and Radiology Divisions at Ospedale Maggiore Policlinico, Milan, Italy). (b–d) Results of a FSI simulation in the ascending and thoracic aorta (MRI images from www.vascularmodel.com/sandbox/doku.php?id=repository). Blood velocity magnitude in the whole domain (b) and on a selected longitudinal section (c), vessel wall displacements (d). All cases refer to the systolic peak. Numerical results were obtained using LifeV, P1-Bubble/P1 finite elements for the fluid problem and P1 finite elements for the vessel wall problem; the backward Euler scheme and the midpoint method was used for the time discretization of the fluid and vessel wall problems, respectively. The implicit Robin–Robin partitioned scheme was used in case (a) and the FaCSI preconditioner in cases (b–d).

procedure, requiring the solution of m steady problems out of the temporal loop and of one unsteady null flow rate problem at each time step. This strategy introduces an error near the section which is smaller than the one based on conjecturing the velocity profile in the original (non-null) flow rate problem: see Section 4.4.1.

For the numerical solution of the control-based approach described in Section 4.4.3, Formaggia *et al.* (2008, 2009*b*) have considered classical iterative methods for the solution of the resulting KKT system.

Recently, a numerical approach based on the *Nitsche method* has been considered to prescribe a flow rate condition. In particular, the original idea of prescribing Dirichlet conditions with a consistent penalization approach (Nitsche 1970/71) has been extended by Zunino (2009) to the case of flow rate boundary conditions. This strategy does not introduce further variables other than those of the original problem, but it does require careful tuning of a penalization parameter. In addition, it deals with non-standard bilinear forms that need *ad hoc* implementation. However, it should be very effective if flow rate conditions are implemented in a DG code. Vergara (2011) has considered a similar approach to fulfilling the mean pressure condition (3.3) and the FSI case; see also Porpora, Zunino, Vergara and Piccinelli (2012).

For a more comprehensive overview of numerical strategies for defective boundary problems, we refer the reader to Formaggia and Vergara (2012).

4.6.5. Numerical methods for the geometric reduced models and multiscale approach

For the numerical solution of the 1D reduced model (4.30), in principle any convenient approximation method for non-linear hyperbolic equations can be used. The peculiar feature of this model, however, is the lack of discontinuous solutions. A common approach relies on the finite element version of the Lax–Wendroff scheme, thanks to its excellent dispersion properties (Formaggia *et al.* 2001). As this scheme is explicit, a CFL-like condition is required to ensure absolute stability. In the presence of a viscoelastic term, the 1D model is usually discretized by means of a splitting procedure where the solution is split into two components, one satisfying the pure elastic problem and the other satisfying the viscoelastic correction (Formaggia *et al.* 2003, Malossi, Blanco and Deparis 2012). High-order methods are suitable for capturing the (physical) reflections at bifurcations induced by the vessel tapering: see *e.g.* Sherwin, Franke, Peiró and Parker (2003*b*) and Sherwin, Formaggia, Peiró and Franke (2003*a*) for a discontinuous Galerkin discretization and Muller and Toro (2014) for a finite volume scheme.

Regarding 0D models, they are in general described by systems of differential and algebraic equations, possibly non-linear due to the presence of diodes to represent the valves (Formaggia *et al.* 2009*a*). Usually, for haemodynamic applications, these systems can be reduced to classical Cauchy problems and solved by classical Runge–Kutta methods.

As for the solution of the 3D/1D coupled problems described in Section 4.5.2, we can in principle identify three different strategies, namely *partitioned schemes*, *monolithic approaches*, and methods based on the solution of an *interface equation*. In partitioned schemes, the 3D and 1D problems are solved separately in an iterative framework. The coupling interface conditions can be enforced in many different ways. For example, we can prescribe the flow rate condition (4.39a) to the 3D problem and the pressure condition (4.42) to the 1D problem. Different algorithms are obtained by switching the role of the interface conditions in the iterative algorithm or by considering other interface conditions (*e.g.* (4.39b)). This is also the case when one of the two interface conditions is replaced by a condition expressing the continuity of the characteristic variable W_1 entering the 1D domain

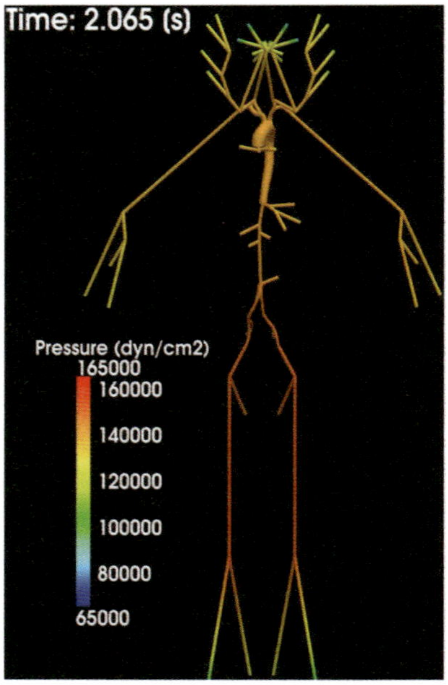

Figure 4.8. Pressure wave propagation in an ascending aorta (3D model) and in the 1D model of the systemic circulation. Numerical results were obtained using LifeV; the Newton method was used for the interface equation. Courtesy of C. Malossi.

(Formaggia *et al.* 2001, Papadakis 2009), that is, according to (4.33),

$$W_1(t)|_{z=0} = \zeta_1\left(|\Gamma_f^t|, \int_{\Gamma_f^t} \boldsymbol{v}\cdot\boldsymbol{n}\,\mathrm{d}\gamma\right).$$

In any case, each of these approaches yields a 3D problem with a defective boundary condition, which could be tackled by one of the strategies described in Section 4.4. Formaggia *et al.* (2007) and Papadakis (2009) have successfully considered explicit algorithms based on the solution of the 3D and 1D problems only once per time step. These algorithms enforce a limitation on Δt, which, however, is milder with respect to that imposed by the numerical scheme adopted for the 1D model. As an alternative, Blanco, Feijóo and Urquiza (2007) and Blanco, Pivello, Urquiza and Feijóo (2009) have introduced iterative methods applied directly to the monolithic linearized system. A different approach to solving the 3D/1D coupled problem relies on writing an interface equation involving only the 3D/1D interface unknowns. We can interpret this equation as the geometric heterogeneous counterpart of the Schur complement equation. For its numerical solution, Leiva, Blanco and Buscaglia (2011), Malossi *et al.* (2013) and Blanco, Deparis and Malossi (2013) have used either the Broyden or the Newton method, in combination with GMRES. Methods relying on the numerical solution of the interface equation are simple to implement in the case of multiple interfaces, such as those arising in complex arterial networks.

In Figure 4.8 we give a numerical result obtained by the coupling between the 3D model of an ascending aorta and a 1D model of the systemic circulation. This result highlights the suitability of the 1D model in providing absorbing conditions for the 3D model and in propagating the pressure wave along the whole network (Malossi 2012).

PART TWO
Heart function

5. Basic facts on quantitative physiology

5.1. Basic anatomy

The heart is a hollow organ that pumps the blood into the arteries of the systemic and pulmonary circulations and collects it after its return through the veins. It is made up of the left and right heart, each consisting of two chambers, an *atrium* and a *ventricle*. The left and right hearts are separated by the *interatrial* and *interventricular septa*, which do not allow the transfer of blood, whereas the atria and the ventricles are connected by the *atrioventricular valves* (*tricuspid valve* in the right heart, *mitral valve* in the left heart) that either allow or prevent the blood transfer from the atria to the ventricles depending on their position (open or closed, respectively): see Figure 5.1. In particular, they open when the atrial pressure is higher than the ventricular pressure and close as soon as the blood flow rate becomes negative, that is, when blood flow, which normally goes from the atrium to the ventricle, comes back into the atrium. The *papillary muscles*, located in the ventricles, attach to the cusps of these valves via the *chordae tendineae*, preventing their inversion (prolapse) during closure.

All four chambers are connected to the circulatory system: the left ventricle through the aorta, the right ventricle through the pulmonary artery, the left atrium through the *pulmonary veins*, and the right atrium through the *superior* and *inferior venae cavae*. The ventricles are separated from the circulatory system by two further valves, the *aortic valve* on the left side and the *pulmonary valve* on the right side, whose opening/closure mechanism is similar to that of the atrioventricular valves, that is, they open when the pressure is higher in the ventricle with respect to the corresponding connected artery, whereas they close when the flow rate becomes negative (*i.e.* going from the artery to the ventricle). No valves are located between the atria and the corresponding terminal veins.

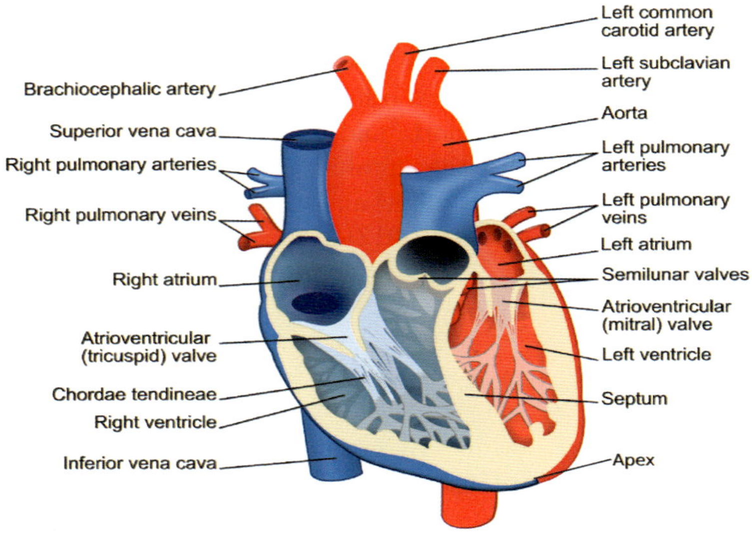

Figure 5.1. Schematic representation of the heart.

The heart wall is made up of three layers: the internal thin *endocardium*, the thick muscular *myocardium* and the external thin *epicardium*. The myocardium of the left ventricle is almost twice as thick as that of the right ventricle. The epicardium is surrounded by the *pericardium*, a serous membrane that isolates the heart from the closest organs, facilitating its movements.

5.2. The cardiac cycle

The main purpose of the heart is to pump the blood in the circulatory system through the aorta and the pulmonary artery. To do this, it needs to exceed the resistance in the arteries where blood has a non-null pressure (about 70 mmHg in the aorta, 10 mmHg in the pulmonary artery) due to the reaction of the elastic vessel wall to the deformation induced by blood inside. Moreover, the heart supplies blood with the energy needed to reach the microvasculature or the lungs. The total work done by the heart per heartbeat is

$$W = QP + \frac{1}{2}mv^2,$$

where Q and m are the blood flow rate and mass ejected, P is the arterial blood pressure, and v is the blood velocity in the aorta or pulmonary artery. The kinetic component of the work is quite negligible with respect to the potential component (about 2% for the left heart and about 5% for the right heart), although it could become more relevant (up to 25%) under physical effort. The flow rate Q in normal conditions is about 5×10^3 cm^3 min^{-1}, so that, assuming 60–90 heartbeats per minute, we have about 55.5–83.3 cm^3 of blood expelled at each heartbeat. The energetic requirements for pumping blood are obtained, as in all organs, by the consumption of oxygen that is provided to the heart by blood in the coronary arteries. The efficiency of the heart, that is, the ratio between the energy W and the total energy consumption, is in physiological cases about 25%.

The cardiac cycle comprises different phases. For its description, we consider the left heart and we start from the situation where the left atrium is filled with oxygenated blood coming from the pulmonary veins, the mitral valve is closed, and the left ventricle has reached its maximum contraction.

(1) *Ventricular filling.* Due to the continuous inflation of blood, when the pressure in the atrium exceeds the ventricular pressure (about 10 mmHg) the mitral valve opens. Blood starts filling the left ventricle, which relaxes. In an initial stage (~ 0.15 s), there is fast inflation of blood due to the pressure gradient between the atrium and the ventricle, where the ventricle probably also exerts suction due to its expansion (phase 1a in Figure 5.2). In any case, this is a passive phase. In a second stage (~ 0.25 s, phase 1b in Figure 5.2), the atrium actively contracts (*atrial systole*), producing a slow ventricular inflation of the remaining blood.

(2) *Isovolumic contraction.* After the atrial systole, the ventricular active contraction starts. This produces an increase in the ventricular pressure causing retrograde flow that accordingly closes the mitral valve. However, the ventricular pressure is still lower than the aortic pressure, so the aortic valve is also closed. Thus, during this phase, there is a continuous and fast increase of ventricular pressure without any change of blood volume due to blood incompressibility (~ 0.05 s, phase 2 in Figure 5.2).

(3) *Ventricular ejection.* As soon as the ventricular pressure reaches the aortic pressure (about 70 mmHg), the aortic valve opens and blood is ejected into the systemic circulation. Since the ventricular contraction carries on after the valve opens, the ventricular pressure continues to increase. Accordingly, the aortic pressure also increases due to the elastic adaptation of the vessel wall. However, at each time, there is a pressure difference between ventricle and aorta that allows blood to accelerate (~ 0.05 s, phase 3a in Figure 5.2). When the ventricle stops its active contraction, this pressure difference reduces and, after a short period (but not immediately due to inertial effects), becomes zero, allowing the flow rate to reach its maximum (~ 0.05 s, phase 3b in Figure 5.2). Then, the pressure difference starts to become negative (higher in the aorta), provoking a deceleration of blood that however continues to enter the aorta. When, due to this deceleration, the flow rate becomes

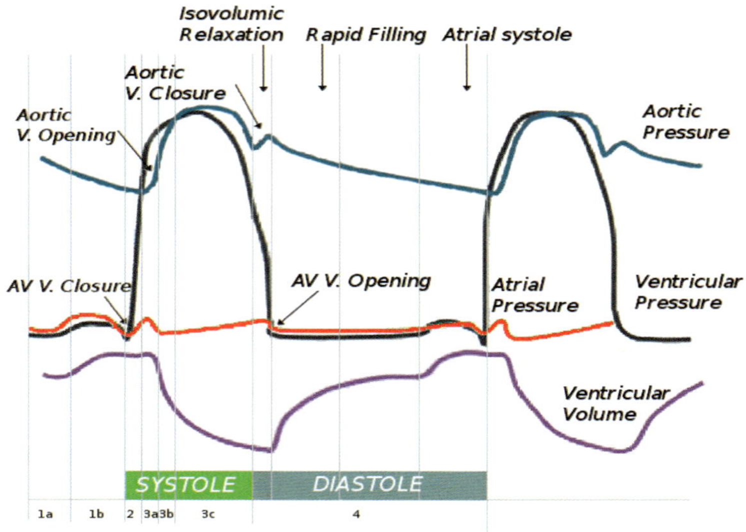

Figure 5.2. Aortic pressure, ventricular pressure, atrial pressure and ventricular volume during a heartbeat.

negative, the aortic valve closes and the blood ejection stops ($\sim$0.15 s, phase 3c in Figure 5.2). Note that, once the active contraction stops, the ventricle starts to relax, releasing the elastic energy accumulated during the contraction. However, this energetic relaxation is not immediately followed by a mechanical relaxation, which should result in an increase in ventricular volume. Indeed, due to inertial effects, there is a 'passive' mechanical contraction which allows ejection of all the blood.

(4) *Isovolumic relaxation.* When the aortic valve closes, the release of energy of the ventricle continues with both valves closed, so that no changes in ventricular volume are produced, resulting in fast decrease of the ventricular pressure ($\sim$0.10 s, phase 4 in Figure 5.2). After the valve closure, the aortic pressure continues to increase (slightly), due to the elastic recoil of the closing valve (*dicrotic wave*: see Figure 5.2).

All these mechanisms apply to the right heart too, the only difference being the pressure values, smaller in this case. This justifies the thicker myocardium of the left heart since higher values of resistances need to be exceeded.

5.3. Electrical propagation

As seen in the previous paragraph, the main agent of blood ejection in the circulatory system is active ventricular contraction. The heart itself is able to produce the electrical impulse that determines this contraction, triggered

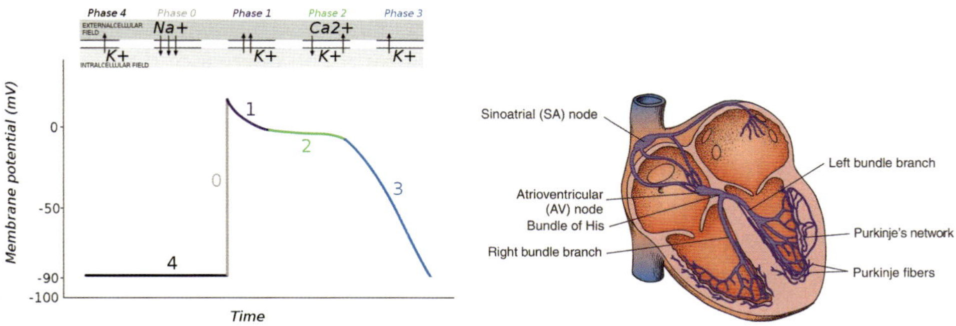

(a) (b)

Figure 5.3. (a) Characteristic action potential of cardiomyocytes and (b) anatomy of the cardiac conduction system (http://medical-dictionary.thefreedictionary.com).

by an electrical potential that propagates along all the myocardium. This is possible owing to the excitability of the heart cells, the *cardiomyocytes*, which, when suitably stimulated, are able to produce a variation in membrane voltage. At rest, the membrane potential is negative (~ -90 mV), whereas when stimulated it reaches a positive value (~ 20 mV) in a very short period (about 2 ms). After this depolarization, a plateau around 0 mV is observed that corresponds to the refractory period (see Figure 5.3). Then, the repolarization phase starts, which brings the potential back to the rest value allowing for a new excitation (see Figure 5.3(a)). This action potential is generated by several ion channels that open and close, and by the resulting currents passing through the membrane. The most important channels are those of calcium, sodium and potassium. In particular, a fast inward sodium current is the main driver of rapid depolarization, a slow inward flux of extra-cellular calcium ions is the main agent behind the characteristic plateau appearing after the depolarization, whereas the outward potassium currents are responsible for the repolarization.

Unlike other cells in the human body, the cardiomyocytes obey the 'all-or-none' law, meaning that if the stimulus is above a suitable threshold, a complete action potential with peak value independent of the stimulus is generated, otherwise no response is provided by the cell. Another characteristic of the heart cells is the presence of a *refractory* period after the generation of an action potential, which inhibits any further stimulus independently of its intensity. Thus, during this period the cell is not excitable at all (*absolute refractory period*, infinite threshold). Afterwards, the cardiomyocytes recover their excitability with a value of the threshold needed to generate the action potential which decreases in time (*relative refractory period*). Finally, once the threshold value reaches its minimum, the cell returns to its excitable state and the threshold remains constant.

The cardiomyocytes act as a syncytium, that is, the stimulation of an individual cell produces the action potential and the corresponding current that results in the excitation of the neighbour cells, and thus of the whole myocardium. This is due to the *gap junctions*, intercellular channels characterized by low resistance and located between cardiomyocytes that permit the electrical potential to travel on the cellular membranes from cell to cell.

In normal conditions, the signal spontaneously originates at the *sinoatrial node*, located in the right atrium at the junction with the superior vena cava. It is the natural pacemaker of the heart, and imposes its rhythm on the entire myocardium (*sinusal rhythm*, $\sim$ 60–90 heartbeats per minute). The impulse generated by the sinoatrial node propagates through all the cardiomyocytes of the atria, activating all their regions and allowing their contraction. The propagation is faster in the direction of the ventricles ($\sim$ 200 cm s^{-1}), allowing it to reach the *atrioventricular node*, located between the atria and the ventricles. When the signal arrives at this node, it is subjected to a delay ($\sim$ 0.12 s) that allows complete contraction of the atria before the propagation in the ventricles starts. Moreover, this node provides a filter limiting possible high frequencies of the atrial signal, induced by atrial fibrillation, for example, protecting the ventricles. This node, when the sinoatrial node loses automatic function, becomes the leading pacemaker and takes on the role of giving the pace to all ventricle stimulation. Then, the electrical signal enters the *bundle of His*, propagating in the ventricles through the two (left and right) *bundle branches* and then through the *Purkinje fibres* (see Figure 5.3(b)). The bundle of His, bundle branches and Purkinje fibres form the *cardiac conduction system* (CCS), a subendocardial specialized network responsible for the fast and coordinated propagation of electrical impulses in the ventricle. The propagation in the CCS is very fast ($\sim$ 350 cm s^{-1}), and its role is to reach the entire endocardium via the dense network of Purkinje fibres, activating it almost simultaneously. Note that the cells of the CCS are dedicated to electrical propagation, so they are not involved in the muscular contraction. Then, the electrical signal enters the myocardium through the *Purkinje muscle junctions* which are the terminal points of the Purkinje network. At the Purkinje muscle junctions, the signal is subjected to a delay ($\sim$ 0.01 s) and then propagates into the ventricular muscle towards the epicardium with reduced velocity ($\sim$ 80 cm s^{-1}).

To better understand how the propagation of the electrical potential spreads into the ventricles, we observe that cardiomyocytes are of cylindrical type. This allows us to define the *fibre direction* of the cell, resulting in a macroscopic fibre direction intended as the average cell orientation in a sufficiently small control volume. A transmural variation of the fibre direction is measured between the epicardium and the endocardium ($\sim -70^\circ$ and $\sim 80^\circ$ with respect to the normal direction to the surface, respectively). The fibres are in turn organized into *sheets* of collagen. The velocity of propagation of the electrical potential is about twice as fast along the fibre direction compared to the directions tangential to the fibres.

5.4. Mechanisms of contraction and cardiac blood fluid dynamics

The propagation of the electrical signal through the cardiomyocytes is responsible for their contraction (*electromechanical coupling*), resulting in the atrial and, more importantly, ventricular contractions. The characteristic connections between the cardiomyocytes resulting in a syncytium allows coordinated contraction of the heart.

At the cellular level, as discussed in the previous paragraph, there is an inward flux of extra-cellular calcium ions just after the depolarization of the cell. Once in the intracellular space, calcium ions bind to *troponin*, which allows *myosin* to bind to *actin* and contraction of the cell to occur.

At the macroscopic level, ventricle contraction results in a longitudinal shortening (from apex to base) of about 15% from the diastolic configuration, in a ventricular wall thickening of about 30%, and torsion around the longitudinal axis of about 15°. This is due to the particular fibre orientation of the cardiomyocytes, which also highly influences the mechanical response of the heart.

The heart obeys the *Frank–Starling law*, stating that an increase (for any reason) of the end diastolic volume (*i.e.* the maximum ventricular expansion) results in an increase in the *stroke volume*, that is, the volume of blood pumped by the left ventricle per heartbeat. This is due to an increase in the load experienced by the cardiomyocytes as a result of the increased volume of blood filling the ventricle. This extra stretching of the cells produces a greater number of actin–myosin bridges, resulting in augmented muscle contraction power. It should be noticed, however, that the Frank–Starling law holds true only up to a certain level of increased stretch: for higher values, the contractility power of the heart, and thus the stroke volume, decreases.

As well as the influence of electrical processes on mechanical behaviour, cardiomyocytes also experience mechano-electrical feedback. This is due to the formation of *stretch-activated* ion channels and to changes in cell conductivity by means of stretching the cell, which changes the membrane shape and the distance between gap junctions. This mechanism seems to be of utmost importance in describing the evolution of spiral waves and thus arrhythmias (Trayanova, Li, Eason and Kohl 2004, Keldermann, Nash and Panfilov 2009).

Blood flow in the cardiac chambers exhibits different characteristics with respect to the vascular flow. The particular challenges are the large-scale motion and complex deformation of the myocardium, the complex interaction with valves leading to specific flow patterns, and the higher Reynolds number (about 4000). For some authors, all of these aspects are responsible for transition to turbulence effects, especially in pathological conditions, even though not all authors agree that turbulence can also occur in normal conditions. Another important aspect of cardiac blood flow is given by the formation of a large vortex ring in the long axis plane of the left ventricle due to the asymmetry of the mitral valve leaflets that forces the blood flow jet to impinge on the posterior wall (Charonko *et al.* 2013). The interaction of this ring with the wall gives rise to a complex flow pattern. Complex vortex rings ejected by the pulmonary veins can also be observed in the left atrium (Mittal *et al.* 2016).

5.5. A brief summary of heart diseases

The principal heart diseases (cardiopathies) are *ischaemic cardiopathy*, the *cardiomyopathies*, *hypertensive cardiopathy* and *valvular pathologies*.

Ischaemic cardiopathy is the most important in terms of frequency and clinical impact. It is caused by a reduced coronary flow rate (due to atherosclerosis of coronary arteries) with consequent malnutrition of the myocardium. When, due to coronary occlusion, a decrease in oxygen supply is abrupt, long-lasting and total, an *infarct of the myocardium* occurs. This is the most classical event of ischaemic cardiopathy, resulting in necrosis of cardiomyocytes and leading, at the end of the process, to the formation of a *scar*.

In *dilated cardiomyopathy* the ventricle wall becomes thinner due to the partial replacement of cardiomyocytes with fibrosis and small scars, leading to dilatation of the ventricle. In *hypertrophic cardiomyopathy* the ventricular wall thickens, with a consequent increase in ventricular blood pressure resulting in decreased filling of the ventricle. As in all cases of ventricular hypertrophy, this produces malnutrition of the myocardium, since the amount of blood supplied by the coronary arteries remains unchanged. Moreover, in this specific hypertrophic cardiomyopathy, the orientation of the fibres is not coordinated as in normal conditions (this is called *electrical disarray*).

In hypertensive cardiopathy there is an increase in blood pressure, resulting in increased work needed by the heart to pump blood, leading again to thickening of the myocardial wall and to malnutrition of the myocardium.

The main valvular pathologies are *stenosis* and *insufficiency*. In aortic valve stenosis, narrowing of the maximum opening of the aortic valve causes the heart to require increased work to pump blood into the aorta properly, again resulting in thickening of the myocardial wall and malnutrition. In mitral valve stenosis, there is an increase in the pressure of the left atrium and in the pulmonary circulation, with consequences for the right heart. Aortic and mitral valve insufficiency is due to the partial reversal of flow (going from the aorta to the left ventricle in aortic insufficiency, and from the left ventricle to the left atrium in mitral insufficiency), which occurs as a consequence of inadequate closure of the valve. In this case the heart dilates in order to supply this retrograde flow, resulting in an increase in its contraction power due to the Frank–Starling law. However, sooner or later the diastolic cell stretching overcomes the threshold of validity of the Frank–Starling law, resulting in decreased contractility power.

The two main events, possibly fatal, that the heart may encounter as a consequence of these cardiopathies are *cardiac arrest*, due to *ventricular fibrillation*, and *heart failure*.

In ventricular fibrillation, the cardiomyocytes are not excited in a coordinated way and thus do not contract homogeneously; instead, groups of cells contract continuously whereas other groups relax. This chaotic excitation, if the fibrillation continues, inhibits the normal functioning of the heart pump, which is no longer able to pump the blood around the circulatory system regularly, leading finally to death. All the cardiopathies described above could yield altered electrical properties (*i.e.* velocity of conduction and refractoriness), which could lead to ventricular fibrillation. For example, in a myocardial infarct, the arrhythmogenic substrate is formed by a matrix of healthy cardiomyocytes, necrotic cardiomyocytes and scars, whereas in hypertrophic cardiomyopathy it is provided by the electrical disarray.

When the blood pumped by the heart at each heartbeat is less than the amount required by the body (but not absent as in ventricular fibrillation), the heart on the one hand increases its frequency so as to guarantee that the blood ejected per minute (*cardiac output*) is almost normal, and on the other hand it increases its diastolic filling to increase the stroke volume. However, the increase in frequency (*tachycardia*) is energetically disadvantageous for the heart, and possible excessive diastolic cell stretching leads to the loss of validity of the Frank–Starling law. When these two compensating mechanisms are no longer effective, cardiac output dramatically decreases, leading to heart failure. In the absence of other causes of death (cardiac arrest or death for non-cardiac reasons), all the cardiopathies described above will sooner or later lead to heart failure.

6. All about data

As discussed in Section 3, the use of patient-specific data is needed for numerical modelling aiming to understand biophysical processes and support clinicians. This also holds true for heart modelling. The latter accounts for three processes that are intimately coupled: electrical propagation, mechanical contraction and relaxation, and the blood fluid dynamics. The first two processes occur in the cardiac tissue (here called the muscle region), whereas the latter occurs in the four hollow regions (chambers) delimited

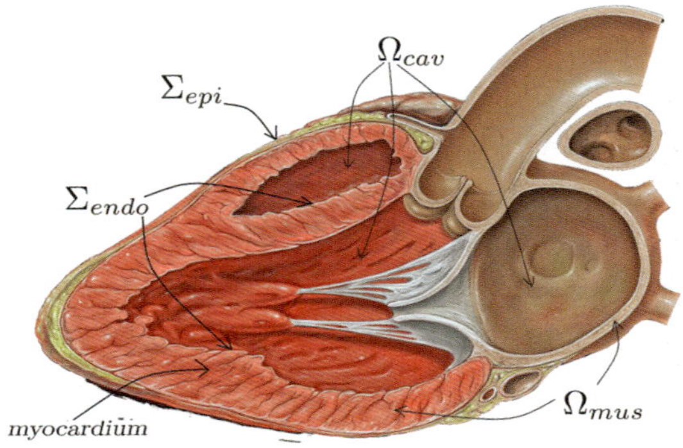

Figure 6.1. Longitudinal section of a complete heart domain.

by the endocardium. As in Section 3 for the circulatory system, in what follows we briefly discuss how to obtain geometric, boundary and biological cardiac data.

6.1. Cardiac geometric data

Referring to Figure 6.1, we need to build two computational domains: the muscle region Ω_{mus}, delimited by the external surface of the epicardium Σ_{epi} and by the internal surface of the endocardium Σ_{endo}, and the cavities Ω_{cav}, referred to below as heart cavities or chambers, delimited by the endocardium. The cavities are easily obtained once the endocardium has been reconstructed, so that in fact the cardiac image reconstruction process relies on identifying the endocardium and epicardium surfaces.

This problem presents several challenging issues. In the vascular case, the external wall surface is usually obtained by extruding the internal surface under the assumption of constant (or in any case known) wall thickness (see Section 3.1.3). This procedure is unsuitable for the reconstruction of the myocardium. On one hand, the thickness changes significantly while moving from apex to base. On the other hand, changes can be dramatic from patient to patient. Moreover, due to the large displacements induced by heart motion, a dynamic acquisition procedure (allowing for the acquisition of several frames per heartbeat) is required to obtain a reconstruction of the heart other than at the end-diastolic phase. Finally, the presence of the papillary muscle and wall irregularities given by *trabeculations* makes the reconstruction of the endocardium very problematic.

Let us briefly review the most common radiological procedures for acquisition of cardiac images. As in the vascular case, MRI and CT play a major role. Often, due to heart motion, temporally resolved acquisitions are performed, allowing us to obtain 20–30 frames per heartbeat.

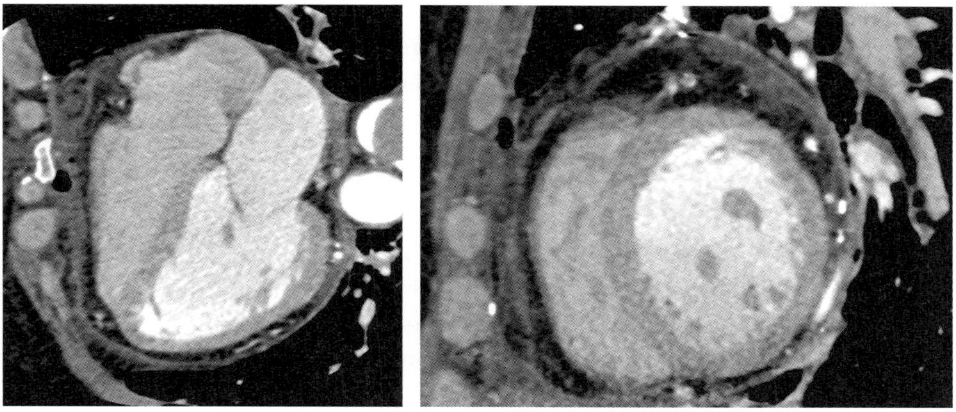

(a) (b)

Figure 6.2. (a) Longitudinal CT slice of the heart. Right atrium (top left), right ventricle (bottom left), left atrium (top right), left ventricle (bottom right). (b) CT slice in the plane orthogonal to the long axis. In both figures, on the right the thick left ventricle myocardium is detectable in darker grey. Radiological images from Ospedale Sacco, Milan, Italy.

The main interest is in the left ventricle, due to its vital importance and its pronounced thickness, ranging between 6 and 16 mm. The shape of its cavity is often approximated by an ellipsoid. In contrast, the right ventricle and the atria are characterized by a thickness that usually does not reach the spatial resolution of the acquisition technologies, so their reconstruction is hard and consequently less studied.

The standard cardiac acquisition plane is orthogonal to the long axis. Blood appears brighter whereas the myocardium and the surrounding tissue are darker: see Figure 6.2. Not all of the slices share the same degree of complexity in their reconstruction. In particular, apical and basal (upward) slice images are more difficult to segment than mid-ventricular slices.

Before the ventricle segmentation starts, a preliminary localization procedure is performed, in order to identify a region of interest and reduce the computational effort. This step is usually performed automatically, taking advantage of the movement of the heart on a fixed background: see *e.g.* Cocosco *et al.* (2004). Another automatic heart location is based on extracting rectangular subwindows from the image and computing specific features for them. Then, based on *a priori* chosen class, the subwindows satisfying specific features are recognized as belonging to the heart.

An initial class of ventricle segmentation methods makes use of little (or even no) *a priori* information. Usually in these methods the endocardium is first segmented by means of thresholding (see Section 3 and Goshtasby and

Turner 1995). Alternatively, *dynamic programming* methods have been considered, where the optimal path in a cost matrix is sought, assigning a low cost to boundaries. Taking advantage of the circular shape of the ventricle, polar coordinates are used and one-dimensional search is performed (Gupta *et al.* 1993). In order to exclude papillary muscles from the segmentation, possible strategies are the computation of the convex hull of the contour (van der Geest, Jansen, Buller and Reiber 1994) or a smoothing of the latter by fitting a parametric curve (Waiter *et al.* 1999). As in the vascular case, deformable models are used for cardiac segmentation (see Section 3). In particular, the functional to be minimized often includes a regularization term to control the smoothness of the reconstructed curves (Pham, Xu and Prince 2000).

Specific automatic methods that have been developed for cardiac image segmentation are based on strong *a priori* information concerning the shape of the ventricles, which is included in the segmentation algorithm by means of statistical models. These strategies are suited to cardiac segmentation because variability in heart shape among patients, as opposed to arteries, is very small in normal conditions. These statistical model-based segmentation strategies rely on identifying an average shape of available geometries forming a training set, and modelling the variability within the latter. This is usually done by means of *principal component analysis* of positions and, if needed, displacements, allowing computation of the eigenvalues and eigenvectors of the covariance matrix related to the training set. These strategies allow automatic segmentation without user intervention, at the expense of needing a training set. For example, deformable models have been extended to this framework by adding a term to the functional to be minimized that penalizes the distance to a reference model (*e.g.* the mean shape of the training set). Another very common statistical model-based strategy is *atlas-guided segmentation*. Given an atlas, that is, an integrated image from multiple segmentations, a registration procedure is performed based on mapping the coordinates of the image under investigation to those of the atlas (Lorenzo-Valdés *et al.* 2004). This transformation is then applied to the atlas obtaining the final segmentation. The registration process could be based on non-rigid transformations that account for elastic deformations. For a recent review of cardiac segmentation methods we refer to Petitjean and Dacher (2011).

Finally, we observe the importance of including the fibre orientation in the reconstructed geometries with respect to modelling electrical propagation and mechanical contraction in the muscle region. Indeed, as discussed in Section 5, the conduction velocity of the action potential propagation assumes different values along the fibres than in the tangential direction. Moreover, the stretching ability of the myocardium is facilitated along the fibre direction.

Diffusion-tensor-MRI is an MRI technology able to identify fibre orientation, but it is not yet used every day in clinical practice, and it is difficult to apply because of heart movement. See Nagler *et al.* (2015) for a preliminary study of the cardiac fibre estimation using synthetic diffusion-tensor-MRI datasets. Analytical representations of the fibres, suitably mapped onto the geometry under investigation, have been proposed to supply the lack of information provided by the imaging. For example, Peskin (1989) described fibres as geodesics, whereas Pravdin *et al.* (1989) represented them by means of spiral surfaces. Other strategies are based on computational generation of the fibre orientation to provide a plausible configuration, for example by means of the solution of a Poisson equation (Bayer, Blake, Plank and Trayanova 2012, Rossi *et al.* 2014, Wong and Kuhl 2014), or by using the unscented Kalman filter (Nagler, Bertoglio, Gee and Wall 2013).

The Purkinje fibres are not detectable via classical radiological acquisition since their thickness falls below the spatial resolution. For this reason, Abboud, Berenfeld and Sadeh (1991), Sebastian, Zimmerman, Romero and Frangi (2011) and Ijiri *et al.* (2008) proposed exploiting the fractal nature of these fibres to generate a realistic Purkinje network, whereas Vergara *et al.* (2014), Palamara *et al.* (2014) and Palamara, Vergara, Faggiano and Nobile (2015) have personalized such fractal networks to the patient at hand by including patient-specific measurements of the activation times (see Section 6.2 for a description of the latter).

For the muscle region mesh generation, the strategies described in Section 3 could be applied to cardiac geometries as well. For ideal ellipsoid geometries, usually considered in numerical experiments to test the performance of the numerical algorithms, structured hexahedral meshes are often used, exploiting the symmetry of the ventricles around the long axis: see *e.g.* Pavarino, Scacchi and Zampini (2015). However, unstructured meshes composed of tetrahedra have also been considered (Goktepe and Kuhl 2010), in particular for real geometries reconstructed by MRI or CT (see *e.g.* Rossi *et al.* 2014, Wong, Goktepe and Kuhl 2013, Vergara *et al.* 2016), or for the atria (Virag *et al.* 2002). Hybrid unstructured meshes composed of tetrahedra and hexahedra have also been successfully considered (Gurev *et al.* 2015). We notice that for the solution of the mechanical problem in the muscle region, no particular requirements are needed for mesh generation, whereas for the electrical propagation problem, due to the very steep front (about 200 μm), the required mesh resolution should be at least of the order of 100 μm in order to spatially resolve this front (Clayton and Panfilov 2008). For mesh generation of the heart chambers, in view of the numerical solution of the fluid dynamics problem, unstructured tetrahedral elements are often considered: see *e.g.* Mittal *et al.* (2016). In this case the mesh resolution needs to be very fine, even smaller than that for CT or MRI technologies ($\simeq$ 0.5 mm), in order to capture the complex flow structures that arise, particularly in the left atrium and left ventricle.

6.2. Cardiac boundary data

The acquisition of boundary data on electrical activation in the muscle region is now possible thanks to specific therapeutic techniques. For example, the NavX system is able to locate accurately any electrode catheter inserted in the heart chambers, providing accurate, real-time recording of the local endocardial electrical activity at a point in contact with the catheter (Eitel *et al.* 2010, Vergara *et al.* 2014). In particular, this allows us to acquire endocardial maps of the activation times, defined as the time at which the action potential in a point reaches an intermediate value between the rest and the plateau potentials. However, this acquisition is performed only for specific therapeutic purposes, such as the ablation of anomalous electrical pathways.

For the mechanical problem involving the muscle region, data that are commonly available include the stresses exerted by the blood on the endocardium of the left ventricle and the endocardial and/or epicardial vessel wall displacements. Stresses are usually obtained from measurements of aortic pressure (see Section 3.2). This is a good approximation; as a matter of fact the ventricular and circulatory pressures are not the same (thus allowing the acceleration and deceleration of blood) but they are very similar: see Figure 5.2. Vessel wall displacements can be obtained from dynamic MRI or CT images, yielding 20–30 frames per heartbeat, providing the position of the endocardium and epicardium at multiple times. After suitable postprocessing, these techniques can provide an estimate of the vessel wall displacement (and thus velocity) by comparing two consecutive frames. The endocardial vessel wall velocity, thanks to a continuity argument, could also be interpreted as the blood velocity at the interface with the endocardium (Khalafvand, Zhong and Ng 2014, Mittal *et al.* 2016).

Another useful measurement that could be provided quite easily by means of Doppler echocardiographic methods or PC-MRA (see Section 3.2) is the flow rate at the mitral and aortic valve orifices. With PC-MRA technology, measurement of blood velocity is possible in principle at any point of the ventricles and atria chambers.

In Section 7 we will see how these data can be used to provide boundary conditions for the different cardiac models we are about to introduce.

6.3. Cardiac biological data

The classical *monodomain* and *bidomain* models, widely used to describe the electrical propagation in the muscle region and Purkinje network (see Section 7.1), would require the following data: the membrane capacitance per unit area C, the membrane surface-to-volume ratio χ, and the conductivities σ (see the next section). Their subject-specific acquisition is generally out of reach. An acceptable range of values for the capacitance C_m in the myocardium is 0.8–1.0 μF cm^{-2} (Roth 1991, Huiskamp 1998); the capacitance C_p in the Purkinje network is higher ($\sim$ 1.3 μF cm^{-2}: Legato 1973). The value of the membrane surface-to-volume ratio in the muscle region χ_m varies in the range 200–3000 cm^{-1} (Niederer *et al.* 2011, Potse *et al.* 2006), whereas for the Purkinje network, a measurement in a pig heart gives $\chi_p = 1467$ cm^{-1} (Stankovičová *et al.* 2003). In the myocardium, we need to distinguish between conductivity along the fibre direction (index f) and conductivity in the direction of sheets (index s). Sometimes, a third direction (orthogonal to sheets) is also considered (if this is the case we will use the index n). Moreover, with the bidomain model in mind, we also distinguish between the intra-cellular (index i) and extra-cellular (index e) conductivities. Following Roth (1997), acceptable ranges used in the numerical experiments are $\sigma_f^i \in (0.17, 0.34)$ S m^{-1}, $\sigma_s^i \in (0.02, 0.06)$ S m^{-1}, $\sigma_f^e \in (0.12, 0.62)$ S m^{-1}, $\sigma_s^e \in (0.08, 0.24)$ S m^{-1}.

Another classical model used to describe the activation of Purkinje and muscle cells is provided by the *eikonal* equation (see Section 7.1). In this case, one needs to prescribe the conduction velocity V explicitly, unlike the monodomain and bidomain models where this is determined by the conductivity and the membrane capacitance. Acceptable ranges of values of the conduction velocity are $V_f \in (0.6, 1.0)$ m s^{-1} (Kerckoffs *et al.* 2003), $V_s \simeq V_f/2$ (Frazier *et al.* 1988), $V_n \simeq V_f/4$ (Clayton *et al.* 2011), and $V_p \in (3.0 - 4.0)$ m s^{-1} (Iaizzo 2009), where again index f refers to the direction along the fibres in the muscle region, s to the direction along sheets, and p to the Purkinje network.

These electrical data are barely measurable *in vivo*, so subject-specific measurements are not usually available. Nevertheless, the use of extra data, such as the activation time at the endocardium provided by the NavX system, could be used to estimate some of these parameters by solving a suitable inverse problem; see *e.g.* Sermesant *et al.* (2012) and Vergara *et al.* (2014) for the case of the Purkinje network.

The parameters involved in the cardiac mechanical model depend on the chosen constitutive law. In general, linearization of the stress–strain curves gives the following values for the corresponding time-varying Young's modulus E (Rossi 2014): during a traction/compression test along the fibre direction, we have $E \simeq 20$ kPa during traction at small deformation and during compression, and $E \simeq 1500$ kPa during traction for deformation of about 15%; during traction the material behaves as transversally isotropic, so the stiffness in the two tangential directions (along sheets and orthogonal to fibres and sheets) is the same, whereas during compression the stiffness in the direction of the sheets is higher than that in the direction orthogonal to fibres and sheets. For the Poisson modulus, a value of 0.45 is usually considered.

Regarding the aortic valve leaflets, measurements in dogs exhibited low stiffness during systole ($\simeq$ 240 kPa) and increased stiffness during diastole ($\simeq$ 5000 kPa). For the mitral valve, different stiffness behaviour is observed for the two leaflets (anterior and posterior). In particular, Kunzelman *et al.* (1993) propose the following values for Young's modulus: for the anterior leaflet, $E = 6200$ kPa and $E = 2100$ kPa in the directions parallel and perpendicular to the annulus, respectively, and for the posterior leaflet, $E = 2300$ kPa and $E = 1900$ kPa, respectively.

Blood in the heart exhibits the same behaviour as in the circulatory systems, so its parameters (density and viscosity) are taken in the ranges reported in Section 3.3.

7. Modelling the heart

In this section we discuss the main mathematical models introduced so far to describe heart function and the related numerical strategies developed for their solution. In particular, Section 7.1 is devoted to the electrical propagation, Section 7.2 to the cardiac mechanics and electromechanical coupling, Section 7.3 to the ventricular fluid dynamics, Section 7.4 to the valve modelling and its interaction with blood fluid dynamics, and finally Section 7.5 to their integration.

7.1. Cardiac electrical activity

7.1.1. The bidomain model

As observed in Section 5.3, the electrical activation of the heart is the result of two processes: at the microscopic scales, the generation of ionic currents through the cellular membrane producing a local action potential, and at the macroscopic scales, the travelling of the action potential from cell to cell allowed by the presence of the gap junctions. The former is a discrete process, in the sense that there is a delay between the depolarization of a cardiomyocyte and its neighbours, whereas the latter can be assimilated to a smooth process (Durrer *et al.* 1970).

At the macroscopic level, the propagation of the potentials is described by means of partial differential equations, suitably coupled with ordinary differential equations modelling the ionic currents in the cells. In particular, the single membrane cell can be modelled as a capacitor separating charges that accumulate at its *intracellular* and *extracellular* surfaces. Moreover, as observed in Section 5, ionic currents cross the membrane through channels which open and close during excitation. A suitable model can therefore

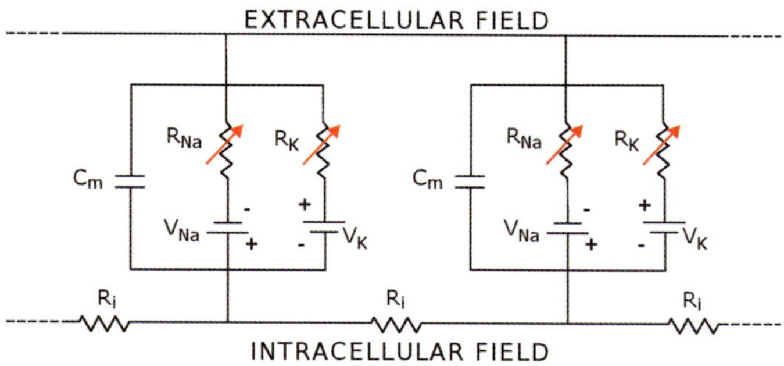

Figure 7.1. Electrical circuit for the sequence of two cardiac cells. Each consists of a capacitor and a series of resistances, one for each ionic current (here only sodium and potassium ionic channels are depicted). In the intracellular region, two adjacent cells are connected by a resistance representing a gap junction. However, the latter is not explicitly modelled at the macroscopic scales: instead its effect is hidden in the conductivity tensor (see the text).

be expressed via the simple electric circuit depicted in Figure 7.1, for which

$$I_m = \chi_m \left(C_m \frac{\partial V_m}{\partial t} + I_{ion} \right),$$

where I_m is the membrane current per unit volume, C_m is the membrane capacitance, χ_m is the surface area-to-volume ratio (see Section 6.3 for a quantification of the latter two), $V_m(t, \boldsymbol{x})$ is the trans-membrane potential, and $I_{ion}(t, \boldsymbol{x})$ are the ionic currents. Due to the conservation of current and charge, this current should equal the divergence of both the intracellular and extracellular current fluxes $\boldsymbol{j}_i$ and $\boldsymbol{j}_e$:

$$\nabla \cdot \boldsymbol{j}_i = -I_m, \quad \nabla \cdot \boldsymbol{j}_e = I_m. \tag{7.1}$$

Ohm's law in the intracellular and extracellular regions gives

$$\boldsymbol{j}_i = -\boldsymbol{\Sigma}_i \nabla \phi_i, \quad \boldsymbol{j}_e = -\boldsymbol{\Sigma}_e \nabla \phi_e, \tag{7.2}$$

where $\boldsymbol{\Sigma}_i, \boldsymbol{\Sigma}_e$ are the conductivity tensors and $\phi_i(t, \boldsymbol{x}), \phi_e(t, \boldsymbol{x})$ are the intracellular and extracellular potentials, so that

$$V_m = \phi_i - \phi_e. \tag{7.3}$$

Note that, due to the anisotropy of the cardiac tissue induced by the presence of fibres and sheets, each conductivity tensor is in general expressed in terms of three scalar quantities representing the conductivities along the fibre direction $\boldsymbol{a}_f(\boldsymbol{x})$, the direction $\boldsymbol{a}_s(\boldsymbol{x})$ orthogonal to $\boldsymbol{a}_f$ and tangential to sheets, and the direction $\boldsymbol{a}_n(\boldsymbol{x})$ orthogonal to sheets, that is,

$$\boldsymbol{\Sigma}_\beta = \sigma_f^\beta \boldsymbol{a}_f \boldsymbol{a}_f^T + \sigma_s^\beta \boldsymbol{a}_s \boldsymbol{a}_s^T + \sigma_n^\beta \boldsymbol{a}_n \boldsymbol{a}_n^T, \quad \beta = i, e. \tag{7.4}$$

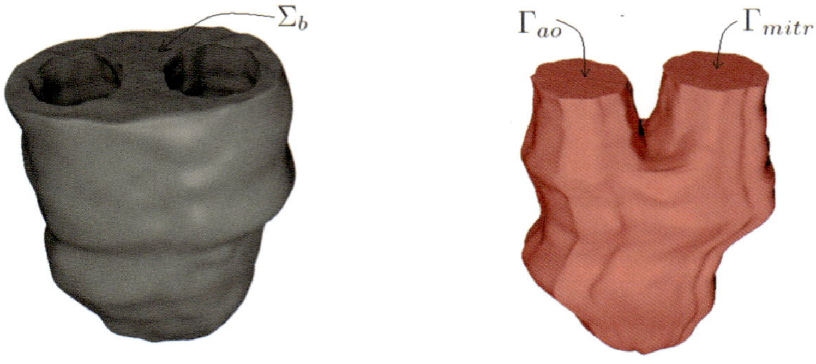

(a) (b)

Figure 7.2. (a) Left ventricular myocardial domain obtained by the cut at the base (corresponding to Σ_b), and (b) corresponding fluid cavity domain.

Putting together all the above equations and using a homogenization procedure (see *e.g.* Colli Franzone *et al.* 2014 for a rigorous derivation), we obtain for each $t > 0$ the following system of two partial differential equations called the *parabolic–parabolic* (PP) *formulation of the bidomain* equations:

$$\chi_m C_m \frac{\partial V_m}{\partial t} - \nabla \cdot (\boldsymbol{\Sigma}_i \nabla \phi_i) + \chi_m I_{ion} = I_i^{ext} \quad \text{in } \Omega_{mus}, \tag{7.5a}$$

$$-\chi_m C_m \frac{\partial V_m}{\partial t} - \nabla \cdot (\boldsymbol{\Sigma}_e \nabla \phi_e) - \chi_m I_{ion} = -I_e^{ext} \quad \text{in } \Omega_{mus}, \tag{7.5b}$$

where $I_i^{ext}(t, \boldsymbol{x})$, $I_e^{ext}(t, \boldsymbol{x})$ are applied currents per unit volume.

Thanks to (7.1), $\nabla \cdot (\boldsymbol{j}_i + \boldsymbol{j}_e) = 0$; thus, using (7.2) and (7.3), we obtain the following *parabolic–elliptic* (PE) *formulation of the bidomain* equations:

$$\chi_m C_m \frac{\partial V_m}{\partial t} - \nabla \cdot (\boldsymbol{\Sigma}_i (\nabla \phi_e + \nabla V_m)) + \chi_m I_{ion} = I_i^{ext} \quad \text{in } \Omega_{mus}, \tag{7.6a}$$

$$-\nabla \cdot (\boldsymbol{\Sigma}_i \nabla V_m) - \nabla \cdot ((\boldsymbol{\Sigma}_i + \boldsymbol{\Sigma}_e) \nabla \phi_e) = I_i^{ext} - I_e^{ext} \quad \text{in } \Omega_{mus}. \tag{7.6b}$$

Due to the homogenization procedure, the effect of the gap junctions, which at the cellular level contributes to determining the current flux $\boldsymbol{j}_i$, is hidden in the conductivity tensor $\boldsymbol{\Sigma}_i$. We also notice that both bidomain problems (7.5) and (7.6) hold in the entire computational domain Ω_{mus} given by the union of the myocardium with the endocardium and epicardium: see Figures 6.1 and 7.2. Indeed, again because of the homogenization procedure, no geometric distinction is made between the intracellular and extracellular regions, even if their different functionality is maintained in the bidomain models.

7.1.2. Cardiac cell models

In order to close equations (7.5) and (7.6), we need to provide a model for the ionic current I_{ion}. In what follows we briefly describe three families of models, featuring different levels of complexity and accuracy.

The first family, the so-called *reduced ionic models*, only provide a description of the action potential and disregard sub-cellular processes:

$$I_{ion} = f(V_m, \boldsymbol{w}), \tag{7.7a}$$

$$\frac{\partial \boldsymbol{w}}{\partial t} = \boldsymbol{g}_w(V_m, \boldsymbol{w}), \tag{7.7b}$$

where $f, \boldsymbol{g}_w$ are suitable functions, while $\boldsymbol{w} : [0,T] \times \Omega_{mus} \to \mathbb{R}^M$ collects the so-called *gating variables* which represent the percentage of open channels per unit area of the membrane. The most celebrated reduced model for ventricular cells is the *FitzHugh–Nagumo* model (FitzHugh 1961), where $f(V_m, w) = -kV_m(V_m - a)(V_m - 1) - w$ and $g_w(V_m, w) = \epsilon(V_m - \gamma w)$, for suitable constant parameters k, a, γ. In this case the gating variable w in fact plays the role of a *recovery* function, which allows us to model the refractoriness of cells. Other, more sophisticated ventricular cell models of this family include the *Rogers–McCulloch* model (Rogers and McCulloch 1994), the *Aliev–Panfilov* model (Aliev and Panfilov 1996), the *Fenton–Karma* model (Fenton and Karma 1998) and the *Bueno-Orovio* model (Bueno-Orovio, Cherry and Fenton 2008). Like FitzHugh–Nagumo, the first two models are characterized by the dynamics of one gating variable and by a cubic non-linear expression of the ionic current. In contrast, the Fenton–Karma model and its Bueno-Orovio variant, specifically applied to human ventricular cells, include two and three gating variables, respectively, and a more complex non-linearity in the ionic current expression. These simple models are very appealing, especially because their parameters have a direct physical interpretation, such as the action potential duration, allowing easy setting of the model properties. For example, the Aliev–Panfilov model has been used successfully in the initial simulations of ventricular fibrillation in a real geometry (Panfilov 1999). However, they are not able to describe any process occurring at the level of the ionic channels or the cell, so they are recommended when one is only interested in electrical activity of the heart.

The second family of ventricular cell models we consider is that of the so-called *first-generation* models. Unlike reduced models, which express the ionic current by means of the sole function f, they allow explicit description of the kinetics of different ionic currents by using several gating variables. They are given by

$$I_{ion} = \sum_{k=1}^{N} I_k(V_m, \boldsymbol{w}), \quad I_k = G_k\left(\prod_{j=1}^{M} w_j^{p_{j_k}}\right)(V_m - V_k), \quad \frac{\partial \boldsymbol{w}}{\partial t} = \boldsymbol{g}_w(V_m, \boldsymbol{w}),$$

where N is the total number of ionic currents, M is the total number of gating variables, V_k is the Nernst potential of the kth ion (a constant value corresponding to the thermodynamic equilibrium of the ion at hand), I_k is the current related to the kth ion, p_{j_k} accounts for the influence of the jth gating variable on the kth ionic current (possibly vanishing) and G_k is the maximal conductance of the kth ion. The M components of $\boldsymbol{g}$ usually have the expression

$$g_{w,j} = (w_j^{\infty}(V_m) - w_j)/\tau_j(V_m),$$

where w_j^{∞} is the equilibrium state and τ_j the characteristic time constant. The most famous model of this family, the *Hodgkin–Huxley* (HH) model (Hodgkin and Huxley 1952), depends on three ionic currents, namely the sodium, potassium and leakage currents, and three gating variables:

$$I_{Na} = G_{Na} w_1^3 w_2 (V_m - V_{Na}), \quad I_K = G_K w_3^4 (V_m - V_K), \quad I_L = G_L (V_M - V_L).$$

Although introduced to describe the action potentials in nerves, the HH model inspired all the following models introduced specifically for the ventricle. Among these, we cite the *Beeler–Reuter* model (Beeler and Reuter 1977), the *Luo–Rudy I* model (Luo and Rudy 1991) and the *ten Tusscher–Panfilov* model (ten Tusscher and Panfilov 2006). These models have been widely used to study specific features of the ventricular electrical activation, such as re-entry and fibrillation (Xie *et al.* 2004).

Finally, we mention the family of *second-generation* ventricular cell models, such as the *Luo–Rudy dynamic* model (Luo and Rudy 1994*a*, 1994*b*), which, unlike first-generation models, provide a detailed description of ion concentration variables $\boldsymbol{c}$ and of many processes allowing for the study of channelopathies and drug action, for example. However, due to their increased complexity, the computational time required is huge for a complete heart model, and tuning of parameters is often very demanding. We refer to Clayton and Panfilov (2008) for a discussion of second-generation models and Rudy and Silva (2006) for a general review of cardiac cell models.

Although most research has focused on ventricular cell models like those mentioned above, specific models have also been introduced for the atrial cells (*e.g.* Hilgemann and Noble 1987) and sinoatrial node cells (*e.g.* Yanagihara, Noma and Irisawa 1980).

All the cardiac cell models belonging to the three families described above, used in combination with the bidomain problem (7.5) or (7.6), lead to a system of two PDEs coupled with two systems of ODEs, that is, the equations for the gating variables and ion concentrations at each point $\boldsymbol{x}$. The general form of this coupled problem is as follows (we only describe the PP case in detail). At each time $t > 0$, find the potentials V_M, ϕ_i and ϕ_e, the gating variable $\boldsymbol{w}$ and the ion concentrations $\boldsymbol{c}$ such that

$$\chi_m C_m \frac{\partial V_m}{\partial t} - \nabla \cdot (\boldsymbol{\Sigma}_i \nabla \phi_i) + \chi_m I_{ion}(V_m, \boldsymbol{w}, \boldsymbol{c}) = I_i^{ext} \quad \text{in } \Omega_{mus}, \tag{7.9a}$$

$$-\chi_m C_m \frac{\partial V_m}{\partial t} - \nabla \cdot (\boldsymbol{\Sigma}_e \nabla \phi_e) - \chi_m I_{ion}(V_m, \boldsymbol{w}, \boldsymbol{c}) = -I_e^{ext} \quad \text{in } \Omega_{mus}, \tag{7.9b}$$

$$I_{ion} = \sum_{k=1}^{N} I_k(V_m, \boldsymbol{w}, \boldsymbol{c}) \quad \text{in } \Omega_{mus}, \tag{7.9c}$$

$$I_k = G_k \left(\prod_{j=1}^{M} w_j^{p_{jk}} \right) (V_m - V_k(\boldsymbol{c})) \quad \text{in } \Omega_{mus}, \tag{7.9d}$$

$$\frac{\partial \boldsymbol{w}}{\partial t} = \boldsymbol{g}_w(V_m, \boldsymbol{w}) \quad \text{in } \Omega_{mus}, \tag{7.9e}$$

$$\frac{\partial \boldsymbol{c}}{\partial t} = \boldsymbol{g}_c(V_m, \boldsymbol{w}, \boldsymbol{c}) \quad \text{in } \Omega_{mus}, \tag{7.9f}$$

where, together with the notation introduced above for the reduced and first-generation models, $\boldsymbol{c} : [0, T] \times \Omega_{mus} \to \mathbb{R}^S$ collects the S ionic concentration variables and $\boldsymbol{g}_c$ is a suitable function: see *e.g.* Colli Franzone *et al.* (2014). We observe, in general, the dependence of the Nernst potential V_k on the variable $\boldsymbol{c}$. Well-posedness results of the previous coupled problem are provided in Colli Franzone and Savaré (2002), where the existence and uniqueness of the solution of the PP formulation coupled with the FitzHugh–Nagumo model is proved, and in Bourgault, Coudière and Pierre (2006), where a Faedo–Galerkin technique is applied to the PE formulation coupled with a general first-generation cell model.

The ODE systems modelling the gating variables and the ionic concentration variables are in general stiff, since the Jacobians $\partial \boldsymbol{g}_w / \partial \boldsymbol{w}$ and $\partial \boldsymbol{g}_c / \partial \boldsymbol{c}$ exhibit a wide range of eigenvalues.

For each $t > 0$, the weak formulation of the bidomain model (7.9) together with homogeneous Neumann conditions and initial conditions (see Section 7.1.4) reads as follows. Given $I_i^{ext}(t)$, $I_e^{ext}(t) \in L^2(\Omega_{mus})$, find $V_m(t), \phi_e(t), \phi_i(t) \in H^1(\Omega_{mus})$, $\boldsymbol{w} \in [L^2(\Omega_{mus})]^M$ and $\boldsymbol{c} \in [L^2(\Omega_{mus})]^S$ such that

$$\chi_m C_m \int_{\Omega_{mus}} \frac{\partial V_m}{\partial t} z \, \mathrm{d}\omega + \int_{\Omega_{mus}} \boldsymbol{\Sigma}_i \nabla \phi_i \cdot \nabla z \, \mathrm{d}\omega + \chi_m \int_{\Omega_{mus}} I_{ion}(V_m, \boldsymbol{w}, \boldsymbol{c}) z \, \mathrm{d}\omega = \int_{\Omega_{mus}} I_i^{ext} z \, \mathrm{d}\omega, \tag{7.10a}$$

$$- \chi_m C_m \int_{\Omega_{mus}} \frac{\partial V_m}{\partial t} z \, d\omega + \int_{\Omega_{mus}} \boldsymbol{\Sigma}_e \nabla \phi_e \cdot \nabla z \, d\omega$$

$$- \chi_m \int_{\Omega_{mus}} I_{ion}(V_m, \boldsymbol{w}, \boldsymbol{c}) z \, d\omega = - \int_{\Omega_{mus}} I_e^{ext} z \, d\omega, \tag{7.10b}$$

$$\int_{\Omega_{mus}} \frac{\partial \boldsymbol{w}}{\partial t} \cdot \boldsymbol{y} \, d\omega = \int_{\Omega_{mus}} \boldsymbol{g}(V_m, \boldsymbol{w}) \cdot \boldsymbol{y} \, d\omega, \tag{7.10c}$$

$$\int_{\Omega_{mus}} \frac{\partial \boldsymbol{c}}{\partial t} \cdot \boldsymbol{\zeta} \, d\omega = \int_{\Omega_{mus}} \boldsymbol{g}_c(V_m, \boldsymbol{w}, \boldsymbol{c}) \cdot \boldsymbol{\zeta} \, d\omega, \tag{7.10d}$$

for all $z \in H^1(\Omega_{mus})$, $\boldsymbol{y} \in [L^2(\Omega_{mus})]^M$ and $\boldsymbol{\zeta} \in [L^2(\Omega_{mus})]^S$, together with (7.9c, 7.9d).

7.1.3. Reduced continuous models: the monodomain and eikonal equations
To reduce the complexity of the bidomain models (7.5) and (7.6), an assumption of proportionality between the intracellular and extracellular conductivities is introduced, that is, $\boldsymbol{\Sigma}_e = \xi \boldsymbol{\Sigma}_i$ for a suitable constant ξ. Substituting this relation in (7.6b), eliminating $\boldsymbol{\Sigma}_e$ and substituting the corresponding relation for $\boldsymbol{\Sigma}_i$ in (7.6a), we obtain the following *monodomain* equation. For each $t > 0$, find the trans-membrane potential V_m such that

$$\chi_m C_m \frac{\partial V_m}{\partial t} - \nabla \cdot (\boldsymbol{\Sigma} \nabla V_m) + \chi_m I_{ion} = I^{ext} \quad \text{in } \Omega_{mus}, \tag{7.11}$$

where

$$\boldsymbol{\Sigma} = \frac{\xi}{1+\xi} \boldsymbol{\Sigma}_i \tag{7.12}$$

is the effective conductivity and

$$I^{ext} = \frac{\xi I_i^{ext} + I_e^{ext}}{1+\xi}.$$

Again, a ventricular cell model is needed to provide the ionic current I_{ion}. The same models discussed above for the coupling with the bidomain problem are used in combination with the monodomain problem too. Once the trans-membrane potential V_m has been computed, the extracellular potential ϕ_e could be computed as a postprocessing by solving the elliptic problem (7.6b).

For each $t > 0$, the weak formulation of the monodomain problem (7.11) together with homogeneous Neumann conditions and initial conditions (see Section 7.1.4) reads as follows. Given $I^{ext}(t) \in L^2(\Omega_{mus})$, find $V_m(t) \in H^1(\Omega_{mus})$, $\boldsymbol{w} \in [L^2(\Omega_{mus})]^M$ and $\boldsymbol{c} \in [L^2(\Omega_{mus})]^S$ such that

$$\chi_m C_m \int_{\Omega_{mus}} \frac{\partial V_m}{\partial t} z \, d\omega + \int_{\Omega_{mus}} \boldsymbol{\Sigma} \nabla V_m \cdot \nabla z \, d\omega$$

$$+ \chi_m \int_{\Omega_{mus}} I_{ion}(V_m, \boldsymbol{w}, \boldsymbol{c}) z \, d\omega = \int_{\Omega_{mus}} I^{ext} z \, d\omega, \tag{7.13a}$$

$$\int_{\Omega_{mus}} \frac{\partial \boldsymbol{w}}{\partial t} \cdot \boldsymbol{y} \, \mathrm{d}\omega = \int_{\Omega_{mus}} \boldsymbol{g}(V_m, \boldsymbol{w}) \cdot \boldsymbol{y} \, \mathrm{d}\omega, \tag{7.13b}$$

$$\int_{\Omega_{mus}} \frac{\partial \boldsymbol{c}}{\partial t} \cdot \boldsymbol{\zeta} \, \mathrm{d}\omega = \int_{\Omega_{mus}} \boldsymbol{g}_c(V_m, \boldsymbol{w}, \boldsymbol{c}) \cdot \boldsymbol{\zeta} \, \mathrm{d}\omega, \tag{7.13c}$$

for all $z \in H^1(\Omega_{mus})$, $\boldsymbol{y} \in [L^2(\Omega_{mus})]^M$ and $\boldsymbol{\zeta} \in [L^2(\Omega_{mus})]^S$, together with (7.9c, 7.9d).

Although the hypothesis underlying the monodomain model, that is, the proportionality between the internal and external conductivities, is not physiological, as shown by experiment, in some cases this model provides a very accurate solution compared to the bidomain model. In particular, this is true when there is no injection of current in the extracellular region (Colli Franzone, Pavarino and Taccardi 2005, Potse *et al.* 2006). In contrast, when an external current is injected such as in defibrillation, the monodomain solution is no longer accurate, and the bidomain model is mandatory since the unequal anisotropy is fundamental to successful description of these scenarios (Trayanova 2006).

A further simplification is provided by the *eikonal* equation. Starting from the bidomain model coupled with a simplified representation of the ionic current, which does not consider any gating variable and allows for the description only of the depolarization phase, Colli Franzone, Guerri and Rovida (1990) derive the following *eikonal-diffusion* equation:

$$c_o \sqrt{\nabla \psi \cdot \boldsymbol{M} \nabla \psi} - \nabla \cdot (\boldsymbol{M} \nabla \psi) = 1, \tag{7.14}$$

where $\psi(\boldsymbol{x})$ is the unknown *activation time* (see Section 6.2), c_o denotes the velocity of the depolarization wave along the fibre direction for a planar wavefront, and $\boldsymbol{M} = \boldsymbol{\Sigma}/(\chi C_m)$. A different derivation has been provided in Keener (1991), leading to the following *eikonal-curvature* equation:

$$c_o \sqrt{\nabla \psi \cdot \boldsymbol{M} \nabla \psi} - \sqrt{\nabla \psi \cdot \boldsymbol{M} \nabla \psi} \nabla \cdot \left(\frac{\boldsymbol{M} \nabla \psi}{\sqrt{\nabla \psi \cdot \boldsymbol{M} \nabla \psi}} \right) = 1. \tag{7.15}$$

These are both steady equations providing information on the activation of each cell. The contours of $\psi(\boldsymbol{x})$ give the position of the wavefront at time $t = \psi$. The eikonal-diffusive model (7.14) is an elliptic equation, where the propagation speed is influenced by the tissue surrounding the wavefront. Once the activation time ψ has been computed, it is possible to obtain an approximate value of the extracellular potential ϕ_e by solving at each time step a suitable elliptic problem: see Colli Franzone and Guerri (1993).

In contrast, the eikonal-curvature model (7.15) is of parabolic type since the 'diffusive' term lacks the second derivative in the direction of propagation. This term is also proportional to an anisotropic generalization of the mean curvature (Tomlinson, Hunter and Pullan 2002). This implies that the propagation is faster when the wavefront is concave. This is in accordance with the diffusion of charge, which allows for faster depolarization in regions close to already depolarized tissues.

The eikonal equations are unsuitable for recovering the action potential and the ionic currents. However, they provide accurate results about the activation of cells even in complex scenarios such as front–front collision: see *e.g.* Colli Franzone and Guerri (1993) for the eikonal-diffusive model.

The eikonal models are, however, very appealing from the computational point of view. First of all, they consist of a single steady PDE. Although non-linear, they do not require coupling with ODE systems. More importantly, the activation time, unlike the trans-membrane potential, does not feature any internal or boundary layer, so no special restriction on the mesh is needed in this case (see Section 6.1).

7.1.4. Boundary conditions and Purkinje network models

We discuss here the initial and boundary conditions of the problems introduced above. The bidomain and monodomain equations and the ODE systems for the gating variables and ionic concentrations need to be equipped with suitable initial conditions, that is,

$$V_m|_{t=0} = V_{m,0}, \quad \boldsymbol{w}|_{t=0} = \boldsymbol{w}_0, \quad \boldsymbol{c}|_{t=0} = \boldsymbol{c}_0 \quad \text{in } \Omega_{mus},$$

for given functions $V_{m,0}(\boldsymbol{x}), \boldsymbol{w}_0(\boldsymbol{x}), \boldsymbol{c}_0(\boldsymbol{x})$.

For the boundary conditions for the bidomain, monodomain and eikonal-diffusion problems, a homogeneous Neumann condition is commonly prescribed at the external surface Σ_{epi} of the epicardium and, for a ventricular domain only, at the base Σ_b (see Figures 6.1 and 7.2(a)) to prescribe null outgoing current fluxes. In particular, the following conditions have to be prescribed on $\Sigma_{epi} \cup \Sigma_b$:

$$\begin{aligned}
&(\boldsymbol{\Sigma}_\beta \nabla \phi_\beta) \cdot \boldsymbol{n} = 0 \quad \beta = i, e && \text{for PP (7.5)},\\
&\left.\begin{aligned}&(\boldsymbol{\Sigma}_i \nabla (V_m + \phi_e)) \cdot \boldsymbol{n} = 0\\ &((\boldsymbol{\Sigma}_i + \boldsymbol{\Sigma}_e)\nabla \phi_e) \cdot \boldsymbol{n} + (\boldsymbol{\Sigma}_i \nabla V_m) \cdot \boldsymbol{n} = 0\end{aligned}\right\} && \text{for PE (7.6)},\\
&(\boldsymbol{\Sigma} \nabla V_m) \cdot \boldsymbol{n} = 0 && \text{for monodomain (7.11)},\\
&(\boldsymbol{M} \nabla \psi) \cdot \boldsymbol{n} = 0 && \text{for eikonal-diffusion (7.14)}.
\end{aligned}$$

Moreover, for the bidomain problems (7.5) and (7.6), they force the following compatibility conditions on the applied external currents:

$$\int_{\Omega_{mus}} I_i^{ext} = \int_{\Omega_{mus}} I_e^{ext}.$$

On the internal surface Σ_{endo} of the endocardium, again Neumann conditions are prescribed. In this case, however, they could be non-homogeneous at specific stimulation points (*e.g.* the atrioventricular node and the points of the bundle of His). For the eikonal problem, Dirichlet data on the activation time could be prescribed at some specific locations if they are available, thanks to the NavX system, for example; see Section 6.2 (Sermesant *et al.* 2012). When redundant (*e.g.* when the electrical problem in the myocardium is coupled with the Purkinje network: see below) these data have been used to solve inverse problems, for example to estimate the conduction velocity in the myocardium (Sermesant *et al.* 2012) or to obtain personalized Purkinje networks (Vergara *et al.* 2014, Palamara *et al.* 2014, Palamara *et al.* 2015).

If the mathematical model accounts for the presence of the Purkinje network, interface conditions on Σ_{endo} describing the continuity of the current and of the potential at the Purkinje muscle junctions (PMJs) are implicitly provided for the bidomain and monodomain problems by the solution of the coupled muscle region/Purkinje network problem (Vigmond and Clements 2007, Vergara *et al.* 2016). For the sake of exposition, we will not detail the bidomain and monodomain models for the Purkinje network, instead referring interested readers to Vigmond and Clements (2007), Bordas *et al.* (2012) and Vergara *et al.* (2016). We will only note that, unlike the muscular case, in the network the gap junctions connecting two consecutive Purkinje cells are often explicitly modelled by means of resistances. Specific Purkinje cell models have also been developed, with the same structure as those developed for the muscular cells: see *e.g.* DiFrancesco and Noble (1985). However, we will describe the mechanisms of coupling; in particular, we will refer to the coupled problem obtained by considering the monodomain problem both in the muscle region and in the Purkinje network (Vergara *et al.* 2016). We consider N Purkinje muscle junctions located at $\boldsymbol{x} = \boldsymbol{s}_j$ and we assume that each of them could be modelled by means of a resistance R_{PMJ}. Then, the *monodomain/monodomain coupled problem* reads as follows. For each $t > 0$, find V_p, V_m, $\boldsymbol{w}_p$, $\boldsymbol{w}$ and γ_j, $j = 1, \ldots, N$, such that

$$P_m\left(V_m, \boldsymbol{w}, \sum_{j=1}^{N} \frac{1}{A_r}\mathcal{I}_{\mathcal{B}_r(\boldsymbol{s}_j)}\gamma_j + I^{ext}\right) = 0, \tag{7.16a}$$

$$P_p(V_p, \boldsymbol{w}_p, \boldsymbol{\gamma}) = 0, \tag{7.16b}$$

$$\gamma_j = \frac{V_p(\boldsymbol{s}_j) - \frac{1}{A_r}\int_{\mathcal{B}_r(\boldsymbol{s}_j)} V_m \,\mathrm{d}\boldsymbol{x}}{R_{PMJ}} \quad j = 1, \ldots, N, \tag{7.16c}$$

where $P_m(V_m, \boldsymbol{w}, F) = 0$ denotes the monodomain problem in the myocardium with source term F, $P_p(V_p, \boldsymbol{w}_p, \boldsymbol{\eta}) = 0$ is the monodomain problem in the Purkinje network with Neumann conditions with data η_j at the Purkinje muscle junctions, V_p and $\boldsymbol{w}_p$ are the trans-membrane potential and the gating variables in the Purkinje network, γ_j are the PMJ currents which are determined by Ohm's laws (7.16c), $\mathcal{I}_Y$ is the characteristic function related to the region $Y \subset \Omega_{mus}$, $\mathcal{B}_r(\boldsymbol{s}_j)$ is the ball of radius r centred at the point $\boldsymbol{s}_j$, and A_r is the volume of this ball. We observe that the two monodomain problems are coupled by means of the PMJ currents γ_j: for the 3D problem the latter act as source terms distributed in balls of radius r, whereas for the network they act as Neumann conditions (Bordas *et al.* 2012). A similar approach could be considered for the bidomain problems as well.

The coupling between eikonal muscular and Purkinje network problems has been addressed by Vergara *et al.* (2014) for normal propagation and Palamara *et al.* (2014) for pathological propagations.

7.1.5. Computing the surface electrocardiogram signals

The computation of the potentials related to electrical propagation in the heart could be used to numerically compute the surface electrocardiogram (ECG) signals, that is, electrical potential on the surface of the body, thus simulating what happens in normal clinical practice. This could be obtained by coupling the bidomain or monodomain problem with the propagation in the torso, modelled by a simple diffusion problem for the extracellular potential at each $t > 0$:

$$\nabla \cdot (\boldsymbol{\Sigma}_T \nabla \phi_T) = 0 \quad \text{in } \Omega_T,$$

where Ω_T is the torso domain that surrounds the heart domain Ω_{mus} (*i.e.* the heart boundary surface $\Sigma_{epi} \cup \Sigma_b$ denotes the interface with the torso), $\Omega_T \cap \Omega_{mus} = \emptyset$, with ϕ_T denoting the extracellular potential in the torso. Homogeneous Neumann conditions have to be applied on the external torso surface Σ_T, whereas the following interface conditions, which replace the boundary conditions for the heart problem, need to be prescribed at the torso/heart interface:

$$\begin{aligned}
&\phi_e = \phi_T && \text{on } \Sigma_{epi} \cup \Sigma_b, \\
&(\boldsymbol{\Sigma}_e \nabla \phi_e) \cdot \boldsymbol{n} = (\boldsymbol{\Sigma}_T \nabla \phi_T) \cdot \boldsymbol{n} && \text{on } \Sigma_{epi} \cup \Sigma_b, \\
&(\boldsymbol{\Sigma}_i \nabla (V_m + \phi_e)) \cdot \boldsymbol{n} = 0 && \text{on } \Sigma_{epi} \cup \Sigma_b.
\end{aligned}$$

7.1.6. Numerical discretization

Numerical solution of the bidomain and monodomain problems is very demanding. Together with the strict constraint on the spatial mesh size due to the propagation of a very steep front (see Section 6.1), a time step of the order of tenths or even hundredths of milliseconds must be used in order to capture the fast dynamics characterizing the propagation, with time constants of the order of 0.1 ms. Moreover, as discussed below, the algebraic solution of the linear systems arising at each time step using finite elements, for example, requires careful treatment, due to the coupled nature of the problem, the singularity of some of the matrices involved, and the presence of the non-linear term due to the ionic currents.

Several time discretization strategies have been considered so far. We can group them into four main categories, *explicit, semi-implicit, implicit* and *operator splitting-based* methods, which are now presented briefly. We start by introducing the finite element matrices and vectors:

$$\begin{aligned}
(M)_{jk} &= \int_{\Omega_{mus}} \psi_k \psi_j \, \mathrm{d}\omega, \\
(A_\beta)_{jk} &= \int_{\Omega_{mus}} \boldsymbol{\Sigma}_\beta \nabla\psi_k \cdot \nabla\psi_j \, \mathrm{d}\omega, \qquad \beta = i, e \\
(\boldsymbol{I}_{ion}(\boldsymbol{V}_m^n, \boldsymbol{W}^r, \boldsymbol{C}^s))_j &= \int_{\Omega_{mus}} I_{ion}(V_m^n, \boldsymbol{w}^r, \boldsymbol{c}^s)\psi_j \, \mathrm{d}\omega, \\
(\boldsymbol{G}(\boldsymbol{V}_m^n, \boldsymbol{W}^r))_j &= \int_{\Omega_{mus}} \boldsymbol{g}(V_m^n, \boldsymbol{w}^r) \cdot \boldsymbol{y}_j \, \mathrm{d}\omega, \\
(\boldsymbol{S}(\boldsymbol{V}_m^n, \boldsymbol{W}^r, \boldsymbol{C}^s))_j &= \int_{\Omega_{mus}} \boldsymbol{g}_c(V_m^n, \boldsymbol{w}^r, \boldsymbol{c}^s) \cdot \boldsymbol{\zeta}_j \, \mathrm{d}\omega,
\end{aligned}$$

where ψ_i, $\boldsymbol{y}_j$ and $\boldsymbol{\zeta}_j$ denote the basis functions of the finite element spaces, and n, r, s are integers denoting discretized times. Vectors $\boldsymbol{V}_m$, $\boldsymbol{\Phi}_e$, $\boldsymbol{\Phi}_i$, $\boldsymbol{W}$ and $\boldsymbol{C}$ denote the unknown coefficients of the finite element solutions associated with the unknowns V_m, ϕ_e, ϕ_i, $\boldsymbol{w}$, $\boldsymbol{c}$. Note that the dimensions of $\boldsymbol{W}$ and $\boldsymbol{C}$ are MK and SK, respectively, where K is the number of degrees of freedom associated with the mesh and the choice of finite elements (*e.g.* the number of vertices for linear finite elements). In all the cases, in order to simplify the notation, we will set $I_e^{ext} = I_i^{ext} = 0$, and we assume that the running temporal index $^{n+1}$ is understood.

Explicit methods. In explicit methods, all the problems in (7.10) are discretized by means of an explicit scheme, for example forward Euler for both the PDE and ODE systems (Puwal and Roth 2007, Ethier and Bourgault 2008). This choice allows us to decouple the four blocks of the bidomain system (the two PDEs and the two ODEs system), involving only the mass matrix in the PP formulation, and thus in principle avoiding the need to solve any linear system provided that mass lumping is performed. In contrast, for the PE formulation, the absence of time derivatives in the second PDE implies that a linear system in the unknown $\boldsymbol{\Phi}_e$ needs to be solved in any case (we detail the forward Euler method):

$$\begin{aligned}
&\chi_m C_m M \frac{\boldsymbol{V}_m - \boldsymbol{V}_m^n}{\Delta t} = -A_i(\boldsymbol{V}_m^n + \boldsymbol{\Phi}_e^n) - \chi_m \boldsymbol{I}_{ion}(\boldsymbol{V}_m^n, \boldsymbol{W}^n, \boldsymbol{C}^n), \\
&A_i \boldsymbol{V}_m + (A_i + A_e)\boldsymbol{\Phi}_e = \boldsymbol{0}, \\
&M \frac{\boldsymbol{W} - \boldsymbol{W}^n}{\Delta t} = \boldsymbol{G}(\boldsymbol{V}_m^n, \boldsymbol{W}^n), \quad M \frac{\boldsymbol{C} - \boldsymbol{C}^n}{\Delta t} = \boldsymbol{S}(\boldsymbol{V}_m^n, \boldsymbol{W}^n, \boldsymbol{C}^n).
\end{aligned}$$

Explicit methods lead to a severe constraint on the time discretization of the type $\Delta t < h^2/\Sigma$, where Σ is a representative value of the conductivity (Puwal and Roth 2007). Dos Santos, Plank, Bauer and Vigmond (2005) showed for a model problem that an explicit method is not absolutely stable for a value of Δt much smaller than that required to capture the front propagation. For this reason and due to the increased CPU availability, explicit methods are no longer used so often.

Semi-implicit methods. ODE systems are usually solved at each time step by means of explicit or semi-implicit methods (in the latter case the dependence on V_m is treated explicitly). This suggests using semi-implicit methods (Keener and Bogar 1998, Pennacchio and Simoncini 2002, Colli Franzone and Pavarino 2004) for the whole coupled PDE/ODE problem. These methods are the most widely used together with operator splitting-based methods (see below). They are based on treating the diffusive term implicitly and the non-linear term explicitly. A possible semi-implicit (first-order) scheme for the PP formulation (7.10) is as follows:

$$
\begin{aligned}
&M\frac{\boldsymbol{W}-\boldsymbol{W}^n}{\Delta t} - \boldsymbol{G}(\boldsymbol{V}_m^n,\boldsymbol{W}) = \boldsymbol{0}, \quad M\frac{\boldsymbol{C}-\boldsymbol{C}^n}{\Delta t} - \boldsymbol{S}(\boldsymbol{V}_m^n,\boldsymbol{W},\boldsymbol{C}) = \boldsymbol{0},\\
&\chi_m C_m M\frac{\boldsymbol{V}_m-\boldsymbol{V}_m^n}{\Delta t} + A_i\boldsymbol{\Phi}_i + \chi_m \boldsymbol{I}_{ion}(\boldsymbol{V}_m^n,\boldsymbol{W},\boldsymbol{C}) = \boldsymbol{0},\\
&-\chi_m C_m M\frac{\boldsymbol{V}_m-\boldsymbol{V}_m^n}{\Delta t} + A_e\boldsymbol{\Phi}_e - \chi_m \boldsymbol{I}_{ion}(\boldsymbol{V}_m^n,\boldsymbol{W},\boldsymbol{C}) = \boldsymbol{0}
\end{aligned}
$$

(see *e.g.* Colli Franzone and Pavarino 2004).

A semi-implicit method like the one reported here has two nice properties. First of all, the two PDEs are decoupled by the ODE systems, greatly simplifying the numerical solution of the entire problem. This decoupling strategy is justified by noticing that the Jacobian matrices of a fully implicit discretization feature dominant values on the diagonal, suggesting weak coupling between potentials and gating/ionic concentration variables (Munteanu, Pavarino and Scacchi 2009). Secondly, it allows for a linearization of the non-linear reaction term given by the ionic currents, thus requiring the solution of a 2×2 block linear system.

Special attention has been paid to evaluation of the ionic current in a finite element context. In particular, two strategies have mainly been considered, namely *ionic current interpolation* (ICI) and *state variable interpolation* (SVI). In the first case, only the nodal values are used to build an interpolation of the ionic current, whereas in the second case the trans-membrane potential and the gating/ionic concentration variables are interpolated within each element and the ionic current is then evaluated by using these interpolated variables (Pathmanathan, Mirams, Southern and Whiteley 2011). The ICI approach is of course more efficient from the computational point of view, and allows expression of the reaction term by means of a mass matrix. However, it has lower accuracy with respect to state variable interpolation; in particular, the computed conduction velocity is generally larger than the true velocity: see Pathmanathan *et al.* (2011).

From the algebraic point of view, the solution of a semi-implicit discretized problem is very demanding, since the matrix of the linear system associated with the PDEs written in terms of the unknowns $\boldsymbol{\Phi}_i$ and $\boldsymbol{\Phi}_e$ is given by

$$B = \frac{\chi_m C_m}{\Delta t} \begin{bmatrix} M & -M \\ -M & M \end{bmatrix} + \begin{bmatrix} A_i & 0 \\ 0 & A_e \end{bmatrix}.$$

Both terms of this matrix are singular, the first due to the degenerate parabolic nature of the PP bidomain formulation and the second since each block A_β is related to a pure Neumann diffusive problem. However, matrix B is positive semidefinite, so the preconditioned conjugate gradient method is often used for its numerical solution (Pennacchio and Simoncini 2002, Colli Franzone and Pavarino 2004). Preconditioning is needed since the matrix B is highly ill-conditioned, due to the block 2×2 mass matrix which, unlike the classical mass matrix, worsens the spectrum of the stiffness matrix. Efficient preconditioners include block Jacobi and Gauss–Seidel (Mardal, Nielsen, Cai and Tveito 2007), block SSOR (Pennacchio and Simoncini 2002), multilevel additive Schwarz (Pavarino and Scacchi 2008), multigrid (Plank *et al.* 2007, Vigmond *et al.* 2008), and a functional block preconditioner obtained by the monodomain solver (Gerardo-Giorda *et al.* 2009).

Semi-implicit methods have also been proposed for the PE formulation: see *e.g.* Fernández and Zemzemi (2010). For example, with respect to the unknowns $\boldsymbol{V}_m$ and $\boldsymbol{\Phi}_e$, we have

$$B = \frac{\chi_m C_m}{\Delta t} \begin{bmatrix} M & 0 \\ 0 & 0 \end{bmatrix} + \begin{bmatrix} A_i & A_i \\ A_i & A_i + A_e \end{bmatrix}.$$

Again the matrix is singular, ill-conditioned, and semidefinite positive. In this context, we cite Pierre (2012) for an incomplete block LU factorization preconditioner, and Vigmond, Aguel and Trayanova (2002), where the two PDEs are decoupled by treating V_m explicitly in (7.6a) and ϕ_e in (7.6b).

In any case, semi-implicit methods are conditionally stable with a bound on Δt which is, however, independent of the mesh size (Fernández and Zemzemi 2010, Colli Franzone *et al.* 2014).

A variant of the semi-implicit method reported above arises from treating the reaction term I_{ion} implicitly. In this case, Newton–Krylov–Schwarz methods are very efficient (Munteanu *et al.* 2009). Second-order semi-implicit schemes have been successfully considered as well: see *e.g.* Ethier and Bourgault (2008).

Operator splitting-based methods. These methods separate the reaction operator from the diffusive operator, in a similar way to what is done in fractional step methods for fluid problems. They were first introduced for the monodomain problem by Qu and Garfinkel (1998). Here we present an operator splitting-based method for the PP formulation of the bidomain problem (Sundnes, Lines and Tveito 2005, Colli Franzone *et al.* 2014).

(1) Given the quantities at time step t^n, solve the reaction problem and the ODE systems in $(t^n, t^n + \theta\Delta t]$:

$$\chi_m C_m M \frac{\widetilde{\boldsymbol{V}}_m^{n+\theta} - \boldsymbol{V}_m^n}{\Delta t} + \chi_m \boldsymbol{I}_{ion}(\boldsymbol{V}_m^*, \boldsymbol{W}^*, \boldsymbol{C}^*) = \boldsymbol{0},$$

$$M \frac{\widetilde{\boldsymbol{W}}^{n+\theta} - \boldsymbol{W}^n}{\Delta t} - \boldsymbol{G}(\boldsymbol{V}_m^*, \boldsymbol{W}^*) = \boldsymbol{0},$$

$$M \frac{\widetilde{\boldsymbol{C}}^{n+\theta} - \boldsymbol{C}^n}{\Delta t} - \boldsymbol{S}(\boldsymbol{V}_m^*, \boldsymbol{W}^*, \boldsymbol{C}^*) = \boldsymbol{0}.$$

(2) Given $\widetilde{\boldsymbol{V}}_m^{n+\theta}$, solve the diffusion problems in $(t^n, t^{n+1}]$:

$$\chi_m C_m M \frac{\widehat{\boldsymbol{V}}_m^{n+1} - \widetilde{\boldsymbol{V}}_m^{n+\theta}}{\Delta t} + A_i \boldsymbol{\Phi}_i^* = \boldsymbol{0},$$

$$-\chi_m C_m M \frac{\widehat{\boldsymbol{V}}_m^{n+1} - \widetilde{\boldsymbol{V}}_m^{n+\theta}}{\Delta t} + A_e \boldsymbol{\Phi}_e^* = \boldsymbol{0}.$$

(3) Given $\widehat{\boldsymbol{V}}_m^{n+1}$, $\widetilde{\boldsymbol{W}}^{n+\theta}$, $\widetilde{\boldsymbol{C}}^{n+\theta}$, solve the reaction problem and the ODE systems in $(t^n + \theta\Delta t, t^{n+1}]$:

$$\chi_m C_m M \frac{\boldsymbol{V}_m^{n+1} - \widehat{\boldsymbol{V}}_m^{n+1}}{\Delta t} + \chi_m \boldsymbol{I}_{ion}(\boldsymbol{V}_m^*, \boldsymbol{W}^*, \boldsymbol{C}^*) = \boldsymbol{0},$$

$$M \frac{\boldsymbol{W}^{n+1} - \widetilde{\boldsymbol{W}}^{n+\theta}}{\Delta t} - \boldsymbol{G}(\boldsymbol{V}_m^*, \boldsymbol{W}^*) = \boldsymbol{0},$$

$$M \frac{\boldsymbol{C}^{n+1} - \widetilde{\boldsymbol{C}}^{n+\theta}}{\Delta t} - \boldsymbol{S}(\boldsymbol{V}_m^*, \boldsymbol{W}^*, \boldsymbol{C}^*) = \boldsymbol{0}.$$

The superscript * means that the quantity at hand could be treated either explicitly or implicitly. The variable θ could assume value $1/2$ or 1. In the latter case, step (3) is unnecessary: we set

$$\boldsymbol{V}_m^{n+1} = \widehat{\boldsymbol{V}}_m^{n+1}, \quad \boldsymbol{W}^{n+1} = \widetilde{\boldsymbol{W}}^{n+1}, \quad \boldsymbol{C}^{n+1} = \widetilde{\boldsymbol{C}}^{n+1},$$

and a first-order method is obtained (*Godunov splitting*). For $\theta = 1/2$, we have a second-order method provided that all the subproblems are solved with a second-order strategy (*Strang splitting*). Note that if step (2) is solved implicitly, then the same preconditioners introduced above for the semi-implicit schemes could be applied as well, since it applies to the same operator.

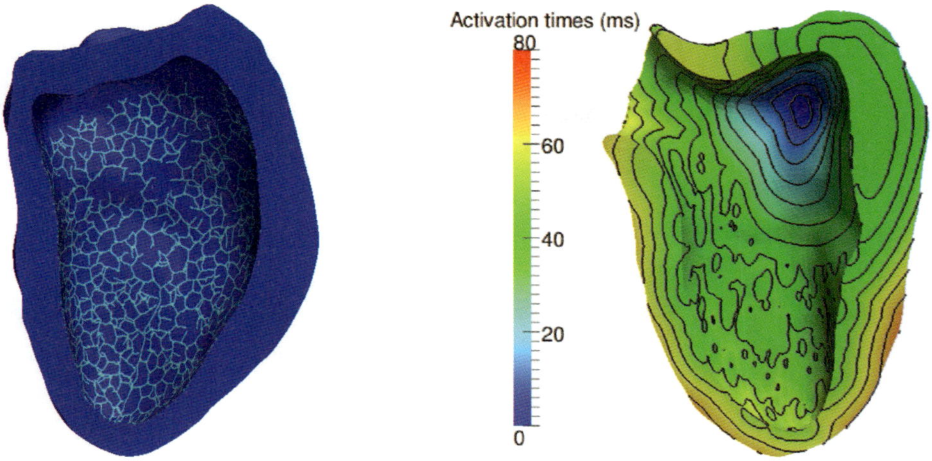

(a) (b)

Figure 7.3. (a) Purkinje network generated by the algorithm proposed in Palamara *et al.* (2015) in the case of a real left ventricle. (b) Map of the activation times computed by means of the eikonal equation. The time marching scheme and P1 finite elements have been used. The source term (dark blue) is located within the myocardium, as typically happens in Wolff–Parkinson–White syndrome. Results were obtained by means of a code implemented in the VMTK environment (www.vmtk.org). CT images from the Cardiology Division at Ospedale S. Maria del Carmine, Rovereto (TN), Italy, and from the Radiology Division of Borgo-Trento (TN), Italy.

Implicit methods. Some authors have considered a fully implicit discretization of the full bidomain problem (7.9): see *e.g.* Bourgault, Ethier and LeBlanc (2003) and Murillo and Cai (2004). In this case the whole Jacobian is built and the Newton method is applied. Due to the small time step required in bidomain simulations to capture the propagating front and due to the excellent stability properties of semi-implicit and operator splitting-based methods, fully implicit method are no longer considered so often.

In Figures 7.3 and 7.4 we give some examples of numerical results related to the solution of the electrical propagation in the myocardium.

7.2. Cardiac mechanics and electromechanical coupling

7.2.1. The continuous mechanics problems

During a physiological contraction, the cardiac cells change in length by up to 20–30%, so finite elasticity models are needed to describe heart contraction and relaxation. In particular, the first Piola–Kirchhoff tensor is written as the sum of two terms (Nash and Panfilov 2004, Niederer and Smith 2008): a *passive component* $\widehat{\boldsymbol{T}}_s^P$ describing the stress required to obtain a given deformation of the passive myocardium (similarly to arteries),

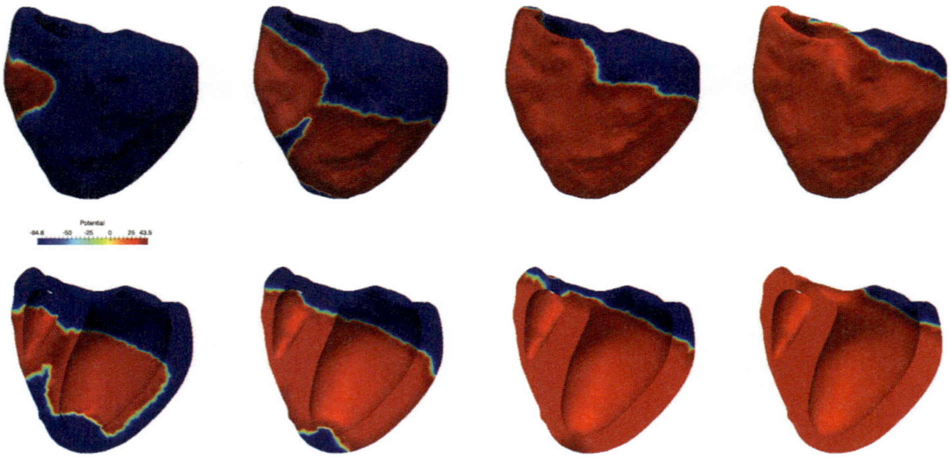

Figure 7.4. Propagation of the trans-membrane potential in the two ventricles at eight different instants during a heartbeat. Monodomain simulation, semi-implicit method, P1 finite elements. Results were obtained using LifeV; the computational mesh was obtained by an open source biventricular geometry segmented from CT images: see Rousseau (2010).

and an *active component* $\widehat{\boldsymbol{T}}_s^A$ denoting the tension generated by the depolarization of the propagating electrical signal that provides the internal active forces responsible for the contraction (see Section 5.4):

$$\widehat{\boldsymbol{T}}_s = \widehat{\boldsymbol{T}}_s^P + \widehat{\boldsymbol{T}}_s^A, \tag{7.23}$$

where, using the notation of Part 1, $\widehat{\ }$ refers to quantities computed in the reference domain.

The passive component of the stress is obtained as the derivative of a suitable strain energy function: see (4.8). The heart's mechanical response (just like the electrical propagation: see Section 7.1) is highly dependent on the presence of fibres and sheets. For this reason, the passive myocardium is modelled as an *orthotropic* material, characterized by two principal directions and with different material responses on three mutually orthogonal planes, identified by these directions. This is in accordance with the shear tests performed on pig hearts, which highlighted an elevated resistance to shear deformations producing an extension along the fibre direction, an intermediate resistance in the direction orthogonal to fibres and tangential to sheets, and the least resistance in the third orthogonal direction (Dokos, Smaill, Young and LeGrice 2002). Letting $\widehat{\boldsymbol{a}}_f$, $\widehat{\boldsymbol{a}}_s$, $\widehat{\boldsymbol{a}}_n$ denote the unit vectors along these directions (see Section 7.1) in the reference configuration, the following strain energy function has been proposed in Holzapfel and

Ogden (2009):

$$\Theta(I_1, I_{4,f}, I_{4,s}, I_{8,fs}) \\ = \frac{a}{2b}\mathrm{e}^{b(I_1-3)} + \sum_{i=f,s} \frac{a_i}{2b_i}\left[\mathrm{e}^{b_i(I_{4,i}-1)^2} - 1\right] + \frac{a_{fs}}{2b_{fs}}\left[\mathrm{e}^{b_{fs}I_{8,fs}^2} - 1\right], \tag{7.24}$$

where, referring to the notation introduced in Section 4.2, $I_1 = \operatorname{tr} \boldsymbol{C}$, $I_{4,i} = \widehat{\boldsymbol{a}}_i \cdot (\boldsymbol{C}\widehat{\boldsymbol{a}}_i)$, $i = f, s$, $I_{8,fs} = \widehat{\boldsymbol{a}}_f \cdot (\boldsymbol{C}\widehat{\boldsymbol{a}}_s)$ are invariants of $\boldsymbol{C}$, and a, b, a_f, b_f, a_s, b_s, a_{fs}, b_{fs} are material parameters. I_1 is the first isotropic invariant and the related term in Θ accounts for the isotropic response of the myocardium at small deformations; $I_{4,f}$ and $I_{4,s}$ are the squares of the stretch of $\widehat{\boldsymbol{a}}_f$ and $\widehat{\boldsymbol{a}}_s$ in the deformed configuration, and the related terms in the energy are associated with the increased stiffness of the material along the two principal directions $\boldsymbol{a}_f$ and $\boldsymbol{a}_s$ for large deformations; finally, $I_{8,fs}$ denotes the angle spanned by the two principal directions in the deformed configuration, and the related term in the energy describes the coupling between the two principal directions. Convexity of this energy is guaranteed for positive parameter values (Holzapfel and Ogden 2009). Other orthotropic models have been proposed, for example by Hunter, Nash and Sands (1997) and Costa, Holmes and McCulloch (2001), whereas *transversally isotropic* models with only one principal direction (along the fibres) were introduced by Humphrey and Yin (1987) and Guccione, McCulloch and Waldman (1991), for example. Some authors model the myocardium as incompressible (Holzapfel and Ogden 2009), in accordance with the experiments reported by Vossoughi, Vaishnav and Patel (1980). In this case the term $p_s(J-1)$ is added to the strain energy functions, as in (4.13).

On the basis of experimental evidence, reported by Dokos *et al.* (2002), for example, highlighting hysteresis under shear deformations, viscoelastic orthotropic models have recently been proposed for the passive myocardium: see Gultekin, Sommer and Holzapfel (2016). The viscoelastic behaviour is probably due to the extracellular fluid that filtrates through the solid part.

The active contribution of the cardiac cells to the contraction is regulated by the opening of calcium channels as a response to the depolarization, with a consequent entry of calcium ions in the cells. As detailed in Section 5.4, this process is responsible for cardiac contraction. Since the latter occurs along the axial direction of the cardiac cells, that is, along the fibre direction, the active part of the stress tensor usually takes the form

$$\widehat{\boldsymbol{T}}_s^A = P^A \boldsymbol{a}_f \otimes \widehat{\boldsymbol{a}}_f \tag{7.25}$$

(Nash and Panfilov 2004), where the scalar function of time and space P^A denotes the pointwise active stress, and should be properly modelled.

In the classical model for the active stress function proposed in Nash and Panfilov (2004), P^A depends only on the trans-membrane potential V_m. In particular, for each spatial point, the following ODE equation is introduced:

$$\frac{\mathrm{d}P^A(t)}{\mathrm{d}t} = \varepsilon(V_m(t))\big[k_P(V_m(t) - V_r) - P^A(t)\big], \tag{7.26}$$

where V_r is the resting potential, k_P is a suitable parameter controlling the amplitude of P^A, and ε controls the delay of the contraction with respect to the action potential. For example, Goktepe and Kuhl (2010) proposed the expression

$$\varepsilon(V_m) = \varepsilon_0 + (\varepsilon_\infty - \varepsilon_0)\mathrm{e}^{-\mathrm{e}^{-\xi(V_m - \bar{V}_m)}},$$

where ε_0, ε_∞, ξ, $\bar{V}_m$ are suitable parameters.

Since cardiac cell contraction is regulated by calcium ion concentration, the active stress function P^A can be assumed to depend directly on this specific concentration, say c_{ca}, rather than on the more general variable V_m. For example, Wong, Goktepe and Kuhl (2013) proposed a system of ODEs of the same structure as (7.26), but using c_{ca} instead of V_m on the right-hand side.

In more sophisticated models, the active stress function also depends on the stretch in the fibre direction $\lambda = (\widehat{\boldsymbol{a}}_f^T C \widehat{\boldsymbol{a}}_f)^{1/2}$ and on the fibre stretch rate $\mathrm{d}\lambda/\mathrm{d}t$ (Niederer, Hunter and Smith 2006, Land *et al.* 2012). In compact form, these models are written as

$$\begin{aligned} \frac{\mathrm{d}\boldsymbol{y}}{\mathrm{d}t} &= \boldsymbol{g}_y\left(P^A, \lambda, \frac{\mathrm{d}\lambda}{\mathrm{d}t}, \boldsymbol{c}, \boldsymbol{y}\right), \\ P^A &= g_{P^A}(\lambda, \boldsymbol{y}), \end{aligned} \tag{7.27}$$

for suitable functions $\boldsymbol{g}_y$ and g_{P^A} and where $\boldsymbol{c}$, as in Section 7.1, is the ionic concentration variable (in particular the calcium one), whereas $\boldsymbol{y}$ collects other myofilament and electrophysiology state variables.

As an alternative to the decomposition (7.23), where the stress tensor is split into a passive and an active component (the *active stress* approach), Cherubini, Filippi, Nardinocchi and Teresi (2008) and Ambrosi, Arioli, Nobile and Quarteroni (2011) have proposed a different strategy based on an *active strain* approach. In this case, the following factorization of the deformation tensor is used,

$$\boldsymbol{F} = \boldsymbol{F}^P \boldsymbol{F}^A,$$

where $\boldsymbol{F}^A$ is the active factor acting at the microscales, describing fibre distortion not preserving geometric compatibility, whereas $\boldsymbol{F}^P$ describes passive macroscale deformation and the deformation needed to restore compatibility. With this choice, quite common when analysing plasticity and the remodelling of living tissues, the microscale information related to the fibre contraction is directly incorporated in the body kinematics, allowing for the inclusion of fibre contraction driven by the depolarization as a prescribed active deformation rather than a further contribution to stress. In this case, the first Piola–Kirchhoff stress tensor is given by

$$\widehat{\boldsymbol{T}}_s = J^A \frac{D\Theta}{D\boldsymbol{F}^P} (\boldsymbol{F}^A)^{-T},$$

and, in analogy with (7.25), a possible expression for $\boldsymbol{F}^A$ is

$$\boldsymbol{F}^A = \boldsymbol{I} - \beta V_m \widehat{\boldsymbol{a}}_f \otimes \widehat{\boldsymbol{a}}_f,$$

for a suitable parameter β (Ambrosi *et al.* 2011). The active strain approach is probably more satisfactory from the modelling point of view, since, unlike active stress, it should provide the expected fibre contraction without any tuning. However, from the numerical point of view it is more problematic, since standard finite element finite elasticity solvers cannot be used directly and should be properly adapted. From now on we refer only to the active stress approach. For a comparison between the two approaches we refer to Ambrosi *et al.* (2011) and Ambrosi and Pezzuto (2012), and for related computational results to Pezzuto (2013) and Rossi *et al.* (2014).

7.2.2. The coupled electromechanics problem

The values of the trans-membrane potential V_m or calcium ion concentration c_{ca}, to be used in (7.26) or (7.27) to compute the active stress function P^A, are sometimes prescribed as given data to the mechanics problem: see *e.g.* Eriksson, Prassl, Plank and Holzapfel (2013). However, in most cases they are obtained from the bidomain or monodomain equations. This leads to a *coupled electromechanical* problem.

The electrical propagation problem needs to be solved in a domain that changes in time, because of the cardiac contraction and relaxation. Under suitable assumptions (Colli Franzone, Pavarino and Scacchi 2016), in an Eulerian framework, these problems assume the form (7.5), (7.6) or (7.11), provided that the conductivity tensors (7.4) or (7.12) are computed by using the deformed unit directions $\boldsymbol{a}_f$, $\boldsymbol{a}_s$, $\boldsymbol{a}_n$: see *e.g.* Keldermann *et al.* (2009). However, for computational simplicity, the bidomain and monodomain equations are usually set in a Lagrangian framework and written in the reference undeformed configuration. Here we report the corresponding monodomain equation,

$$J\chi_m C_m \left(\frac{\partial \widehat{V}_m}{\partial t} - \boldsymbol{F}^{-T} \nabla \widehat{V}_m \cdot \frac{\partial \boldsymbol{\psi}}{\partial t} \right) - \nabla \cdot \left(J\boldsymbol{F}^{-1} \boldsymbol{\Sigma} \boldsymbol{F}^{-T} \nabla \widehat{V}_m \right) + J\chi_m \widetilde{I}_{ion}\left(\widehat{V}_m, \widehat{\boldsymbol{w}}, \widehat{\boldsymbol{c}}, \lambda\right) = J\widehat{I}^{ext} \quad \text{in } \Omega_{mus} \tag{7.28}$$

(using the notation of Part 1, we set $\Omega_{mus} = \Omega^0_{mus}$), coupled with the ODE systems (7.9e, 7.9f) (similar arguments hold true for the bidomain equations as well). The spatial derivatives have to be taken with respect to the undeformed domain, $\boldsymbol{\psi}$ is the deformation map between Ω^0_{mus} and Ω^t_{mus} and λ is the stretch in the fibre direction introduced above.

From the above equation, we observe that there are three sources of *mechano-electrical feedback*, that is, three ways in which the mechanics problem influences the electrical problem:

(i) the effective conductivity tensor $\widehat{\boldsymbol{\Sigma}} = J\boldsymbol{F}^{-1}\boldsymbol{\Sigma}\boldsymbol{F}^{-T}$ depends on the deformation gradient $\boldsymbol{F}$,

(ii) the ionic current term I_{ion} depends on the stretch in the fibre direction λ,

(iii) the relation between the spatial and material time derivatives introduces the advection term $-\boldsymbol{F}^{-T}\nabla\widehat{V}_m \cdot \partial\boldsymbol{\psi}/\partial t$, which depends on the solution of the mechanics problem (Colli Franzone *et al.* 2016).

Terms (i) and (iii) are due to the pull-back of the monodomain equation into the undeformed domain (geometric feedback). Term (ii) is due to a well-known physical process consisting of the opening of ion channels under the action of deformation (stretch-activated channels: see Section 5.4) (Kohl, Hunter and Noble 1999, Kohl and Sachs 2001). Accordingly, the ionic current term is written as the sum of two contributions,

$$\widetilde{I}_{ion} = I_{ion}(\widehat{V}_m, \widehat{\boldsymbol{w}}, \widehat{\boldsymbol{c}}) + I^{SAC}(\widehat{V}_m, \lambda), \tag{7.29}$$

where I_{ion} denotes one of the classical independent-stretch models described in Section 7.1.2, and I^{SAC} is the current activated by the deformation. A fairly general expression for the latter term is given by

$$I^{SAC} = \sum_i K_i(\widehat{V}_m)(\lambda - 1)(\widehat{V}_m - V_i)H(\lambda - 1), \tag{7.30}$$

where the ith term of the sum represents the stretch-activated currents related to the ith ion (usually sodium and potassium), K_i is a suitable function, and V_i is the Nerst potential introduced in Section 7.1.2. The Heaviside function $H(\cdot)$ guarantees that the stretch-activated ion channels open only under fibre tension ($\lambda > 1$) (Niederer and Smith 2007). Colli Franzone *et al.* (2016) showed (computationally) that these terms do not significantly alter the morphology of the action potential, but they strongly influence the action potential duration.

Stretch-activated ionic currents together with the active stress component of the stress tensor (or equivalently the active strain component of the deformation gradient) make the electromechanics problem a highly coupled system, which can be summarized as follows. Find the muscle displacement $\boldsymbol{d}$, the trans-membrane potential V_m, the gating variables $\boldsymbol{w}$, and the ionic concentration variables $\boldsymbol{c}$, such that

$$\nabla \cdot \left(\widehat{\boldsymbol{T}}_s^P(\widehat{\boldsymbol{d}}) + \widehat{\boldsymbol{T}}_s^A\left(\widehat{\boldsymbol{c}}, \widehat{\boldsymbol{d}}, \frac{\mathrm{d}\widehat{\boldsymbol{d}}}{\mathrm{d}t}\right)\right) = \boldsymbol{0} \qquad \text{in } \Omega_{mus}, \tag{7.31a}$$

$$\widehat{\boldsymbol{T}}_s^A = P^A \boldsymbol{a}_f \otimes \widehat{\boldsymbol{a}}_f, \quad P^A = \mathcal{A}\left(\widehat{\boldsymbol{c}}, \widehat{\boldsymbol{d}}, \frac{\mathrm{d}\widehat{\boldsymbol{d}}}{\mathrm{d}t}\right) \qquad \text{in } \Omega_{mus}, \tag{7.31b}$$

$$\widehat{V}_m = \mathcal{M}(\widehat{V}_m; \widehat{\boldsymbol{d}}, \widehat{\boldsymbol{w}}, \widehat{\boldsymbol{c}}) \qquad \text{in } \Omega_{mus}, \tag{7.31c}$$

$$(\widehat{\boldsymbol{w}}, \widehat{\boldsymbol{c}}) = \mathcal{G}(\widehat{\boldsymbol{w}}, \widehat{\boldsymbol{c}}; \widehat{V}_m) \qquad \text{in } \Omega_{mus}. \tag{7.31d}$$

Here, equation (7.31a) stands for the mechanical problem. Note that, unlike for the vessel case (see (4.7)), the time-dependent inertial term has been neglected, as is commonly done for heart mechanics. Moreover, $\widehat{\boldsymbol{T}}_s^P = \partial\Theta/\partial\boldsymbol{F}$, where Θ is given by the orthotropic law (7.24), for example, and P^A is given by either (7.26) or (7.27). Equation (7.31c) represents the monodomain problem (7.28) in compact form, together with the ionic current expression given by (7.29, 7.9c, 7.9d, 7.30). Finally, (7.31d) is shorthand for the ODE systems (7.9e, 7.9f) for the gating and ionic concentration variables characterizing the cardiac cell model. The symbols $\mathcal{M}$ and $\mathcal{G}$ are abstract notations to identify the underlying PDE and ODEs, respectively. Note also that in writing system (7.31), we have exploited the fact that λ and $\boldsymbol{F}$ could be written in terms of $\boldsymbol{d}$, allowing us to indicate the dependences of the monodomain problem on λ and $\boldsymbol{F}$ and of $\widehat{\boldsymbol{T}}_s^A$ on λ in compact form through $\boldsymbol{d}$.

Regarding the well-posedness of the electromechanical coupled problem (7.31), very few results have been obtained so far. We mention Pathmanathan, Chapman, Gavaghan and Whiteley (2010), who noticed that for the general active stress model (7.27), the equilibrium equation (7.31a) is not even elliptic when there is explicit dependence on the rate of stretch $\mathrm{d}\lambda/\mathrm{d}t$, and Andreianov, Bendahmane, Quarteroni and Ruiz-Baier (2015), who proved the existence of a weak solution for the case of a linearized elasticity equation in the active strain formulation coupled with the bidomain equations including the geometric feedback affecting the conductivity tensors.

7.2.3. The issue of boundary conditions for the mechanics problem

For the mechanical contraction problem (7.31a), proper boundary conditions should be prescribed at both the external epicardium and internal endocardium surfaces. For the former, the presence of both the pericardium and the surrounding tissue has to be accounted for, because of their effect on heart movement. Fritz *et al.* (2014) have proposed a sophisticated model of interaction with the pericardium accounting for a frictionless contact. More commonly, and similarly to the vascular case, a Robin condition such as (3.1) is prescribed at Σ_{epi}.

At the endocardium internal surface Σ_{endo}, in the presence of a fluid model in the atrial and ventricular cavities, the usual dynamic and kinematic conditions (4.19c, 4.19d), arising from the fluid–structure interaction model, implicitly provide both the endocardial displacement and normal stresses (see Section 7.5). When the fluid in the ventricular cavity is not modelled, suitable strategies to prescribe boundary conditions at Σ_{endo} are mandatory. A common strategy relies on prescribing a Neumann condition of the form

$$\widehat{\boldsymbol{T}}_s\widehat{\boldsymbol{n}} = P\widehat{\boldsymbol{n}}, \tag{7.32}$$

where P is a suitable estimate of the blood pressure distribution at the endocardium.

During ventricular filling, condition (7.32) is prescribed both for atria and ventricles by means of selected values of blood pressures taken from the literature (see Figure 5.2).

During isovolumic contraction, the ventricular volume is kept constant by means of fixed-point iterations, for example, where the value of the endocardial pressure to be prescribed in (7.32) at each time step is updated until satisfaction of the constraint given by the unchanged volume:

$$P_{(k+1)} = P_{(k)} + (V - V^n)/C_k$$

(Usyk, LeGrice and McCulloch 2002, Eriksson *et al.* 2013), where V is the cavity volume, k is the sub-iteration index and C_k is a penalty parameter, and where the current temporal index $^{n+1}$ is as usual understood. Alternatively, a Lagrange multiplier approach could be employed (Gurev *et al.* 2015). During this phase the atrial pressure is kept constant to the values reached when the mitral and tricuspid valves close.

For ventricular ejection, a reduced 0D model (based on the analogy with electrical circuits: see Section 4.5.1) for the systemic (or pulmonary) circulation is usually coupled with the cardiac mechanics problem, assuming that the ventricular pressure equals the pressure in the aorta (or in the pulmonary artery) (Usyk *et al.* 2002, Eriksson *et al.* 2013). This is a good approximation since, although the ventricular and circulatory pressures are different (thus allowing for the acceleration and deceleration of blood: see Figure 5.2), they are very similar. In this case we have a coupled problem between the cardiac mechanics and the lumped parameter model where the two subproblems exchange suitable interface conditions (*e.g.* in an iterative framework). For example, the 0D model could provide the pressure to be prescribed to the mechanics problem by means of (7.32), whereas the flow rate $Q = \rho_f(\mathrm{d}V/\mathrm{d}t)$ is prescribed to the 0D model. During the ejection phase the atrial pressure is determined by the venous pressure, which could be obtained with a 0D model or by values from the literature.

Finally, during isovolumic relaxation the endocardial pressure is decreased according to values from the literature until it reaches the pressure atrial value. Also during this phase the atrial pressure is determined by the venous pressure.

When only the ventricular mechanics is modelled (no atria), a truncated computational domain is considered, like the one depicted in Figure 7.2(a). In this case we have to provide suitable boundary conditions for the ventricular base Σ_b. Often, this surface is kept fixed by enforcing homogeneous Dirichlet conditions. In other cases, the tangential displacement is allowed by prescribing a homogeneous Neumann condition along the tangential directions.

7.2.4. Numerical approximation

The coupled electromechanical problem (7.31) is composed of four different blocks, namely the two PDEs (7.31a) and (7.31c) describing the heart mechanics and electrical propagation, respectively, the ODE system (7.31b) providing the active stress function, and the ODE systems (7.31d) modelling gating and ionic concentration variables. Moreover, the algebraic source of coupling (7.29, 7.30) needs to be accounted for to determine the stretch-activated ion currents. Their influence on the electrical problem has been included directly in the right-hand side of (7.31c).

A common numerical solution strategy for problem (7.31) addresses the two PDEs (7.31a) and (7.31c) separately by relying (when possible) on pre-existing mechanics and electrical codes: for instance, the mechanics subproblem is solved by Pavarino *et al.* (2015) using efficient Newton–Krylov iterations, while the electrical subproblem (monodomain+cell ODE model) is solved by means of one of the strategies described in Section 7.1.6 (see also Sundnes *et al.* 2014 for an operator splitting method).

In this context, at each time step the electrical subproblem is solved first and the mechanics problem later (Nash and Panfilov 2004, Gurev *et al.* 2015, Sundnes *et al.* 2014). In particular, after time discretization, the following scheme is employed (as usual the current temporal index $^{n+1}$ is understood and * means that the related term could be treated either implicitly or explicitly, depending on the temporal scheme used).

Algorithm EM1

At each time step:

(1) Solve the monodomain problem together with the cell model:

$$\widehat{V}_m = \mathcal{M}(\widehat{V}_m^*; \widehat{\boldsymbol{d}}^n, \widehat{\boldsymbol{w}}^*, \widehat{\boldsymbol{c}}^*) \quad \text{in } \Omega_{mus}, \tag{7.33a}$$

$$(\widehat{\boldsymbol{w}}, \widehat{\boldsymbol{c}}) = \mathcal{G}(\widehat{\boldsymbol{w}}^*, \widehat{\boldsymbol{c}}^*; \widehat{V}_m^*) \qquad \text{in } \Omega_{mus}. \tag{7.33b}$$

(2) Then, solve the following mechanics problem.

(2a) Update the active stress contribution

$$\widehat{\boldsymbol{T}}_s^A = P^A \boldsymbol{a}_f \otimes \widehat{\boldsymbol{a}}_f, \quad P^A = \mathcal{A}\left(\widehat{\boldsymbol{c}}, \widehat{\boldsymbol{d}}^n, \frac{\widehat{\boldsymbol{d}}^n - \widehat{\boldsymbol{d}}^{n-1}}{\Delta t}\right) \quad \text{in } \Omega_{mus}.$$

(2b) Solve the equilibrium equation by means of the following Newton iterations. At each iteration k solve

$$\begin{aligned}
&\nabla \cdot \big(D_F \widehat{\boldsymbol{T}}_s^P(\widehat{\boldsymbol{d}}_{(k-1)})\big)\delta\widehat{\boldsymbol{d}}_{(k)} \\
&\quad = -\nabla \cdot \left(\widehat{\boldsymbol{T}}_s^P(\widehat{\boldsymbol{d}}_{(k-1)}) + \widehat{\boldsymbol{T}}_s^A\left(\widehat{\boldsymbol{c}}, \widehat{\boldsymbol{d}}^n, \frac{\widehat{\boldsymbol{d}}^n - \widehat{\boldsymbol{d}}^{n-1}}{\Delta t}\right)\right) \quad \text{in } \Omega_{mus}.
\end{aligned}$$

Using the notation introduced in Section 4.6.2, D_F is the Gâteaux derivative with respect to $\boldsymbol{F}$, $\delta\widehat{\boldsymbol{d}}_{(k)} = \widehat{\boldsymbol{d}}_{(k)} - \widehat{\boldsymbol{d}}_{(k-1)}$, and, for simplicity, we have considered a Forward Euler approximation of $\mathrm{d}w/\mathrm{d}t|_{t^n}$. The active stress contribution is treated explicitly, that is, it is updated once per time step. This choice may lead to numerical instabilities, as reported computationally in Niederer and Smith (2008). Pathmanathan *et al.* (2010) speculate that such instability could be ascribed to the (necessarily) explicit time discretization of the explicit stretch rate term $\mathrm{d}\lambda/\mathrm{d}t|_{t^n}$ (instead of $\mathrm{d}\lambda/\mathrm{d}t|_{t^{n+1}}$) in the active stress function solution.

To overcome these instabilities, Niederer and Smith (2008) propose updating the active stress function at each Newton step, that is, to replace step (2) in Algorithm EM1 with the following one.

Algorithm EM1′

(2′) Solve the mechanics problem by means of the following Newton iterations. At each iteration k:

(2′a) Update the active stress contribution:

$$\widehat{\boldsymbol{T}}_s^A = P^A \boldsymbol{a}_f \otimes \widehat{\boldsymbol{a}}_f, \quad P^A = \mathcal{A}\left(\widehat{\boldsymbol{c}}, \widehat{\boldsymbol{d}}_{(k-1)}, \frac{\widehat{\boldsymbol{d}}_{(k-1)} - \widehat{\boldsymbol{d}}^n}{\Delta t}\right) \quad \text{in } \Omega_{mus}.$$

(2′b) Solve the equilibrium equation:

$$\begin{aligned}
&\nabla \cdot \big(D_F \widehat{\boldsymbol{T}}_s^P(\widehat{\boldsymbol{d}}_{k-1})\big)\delta\widehat{\boldsymbol{d}}_k \\
&\quad = -\nabla \cdot \left(\widehat{\boldsymbol{T}}_s^P(\widehat{\boldsymbol{d}}_{k-1}) + \widehat{\boldsymbol{T}}_s^A\left(\widehat{\boldsymbol{c}}, \widehat{\boldsymbol{d}}_{(k-1)}, \frac{\widehat{\boldsymbol{d}}_{(k-1)} - \widehat{\boldsymbol{d}}^n}{\Delta t}\right)\right) \quad \text{in } \Omega_{mus}.
\end{aligned}$$

Note that step (2′) does not coincide with a Newton iteration for (7.31a, 7.31b). Although at the expense of a higher number of iterations, this scheme provides stable solutions and allows the use of a standard solver for the mechanics part.

A Newton method for the solution of the mechanics problem (7.31a, 7.31b), used in combination with a different explicit decoupled scheme, has been successfully proposed by Pavarino *et al.* (2015), who split the electrical problem (7.31c, 7.31d) into the cellular and macroscopic problems (semi-implicit treatment), and solved the mechanics problem in between, as follows.

Algorithm EM2
At each time step:

(1) Solve the ODE system for the gating and ionic concentration variables

$$(\widehat{\boldsymbol{w}}, \widehat{\boldsymbol{c}}) = \mathcal{G}(\widehat{\boldsymbol{w}}^*, \widehat{\boldsymbol{c}}^*; \widehat{V}_m^*) \quad \text{in } \Omega_{mus}.$$

(2) Then, solve the mechanics problem by means of the following Newton method. At each iteration k solve:

(2a) Update the active stress contribution:

$$\widehat{\boldsymbol{T}}_s^A = P^A \boldsymbol{a}_f \otimes \widehat{\boldsymbol{a}}_f, \quad P^A = \mathcal{A}\left(\widehat{\boldsymbol{c}}, \widehat{\boldsymbol{d}}_{(k-1)}, \frac{\widehat{\boldsymbol{d}}_{(k-1)} - \widehat{\boldsymbol{d}}^n}{\Delta t}\right) \quad \text{in } \Omega_{mus}.$$

(2b) Solve the equilibrium equation:

$$\begin{aligned} \nabla \cdot \left(D_F \widehat{\boldsymbol{T}}_s^P (\widehat{\boldsymbol{d}}_{(k-1)}) + D_F \widehat{\boldsymbol{T}}_s^A \left(\widehat{\boldsymbol{c}}, \widehat{\boldsymbol{d}}_{(k-1)}, \frac{\widehat{\boldsymbol{d}}_{(k-1)} - \widehat{\boldsymbol{d}}^n}{\Delta t} \right) \right) \delta \widehat{\boldsymbol{d}}_{(k)} \\ = -\nabla \cdot \left(\widehat{\boldsymbol{T}}_s^P (\widehat{\boldsymbol{d}}_{(k-1)}) + \widehat{\boldsymbol{T}}_s^A \left(\widehat{\boldsymbol{c}}, \widehat{\boldsymbol{d}}_{(k-1)}, \frac{\widehat{\boldsymbol{d}}_{(k-1)} - \widehat{\boldsymbol{d}}^n}{\Delta t} \right) \right) \quad \text{in } \Omega_{mus}. \end{aligned}$$

(3) Solve the monodomain problem:

$$\widehat{V}_m = \mathcal{M}(\widehat{V}_m^*; \widehat{\boldsymbol{d}}, \widehat{\boldsymbol{w}}, \widehat{\boldsymbol{c}}) \quad \text{in } \Omega_{mus}.$$

Andreianov *et al.* (2015) have proposed a simplified variant of the above decoupled algorithms for the active strain formulation.[2] We provide the same variant below for the active stress formulation; in the second equation of (7.31b), however, the active stress function $P^A = \mathcal{A}(\widehat{\boldsymbol{c}})$ does not depend on the stretch and stretch rate.

Algorithm EM3
At each time step:

(1) Update the active stress contribution:

$$\widehat{\boldsymbol{T}}_s^A = P^A \boldsymbol{a}_f \otimes \widehat{\boldsymbol{a}}_f, \quad P^A = \mathcal{A}(\widehat{\boldsymbol{c}}^{n-1}) \quad \text{in } \Omega_{mus}.$$

[2] This is one of the very few papers that contain a convergence result for the finite element solution to the differential problem (7.31).

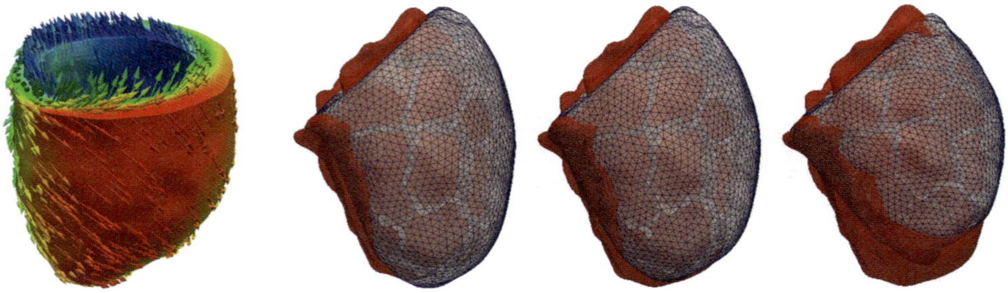

(a) (b)

Figure 7.5. (a) Fibre orientation in a real left ventricle obtained with the method proposed in Rossi *et al.* (2014). CT images from the Cardiology Division of Ospedale S. Maria del Carmine, Rovereto (TN), Italy, and from the Radiology Division of Borgo-Trento (TN), Italy. (b) Displacement configuration of a real left ventricle during the contraction phase at three different times. Orthotropic model of activation: see Barbarotta (2014). Numerical results were obtained using LifeV and taken from Barbarotta (2014); CT images from the Cardio Surgery and Radiology Divisions at Ospedale Sacco, Milan, Italy.

(2) Solve the equilibrium mechanics problem via the following Newton method. At each iteration k solve

$$\nabla \cdot \big(D_F \widehat{\boldsymbol{T}}_s^P(\widehat{\boldsymbol{d}}_{k-1})\big)\delta\widehat{\boldsymbol{d}}_k = -\nabla \cdot \big(\widehat{\boldsymbol{T}}_s^P(\widehat{\boldsymbol{d}}_{k-1}) + \widehat{\boldsymbol{T}}_s^A(\widehat{\boldsymbol{c}}^{n-1})\big) \quad \text{in } \Omega_{mus}.$$

(3) Then, solve the electrical problem:

$$\begin{aligned} \widehat{V}_m &= \mathcal{M}\big(\widehat{V}_m^*; \widehat{\boldsymbol{d}}^{n-1}, \widehat{\boldsymbol{w}}^*, \widehat{\boldsymbol{c}}^*\big) && \text{in } \Omega_{mus}, \\ (\widehat{\boldsymbol{w}}, \widehat{\boldsymbol{c}}) &= \mathcal{G}(\widehat{\boldsymbol{w}}^*, \widehat{\boldsymbol{c}}^*; \widehat{V}_m^*) && \text{in } \Omega_{mus}. \end{aligned}$$

Although segregated algorithms such as those reported above are the most widely used for numerical solution of the coupled electromechanics problem (7.31), a monolithic approach has been used by Goktepe and Kuhl (2010), who have successfully applied a Newton method to the whole coupled problem with a FitzHugh–Nagumo cardiac cell model (*i.e.* without the ODE systems (7.31d)).

In Figure 7.5 we show an example of fibre configuration in the left ventricle, and in the same figure and Figure 7.6 we give the results of an electromechanical simulation.

7.3. The ventricular fluid dynamics

Blood flow in heart cavities Ω_{cav}, shown in Figure 6.1, can be regarded as homogeneous, Newtonian and incompressible (Vierendeels, Riemslagh, Dick and Verdonck 2000, Watanabe *et al.* 2002). The displacements induced by the interaction with the myocardium can be larger than 30%, so

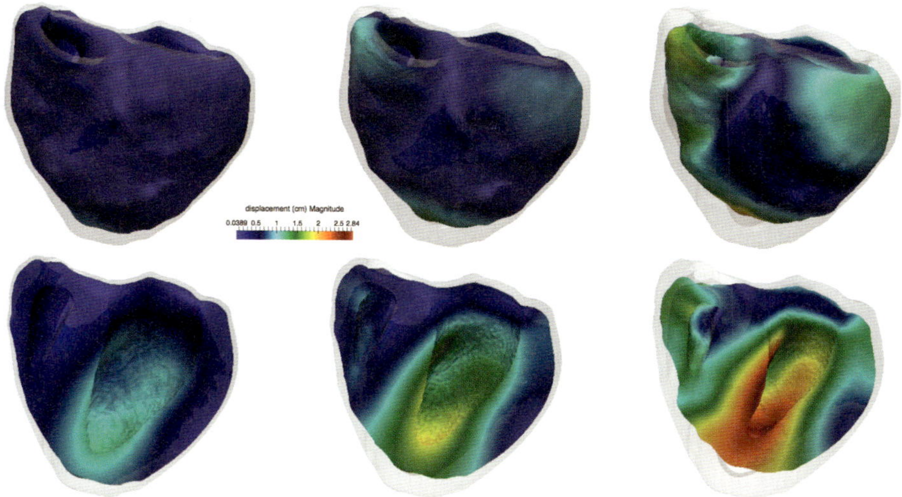

Figure 7.6. Displacements of the two ventricles at six different times during systolic contraction. Electromechanical coupled simulation, P1 finite elements. Numerical results were obtained using LifeV; the computational mesh was obtained from an open source biventricular geometry segmented from CT images: see Rousseau (2010).

the solution of fluid–structure interaction problems is mandatory. Despite the large displacements, the ALE technique introduced in Section 4.6.3 is suitable in this context, as shown by several authors: see *e.g.* Cheng, Oertel and Schenkel (2005) and Nordsletten *et al.* (2011*a*). Alternatively, the immersed boundary method (Peskin 1972), which does not require remeshing or induced mesh deformation, has often been used since the celebrated paper by Peskin (Mittal *et al.* 2016, Choi *et al.* 2015); see Section 7.4 for a mathematical description.

One of the main features of the intraventricular fluid dynamics is represented by the formation of a vortex ring during diastole (*i.e.* the filling phase) just downstream of the mitral valve. This ring is generated during the peak flow due to a pressure gradient between left atrium and ventricle, and it is highly distorted due to the asymmetry of the mitral valve leaflets (Seo *et al.* 2014). At the end of this phase, the vortex ring is broken up into small-scale structures that propagate towards the posterior wall and the apex. After the second mitral peak flow due to the atrium contraction, additional vortex rings pop up. However, due to the small duration of this phase, they do not propagate far into the chamber (Le and Sotiropoulos 2013). Similar vortex rings are generated in the left atrium from the flow ejected by the pulmonary vein (Mittal *et al.* 2016). This complex and disturbed fluid dynamics, in the presence of a fairly high Reynolds number ($\simeq 4000$), leads to transition to turbulence effects even in healthy cases (Querzoli, Fortini and

Cenedese 2010). For this reason, it is common practice to include a suitable turbulence model in heart simulations; see *e.g.* Chnafa, Mendez and Nicoud (2014) for LES models and Tagliabue, Dede' and Quarteroni (2015) for the variational multiscale (VMS) formulation.

Regarding boundary conditions, on the endocardium Σ_{endo} (Figure 6.1) the kinematic and dynamic conditions are implicitly determined by the coupling with the structural problem holding in the myocardium, leading to a fluid–structure interaction problem (see Section 7.5). The solution of a coupled fluid–structure interaction (FSI) problem can be avoided in those cases where the wall velocity can be derived from dynamic MRI or CT images. However, this requires us to have at our disposal several (20–30) wall displacement fields at different time instants during every heartbeat. The wall myocardium velocity (which provides the Dirichlet boundary condition for the fluid equations in the ventricle) can be generated by numerically differentiating the displacement field (Khalafvand *et al.* 2014, Mittal *et al.* 2016). Alternatively, the wall myocardium velocity could be provided by the solution of the electromechanics model if the latter were segregated from the fluid model (Choi *et al.* 2015).

The heart also has four (virtual) sections that delimit it from the circulatory system, namely the aortic and pulmonary valve orifices (outlets) and the interfaces with the venae cavae and the pulmonary vein (inlets). In these cases, suitable conditions could be obtained by coupling the heart fluid dynamics with the aorta, the pulmonary artery, and the above-mentioned veins. These can be modelled by means of 3D, 1D or 0D models.

In the case of simulation in the left ventricle alone (see Figure 7.2(b)) we have to prescribe suitable boundary conditions at the mitral valve orifice Γ_{mitr} as well. For example, Khalafvand *et al.* (2014) propose prescribing the flow rates at Γ_{mitr} and at the aortic valve outlet Γ_{ao} obtained by MRI measurements of the volume changes of the left ventricle chamber, noticing that the two valves are never open simultaneously. Alternatively, flow rate curves taken from the literature could be applied: see *e.g.* Nordsletten *et al.* (2011*a*).

7.4. Modelling the valves

Heart valves are made of thin leaflets. They do not cause any resistance to the blood during systole, and sustain large pressure gradients during diastole. Moreover, unlike the vascular vessel wall, they are subjected to very large displacements. Because of these features, the mathematical and numerical modelling of the valve mechanics and their interaction with blood flow is very demanding, requiring *ad hoc* techniques for their description. Our main focus will be on the aortic and mitral valves of the left ventricle, the most studied from the mathematical point of view. For recent reviews on valve modelling, see Marom (2015) and Votta *et al.* (2013), for example.

7.4.1. Mechanical models of the leaflets

Due to fibre alignment, which occurs mostly in the circumferential direction $\boldsymbol{a}$, and to the presence of collagen, most of the valve mechanical models proposed so far are of the non-linear transversally isotropic type. In particular, they depend on the two invariants $I_1 = \operatorname{tr} \boldsymbol{C}$ and $I_4 = \widehat{\boldsymbol{a}} \cdot (\boldsymbol{C}\widehat{\boldsymbol{a}})$. For example, May-Newman, Lam and Yin (2009) and Prot, Skallerud and Holzapfel (2007) use the strain energy function

$$\Theta(I_1, I_4) = c\big[\mathrm{e}^{b_1(I_1-3)^2+b_4(I_4-1)^2} - 1\big],$$

for suitable constants c, b_1, b_4. Variants of the above energy have been considered, for example by Humphrey and Yin (1987), May-Newman and Yin (1998) and Weinberg and Kaazempur-Mofrad (2006). For a comparison of numerical results obtained with these constitutive laws, see Auricchio, Ferrara and Morganti (2012). More sophisticated energy functions include the microstructure of the leaflet tissue: see *e.g.* Einstein *et al.* (2005).

Due to their thinness, heart valves are often modelled as membrane shells by neglecting the shear bending forces, which are an order of magnitude smaller than the in-plane stresses (Merryman, Huang, Schoen and Sacks 2006, Kim, Lu, Sacks and Chandran 2008, Hsu *et al.* 2014).

Some works on the mitral valve also consider the presence of the chordae tendinae (see Section 5.1). These comprise independent ring models, where the chordae are modelled as non-linear springs with zero elasticity during compression and exponential-like stiffness during traction (Kunzelman and Cochran 1990, Mansi *et al.* 2012), and one-dimensional models characterized by a non-linear energy function (Einstein *et al.* 2005, Votta *et al.* 2007, Padala *et al.* 2010).

Another difficulty arises when modelling the contact among the leaflets during closure. A common strategy to handle this process is a penalty method (Hsu *et al.* 2014, Marom 2015, Mansi *et al.* 2012). This is based on measuring the distance between each vertex and the closest leaflet, and on locating a spring between the vertex and the collision point when the distance is below a critical value. Astorino, Gerbeau, Pantz and Traoré (2010) propose a more sophisticated algorithm, based on the introduction of sub-iterations that guarantee the satisfaction of the contact constraint via the introduction of Lagrange multipliers that act as a force of repulsion between the leaflets.

The numerical simulation of valve mechanics has usually been obtained by prescribing a pressure difference between the two sides of the leaflets, mimicking the effect of the fluid (Gnyaneshwar, Kumar and Balakrishnan 2002, Votta *et al.* 2013). However, more realistic models are obtained by considering the FSI problem arising between the blood flow and the leaflets. This is the topic of the next two subsections.

7.4.2. Reduced models for fluid–valve interaction

As noticed, the simulation of the fluid–structure interaction between blood and valve leaflets is computationally demanding. However, if one is interested only in the fluid-dynamic quantities and not in the internal leaflet stresses, reduced models could be considered.

The first family does not explicitly represent the leaflets; rather, only the opening/closure mechanism of the valve is modelled through the description of the valve orifice. The simplest strategy consists in *a priori* operating the opening and closing of the valve on the basis of a flow rate or pressure profile obtained from the literature or from clinical data: see *e.g.* Faggiano *et al.* (2013). For an example, see Bazilevs *et al.* (2009), Vergara *et al.* (2012) and Nestola *et al.* (2017) for the flow simulation in the aorta and Khalafvand *et al.* (2014) for that in the left ventricle. A more realistic situation is obtained by simulating the opening/closure mechanism of the aortic valve orifice by means of the following conditions prescribed to the fluid problem:

$$\begin{aligned} &\text{if } P_u > P_d \quad \text{then the valve opens,} \\ &\text{if } Q_d < 0 \quad \text{then the valve closes,} \end{aligned} \tag{7.34}$$

where P_u is the upstream pressure and P_d and Q_d are the downstream pressure and flow rate (Formaggia *et al.* 1999). The quantities P_u, P_d, Q_d could be prescribed as data or else be the results of the numerical computation of the flow field. For example, in a fluid-dynamic simulation in the aorta, P_u is the prescribed left ventricular pressure, and P_d and Q_d are the unknown aortic pressure and flow rate. For a left ventricular flow simulation, the situation is more involved as we have to distinguish between the mitral and the aortic valve: for the former, P_u is the prescribed atrial pressure and P_d and Q_d are the unknown left ventricular pressure and flow rate; for the latter, P_u is the unknown left ventricular pressure, and P_d and Q_d the prescribed aortic pressure and flow rate. Many works have considered a zero-dimensional model of the systemic circulation to provide the latter quantities (Wenk *et al.* 2013). In any case, the opening/closure mechanism is not prescribed *a priori*. This leads to a non-linear boundary condition for the fluid problem at the valve orifice which, in the electric circuit analogy of zero-dimensional models (see Section 4.5.1), is represented by an ideal diode. Similar models have been proposed by Korakianitis and Shi (2006*b*) and Blanco and Feijóo (2013), to account for diseased valves, by introducing a suitable non-ideal law for the diode. More refined reduced models consider the opening/closure mechanism not simply in an on–off mode: the open part of the orifice dynamically changes continuously by projecting the supposed leaflet position on the orifice plane. For example, in Sundaram,

Balakrishnan and Kumar (2015) the dynamics of the valve orifice opening/closure is given by a prescribed law, whereas in Korakianitis and Shi (2006*a*) it is given by

$$A_V = \frac{(1-\cos\theta)^2}{(1-\cos\theta_{\max})^2},$$

where $A_V \in [0,1]$ is the percentage of open orifice area, θ is the opening leaflet angle ($\theta = 0$ means closed valve) and $\theta_{\max}$ is the maximum opening angle. The value of θ is determined by the ODE

$$\frac{\mathrm{d}^2\theta}{\mathrm{d}t^2} + k_1\frac{\mathrm{d}\theta}{\mathrm{d}t} + k_2(P_d - P_u)\cos\theta = 0, \tag{7.35}$$

for suitable parameters k_1 and k_2 and appropriate initial conditions on θ and $\mathrm{d}\theta/\mathrm{d}t$.

For all these models, which do not explicitly include the leaflets, there might be a need to switch between Dirichlet and Neumann boundary conditions (and *vice versa*) along a single heartbeat in a fluid problem. Indeed, a Dirichlet condition is usually prescribed when the valve is closed (*e.g.* homogeneous in the physiological case), whereas a Neumann or a resistance condition might be preferred when the valve is open. This might be problematic at the numerical level, particularly in view of the implementation. Tagliabue, Dede' and Quarteroni (2015) propose a new way to overcome this problem, based on a Robin-like condition implemented by means of the extended Nitsche's method proposed in Juntunen and Stenberg (2009). For simplicity we detail the case of a Neumann-like condition for the open valve, and we refer the reader to Tagliabue *et al.* (2015) for the more physiological case of a resistance condition. Let Γ be the valve orifice at hand, and suppose that one wants to prescribe at each time the following conditions:

$$\boldsymbol{u} = \boldsymbol{g} \quad \text{if the valve is closed,} \tag{7.36a}$$

$$\boldsymbol{T}_f\boldsymbol{n} = \boldsymbol{h} \quad \text{if the valve is open.} \tag{7.36b}$$

For example, $\boldsymbol{g}$ might be obtained via a prescribed flow rate (null in the physiological case) by assuming *a priori* the shape of the velocity profile (*e.g.* flat), whereas usually $\boldsymbol{h} = P\boldsymbol{n}$, where P is a prescribed mean pressure. Thus, the following term is added to the weak formulation (4.6) of the fluid problem:

$$\begin{aligned}
&-\int_\Gamma \frac{\gamma h}{\xi+\gamma h}\boldsymbol{T}_f(\boldsymbol{v},p)\boldsymbol{n}\cdot\boldsymbol{w}\,\mathrm{d}\omega \\
&\quad -\int_\Gamma \frac{\gamma h}{\xi+\gamma h}(\boldsymbol{v}-\boldsymbol{g})\cdot(\boldsymbol{T}_f(\boldsymbol{w},q)\boldsymbol{n})\,\mathrm{d}\omega + \int_\Gamma \frac{1}{\xi+\gamma h}(\boldsymbol{v}-\boldsymbol{g})\cdot\boldsymbol{w}\,\mathrm{d}\omega \\
&\quad -\int_\Gamma \frac{\xi\gamma h}{\xi+\gamma h}(\boldsymbol{T}_f(\boldsymbol{v},p)\boldsymbol{n}-\boldsymbol{h})\cdot(\boldsymbol{T}_f(\boldsymbol{w},q)\boldsymbol{n})\,\mathrm{d}\omega + \int_\Gamma \frac{\xi}{\xi+\gamma h}\boldsymbol{h}\cdot\boldsymbol{w}\,\mathrm{d}\omega,
\end{aligned}$$

where $\xi \in [0, +\infty)$ and $\gamma > 0$ are two parameters and h is the mesh size (possibly not constant). This discrete problem preserves the usual finite element accuracy. Note that for $\xi = 0$ we recover the classical Nitsche method to prescribe the Dirichlet condition (7.36a), whereas for $\xi \to +\infty$ the formulation is consistent with the Neumann condition (7.36b). Thus, the idea is to change the value of ξ (0 or $+\infty$) allowing for the switch between Dirichlet and Neumann conditions. Alternatively, the switch between a flow rate and mean pressure condition could be prescribed via the defective version of the extended Nitsche method proposed in Vergara (2011).

A second family of reduced strategies includes a reduced model of the leaflets without solving the full 3D mechanical problem. These methods are very useful when one wants to determine with a good accuracy the influence of the leaflets on the direction of the fluid jet. A first simple model is based on including the position of the open leaflets obtained by a preliminary mechanical simulation and considering an on–off opening/closure mode: see *e.g.* Bonomi *et al.* (2015). A similar model has been proposed by Astorino, Hamers, Shadden and Gerbeau (2012), who consider an immersed resistive approach to switch between the open and closed configuration. In particular, the following term is added to the variational formulation of the fluid problem:

$$\int_{\Gamma_o} R_o \boldsymbol{v} \cdot \boldsymbol{w} \,\mathrm{d}\gamma + \int_{\Gamma_c} R_c \boldsymbol{v} \cdot \boldsymbol{w} \,\mathrm{d}\gamma,$$

where Γ_o and Γ_c are the surfaces representing the open and closed leaflet configurations, respectively, and R_o and R_c are the corresponding resistance, which act as penalization terms to enforce the no-slip condition $\boldsymbol{v} = \mathbf{0}$ at the leaflet. Thus, when the value of the resistance is high, the corresponding surface configuration is physically included in the model, the velocity is zero at the leaflet and, accordingly, a pressure drop is generated across the leaflet; in contrast, when the value of the resistance is zero, the leaflet is invisible to the model, no constraint is prescribed on the velocity, and no pressure drop is generated. The switch between large and vanishing values of the resistances is determined by (7.34).

More sophisticated models account not only for the open and closed configurations of the leaflets, but also provide a simplified dynamics of the entire valve opening/closure mechanism. For example, Laadhari and Quarteroni (2016) and Fedele, Faggiano, Dede' and Quarteroni (2016) represent a leaflet Γ as a surface described by two embedded level set functions which depend on a single scalar function of time, that is, the opening angle θ. The latter is determined by a relation very similar to (7.35). At each time step, the term

$$\int_{\Gamma} R(\boldsymbol{v} - \boldsymbol{v}_{leaf}) \cdot \boldsymbol{w} \,\mathrm{d}\gamma$$

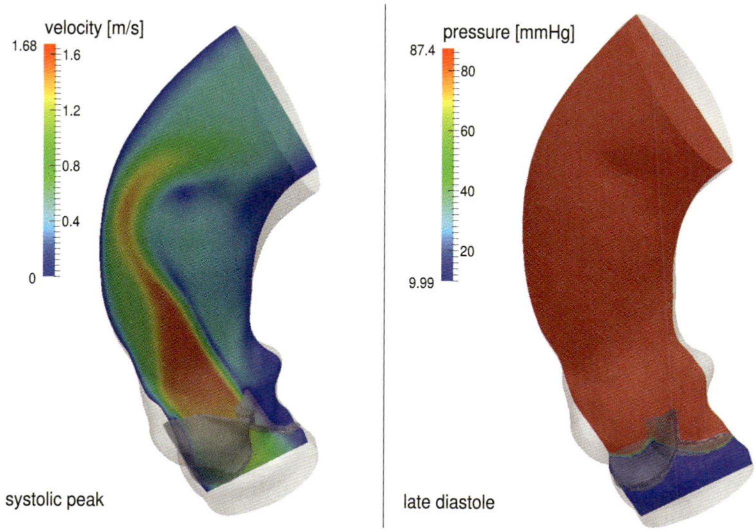

Figure 7.7. Fluid dynamics in the ascending aorta with a patient-specific aortic valve. Numerical results were obtained using LifeV. See Fedele *et al.* (2016) for a complete overview of the results.

is added to the weak formulation of the fluid problem, in order to guarantee a no-slip condition at the leaflet ($\boldsymbol{v} = \boldsymbol{v}_{leaf}$, where $\boldsymbol{v}_{leaf}$ is the leaflet velocity determined from θ) by selecting a sufficiently large resistance R. A similar approach has recently been adopted in Auricchio, Lefieux, Reali and Veneziani (2016), where however the leaflet is represented in a more simplified way by means of the opening angle alone (*i.e.* no level set functions are involved).

In Figure 7.7 we give an example of the fluid dynamics across the aortic valve obtained by means of the method proposed in Fedele *et al.* (2016).

7.4.3. Interaction between blood and leaflets

To accurately model the interaction between blood and valve leaflets, aimed, for example, at detailed computation of the internal leaflet stresses, the solution of a full FSI problem is required.

Unlike for the blood vessel and ventricular simulations, the arbitrary Lagrangian–Eulerian (ALE) formulation, described in Section 4.6, is not suited to handling the FSI problem arising between blood and heart valve leaflets. Indeed, due to the large displacements of the leaflets, the fluid mesh becomes highly distorted, producing severely stretched mesh elements and thus requiring a frequent remeshing of the grid. For this reason, specific FSI techniques have been developed specifically for the numerical solution of this problem. The most successful are the *immersed boundary* method, the *fictitious domain* approach and the *cut-FEM* method, together with their numerous variants. All these methods are characterized by a fixed fluid mesh and enable us to treat non-conforming interface meshes.

Immersed boundary method. The celebrated immersed boundary method (IB) was specifically developed by Peskin (1972) for the fluid dynamics in the heart. The leaflets are represented by membranes in a Lagrangian framework regarded as part of the fluid, represented in an Eulerian framework, and exerting internal localized forces on the latter. This is achieved by introducing a Dirac *delta* distribution, δ, on the membrane.

The IB method was originally developed for finite difference spatial discretization: see *e.g.* Peskin (1972), Mittal and Iaccarino (2005) and Peskin (2002). In this case, in order to avoid leaks, the Lagrangian grid should be sufficiently fine for the distance between two adjacent points to be less than the Eulerian mesh size. Then, the structural forces are interpolated into the fixed nodes of the fluid mesh. The major issue when finite differences are used is the approximation of δ. This is usually obtained via a function δ_h, which is non-singular for each h and tends to δ for $h \to 0$ (Peskin 2002).

Here we report the variational formulation of the IB method proposed in Boffi and Gastaldi (2003), which is useful for a finite element approximation and does not require any specific approximation of the delta distribution. Given a two-dimensional fluid domain Ω_f, let Γ_s be the immersed (one-dimensional) structure, whose material points are located at each t in $\boldsymbol{X}(s,t)$, where $s \in [0, L]$ is the Lagrangian coordinate. We consider as an example the case of a massless linear elastic structure with elasticity constant κ. Thus, referring to the notation introduced in Section 4.1 and assuming homogeneous boundary conditions, the weak formulation of the IB method is as follows (Boffi, Gastaldi and Heltai 2007). For each $t > 0$, find $\boldsymbol{v} \in \boldsymbol{V}$, $\boldsymbol{v} = \boldsymbol{v}_0$ for $t = 0$, $p \in L^2(\Omega_f)$, and $\boldsymbol{X} = \boldsymbol{X}(t)$, $\boldsymbol{X} = \boldsymbol{X}_0$ for $t = 0$, such that

$$\rho_f \int_{\Omega_f} \frac{\partial \boldsymbol{v}}{\partial t} \cdot \boldsymbol{w} \, \mathrm{d}\omega + \mathcal{A}_f(\boldsymbol{v}, \boldsymbol{v}, \boldsymbol{w}) + \mathcal{B}(p, \boldsymbol{w}) = \int_0^L \kappa \frac{\partial^2 \boldsymbol{X}(s,t)}{\partial s^2} \boldsymbol{w}(\boldsymbol{X}(s,t)) \, \mathrm{d}s, \tag{7.37a}$$

$$\mathcal{B}(q, \boldsymbol{v}) = 0, \tag{7.37b}$$

$$\frac{\partial \boldsymbol{X}}{\partial t}(s,t) = \boldsymbol{v}(\boldsymbol{X}(s,t), t), \tag{7.37c}$$

for all $\boldsymbol{w} \in \boldsymbol{V}$ and $q \in L^2(\Omega_f)$. In fact, the right-hand side of (7.37a) guarantees the satisfaction of the normal stress continuity at the membrane, whereas (7.37c) is the velocity continuity.

For the finite element approximation of the above problem, the forcing term is usually integrated by parts. For example, in the case of linear finite elements for the structure, for each $t > 0$ it becomes

$$\sum_{i=0}^{M-1} \kappa\left(\frac{\partial \boldsymbol{X}_h^{i+1}}{\partial s} - \frac{\partial \boldsymbol{X}_h^{i}}{\partial s}\right)\boldsymbol{w}(\boldsymbol{X}_h^i)$$

(Boffi *et al.* 2007), where $\boldsymbol{X}_h^i = \boldsymbol{X}_h(s_i)$, $i = 0, \ldots, M$, is the finite element approximation of $\boldsymbol{X}$ and s_i are the nodes of the one-dimensional mesh.

Regarding the time discretization, a first approach is to consider a fully implicit scheme. For the backward Euler approximation, unconditional stability is guaranteed (Tu and Peskin 1992). However, this scheme requires the introduction of sub-iterations to handle the coupling between fluid and structure in (7.37). A more useful approach relies on treating the right-hand side of (7.37a) explicitly and the remaining part of the equation implicitly. The position occupied by the structure is then updated in a second step by means of a suitable implicit approximation of (7.37c). This scheme has been proved to be stable under a CFL-like condition (Boffi *et al.* 2007).

Several extensions and applications of the IB method have been provided. Among them, we mention the following: the variational formulation for the case of a thick immersed structure proposed in Boffi, Gastaldi, Heltai and Peskin (2008), where the right-hand side of (7.37a) becomes

$$\int_{\Omega_s} \widehat{\boldsymbol{T}}_s(\widehat{\boldsymbol{d}}(\boldsymbol{s},t)) : \nabla \boldsymbol{w}(\boldsymbol{X}(\boldsymbol{s},t))\,\mathrm{d}\boldsymbol{s},$$

where $\boldsymbol{s}$ in this case is a vectorial Lagrangian coordinate; a mesh-adaptive approach used in combination with a second-order scheme (see Griffith, Hornung, McQueen and Peskin 2007); the use of a curvilinear fixed fluid mesh to improve flexibility and efficiency (the CURVIB approach: see Borazjani, Ge and Sotiropoulos 2008); the application to realistic three-dimensional mitral and aortic valves successfully addressed in Griffith, Luo, McQueen and Peskin (2009) and Yin, Luo, Wang and Watton (2009).

Fictitious domain approach. The fictitious domain (FD) approach was first introduced by Glowinski, Pan and Periaux (1997) and Bertrand, Tanguy and Thibault (1997), and then used by de Hart, Peters, Schreurs and Baaijens (2000) and de Hart, Baaijens, Peters and Schreurs (2003) in the context of heart valves. The FD approach can be regarded as the dual of the IB method, in the sense that in the latter a weak enforcement of the normal stress continuity at the FS interface is added to the weak formulation of fluid problem, whereas in the FD approach the velocity continuity is weakly enforced at the FS interface. This is achieved by means of Lagrange multipliers, as detailed in what follows (we refer to Sections 4.1, 4.2 and 4.3 for the notation; in particular, variables with $\widehat{\ }$ are referred to the reference domain, and we assume homogeneous boundary conditions). Find

$$\boldsymbol{v}(t) \in \boldsymbol{V}^t, \quad \widehat{\boldsymbol{d}}(t) \in \boldsymbol{D}, \quad \boldsymbol{\lambda}(t) \in \boldsymbol{H}^{-1/2}(\Sigma^t) \quad \text{and} \quad p(t) \in L^2(\Omega_f^t),$$

such that

$$\begin{aligned}
&\rho_f \int_{\Omega_f^t} \frac{\partial \boldsymbol{v}}{\partial t} \cdot \boldsymbol{w}\,\mathrm{d}\omega + \mathcal{A}_f(\boldsymbol{v},\boldsymbol{v},\boldsymbol{w}) + \mathcal{B}(p,\boldsymbol{w}) \\
&\qquad + \rho_s \int_{\Omega_s} \frac{\partial^2 \widehat{\boldsymbol{d}}}{\partial t^2} \cdot \widehat{\boldsymbol{e}}\,\mathrm{d}\omega + \int_{\Omega_s} \widehat{\boldsymbol{T}}_s(\widehat{\boldsymbol{d}}) : \nabla \widehat{\boldsymbol{e}}\,\mathrm{d}\omega = \int_{\Sigma^t} \boldsymbol{\lambda} \cdot (\boldsymbol{w} - \boldsymbol{e})\,\mathrm{d}\sigma, \\
&\mathcal{B}(q,\boldsymbol{v}) = 0, \\
&\int_{\Sigma^t} \boldsymbol{\mu} \cdot \left(\boldsymbol{v} - \frac{\partial \boldsymbol{d}}{\partial t}\right) \mathrm{d}\sigma = 0, \\
&\boldsymbol{d}_f = \boldsymbol{d} \quad \text{at } \Sigma^t,
\end{aligned} \tag{7.38}$$

for all $\boldsymbol{w} \in \boldsymbol{V}$, $\widehat{\boldsymbol{e}} \in \boldsymbol{D}$, $\boldsymbol{\mu} \in \boldsymbol{H}^{-1/2}(\Sigma^t)$ and $q \in L^2(\Omega_f)$. For the well-posedness of the time-discretized version of problem (7.38) we refer to Formaggia *et al.* (2009*b*). We notice that the use of the same Lagrange multiplier $\boldsymbol{\lambda}$ from both the fluid and structure sides also guarantees the continuity of normal stresses (Formaggia *et al.* 2009*b*).

Stijnen, de Hart, Bovendeerd and van de Vosse (2004) have successfully reported a validation with experimental measurements, and Astorino *et al.* (2010) have reported an application of the FD approach to three-dimensional problems in combination with a contact model for the leaflets.

Cut-FEM approach. A natural strategy for handling an internal interface cutting the mesh in an arbitrary way is to write two weak formulations of the problem at hand, one for each of the two subdomains generated by the presence of the interface, and then sum them up. In this case, the meshes of the two subdomains are fixed (background meshes). Since some of the mesh elements are cut by the interface, their contribution to the matrices is split into two parts: see Figure 7.8(a). This method is known as cut-FEM (Hansbo and Hansbo 2002, Hansbo, Hansbo and Larson 2003). Here, the jump between the normal stresses at the interface is determined by the physical interaction with the interface, as happens for the blood/valve interaction.

A common strategy used to guarantee the satisfaction of the interface continuity conditions is *discontinuous Galerkin* (DG) mortaring (often referred to as *Nitsche* mortaring). This is a very effective choice since the DG method prescribes interface conditions in a weak sense, thus allowing a great degree of flexibility of the solution at the interface (Hansbo and Hansbo 2002, Hansbo *et al.* 2003).

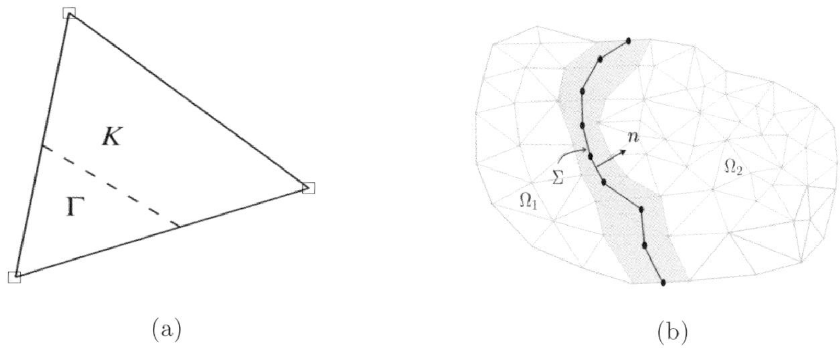

Figure 7.8. Element mesh K cut by the interface Γ (a) and FS domain (b). For the latter, we notice that Ω_f^1 and Ω_f^2 are two non-overlapping subdomains, whereas the related computational meshes have an overlap (in grey) in view of the X-FEM approach.

Finally, we notice that the implementation of the cut-FEM method is not standard since in the cut elements we may have polygonal elements: see for example Figure 7.8(a), where the original triangle is split into a triangle and a trapezoid. A solution is offered by the *extended finite element* method (X-FEM), where the finite element space is suitably enriched in order to make the treatment of the cut elements easy. In particular, the basic idea is to duplicate the variables in the cut elements and to use the basis functions of the original triangle in both subdomains, to represent the finite element solution and to compute the integrals (Hansbo and Hansbo 2002).

To provide a concrete example of the cut-FEM method, we will give the weak formulation introduced and analysed by Alauzet, Fabrèges, Fernández and Landajuela (2016), related to the case of a valve embedded in blood and represented by

$$\rho_s H_s \frac{\partial \boldsymbol{d}}{\partial t} + \boldsymbol{L}\boldsymbol{d} = -[\![\boldsymbol{T}_f \boldsymbol{n}]\!] \quad \text{in } \Sigma,$$

where Σ is the embedded membrane, $\boldsymbol{L}$ is an elliptic operator, and $[\![\cdot]\!]$ denotes the jump across the membrane. In particular, referring to Figure 7.8(b), let Ω_f^1 and Ω_f^2 be two non-overlapping subdomains separated by Σ, and $\Omega_{f,h}^1$ and $\Omega_{f,h}^2$ two meshes containing Ω_f^1 and Ω_f^2, respectively, with an overlap region (in grey in the figure) containing the membrane. Moreover, let $\boldsymbol{V}_h^i$ and Q_h^i be velocity and pressure finite element spaces related to $\Omega_{f,h}^i$ and let $\mathcal{C}^i$ be one of the bilinear or trilinear forms $\mathcal{C}$ introduced in Section 2 restricted to $\Omega^i_{f,h}$. Thus we have

$$\sum_{i=1}^{2}\left[\rho_f\int_{\Omega_f^i}\frac{\partial \boldsymbol{v}_h^i}{\partial t}\cdot\boldsymbol{w}\,\mathrm{d}\omega+\mathcal{A}_f^i(\boldsymbol{v}_h^i,\boldsymbol{v}_h^i,\boldsymbol{w})+\mathcal{B}^i(p_h^i,\boldsymbol{w})-\mathcal{B}^i(q,\boldsymbol{v}_h^i)\right]$$
$$+s_h(\boldsymbol{v}_h,\boldsymbol{w})+\rho_s H_s\int_\Sigma\frac{\partial \boldsymbol{d}_h}{\partial t}\cdot\boldsymbol{e}\,\mathrm{d}\sigma+\int_\Sigma \boldsymbol{L}\boldsymbol{d}_h\cdot\boldsymbol{e}\,\mathrm{d}\sigma$$
$$+\sum_{i=1}^{2}\left[-\int_\Sigma \boldsymbol{T}_f(\boldsymbol{v}_h^i,p_h^i)\boldsymbol{n}^i\cdot(\boldsymbol{w}-\boldsymbol{e})\,\mathrm{d}\sigma-\int_\Sigma\boldsymbol{T}_f(\boldsymbol{w},q)\boldsymbol{n}^i\cdot\left(\boldsymbol{v}_h^i-\frac{\partial \boldsymbol{d}_h}{\partial t}\right)\mathrm{d}\sigma\right.$$
$$\left.+\frac{\gamma}{h}\int_\Sigma\left(\boldsymbol{v}_h^i-\frac{\partial \boldsymbol{d}_h}{\partial t}\right)\cdot(\boldsymbol{w}-\boldsymbol{e})\,\mathrm{d}\sigma\right]=0.$$

The terms in the third and fourth rows above are simply the usual DG terms guaranteeing consistency, symmetry and stability of the method, whereas $s_h(\cdot,\cdot)$ accounts for the *ghost node stabilization* (Burman *et al.* 2015), which guarantees an optimal convergence order. This formulation was introduced for infinitesimal displacements. However, the authors also treat the case of a moving interface: for details see Alauzet *et al.* (2016).

Finally, we notice that the cut-FEM approach has been also considered for the case of a thick structure: see Burman and Fernández (2014) and Zonca, Formaggia and Vergara (2016).

7.5. Modelling the entire heart function

We conclude this section by providing some hints on the modelling of the complete heart function, a challenging and far-reaching endeavour. The latter is very complex and requires us to merge all the mathematical, numerical and computational issues highlighted previously in this section: see Lee *et al.* (2016), Chabiniok *et al.* (2016), Nordsletten *et al.* (2011*b*) and Quarteroni *et al.* (2017).

Basically, an integrated model of the heart involves the coupling of the electromechanical problem (7.31), the blood fluid dynamics (see Section 7.3), and the valve functioning (see Section 7.4). The coupling between the first two subproblems occurs at the endocardium Σ_{endo}: see Figure 6.1. In particular, this is determined by the classical fluid–structure interaction coupling conditions, *i.e.*, the kinematic condition (4.19c) and the dynamic condition (4.19d). In particular, referring to the notation of Sections 4.1 and 7.2 and to Figure 6.1, we have the following problem. Find the blood velocity $\boldsymbol{v}$ and pressure p, the muscle displacement $\boldsymbol{d}$, the trans-membrane potential V_m, the gating variables $\boldsymbol{w}$, and the ionic concentration variables $\boldsymbol{c}$, such that

$$\rho_f\left(\frac{\partial \boldsymbol{v}}{\partial t}+\rho_f(\boldsymbol{v}\cdot\nabla)\boldsymbol{v}\right)-\nabla\cdot\boldsymbol{T}_f(\boldsymbol{v},p)=\boldsymbol{0}\quad\text{in }\Omega^t_{cav},\tag{7.39a}$$

$$\nabla\cdot\boldsymbol{v}=0\quad\text{in }\Omega^t_{cav},\tag{7.39b}$$

$$\boldsymbol{v} = \frac{\partial \boldsymbol{d}}{\partial t} \qquad \text{at } \Sigma^t_{endo}, \tag{7.39c}$$

$$\left(\boldsymbol{T}^P_s(\boldsymbol{d}) + \boldsymbol{T}^A_s\left(\boldsymbol{c}, \boldsymbol{d}, \frac{\mathrm{d}\boldsymbol{d}}{\mathrm{d}t}\right)\right)\boldsymbol{n} = \boldsymbol{T}_f(\boldsymbol{v}, p)\,\boldsymbol{n} \quad \text{at } \Sigma^t_{endo}, \tag{7.39d}$$

$$\nabla \cdot \left(\widehat{\boldsymbol{T}}^P_s(\widehat{\boldsymbol{d}}) + \widehat{\boldsymbol{T}}^A_s\left(\widehat{\boldsymbol{c}}, \widehat{\boldsymbol{d}}, \frac{\mathrm{d}\widehat{\boldsymbol{d}}}{\mathrm{d}t}\right)\right) = \boldsymbol{0} \qquad \text{in } \Omega_{mus}, \tag{7.39e}$$

$$\widehat{\boldsymbol{T}}^A_s = P^A \boldsymbol{a}_f \otimes \widehat{\boldsymbol{a}}_f, \quad P^A = \mathcal{A}\left(\widehat{\boldsymbol{c}}, \widehat{\boldsymbol{d}}, \frac{\mathrm{d}\widehat{\boldsymbol{d}}}{\mathrm{d}t}\right) \qquad \text{in } \Omega_{mus}, \tag{7.39f}$$

$$\widehat{V}_m = \mathcal{M}(\widehat{V}_m; \widehat{\boldsymbol{d}}, \widehat{\boldsymbol{w}}, \widehat{\boldsymbol{c}}) \qquad \text{in } \Omega_{mus}, \tag{7.39g}$$

$$(\widehat{\boldsymbol{w}}, \widehat{\boldsymbol{c}}) = \mathcal{G}(\widehat{\boldsymbol{w}}, \widehat{\boldsymbol{c}}; \widehat{V}_m) \qquad \text{in } \Omega_{mus}. \tag{7.39h}$$

After numerical discretization in time and space, the corresponding nonlinear algebraic system is tremendously stiff and can have a very high numerical dimension. Devising efficient numerical strategies for its solution is a very active research area.

An effective strategy to solve the entire heart coupled problem numerically is given by the iterative solution at each time step of the electromechanical and fluid subproblems and based on the exchange of the interface conditions (7.39c, 7.39d). In particular, one of the partitioned strategies described in Section 4.6.3 can be adapted and used for problem (7.39) as well. Of course, at each iteration of the partitioned algorithm, the electromechanical problem could be solved by means of one of the strategies reported in Section 7.2.

Another partitioned algorithm for the solution of problem (7.39) is obtained by considering the solution at each time step of the electrical problem first and the FSI problem later (Nordsletten *et al.* 2011*b*). This is simply an extension of Algorithms EM1 and EM1′ reported in Section 7.2, where steps (2b) and (2′b) are replaced by a Newton iteration over the FSI problem.

A different approach is based on the monolithic solution of the integrated heart problem (7.39) by means of a full Newton method. In particular, Quarteroni *et al.* (2017) describe fully implicit and semi-implicit methods. The latter is based on updating only a part of the Jacobian as a consequence of an operator splitting technique. By means of numerical experiments, the authors emphasized that particular choices of the splitting and of the temporal scheme could lead to numerical instabilities. Whatever strategy is adopted, the tangent problem at each Newton iteration is solved by means of a preconditioned iterative method, which in this case is particularly suitable due to the block structure of the Jacobian: see Quarteroni *et al.* (2017).

Stability and convergence analysis of the different approaches is a field of current investigation.

PART THREE
Optimization, control, uncertainty and complexity reduction

8. Beyond direct simulation

In all the problems considered so far, for a given set of inputs (*e.g.* geometry, source terms, boundary and initial conditions) the goal has been to solve a forward PDE problem numerically, and, possibly, to evaluate some physical indices of clinical interest, such as the flow rate across a vessel section, the wall shear stress over a portion of the lumen boundary, or the oscillatory shear index. On the other hand, the efficient numerical algorithms and computational resources now available enable us to tackle several additional kinds of problem.

When simulating blood flows one has to deal with the variability of input data, such as geometric features of the vessel, boundary conditions and physical coefficients (*e.g.* the Reynolds number, the Womersley number, the structural model or material parameters related to the vessel wall). Very often, these data vary within a broad range and are almost impossible to characterize precisely. All these inputs affect the solution of the problem under investigation, as well as the outcomes of clinical interest. In the clinical context it is thus important to be able to characterize input/output relationships efficiently, in order to investigate both *intra-patient* and *inter-patient* variability. For the former case we include all those effects due to variations affecting a single patient, for example over time, or before/after clinical intervention. For the latter case, we mention for instance the morphological variability among vessel shapes due to age, size or pathological factors (see *e.g.* Xiong and Chong 2008 for the case of distal coronary anastomoses). It is thus of paramount importance to develop mathematical techniques capable of detecting the most relevant parameters, and then address the impact of their variation on the outputs of interest. This requires the solution of *many queries* to the *forward* problem. In this context, three classical situations we may face are (i) optimal control and optimal design, (ii) parameter identification and data assimilation, and (iii) uncertainty quantification.

(i) *Optimal control and optimal design problems.* When we pursue a suitable objective via an optimal strategy to be determined, we act on some *inputs* of the given partial differential equation, the *state* problem, such as boundary data, physical coefficients, or the shape of the domain itself, and the *control* variables, so that the solution of the corresponding state problem could meet the prescribed goal. This yields a PDE-constrained optimization problem; its numerical solution usually poses severe computational challenges, as the state problem needs to be solved several times while searching for the optimal control in an iterative way. Exploiting numerical optimization procedures for blood flow simulations is intended to (a) improve the design of prosthetic devices (such as stents, grafts, ventricular assist devices) with the aim of fulfilling a specific target, or (b) customize treatments or surgical procedures, provided that patient-specific information can be acquired. Examples include the optimal placement of electrodes in a defibrillation stage to damp the electrical activity in the myocardium in a desired way when dealing with the therapeutic treatment of cardiac arrhythmias (Nagaiah, Kunisch and Plank 2013*b*), the optimization of the shape of a cannula in order to maximize the flow rate through a ventricular assist device (Marsden 2014), and the improvement of the shape of a coronary bypass graft in order to possibly avoid vessel reocclusion (Lei, Archie and Kleinstreuer 1997, Dur *et al.* 2011). The first example mentioned above is an optimal control problem, where the control variable is a current source for the monodomain (or bidomain) equation (see Section 7.1), that is, one of its data. The two other examples address a *shape optimization* or *optimal design* problem to be solved, the control variable being the *shape* of the domain where the state problem is set.

(ii) *Parameter identification and data assimilation.* In principle, cardiovascular models necessitate a huge amount of data, for example the patient's radiological images and measurements. In general, however, some of them may be missing; remember for instance the issue of missing or defective boundary conditions discussed in Section 4.4, or the lack of biological parameters characterizing the tissue properties. For example, it is very hard to estimate the electrical conductivity of the myocardium for electrophysiology models, whereas (pointwise) measurements of the electrical potential – whose mathematical modelling can be characterized by a PDE model requiring electrical conductivities as inputs – can be easier to acquire. By solving *inverse/identification problems* in cardiovascular modelling, we aim to identify those inputs which are unknown or affected by uncertainty. For that, we rely on suitable quantities which are (a) acquired from measurements, such as displacements, pressures or flow rates, and (b) obtained as the outcome of a numerical simulation, and match these two sets of values. In other words, to identify those input values yielding the acquired measurements, we need to drive the PDE outcome as near as possible to the measured quantity. In general, vastly different inputs may have produced the observed outcome: this is why, instead of finding the most likely input configuration resulting in the observation performed, we instead rely on statistical inversion theory, in order to incorporate all possible information about the unknown inputs we may have prior to the measurement. This requires us to reformulate inverse problems as problems of statistical inference, for example by means of Bayesian statistics. A second issue is *data assimilation*. This rather generic term encompasses a wide range of techniques exploited to merge measurements and images into the mathematical model in order to improve numerical simulations. Furthermore, a filtering effect induced by the numerical simulation can also be exploited in view of noise reduction in image and measurement acquisition.

(iii) *Uncertainty quantification and propagation.* Confidence in the outcome of a cardiovascular simulation depends directly on the level of accuracy and certainty at which inputs can be determined. In fact, although outcomes are computed from inputs via a deterministic process, input data are often contaminated by experimental noise, or cannot be fully ascertained. Common sources of uncertainty in cardiovascular simulations include (a) boundary conditions, (b) anatomical models, where each geometric model is contaminated by image noise, (c) flow split, since very often there is a lack of clinical data to determine flow distribution to multiple distal branches, and (d) material properties, related to vessel walls or blood (Sankaran and Marsden 2010). Due to uncertainty, computational simulations have to be performed for a set of different parameter configurations and then merged in order to determine how robust simulation outcomes are with respect to variations of uncertain inputs (Sankaran and Marsden 2011). This is indeed strictly related to the task of parametric studies and sensitivity analyses, and can be seen as a *forward uncertainty quantification* (UQ) *problem*. Investigating the propagation of input uncertainties via computed outputs means evaluating suitable statistics of the outputs (such as expected values, moments, confidence bands), which are functions of the parameters affected by uncertainty. On the other hand, the solution of optimal control and inverse identification problems also depends on the experimental noise affecting observations and measurements used during the identification process, or the set-up of a desired target. Evaluating uncertainties in this case, providing suitable *confidence* intervals for the estimated quantities (not simply point estimates) and characterizing the statistical distribution of the unknown parameters are all *inverse UQ problems*. In this second case, quantifying uncertainties is even more important because an inverse problem is intrinsically ill-posed.

In the following sections we hone our understanding of these three classes of problems, showing relevant examples in cardiovascular modelling.

9. Control and optimization

Solving a PDE-constrained optimization problem consists of reaching a target by acting on a system governed by PDEs. The goal is to act on a *state problem* $\mathcal{E}(y, u) = 0$ given by a PDE system modelling the behaviour of a state variable y, through a control variable u in order to reach a desirable target. This is usually done by minimizing (or maximizing) a cost functional $J = J(y, u)$ expressing some physical index of interest encoding the desired objective – for instance, the squared distance from a target state. For a comprehensive presentation of the functional setting and well-posedness analysis of PDE-constrained optimization, see the monographs by Hinze, Pinnau, Ulbrich and Ulbrich (2009) and Gunzburger (2003). Other classes of problems of interest in cardiovascular applications, such as those related to *parameter estimation*, can also be cast in a PDE-constrained optimization framework; see *e.g.* Section 10.1 for further insights.

Generally speaking, a PDE-constrained optimization problem reads

$$\min_{y,u} J(y, u) \quad \text{subject to} \quad \mathcal{E}(y, u) = 0, \quad y \in V, \quad u \in \mathcal{U}_{ad} \subseteq \mathcal{U}. \tag{9.1}$$

Here V and $\mathcal{U}$ denote the state and the control space, whereas $\mathcal{E} : V \times \mathcal{U} \to V^*$ and $J : V \times \mathcal{U} \to \mathbb{R}$ denote the state equation and the cost functional, respectively; V^* is the dual space of V, and $\langle \cdot\,,\ \cdot \rangle_{V^*,V}$ denotes the duality between two elements in V^* and V; similarly, $\langle \cdot\,,\ \cdot \rangle_{\mathcal{U}^*,\mathcal{U}}$ indicates the duality between two elements in $\mathcal{U}^*$ and $\mathcal{U}$. Additional constraints, depending on the problem at hand, can also be imposed, for example in the form of inequalities; we express this fact by saying that we seek the optimal control in a closed subset $\mathcal{U}_{ad} \subseteq \mathcal{U}$ of admissible controls.

We assume that the state equation $\mathcal{E}(y, u) = 0$ has a unique solution $y = y(u) \in V$, and that the Fréchet derivative (with respect to y) $\mathcal{E}_y(y(u), u)$ has a bounded inverse for each $u \in \mathcal{U}$. Under these assumptions, the solution operator of the state equation is continuously differentiable – we let $y'(u)$ denote its derivative – and the *reduced formulation*

$$\min_{u} \tilde{J}(u) = J(y(u), u), \quad u \in \mathcal{U}_{ad} \tag{9.2}$$

is equivalent to (9.1). Note that this is a convenient framework to embrace – upon defining suitable functional spaces and operators – both stationary and time-dependent state problems; in the following we will make it clear whether the proposed methods can be used to tackle both classes of problems.

9.1. Optimality conditions

The solution $(\hat{y}, \hat{u})$ of a PDE-constrained optimization problem such as (9.1) requires a system of optimality conditions to be fulfilled, including the state equation and an optimality condition which involves the gradient of the cost functional. Indeed, if $\hat{u} \in U_{ad}$ is a local minimum of $\tilde{J}$, then it satisfies the variational inequality (also referred to as the *optimality condition* or *minimum principle*)

$$\langle \tilde{J}'(\hat{u}), v - \hat{u} \rangle_{\mathcal{U}^*,\mathcal{U}} \geq 0 \quad \text{for all } v \in \mathcal{U}_{ad}. \tag{9.3}$$

The quantity appearing on the left-hand side is the so-called Gâteaux derivative of $\tilde{J}$, evaluated at $\hat{u}$, in the generic, admissible direction $v - \hat{u}$ (see *e.g.* Tröltzsch 2010); $\tilde{J}'(\hat{u})$ denotes the gradient of $\tilde{J}$ at $\hat{u}$. To express this latter quantity in terms of the state solution y, we can use either (i) the sensitivity approach or (ii) the adjoint approach. As we will see below, in both cases at least a second PDE problem has to be solved in order to evaluate $\tilde{J}'(u)$.

9.1.1. Sensitivity approach

Computing *sensitivities* requires the evaluation of directional derivatives of both the cost functional and the state solution. For any $u, v \in \mathcal{U}$, we can write

$$\langle \tilde{J}'(u), v \rangle_{\mathcal{U}^*,\mathcal{U}} = \langle J_y(y(u), u), y'(u)v \rangle_{V^*,V} + \langle J_u(y(u), u), v \rangle_{\mathcal{U}^*,\mathcal{U}}. \tag{9.4}$$

The quantity $\delta_v y = y'(u)v$ denotes the sensitivity of the state with respect to the control, evaluated at u, for a given variation v, and can be obtained by solving a further PDE. Indeed, differentiating the state equation $\mathcal{E}(y(u), u) = 0$ along the direction v, we obtain

$$\mathcal{E}_y(y(u), u)y'(u)v + \mathcal{E}_u(y(u), u)v = 0, \tag{9.5}$$

where $\mathcal{E}_y(y, u), \mathcal{E}_u(y, u)$ are the Fréchet derivatives of the state operator with respect to y and u, respectively; see *e.g.* Tröltzsch (2010). Then $\delta_v y = y'(u)v$ is given by the solution of the linearized state equation

$$\mathcal{E}_y(y(u), u)\delta_v y = -\mathcal{E}_u(y(u), u)v. \tag{9.6}$$

Note that $y'(u) : \mathcal{U} \to V$ is a linear operator so that, for any admissible $v \in \mathcal{U}$, a new problem has to be solved to evaluate $\delta_v y$. Numerically, this approach is very demanding if the whole gradient $\tilde{J}'(u)$ is required: in this case, the number of linearized state equations (9.6) to be solved is equal to the dimension of the control space $\mathcal{U}$, a task that becomes out of reach as soon as the control has dimension larger than $O(10)$.

9.1.2. Adjoint approach and Lagrange multipliers

A convenient alternative is based on the so-called *adjoint approach*. By rewriting (9.4) as

$$\langle \tilde{J}'(u), v \rangle_{\mathcal{U}^*,\mathcal{U}} = \langle (y'(u))^* J_y(y(u), u) + J_u(y(u), u), v \rangle_{\mathcal{U}^*,\mathcal{U}} \tag{9.7}$$

and exploiting problem (9.5), we can express the first term on the right-hand side of (9.7) as

$$(y'(u))^* J_y(y(u), u) = -\mathcal{E}_u^*(y(u), u)((\mathcal{E}^*{}_y(y(u), u)))^{-1} J_y(y(u), u),$$

where $\mathcal{E}_u^*(y(u), u)$ is the adjoint operator[3] of $\mathcal{E}_u(y(u), u)$. Let us introduce an additional variable $\lambda = \lambda(u) \in V$, called the *adjoint state*, the solution of

$$\mathcal{E}_y(y(u), u))^* \lambda = -J_y(y(u), u). \tag{9.8}$$

Then $(y'(u))^* J_y(y(u), u) = \mathcal{E}_u^*(y(u), u)\lambda$; moreover, owing to (9.7), the evaluation of

$$\tilde{J}'(u) = \mathcal{E}_u^*(y(u), u)\lambda(u) + J_u(y(u), u) \tag{9.9}$$

simply requires us to solve the state problem and a further PDE problem, regardless of the dimension of the control space $\mathcal{U}$. Note that the adjoint problem is always a linear PDE.

The adjoint-based expression of the gradient of the cost functional (and more generally a system of first-order optimality conditions) can also be obtained, in a more straightforward way, by relying on the Lagrange multiplier method (Hinze *et al.* 2009). Let us define the Lagrangian functional $\mathcal{L} : V \times \mathcal{U} \times V \to \mathbb{R}$,

$$\mathcal{L}(y, u, \lambda) = J(y, u) + \langle \lambda, \mathcal{E}(y, u)\rangle_{V,V^*},$$

where $\lambda \in V$ denotes a Lagrange multiplier enforcing the PDE constraint (playing the role of adjoint variable); note that in this case the three variables are independent.

By deriving $\mathcal{L}$ with respect to λ and imposing that the derivative is equal to zero, we recover the state problem,

$$\langle \mathcal{L}_\lambda(y, u, p), \varphi\rangle_{V^*,V} = 0 \quad \text{for all } \varphi \in V \;\Leftrightarrow\; \mathcal{E}(y, u) = 0.$$

Similarly, by deriving $\mathcal{L}$ with respect to y, we obtain the expression of the adjoint problem (9.8),

$$\langle \mathcal{L}_y(y, u, p), \psi\rangle_{V^*,V} = 0 \quad \text{for all } \psi \in V \;\Leftrightarrow\; J_y(y, u) + \mathcal{E}_y^*(y, u) = 0. \tag{9.10}$$

Finally, at the optimum we impose

$$\begin{aligned} &\langle \mathcal{L}_u(y, u, \lambda), v - u\rangle_{\mathcal{U}^*,\mathcal{U}} = 0 \quad \text{for all } v \in \mathcal{U}_{ad} \\ &\quad\Leftrightarrow\; \langle J_u(y, u) + \mathcal{E}_u(y, u)^*\lambda, v - u\rangle_{\mathcal{U}^*,\mathcal{U}} \geq 0 \quad \text{for all } v \in \mathcal{U}_{ad}. \end{aligned}$$

In this way, a system of first-order necessary optimality conditions to be

[3] The adjoint operator of $\mathcal{E}$, denoted by $\mathcal{E}^*$, is given by the relation $(\mathcal{E}^*\varphi, \psi)_{L^2(\Omega)} = (\varphi, \mathcal{E}\psi)_{L^2(\Omega)}$ for any $\varphi, \psi \in C_0^\infty(\Omega)$; note that no boundary condition is involved in its definition. For this reason, $\mathcal{E}^*$ is also referred to as the *formal* adjoint operator.

fulfilled by the optimal solution (y,u) and the corresponding adjoint state λ is given by the following Karush–Kuhn–Tucker (KKT) system:

$$\mathcal{E}(y,u) = 0, \tag{9.11a}$$

$$\mathcal{E}_y(y,u)^*\hat{\lambda} = -J_y(y,u), \tag{9.11b}$$

$$\langle J_u(y,u) + \mathcal{E}_u(y,u)^*\lambda, v-u\rangle_{\mathcal{U}^*,\mathcal{U}} \geq 0 \quad \text{for all } v \in \mathcal{U}_{ad}. \tag{9.11c}$$

In the unconstrained case $\mathcal{U}_{ad} \equiv \mathcal{U}$, the variational inequality reduces to the equation

$$\langle J_u(y,u) + \mathcal{E}_u(y,u)^*\lambda, v\rangle_{\mathcal{U}^*,\mathcal{U}} = 0 \quad \text{for all } v \in \mathcal{U},$$

so (9.11) can be viewed as the Euler–Lagrange system for the Lagrangian functional, that is, the solutions of (9.11) are the stationary points of $\mathcal{L}(\cdot,\cdot,\cdot)$:

$$\nabla\mathcal{L}(y,u,\lambda)[w,v,\eta] = 0 \quad \text{for all } (\phi,v,\psi) \in Y \times \mathcal{U} \times V.$$

The third inequality of system (9.11) allows us to recover the expression of the gradient $\tilde{J}'(u)$. Indeed, since $\mathcal{E}(y(u),u) = 0$, it holds that $\mathcal{L}(y(u),u,\lambda) = \tilde{J}(u)$ for any arbitrary $\lambda \in V$, so that, by differentiating this latter equality with respect to u, we obtain

$$\begin{aligned}\langle \tilde{J}'(u), v\rangle_{\mathcal{U}^*,\mathcal{U}} &= \langle \mathcal{L}_y(y(u),u,\lambda), y'(u)v\rangle_{V^*,V} + \langle \mathcal{L}_u(y(u),u,\lambda), v\rangle_{\mathcal{U}^*,\mathcal{U}} \\ &= \langle \mathcal{L}_u(y(u),u,\lambda), v\rangle_{\mathcal{U}^*,\mathcal{U}}.\end{aligned}$$

Hence, $\tilde{J}'(u) = \mathcal{L}_u(y(u),u,\lambda(u))$, since $\langle \mathcal{L}_y(y(u),u,\lambda),\psi\rangle_{V^*,V} = 0$ for any $\psi \in V$, thanks to (9.10).

For illustration we discuss a specific example, by considering the problem of minimizing the energy

$$J(\boldsymbol{v},\boldsymbol{u}) = \frac{1}{2}\int_{\Omega_{obs}} |\boldsymbol{v}-\boldsymbol{v}_d|^2 \,\mathrm{d}\Omega + \frac{\alpha}{2}\int_{\Gamma_{con}} |\nabla\boldsymbol{u}|^2 \,\mathrm{d}\Gamma,$$

of a fluid flow in a tract of blood vessel where a bypass is inserted. As a matter of fact, the bypass is not simulated. Its action is represented via a velocity control $\boldsymbol{u}$ acting on the boundary $\Gamma_{con} \subset \partial\Omega_f$, the interface where the bypass and the vessel meet: see Figure 9.1. The goal is to regularize the velocity pattern in a suitable observation region $\Omega_{obs} \subseteq \Omega_f$ by requiring $\boldsymbol{v}$ to be as close as possible to a desired distribution $\boldsymbol{v}_d$; see the related discussion in the next section. Referring to the notation in Section 4, for simplicity, we consider a steady version of the Navier–Stokes equations (4.1a, 4.1b), for which the velocity/pressure couple $y = (\boldsymbol{v},p)$ solves the state problem

$$\begin{aligned}&-\nabla\cdot\mathbf{T}_f(\boldsymbol{v},p) + \rho_f(\boldsymbol{v}\cdot\nabla)\boldsymbol{v} = \mathbf{0} && \text{in } \Omega_f,\\ &\nabla\cdot\boldsymbol{v} = 0 && \text{in } \Omega_f,\\ &\boldsymbol{v} = \boldsymbol{v}_{in} && \text{on } \Gamma_{in},\\ &\boldsymbol{v} = \mathbf{0} && \text{on } \Gamma_w,\end{aligned}$$

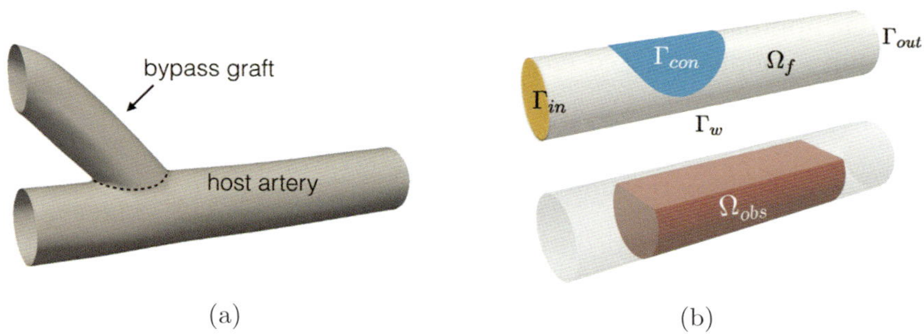

Figure 9.1. (a) Schematic representation of a bypass graft. (b) Domain, boundary portion and observation region for the bypass model problem.

$$\begin{aligned} \boldsymbol{v} &= \boldsymbol{u} && \text{on } \Gamma_{con}, \\ \boldsymbol{T}_f(\boldsymbol{v}, p)\boldsymbol{n} &= \boldsymbol{0} && \text{on } \Gamma_{out}, \end{aligned}$$

where the control variable is the velocity $\boldsymbol{u}$ imposed on the boundary Γ_{con}. Here $\alpha > 0$ is a parameter penalizing the control magnitude (or cost); this can also be seen as a regularization term, ensuring the convexity of the cost functional. The fluid Cauchy stress tensor $\mathbf{T}_f(\boldsymbol{v}, p)$ has been defined in (4.2).

Following the Lagrangian approach, we can derive a system of first-order optimality conditions, where the adjoint problem for the adjoint variables $\lambda = (\boldsymbol{z}, q)$ is given by

$$\begin{aligned} & -\nabla \cdot \mathbf{T}_f(\boldsymbol{z}, q) + \rho_f(\nabla^T \boldsymbol{v})\boldsymbol{z} - \rho_f(\boldsymbol{v} \cdot \nabla)\boldsymbol{z} \\ & \qquad\qquad = (\boldsymbol{v} - \boldsymbol{v}_d)\mathcal{I}_{\Omega_{obs}} && \text{in } \Omega_f, \\ & \nabla \cdot \boldsymbol{z} = 0 && \text{in } \Omega_f, \\ & \boldsymbol{z} = \boldsymbol{0} && \text{on } \Gamma_{in} \cup \Gamma_w \cup \Gamma_{con}, \\ & \boldsymbol{T}_f(\boldsymbol{z}, q)\boldsymbol{n} = \boldsymbol{0} && \text{on } \Gamma_{out}, \end{aligned}$$

where $\mathcal{I}_{\Omega_{obs}} = \mathcal{I}_{\Omega_{obs}}(\mathbf{x})$ is the characteristic function of the region Ω_{obs}. Note that the adjoint problem is linear in $(\boldsymbol{z}, q)$, and comes from the linearization of the Navier–Stokes equations around the state solution; the optimality condition instead reads

$$\alpha \boldsymbol{u} + \boldsymbol{z} = \boldsymbol{0} \quad \text{on } \Gamma_{con}.$$

Remark 9.1. In the case of time-dependent state problems, the adjoint problem is backward-in-time. Depending on the observation appearing in the cost functional – which can be either on the whole time interval $(0, T)$ or at the final time T only – the dependence of the adjoint problem on the state is only at $t = T$ (thus, as initial condition) or on the whole interval

$(0, T)$. The approach followed so far can still be employed to derive a system of optimality conditions: see *e.g.* Section 9.3.2 for further details of a case of interest.

9.2. Numerical approximation

Solving a PDE-constrained optimization problem entails a computational effort larger than that required for the solution of a forward (state) problem. The two formulations (9.1) and (9.2) yield two different paradigms for the approximation of such a problem. In the former case, both state and control variables are optimization variables and PDE constraints are explicitly specified; in the latter, only the control variable u is an optimization variable, whereas the state variable y is considered to be an implicit function of u via the PDE constraint.[4] In this latter case, the solution of the state problem is *nested* in the evaluation of the gradient $\tilde{J}'(u)$ of the reduced cost functional.

Algorithms for solving PDE-constrained optimization problems can be sorted according to several criteria. An initial classification criterion is between *iterative methods*, rooted in iterative minimization algorithms for the reduced cost functional, and *all-at-once methods*, where the PDE constraint is kept explicitly and the three equations forming (9.11) are solved simultaneously. Another criterion is concerned with the highest order of derivatives exploited by the algorithm, yielding derivative-free methods, gradient-based methods and Hessian-based methods.

A different perspective is taken when addressing the interplay between *optimization* and *discretization*: numerical discretization can be performed before or after the derivation of a system of optimality conditions. More precisely, in the so-called *optimize-then-discretize* approach, optimization is carried out at the continuous level (*e.g.* to find system (9.11)) and then the discretization is operated on the resulting optimality system. Alternatively, using the *discretize-then-optimize* approach, we first approximate the state equation (and the cost functional) and then carry out the optimization at the discrete level. Here we address the former approach; further details are given at the end of this section. For brevity, we recall the main features of iterative and all-at-once methods in the case of *unconstrained* problems, that is, problems without further equality/inequality constraints or, equivalently, for which $\mathcal{U}_{ad} \equiv \mathcal{U}$. In particular, $\boldsymbol{Y} \in \mathbb{R}^{n_y}$, $\boldsymbol{U} \in \mathbb{R}^{n_u}$ denote the discrete representation of the state and the control variable, respectively, whereas $\boldsymbol{\lambda} \in \mathbb{R}^{n_\lambda}$ is the discrete adjoint variable.

[4] From the numerical standpoint, the former approach is often given the name of *simultaneous analysis and design* (SAND) and the latter is referred to as *nested analysis and design* (NAND).

9.2.1. Iterative methods

Iterative (also referred to as *black-box*) methods treat the reduced problem

$$\min_{\boldsymbol{U}} \tilde{J}(\boldsymbol{U}) = J(\boldsymbol{Y}(\boldsymbol{U}), \boldsymbol{U}),$$

once $\boldsymbol{U}$ is known; $\boldsymbol{Y}(\boldsymbol{U})$ is obtained as the solution of the state equation. An existing algorithm for the solution of the state equation therefore has to be embedded into an optimization loop, and any available PDE and optimization routines can be freely combined. In particular, iterative methods are a popular choice when dealing with optimal control problems by extending an existing code for the state problem. Within this class, a notable difference exists between gradient-based and non-gradient-based algorithms.

Non-gradient-based (or *derivative-free*) algorithms, such as the popular *Nelder–Mead algorithm*, exploit either comparisons between function evaluations in different directions at each step, or low-order local approximants of $\tilde{J}$, in order to assess its local behaviour and find the minimizer: see *e.g.* Marsden, Feinstein and Taylor (2008). Employing a finite difference approximation of the gradient is an attractive alternative due to its ease of implementation, but it may suffer from limited accuracy and large costs in the presence of many design variables. Hence, these methods are feasible only in the case where the control space has very small dimension, for example if the control is expressed in terms of a vector of $n_u = O(10)$ design variables.

Gradient-based algorithms exploit the gradient $\tilde{J}'$ to iteratively update the control until a suitable convergence criterion is fulfilled. Notable instances are descent methods, such as the gradient, (non-linear) conjugate gradient, quasi-Newton or Newton methods. In the simplest case of a gradient method, starting from an initial guess $\boldsymbol{U}^{(0)}$ we iteratively generate a sequence

$$\boldsymbol{U}^{(k+1)} = \boldsymbol{U}^{(k)} - \tau_k \tilde{J}'(\boldsymbol{U}^{(k)}), \quad k = 0, 1, \ldots,$$

where $\tau_k > 0$ is a step size, until *e.g.* $\|\tilde{J}'(\boldsymbol{U}^{(k)})\| < \varepsilon$, for a given tolerance $\varepsilon > 0$. Further details can be found in Nocedal (1992) and Kelley (1999), for example. The solver for the state equation has to be augmented with a routine which provides the gradient of the state with respect to the optimization variables, and hence the solution of the adjoint problem. Optimization algorithms with faster convergence rates are needed to speed up the execution of the whole algorithm. Although straightforward to implement, the gradient method suffers from a poor rate of convergence when dealing with the numerical solutions of PDE-constrained optimization problems. More efficient methods are more typically employed, such as quasi-Newton methods: see *e.g.* Borzì and Schulz (2011) for further details.

9.2.2. All-at-once methods

By treating both the control and the state variables as independent optimization variables, coupled via the PDE constraint, we deal with an equality constrained non-linear optimization problem, the state equation now playing the role of equality constraint. The goal of all-at-once (also referred to as *one-shot*) methods is to tackle the (possibly) non-linear optimality system (9.11) as a whole problem to be solved. After numerical discretization, in the unconstrained case we obtain the algebraic system

$$J_{\boldsymbol{Y}}(\boldsymbol{Y},\boldsymbol{U}) + \mathbf{e}_{\boldsymbol{Y}}^T(\boldsymbol{Y},\boldsymbol{U})\boldsymbol{\lambda} = \mathbf{0}, \tag{9.12a}$$

$$J_{\boldsymbol{U}}(\boldsymbol{Y},\boldsymbol{U}) + \mathbf{e}_{\boldsymbol{U}}^T(\boldsymbol{Y},\boldsymbol{U})\boldsymbol{\lambda} = \mathbf{0}, \tag{9.12b}$$

$$\mathbf{e}(\boldsymbol{Y},\boldsymbol{U}) = \mathbf{0}. \tag{9.12c}$$

Here $\mathbf{e}(\boldsymbol{Y},\boldsymbol{U})$ denotes the discrete state operator, $J_{\boldsymbol{Y}}$ and $J_{\boldsymbol{U}}$ are the gradients of J with respect to state and control variables, respectively, whereas $\mathbf{e}_{\boldsymbol{Y}}$ and $\mathbf{e}_{\boldsymbol{U}}$ are the Jacobians of the state equations with respect to state and control variables, respectively. The three equations of system (9.12) can also be seen as the conditions obtained by requiring that the gradient of the discrete Lagrangian $\mathcal{L}(\boldsymbol{Y},\boldsymbol{U},\boldsymbol{\lambda}) = J(\boldsymbol{Y},\boldsymbol{U}) - \boldsymbol{\lambda}^T\mathbf{e}(\boldsymbol{Y},\boldsymbol{U})$ vanishes.

The strategy above is well suited to PDE-constrained optimization problems involving stationary state systems, but it is more computationally involved in the time-dependent case. If $J(\boldsymbol{Y},\boldsymbol{U})$ is quadratic and $\mathbf{e}(\boldsymbol{Y},\boldsymbol{U})$ is linear in $\boldsymbol{Y}$ and $\boldsymbol{U}$, (9.12) is a linear system of equations in saddle-point form, such as those arising from quadratic programming. In this case, preconditioned iterative methods for linear systems could be employed, such as those based on Krylov subspaces. In this respect, several preconditioners have been proposed in the past decade, in which multigrid schemes are exploited as inner solvers (or preconditioners) for some blocks of the KKT matrix within an outer iterative solver; see *e.g.* Benzi *et al.* (2005) and Rees, Dollar and Wathen (2010). More recent extensions to constrained problems have been addressed, for example in Borzì and Schulz (2011) and the references therein.

If the state problem is non-linear, the optimality system has to be solved via appropriate linearization procedures (*e.g.* sequential quadratic programming methods) or modern penalty methods (*e.g.* augmented Lagrangian methods). Indeed, the equations of system (9.12) are still linear in $\boldsymbol{\lambda}$ but non-linear in $(\boldsymbol{Y},\boldsymbol{U})$. When a Newton-type method is applied to (9.12), each iteration on the KKT system entails the solution of the linear system

$$\begin{pmatrix} \mathcal{L}_{\boldsymbol{YY}} & \mathcal{L}_{\boldsymbol{YU}} & \mathbf{e}_{\boldsymbol{Y}}^T \\ \mathcal{L}_{\boldsymbol{UY}} & \mathcal{L}_{\boldsymbol{UU}} & \mathbf{e}_{\boldsymbol{U}}^T \\ \mathbf{e}_{\boldsymbol{Y}} & \mathbf{e}_{\boldsymbol{U}} & 0 \end{pmatrix} \begin{pmatrix} \delta\boldsymbol{Y} \\ \delta\boldsymbol{U} \\ \boldsymbol{\lambda}^{(k+1)} \end{pmatrix} = - \begin{pmatrix} J_{\boldsymbol{Y}} \\ J_{\boldsymbol{U}} \\ \mathbf{e} \end{pmatrix} \tag{9.13}$$

for the update $\delta\boldsymbol{X}^T = (\delta\boldsymbol{Y},\delta\boldsymbol{U})^T$ of the optimization variable and the new value $\boldsymbol{\lambda}^{(k+1)}$ of the adjoint variable. Then we set $(\boldsymbol{Y}^{(k+1)}, \boldsymbol{U}^{(k+1)})^T = (\boldsymbol{Y}^{(k)}, \boldsymbol{U}^{(k)})^T + (\delta \boldsymbol{Y}, \delta \boldsymbol{U})^T$. Here

$$\mathcal{H} = \begin{pmatrix} \mathcal{L}_{\boldsymbol{YY}} & \mathcal{L}_{\boldsymbol{YU}} \\ \mathcal{L}_{\boldsymbol{UY}} & \mathcal{L}_{\boldsymbol{UU}} \end{pmatrix}$$

denotes the Hessian of the Lagrangian functional, to be evaluated at $(\boldsymbol{Y}^{(k)}, \boldsymbol{U}^{(k)}, \boldsymbol{\lambda}^{(k)})^T$. Note the saddle-point structure of system (9.13), where the adjoint variable indeed plays the role of multiplier for the constraint expressed by the (linearized) state equation. The system (9.13) can be equivalently obtained as the KKT system for the linear-quadratic optimization problem

$$\min_{\delta \boldsymbol{X}} \left[\frac{1}{2} \delta \boldsymbol{X}^T \mathcal{H} \delta \boldsymbol{X} + \begin{pmatrix} J_{\boldsymbol{Y}} \\ J_{\boldsymbol{U}} \end{pmatrix}^T \delta \boldsymbol{X} \right] \quad \text{subject to} \quad \begin{pmatrix} \mathbf{e}_{\boldsymbol{Y}} \\ \mathbf{e}_{\boldsymbol{U}} \end{pmatrix} \delta \boldsymbol{X} + \mathbf{e} = 0$$

(*e.g.* Hinze *et al.* 2009, Borzì and Schulz 2011), whence the name *sequential quadratic programming* method, which is commonly used to refer to the Newton iterations on system (9.13). Indeed, a quadratic programming problem has to be solved at each step, for example by means of a preconditioned Krylov method such as MINRES, until a suitable convergence criterion is fulfilled. Suitable approximations of the Hessian, based for example on quasi-Newton methods, are required to make this algorithm more computationally attractive: see *e.g.* Borzì and Schulz (2011).

Remark 9.2. Constraints on the control and/or state variables add nonlinearity to the optimization problem. A first option is to treat inequality constraints in an outer loop, via *penalty methods* which allow us to convert them into additional terms in the cost functional. For instance, if $u \leq b$ is a pointwise control constraint, the term $\frac{c}{2} \| \max\{0, u - b\} \|_{\mathcal{U}}^2$ can be added, where $c > 0$ is a penalty parameter to be properly selected, and then an iterative method can be used in the inner loop. Another option in iterative methods is to perform a projection over the space of admissible controls at each step. More efficient strategies to tackle constrained problems are usually obtained when dealing with constraints in the main optimization loop, such as in the case of *primal–dual active set* strategies: see *e.g.* the monographs by Borzì and Schulz (2011) and Hinze *et al.* (2009).

We finally point out that in this section we have opted for the optimize-then-discretize approach: that is, we have shown how to recover a system of optimality conditions and then proceed to its numerical discretization. The opposite strategy (discretize-then-optimize) would have led to substantially similar numerical methods, by choosing either an iterative or an all-at-once method for the system of optimality conditions derived once the original state system had been discretized. The two approaches do not generally yield identical solutions; see *e.g.* Hinze *et al.* (2009) for a discussion.

9.3. Applications to cardiovascular modelling

9.3.1. Optimal design of bypass grafts

PDE-constrained optimization problems are obtained when looking for the optimal design of prosthetic devices, such as ventricular assist devices or bypass grafts. For instance, coronary artery bypass grafting (CABG) is a standard surgical procedure to restore blood perfusion to the cardiac muscle by redirecting blood from the aorta through a graft vessel downstream of a coronary artery affected by stenosis or occlusion. The ability to design the graft-vessel connection (the so-called anastomosis, which we refer to as the domain Ω) in an efficient way is a potentially critical factor in preventing post-operative recurrence of re-stenosis.

Today it is accepted that intimal wall thickening, caused by the accelerated growth of smooth muscle cells and the surrounding matrix, is one of the leading causes of long-term failure of end-to-side vascular grafts (Haruguchi and Teraoka 2003). Low and/or highly oscillatory patterns of wall shear stress (WSS), as well as strong vorticity and recirculations, cause intimal wall thickening (Ethier *et al.* 1998, Keynton *et al.* 2001) at sites where curvatures, bifurcations, tortuosity and branching occur, and, more generally, where flow departs from unidirectional patterns (Giordana *et al.* 2005, Loth, Fischer and Bassiouny 2008). In mathematical terms, an *optimal* graft is one that minimizes suitable cost functionals involving the area of low WSS, the spatial WSS gradient (Lei *et al.* 1997) or the vorticity (Quarteroni and Rozza 2003, Manzoni, Quarteroni and Rozza 2012*b*); see *e.g.* Kleinstreuer (2006) for a detailed review. Taking blood velocity $\mathbf{v}$ and pressure p as state variables, and $u = \Omega$ (*i.e.* the shape of the domain itself) as the control variable, the goal is thus to find the optimal shape $\hat{\Omega}$ of the graft by minimizing the cost functional

$$J(\boldsymbol{v}, \Omega) = \int_0^T \int_{\Omega_{obs}} |\nabla \times \boldsymbol{v}|^2 \, \mathrm{d}\omega \, \mathrm{d}t,$$

where $\Omega_{obs} \subset \Omega$ is a given observation region in the artery portion immediately after the anastomosis. Indeed, high downstream vorticity may lead to strong flow recirculation, yielding similar effects in terms of intimal thickening.

Other cost functionals that can be employed include the following.

- A tracking-type functional, in order to drive the blood velocity (and pressure, if $\delta > 0$) towards a specified velocity (and pressure) target state $\boldsymbol{v}_d$, p_d, featuring a regular pattern:

$$J(\boldsymbol{v}, \Omega) = \frac{1}{2} \int_0^T \int_{\Omega_{obs}} |\boldsymbol{v} - \boldsymbol{v}_d|^2 \, \mathrm{d}\Omega \, \mathrm{d}t + \frac{\delta}{2} \int_0^T \int_{\Omega_{obs}} |p - p_d|^2 \, \mathrm{d}\omega \, \mathrm{d}t.$$

A typical choice of $(\boldsymbol{v}_d, p_d)$ is provided by the Stokes (or a low-Reynolds Navier–Stokes) flow in the same domain.

- In two dimensions, a *Galilean invariant* vortex measure identifies a region as a vortex if $\nabla \boldsymbol{v}$ has complex eigenvalues, that is, if $\det(\nabla \boldsymbol{v}) > 0$. Following Hintermüller, Kunisch, Spasov and Volkwein (2004), Kunisch and Vexler (2007) and Lassila, Manzoni, Quarteroni and Rozza (2013*a*), then

$$J(\boldsymbol{v}, \Omega) = \int_0^T \int_{\Omega_{obs}} \max(0, \det(\nabla \boldsymbol{v})) \, d\omega \, dt$$

can be used when dealing with vortex suppression.

- A WSS gradient-based functional is given by

$$J(\boldsymbol{v}, \Omega) = \int_0^T \int_{\Gamma_{obs}} WSSG(t) \, d\gamma \, dt,$$
$$WSSG(t) = \left(\left(\frac{\partial \boldsymbol{w}_p}{\partial \boldsymbol{\tau}_p} \right)^2 + \left(\frac{\partial \boldsymbol{w}_n}{\partial \boldsymbol{\tau}_n} \right)^2 \right),$$

that is, by a time-averaged WSS gradient, measured over the portion $\Gamma_{obs} \subset \partial\Omega$ of the boundary in the anastomosis region. Here, *WSSG* denotes the WSS gradient, where $\boldsymbol{w} = \boldsymbol{w}(t, \boldsymbol{x})$ is the WSS vector of components $w^{(j)} = \mu \nabla \boldsymbol{v} \boldsymbol{n}(t, \boldsymbol{x}) \cdot \boldsymbol{\tau}^{(j)}$, and $\boldsymbol{\tau}_p$ and $\boldsymbol{\tau}_n$ are the unit vectors parallel and normal to the direction of the time-averaged WSS vector $\boldsymbol{w}$, respectively. Only a linear combination of the *normal* components $\partial \boldsymbol{x}_j / \partial \boldsymbol{\tau}_j$, $j = p, n$ is considered as an index to quantify the tension yielding intimal thickening. Moreover, the time-averaged WSS is relatively insensitive to changes in the anastomosis configuration, whereas the time-averaged WSS gradient is highly sensitive, and has been linked to localized mechanobiological responses in tissues (see Lei *et al.* 1997 and the discussion therein). Using *WSSG* rather than *WSS* can be understood as filtering the WSS by removing its component induced by the steady mean flow and considering only the spatially fluctuating term as part of the indicator. For the sake of numerical efficiency of the whole optimization process, a steady flow simulation (*e.g.* corresponding to the systolic peak) can be considered, and the WSS gradient of the steady flow can be taken as an approximation of the time-averaged WSS gradient. The interested reader can refer to Kleinstreuer (2006), for example, for further details.

Many works of the past few decades have focused on the optimal shape design of end-to-side anastomoses, typically by acting on the wall shape near the anastomosis by local shape variation. If a simpler parametric optimization approach is considered, the three most significant design variables (see Loth *et al.* 2008) are the anastomosis angle, the graft-to-host diameter ratio (Keynton *et al.* 2001, Do, Owida, Yang and Morsi 2011) and the toe shape, that is, the profile of the bypass junction. Optimizing a complex configuration in terms of few design variables is more advantageous since it is known that grafts usually have an extended patch or cuff that effectively provides a hood to ease the flow transition from the graft to the artery. Further, improved anastomosis design must comply with the constraints set by surgical realization. For a more comprehensive review of bypass graft design results, we refer to Migliavacca and Dubini (2005), Loth *et al.* (2008), Owida, Do and Morsi (2012) and Marsden (2014).

Concerning the objective to be minimized, each cost functional involves suitable observations, obtained as functions of the state variables over the entire domain or some boundary portions, and over a given time interval and, possibly, data acquired from measurements, as in the case of a tracking-type functional. There are of course other parameters affecting the cost functional, such as the Reynolds number, or the flow split between the proximal host artery and the graft, *i.e.* the ratio of the corresponding flow rates (Giordana *et al.* 2005). All these parameters should be considered as uncertainties within a robust optimization framework, in order to characterize the optimal shape of the graft in a range of possible operating conditions. Examples of optimal design in the presence of uncertainty in cardiovascular applications have been considered by Sankaran and Marsden (2010).

As a concluding remark, we point out that the coupling of optimization algorithms to blood flow simulations is computationally challenging, since each evaluation of the cost functional requires the solution of an unsteady, three-dimensional Navier–Stokes problem. If relying on the adjoint approach to characterize the gradient of the cost functional, a further linearized Navier–Stokes (Oseen) problem has to be solved at every iteration. For this reason, the majority of works related to applications in surgery and device optimization have focused, so far, on small-scale and/or two-dimensional problems, or on steady-flow problems, usually relying on gradient-free methods.

For the sake of illustration, we report some numerical results for the optimal design of a femoropopliteal bypass graft. This surgery is used to bypass diseased blood vessels above or below the knee, and is one of the most common surgical treatments of chronic lower-extremity ischaemia. We model the blood flow across a bypass graft via a steady Navier–Stokes model, and consider a vorticity cost functional

$$J(\boldsymbol{v}, \Omega) = \int_{\Omega_{obs}} |\nabla \times \boldsymbol{v}|^2 \, \mathrm{d}\omega,$$

where $\Omega_{obs} \subset \Omega$ is the observation region. Initial and optimal shapes are shown in Figure 9.2 for the case of a stenosed host artery; its occlusion,

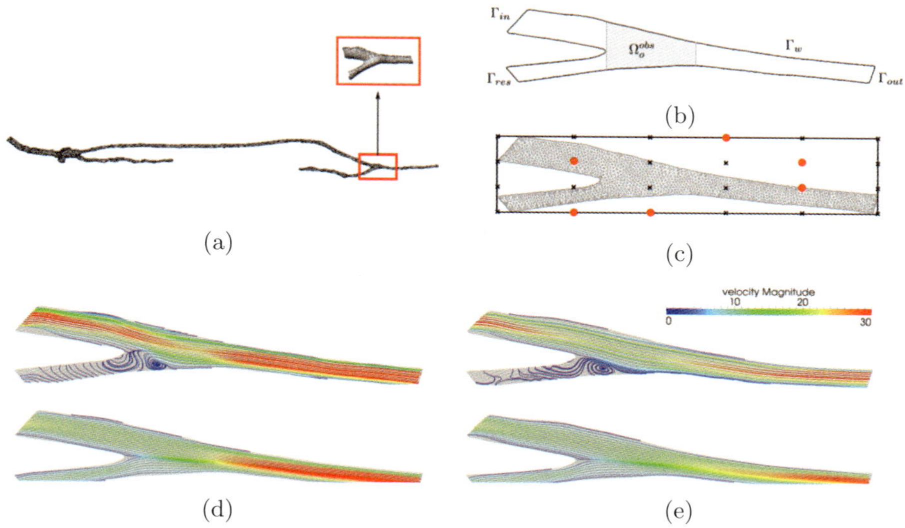

Figure 9.2. Optimal design of bypass grafts. (a) A tract of femoral artery with a bypass graft. (b) Computational domain, boundaries and observation region; (c) FFD shape parametrization used to generate admissible shapes. Global shape deformations are induced by the displacement of a few selected control points (shown in red) in the 6×4 FFD lattice. These control points are selected by a preliminary screening procedure based on sensitivity analysis. (d) Initial and (e) optimal bypass configurations in the case of total (above) or partial (below) occlusion. Numerical results were obtained using the MATLAB finite element library MLife.

either total or partial, is expressed via a Dirichlet boundary condition on the incoming velocity field on Γ_{res}, homogeneous for the completely occluded case and non-homogeneous otherwise. See Figure 9.2(b) for the definition of Ω_{obs} and Γ_{ref}. A shape parametrization technique based on free-form deformations (FFDs) is very suitable for describing admissible shapes via deformation of a reference configuration by acting on a small set of control points: see *e.g.* Manzoni (2012) and Manzoni *et al.* (2012*b*).

9.3.2. Optimal control of electrical defibrillation

Whereas in healthy conditions the electrical activation of the heart is an extremely organized (and efficient) process, some disturbances in the formation and/or propagation of electrical signals may induce re-entrant activation patterns which lead to tachycardia, that is, a noticeable increase in the heart's activation rate. In the worst cases, this may turn into an even less organized activation pattern, called fibrillation. A common therapy to terminate fibrillation and restore regular cardiac rhythm is electrical defibrillation, consisting in the delivery of a strong electrical shock by injecting external currents through a set of electrodes. This restores a spatially uniform activation pattern, recovering an extracellular potential distribution showing damped voltage gradients. Today, electrical defibrillation is carried out by implanting devices (so-called cardioverter defibrillators) able to monitor the heart rhythm and then deliver electrical discharges when needed.

This process can be modelled by considering the monodomain equation (7.11) as the state system, for which the extracellular stimulation current $I_e = I_e(t)$ to be applied plays the role of a distributed control function, and has to be determined in order to minimize the cost functional

$$J(V_m, I_e) = \frac{1}{2}\int_0^T \int_{\Omega_{obs}} |V_m - v_d|^2 \,\mathrm{d}\Omega\,\mathrm{d}t + \frac{\alpha}{2}\int_0^T \int_{\Omega_{con}} |I_e|^2 \,\mathrm{d}\Omega\,\mathrm{d}t. \quad (9.14)$$

Here V_m is the trans-membrane potential, $\Omega_{obs} \subseteq \Omega_{mus}$ is the observation domain, $\Omega_{con} \subseteq \Omega_{mus}$ is the control domain, and v_d is the target potential distribution. For instance, if $v_d = 0$, the minimum of $J(V_m, I_e)$ corresponds to the case of an excitation wave which is suppressed in the region Ω_{obs}: see *e.g.* Nagaiah, Kunisch and Plank (2011). The trans-membrane potential V_m can be obtained by solving the monodomain equations

$$\chi_m C_m \frac{\partial V_m}{\partial t} - \nabla \cdot (\mathbf{\Sigma} \nabla V_m) + \chi_m I_{ion} = I_e \quad \text{in } \Omega_{mus} \times (0, T), \quad (9.15a)$$

$$\frac{\partial w}{\partial t} = g(V_m, w) \quad \text{in } \Omega_{mus} \times (0, T), \quad (9.15b)$$

$$V_m|_{t=0} = V_{m,0} \quad \text{in } \Omega_{mus}, \quad (9.15c)$$

$$w|_{t=0} = w_0 \quad \text{in } \Omega_{mus}, \quad (9.15d)$$

$$\mathbf{\Sigma} \nabla V_m \cdot \mathbf{n} = 0 \quad \text{on } \Sigma_{epi} \cup \Sigma_{endo} \quad (9.15e)$$

(see Section 7.1.3), where I_{ion} is provided, for example, by the FitzHugh–Nagumo model (see Section 7.1.2):

$$I_{ion} = f(V_m, w) = -kV_m(V_m - a)(V_m - 1) - w, \quad g(V_m, w) = \epsilon(V_m - \gamma w).$$

An analysis of this optimal control problem can be found in Nagaiah *et al.* (2011), for example. We can exploit the Lagrangian approach to derive a system of first-order optimality conditions, by introducing the Lagrangian

$$\begin{aligned}
&\mathcal{L}(V_m, w, I_e, z, q) \\
&\quad = J(V_m, I_e) + \langle \mathcal{E}(V_m, w, I_e), (z, q) \rangle_{V,V^*} \\
&\quad = J(V_m, I_e) \\
&\qquad + \int_0^T \int_{\Omega_{mus}} \left(\chi_m C_m \frac{\partial V_m}{\partial t} - \nabla \cdot (\mathbf{\Sigma} \nabla V_m) + \chi_m I_{ion} - I_e \right) z \,\mathrm{d}\Omega\,\mathrm{d}t \\
&\qquad + \int_0^T \int_{\Omega_{mus}} \left(\frac{\partial w}{\partial t} - g(V_m, w) \right) q \,\mathrm{d}\Omega\,\mathrm{d}t,
\end{aligned}$$

where (z, q) denote the dual variables of V_m, w, respectively. Here $(V_m, w) \in$

$V = L^2(0,T;\mathcal{V}) \times W^{1,2}(0,T;\mathcal{H})$, where $\mathcal{V} = H^1(\Omega_{mus})$ and $\mathcal{H} = L^2(\Omega_{mus})$; the control space can be chosen as $U = L^2(0,T;L^2(\Omega))$, and the initial conditions can be kept as explicit constraints. By setting the partial derivatives of $\mathcal{L}$ equal to zero, we find the following expression for the adjoint problem:

$$\begin{aligned}
&-\chi_m C_m \frac{\partial z}{\partial t} - \nabla\cdot(\boldsymbol{\Sigma}\nabla z) + \chi_m \frac{\partial I_{ion}}{\partial V_m} z - \frac{\partial g}{\partial V_m} q \\
&\qquad\qquad\qquad\qquad = v_d - V_m && \text{in } \Omega_{mus}\times(0,T), && (9.16a)\\
&-\frac{\partial q}{\partial t} - \frac{\partial g}{\partial w} q + \frac{\partial I_{ion}}{\partial w} z = 0 && \text{in } \Omega_{mus}\times(0,T), && (9.16b)\\
&z|_{t=T} = 0 && \text{in } \Omega_{mus}, && (9.16c)\\
&q|_{t=T} = 0 && \text{in } \Omega_{mus}, && (9.16d)\\
&\boldsymbol{\Sigma}\nabla z\cdot\mathbf{n} = 0 && \text{on } \Sigma_{epi}\cup\Sigma_{endo} && (9.16e)
\end{aligned}$$

and the optimality condition

$$z + \alpha I_e = 0 \quad \text{on } \Omega_{con}. \tag{9.17}$$

The optimal control problem (9.14, 9.15) can be solved by an *optimize-then-discretize* strategy, where both spatial and temporal discretizations are required because of the time-dependent nature of the problem. After discretization, we can employ an iterative method, by computing at each step the solution of the (coupled) state problem (9.15) over $(0,T)$, and the solution of the adjoint problem (9.16), which is a linear problem, backward in time, where the adjoint variables are coupled similarly to (V_m, w) in the state problem. Note that the data of the adjoint problem are related to the Fréchet derivative of the cost functional with respect to the state variables, and that the adjoint problem depends on the control function only through the state variable. Moreover, the adjoint operator calls into play the linearization of the state operator around the computed state solution; that is, the derivatives $\partial I_{ion}/\partial V_m$, $\partial g/\partial V_m$, $\partial I_{ion}/\partial w$, $\partial g/\partial w$ have to be evaluated, at each step, around the computed solution of the state system. The optimality condition (9.17) then allows us to determine the gradient $\tilde{J}'(I_e)$ of the cost functional $\tilde{J}(I_e) = J(V_m(I_e), I_e)$, required to update the control function at each step.

To simplify the optimal control problem, we can for instance look for control functions of the form

$$I_e(t,\mathbf{x}) = \sum_{k=1}^{N_{el}} u_k(t)\mathcal{I}_{\Omega_{con,k}}(\mathbf{x})\mathcal{I}_{(0,T_{def})}(t),$$

where the location of N_{el} electrodes through which the current is delivered is prescribed, and only its time intensity has to be controlled over time. Here $\mathcal{I}_{\Omega_{con,k}}(\mathbf{x})$ denotes the indicator function of the region $\Omega_{con,k}$ where the kth electrode is located. Additional inequality constraints of the form $u_{\min} \leq u_k(t) \leq u_{\max}$ can also be taken into account in order to limit current amplitude.

Further extensions of this framework consider, for example, the bidomain model (Nagaiah *et al.* 2013*b*), the case of boundary controls (Nagaiah, Kunisch and Plank 2013*a*), and a different Mitchell–Schaeffer model to describe ionic currents, together with the presence of a conductive bath medium outside the heart effects and experimentally based fibre and sheet orientations (Nagaiah, Kunisch and Plank 2016). Moreover, the case where the shock duration itself is also optimized has been considered by Kunisch and Rund (2015). The reason to consider this further effect is that after applying a defibrillation shock, the muscle tissue needs a suitable amount of time to reach a non-fibrillatory state, so that a successful defibrillation can only be assessed at a time $T \gg T_{def}$, where T_{def} is the end time of the defibrillation shock and T the final observation time, at which the tissue should have reached a non-excited state. In this context, the cost functional

$$J(V_m, u, T_{def}) = T_{def} + \frac{\mu}{2} \int_{\Omega_{obs}} |V_m(\cdot, T)|^2 \, \mathrm{d}\Omega + \frac{\alpha}{2} \sum_{k=1}^{N_{el}} \int_0^T |u_k|^2 \, \mathrm{d}t$$

can be minimized, accounting for (i) a minimum time term, (ii) a final time observation at $t = T$, and (iii) the amplitude of the applied currents, subject to the state system (9.15) to be solved over $(0, T)$. Indeed, the shock duration and its energy have to be minimized to avoid negative side effects of the applied shock, while the final time observation term is taken into account to quantify defibrillation, thus requiring that the tissue reaches a non-excited state at $T \gg T_{def}$.

For illustration, we consider the case of an axially symmetric domain $\Omega_{mus} = (0, 2) \times (0, 0.8)$, where a constant pulse of intensity u is delivered in the control domain $\Omega_{con} = (0, 0.25) \times (0.3, 0.55) \cup (1.75, 2) \times (0.3, 0.55)$, until time T_{def}. Here $\boldsymbol{U} = (u, T_{def})^T$ is the control variable, with $1 \leq u \leq 100$ mV and $0.1 \leq T_{def} \leq 4$ ms; the final time is $T = 64$ ms. The initial condition $(V_{m,0}, w_0)^T$ describes a re-entry wave of the 'figure of eight' type, obtained following the procedure described by Kunisch and Rund (2015), who also take into account more general optimal control problems. For the case at hand, a planar wavefront travelling from the bottom to the top can be damped by imposing an optimal control of intensity $u \approx 95$ mV until $T_{def} \approx 1.2$ ms on the control region. The successful defibrillation – resulting from a trade-off between a large intensity and a short duration of the pulse – is clearly visible in Figure 9.3, where in the controlled case at the final time the tissue is almost completely unexcited. Indeed, the pulse acts on the excitable region of the tissue adjacent to the wavefront, bringing it to a non-excitable state.

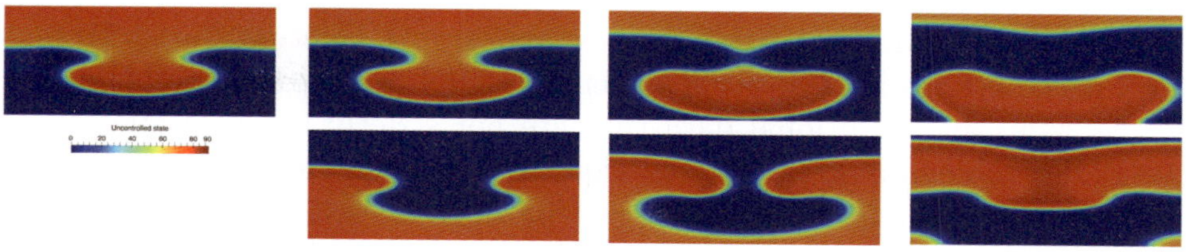

(a)

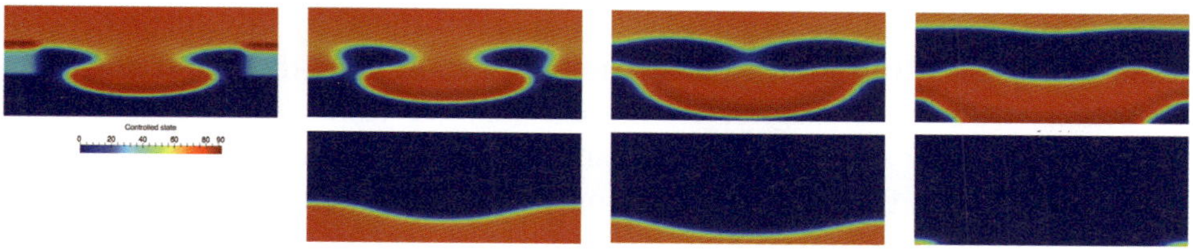

(b)

Figure 9.3. Electrical potential at times $t = 0, 4, 12, 20, 40, 52, 64$ ms in the uncontrolled case (a) and in the controlled case (b). The re-entry wave appearing in the uncontrolled case is damped by the control acting on Ω_{con}.

Remark 9.3. We point out that choosing the cost functional and, if necessary, imposing suitable constraints on the control and/or the state variable, are two extremely hard, problem-dependent tasks. Moreover, very often control functions are described in terms of (possibly few) relevant parameters, which play the role of design variables. However, in the case of complex, patient-dependent geometries, for example, their automatic selection can result in a fairly involved procedure. Last but not least, in the case where a target state to be reached depends on acquired data, the effect of uncertainty has to be considered in the formulation (and then solution) of the problem. This leads to robust optimization problems, for example, or PDE-constrained optimization problems under uncertainty, an active research field, where very few applications to cardiovascular modelling have yet been considered.

10. Parameter estimation from clinical data

When dealing with the mathematical and numerical modelling of the circulatory system, initial conditions, boundary conditions or physical coefficients, for example tissue properties, might be (partially) unknown (Veneziani and Vergara 2013): see Sections 3 and 6. From here on we refer to any of these quantities as input parameters, independent of their mathematical nature (they could be scalars or vectors, or even parametric fields varying in space) and we refer to the issue of making inferences about unknown parameters from data as *parameter estimation* (or *identification*). To perform parameter estimation for a system of PDEs, we need to combine state observations and additional data that are not strictly required for solving the PDE problem, and have to be acquired from measurements.

Parameter estimation for PDEs can be done according to several techniques, which can be roughly classified depending on their outcome (Tarantola 2004, Le Dimet and Talagrand 1986). *Point estimates* rely on either *variational* or *sequential* methods (Ide, Courtier, Ghil and Lorenc 1997). Both methods provide optimal least-squares estimates by minimizing a cost functional accounting for the misfit between measured data and state observations. A second class of techniques instead yields *confidence regions* or, more generally speaking, the possibility of characterizing the probability distribution of the unknown parameters provided they are described in terms of random variables; this is the goal of statistical inversion theory relying for example on *Bayesian inference*, which will be the main focus of Section 11.2.

In the context of cardiovascular modelling, parameter estimation is necessary for *model calibration/personalization*, for the purpose of diagnosis or treatment. Indeed, parameters that are not directly measurable (*e.g.* for tissue conductivity or vessel compliance) are tuned so that the outcome of the numerical model is able to reproduce patient-specific data (Krishnamurthy *et al.* 2013). Difficulties arise because of data sparsity (Konukoglu *et al.* 2011). Indeed, spatial resolution and temporal frequency are typically undersampled; experimental data, on the other hand, are always contaminated by measurement noise.

Remark 10.1. Data assimilation (DA) is the process by which a numerical model of a given system, usually affected by noise or model uncertainties, is improved by incorporating system observations. Although data assimilation relies on the same variational or filtering approaches addressed in this section, in the case of geophysical fluids, for example, its main goal often goes beyond parameter estimation; indeed, it is more often related to *state estimation*, namely, to improving the outcome of the numerical model and of its initial state to correctly initialize forecasts, by assimilating available measurements into the numerical model itself. Data assimilation is intrinsically related to time-varying phenomena and deals with highly nonlinear dynamical systems, very often far from being periodical (such as in the case of meteorological models) and ill-posed: see *e.g.* Blum, Le Dimet and Navon (2009), Voss, Timmer and Kurths (2004) and Le Dimet and Talagrand (1986) for a detailed discussion. In the past decade several works dealing with cardiovascular applications have focused on *data assimilation* (Sermesant *et al.* 2006, Delingette *et al.* 2012, Bertagna, D'Elia, Perego and Veneziani 2014, Lal, Mohammadi and Nicoud 2016), which has been considered, in many cases, as synonymous with *parameter estimation.*

In this section we provide an overview of both variational and sequential approaches for parameter estimation in time-dependent systems. Parameter estimation problems dealing with stationary systems in cardiovascular applications have also been solved: see *e.g.* D'Elia, Perego and Veneziani (2012), Bertagna *et al.* (2014), MacLachlan, Nielsen, Lysaker and Tveito (2006), Nielsen, Cai and Lykaser (2007*a*), Lassila, Manzoni, Quarteroni and Rozza (2013*b*) and Manzoni, Lassila, Quarteroni and Rozza (2014). For the sake of space and relevance, we will focus on time-dependent problems.

10.1. Variational approach: PDE-constrained optimization

The variational approach recasts parameter estimation in the framework of PDE-constrained optimization, by considering the equations governing the problem at hand as the state system, and the discrepancy between the observation of the state and the measured data as the cost functional to be minimized (Banks and Kunisch 1989, Chavent 2010). The parameters to be estimated (often involving the initial condition) play the role of optimization variables, just like control variables in the case of optimal control problems; however, unlike the optimal control case, the parameters to be estimated are quantities which no one can actually control. They are often coefficients of the operator appearing in the PDE problem.

Here we provide an abstract formulation of the variational approach, following Bertoglio, Moireau and Gerbeau (2012) and Blum *et al.* (2009); for more details see *e.g.* Nichols (2010), Chapelle, Fragu, Mallet and Moireau (2013*a*) and Sermesant *et al.* (2006). We assume that the state problem has already been discretized in space. We let $\boldsymbol{X}(t) \in \mathbb{R}^{n_x}$ denote the semi-discrete state, $\boldsymbol{\theta} \in \mathbb{R}^p$ the parameters of the model to be estimated, and $\boldsymbol{A}(t, \boldsymbol{X}(t), \boldsymbol{\theta})$ the (semi-discretized in space) state operator; note that usually $p \ll n_x$. The state variable then solves the dynamical system

$$\dot{\boldsymbol{X}}(t) = \boldsymbol{A}(t, \boldsymbol{X}(t), \boldsymbol{\theta}), \quad t \in (0, T), \tag{10.1a}$$

$$\boldsymbol{X}(0) = \boldsymbol{G}. \tag{10.1b}$$

For instance, in the case of the FSI system (4.19), $\boldsymbol{X} = (\boldsymbol{v}, p, \boldsymbol{d})$ contains fluid velocity and pressure, and structure displacement, and $\boldsymbol{\theta}$ may contain the value of the Young's modulus (here assumed to be piecewise constant) in different patches of the arterial wall.

We consider the case where the parameter vector $\boldsymbol{\theta}$ is unknown, and for which the estimation problem consists in finding $\widehat{\boldsymbol{\theta}}$ such that the discrepancy between the observation and a set of measurements $\boldsymbol{Z}(t) \in \mathbb{R}^{n_z}$, $t \in (0, T)$, is minimized, for example in a least-squares sense. The case of unknown initial data $\boldsymbol{G}$ to be estimated can be treated in essentially the same way.

We usually assume that measurements $\boldsymbol{Z}(t)$ are related to observations of the true state through an *additive noise* model, that is,

$$\boldsymbol{Z}(t) = H(t)\boldsymbol{X}(t) + \boldsymbol{\varepsilon}(t), \quad t \in (0,T),$$

where $H = H(t) \in R^{n_z \times n_x}$ is an observation operator which maps the state space into the observation space R^{n_z}, and $\boldsymbol{\varepsilon} = \boldsymbol{\varepsilon}(t)$ is a *noise* term accounting for measurement errors. The following minimization problem is then solved:

$$J(\boldsymbol{X},\boldsymbol{\theta}) = \frac{1}{2}\int_0^T \|\boldsymbol{Z}(t) - H(t)\boldsymbol{X}(t)\|_M^2 \,\mathrm{d}t + \frac{\alpha_\theta}{2}\|\boldsymbol{\theta} - \boldsymbol{\theta}_0\|^2_{P_\theta^{-1}} \to \min_{\boldsymbol{\theta}\in\mathcal{P}}, \tag{10.2}$$

where $\boldsymbol{X} = \boldsymbol{X}(t)$ is the solution of (10.1). Here $\mathcal{P} \subset R^p \times R^{n_x}$ denotes the set of admissible parameters. Additional information is usually added to the least-squares objective expressed by the first term in (10.2), via a background estimate $\boldsymbol{\theta}_0$ of $\boldsymbol{\theta}$; M and P_θ^{-1} are suitable symmetric positive definite matrices (the reason why we consider an inverse matrix to define this latter norm will be clarified in the following). This procedure goes by the name of *Levenberg–Marquardt–Tikhonov regularization.* For a discussion of classical regularization methods for inverse problems, see Kaipio and Somersalo (2005); note that the usual penalization coefficients are embedded in the definition of the matrix P_θ^{-1}.

The minimization problem (10.2) can be solved by an optimization algorithm based on the evaluation of the gradient of J with respect to $\boldsymbol{\theta}$. As shown in Section 9.1, this latter can be computed by relying on the solution of a suitable *adjoint problem.* This is a *four-dimensional variational* (4D-Var) assimilation; a three-dimensional variational (3D-Var) assimilation would arise in the case of steady-state systems. See Section 10.3.1 for further details on a relevant example in cardiovascular modelling, and Blum *et al.* (2009) and Chapelle *et al.* (2013*a*) for more on 4D-Var assimilation problems.

Since measurements $\boldsymbol{Z}$ are only available at a discrete number of times $\tau^1, \ldots, \tau^K$, we formulate the identification problem by replacing the dynamical system (10.1) with its time-discretized version:

$$\boldsymbol{X}^{k+1} = \boldsymbol{A}_{k|k+1}(\boldsymbol{X}^k, \boldsymbol{\theta}), \quad k = 0, \ldots, K-1, \tag{10.3a}$$

$$\boldsymbol{X}^0 = \boldsymbol{G} \tag{10.3b}$$

where $\boldsymbol{A}_{k|k+1}$ is a non-linear function describing the evolution of the state from time τ^k to time τ^{k+1} and $\boldsymbol{X}^k \approx \boldsymbol{X}(k\Delta\tau)$ denotes the state vector at time τ^k. Note that the length $\Delta\tau = \tau^{k+1} - \tau^k$ of the time window between two subsequent measurements is usually larger than the time step Δt used for time discretization, and that $\boldsymbol{\theta}$ does not depend on k. From here on k will denote the temporal index of system evolution, thus using a different notation from that introduced in Parts 1 and 2 (where the temporal index was denoted by n).

We then formulate an optimal time-discretized minimization criterion and, finally, determine the corresponding adjoint problem, rather than discretizing the adjoint problem in time. This yields the minimization problem

$$J_K(\boldsymbol{X},\boldsymbol{\theta}) = \frac{1}{2}\sum_{k=1}^{K} \|\boldsymbol{Z}^k - H\boldsymbol{X}^k\|^2_{M_k} + \frac{1}{2}\|\boldsymbol{\theta} - \boldsymbol{\theta}_0\|^2_{P_\theta^{-1}} \to \min_{\boldsymbol{\theta}\in\mathcal{P}}, \tag{10.4}$$

where $\boldsymbol{X} = (\boldsymbol{X}^1, \ldots, \boldsymbol{X}^k)$, and we set

$$\boldsymbol{Z}^k = H_k\boldsymbol{X}^k + \boldsymbol{\varepsilon}^k. \tag{10.5}$$

Here $\boldsymbol{\varepsilon}^k$ denotes the noise of the measurement device at $\tau^k = k\Delta\tau$; a possible choice for M_k is $M_k = \Delta\tau M$, whereas $H_k \approx H(k\Delta\tau)$. Also in this case a gradient-based optimization procedure can be used to solve the constrained optimization problem (10.3, 10.4) with gradients evaluated by introducing a suitable adjoint problem.

10.2. Sequential approach: Kalman filter and extensions

A drawback of the variational approach is the need to wait until the whole set of measurements has been acquired in order to perform an optimization step. A sequential approach instead performs the assimilation of acquired measurements *on the fly*, and updates the estimate of the unknown quantities accordingly.

A numerical milestone for the solution of sequential estimation problems, the Kalman filter (KF) (Kalman 1960), was introduced as a recursive filter for the estimation of the state of a noisy dynamical system from a set of measurements, that is, to improve the prediction of the state dynamics by taking into account additional data. An augmented form of the Kalman filter can be easily adapted to the problem of estimating unknown parameters, as we will see. Originally designed for linear dynamical systems, the Kalman filter has severe memory requirements. To mitigate these two limitations, several improvements and extensions have been proposed in the past few decades, most notably the extended Kalman filter (EKF), the unscented Kalman filter (UKF) and the ensemble Kalman filter (EnKF); detailed reviews can be found, for example, in Kaipio and Somersalo (2005), Simon (2006), Humpherys, Redd and West (2012) and Asch, Bocquet and Nodet (2017). In this section we recall the formulation of the basic Kalman filter and provide some hints about its extensions, with a special focus on the field of cardiovascular modelling where these techniques have been applied.

The literature offers many possible derivations of the Kalman filter; following Humpherys *et al.* (2012) and Chapelle *et al.* (2013*a*), we will exploit the analogy with the solution of a recursive least-squares problem yielding the best linear unbiased estimator for a linear model; alternative derivations can be obtained, for example, by relying on the so-called push-forward and subsequent conditioning of Gaussian measures (Sullivan 2015) or a sequential Bayesian estimation framework (Law, Stuart and Zygalakis 2015).

10.2.1. The Kalman filter algorithm

The KF algorithm sequentially generates an estimate of the unknown quantity by means of a linear combination of the current estimate and the acquired measurement. Let us first consider the case where there is no dynamics, and data are generated by the linear model

$$\boldsymbol{Z} = H\boldsymbol{X} + \boldsymbol{\varepsilon}, \tag{10.6}$$

where H is a given $n_z \times n_x$ matrix of rank n_x, $\boldsymbol{\varepsilon}$ is an n_z-dimensional random variable with zero mean and known positive definite covariance $Q = \mathbb{E}[\boldsymbol{\varepsilon}\boldsymbol{\varepsilon}^T] > 0$, and $\boldsymbol{Z}$ denotes known, but inexact, measurements with errors given by $\boldsymbol{\varepsilon}$. The vector $\boldsymbol{X} \in R^{n_x}$ is the quantity to be estimated from the observation $\boldsymbol{Z}$; $\mathbb{E}[\cdot]$ denotes the expected value.

Among all linear estimators of $\boldsymbol{X}$, that is, estimators of the form $\widehat{\boldsymbol{X}} = K\boldsymbol{Z}$ for some matrix $K \in R^{n_x \times n_z}$, which are unbiased (*i.e.* $\mathbb{E}[\widehat{\boldsymbol{X}}] = \boldsymbol{X}$), the best choice is the one that minimizes the mean-square error $\mathbb{E}[(\widehat{\boldsymbol{X}} - \boldsymbol{X})^T(\widehat{\boldsymbol{X}} - \boldsymbol{X})]$; by the Gauss–Markov theorem (see *e.g.* Sullivan 2015, Chapter 7), the best or minimum variance linear unbiased estimator for (10.6) is given by

$$\widehat{\boldsymbol{X}} = (H^T Q^{-1} H)^{-1} H^T Q^{-1} \boldsymbol{Z}. \tag{10.7}$$

In that case, $E[(\widehat{\boldsymbol{X}} - \boldsymbol{X})(\widehat{\boldsymbol{X}} - \boldsymbol{X})^T] = (H^T Q^{-1} H)^{-1}$. Equivalently, (10.7) can also be obtained by solving the weighted least-squares problem

$$\widehat{\boldsymbol{X}} = \frac{1}{2}\|H\boldsymbol{X} - \boldsymbol{Z}\|^2_{Q^{-1}} \to \min_{\boldsymbol{X}}, \tag{10.8}$$

again resorting to a variational argument. In the slightly different case where we want to combine the observation and a background estimate $\boldsymbol{X}_0$ of $\boldsymbol{X}$ with covariance P^-, (10.8) becomes

$$\widehat{\boldsymbol{X}} = \frac{1}{2}\|H\boldsymbol{X} - \boldsymbol{Z}\|^2_{Q^{-1}} + \frac{1}{2}\|\boldsymbol{X} - \boldsymbol{X}_0\|^2_{(P^-)^{-1}} \to \min_{\boldsymbol{X}},$$

and instead of (10.7) we find

$$\widehat{\boldsymbol{X}} = \boldsymbol{X}_0 + K(\boldsymbol{Z} - H\boldsymbol{X}_0), \quad K = P^+ H^T Q^{-1}, \tag{10.9}$$

upon defining the matrix playing the role of *updated* covariance as

$$P^+ = ((P^-)^{-1} + H^T Q^{-1} H)^{-1}. \tag{10.10}$$

Note that this estimate is given by a linear combination of the background estimate $\boldsymbol{X}_0$ and the so-called *innovation* $\boldsymbol{Z} - H\boldsymbol{X}_0$. K is usually referred to as the *Kalman gain matrix*; it can also be evaluated by exploiting the prior covariance P^- instead of the updated covariance P^+, according to the equivalence

$$P^+ H^T Q^{-1} = P^- H^T (HP^- H^T + Q)^{-1}. \tag{10.11}$$

We now consider the case of a time-discretized linear system, with data acquired over a time interval. Then the model (10.6) is replaced by

$$\begin{aligned} \boldsymbol{X}^k &= A_{k-1|k}\boldsymbol{X}^{k-1} + \boldsymbol{v}^k, \quad k = 1, \dots, K, \ \boldsymbol{X}^0 = \boldsymbol{G}, \\ \boldsymbol{Z}^k &= H_k \boldsymbol{X}^k + \boldsymbol{w}^k, \end{aligned} \tag{10.12}$$

where $\boldsymbol{X}^k \in \mathbb{R}^{n_x}$ denotes the state and $\boldsymbol{Z}^k \in \mathbb{R}^{n_z}$ are the measurements; $\boldsymbol{v}_k$ and $\boldsymbol{w}_k$ are uncorrelated zero-mean random noise processes with positive definite covariances Q_k and R_k, modelling the uncertainty of the model and the additive noise in the observation, respectively. The state estimation problem is the problem of finding the state $\boldsymbol{X}^k$ given k known observations $\boldsymbol{Z}^1, \dots, \boldsymbol{Z}^k$.

The Kalman filter is a recursive algorithm that provides the best linear unbiased estimate $\boldsymbol{X}_a^k$ of $\boldsymbol{X}^k$ in terms of both the previous estimate $\boldsymbol{X}_a^{k-1}$ and the latest data $\boldsymbol{Z}^k$ up to that point in time. It is based on a *predictor-corrector* strategy, consisting of the following steps.

(1) A *prediction* step (called the forecast or the time update) consists in letting the system dynamics evolve from $\boldsymbol{X}_a^{k-1}$ without taking into account the observations, yielding the *forecast* state $\boldsymbol{X}_f^k$.

(2) A *correction* step (called the analysis or the measurement update) updates the forecast state $\boldsymbol{X}_f^k$ by *assimilating* the measurements into the model, yielding the *assimilated* state $\boldsymbol{X}_a^k$.

To derive the expression of the correction step, let us suppose that the current prediction based on observations $\boldsymbol{Z}^1, \dots, \boldsymbol{Z}^{k-1}$ is $\boldsymbol{X}_f^k$, with covariance matrix P_k^f. If the true state is $\boldsymbol{X}_k$, model (10.6) becomes

$$\begin{bmatrix} \boldsymbol{X}_f^k \\ \boldsymbol{Z}^k \end{bmatrix} = \begin{bmatrix} I \\ H_k \end{bmatrix} \boldsymbol{X}^k + \boldsymbol{\varepsilon},$$

where $\boldsymbol{\varepsilon}$ is an $(n_x + n_z)$-dimensional random variable with zero mean and covariance $Q_k = \operatorname{diag}(P_k^f, R_k)$. The best linear unbiased estimator of this system (see (10.7)) is given by

$$\boldsymbol{X}_a^k = P_k^a \begin{bmatrix} I & H_k^T \end{bmatrix} \begin{bmatrix} (P_k^f)^{-1} & 0 \\ 0 & R_k^{-1} \end{bmatrix} \begin{bmatrix} \boldsymbol{X}_f^k \\ \boldsymbol{Z}^k \end{bmatrix} = P_k^a \big((P_k^f)^{-1} \boldsymbol{X}_f^k + H_k^T R_k^{-1} \boldsymbol{Z}^k \big), \tag{10.13}$$

and results from the linear combination of the current estimate and the *last* observation $\boldsymbol{Z}^k$, where

$$P_k^a = \left(\begin{bmatrix} I & H_k^T \end{bmatrix} \begin{bmatrix} (P_k^f)^{-1} & 0 \\ 0 & R_k^{-1} \end{bmatrix} \begin{bmatrix} I \\ H_k \end{bmatrix} \right)^{-1} = ((P_k^f)^{-1} + H_k^T R_k^{-1} H_k)^{-1}.$$

Note the formal analogy with (10.10), where P_k^f and P_k^a now play the role of P^- and P^+, respectively. Equation (10.13) can be written in the more convenient form

$$\begin{aligned} \boldsymbol{X}_a^k &= P_k^a \left[(P_k^a)^{-1} - H_k^T R_k^{-1} H_k)) \boldsymbol{X}_f^k + H_k^T R_k^{-1} \boldsymbol{Z}^k \right] \\ &= \boldsymbol{X}_f^k + P_k^a H_k^T R_k^{-1} (\boldsymbol{Z}^k - H_k \boldsymbol{X}_f^k) = \boldsymbol{X}_f^k + K_k (\boldsymbol{Z}^k - H_k \boldsymbol{X}_f^k) \end{aligned}$$

as a function of the innovation $\boldsymbol{Z}^k - H_k \boldsymbol{X}_f^k$. Similarly to (10.11), we obtain

$$K_k = P_k^a H_k^T R_k^{-1} = P_k^f H_k^T (H_k P_k^f H^T + R_k)^{-1}, \tag{10.14}$$

so it is possible to evaluate the Kalman gain matrix K_k as a function of P_k^f, and then correct the covariance:

$$\begin{aligned} P_k^a &= ((P_k^f)^{-1} + H_k^T R_k^{-1} H_k)^{-1} \\ &= (I - K_k H_k) P_k^f (I - K_k H_k)^T + K_k R_k K_k^T = (I - K_k H_k) P_k^f. \end{aligned}$$

The prediction step instead exploits the dynamical system to propagate the state, yielding

$$\boldsymbol{X}_f^{k+1} = A_{k|k+1} \boldsymbol{X}_a^k$$

for the time update of the state, and

$$\begin{aligned} P_{k+1}^f &= \mathbb{E}[(\boldsymbol{X}_f^{k+1} - \boldsymbol{X}^{k+1})(\boldsymbol{X}_f^{k+1} - \boldsymbol{X}^{k+1})^T] \\ &= \mathbb{E}[(A_{k|k+1} \boldsymbol{X}_a^k - A_{k|k+1} \boldsymbol{X}_k - \boldsymbol{w}_k)(A_{k|k+1} \widehat{\boldsymbol{X}}_a^k - A_{k|k+1} \boldsymbol{X}_k - \boldsymbol{w}_k)^T] \\ &= A_{k|k+1} P_k^a A_{k|k+1}^T + Q_{k+1} \end{aligned}$$

for the time update of the covariance.

Grouping the prediction and the correction steps together, we finally obtain the kth step of the KF algorithm:

$$\boldsymbol{X}_f^k = A_{k-1|k} \boldsymbol{X}_a^{k-1} \qquad \text{state prediction,} \tag{10.15a}$$

$$P_k^f = A_{k-1|k} P_{k-1}^a A_{k-1|k}^T + Q_k \qquad \text{error covariance prediction,} \tag{10.15b}$$

$$K_k = P_f^k H_k^T (H_k P_k^f H_k^T + R_k)^{-1} \qquad \text{Kalman gain evaluation,} \tag{10.15c}$$

$$\boldsymbol{X}_a^k = \boldsymbol{X}_f^k + K_k (\boldsymbol{Z}^k - H_k \boldsymbol{X}_f^k) \qquad \text{state correction,} \tag{10.15d}$$

$$P_k^a = (I - K_k H_k) P_k^f \qquad \text{error covariance correction.} \tag{10.15e}$$

Note that from (10.15d) only, the estimated state from the above step and the current measurement are needed to compute the estimate of the current state. The two prediction and correction steps alternate: the prediction advances the state until the next measurement is acquired, and then the correction incorporates this measurement.

Remark 10.2. The Kalman gain K_k defined in (10.14) can also be expressed as $K_k = P_k^{X_f Z}(P_k^Z)^{-1}$. Here

$$P_k^{X_f Z} = \mathbb{E}[\boldsymbol{X}_f^k(\boldsymbol{Z}^k - H_k\boldsymbol{X}_f^k)^T] = P_k^f H_k^T$$

is the cross-covariance between $\boldsymbol{X}_f^k$ and the innovation $\boldsymbol{Z}^k - H_k\boldsymbol{X}_f^k$, and

$$P_k^Z = \mathbb{E}[(\boldsymbol{Z}^k - H_k\boldsymbol{X}_f^k)(\boldsymbol{Z}^k - H_k\boldsymbol{X}_f^k)^T] = H_k P_k^f H_k + R_k$$

is the innovation covariance. Similarly, (10.15e) can be rewritten as $P_k^a = P_k^f - K_k P_k^Z K_k^T$. This interpretation is useful when dealing with the unscented Kalman filter (see Section 10.2.3).

Remark 10.3. In the linear case, the variational and the sequential approaches yield the same result at the end of a time window, provided the following assumptions are made: the same background estimation and the same covariance matrices are used, and the same measurements are acquired – that is, both algorithms are optimal from a least-squares or minimum variance standpoint.

Let us now return to our problem of estimating the parameter $\boldsymbol{\theta}$. We apply the KF algorithm to the system

$$\begin{aligned}
\boldsymbol{X}^k &= A_{k-1|k}\boldsymbol{X}^{k-1} + B_k\boldsymbol{\theta}^k + \boldsymbol{v}^k, \quad k = 1, \dots, K, \quad \boldsymbol{X}^0 = \boldsymbol{G}, \\
\boldsymbol{\theta}^k &= \boldsymbol{\theta}^{k-1}
\end{aligned}$$

with observations

$$\boldsymbol{Z}^k = H_k\boldsymbol{X}^k + \boldsymbol{w}^k, \quad k = 1, \dots, K.$$

Here $\boldsymbol{\theta}_k \in \mathbb{R}^p$ denotes the parameter vector and (under the linearity assumption) $B_k \in \mathbb{R}^{n_x \times p}$, and we assume that no random error is associated with model parameters. This is the so-called *state augmentation* technique. In order to exploit the KF algorithm, we consider $\widetilde{\boldsymbol{X}}^k = (\boldsymbol{X}^k, \boldsymbol{\theta}^k)^T$ as state vector instead of $\boldsymbol{X}^k$, thus yielding the *augmented* KF algorithm

$$\widetilde{\boldsymbol{X}}_f^k = \widetilde{A}_{k-1|k}\widetilde{\boldsymbol{X}}_a^{k-1} \qquad \text{state prediction,} \tag{10.16a}$$

$$\widetilde{P}_k^f = \widetilde{A}_{k-1|k}\widetilde{P}_{k-1}^a\widetilde{A}_{k-1|k}^T + \widetilde{Q}_k \qquad \text{error covariance prediction,} \tag{10.16b}$$

$$K_k = \widetilde{P}_f^k\widetilde{H}_k^T(\widetilde{H}_k\widetilde{P}_k^f H_k^T + \widetilde{R}_k)^{-1} \qquad \text{Kalman gain evaluation,} \tag{10.16c}$$

$$\widetilde{\boldsymbol{X}}_a^k = \widetilde{\boldsymbol{X}}_f^k + K_k(\widetilde{\boldsymbol{Z}}^k - \widetilde{H}_k\widetilde{\boldsymbol{X}}_f^k) \qquad \text{state correction,} \tag{10.16d}$$

$$\widetilde{P}_k^a = (I - K_k\widetilde{H}_k)\widetilde{P}_k^f \qquad \text{error covariance correction,} \tag{10.16e}$$

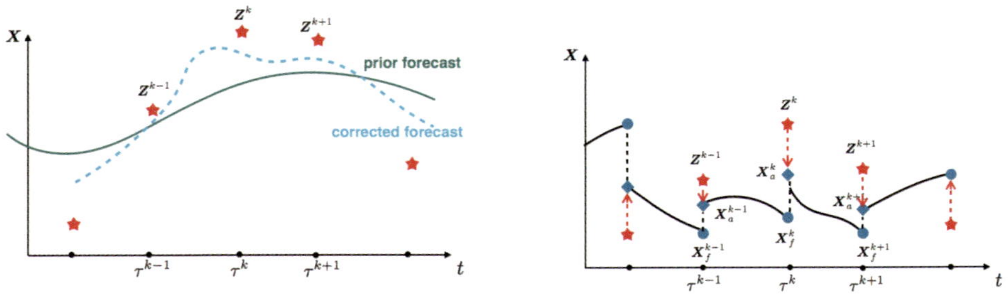

Figure 10.1. Variational approach (a) versus Kalman filter (b) approach: in the former, at each optimization stage the whole state dynamics has to be computed, whereas in the latter each measurement is sequentially used for the state (and parameter) correction.

where

$$\widetilde{A}_{k-1|k} = \begin{bmatrix} A_{k-1|k} & B_k \\ 0 & I \end{bmatrix}, \quad \widetilde{\boldsymbol{Z}}^k = \begin{bmatrix} \boldsymbol{Z}^k \\ 0 \end{bmatrix},$$
$$\widetilde{H}_k = \begin{bmatrix} H_k & 0 \\ 0 & 0 \end{bmatrix}, \quad \widetilde{Q}_k = \begin{bmatrix} Q_k \\ 0 \end{bmatrix}, \quad \widetilde{R}_k = \begin{bmatrix} R_k & 0 \\ 0 & 0 \end{bmatrix}.$$

We point out that, by construction of the filtering procedure, the estimated parameter values evolve along the simulation period and the actual estimation is achieved with the final values, that is, the estimated parameter vector is $\widehat{\boldsymbol{\theta}} = \boldsymbol{\theta}_a^K$. (Note that in the current formulation θ is independent of time.) Hence, we expect these estimation *trajectories* to fluctuate less and less during the course of the simulation (see *e.g.* Figure 11.3 in Section 11.3.2); the non-converging case would therefore denote the presence of persistent modelling errors.

We conclude this section by pointing out that when a sequential approach like the Kalman filter is used for parameter estimation, the dynamical system has to be solved only once, by updating the parameter value after each assimilation of new measurements. On the other hand, a variational approach would require us to solve the dynamical system on the whole time interval several times, assuming that an iterative approach is used to perform the optimization: see the sketch in Figure 10.1.

10.2.2. The extended Kalman filter

The classical Kalman filter formulation is well suited to low-dimensional, linear dynamical systems, although in real applications these assumptions are seldom verified. The extended Kalman filter (EKF) was introduced for non-linear dynamical systems, where (10.12) is replaced by

$$\boldsymbol{X}^k = \mathbf{f}(\boldsymbol{X}^{k-1}, \tau^k) + \boldsymbol{\varepsilon}^k, \quad \boldsymbol{Z}^k = \mathbf{h}(\boldsymbol{X}^k, \tau^k) + \boldsymbol{\eta}^k, \tag{10.17}$$

where **f** and **h** are two non-linear functions. Non-linearity here involves both the system dynamics and the observation model. The EKF consists in applying the KF algorithm (10.15) to a linearized version of (10.17) around the previous state, so that at each step we set

$$A_{k-1|k} = \left.\frac{\partial \mathbf{f}}{\partial \boldsymbol{X}}\right|_{\boldsymbol{X}_a^{k-1},\boldsymbol{\theta}^k}, \qquad H_k = \left.\frac{\partial \mathbf{h}}{\partial \boldsymbol{X}}\right|_{\boldsymbol{X}_f^k}.$$

At each step two Jacobian matrices have to be evaluated at the current predicted state/parameters. A similar extension of the algorithm (10.16) provides the EKF for parameter estimation. Although feasible in principle, the EKF suffers from several drawbacks. For instance, it entails prohibitive computational costs to invert large matrices and to propagate the covariance matrix in time. Even more importantly, the EKF may lack stability, meaning that as the estimated state deviates from the true state, the linearized model becomes inaccurate, which may lead to an even larger error in state estimation. To mitigate these shortcomings, several strategies have been put in place: low-rank approximation of the covariance matrices have been considered, and other extensions of the original Kalman filter such as the UKF and EnKF have been introduced.

10.2.3. The unscented Kalman filter

While the EKF exploits the differentiation of non-linear operators, modelling both the state dynamics and the observation process, to evaluate the propagation of means and covariances, the unscented Kalman filter (UKF), introduced by Julier, Uhlmann and Durrant-Whyte (1995), relies on a set of well-chosen deterministic points (or *sigma points*) whose propagation via the non-linear operators yields the empirical means and covariances required in the Kalman prediction-correction formulas. The rationale goes by the name of *unscented transformation*, whose goal is to map a set of points so that their sample distribution approximates the true distribution (Julier, Uhlmann and Durrant-Whyte 2000).

Supposing that we know the mean $\mathbb{E}[\boldsymbol{X}]$ and the covariance Σ of a random vector $\boldsymbol{X} \in \mathbb{R}^{n_x}$, the simplest choice is to select $2n_x$ (symmetric) sigma points $x_{(i)}$ as

$$x_{(i)} = \mathbb{E}[\boldsymbol{X}] + \left(\sqrt{n_x\Sigma}\right)_i, \quad x_{(n_x+i)} = \mathbb{E}[\boldsymbol{X}] - \left(\sqrt{n_x\Sigma}\right)_i, \quad i = 1,\dots,n_x,$$

where $\left(\sqrt{n_x\Sigma}\right)_i$ is the ith column of the Cholesky factor of $n_x\Sigma$. Different options for the sigma points selection are however possible: see *e.g.* Julier and Uhlmann (2004).

The prediction-correction strategy of the Kalman filter is then performed. Referring to the same notation used in Section 10.2.1 and supposing that the current forecast based on $\boldsymbol{Z}^1, \dots, \boldsymbol{Z}^{k-1}$ is $\widehat{\boldsymbol{X}}_f^k$, with covariance matrix P_k^f, the following *correction step* is performed.

- From the estimated mean $\widehat{\boldsymbol{X}}_f^k$ and covariance P_k^f at time τ^k, select a set of $2n_x$ sigma points

$$x_{f,(i)}^k = \widehat{\boldsymbol{X}}_f^k + \left(\sqrt{n_x P_k^f}\right)_i, \quad x_{f,(n_x+i)}^k = \widehat{\boldsymbol{X}}_f^k - \left(\sqrt{n_x P_k^f}\right)_i, \quad i = 1, \dots, n_x,$$

  centred around $\widehat{\boldsymbol{X}}_f^k$ at a distance given by the standard deviation extracted from the covariance matrix, and obtain the predicted measurement

$$\widehat{\boldsymbol{Z}}^k = \frac{1}{2n_x} \sum_{i=1}^{2n_x} \widehat{\boldsymbol{Z}}_{(i)}^k, \quad \text{where } \widehat{\boldsymbol{Z}}_{(i)}^k = \mathbf{h}(x_{f,(i)}^k, \tau^k), \quad i = 1, \dots, 2n_x.$$

- Estimate the covariance of the predicted measurement,

$$P_k^Z = \frac{1}{2n_x} \sum_{i=1}^{2n_x} \left(\widehat{\boldsymbol{Z}}_{(i)}^k - \widehat{\boldsymbol{Z}}^k\right)\left(\widehat{\boldsymbol{Z}}_{(i)}^k - \widehat{\boldsymbol{Z}}^k\right)^T + R_k,$$

  and the cross-covariance between $\boldsymbol{X}_f^k$ and $\widehat{\boldsymbol{Z}}^k$:

$$P_k^{X_f Z} = \frac{1}{2n_x} \sum_{i=1}^{2n_x} \left(\widehat{\boldsymbol{X}}_f^k - x_{f,(i)}^k\right)\left(\widehat{\boldsymbol{Z}}^k - \widehat{\boldsymbol{Z}}_{(i)}^k\right)^T.$$

- Perform the Kalman gain evaluation, state correction and the error covariance correction similarly to (10.16c–10.16e) (recall Remark 10.2),

$$\begin{aligned} K_k &= P_k^{X_f Z} (P_k^Z)^{-1}, \\ \widehat{\boldsymbol{X}}_a^k &= \widehat{\boldsymbol{X}}_f^k + K_k(\boldsymbol{Z}^k - \widehat{\boldsymbol{Z}}^k), \\ P_k^a &= P_k^f - K_k P_k^Z K_k^T. \end{aligned}$$

In the *prediction step*, a set of sigma points is selected as

$$x_{a,(i)}^k = \widehat{\boldsymbol{X}}_a^k + \left(\sqrt{n_x P_k^a}\right)_i, \quad x_{a,(n_x+i)}^k = \widehat{\boldsymbol{X}}_a^k - \left(\sqrt{n_x P_k^a}\right)_i, \quad i = 1, \dots, n_x$$

and a forward computation of one time window is performed to get the state prediction at step $k+1$:

$$\widehat{\boldsymbol{X}}_f^{k+1} = \frac{1}{2n_x} \sum_{i=1}^{2n_x} x_{f,(i)}^{k+1}, \quad \text{where } x_{f,(i)}^{k+1} = \mathbf{f}(x_{a,(i)}^k, \tau^{k+1})$$

(note that each sigma point is treated independently). Finally, the covariance is updated as

$$P_{k+1}^f = \frac{1}{2n_x} \sum_{i=1}^{2n_x} \left(\widehat{\boldsymbol{X}}_f^{k+1} - x_{f,(i+1)}^k\right)\left(\widehat{\boldsymbol{X}}_f^{k+1} - x_{f,(i+1)}^k\right)^T + Q_{k+1}.$$

Provided the analysis step is performed in parallel, the cost of the whole estimation procedure is comparable to that of a simple forward simulation. By contrast, a variational estimation would require a significant number of successive iterations (typically, of the order of hundreds) of both forward and adjoint simulations.

An augmented formulation of the UKF can be easily obtained, similarly to (10.16), in order to treat the case where the system depends on a set of p parameters affected by uncertainty, which need to be estimated. Nevertheless, the very large dimension of the state vector (depending on the spatial discretization of the state variable) makes this filter intractable in practice. In those cases where uncertainty only affects the parameters, and $p \ll n_x$, a much more feasible version yielding a reduced-order UKF limits the computations of the filter operator to a subspace of small dimension and is much more efficient. The reduction to the parametric space as regarding the choice of the sigma-points for the sake of parameter estimation was originally introduced by Pham (2001) and Hoteit, Pham and Blum (2002); an initial application to the estimation of electrophysiology parameters can be found in Wang, Zhang, Wong and Shi (2009) and Wang *et al.* (2011). A detailed analysis of the reduced-order UKF can be found in Moireau and Chapelle (2011), for example. A possible alternative, recently explored by Pagani, Manzoni and Quarteroni (2016) and Pagani (2016), is to perform a *state reduction* (rather than a reduction of the parameter space) relying on a reduced-order model, and then consider the problem of simultaneous state/parameter estimation.

10.3. Applications to cardiovascular modelling

In the past decade, parameter identification problems have been considered in several applications to cardiovascular modelling. In this section we report a list of relevant contributions; two substantive examples will be more specifically discussed in Sections 10.3.1 and 10.3.2.

The problem of identifying a set of parameters of one-dimensional models for the arterial circulation was first considered by Martin, Clément, Decoene and Gerbeau (2005), who considered a non-linear least-squares approach based on the optimization of a cost function and the introduction of a suitable adjoint problem. A relevant application in this context is the estimation of the elastic coefficient of a subject-specific vessel based on measurements of its displacement recovered from medical images. A *variational approach* based on the minimization of suitable functionals has been proposed by Perego, Veneziani and Vergara (2011) and, more recently, by Bertagna and Veneziani (2014); we present this technique in Section 10.3.1. A different approach based on an unscented Kalman filter has been proposed by Bertoglio *et al.* (2012) and further explored by Moireau *et al.*

(2013) to estimate modelling parameters pertaining to vessel wall boundary conditions. Similar strategies based on the unscented Kalman filter have been considered by Lombardi (2014), for example, to solve inverse problems in large one-dimensional arterial networks, and by Pant, Fabrèges, Gerbeau and Vignon-Clementel (2014) to estimate lumped (*e.g.* windkessel) model parameters by using the pressure curve in the ascending aorta.

Other approaches have been considered, for example in Blanco, Watanabe and Feijóo (2012), to characterize the terminal peripheral resistances in a network describing one-dimensional arterial blood flow, or in Spilker and Taylor (2010) to adjust the parameters of windkessel outflow boundary conditions of three-dimensional blood flow models, in order to match some desired features of pressure and flow waveforms.

Several recent works have focused on parameter identification and, more generally speaking, on the solution of inverse problems in heart modelling. The classical *inverse problem in electrocardiology* has been considered by many authors in the past few decades. It consists in recovering the electrical potential at the epicardial surface by using a number of remote, non-invasive or minimally invasive potential recordings, such as those acquired along the body surface (Rudy and Messinger-Rapport 1988, Pullan *et al.* 2001, Cheng, Bodley and Pullan 2003), usually referred to as *body surface-potential data*. In its original formulation, this problem involves the pure (linear) diffusion model for the torso as a direct problem (Colli Franzone, Taccardi and Viganotti 1978, Colli Franzone, Guerri, Viganotti and Taccardi 1985, Yamashita 1982): see Section 7.1.5. This inverse problem is mathematically ill-posed, and several regularization strategies have been proposed; see *e.g.* the reviews by Pullan *et al.* (2010) and Colli Franzone *et al.* (2014). Another method to solve the inverse electrocardiology problem relies on measurements of the electrical potential within one of the heart chambers (ventricles and atria) by means of non-contact multi-electrode catheter arrays, and tries to estimate the endocardial surface potential from them. Since the measurements are recorded closer to the endocardial surface than those acquired on the body surface, this problem is less ill-conditioned than the former. If the problem is recast in the form of a parameter estimation problem, that is, where the epicardial potential distribution is described in terms of a set of parameters, the inverse problem can be more easily tackled.

More generally speaking, inverse and parameter estimation problems have been considered in several works in order to approximately reconstruct the cardiac electrical activity in the myocardium, for example, and to provide indications of the presence of ischaemic or infarcted zones by estimating conductibility parameters. Concerning *variational approaches*, we can mention a level set framework for identifying heart infarctions (Lykaser and Nielsen 2006) by relying on a least-squares formulation and an adjoint problem to determine the gradient of the cost functional. In this case, the infarcted region in a simplified two-dimensional domain has been described in terms of a discrete level set function, involving a set of parameters to be identified from synthetic ECG boundary measurements, and a Tikhonov regularization procedure. MacLachlan *et al.* (2006) and Nielsen *et al.* (2007*a*) considered the solution of a parameter identification problem to locate (in terms of size and position) ischaemic regions, where a simplified state elliptic system models the electrical potential in both the heart and the torso in the resting phase. In all these papers, the PDE-constrained optimization problem was solved using an iterative method, following an optimize-then-discretize approach. Nielsen, Lykaser and Tveito (2007*b*) instead considered an all-at-once approach for a similar problem, also taking into account anisotropic cardiac conductivities and fibre orientation. A more involved version of this problem, still focusing on the inverse electrocardiographic source localization of ischaemias, has been considered more recently by Wang, Kirby, MacLeod and Johnson (2013).

The personalization of an eikonal model, via the identification of local conduction velocities and parameters related to the action potential duration restitution curve, has been addressed by Chinchapatnam *et al.* (2008) and Relan *et al.* (2011). More recently, Yang and Veneziani (2015) proposed a variational approach for the estimation of cardiac conductivities in a bidomain model, from measurements of the trans-membrane and extracellular potentials available at some locations in the tissue. Boulakia, Fernández, Gerbeau and Zemzemi (2008) have considered parameter estimation in a problem where the bidomain model for the heart is coupled with the Laplace equation for the passive conduction in the torso, in order to estimate the torso conductivity parameters.

The unscented Kalman filter was exploited by Wang *et al.* (2011) and by Marchesseau *et al.* (2013) and Talbot *et al.* (2015) to identify scar locations and their size from body surface-potential and MRI data, using a monodomain two-variable Aliev–Panfilov model. A reduced-order UKF was used by Corrado, Gerbeau and Moireau (2015) to estimate electrical parameters within a coupled electromechanical model, by taking advantage of observations from both electrocardiograms and myocardium displacements.

Another inverse problem in electrophysiology has been studied by Vergara *et al.* (2014) and Palamara *et al.* (2015). Measurements of the electrical activation time on the endocardium, acquired via NavX (see Section 6.2), were used to find the optimal Purkinje network configuration. For this problem, the parameters to be estimated are the coordinates of the network. In particular, starting from an initial network with a fractal shape, a functional accounting for the discrepancy between measured and computed activation times is minimized to find the patient-specific location of the Purkinje muscle junctions (see Section 5.3). This methodology has been successfully applied to real pathological scenarios: see Palamara *et al.* (2014).

Regarding cardiac biomechanics, Moireau, Chapelle and Le Tallec (2009) and Moireau and Chapelle (2011) addressed the estimation of contractility parameters using the unscented Kalman filter, where the degree of damage in cardiac tissues caused by an infarct is estimated using velocity measurements from tagged MRI; see also Moireau, Chapelle and Le Tallec (2008). A reduced-order UKF was used by Xi *et al.* (2011) to identify material parameters in a transversally isotropic mechanical model; further details concerning this application will be provided in Section 10.3.2. Subsequently, the same technique was exploited by Chabiniok *et al.* (2012) to estimate contractility values in a more complex cardiac model, where the tissue is described by combining an active constitutive law in the muscle fibre direction and a visco-hyperelastic material, using actual clinical data consisting of *in vivo* cine-MRI image sequences and pressure measurements.

Finally, for the coupled electromechanical problem, very few numerical results are now available. The personalization of a 3D electromechanical model has been tackled by variational approaches, for example by Sermesant *et al.* (2006), who estimated local ventricular myocardium contractility using MRI in an electromechanical model. Similarly, Delingette *et al.* (2012) estimated both the parameters of a Mitchell–Schaeffer model and cardiac contractilities from catheterized electrophysiology data and cine-MRI images, respectively. Marchesseau, Delingette, Sermesant and Ayache (2012) performed the calibration of mechanical parameters of a complete electromechanical model of the heart involving the eikonal model for electrophysiology and an isotropic Mooney–Rivlin material model for cardiac mechanics, relying on the unscented Kalman filter.

Although several works have made a big step forward, the solution of parameter estimation problems remains an open computational challenge if complex, coupled models and patient-specific data are taken into account. Further, when UKF and EnKF techniques are exploited – which exhibit natural parallelism to a high degree – the need to evaluate the state dynamics for several different scenarios (given by the elements of the ensemble, for example) makes the computational effort exorbitant. For this reason, simplified physical models have been considered in many contexts; on the other hand, reduced-order strategies, such as the reduced-order unscented Kalman filter, have recently been proposed as a possible way to overcome the computational complexity arising from these problems. Dramatic progress is expected in this field in the next few years.

10.3.1. A variational approach for estimating the tissue compliance

In this section we illustrate a variational approach introduced by Perego *et al.* (2011) to estimate the Young's modulus E of a cardiovascular tissue from displacement data. Indeed, this parameter – and, more generally speaking, the deformability of a soft tissue – is an important index for detecting diseases: low compliance may indicate atherosclerosis or hypertension in the case of an artery, or a marker of diastolic dysfunction in the case of the left ventricle wall. The basic steps are (i) to retrieve the vessel displacement $\boldsymbol{d}_{meas}$ by image registration procedures on time frames of the vessel of interest, and then (ii) to minimize the difference between $\boldsymbol{d}_{meas}$ and the displacement computed by solving the coupled 3D blood vessel problem, in order to estimate Young's modulus. Here the structure is assumed to be linearly elastic, and the only parameter to be estimated is E, which is generally a function of space (*e.g.* an atherosclerotic plaque has a different modulus to healthy tissue) but not of time.

The former step consists of data acquisition, image segmentation and reconstruction, and finally registration; see Perego *et al.* (2011) for further details. The latter, which we will address, is an example of an *inverse fluid–structure interaction* (IFSI) problem, and results in a constrained optimization problem, where the constraint is the FSI problem and the functional to be minimized is a measure of the mismatch between the data and the computed solution.

An initial formulation of the IFSI problem is obtained by assuming that the displacement data $\boldsymbol{d}_{meas}(\boldsymbol{x}, \tau_k)$ for $\boldsymbol{x} \in \Sigma^{\tau_k}$ retrieved from the image registration process are available within the interval $[0, T]$ in some instants denoted by τ^k, $k = 1, \ldots, K$. Here K is the number of instants when image registration is performed, $\Delta\tau$ is the time step between two measurements, driven by the sampling frequency of the image devices, and Σ^{τ_k} denotes the interface between the fluid and the structure domain. Following the formulation of problem (10.4), we consider the functional

$$\begin{aligned}\mathcal{J}_K(\boldsymbol{d}, E) = \frac{1}{2}\sum_{k=1}^{K} \int_{\Sigma^{\tau_k}} \left(\boldsymbol{d}_{meas}(\boldsymbol{x}, \tau^k) - \boldsymbol{d}(\boldsymbol{x}, \tau^k)\right)^2 \mathrm{d}\sigma \\ + \frac{\alpha_E}{2}\sum_{k=1}^{K} \int_{\Omega_s} \left(E(\boldsymbol{x}, \tau^k) - E_{ref}\right)^2 \mathrm{d}\boldsymbol{x},\end{aligned} \tag{10.18}$$

where $\boldsymbol{d}(\boldsymbol{x}, \tau^k)$ denotes the solution of the FSI system (4.44) in the ALE formulation, at $t = \tau^k$ and the second term is a non-negative *Tikhonov* regularization term. Here E_{ref} is a prior estimate (available from *ex vivo* specimens, for example), so that the regularization forces E to be close to its reference value. Given an admissible set $\mathcal{E}_{ad}$ where we seek the parameter E, a possible formulation of the IFSI problem reads as follows. For $t > 0, \boldsymbol{x} \in \Omega_s$, find $E = E(\boldsymbol{x}, t) \in \mathcal{P}$ that minimizes (10.18) under the constraint (4.44).

A possible choice for $\mathcal{P}$ is

$$\mathcal{P} = \{E : E \in L^\infty(\Omega_s),\ 0 < E_{\min} \leq E \leq E_{\max}, \text{with } E_{\min}, E_{\max} \in \mathbb{R}\}. \tag{10.19}$$

Such a problem entails the solution of a time-dependent minimization problem, for which a classical KKT system can be obtained using the Lagrange multiplier method. In this setting the parameter E plays the role of the control variable. However, this approach is rather involved, since the adjoint problem results in a final value problem which would involve differentiation with respect to the deformable domain as well: the so-called *shape derivative*. Moreover, it requires a massive memory occupancy as the solution at all the time steps needs to be stored due to the back-in-time nature of the adjoint problem.

A more convenient alternative numerical approach proposed by Perego *et al.* (2011) first considers the time-discretization of the forward problem, and formulates a minimization problem at each time step. For simplicity a constant time step $\Delta t = \Delta\tau/m$ is assumed, for a suitable $m \in \mathbb{N}$, $m \geq 1$, that is, the instants τ^k in which measurements are acquired are a subset of the time discretization of (4.44); for simplicity, we will consider $m = 1$. For the time discretization of (4.44), we consider implicit methods with a semi-implicit treatment of the convective term and of the fluid domain, whereas the fluid viscous term is treated implicitly. As in Section 4.6.1, we use the superscript $*$ to identify fluid quantities extrapolated from corresponding quantities at previous time steps. Moreover, we introduce the spaces

$$\boldsymbol{V}^* = \{\boldsymbol{v} \in \boldsymbol{H}^1(\Omega_f^*) : \boldsymbol{v}|_{\Gamma^*_{D,f}} = \boldsymbol{0}\}, \quad Q^* = L^2(\Omega_f^*),$$
$$\boldsymbol{W} = \{\boldsymbol{\psi} \in \boldsymbol{H}^1(\Omega_s) : \boldsymbol{\psi}_{\Gamma_{D,s}} = \boldsymbol{0}\},$$

where $\Gamma^*_{D,f}$ and $\Gamma_{D,s}$ are the portions of the boundary where a Dirichlet condition is prescribed. Further, we let

$$\boldsymbol{Z}^* = \{(\boldsymbol{v}, \boldsymbol{\psi}) \in \boldsymbol{V}^* \times \boldsymbol{W} : \boldsymbol{v}|_{\Sigma^*} - \widehat{\boldsymbol{\psi}}|_{\Sigma^*}/\Delta t = \boldsymbol{0}\},$$

and introduce the following bilinear forms:

$$\begin{aligned} a(\boldsymbol{v}, \boldsymbol{d}; \boldsymbol{w}, \boldsymbol{\psi})^* &= \frac{\rho_f}{\Delta t}(\boldsymbol{v}, \boldsymbol{w})_f^* + (\boldsymbol{T}_f(\boldsymbol{v}, p), \nabla \boldsymbol{w})_f^* \\ &\qquad + \rho_f(((\boldsymbol{v}^* - \boldsymbol{v}_f^*) \cdot \nabla)\boldsymbol{v}, \boldsymbol{w})_f^* + \rho_s\left(\frac{\boldsymbol{d}}{\Delta t^2}, \frac{\boldsymbol{\psi}}{\Delta t}\right)_s, \\ b(q; \boldsymbol{w})^* &= -(q, \nabla \cdot \boldsymbol{w})_f^*, \end{aligned}$$

where

$$(\boldsymbol{v}, \boldsymbol{w})_f^* = \int_{\Omega_f^*} \boldsymbol{v} \cdot \boldsymbol{w} \,\mathrm{d}\boldsymbol{x} \quad \text{and} \quad (\boldsymbol{\psi}, \boldsymbol{\chi})_s = \int_{\Omega_s} \boldsymbol{\psi} \cdot \boldsymbol{\chi} \,\mathrm{d}\boldsymbol{x}.$$

Then, for any given $E \in \mathcal{E}_{ad}$, at each time t^{n+1}, the time-discretized forward FSI problem consists of the following steps.

(1) Compute extrapolations Ω_f^*, $\boldsymbol{v}^*$ and $\boldsymbol{v}_f^*$ to approximate Ω_f^{n+1}, $\boldsymbol{v}^{n+1}$ and $\boldsymbol{v}_f^{n+1}$.

(2) Given $\boldsymbol{f}_f^{n+1} \in \boldsymbol{L}^2(\Omega_f^*)$ and $\boldsymbol{f}_s^{n+1} \in \boldsymbol{L}^2(\Omega_s)$, find $(\boldsymbol{v}^{n+1}, \boldsymbol{d}^{n+1}) \in \boldsymbol{Z}^*$ and $p^{n+1} \in Q^*$ such that

$$\begin{aligned}
&a(\boldsymbol{v}^{n+1}, \boldsymbol{d}^{n+1}; \boldsymbol{w}, \boldsymbol{\psi})^* + \left(E\, \boldsymbol{S}_s(\boldsymbol{d}^{n+1}), \frac{1}{\Delta t}\nabla\boldsymbol{\psi}\right)_s + b(p^{n+1}; \boldsymbol{v})^* \\
&\quad = F_f^{n+1}(\boldsymbol{w}) + F_s^{n+1}\left(\frac{\boldsymbol{\psi}}{\Delta t}\right) - a(\boldsymbol{R}_f^{n+1}, \boldsymbol{0}; \boldsymbol{w}, \boldsymbol{\psi})^* \quad \text{for all } (\boldsymbol{w}, \boldsymbol{\psi}) \in \boldsymbol{Z}^* \\
&b(q; \boldsymbol{v}^{n+1})^* = 0 \quad \text{for all } q \in Q^*.
\end{aligned} \tag{10.20}$$

(3) Update the fluid domain to obtain Ω_f^{n+1}.

For notation we have set

$$\boldsymbol{S}_s(\boldsymbol{d}) = \frac{1}{2(1+\nu)}(\nabla\boldsymbol{d} + (\nabla\boldsymbol{d})^T) + \frac{\nu}{(1+\nu)(1-2\nu)}(\nabla\cdot\boldsymbol{d})\boldsymbol{I}$$

so that we can write $\boldsymbol{T}_s = E\,\boldsymbol{S}_s$. Moreover, given $\boldsymbol{g} \in \boldsymbol{H}^{1/2}(\Sigma^*)$, we let

$$\boldsymbol{R}_f(\boldsymbol{g}) \in \boldsymbol{H}^{div}(\Omega_f^*) = \{\boldsymbol{v} \in \boldsymbol{H}^1(\Omega_f^*) : \nabla\cdot\boldsymbol{v} = 0\} \quad \text{and} \quad \boldsymbol{R}_s(\boldsymbol{g}) \in \boldsymbol{H}^1(\Omega_s)$$

denote two lifting functions for the fluid–structure interface continuity, given by

$$\boldsymbol{R}_f(\boldsymbol{g}) - \boldsymbol{R}_s(\boldsymbol{g}) = -\boldsymbol{g}/\Delta t \quad \text{on } \Sigma^*.$$

Due to the arbitrariness of one of these functions, in what follows we set $\boldsymbol{R}_s(\boldsymbol{g}) = \boldsymbol{0}$. Moreover, when applied to function $\boldsymbol{d}^{n-1}$, it is possible to set $\boldsymbol{R}_f^n = \boldsymbol{R}_f(\boldsymbol{d}^{n-1})$. Finally, the functionals F_f^{n+1} and F_s^{n+1} in (10.20) account for forcing terms, boundary data on $\partial\Omega_f^* \setminus \Sigma^*$ and $\partial\Omega_s \setminus \Sigma$ and terms coming from the time discretization at previous time steps.

Regarding the parameter estimation problem, once the problem has been discretized in time, the minimization of the cost functional

$$\widetilde{\mathcal{J}}_k(\boldsymbol{d}, E) = \int_\Sigma (\boldsymbol{d}_{meas}(\boldsymbol{x}, \tau^k) - \boldsymbol{d}(\boldsymbol{x}, \tau^k))^2 \,\mathrm{d}\sigma + \frac{\alpha_E}{2}\int_{\Omega_s} (E(\boldsymbol{x}, \tau^k) - E_{ref})^2 \,\mathrm{d}\boldsymbol{x}$$

for each $k = 1, 2, \ldots, K$ is considered by Perego *et al.* (2011), under the constraint (10.20). Hence, a system of KKT conditions can be derived relying on the standard Lagrange multiplier approach (see Section 9.1.2).

With this aim we introduce the Lagrangian functional at time τ^k:

$$\mathcal{L}(\boldsymbol{v}, p, \boldsymbol{d}; \boldsymbol{\lambda}_v, \lambda_p, \boldsymbol{\lambda}_d; E) = \widetilde{\mathcal{J}}_k(\boldsymbol{d}, E) + a(\boldsymbol{v}, \boldsymbol{d}; \boldsymbol{\lambda}_v, \boldsymbol{\lambda}_d)^* + \left(E\, \boldsymbol{S}_s(\boldsymbol{d}), \frac{1}{\Delta t} \nabla \boldsymbol{\lambda}_d\right)_s$$
$$+ b(p; \boldsymbol{\lambda}_v)^* + b(\lambda_p; \boldsymbol{v})^* - F_f(\boldsymbol{\lambda}_v) - F_s\left(\frac{\boldsymbol{\lambda}_d}{\Delta t}\right) + a(\boldsymbol{R}_f, \mathbf{0}; \boldsymbol{\lambda}_v, \boldsymbol{\lambda}_d)^*.$$

The dependence on the superscript k is omitted. By the requirement that the gradient of $\mathcal{L}$ vanishes, we obtain the following.

- The *adjoint problem*, by forcing to zero the (Gâteaux) derivatives of the Lagrangian functional with respect to $(\boldsymbol{v}, p, \boldsymbol{d})$, which in fact is a time-discretized FSI problem (see below for the interface condition). Find $(\boldsymbol{\lambda}_v, \boldsymbol{\lambda}_d) \in \boldsymbol{Z}^*$, $\lambda_p \in Q^*$ such that

$$\begin{aligned} &a(\boldsymbol{v}, \boldsymbol{\psi}; \boldsymbol{\lambda}_v, \boldsymbol{\lambda}_d)^* + \left(E\, \boldsymbol{S}_s\left(\frac{\boldsymbol{\psi}}{\Delta t}\right), \nabla \boldsymbol{\lambda}_d\right)_s + b(\lambda_p; \boldsymbol{w})^* \\ &\quad + \int_\Sigma (\boldsymbol{d} - \boldsymbol{d}_{meas}) \cdot \boldsymbol{\psi}\, \mathrm{d}\sigma = 0, \\ &b(q; \boldsymbol{\lambda}_v)^* = 0 \quad \text{for all } (\boldsymbol{w}, \boldsymbol{\psi}) \in \boldsymbol{Z}^*, q \in Q^*. \end{aligned} \tag{10.21}$$

- The *optimality condition*

$$(\varphi\, \boldsymbol{S}_s(\boldsymbol{d}), \nabla \boldsymbol{\lambda}_d)_s = 0 \quad \text{for all } \varphi \in L^\infty(\Omega_s),$$

  by forcing to zero the derivative with respect to E.

- We also require the following state problem. Find $(\boldsymbol{v}, \boldsymbol{d}) \in \boldsymbol{Z}^*$, $p \in Q^*$ such that

$$\begin{aligned} &a(\boldsymbol{v}, \boldsymbol{d}; \boldsymbol{w}, \boldsymbol{\psi})^* + \left(E\, \boldsymbol{S}_s(\boldsymbol{d}), \frac{1}{\Delta t} \nabla \boldsymbol{\psi}\right)_s + b(p; \boldsymbol{w})^* \\ &\quad = F_f(\boldsymbol{w}) + F_s\left(\frac{\boldsymbol{\psi}}{\Delta t}\right) - a(\boldsymbol{R}_f, \mathbf{0}; \boldsymbol{w}, \boldsymbol{\psi})^* \\ &b(q; \boldsymbol{v})^* = 0 \quad \text{for all } (\boldsymbol{w}, \boldsymbol{\psi}) \in \boldsymbol{Z}^*, q \in Q^*, \end{aligned}$$

  formally obtained by deriving the Lagrangian with respect to $(\boldsymbol{\lambda}_v, \lambda_p, \boldsymbol{\lambda}_d)$.

The adjoint problem, the optimality condition and the state problem yield the system of KKT conditions at each τ^k, which couples two linearized FSI problems and a scalar equation.

In particular, for the adjoint problem the interface velocity condition reads

$$\boldsymbol{\lambda}_v = \frac{\boldsymbol{\lambda}_d}{\Delta t} \quad \text{on } \Sigma^*,$$

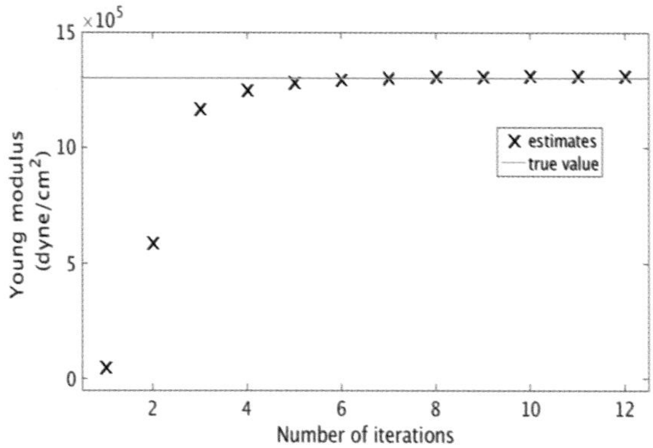

Figure 10.2. Convergence history for the estimation of Young's modulus by means of the algorithm proposed by Perego *et al.* (2011). Rectangular fluid and structure domains are used, with synthetic measurements generated by means of forward FSI simulations. Numerical results were obtained using the MATLAB finite element library MLife.

whereas the interface stress condition is

$$\boldsymbol{T}_s(\boldsymbol{\lambda}_d)\,\boldsymbol{n} - \boldsymbol{T}_f(\boldsymbol{\lambda}_v, \lambda_p)\,\boldsymbol{n} = -(\boldsymbol{d} - \boldsymbol{d}_{meas}) \quad \text{on } \Sigma^*, \tag{10.22}$$

taking into account the mismatch between the data and the solution, and modifying the homogeneous interface stress condition (4.19d) accordingly.

The same strategies described in Section 9.2 (*e.g.* gradient-based methods) can be exploited to solve the KKT system numerically. To take into account the constraint $E > 0$ in the case at hand, it is possible to transform the parameter as $\psi = \log(E)$, so that $E = \exp(\psi) > 0$ for every $\psi \in L^\infty(\Omega_s)$, and then optimize with respect to ψ. Finally, a finite element discretization in space is required to solve the state and the adjoint problem numerically.

By construction, this approach provides an estimate for E at each time; a possible option is then to average them to obtain a unique estimate, although this may suffer from instabilities in the presence of highly noisy data. A more robust approach is to solve the minimization problem over time. In this case, however, the complexity of the problem would increase, due to the inclusion of the shape derivatives in the minimization problem. In Figure 10.2 we show the number of iterations and the convergence history for the algorithm described above for the estimation of Young's modulus.

10.3.2. A Kalman filter approach for estimating material parameters in cardiac mechanics

In this section we illustrate a Kalman filter approach introduced by Xi *et al.* (2011) to estimate material parameters occurring in the mechanical model describing myocardial contractility. Only passive mechanics is modelled, making a quasi-static assumption, and without taking into account the electromechanical coupling. A transversely isotropic Guccione's constitutive law (Guccione *et al.* 1991) is used, and four parameters are estimated *in silico* from noisy displacement measurements of material points located on the myocardial surface. In principle, a similar approach can also be used, for example, to identify the material parameters affecting the orthotropic model (7.24).

The transversely isotropic strain energy function in the Guccione law can be expressed as

$$\Theta(\boldsymbol{E}) = \frac{1}{2}\theta_1(\mathrm{e}^Q - 1),$$

where

$$Q = \theta_2 E_{ff}^2 + \theta_3(E_{ss}^2 + E_{nn}^2 + 2E_{sn}^2) + \theta_4(2E_{fs}^2 + 2E_{fn}^2).$$

Here $\boldsymbol{\theta} = (\theta_1, \dots, \theta_4) \in \mathbb{R}^p$ denotes a vector of $p = 4$ material parameters, and E_{ij} is the (i, j)th component of the Green–Lagrange strain tensor $\boldsymbol{E} = \frac{1}{2}(\boldsymbol{F}^T\boldsymbol{F} - I) = \frac{1}{2}(\boldsymbol{C} - I)$, with $i, j \in \{f, s, n\}$ denoting fibre, sheet and sheet-normal directions, respectively. After performing spatial discretization, at each time step the augmented state $\widetilde{\boldsymbol{X}}^k = (\boldsymbol{X}^k, \boldsymbol{\theta}^k)^T \in \mathbb{R}^{n_x+p}$, made up of the deformed configuration $\boldsymbol{X}^k$ and the material parameter vector $\boldsymbol{\theta}^k$, has to be determined by solving a non-linear problem of the form

$$\widetilde{\boldsymbol{X}}^k = \mathbf{f}(\widetilde{\boldsymbol{X}}^{k-1}, \boldsymbol{u}^{k-1}) + \boldsymbol{\varepsilon}^k, \quad \mathbf{f}(\widetilde{\boldsymbol{X}}^{k-1}, \boldsymbol{u}^{k-1}) = \begin{bmatrix} G(\boldsymbol{\theta}^{k-1}, \boldsymbol{u}^{k-1}) \\ \boldsymbol{\theta}^{k-1} \end{bmatrix},$$

where the input vector $\boldsymbol{u}^{k-1}$ at time $k-1$ is given by the external forces, with observations

$$\boldsymbol{Z}^k = \boldsymbol{h}(\widetilde{\boldsymbol{X}}^k + \boldsymbol{w}^k), \quad k = 1, \dots, K.$$

Xi *et al.* (2011) take $\boldsymbol{h} : \mathbb{R}^{n_x+p} \to \mathbb{R}^{n_z}$ to be a linear (interpolation) matrix $H \in \mathbb{R}^{(n_x+p)\times n_z}$, mapping the augmented state vector to the coordinates of the points where measurements are acquired. These measurements are assumed to be contaminated by a noise term $\boldsymbol{w}^k$ with zero mean and covariance R_k; note that the material parameter has no explicit time dependence. Nevertheless, as already pointed out, the filter will provide recursively updated estimates of $\boldsymbol{\theta}^k$. The goal is thus to compute the estimate $\boldsymbol{\theta}^K$ at the final time $t = t^K$.

Moreover, since the augmented state vector $\widetilde{\boldsymbol{X}}^k$ does not depend on the deformed configuration $\widetilde{\boldsymbol{X}}^{k-1}$, the rank of the error covariance matrix $\widetilde{P}_k$ at each step will be equal to p. This is a crucial point in order to devise a reduced-order UKF, making it possible to (i) store a covariance matrix of dimension $(n_x + p) \times p$, (ii) perform inversion of $p \times p$ matrices and, even more importantly, (iii) decrease the number of model evaluations required by the filter at each time step from $O(n_x + p)$ to $O(p)$.

Slight differences arise with respect to the UKF formulation addressed in Section 10.2.3, due to the reduction of the estimation process to the subset of $\widetilde{\boldsymbol{X}}^k$ corresponding to the p parameter components. The filtering algorithm consists of three steps.

- *Initialization.* A singular value decomposition $P_0^a = L_0^a \Lambda_0^a (L_0^a)^T$ of the initial error covariance $P_0^a \in \mathbb{R}^{(n_x+p)\times(n_x+p)}$ is performed. For instance, P_0^a can be given by a diagonal matrix with p non-zero entries representing the variances of the initial (or background) parameter estimates. Letting $L_{0,p}^a$ be the matrix whose columns are the first p singular vectors of L_0^a and $\Lambda_{0,p}^a$, the $p \times p$ diagonal matrix of the singular values, the reduced-rank square-root approximation $S_0^a \in \mathbb{R}^{(n_x+p)\times p}$ can be obtained as
$$S_0^a = L_{0,p}^a \sqrt{\Lambda_{0,p}^a}.$$
- *Prediction* (*or time update*). At each step $k = 1, \ldots,$ starting from the assimilated state $\widehat{\widetilde{\boldsymbol{X}}}_a^{k-1}$ at step $k-1$, a set of sigma points is selected as
$$\widetilde{x}_{a,(i)}^{k-1} = \widehat{\widetilde{\boldsymbol{X}}}_a^{k-1} + (S_{k-1}^a)_i, \quad \widetilde{x}_{a,(p+i)}^{k-1} = \widehat{\widetilde{\boldsymbol{X}}}_a^{k-1} - (S_{k-1}^a)_i, \quad i = 1, \ldots, p.$$
The sigma points are then transformed via the state dynamics to obtain
$$\widetilde{x}_{f,(i)}^k = \mathbf{f}\big(\widetilde{x}_{a,(i)}^{k-1}, \boldsymbol{u}^{k-1}\big), \quad i = 1, \ldots, 2p,$$
and the mean $\widehat{\widetilde{\boldsymbol{X}}}_f^k$ and the error covariance P_k^f are estimated as
$$\widehat{\widetilde{\boldsymbol{X}}}_f^k = \sum_{i=1}^{2p} \widetilde{x}_{f,(i)}^k, \quad P_k^f = \sum_{i=1}^{2p} (\widehat{\widetilde{\boldsymbol{X}}}_f^k - \widetilde{x}_{f,(i)}^k)(\widehat{\widetilde{\boldsymbol{X}}}_f^k - \widetilde{x}_{f,(i)}^k)^T.$$
To preserve the low-rank structure of the correlation matrix P_k^f, a singular value decomposition $P_k^f = L_k^f \Lambda_0^f (L_k^f)^T$ is performed, thus yielding
$$P_k^f = S_k^f (S_k^f)^T, \quad S_k^f = L_{k,p}^f \sqrt{\Lambda_{k,p}^f}.$$
- *Correction* (*or measurement update*). At each step $k = 1, \ldots,$ the observation $\boldsymbol{Z}^k$ is assimilated into the forecast $\widehat{\widetilde{X}}_f^k$ yielding the corrected state
$$\widehat{\widetilde{\boldsymbol{X}}}_a^k = \widehat{\widetilde{\boldsymbol{X}}}_f^k + K_k(\boldsymbol{Z}^k - H\widehat{\widetilde{\boldsymbol{X}}}_f^k).$$
In this case the Kalman gain can be shown to be
$$K_k = S_k^f (I + (HS_k^f)^T R_k^{-1} H S_k^f)^{-1} (HS_k^f)^T R_k^{-1}$$

along with the square-root of the error covariance, given by

$$S_k^a = S_k^f (I + (HS_k^f)^T R_k^{-1} HS_k^f)^{-1/2}.$$

Note that here, compared to the UKF formulation of Section 10.2.3, update formulas for the Kalman gain and the error covariance are directly obtained for the square-root of the covariance matrix. For further details see Simon (2006) or Xi *et al.* (2011).

Xi *et al.* (2011) show that such a reduced-order UKF is able to estimate $p = 4$ (and subsequently $p = 12$) material parameters, starting from a set of synthetic measurements generated by adding Gaussian noise to a set of noise-free measurements, obtained via model simulation for a chosen ground truth parameter vector. Myocardial dilatation is considered by loading the structure with a given ventricular pressure on the endocardium, and $n_z = 600$ material points located on the epicardium and endocardium.

We have chosen to report this example to illustrate how to take advantage of a Kalman filter strategy in a *relatively* simple cardiovascular application; further details and more involved cases can be found in Moireau and Chapelle (2011), Bertoglio *et al.* (2012), Chabiniok *et al.* (2012), Moireau *et al.* (2013) and Corrado *et al.* (2015), for example. Numerical results related to parameter identification in electrophysiology by means of an ensemble Kalman filter – a technique that indeed shares several similarities with the UKF – will be shown in Section 11.2.2.

11. Including uncertainty

Moving towards model personalization – that is, the adaptation of model inputs to subject-specific conditions – the question immediately arises as to the sensitivity of model predictions to errors and uncertainties in the model inputs. The inputs to be personalized may include the computational domain (*e.g.* vascular networks), physical parameters (*e.g.* vascular material properties) and boundary conditions. Because of noise in input measurements, as well as their large biological variability, model inputs are inevitably hampered by uncertainty. Furthermore, as already seen in Section 10, not all model inputs are directly measurable, as in the case of the local mechanical properties of the arterial wall, thus calling into play the need to solve parameter estimation problems. The uncertainties carried by patient-specific features should then be incorporated into the computational model, to quantify their impact on the computed results and to obtain more reliable predictions or best/worst-case scenarios (Eck *et al.* 2016). These are the main reasons behind the very rapid growth of applications of sensitivity analysis and uncertainty quantification (UQ) to cardiovascular problems in the past decade. Since UQ is a very active field (and less mature than

PDE-constrained optimization or parameter estimation) from both a mathematical and computational viewpoint, in this section we only touch on some basic, yet relevant, issues related to UQ techniques; see *e.g.* Sullivan (2015), Kaipio and Somersalo (2005) and Le Maître and Knio (2010) for a more in-depth overview of numerical aspects of UQ.

Typically, UQ problems involve a mathematical model for a process of interest, subject to some uncertainty about the correct form of this model or, more frequently, about some of its parameters. Although exhibiting substantial overlap with the field of parameter estimation and data assimilation, UQ problems also involve the propagation of uncertainty on outputs of interest, reliability or certification problems, prediction problems and, very often, consist of several of these aspects coupled together. For instance, after estimating some model parameters, we may use them to forward-propagate other uncertainties. Often, though not always, uncertainties are treated probabilistically, thus calling into play probabilistic and statistical methods. Here we focus on basic aspects related to *forward propagation* and *backward propagation* of uncertainty, for the latter case focusing on statistical inversion methods within a Bayesian framework.

11.1. Forward uncertainty quantification

The goal of forward UQ is to derive information about the uncertainties in system outputs of interest, given information about the uncertainties in the system inputs. The goal is to obtain the probability density function (PDF) of some output of interest $\boldsymbol{Z} = \boldsymbol{f}(\boldsymbol{\theta}) \in \mathbb{R}^{n_z}$ given the probability distribution of the input $\boldsymbol{\theta} \in \mathbb{R}^p$, or evaluate moments (*e.g.* the expected value and the variance), correlation functions, confidence regions or quantiles.[5] Here $\boldsymbol{f} : \mathbb{R}^p \to \mathbb{R}^{n_z}$ denotes an input/output map. For instance, Sankaran and Marsden (2011) take the radius of an abdominal aortic aneurysm, the radii and the inflow velocity of the carotid artery bifurcation, and the flow split of the left and right pulmonary arteries as random variables to account for the uncertainty impact on blood flows modelled by three-dimensional Navier–Stokes equations with rigid arterial walls, by considering as outputs blood velocity and wall shear stresses.

Once a probabilistic description of the random inputs has been provided, a suitable strategy is needed to propagate uncertainties through the model. The simplest case is one where the PDE system depends on a set of random inputs that are constant with respect to space and time. In this case,

[5] Sensitivity analysis (see *e.g.* Saltelli *et al.* 2008 for a detailed review) may be seen as a precursor of forward UQ, sharing the need to provide a quantitative description of the dependence of the solution of a model on input parameters. See Donders, Huberts, van de Vosse and Delhaas (2015) for the application of sensitivity analysis to a pulse wave propagation model of arterial flows.

they can be described by means of a finite-dimensional random vector, with a given probability distribution. A more involved case is one where input data may vary randomly from one point of the physical domain to another (and possibly from one time instant to another); in this case, they are described in terms of random fields. Two popular strategies to describe correlated random fields include Karhunen–Loève expansions, and expansions in terms of global orthogonal polynomials (Ghanem and Spanos 2003, Xiu and Karniadakis 2002*b*). Made up of infinitely many terms, these expansions are usually truncated to approximate random fields: the milder the variations in space and time of the random field realizations, the fewer the terms retained in the expansion.

Several approaches are available for solving PDE problems with random input data. Monte Carlo (MC) methods are the most popular. These are based on independent realizations $\boldsymbol{\theta}_1, \dots, \boldsymbol{\theta}_M$ of the random variable $\boldsymbol{\theta}$ (whose probability distribution is denoted by μ) yielding approximations of the expectation by averaging over the corresponding realizations of that quantity, that is,

$$\mathbb{E}[\boldsymbol{f}(\boldsymbol{\theta})] = \int_{\mathbb{R}^p} f(\boldsymbol{\theta})\mu(\mathrm{d}\boldsymbol{\theta}) \approx \frac{1}{M}\sum_{i=1}^{M} \boldsymbol{f}(\boldsymbol{\theta}_i), \tag{11.1}$$

where the sample $\boldsymbol{\theta}_1, \dots, \boldsymbol{\theta}_M$ is generated randomly from the probability distribution of $\boldsymbol{\theta}$. Such a method requires a deterministic PDE query for each realization and a very large number of queries to achieve a small error, this latter being proportional to $1/\sqrt{M}$. This entails an excessive computational burden, especially for systems which are already computationally expensive in their deterministic setting. See, for example, Fishman (1996), Robert and Casella (2004) and Voss (2013) for more on Monte Carlo methods, and Dick, Kuo and Sloan (2013) and Giles (2015) for quasi-Monte Carlo and multilevel Monte Carlo methods, respectively.

When the solution depends analytically on the random input parameters, other approaches exhibit much faster convergence rates. These include spectral (global) stochastic Galerkin (SG) methods (Ghanem and Spanos 2003, Xiu and Karniadakis 2002*a*, Babuška, Tempone and Zouraris 2004) and stochastic collocation (SC) methods (Babuška, Nobile and Tempone 2007, Xiu and Hesthaven 2005, Nobile, Tempone and Webster 2008); for a mathematical and numerical discussion, see the detailed review by Gunzburger, Webster and Zhang (2014) and the recent books by Le Maître and Knio (2010) and Sullivan (2015). These methods are based on the discretization of a PDE system, not only with respect to spatial variables but also with respect to the random inputs. Stochastic Galerkin methods are intrusive approaches since the physical and the probabilistic variables are coupled; stochastic sampling (SS) and interpolatory-type stochastic collocation methods are instead non-intrusive. Further, stochastic Galerkin methods require the solution of discrete systems that couple all spatial and probabilistic variables, whereas stochastic collocation methods can combine standard approximations in physical space with globally defined polynomial approximation in the probability domain, either by full polynomial spaces, tensor product polynomial spaces or sparse tensor product polynomials. The stochastic space can then be queried at any point in order to construct the PDF of the output, the interpolation in the stochastic space being built on top, and independently, of the PDE solver.

By way of illustration, Figure 11.1 gives the activation times obtained using the monodomain model coupled with the Aliev–Panfilov cell model. This computation is carried out on a patient-specific left ventricle geometry, obtained by considering different levels of tissue damage (or ischaemia), caused by an insufficient blood perfusion of the myocardium. In this case, each component of the conductivity $\boldsymbol{\Sigma}$ is multiplied by a spatial field $\sigma(\boldsymbol{x};\boldsymbol{\theta})$ ranging from 0 (lack of conductivity, damaged tissue) to 1 (regular conductivity, healthy tissue), which we refer to as relative conductivity; as soon as the position and size of the ischaemia are unknown, σ is a random field, depending on (a finite number p of) random inputs $\theta_1, \ldots, \theta_p$. We emphasize that in this case input uncertainty yields significant variability in the output, thus making uncertainty propagation a problem of interest in this context.

11.2. Inverse uncertainty quantification

Inverse UQ (or backward uncertainty propagation) problems refer to those situations in which input quantities are inferred starting from observed system outputs. Variational or sequential methods described in Sections 10.1 and 10.2 provide point estimates of the quantities of interest, given a set of observations. The *statistical inversion* approach provides a well-established framework to better characterize the uncertainty in the data and the resulting uncertainty in the computed estimates, adopting a Bayesian formulation. In this setting, all the model inputs included in the model are described by random variables, where randomness accounts for the available degree of information concerning their realizations, and is expressed in terms of probability distributions; as a result, the solution of the inverse UQ problem is the *posterior* PDF of the unknown inputs. Compared to the variational methods of Section 10.1, where classical regularization methods yield point estimates by curing the ill-posedness of the problem, statistical inversion aims at removing ill-posedness by recasting the inverse problem in a larger space of probability distributions (Kaipio and Somersalo 2005). This strategy also enables better characterization of the prior information contained in the regularization terms in (10.2), in the form of a *prior* PDF

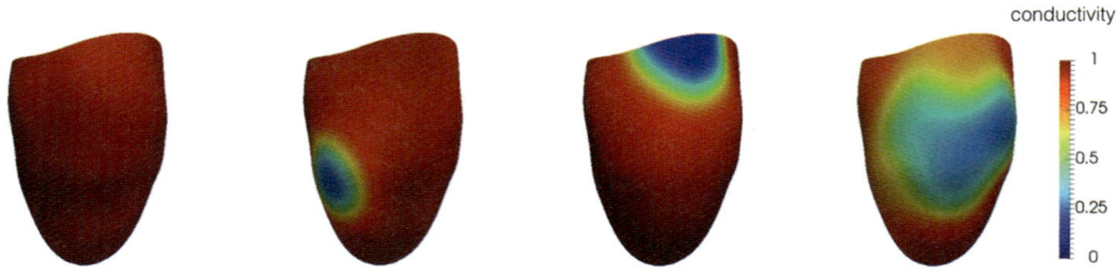

(a)

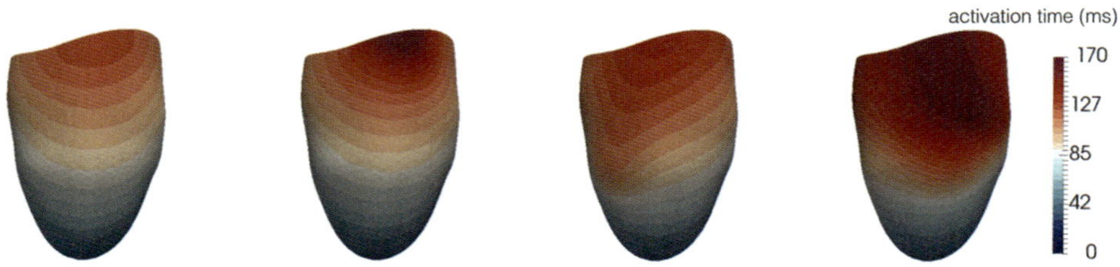

(b)

Figure 11.1. Different degrees of tissue damage in terms of relative conductivity (a) and activation times in milliseconds (b) for a healthy case (left) and different ischaemic regions on the myocardium. The patient-specific geometry of the left ventricle has been reconstructed using the semi-automatic segmentation method proposed by Fedele *et al.* (2015). Numerical results were obtained using the finite element library redbKIT v2.1 (github.com/redbKIT/redbKIT/releases).

of the unknown inputs. The task of Bayesian inversion is to improve the knowledge of the unknown system features starting from the prior belief and exploiting information from a set of model realizations. For more on this topic, see Stuart (2010) and Sullivan (2015), for example.

11.2.1. Static problems

Assume that we are measuring an output $\boldsymbol{z} = \boldsymbol{f}(\boldsymbol{\theta}, \boldsymbol{\varepsilon}) \in \mathbb{R}^{n_z}$ in order to get information on the unknown input $\boldsymbol{\theta} \in \mathbb{R}^p$, and that the measured quantity is contaminated by noise $\boldsymbol{\varepsilon}$. By treating input and output as random quantities, let us denote by $\boldsymbol{Z} \in \mathbb{R}^{n_z}$ and $\boldsymbol{\Theta} \in \mathbb{R}^p$ two random variables, of which $\boldsymbol{z}$ and $\boldsymbol{\theta}$ denote the corresponding realizations. Very often, the noise is modelled as additive and mutually independent of $\boldsymbol{\Theta}$, so that the input/output map takes the form

$$\boldsymbol{Z} = \boldsymbol{f}(\boldsymbol{\Theta}) + \boldsymbol{\varepsilon},$$

where $\boldsymbol{Z}, \boldsymbol{\varepsilon} \in \mathbb{R}^{n_z}$, $\boldsymbol{\Theta} \in \mathbb{R}^p$ are random variables. Here $\boldsymbol{Z}$ is called the *measurement*, and its realization $\boldsymbol{Z} = \boldsymbol{z}$ in the actual measurement process is called the *data*. Moreover, let $\pi_{noise}(\boldsymbol{\varepsilon})$ denote the PDF of the noise $\boldsymbol{\varepsilon}$, usually encoding experimental errors. Before performing output measurements, all the information (*e.g.* structure or regularity) about the distribution of the input $\boldsymbol{\Theta}$ are encapsulated in the *prior* PDF $\pi_{prior}(\boldsymbol{\theta})$, to be selected according to problem-specific considerations.

The conditional probability $\pi(\boldsymbol{z} \mid \boldsymbol{\theta})$ of $\boldsymbol{Z} \mid \boldsymbol{\Theta} = \boldsymbol{\theta}$ (*i.e.* of $\boldsymbol{Z}$ conditioned on $\boldsymbol{\Theta} = \boldsymbol{\theta}$) is the so-called (*conditional*) *likelihood function*, and expresses the likelihood of different measurement outcomes $\boldsymbol{z}$ given $\boldsymbol{\Theta} = \boldsymbol{\theta}$. In our case the input/output map $\boldsymbol{f} : \mathbb{R}^p \to \mathbb{R}^{n_z}$ is the solution of a PDE problem, and the evaluation of the output $\boldsymbol{f}(\boldsymbol{\theta})$. Thanks to the assumption of mutual independence of $\boldsymbol{\Theta}$ and ε, $\boldsymbol{Z} \mid \boldsymbol{\Theta} = \boldsymbol{\theta}$ is distributed like ε, that is, the likelihood function is

$$\pi(\boldsymbol{z} \mid \boldsymbol{\theta}) = \pi_{noise}(\boldsymbol{z} - \boldsymbol{f}(\boldsymbol{\theta})).$$

Assuming that the measurement data $\boldsymbol{Z} = \boldsymbol{z}$ is given, in the Bayesian framework the inverse problem is to find the conditional PDF $\pi(\boldsymbol{\theta} \mid \boldsymbol{z})$ of $\boldsymbol{\Theta}$. This latter is the *posterior* PDF of $\boldsymbol{\Theta}$ given the data $\boldsymbol{Z} = \boldsymbol{z}$ and can be expressed via Bayes' theorem as

$$\pi(\boldsymbol{\theta} \mid \boldsymbol{z}) = \frac{\pi_{prior}(\boldsymbol{\theta})\pi(\boldsymbol{z} \mid \boldsymbol{\theta})}{\pi(\boldsymbol{z})},$$

where

$$\pi(\boldsymbol{z}) = \int_{\mathbb{R}^{n_z}} \pi(\boldsymbol{z} \mid \boldsymbol{\theta})\pi_{prior}(\boldsymbol{\theta})\, \mathrm{d}\boldsymbol{\theta}$$

plays the role of a normalization constant, and often has little importance from a computational standpoint.

Solving an inverse UQ problem in the static case – alternatively, in the literature such a problem is referred to as the *stationary inverse problem* – thus consists in finding a prior PDF $\pi_{prior}(\boldsymbol{\theta})$, expressing the likelihood function $\pi(\boldsymbol{z} \mid \boldsymbol{\theta})$ using the interplay between the observation and the unknown, and finally developing suitable numerical techniques to explore the posterior PDF. Each of these tasks is a challenging problem from a computational standpoint; here we provide some hints on how to use this framework for a wide range of applications related to backward uncertainty quantification in cardiovascular modelling.

In the case where the unknown is a random variable with few components, the posterior PDF can also be visualized in the form of a non-negative function of these variables. Most applications, however, yield larger-scale inverse UQ problems, and resulting PDFs in high-dimensional spaces, for which it is much more effective to evaluate suitable *point estimators*, such as the *maximum a posteriori* estimator

$$\boldsymbol{\theta}_{MAP} = \arg\max_{\boldsymbol{\theta} \in \mathbb{R}^p} \pi(\boldsymbol{\theta} \mid \boldsymbol{z})$$

or the *conditional mean*

$$\boldsymbol{\theta}_{CM} = \mathbb{E}[\boldsymbol{\theta} \mid \boldsymbol{z}] = \int_{\mathbb{R}^p} \boldsymbol{\theta}\, \pi(\boldsymbol{\theta} \mid \boldsymbol{z})\, \mathrm{d}\theta.$$

Evaluating the former requires the solution of an optimization problem, for example using iterative, gradient-based methods; computing the latter involves a numerical quadrature problem in high-dimensional spaces. The evaluation of variability estimators such as the *conditional covariance*

$$\mathrm{Cov}\,(\boldsymbol{\theta} \mid \boldsymbol{z}) = \int_{\mathbb{R}^p} (\boldsymbol{\theta} - \boldsymbol{\theta}_{CM})(\boldsymbol{\theta} - \boldsymbol{\theta}_{CM})^T\, \pi(\boldsymbol{\theta} \mid \boldsymbol{z})\, \mathrm{d}\theta \in \mathbb{R}^{p\times p},$$

or *confidence regions*, also provides further indicators for uncertainty quantification.

The simplest probabilistic model that can be used to describe experimental uncertainties is the Gaussian model, for which the noise $\boldsymbol{\varepsilon} \sim \mathcal{N}(\mathbf{0}, \Sigma_\varepsilon)$ is normally distributed, with mean $\mathbf{0}$ and covariance matrix Σ_ε. In this case, the likelihood function is

$$\pi(\boldsymbol{z} \mid \boldsymbol{\theta}) \propto \exp\left(-\frac{1}{2}\|\boldsymbol{z} - \boldsymbol{f}(\boldsymbol{\theta})\|^2_{\Sigma_\varepsilon^{-1}}\right).$$

If we can also assume a Gaussian model on the *prior* knowledge of the parameter distributions, that is, $\pi_{prior} \sim \mathcal{N}(\boldsymbol{\theta}_p, \Sigma_p)$, then the posterior PDF will be normally distributed as well, that is,

$$\pi(\boldsymbol{\theta} \mid \boldsymbol{z}) \propto \exp\left(-\frac{1}{2}\|\boldsymbol{z} - \boldsymbol{f}(\boldsymbol{\theta})\|^2_{\Sigma_\varepsilon^{-1}} - \frac{1}{2}\|\boldsymbol{\theta} - \boldsymbol{\theta}_p\|^2_{\Sigma_p^{-1}}\right). \tag{11.2}$$

In this case, the maximum *a posteriori* estimator is

$$\boldsymbol{\theta}_{MAP} = \arg\min_{\boldsymbol{\theta}\in\mathbb{R}^p} \left(\frac{1}{2}\|\boldsymbol{z} - \boldsymbol{f}(\boldsymbol{\theta})\|^2_{\Sigma_\varepsilon^{-1}} + \frac{1}{2}\|\boldsymbol{\theta} - \boldsymbol{\theta}_p\|^2_{\Sigma_p^{-1}}\right),$$

that is, it coincides with the estimator obtained by solving the (static version, with $K = 1$ of the) regularized least-squares problem (10.4), as long as we choose $\boldsymbol{\theta}_0 = \boldsymbol{\theta}_p$, $P_\theta = \Sigma_p$ and $M = \Sigma_\varepsilon^{-1}$. If we assume instead that no information is available about the parameter distribution except that it resides in a subset $\mathcal{D}$ of the parameter space $\mathbb{R}^p$, then $\pi_{prior}(\boldsymbol{\theta}) \sim \mathcal{U}(\mathcal{D})$ is a uniform distribution over $\mathcal{D}$.

Remark 11.1. Note that the parameter estimation techniques of Section 10.1 can be seen as strategies yielding point estimates without any reference to underlying statistical models. However, when a Gaussian assumption is made in the Bayesian framework, the regularization term and the norms $\|\cdot\|_{\Sigma_\varepsilon^{-1}}$, $\|\cdot\|_{\Sigma_p^{-1}}$ and the value $\boldsymbol{\theta}_p$ have a clear interpretation in terms of measurement noise and prior probability distribution. Hence, a prior that carries sufficient information about the true underlying structure of the parameters often provides more meaningful estimates and regularizes the inverse problem in a more natural way than relying on abstract regularization terms, as in (10.4), that might not have any interpretation. The benefit of casting parameter estimation problems in the form of inverse UQ problems is that one is also able to characterize the variance of the prediction due to measurement and model errors more precisely than from the single-point estimates.

Exploring the posterior distribution in the case where the input space $\mathcal{D}$ has a larger dimension than $p = 2$ or 3 calls into play suitable *sampling methods*, among which the Markov chain Monte Carlo (MCMC) techniques are the most relevant example. Instead of evaluating the posterior PDF at a single point, an MCMC technique is a systematic way to generate a sample which can be used to *explore* the distribution, as well as to perform Monte Carlo integration as in (11.1) in order to compute the conditional mean or conditional covariance, for example. In the former case, the posterior $\pi(\boldsymbol{\theta} \mid \boldsymbol{z})$ plays the role of target probability distribution that we want to explore, and is obtained as a realization of a Markov chain by relying on the following *Metropolis–Hastings algorithm*, for example.

(1) Pick an initial $\boldsymbol{\theta}_1$.

(2) For $m = 1, \dots, M$:

  (a) Compute $\pi(\boldsymbol{\theta}^m \mid \boldsymbol{z})$.

  (b) Draw $\boldsymbol{w} \sim \mathcal{N}(\boldsymbol{0}, \Sigma_\varepsilon)$ and take a random step to find the next candidate $\widehat{\boldsymbol{\theta}} = \boldsymbol{\theta}_k + \boldsymbol{w}$.

  (c) Compute $\pi(\widehat{\boldsymbol{\theta}} \mid \boldsymbol{z})$.

  (d) Define the acceptance ratio

$$\alpha_m = \min\left\{1, \frac{\pi(\widehat{\boldsymbol{\theta}} \mid \boldsymbol{z})}{\pi(\boldsymbol{\theta}_m \mid \boldsymbol{z})}\right\}.$$

  (e) Let $u \sim \mathcal{U}([0,1])$. If $u \leq \alpha_m$, accept, set $\boldsymbol{\theta}_{m+1} = \widehat{\boldsymbol{\theta}}$, and add to the set of samples Ξ; otherwise reject, and keep $\boldsymbol{\theta}_{m+1} = \boldsymbol{\theta}_m$.

After M steps of the algorithm, the set Ξ of samples contains realizations of the probability distribution $\pi(\boldsymbol{\theta} \mid \boldsymbol{z})$. The random step in point (b) from the current sample to the next candidate is distributed as white noise; its covariance should be chosen as large as possible while still maintaining a reasonable acceptance rate. Moreover, the initial points of the generated set usually represent the distribution to be explored poorly, and are then removed from the sample. To learn more on MCMC algorithms we refer to

Kaipio and Somersalo (2005), Robert and Casella (2004) and Voss (2013), and references therein.

11.2.2. Dynamical problems

Backward UQ problems whose forward system is stationary, or with observations that are not acquired sequentially in time, can be cast in the form of a Bayesian inverse problem, as shown in the previous section. In these cases, backward UQ can be formulated in the Bayesian framework by means of *Bayesian filtering methods*, among which the Kalman filter can be seen as a particular instance. These problems are also referred to as *non-stationary inverse problems*. As in Section 10.2, depending on the quantities that have to be estimated, the problem can be formulated as a state estimation or a joint state/parameter estimation problem. We will treat both these cases but will limit our discussion to finite-dimensional models (which usually arise from space and time discretization of unsteady PDEs) and using time-discretized evolution models. Further discussion can be found in Kaipio and Somersalo (2005), Särkkä (2013), Sullivan (2015) and Houtemaker and Zhang (2016), for example.

Let $\{\boldsymbol{X}^k\}_{k=0}^K$ and $\{\boldsymbol{Z}^k\}_{k=1}^K$ denote two stochastic processes; the former is related to the quantity we are interested in, whereas the latter represents the measurement. In particular, the random vector $\boldsymbol{X}^k \in \mathbb{R}^{n_x}$ is referred to as the *state vector*, whereas the random vector $\boldsymbol{Z}^k \in \mathbb{R}^{n_z}$ is referred to as the *observation*, both considered at the kth time τ^k. From a Bayesian standpoint, the goal is to use the observations until time k to get information about the state $\boldsymbol{X}^k$ and quantify the uncertainty related to this estimate. To frame this problem in the Bayesian setting, we assume that $\{\boldsymbol{X}^k\}_{k=0}^K$ and $\{\boldsymbol{Z}^k\}_{k=1}^K$ are an *evolution/observation model*, as follows.

(1) $\{\boldsymbol{X}^k\}_{k=0}^K$ and $\{\boldsymbol{Z}^k\}_{k=1}^K$ are Markov processes, that is,

$$\begin{aligned}\pi(\boldsymbol{x}^{k+1} \mid \boldsymbol{x}^0, \boldsymbol{x}^1, \dots, \boldsymbol{x}^k) &= \pi(\boldsymbol{x}^{k+1} \mid \boldsymbol{x}^k), \quad && k = 0, 1, \dots, \\ \pi(\boldsymbol{z}^k \mid \boldsymbol{x}^0, \boldsymbol{x}^1, \dots, \boldsymbol{x}^k) &= \pi(\boldsymbol{z}^k \mid \boldsymbol{x}^k), && k = 1, 2, \dots.\end{aligned}$$

(2) $\{\boldsymbol{X}^k\}_{k=0}^K$ depends on the past observations only through its own history, that is,

$$\pi(\boldsymbol{x}^{k+1} \mid \boldsymbol{x}^k, \boldsymbol{z}^1, \dots, \boldsymbol{z}^k) = \pi(\boldsymbol{x}^{k+1} \mid \boldsymbol{z}^k), \quad k = 0, 1, \dots.$$

Here $\boldsymbol{x}^k$, $\boldsymbol{z}^k$ denote the realizations of the processes $\{\boldsymbol{X}^k\}_{k=0}^K$ and $\{\boldsymbol{Z}^k\}_{k=1}^K$, respectively. In order to characterize such a model, we need to specify the PDF of the initial state $\boldsymbol{X}^0$, $\pi_{prior}(\boldsymbol{x}^0)$, the so-called *transition kernel* $\pi(\boldsymbol{x}^{k+1} \mid \boldsymbol{x}^k)$, $k = 0, 1, \dots$ and the conditional probability $\pi(\boldsymbol{z}^k \mid \boldsymbol{x}^k)$, $k = 1, 2, \dots$, the so-called (conditional) likelihood function. We assume a *state evolution equation* of the form

$$\boldsymbol{X}^{k+1} = \boldsymbol{f}^{k+1}(\boldsymbol{X}^k, \boldsymbol{V}^{k+1}), \quad k = 0, 1, \dots \tag{11.3}$$

and an *observation equation* of the form

$$Z^k = h^k(X^k, W^k), \quad k = 1, 2, \ldots. \tag{11.4}$$

Here f^{k+1} and h^k are known functions, whereas $V^{k+1} \in \mathbb{R}^{n_x}$ and $W^k \in \mathbb{R}^{n_z}$ denote the *state noise* and the *observation noise*, respectively. We want to determine the conditional PDF $\pi(x^k \mid D^k)$ of the state at the kth time instant given the observations $D^k = (z^1, \ldots, z^k)$ up to the same time instant; this procedure is usually referred to as *filtering problem.* By recursive application of Bayes' theorem, we have the following.

- The *time evolution updating, i.e.*, the problem of determining $\pi(x^{k+1} \mid D^k)$ given $\pi(x^k \mid D^k)$ and the transition kernel $\pi(x^{k+1} \mid x^k)$, provides

$$\pi(x^{k+1} \mid D^k) = \int_{\mathbb{R}^{n_x}} \pi(x^{k+1} \mid x^k)\pi(x^k \mid D^k)\, \mathrm{d}x^k. \tag{11.5}$$

- The *observation updating, i.e.*, the problem of determining the posterior distribution $\pi(x^{k+1} \mid D^{k+1})$ of $X^k \mid D^k$ based on the new observation Z^{k+1} given $\pi(x^{k+1} \mid D^k)$ and the likelihood function $\pi(z^{k+1} \mid x^{k+1})$, provides

$$\pi(x^{k+1} \mid D^{k+1}) = \frac{\pi(z^{k+1} \mid x^{k+1})\pi(x^{k+1} \mid D^k)}{\pi(z^{k+1} \mid D^k)}, \tag{11.6}$$

  where

$$\pi(z^{k+1} \mid D^k) = \int_{\mathbb{R}^{n_x}} \pi(z^{k+1} \mid x^{k+1})\pi(x^{k+1} \mid D^k)\, \mathrm{d}x^{k+1}.$$

  Formula (11.6) is Bayes' formula, where $\pi(x^{k+1} \mid D^k)$ is considered as the prior distribution for x^{k+1}.

The Kalman filter introduced in Section 10.2.1 is a remarkable instance of Bayesian filter method. Indeed, let us assume that the state and the observation equations are linear with additive noise processes, that is,

$$f^{k+1}(x^k, v^k) = A_{k|k+1}x^k + v^k, \quad h^k = H_k x^k + w^k,$$

for given matrices $A_{k|k+1}$, H_k, that the noise vectors V^{k+1} and W^k are mutually independent, Gaussian, with zero mean and known covariances Q_{k+1} and R_k, respectively, and that the prior PDF of X^0 is Gaussian with mean m^0 and covariance P^0. Under these assumptions, the time evolution and the observation updating formulas (11.3, 11.4) involve Gaussian distributions, whose means and covariances can be updated at each step according to prediction formulas (10.15a, 10.15b) and correction formulas (10.15d, 10.15e), respectively. In particular, we have that

$$\pi(x^{k+1} \mid x^k) \sim N(A_{k|k+1}x^k, Q^k),$$
$$\pi(z^k \mid x^k) \sim N(H_k x^k, R_k).$$

The Bayesian filtering equations can be evaluated in closed form, yielding the following Gaussian distributions:

$$\pi(\boldsymbol{x}^k \mid \boldsymbol{D}^{k-1}) \sim \mathcal{N}(\boldsymbol{m}_f^k, P_k^f),$$
$$\pi(\boldsymbol{x}^k \mid \boldsymbol{D}^{k}) \sim \mathcal{N}(\boldsymbol{m}_a^k, P_k^a),$$
$$\pi(\boldsymbol{z}^k \mid \boldsymbol{D}^{k-1}) \sim \mathcal{N}(H_k\boldsymbol{m}_f^k, H_k P_k^f H_k^T + R_k),$$

where the means and the variances can be computed with the Kalman filter prediction step

$$\begin{aligned} \boldsymbol{m}_f^k &= A_{k-1|k}\boldsymbol{m}^{k-1}, \\ P_k^f &= A_{k-1|k}P_{k-1}^a A_{k-1|k}^T + Q_{k-1} \end{aligned} \tag{11.7}$$

and the consequent correction step

$$\begin{aligned} K_k &= P_k^f H_k^T (H_k P_k^f H_k^T + R_k)^{-1}, \\ \boldsymbol{m}_a^k &= \boldsymbol{m}_f^k + F_k(\boldsymbol{z}^k - H_k\boldsymbol{m}_f^k), \\ P_k^a &= (I - K_k H_k)P_k^f. \end{aligned} \tag{11.8}$$

In other words, under the Gaussian assumption, the density is updated only through the mean and the covariance. A similar interpretation also holds for the EKF, as soon as a Gaussian approximation of the densities is considered, and the evolution of these densities is taken into account. In this respect, Bayesian filtering can be seen as a generalization of deterministic filters, such as the KF, EKF and UKF introduced in Section 10.2.

As already remarked in Section 10.2.2, when the evolution model is fully non-linear, the EKF, which can be seen as a particular instance of an approximate Gaussian filter, may perform badly: this can be explained by considering that the push-forward of the previous state estimate (which has a Gaussian distribution) by a non-linear map is poorly approximated by a Gaussian distribution. To avoid the linearization of the evolution and the observation models, one can rely on Monte Carlo methods to simulate the distributions by random samples, similarly to what has been done in the static case. This strategy yields the so-called *particle filters* (also referred to as *sequential Monte Carlo methods*), now very popular for complex backward UQ problems.

The goal of a particle filter is to produce an ensemble $\{\boldsymbol{x}_1^{k|k}, \ldots, \boldsymbol{x}_{N_e}^{k|k}\}$ of N_e *particles* sequentially, that is, a random sample distributed according to the conditional probability distribution $\pi(\boldsymbol{x}^k \mid \boldsymbol{D}^k)$. The ensemble Kalman filter (EnKF), introduced by Evensen (1994, 2003), is a particle filter exploiting the idea of approximating the means and the covariances of the current estimate involved in the Kalman filter prediction-correction strategy by a set of particles sampled from the distribution. Unlike the

Kalman filter itself, we evaluate the error covariance predictions and corrections by the ensemble covariance matrices around the corresponding ensemble mean, instead of classical covariance equations (10.15b–10.15e) given in the KF algorithm. The covariance matrices of the state vector $\boldsymbol{X}$ need not be evolved, thus eliminating the costs associated with storing, multiplying and inverting the matrices appearing in the equations (10.15b–10.15e).

The ensemble is initialized by drawing N_e independent particles from, say, a Gaussian distribution with mean $\boldsymbol{m}^0$ and covariance P_0.

- At each *prediction step*, each particle is evolved using the Kalman filter prediction step,

$$\boldsymbol{x}_e^{k|k-1} = A_{k-1|k}\boldsymbol{x}_e^{k-1|k-1} + \boldsymbol{v}^{k-1}$$

  if the system is linear, or

$$\boldsymbol{x}_e^{k|k-1} = \boldsymbol{f}^k(\boldsymbol{x}_e^{k-1|k-1}, \boldsymbol{v}^{k-1})$$

  if the system is non-linear.

- At each *correction step*, the observation $\boldsymbol{z}^k$ is replicated N_e times, obtaining

$$\boldsymbol{d}_e^k = \boldsymbol{z}^k + \boldsymbol{\eta}_e^k, \quad \boldsymbol{\eta}_e^k \sim N(\boldsymbol{0}, R_k).$$

  Then, the empirical mean

$$\bar{\boldsymbol{x}}_e^{k|k-1} = \frac{1}{N_e}\sum_{e=1}^{N_e} \boldsymbol{x}_e^{k|k-1}$$

  and the empirical covariance

$$C_{k|k-1}^E = \frac{1}{N_e - 1}\sum_{e=1}^{N_e} \big(\boldsymbol{x}_e^{k|k-1} - \bar{\boldsymbol{x}}_e^{k|k-1}\big)\big(\boldsymbol{x}_e^{k|k-1} - \bar{\boldsymbol{x}}_e^{k|k-1}\big)^T$$

  of the particles set $\{\boldsymbol{x}_e^{k|k-1}\}_{e=1}^{N_e}$ are computed. The exact Kalman gain is then approximated by

$$K_k^E = C_{k|k-1}^E H_k^T (H_k C_{k|k-1}^E H_k^T + R_k)^{-1}$$

  and, finally, the state correction is obtained by applying the formula (10.15d) to each particle, that is,

$$\boldsymbol{x}_e^{k|k} = \boldsymbol{x}_e^{k|k-1} + K_k^E(\boldsymbol{d}_e^k - H_k\boldsymbol{x}_e^{k|k-1}).$$

Several alternative implementations can be found, for example in Evensen (1994, 2009); for more on the ensemble Kalman filter, see also Iglesias, Law and Stuart (2013), Kelly, Law and Stuart (2014), Ernst, Sprungk and Starkloff (2014) and Houtemaker and Zhang (2016).

If the backward UQ problem also involves random inputs (as in the case of model parameters $\boldsymbol{\theta}$, which can also be time-varying), the problem of state estimation and parameter estimation simultaneously arises. Generally speaking, there is no unique optimal solution for this problem. Similarly to the *state augmentation* technique presented in Section 10.2, a possible way to face this problem is to treat the unknown parameters $\boldsymbol{\theta}$ as part of the state, and use conventional filtering technique to infer the parameter and state simultaneously. This strategy goes by the name of joint estimation: see *e.g.* Moradkhani, Sorooshian, Gupta and Houser (2005), Ching, Beck and Porter (2006) and Evensen (2009) for more details. An example related to cardiovascular modelling where an EnKF can be exploited to deal with state and parameter estimation is addressed in Section 11.3.2.

11.3. Applications to cardiovascular modelling

Until recently, UQ has not been a priority for cardiovascular modelling. Today, a growing number of works focus on both forward and inverse UQ problems, taking into account uncertainties related to (i) measurement errors in experimental data (also referred to as observational uncertainty), (ii) model parameters, which may result from observational uncertainty as well as from variability, or lack of information, (iii) boundary and/or initial conditions and, possibly, (iv) the computational model itself, because of model limitations (model uncertainty) or systematic approximation errors (which can be seen as a form of *epistemic uncertainty*) introduced, for example, when the original high-fidelity model is replaced by a cheaper surrogate model or a reduced-order model. The list of contributions we mention is inevitably incomplete.

The effect of uncertain parameters in one-dimensional models of the arterial network was first considered by Xiu and Sherwin (2007), who exploited a high-order stochastic collocation method based on the generalized polynomial chaos expansion, combined with a discontinuous Galerkin spectral/hp element discretization in physical space, in order to analyse the effects of parametric uncertainties related to material properties and the initial cross-sectional area of the arterial vessel in pulse wave propagation, in a network of up to 37 vessels connected via 16 bifurcations. Chen, Quarteroni and Rozza (2013) have considered a more general setting, taking into account a wider network and many more sources of parametric uncertainties, including parameter-dependent boundary conditions in each distal boundary site and geometric parameters describing the cross-section area in each arterial segment. Similar problems, involving both sensitivity analysis and uncertainty propagation for networks built over patient-specific datasets, can be found in Leguy *et al.* (2011), Huberts *et al.* (2013*a*, 2013*b*) and Brault, Dumas and Lucor (2017).

The use of an adaptive stochastic collocation framework to deal with UQ in haemodynamic simulations has been considered by Sankaran and Marsden (2011), who extract relevant haemodynamic features – and quantify their uncertainty – for two idealized problems, namely an abdominal aortic aneurysm and a carotid artery bifurcation, and for a patient-specific problem, a Fontan procedure for congenital heart defects. In the former case, for instance, the radius of the abdominal aortic aneurysm, the radius and inflow velocity of the carotid artery bifurcation, and the flow split of the left and right pulmonary arteries are treated as random variables, following either Gaussian or uniform distributions, to account for the uncertainty impact on blood flows described by three-dimensional Navier–Stokes equations with rigid walls in small arterial portions. More detailed versions of this methodology have recently been exploited by Sankaran, Grady and Taylor (2015) and Sankaran, Kim, Choi and Taylor (2016) to analyse the impact of geometric uncertainties (*i.e.* anatomic uncertainties resulting in errors in the reconstructed geometry) and physiological uncertainties (yielding errors in boundary conditions or blood viscosity) for the blood flow and pressures in the coronary arteries.

The problem of calibrating outflow boundary conditions of blood flow simulations in truncated arterial domains has been cast in a Bayesian framework by D'Elia and Veneziani (2013) and Perdikaris and Karniadakis (2015), with the goal of quantifying the uncertainty affecting velocity and flow-related variables of interest, all treated as random variables.

A complete uncertainty propagation pipeline from clinical data to computational results has been considered by Schiavazzi *et al.* (2016), focusing on an example of single ventricle palliation surgery. After determining the probability density functions of right pulmonary artery flow split and average pulmonary pressures from clinical measurements, Bayesian parameter estimation is carried out in order to characterize the distributions of boundary conditions yielding the observed flow splits and average pressure distributions. Then, uncertainties in the boundary conditions are propagated to simulation predictions by employing sparse grid stochastic collocation to statistically characterize model predictions of post-operative haemodynamics in models with and without pulmonary artery stenosis, in order to quantify the statistical variability in virtual surgery predictions. Finally, the impact of uncertainty on the optimal design of idealized bypass graft models has been considered by Sankaran and Marsden (2010) and Lassila *et al.* (2013*a*, 2013*b*), for example.

Regarding cardiac electrophysiology, a systematic application of UQ techniques to the forward problem of electrocardiography (*i.e.* the characterization of the torso potential given the electrical conductivity inside the heart and the torso) has been considered by Geneser, Kirby and MacLeod (2008), focusing on a simplified 2D configuration representing a geometric model of a slice through the human thorax. The forward UQ propagation problem yielding the standard deviation of the resulting stochastic torso potentials depending on several uncertainties affecting, for example, the electrical conductivities of the organs and the representation of cardiac sources, has been tackled by a stochastic Galerkin method based on the generalized polynomial chaos expansion. Model personalization in cardiac electrophysiology has been considered by Konukoglu *et al.* (2011), who applied an efficient Bayesian method exploiting polynomial chaos and compressed sensing to an eikonal-diffusion model involving a large (> 10) number of parameters, by integrating uncertainty on data and parameters.

Johnstone *et al.* (2016) have addressed a UQ framework for cardiac action potential models, focusing on (i) the inverse UQ problem of inferring the maximal conductance of ionic channels from noisy experimental recordings, and (ii) the forward UQ problem of propagating the uncertainty in maximal ion channel conductances to suitable outputs of interest, such as the action potential duration. To solve the inverse UQ problem a Bayesian framework, such as that described in Section 11.2, has been successfully employed, whereas the UQ propagation problem has been tackled by means of a Monte Carlo approach exploiting suitable surrogate models, such as Gaussian process emulators, to speed up the evaluation of the system model under analysis. A detailed analysis of the mechanisms underlying physiological variability in cardiac electrophysiology and pro-arrhythmic risks under a variety of conditions can be found in Muszkiewicz *et al.* (2016).

Concerning heart electromechanics, Osnes and Sundnes (2012) have addressed the quantification of the effect of uncertainties in the fibre orientation and the elasticity parameters included in the strain energy function on global output quantities such as the increase in cavity volume, the elongation of the ventricle, the increase in the inner radius, the decrease in wall thickness, and the rotation at the apex during the passive filling phase. Wallman, Smith and Rodriguez (2014) have estimated cardiac tissue properties by integrating structural information with electrophysiological data from electro-anatomical mapping systems, for example. By means of Bayesian inference tools, they obtained a simultaneous description of clinically relevant electrophysiological conduction properties and their associated uncertainty for various levels of noise, together with suitable design strategies to optimize the location and number of measurements required to maximize information and reduce uncertainty. A relevant application of UQ techniques to nonlinear biomechanics has been considered by Biehler, Gee and Wall (2015), where parametric uncertainties related to the constitutive law for the artery wall of an AAA are modelled as random fields. In particular, Biehler *et al.* use a lognormal three-dimensional random field to describe the inter- and intra-patient variations of one constitutive parameter of a hyperelastic constitutive model, and sampling-based approaches such as Monte Carlo to solve the resulting stochastic mechanical problem. A multi-fidelity Bayesian framework incorporating information from different low-fidelity models has been developed to speed up the intensive approximation of the resulting problem, following some general ideas reported by Kennedy and O'Hagan (2000) and Koutsourelakis (2009). Employment of surrogate models or more reliable reduced-order models (for more on this subject, see Section 12) to speed up the numerical solution of direct and inverse UQ problems is indeed a powerful alternative to stochastic collocation and stochastic Galerkin methods. However, they have only been partially touched on in the case of simple problems, for example in the works of Cui, Marzouk and Willcox (2015), Chen and Schwab (2015), Dihlmann and Haasdonk (2016), Manzoni, Pagani and Lassila (2016) and Pagani *et al.* (2016).

11.3.1. Backward uncertainty propagation in a simplified blood flow model
We want to compare the solution of a parameter estimation problem in a variational framework and in a Bayesian setting, on a simplified blood flow problem. We consider a two-dimensional section of a carotid artery bifurcation where the diameters d_c, d_b of the common carotid artery (CCA) at the bifurcation and of the mid-sinus level of the internal carotid artery (ICA) are the input parameters, assumed to be uncertain. In this idealized setting, our goal is to identify $\boldsymbol{\theta} = (d_c, d_b)^T$ from the observation of the mean pressure drop

$$\boldsymbol{z} = \frac{1}{|\Gamma_{in}|}\int_{\Gamma_{in}} p \,\mathrm{d}\gamma - \frac{1}{|\Gamma_{ICA}|}\int_{\Gamma_{ICA}} p \,\mathrm{d}\gamma$$

between the internal carotid outflow Γ_{ICA} and the inflow Γ_{in}: see Figure 11.2(a). For simplicity, a steady incompressible Navier–Stokes model is employed to model the blood flow, although the same approach concerning the solution of the inverse UQ problem can be applied to the case of an unsteady fluid model, as soon as a peak (or time average) pressure drop is evaluated. Although presented here in a very simplified way for the sake of illustration, the problem of recovering information about vessel features by evaluating physical indices related to flow variables is of general interest for diagnostic purposes too. For instance, fractional flow reserve is a procedure exploited in coronary catheterization to measure pressure differences across a coronary artery stenosis in order to determine the degree of stenosis.

The parameter estimation problem is first solved by minimizing a least-squares functional for different observations of the pressure drop, $z = -1400$ and $z = -2200$, by assuming 5% relative additive noise in the measurements. The results of the inverse identification problem are represented in Figure 11.2(b) for 100 realizations of random noise in both cases. The recovered diameter values are shown to be quite sensitive to small noise variations. This is due to the fact that several geometric configurations – in

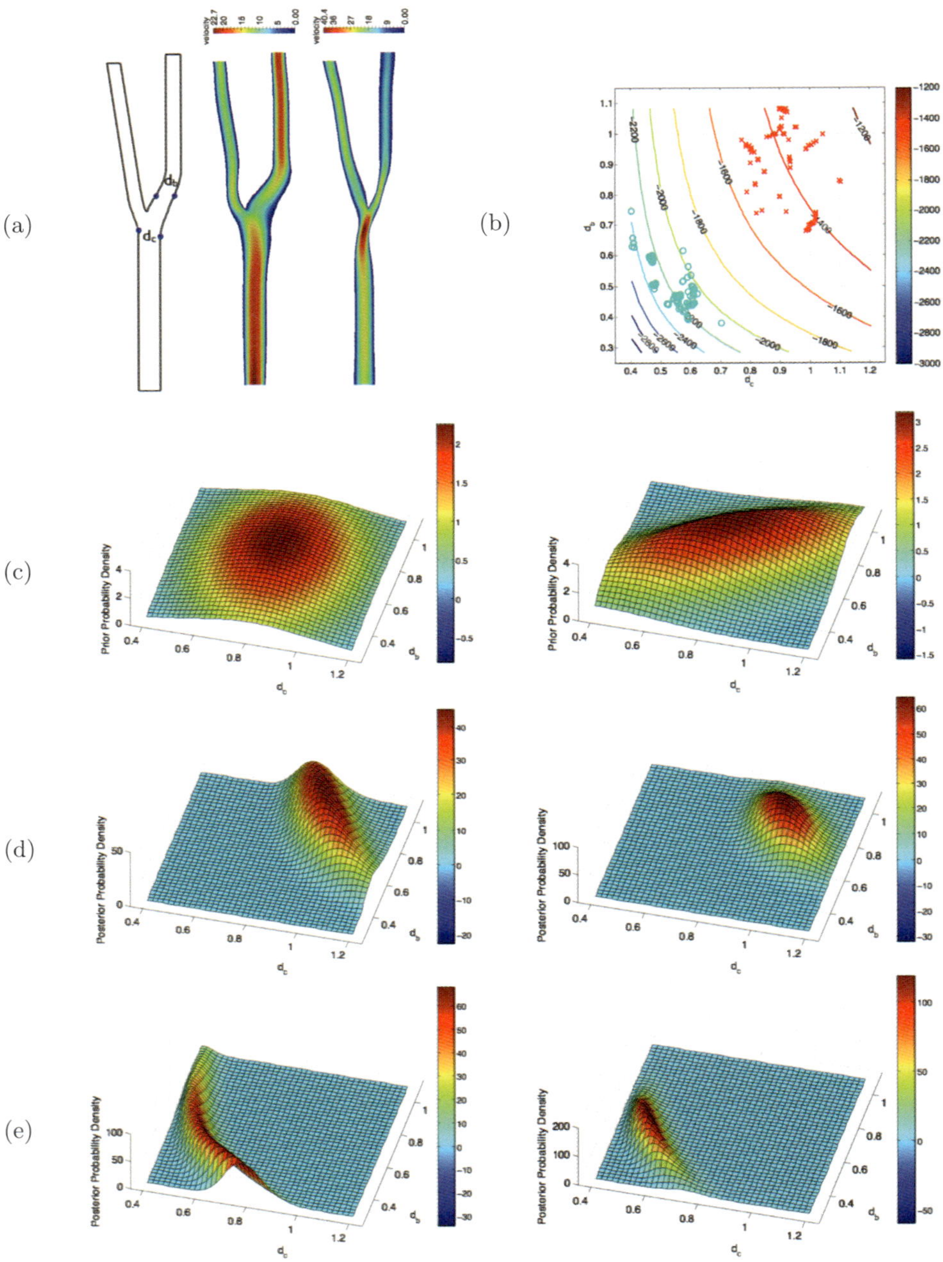

Figure 11.2. (a) Velocity profiles (cm s^{-1}) for different carotid bifurcations parametrized with respect to the diameters d_c, d_b. (b) Variational parameter estimation and isolines of the pressure drop. (c) Two different choices of the prior distribution on diameters $\boldsymbol{\theta} = (d_c, d_b)^T$. (d, e) Results of the backward UQ problem obtained with the priors in (c) with observed pressure drop $\boldsymbol{z}_{obs} = -1400$ and $\boldsymbol{z}_{obs} = -2200$ (dyn cm^{-2}), respectively. Numerical results were obtained using the MATLAB finite element library MLife.

terms of diameters (d_c, d_b) – may correspond to the same output observation. The backward UQ propagation problem is then solved in a Bayesian setting, characterizing the posterior PDF of $\boldsymbol{\theta} = (d_c, d_b)$ starting from a Gaussian prior and assuming a Gaussian model to describe experimental noise. In particular, we can assume that the two diameters are either *a priori* independent (case (c), left), or correlated (case (c), right)). The two corresponding posterior distributions, obtained for the observations $z = -1400$ and $z = -2200$ of the pressure drop, are shown in Figure 11.2(d, e).

11.3.2. Backward uncertainty propagation in cardiac electrophysiology

We finally consider the problem of identifying the size and the position of an ischaemic region in the myocardial tissue, as a remarkable instance of a backward UQ problem dealing with cardiovascular applications. Such a problem case can be cast in the form of a joint state/parameter estimation problem, and can be tackled efficiently by an EnKF technique.

The state evolution equation is given by the (space and time discretization of the) monodomain equation for the trans-membrane potential V_m, coupled with an Aliev–Panfilov cellular model involving a single gating variable w; the state variable is thus $\boldsymbol{u} = (V_m, w)^T$. Similarly to the example in Section 11.1, to represent lack of conductivity in the tissue we introduce the relative conductivity $\sigma = \sigma(\boldsymbol{x}; \boldsymbol{\theta}) \in [0, 1]$; $\sigma = 1$ and $\sigma = 0$ correspond to proper electrical conduction or total lack of conductivity, respectively.

Since the position and the extension of the ischaemic region is unknown, $\sigma(\boldsymbol{x}; \boldsymbol{\theta})$ is a random field. To make its representation low-dimensional, we parametrize the field $\sigma(\boldsymbol{x}; \boldsymbol{\theta})$ in terms of $p = 20$ random inputs $\boldsymbol{\theta} = (\theta_1, \ldots, \theta_{20})^T$ entering into a linear combination of $p = 20$ radial basis functions (RBFs), around p given centres $\{\boldsymbol{x}_i\}_{i=1}^{20}$,

$$\sigma(\boldsymbol{x}; \boldsymbol{\theta}) = \frac{1}{\nu(\boldsymbol{x})} \sum_{i=1}^{p} \theta_i \exp\left(-\frac{\|\boldsymbol{x} - \boldsymbol{x}_i\|^2}{\lambda^2} \right), \quad \nu(\boldsymbol{x}) = \sum_{i=1}^{p} \exp\left(-\frac{\|\boldsymbol{x} - \boldsymbol{x}_i\|^2}{\lambda^2} \right),$$

where $\nu(\boldsymbol{x})$ is a normalization factor so that $\sigma(\boldsymbol{x}; \boldsymbol{\theta}) \in [0, 1]$. We consider a patient-specific left-ventricle geometry, and assume that a dataset $\boldsymbol{z}^1, \ldots, \boldsymbol{z}^K$ is acquired sequentially over a time interval. In particular, at each time τ^k, $k = 1, \ldots, K$, data are given by the trans-membrane potential values computed at a set of nine points located on the endocardium, contaminated by Gaussian noise with zero mean and known covariance, to simulate the effect of experimental noise while evaluating the so-called *simultaneous endocardial mapping* in the human left ventricle; these invasive measurements are usually obtained using a non-contact catheter (see *e.g.* Schilling, Peters and Davies 1998, Álvarez *et al.* 2012). For the sake of the computational experiment, we assume that the data $\boldsymbol{z}^k$, $k = 1, \ldots, K$, are generated by the monodomain model with relative conductivity $\sigma^* = \sigma^*(\boldsymbol{x}; \boldsymbol{\theta}^*)$ for a particular choice $\boldsymbol{\theta} = \boldsymbol{\theta}^*$ of the input vector, that is,

$$\boldsymbol{z}^k = \boldsymbol{h}(\boldsymbol{u}(\boldsymbol{\theta}^*)) + \boldsymbol{\eta}^k, \quad \boldsymbol{\eta}^k \sim N(\mathbf{0}, R_k).$$

The goal is thus to recover the (posterior) PDF of $\boldsymbol{\theta}^K \mid \boldsymbol{D}^k$ given the observations $\boldsymbol{D}^k = (\boldsymbol{z}^1, \dots, \boldsymbol{z}^k)$, evaluate the conditional mean $\boldsymbol{\theta}^K_{CM} = \mathbb{E}[\boldsymbol{\theta}^K \mid \boldsymbol{D}^k]$, and finally estimate the conductivity field as $\sigma(\boldsymbol{x}; \boldsymbol{\theta}^K_{CM})$.

Starting from a prior distribution for the input vector $\boldsymbol{\theta}^0 \sim \mathcal{U}([0,1]^{20})$, reflecting the lack of information about the possible presence and position of the ischaemic region, we rely on the EnKF to produce sequentially an ensemble $\{\boldsymbol{\theta}^k_e\}_{e=1}^{N_e}$ of N_e particles distributed according to the conditional distribution $\pi(\boldsymbol{\theta}^k \mid \boldsymbol{D}^k)$, and the associated ensemble of N_e states $\{\boldsymbol{u}^{k|k}(\boldsymbol{\theta}^k_e)\}_{e=1}^{N_e}$. Hence, starting from the initial ensemble $\{\mathcal{P}^{(0)}, \mathcal{U}^{(0)}\}$ sampled from the prior distribution, the *prediction-analysis* procedure of the EnKF is given by the following two-stage recursion.

(1) At each *prediction step*, compute the solution $\boldsymbol{u}^{k|k-1}(\boldsymbol{\theta}^{k-1}_e)$ of the state system over $[\tau^{k-1}, \tau^k)$ with initial datum $\boldsymbol{u}^{k|k-1}(\boldsymbol{\theta}^{k-1}_e)$.

(2) At each *correction step*, the observation $\boldsymbol{z}^k$ is replicated N_e times, obtaining

$$\boldsymbol{d}^k_e = \boldsymbol{z}^k + \boldsymbol{\eta}^k_e, \quad \boldsymbol{\eta}^k_e \sim N(\mathbf{0}, R_k).$$

Then, compute the sample means

$$\begin{aligned}
\overline{\boldsymbol{u}}^{k|k-1} &= \frac{1}{N_e}\sum_{e=1}^{N_e} \boldsymbol{u}^{k|k-1}(\boldsymbol{\theta}^{k-1}_e), \\
\overline{\boldsymbol{s}}^{k|k-1}_e &= \frac{1}{N_e}\sum_{e=1}^{N_e} \boldsymbol{h}(\boldsymbol{u}^{k|k-1}(\boldsymbol{\theta}^{k-1}_e)), \\
\overline{\boldsymbol{\theta}}^{k-1}_e &= \frac{1}{N_e}\sum_{e=1}^{N_e} \boldsymbol{\theta}^{k-1}_e,
\end{aligned} \tag{11.9}$$

and the sample covariances

$$\mathbf{C}^k_{ss} = \frac{1}{N_e - 1}\sum_{e=1}^{N_e} \boldsymbol{v}_e \boldsymbol{v}_e^T \in \mathbb{R}^{n_z \times n_z}, \tag{11.10}$$

$$\mathbf{C}^k_{\theta s} = \frac{1}{N_e - 1}\sum_{e=1}^{N_e} (\boldsymbol{\theta}^{k-1}_e - \overline{\boldsymbol{\theta}}^{k-1}_e)\boldsymbol{v}_e^T \in \mathbb{R}^{p \times n_z}, \tag{11.11}$$

$$\mathbf{C}^k_{us} = \frac{1}{N_e - 1}\sum_{e=1}^{N_e} (\boldsymbol{u}^{k|k-1}(\boldsymbol{\theta}^{k-1}_e) - \overline{\boldsymbol{u}}^{k|k-1})\boldsymbol{v}_e^T \in \mathbb{R}^{n_x \times n_z}, \tag{11.12}$$

where $\boldsymbol{v}_e = \boldsymbol{h}(\boldsymbol{u}^{k|k-1}(\boldsymbol{\theta}^{k-1}_e))$. Finally, the state/parameter ensembles

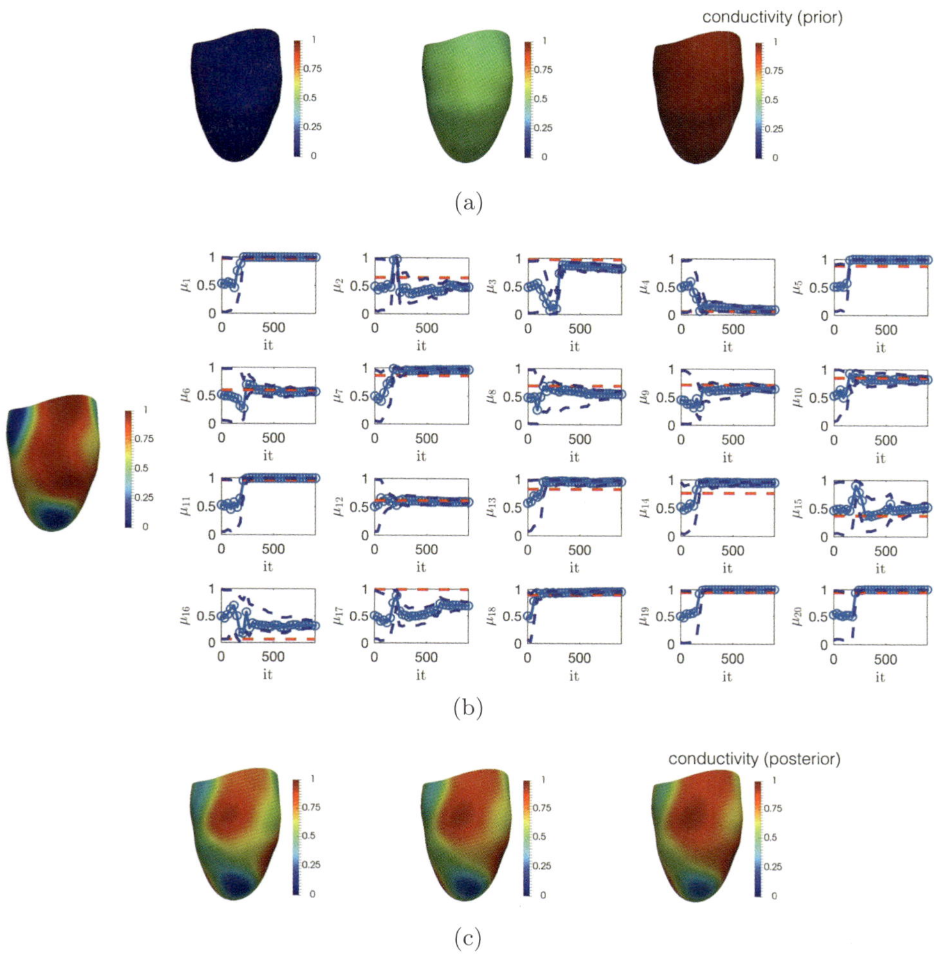

Figure 11.3. (a) 5% quantile, mean and 95% quantile of the uniform prior distribution of $\boldsymbol{\theta}^0$. (b) Identification of $p = 20$ parameters via the EnKF algorithm. Reference values θ_i^*, $i = 1, \ldots, 20$, estimates $(\bar{\theta}_e^k)_i$ and confidence intervals are reported in dotted red, solid blue and dotted blue lines, respectively. (c) 5% quantile, conditional mean and 95% quantile of the posterior distribution of $\boldsymbol{\theta}^K$. Numerical results were obtained using the finite element library redbKIT v2.1 (github.com/redbKIT/redbKIT/releases).

are updated via the *Kalman formula*

$$\begin{bmatrix} \boldsymbol{\theta}_e^k \\ \boldsymbol{u}^{k|k}(\boldsymbol{\theta}_e^k) \end{bmatrix} = \begin{bmatrix} \boldsymbol{\theta}_e^{k-1} \\ \boldsymbol{u}^k(\boldsymbol{\theta}_e^{k-1}) \end{bmatrix} + \begin{bmatrix} \mathbf{C}_{\theta z}^k \\ \mathbf{C}_{uz}^k \end{bmatrix} (R_k + \mathbf{C}_{zz}^k)^{-1} (\boldsymbol{d}_e^k - \boldsymbol{h}(\boldsymbol{u}^{k|k-1}(\boldsymbol{\theta}_e^{k-1}))), \tag{11.13}$$

for $e = 1, \ldots, N_e$.

The numerical results obtained by this procedure are shown in Figure 11.3.

The data were generated by the monodomain model with the relative conductivity field $\sigma^* = \sigma^*(\boldsymbol{x};\boldsymbol{\theta}^*)$ shown in Figure 11.3(a); a uniform prior distribution $\boldsymbol{\theta}^0 \sim \mathcal{U}([0,1]^{20})$ is assumed. After executing the EnKF algorithm, we obtain the posterior distribution whose mean and 5%, 95% quantiles are shown in Figure 11.3(c). The uncertainty in the input parameters is greatly reduced, while the conditional mean of the posterior distribution correctly approximates the realization of the random field reported on the left, by which data exploited for the sake of parameter estimation have been generated.

12. Reduced-order modelling

As seen in Sections 9–11, several numerical strategies used to tackle PDE-constrained optimization, parameter estimation and uncertainty quantification problems arising in cardiovascular modelling involve the approximation of PDE systems for several input parameter values. On the other hand, the repeated evaluation of input/output maps to characterize different scenarios requires many queries to numerical models too. Further, if quantitative outputs are meant to support clinicians and medical doctors in their decisions, each new numerical simulation should be carried out very rapidly (order of minutes, say) on deployed platforms rather than on huge parallel hardware architectures, possibly requiring limited data storage and memory capacity. Meeting all these requirements is a challenging task, with the result that traditional *high-fidelity*, or *full-order*, techniques (*e.g.* the finite element method) are ill-suited, despite the constant growth of computer resources available.

Reduced-order models (ROMs) are emerging methodologies aimed at reducing the computational complexity and costs entailed by the repeated solution of PDE problems (Antoulas 2005, Benner, Gugercin and Willcox 2015, Quarteroni and Rozza 2014). In the case of parametrized PDEs (*i.e.* PDEs depending on a vector of parameters $\boldsymbol{\mu} \in \mathcal{P} \subset \mathbb{R}^p$), the *reduced basis* (RB) method is a remarkable example of a ROM that enables dramatic reduction of the dimension of the discrete problems arising from numerical approximation – from millions to hundreds, or thousands at most, of variables.

Here, $\mathbf{y}(t;\boldsymbol{\mu})$ denote the solution of a time-dependent non-linear problem ($\boldsymbol{\mu} \in \mathcal{P} \subset \mathbb{R}^p$ denotes a set of input parameters) of the form

$$\begin{aligned} M(\boldsymbol{\mu})\frac{\mathrm{d}\mathbf{y}(t;\boldsymbol{\mu})}{\mathrm{d}t} + A(\boldsymbol{\mu})\mathbf{y}(t;\boldsymbol{\mu}) + F(\mathbf{y}(t;\boldsymbol{\mu})) = \mathbf{f}(t;\boldsymbol{\mu}), \quad t \in (0,T], \\ \mathbf{y}(0;\boldsymbol{\mu}) = \mathbf{y}_0(\boldsymbol{\mu}), \end{aligned} \tag{12.1}$$

stemming from the (*e.g.* finite element) discretization of a parametrized PDE, where $A(\boldsymbol{\mu}) \in \mathbb{R}^{N_h \times N_h}$, $M(\boldsymbol{\mu}) \in \mathbb{R}^{N_h \times N_h}$ and $F(\boldsymbol{\mu}) : \mathbb{R}^{N_h} \to \mathbb{R}^{N_h}$ are three operators corresponding to the linear, mass and non-linear terms of the PDE.

The dimensional reduction is made possible by exploiting the parametric dependence of the solution manifold, that is, the set $\mathcal{M}_h = \{\mathbf{y}(t;\boldsymbol{\mu}) : t \in (0,T], \boldsymbol{\mu} \in \mathcal{P}\}$, thanks to the evaluation of a database of solutions, or *snapshots*, for selected parameter values, and to a (Petrov–) Galerkin projection onto the RB space spanned by a set of RB functions. In the case of a stationary problem, these latter are the snapshots themselves if a *greedy* algorithm is used, or the first singular vectors of the snapshot matrix if *proper orthogonal decomposition* (POD) is used: see *e.g.* Quarteroni, Manzoni and Negri (2016), Hesthaven, Rozza and Stamm (2016) and Cohen and DeVore (2015). For time-dependent problems, the parameter space can still be sampled by relying on one of the two techniques mentioned, whereas POD is usually exploited to reduce trajectories of the system over the time interval.

Hence, the RB method seeks an approximation $V_N\mathbf{y}_N(t;\boldsymbol{\mu}) \approx \mathbf{y}(t;\boldsymbol{\mu})$ to the solution of (12.1), where $\mathbf{y}_N(t;\boldsymbol{\mu})$ is the reduced state vector and $V_N = [\mathbf{v}_1,\ldots,\mathbf{v}_N] \in \mathbb{R}^{N_h \times N}$ is a matrix stacking the RB functions by columns. One possible RB approximation of (12.1) might be

$$\begin{aligned} M_N(\boldsymbol{\mu})\frac{\mathrm{d}\mathbf{y}_N(t;\boldsymbol{\mu})}{\mathrm{d}t} &+ A_N(\boldsymbol{\mu})\mathbf{y}_N(t;\boldsymbol{\mu}) \\ &+V_N^T\mathbf{F}(V_N\mathbf{y}_N(t;\boldsymbol{\mu})) = \mathbf{f}_N(t;\boldsymbol{\mu}), \quad t \in (0,T] \\ &\mathbf{y}_N(0;\boldsymbol{\mu}) = \mathbf{y}_{N,0}(\boldsymbol{\mu}), \end{aligned} \tag{12.2}$$

where

$$A_N(\boldsymbol{\mu}) = V_N^T A(\boldsymbol{\mu})V_N, \quad M_N(\boldsymbol{\mu}) = V_N^T M(\boldsymbol{\mu})V_N, \quad \mathbf{f}_N(t;\boldsymbol{\mu}) = V_N^T\mathbf{f}(t;\boldsymbol{\mu}).$$

The arrays appearing in (12.2) can be efficiently assembled in a rapid online phase by combining parameter-independent quantities stored during a more expensive offline phase. Suitable *hyper-reduction* techniques are instead required to manage non-linear terms, in order to make their assembly independent of the dimension N_h of the high-fidelity problem.

We emphasize that a RB method requires the solution of some full-order – and therefore very expensive – discrete equations. The key idea is that these demanding calculations can be done offline, before the optimization with respect to the control parameters, or the parameter estimation, is attempted. In fact, the cost of each optimization step performed online is much smaller than that involving the full-order state approximation; similarly, computing the evolution of a particle set in the ensemble Kalman filter algorithm by querying the ROM can thus be performed in a substantially inexpensive way.

Describing the mathematical principles and numerical algorithms underlying the RB method would take us a long way; interested readers can find a detailed presentation in Quarteroni *et al.* (2016), for example. We limit ourselves to citing some notable applications of the RB method to problems of interest in cardiovascular modelling (this list is of course incomplete). Parametrized blood flows in idealized cardiovascular geometries have been considered by Lassila, Quarteroni and Rozza (2012), Manzoni, Quarteroni and Rozza (2012*a*), Ballarin *et al.* (2016), Ballarin *et al.* (2017), Colciago *et al.* (2014) and Negri (2016), taking into account more complex (and computationally challenging) patient-specific configurations; in all these cases, solutions of Navier–Stokes equations are computed with respect to inflow and/or geometric parameters. Applications to PDE-constrained optimization problems arising in the context of optimal design of prosthetic devices can be found, for example, in Manzoni *et al.* (2012*b*) and Lassila *et al.* (2013*a*). A reduced-order model based on POD is proposed for the bidomain equations of cardiac electrophysiology in Boulakia, Schenone and Gerbeau (2012), yielding the efficient approximation of a restitution curve and the estimation of ionic parameters and infarction locations from synthetic electrocardiograms with an evolutionary algorithm. More recently, Gerbeau, Lombardi and Schenone (2015) have proposed alternative options in this respect. Applications of POD to parameter estimation problems with sequential filtering techniques can be found in Chapelle, Gariah, Moireau and Sainte-Marie (2013*b*), and an application of the RB method to the computational speed-up of Bayesian inverse problems related to blood flows modelling has been reported by Lassila *et al.* (2013*b*) and Manzoni *et al.* (2014).

We emphasize that while physical coefficients, boundary and/or initial conditions, as well as source terms, can usually be described in terms of input parameters in a straightforward way, the task of dealing with geometries of varying shape is much more involved. In this latter case, additional techniques providing flexible descriptions of complex shapes are required, possibly involving few parameters. Notable examples are given by *volume-based* representations, which operate on a control volume and define parametric maps by introducing a set of control points over the control volume; control point displacements, actually inducing a shape deformation, can thus be treated as input parameters. Within this class, free-form deformation techniques and interpolants constructed over a family of radial basis functions have been successfully employed; for further details see Manzoni *et al.* (2012*a*, 2012*b*) and Ballarin *et al.* (2016). We also point out that the need to derive flexible and low-dimensional parametrizations is not confined to the realm of RB methods for parametrized PDEs. Indeed, it also arises when dealing with random inputs or fields; in this latter case, input uncertainties are usually parametrized with respect to a finite number of random variables, corresponding to the retained terms after truncating Karhunen–Loève or polynomial chaos expansions.

The analysis, development and application of reduced-order modelling techniques is a very active field in the numerical approximation of PDEs. Without any doubt, cardiovascular applications are one of the most relevant testing environments.

Disclaimers

Despite being 590 pages long,[6] several topics related to the cardiovascular system have not been addressed in this review paper. Among others, we mention the *venous system* (essential if one wants to consider a closed-loop model of the cardiovascular system, and playing a crucial role in some specific pathologies: see *e.g.* Toro 2016), the *metabolic system* (D'Angelo 2007), the *respiratory system* (Maury 2013, Wall, Wiechert, Comerford and Rausch 2010, Trenhago *et al.* 2016), the *cerebro-spinal fluid circulation* (Fin and Grebe 2003), the *nervous system* (Liang and Liu 2006), and the *lymphatic system* (Margaris and Black 2012). For some of them (*e.g.* the venous and respiratory systems) research has made remarkable progress in recent years. Nonetheless, the mathematical investigation of these systems is still in its infancy; in particular, their coupling with the cardiovascular system is almost absent. Many research avenues are open to the contribution of both pure and applied mathematicians, with the dream of enabling mathematical achievements to play a decisive role in everyday clinical practice.

Acknowledgements

The authors would like to thank P. Biondetti, M. Domanin, L. Forzenigo (Fondazione IRCSS Ca' Granda, Ospedale Maggiore Policlinico, Milan, Italy), S. Ippolito, R. Scrofani (Ospedale Sacco, Milan, Italy), D. Catanzariti, M. Centonze (Ospedale S. Maria del Carmine, Rovereto (TN), Italy), for providing the radiological images; G. Aloe (Politecnico di Milano) for his help in preparing the figures; D. Bonomi, B. Guerciotti, R.M. Lancellotti, S. Pagani (Politecnico di Milano), D. Forti, F. Negri (EPFL, Lausanne, Switzerland), L. Barbarotta (Technische Universiteit Eindhoven), E. Faggiano (University of Pavia) for their help in preparing the plots of some numerical results; E. Faggiano for her fruitful suggestions. AQ acknowledges the project 'Integrative HPC Framework for Coupled Cardiac Simulations' (IFCCS) within the PASC (Platform for Advanced Scientific Computing) network 'Life Sciences Across Scales' and the Swiss National Supercomputing Centre (CSCS), project ID s635. CV was partially supported by INDAM-GNCS and by the Italian MIUR PRIN12 project 201289 A4XL. Last but not least, the authors warmly thank Glennis Starling for her fantastic editorial job.

[6] By slightly rephrasing Blaise Pascal's quotation, we can state that '*we were not good enough* to make it shorter.'

REFERENCES[7]

S. Abboud, O. Berenfeld and D. Sadeh (1991), 'Simulation of high-resolution QRS complex using a ventricular model with a fractal conduction system: Effects of ischemia on high-frequency QRS potentials', *Circ. Res.* **68**, 1751–1760.

S. Ahmed and D. Giddens (1984), 'Pulsatile poststenotic flow studies with laser Doppler anemometry', *J. Biomech.* **17**, 695–705.

I. Akkerman, Y. Bazilevs, V. Calo, T. Hughes and S. Hulshoff (2008), 'The role of continuity in residual-based variational multiscale modeling of turbulence', *Comput. Mech.* **41**, 371–378.

F. Alauzet, B. Fabrèges, M. Fernández and M. Landajuela (2016), 'Nitsche-XFEM for the coupling of an incompressible fluid with immersed thin-walled structures', *Comput. Methods Appl. Mech. Engrg* **301**, 300–335.

R. Aliev and A. Panfilov (1996), 'A simple two-variable model of cardiac excitation', *Chaos Solitons Fract.* **7**, 293–301.

A. Álvarez, F. Alonso-Atienza, J. Rojo-Álvarez, A. Garcia-Alberola and M. Moscoso (2012), 'Shape reconstruction of cardiac ischemia from non-contact intracardiac recordings: A model study', *Math. Comput. Model.* **55**, 1770–1781.

D. Ambrosi and S. Pezzuto (2012), 'Active stress vs. active strain in mechanobiology: Constitutive issues', *J. Elasticity* **2**, 199–212.

D. Ambrosi, G. Arioli, F. Nobile and A. Quarteroni (2011), 'Electromechanical coupling in cardiac dynamics: The active strain approach', *SIAM J. Appl. Math.* **2**, 605–621.

B. Andreianov, M. Bendahmane, A. Quarteroni and R. Ruiz-Baier (2015), 'Solvability analysis and numerical approximation of linearized cardiac electromechanics', *Math. Models Methods Appl. Sci.* **25**, 959–993.

L. Antiga, J. Peiró and D. Steinman (2009), From image data to computational domains. Chapter 4 in *Cardiovascular Mathematics* (L. Formaggia, A. Quarteroni and A. Veneziani, eds), Springer, pp. 123–175.

L. Antiga, M. Piccinelli, L. Botti, B. Ene-Iordache, A. Remuzzi and D. Steinman (2008), 'An image-based modeling framework for patient-specific computational hemodynamics', *Med. Biol. Engrg Comput.* **46**, 1097–1112.

A. Antoulas (2005), *Approximation of Large-Scale Dynamical Systems*, SIAM.

M. Asch, M. Bocquet and M. Nodet (2017), *Data Assimilation: Methods, Algorithms, and Applications*, SIAM.

M. Astorino and C. Grandmont (2009), 'Convergence analysis of a projection semi-implicit coupling scheme for fluid–structure interaction problems', *Numer. Math.* **45–46**, 3603–3612.

M. Astorino, F. Chouly and M. Fernández (2009), 'Robin based semi-implicit coupling in fluid–structure interaction: Stability analysis and numerics', *SIAM J. Sci. Comput.* **31**, 4041–4065.

M. Astorino, J.-F. Gerbeau, O. Pantz and K. Traoré (2010), 'Fluid–structure interaction and multi-body contact: Application to the aortic valves', *Comput. Methods Appl. Mech. Engrg* **116**, 721–767.

[7] The URLs cited in this work were correct at the time of going to press, but the publisher and the authors make no undertaking that the citations remain live or are accurate or appropriate.

M. Astorino, J. Hamers, C. Shadden and J.-F. Gerbeau (2012), 'A robust and efficient valve model based on resistive immersed surfaces', *Int. J. Numer. Methods Biomed. Engrg* **28**, 937–959.

C. Augustin, G. Holzapfel and O. Steinbach (2014), 'Classical and all-floating FETI methods for the simulation of arterial tissues', *Int. J. Numer. Methods Engrg* **99**, 290–312.

F. Auricchio, A. Ferrara and S. Morganti (2012), 'Comparison and critical analysis of invariant-based models with respect to their ability in fitting human aortic valve data', *Ann. Solid Struct. Mech.* **4**, 1–14.

F. Auricchio, F. Lefieux, A. Reali and A. Veneziani (2016), 'A locally anisotropic fluid–structure interaction remeshing strategy for thin structures with application to a hinged rigid leaflet', *Int. J. Numer. Methods Engrg* **107**, 155–180.

A. Avolio (1980), 'Multi-branched model of the human arterial system', *Med. Biol. Engrg Comput.* **18**, 709–718.

I. Babuška, F. Nobile and R. Tempone (2007), 'A stochastic collocation method for elliptic partial differential equations with random input data', *SIAM J. Numer. Anal.* **45**, 1005–1034.

I. Babuška, R. Tempone and G. Zouraris (2004), 'Galerkin finite element approximations of stochastic elliptic partial differential equations', *SIAM J. Numer. Anal.* **42**, 800–825.

S. Badia, F. Nobile and C. Vergara (2008*a*), 'Fluid–structure partitioned procedures based on Robin transmission conditions', *J. Comput. Phys.* **227**, 7027–7051.

S. Badia, F. Nobile and C. Vergara (2009), 'Robin–Robin preconditioned Krylov methods for fluid–structure interaction problems', *Comput. Methods Appl. Mech. Engrg* **198**, 2768–2784.

S. Badia, A. Quaini and A. Quarteroni (2008*b*), 'Modular vs. non-modular preconditioners for fluid–structure systems with large added-mass effect', *Comput. Methods Appl. Mech. Engrg* **197**, 4216–4232.

S. Badia, A. Quaini and A. Quarteroni (2008*c*), 'Splitting methods based on algebraic factorization for fluid–structure interaction', *SIAM J. Sci. Comput.* **30**, 1778–1805.

E. Bagci, Y. Vodovotz, T. Billiar, B. Ermentrout and I. Bahar (2008), 'Computational insights on the competing effects of nitric oxide in regulating apoptosis', *PLOS One* **3**, e2249.

J. Ball (1976), 'Convexity conditions and existence theorems in nonlinear elasticity', *Arch. Rat. Mech. Anal.* **63**, 337–403.

J. Ball (1977), Constitutive inequalities and existence theorems in nonlinear elastostatics. In *Nonlinear Analysis and Mechanics*, Vol. I (R. Knops, ed.), Pitman, pp. 187–241.

F. Ballarin, E. Faggiano, S. Ippolito, A. Manzoni, A. Quarteroni, G. Rozza and R. Scrofani (2016), 'Fast simulations of patient-specific haemodynamics of coronary artery bypass grafts based on a POD–Galerkin method and a vascular shape parametrization', *J. Comput. Phys.* **315**, 609–628.

F. Ballarin, E. Faggiano, A. Manzoni, A. Quarteroni, G. Rozza, S. Ippolito, C. Antona and R. Scrofani (2017), 'Numerical modeling of hemodynamics scenarios of patient-specific coronary artery bypass grafts', *Biomech. Model. Mechanobiol.* https://doi.org/10.1007/s10237-017-0893-7

D. Balzani, D. Brands, A. Klawonn, O. Rheinbach and J. Schroder (2010), 'On the mechanical modeling of anisotropic biological soft tissue and iterative parallel solution strategies', *Arch. Appl. Mech.* **80**, 479–488.

D. Balzani, P. Neff, J. Schroder and G. Holzapfel (2006), 'A polyconvex framework for soft biological tissues: Adjustment to experimental data', *Int. J. Solids Struct.* **43**, 6052–6070.

H. Banks and K. Kunisch (1989), *Estimation Techniques for Distributed Parameter Systems*, Systems & Control: Foundations & Applications, Birkhäuser.

J. Banks, W. Henshaw and D. Schwendeman (2014), 'An analysis of a new stable partitioned algorithm for FSI problems, I: Incompressible flow and elastic solids', *J. Comput. Phys.* **269**, 108–137.

L. Barbarotta (2014), A mathematical and numerical study of the left ventricular contraction based on the reconstruction of a patient specific geometry. MSc thesis, Mathematical Engineering, Politecnico di Milano.

A. Barker and X. Cai (2010*a*), 'Scalable parallel methods for monolithic coupling in fluid–structure interaction with application to blood flow modeling', *J. Comput. Phys.* **229**, 642–659.

A. Barker and X. Cai (2010*b*), 'Two-level Newton and hybrid Schwarz preconditioners for fluid–structure interaction', *SIAM J. Sci. Comput.* **32**, 2395–2417.

A. Barnard, W. Hunt, W. Timlake and E. Varley (1966), 'A theory of fluid flow in compliant tubes', *Biophys. J.* **6**, 717–724.

J. Bayer, R. Blake, G. Plank and N. Trayanova (2012), 'A novel rule-based algorithm for assigning myocardial fiber orientation to computational heart models', *Ann. Biomed. Engrg* **40**, 2243–2254.

Y. Bazilevs, V. Calo, Y. Zhang and T. Hughes (2006), 'Isogeometric fluid–structure interaction analysis with applications to arterial blood flow', *Comput. Mech.* **38**, 310–322.

Y. Bazilevs, J. Gohean, T. Hughes, R. Moser and Y. Zhang (2009), 'Patient-specific isogeometric fluid–structure interaction analysis of thoracic aortic blood flow due to implantation of the Jarvik 2000 left ventricular assist device', *Comput. Methods Appl. Mech. Engrg* **198**, 3534–3550.

Y. Bazilevs, K. Takizawa and T. Tezduyar (2012), *Computational Fluid–Structure Interaction: Methods and Applications*, Wiley.

G. Beeler and H. Reuter (1977), 'Reconstruction of the action potential of ventricular myocardial fibres', *J. Physiol.* **268**, 177–210.

H. Beirão da Veiga (2004), 'On the existence of strong solutions to a coupled fluid–structure evolution problem', *J. Math. Fluid Mech.* **6**, 21–52.

P. Benner, S. Gugercin and K. Willcox (2015), 'A survey of model reduction methods for parametric dynamical systems', *SIAM Review* **57**, 483–531.

J. Bentley and J. Friedman (1979), 'Data structures for range searching', *ACM Comput. Surv.* **11**, 397–409.

M. Benzi, G. Golub and J. Liesen (2005), Numerical solution of saddle point problems. In *Acta Numerica*, Vol. 14, Cambridge University Press, pp. 1–137.

L. Bertagna and A. Veneziani (2014), 'A model reduction approach for the variational estimation of vascular compliance by solving an inverse fluid–structure interaction problem', *Inverse Problems* **30**, 055006.

L. Bertagna, M. D'Elia, M. Perego and A. Veneziani (2014), Data assimilation in cardiovascular fluid–structure interaction problems: An introduction. In *Fluid–Structure Interaction and Biomedical Applications* (T. Bodnár, P. Galdi and Š. Nečasová, eds), Springer, pp. 395–481.

C. Bertoglio, P. Moireau and J.-F. Gerbeau (2012), 'Sequential parameter estimation for fluid–structure problems: Application to hemodynamics', *Int. J. Numer. Methods Biomed. Engrg* **28**, 434–455.

F. Bertrand, P. Tanguy and F. Thibault (1997), 'A three-dimensional fictitious domain method for incompressible fluid flow problems', *Int. J. Numer. Methods Fluids* **25**, 719–736.

R. Bevan, P. Nithiarasu, R. van Loon, I. Sazonov, H. Luckraz and A. Garnham (2010), 'Application of a locally conservative Galerkin (LCG) method for modelling blood flow through a patient-specific carotid bifurcation', *Int. J. Numer. Methods Fluids* **64**, 1274–1295.

J. Biehler, M. Gee and W. Wall (2015), 'Towards efficient uncertainty quantification in complex and large-scale biomechanical problems based on a Bayesian multi-fidelity scheme', *Biomech. Model. Mechanobiol.* **14**, 489–513.

P. Blanco and R. Feijóo (2013), 'A dimensionally-heterogeneous closed-loop model for the cardiovascular system and its applications', *Med. Engrg Phys.* **35**, 652–667.

P. Blanco, S. Deparis and A. Malossi (2013), 'On the continuity of mean total normal stress in geometrical multiscale cardiovascular problems', *J. Comput. Phys.* **51**, 136–155.

P. Blanco, R. Feijóo and S. Urquiza (2007), 'A unified variational approach for coupling 3D–1D models and its blood flow applications', *Comput. Methods Appl. Mech. Engrg* **196**, 4391–4410.

P. Blanco, M. Pivello, S. Urquiza and R. Feijóo (2009), 'On the potentialities of 3D–1D coupled models in hemodynamics simulations', *J. Biomech.* **42**, 919–930.

P. Blanco, S. Watanabe and R. Feijóo (2012), 'Identification of vascular territory resistances in one-dimensional hemodynamics simulations', *J. Biomech.* **45**, 2066–2073.

P. Blanco, S. Watanabe, M. Passos, P. Lemos, R. Feijóo *et al.* (2015), 'An anatomically detailed arterial network model for one-dimensional computational hemodynamics', *IEEE Trans. Biomed. Engrg* **62**, 736–753.

J. Blum, F.-X. Le Dimet and I. Navon (2009), Data assimilation for geophysical fluids. In *Handbook of Numerical Analysis*, Vol. 14: *Computational Methods for the Atmosphere and the Oceans* (R. Temam and J. Tribbia, eds), Elsevier, pp. 385–441.

T. Bodnár, G. Galdi and Š. Nečasová (2014), *Fluid–Structure Interaction and Biomedical Applications*, Springer.

J. Boese, M. Bock, S. Schoenberg and L. Schad (2000), 'Estimation of aortic compliance using magnetic resonance pulse wave velocity measurement', *Phys. Med. Biol.* **45**, 1703–1713.

D. Boffi and L. Gastaldi (2003), 'A finite element approach for the immersed boundary method', *Comput. Struct.* **81**, 491–501.

D. Boffi, F. Brezzi and M. Fortin (2013), *Mixed Finite Element Methods and Applications*, Springer.

D. Boffi, L. Gastaldi and L. Heltai (2007), 'Numerical stability of the finite element immersed boundary method', *Math. Models Methods Appl. Sci.* **17**, 1479–1505.

D. Boffi, L. Gastaldi, L. Heltai and C. Peskin (2008), 'On the hyper-elastic formulation of the immersed boundary method', *Comput. Methods Appl. Mech. Engrg* **197**, 2210–2231.

E. Boileau, P. Nithiarasu, P. Blanco, L. Muller, F. Fossan, L. Hellevik, W. Donders, W. Huberts, M. Willemet and J. Alastruey (2015), 'A benchmark study of numerical schemes for one-dimensional arterial blood flow modelling', *Int. J. Numer. Methods Biomed. Engrg* **31**, e02732.

C. Boldak, Y. Rolland and C. Toumoulin (2003), 'An improved model-based vessel tracking algorithm with application to computed tomography angiography', *Biocybern. Biomed.* **23**, 41–63.

D. Bonomi, C. Vergara, E. Faggiano, M. Stevanella, C. Conti, A. Redaelli, G. Puppini, G. Faggian, L. Formaggia and G. Luciani (2015), 'Influence of the aortic valve leaflets on the fluid-dynamics in aorta in presence of a normally functioning bicuspid valve', *Biomech. Model. Mechanobiol.* **6**, 1349–1361.

I. Borazjani, L. Ge and F. Sotiropoulos (2008), 'Curvilinear immersed boundary method for simulating fluid structure interaction with complex 3D rigid bodies', *J. Comput. Phys.* **227**, 7587–7620.

R. Bordas, K. Gillow, D. Gavaghan, B. Rodríguez and D. Kay (2012), 'A bidomain model of the ventricular specialized conduction system of the heart', *SIAM J. Appl. Math.* **72**, 1618–1643.

A. Borzì and V. Schulz (2011), *Computational Optimization of Systems Governed by Partial Differential Equations*, SIAM.

M. Boulakia, M. Fernández, J.-F. Gerbeau and N. Zemzemi (2008), 'Direct and inverse problems in electrocardiography', *AIP Conference Proceedings* **1048**, 113–117.

M. Boulakia, E. Schenone and J.-F. Gerbeau (2012), 'Reduced-order modeling for cardiac electrophysiology. application to parameter identification', *Int. J. Numer. Meth. Biomed. Engrg* **28**, 727–744.

Y. Bourgault, Y. Coudière and C. Pierre (2006), 'Existence and uniqueness of the solution for the bidomain model used in cardiac electrophysiology', *Nonlinear Analysis: Real World Applications* **10**, 458–482.

Y. Bourgault, M. Ethier and V. LeBlanc (2003), 'Simulation of electrophysiological waves with an unstructured finite element method', *ESAIM Math. Model. Numer. Anal.* **37**, 649–661.

A. Brault, L. Dumas and D. Lucor (2016), 'Uncertainty quantification of inflow boundary condition and proximal arterial stiffness-coupled effect on pulse wave propagation in a vascular network', *Int. J. Numer. Methods Biomed. Engrg.* https://doi.org/10.1002/cnm.2859

A. Bueno-Orovio, E. Cherry and F. Fenton (2008), 'Minimal model for human ventricular action potentials in tissue', *J. Theoret. Biol.* **3**, 544–560.

E. Burman and M. Fernández (2009), 'Stabilization of explicit coupling in fluid–structure interaction involving fluid incompressibility', *Comput. Methods Appl. Mech. Engrg* **198**, 766–784.

E. Burman and M. Fernández (2014), 'An unfitted Nitsche method for incompressible fluid–structure interaction using overlapping meshes', *Comput. Methods Appl. Mech. Engrg* **279**, 497–514.

E. Burman, S. Claus, P. Hansbo, M. Larson and A. Massing (2015), 'CutFEM: Discretizing geometry and partial differential equations', *Int. J. Numer. Methods Engrg* **104**, 472–501.

I. Campbell, J. Ries, S. Dhawan, A. Quyyumi, W. Taylor and J. Oshinski (2012), 'Effect of inlet velocity profiles on patient-specific computational fluid dynamics simulations of the carotid bifurcation', *J. Biomech. Engrg* **134**, 051001.

T. Carew, R. Vaishnav and D. Patel (1968), 'Compressibility of the arterial wall', *Circ. Res.* **23**, 61–68.

J. Carr, W. Fright and R. Beatson (1997), 'Surface interpolation with radial basis functions for medical imaging', *IEEE Trans. Med. Imaging* **16**, 96–107.

P. Causin, J.-F. Gerbeau and F. Nobile (2005), 'Added-mass effect in the design of partitioned algorithms for fluid–structure problems', *Comput. Methods Appl. Mech. Engrg* **194**, 4506–4527.

I. Celik, U. Ghia, P. Roache, C. Freitas, H. Coleman and P. Raad (2008), 'Procedure for estimation and reporting of uncertainty due to discretization in CFD applications', *J. Fluids Engrg: Trans. ASME* **130**, 078001.

R. Chabiniok, P. Moireau, P.-F. Lesault, A. Rahmouni, J.-F. Deux and D. Chapelle (2012), 'Estimation of tissue contractility from cardiac cine-MRI using a biomechanical heart model', *Biomech. Model. Mechanobiol.* **11**, 609–630.

R. Chabiniok, V. Wang, M. Hadjicharalambous, L. Asner, J. Lee, M. Sermesant, E. Kuhl, A. Young, P. Moireau, M. Nash, D. Chapelle and D. Nordsletten (2016), 'Multiphysics and multiscale modelling, data–model fusion and integration of organ physiology in the clinic: Ventricular cardiac mechanics', *Interface Focus* **6**, 20150083.

D. Chapelle, M. Fragu, V. Mallet and P. Moireau (2013*a*), 'Fundamental principles of data assimilation underlying the Verdandi library: Applications to biophysical model personalization within euHeart', *Med. Biol. Engrg Comput.* **51**, 1221–1233.

D. Chapelle, A. Gariah, P. Moireau and J. Sainte-Marie (2013*b*), 'A Galerkin strategy with proper orthogonal decomposition for parameter-dependent problems: Analysis, assessments and applications to parameter estimation', *ESAIM Math. Model. Numer. Anal.* **47**, 1821–1843.

J. Charonko, R. Kumar, K. Stewart, W. Little and P. Vlachos (2013), 'Vortices formed on the mitral valve tips aid normal left ventricular filling', *Ann. Biomed. Engrg* **41**, 1049–1061.

G. Chavent (2010), *Nonlinear Least Squares for Inverse Problems: Theoretical Foundations and Step-by-Step Guide for Applications*, Scientific Computation, Springer.

J. Chen, X.-Y. Lu and W. Wang (2006), 'Non-Newtonian effects of blood flow on hemodynamics in distal vascular graft anastomoses', *J. Biomech.* **39**, 1983–1995.

P. Chen and C. Schwab (2015), 'Sparse-grid, reduced-basis Bayesian inversion', *Comput. Methods Appl. Mech. Engrg* **297**, 84–115.

P. Chen, A. Quarteroni and G. Rozza (2013), 'Simulation-based uncertainty quantification of human arterial network hemodynamics', *Int. J. Numer. Methods Biomed. Engrg* **29**, 698–721.

L. Cheng, J. Bodley and A. Pullan (2003), 'Comparison of potential- and activation-based formulations for the inverse problem of electrocardiology', *IEEE Trans. Biomed. Engrg* **50**, 11–22.

Y. Cheng, H. Oertel and T. Schenkel (2005), 'Fluid–structure coupled CFD simulation of the left ventricular flow during filling phase', *Ann. Biomed. Engrg* **5**, 567–576.

C. Cherubini, S. Filippi, P. Nardinocchi and L. Teresi (2008), 'An electromechanical model of cardiac tissue: Constitutive issues and electrophysiological effects', *Prog. Biophys. Molec. Biol.* **2–3**, 562–573.

S. Cheung, K. K. L. Wong, G. H. Yeoh, W. Yang, J. Tu, R. Beare and T. Phan (2010), 'Experimental and numerical study on the hemodynamics of stenosed carotid bifurcation', *Australas. Phys. Engrg Sci. Med.* **33**, 319–328.

P. Chinchapatnam, K. Rhode, M. Ginks, C. Rinaldi, P. Lambiase, R. Razavi, S. Arridge and M. Sermesant (2008), 'Model-based imaging of cardiac apparent conductivity and local conduction velocity for diagnosis and planning of therapy', *IEEE Trans. Med. Imaging* **27**, 1631–1642.

J. Ching, J. Beck and K. Porter (2006), 'Bayesian state and parameter estimation of uncertain dynamical systems', *Probab. Engrg Mech.* **21**, 81–96.

C. Chnafa, S. Mendez and F. Nicoud (2014), 'Image-based large-eddy simulation in a realistic left heart', *Comput. Fluids* **94**, 173–187.

Y. Choi, J. Constantino, V. Vedula, N. Trayanova and R. Mittal (2015), 'A new MRI-based model of heart function with coupled hemodynamics and application to normal and diseased canine left ventricles', *Front. Bioeng. Biotechnol.* **3**, 140.

A. Chorin (1968), 'Numerical solution of the Navier–Stokes equations', *Math. Comp.* **22**, 745–762.

J. Chung and G. Hulbert (1993), 'A time integration algorithm for structural dynamics with improved numerical dissipation: The generalized-α method', *Trans. ASME J. Appl. Mech.* **60**, 371–375.

P. Ciarlet (1988), *Mathematical Elasticity*, Vol. 1: *Three Dimensional Elasticity*, Elsevier Science.

P. Ciarlet and J. Necas (1985), 'Unilateral problems in nonlinear, three-dimensional elasticity', *Arch. Rat. Mech. Anal.* **87**, 319–338.

R. Clayton and A. Panfilov (2008), 'A guide to modelling cardiac electrical activity in anatomically detailed ventricles', *Prog. Biophys. Molec. Biol.* **96**, 19–43.

R. Clayton, O. Bernus, E. Cherry, H. Dierckx, F. Fenton, L. Mirabella, A. Panfilov, F. Sachse, G. Seemann and H. Zhang (2011), 'Models of cardiac tissue electrophysiology: Progress, challenges and open questions', *Prog. Biophys. Molec. Biol.* **104**, 22–48.

C. Cocosco, T. Netsch, J. Séngas, D. Bystrov, W. Niessen and M. Viergever (2004), Automatic cardiac region-of-interest computation in cine 3D structural MRI.

In *Computer Assisted Radiology and Surgery: Proceedings of the 18th International Congress and Exhibition*, Vol. 1268 of International Congress Series, Elsevier, pp. 1126–1131.

R. Codina and S. Badia (2006), 'On some pressure segregation methods of fractional-step type for the finite element approximation of incompressible flow problems', *Comput. Methods Appl. Mech. Engrg* **195**, 2900–2918.

A. Cohen and R. DeVore (2015), Approximation of high-dimensional parametric PDEs. In *Acta Numerica*, Vol. 24, Cambridge University Press, pp. 1–159.

C. Colciago, S. Deparis and A. Quarteroni (2014), 'Comparisons between reduced order models and full 3D models for fluid–structure interaction problems in haemodynamics', *J. Comput. Appl. Math.* **2754**, 120–138.

P. Colli Franzone and L. Guerri (1993), 'Spreading excitation in 3-D models of the anisotropic cardiac tissue I: Validation of the Eikonal model', *Math. Biosci.* **113**, 145–209.

P. Colli Franzone and L. Pavarino (2004), 'A parallel solver for reaction-diffusion systems in computational electrocardiology', *Math. Models Methods Appl. Sci.* **14**, 883–911.

P. Colli Franzone and G. Savaré (2002), Degenerate evolution systems modeling the cardiac electric field at micro- and macroscopic level. In *Evolution Equations, Semigroups and Functional Analysis: In Memory of Brunello Terreni* (A. Lorenzi and B. Ruf, eds), Birkhäuser, pp. 49–78.

P. Colli Franzone, L. Guerri and S. Rovida (1990), 'Wavefront propagation in an activation model of the anisotropic cardiac tissue: Asymptotic analysis and numerical simulations', *J. Math. Biol.* **28**, 121–176.

P. Colli Franzone, L. Guerri, C. Viganotti and B. Taccardi (1985), 'Finite element approximation of regularized solution of the inverse potential problem of electrocardiography and applications to experimental data', *Calcolo* **12**, 91–186.

P. Colli Franzone, L. Pavarino and S. Scacchi (2014), *Mathematical Cardiac Electrophysiology*, Springer.

P. Colli Franzone, L. Pavarino and S. Scacchi (2016), 'Bioelectrical effects of mechanical feedbacks in a strongly coupled cardiac electro-mechanical model', *Math. Models Methods Appl. Sci.* **26**, 27–57.

P. Colli Franzone, L. Pavarino and B. Taccardi (2005), 'Simulating patterns of excitation, repolarization and action potential duration with cardiac bidomain and monodomain models', *Math. Biosci.* **197**, 35–66.

P. Colli Franzone, B. Taccardi and C. Viganotti (1978), 'An approach to inverse calculation of epicardial potentials from body surface maps', *Adv. Cardiol.* **21**, 50–54.

C. Corrado, J.-F. Gerbeau and P. Moireau (2015), 'Identification of weakly coupled multiphysics problems: Application to the inverse problem of electrocardiography', *J. Comput. Phys.* **283**, 271–298.

K. Costa, J. Holmes and A. McCulloch (2001), 'Modelling cardiac mechanical properties in three dimensions', *Phil. Trans. Royal Soc.* A **359**, 1233–1250.

P. Crosetto, S. Deparis, G. Fourestey and A. Quarteroni (2011), 'Parallel algorithms for fluid–structure interaction problems in haemodynamics', *SIAM J. Sci. Comput.* **33**, 1598–1622.

T. Cui, Y. Marzouk and K. Willcox (2015), 'Data-driven model reduction for the Bayesian solution of inverse problems', *Int. J. Numer. Methods Engrg* **102**, 966–990.

B. Dacorogna (2000), *Direct Methods in the Calculus of Variations*, Vol. 78 of Applied Mathematical Sciences, Springer.

C. D'Angelo (2007), Multiscale modelling of metabolism and transport phenomena in living tissues. PhD thesis, Ecole Polytechnique Fédérale de Lausanne.

J. Degroote (2011), 'On the similarity between Dirichlet–Neumann with interface artificial compressibility and Robin–Neumann schemes for the solution of fluid–structure interaction problems', *J. Comput. Phys.* **230**, 6399–6403.

J. Degroote and J. Vierendeels (2011), 'Multi-solver algorithms for the partitioned simulation of fluid–structure interaction', *Comput. Methods Appl. Mech. Engrg* **25–28**, 2195–2210.

J. Degroote, A. Swillens, P. Bruggeman, Haelterman, P. Segers and J. Vierendeels (2010), 'Simulation of fluid–structure interaction with the interface artificial compressibility method', *Int. J. Numer. Methods Biomed. Engrg* **26**, 276–289.

M. D'Elia and A. Veneziani (2013), 'Uncertainty quantification for data assimilation in a steady incompressible Navier–Stokes problem', *ESAIM Math. Model. Numer. Anal.* **47**, 1037–1057.

M. D'Elia, M. Perego and A. Veneziani (2012), 'A variational data assimilation procedure for the incompressible Navier–Stokes equations in hemodynamics', *J. Sci. Comput.* **52**, 340–359.

H. Delingette, F. Billet, K. Wong, M. Sermesant, K. Rhode, M. Ginks, C. Rinaldi, R. Razavi and N. Ayache (2012), 'Personalization of cardiac motion and contractility from images using variational data assimilation', *IEEE Trans. Biomed. Engrg* **59**, 20–24.

S. Deparis (2004), Numerical analysis of axisymmetric flows and methods for fluid–structure interaction arising in blood flow simulation. PhD thesis, Ecole Polytechnique Fédérale de Lausanne.

S. Deparis, M. Discacciati, G. Fourestey and A. Quarteroni (2006), 'Fluid–structure algorithms based on Steklov–Poincaré operators', *Comput. Methods Appl. Mech. Engrg* **195**, 5797–5812.

S. Deparis, D. Forti, G. Grandperrin and A. Quarteroni (2016), 'FaCSI: A block parallel preconditioner for fluid–structure interaction in hemodynamics', *J. Comput. Phys.*, 700–718.

S. Deparis, G. Grandperrin and A. Quarteroni (2014), 'Parallel preconditioners for the unsteady Navier–Stokes equations and applications to hemodynamics simulations', *Comput. Fluids* **92**, 253–273.

J. Dick, F. Kuo and I. Sloan (2013), High-dimensional integration: The quasi-Monte Carlo way. In *Acta Numerica*, Vol. 22, Cambridge University Press, pp. 133–288.

D. DiFrancesco and D. Noble (1985), 'A model of cardiac electrical activity incorporating ionic pumps and concentration changes', *Phil. Trans. Royal Soc. B* **307**, 353–398.

M. Dihlmann and B. Haasdonk (2016), 'A reduced basis Kalman filter for parametrized partial differential equations', *ESAIM Control Optim. Calc. Var.* **22**, 625–669.

H. Do, A. A. Owida, W. Yang and Y. Morsi (2011), 'Numerical simulation of the haemodynamics in end-to-side anastomoses', *Int. J. Numer. Methods Fluids* **67**, 638–650.

C. Dohrmann and O. Widlund (2009), 'An overlapping Schwarz algorithm for almost incompressible elasticity', *SIAM J. Numer. Anal.* **4**, 8811–8823.

S. Dokos, B. Smaill, A. Young and I. LeGrice (2002), 'Shear properties of passive ventricular myocardium', *Amer. J. Physiol.* **283**, H2650–H2659.

W. Donders, W. Huberts, F. van de Vosse and T. Delhaas (2015), 'Personalization of models with many model parameters: An efficient sensitivity analysis approach', *Int. J. Numer. Methods Biomed. Engrg* **31**, e02727.

J. Donea (1982), 'An arbitrary Lagrangian–Eulerian finite element method for transient dynamic fluid–structure interaction', *Comput. Methods Appl. Mech. Engrg* **33**, 689–723.

O. Dur, S. Coskun, K. Coskun, D. Frakes, L. Kara and K. Pekkan (2011), 'Computer-aided patient-specific coronary artery graft design improvements using CFD coupled shape optimizer', *Cardiovasc. Engr. Tech.* pp. 1–13.

D. Durrer, R. van Dam, G. Freud, M. Janse, F. Meijler and R. Arzbaecher (1970), 'Total excitation of the isolated human heart', *Circulation* **41**, 899–912.

V. Eck, W. Donders, J. Sturdy, J. Feinberg, T. Delhaas, L. Hellevik and W. Huberts (2016), 'A guide to uncertainty quantification and sensitivity analysis for cardiovascular applications', *Int. J. Numer. Methods Biomed. Engrg* **32**, e02755.

D. Einstein, K. Kunzelman, P. Reinhall, M. Nicosia and R. Cochran (2005), 'The effects of cellular contraction on aortic valve leaflet flexural stiffness', *J. Heart Valve Disease* **14**, 376–385.

C. Eitel, G. Hindricks, N. Dagres, P. Sommer and C. Piorkowski (2010), 'EnSite Velocity^{TM} cardiac mapping system: A new platform for 3D mapping of cardiac arrhythmias', *Expert Rev. Med. Devices* **7**, 185–192.

H. Elman and D. Silvester (1996), 'Fast nonsymmetric iterations and preconditioning for Navier–Stokes equations', *SIAM J. Sci. Comput.* **17**, 33–46.

H. Elman, D. Silvester and A. Wathen (2005), *Finite Elements and Fast Iterative Solvers*, Oxford Science Publications.

G. Enden and A. Popel (1992), 'A numerical study of the shape of the surface separating flow into branches in microvascular bifurcations', *J. Biomech. Engrg* **114**, 398–405.

T. Eriksson, A. Prassl, G. Plank and G. Holzapfel (2013), 'Modeling the dispersion in electromechanically coupled myocardium', *Int. J. Numer. Methods Biomed. Engrg* **29**, 1267–1284.

O. Ernst, B. Sprungk and H.-J. Starkloff (2014), Bayesian inverse problems and Kalman filters. In *Extraction of Quantifiable Information from Complex Systems*, Vol. 102 of Lecture Notes in Computational Science and Engineering, Springer, pp. 133–159.

V. Ervin and H. Lee (2007), 'Numerical approximation of a quasi-Newtonian Stokes flow problem with defective boundary conditions', *SIAM J. Numer. Anal.* **45**, 2120–2140.

C. Ethier, D. Steinman, X. Zhang, S. Karpik and M. Ojha (1998), 'Flow waveform effects on end-to-side anastomotic flow patterns', *J. Biomech.* **31**, 609–617.

M. Ethier and Y. Bourgault (2008), 'Semi-implicit time-discretization schemes for the bidomain model', *SIAM J. Numer. Anal.* **5**, 2443–2468.

L. Euler (1775), Principia pro motu sanguinis per arterias determinando. In *Opera posthuma mathematica et physica anno 1844 detecta, editerunt P. H. Fuss et N. Fuss, Petropoli, apud Eggers et socios*, Vol. 1, pp. 814–823.

G. Evensen (1994), 'Sequential data assimilation with a nonlinear quasi-geostrophic model using Monte Carlo methods to forecast error statistics', *J. Geophys. Res.* **9**, 10143–10162.

G. Evensen (2003), 'The ensemble Kalman filter: Theoretical formulation and practical implementation', *Ocean Dynamics* **53**, 343–367.

G. Evensen (2009), 'The ensemble Kalman filter for combined state and parameter estimation', *IEEE Control Syst. Mag.* **29**, 83–104.

E. Faggiano, A. Antiga, G. Puppini, A. Quarteroni, G. Luciani and C. Vergara (2013), 'Helical flows and asymmetry of blood jet in dilated ascending aorta with normally functioning bicuspid valve', *Biomech. Model. Mechanobiol.* **4**, 801–813.

E. Faggiano, T. Lorenzi and A. Quarteroni (2014), 'Metal artefact reduction in computed tomography images by a fourth-order total variation flow', *Comput. Methods Biomech. Biomed. Engrg: Imaging & Visualization* **3-4**, 202–213.

C. Farhat and F. Roux (1991), 'A method of finite element tearing and interconnecting and its parallel solution algorithm', *Int. J. Numer. Methods Engrg* **32**, 1205–1227.

A. Fasano, R. Santos and A. Sequeira (2012), Blood coagulation: A puzzle for biologists, a maze for mathematicians. Chapter 3 in *Modeling of Physiological Flows* (D. Ambrosi, A. Quarteroni and G. Rozza, eds), Springer, pp. 41–75.

M. Fedele, E. Faggiano, L. Barbarotta, F. Cremonesi, L. Formaggia and S. Perotto (2015), Semi-automatic three-dimensional vessel segmentation using a connected component localization of the region-scalable fitting energy. In *2015 9th International Symposium on Image and Signal Processing and Analysis (ISPA)*, IEEE, pp. 72–77.

M. Fedele, E. Faggiano, L. Dede' and A. Quarteroni (2016), A patient-specific aortic valve model based on moving resistive immersed implicit surfaces. MOX-Report 23-2016, Department of Mathematics, Politecnico di Milano.

F. Fenton and A. Karma (1998), 'Vortex dynamics in three-dimensional continuous myocardium with fiber rotation: Filament instability and fibrillation', *Chaos* **8**, 20–47.

M. Fernández (2013), 'Incremental displacement-correction schemes for incompressible fluid–structure interaction: Stability and convergence analysis', *Numer. Math.* **123**, 21–65.

M. Fernández and M. Moubachir (2005), 'A Newton method using exact Jacobians for solving fluid–structure coupling', *Comput. Struct.* **83**, 127–142.

M. Fernández and N. Zemzemi (2010), 'Decoupled time-marching schemes in computational cardiac electrophysiology and ECG numerical simulation', *Math. Biosci.* **226**, 58–75.

M. Fernández, J.-F. Gerbeau and C. Grandmont (2007), 'A projection semi-implicit scheme for the coupling of an elastic structure with an incompressible fluid', *Int. J. Numer. Methods Engrg* **69**, 794–821.

M. Fernández, V. Milisic and A. Quarteroni (2005), 'Analysis of a geometrical multiscale blood flow model based on the coupling of ODEs and hyperbolic PDEs', *Multiscale Model. Simul.* **4**, 215–236.

C. Figueroa, I. Vignon-Clementel, K. Jansen, T. Hughes and C. Taylor (2006), 'A coupled momentum method for modeling blood flow in three-dimensional deformable arteries', *Comput. Methods Appl. Mech. Engrg* **195**, 5685–5706.

L. Fin and R. Grebe (2003), 'Three dimensional modeling of the cerebrospinal fluid dynamics and brain interactions in the aqueduct of sylvius', *Comput. Methods Biomech. Biomed. Engrg* **3**, 163–170.

P. Fischer, F. Loth, S. Lee, S. Lee, D. Smith and H. Bassiouny (2007), 'Simulation of high-Reynolds number vascular flows', *Comput. Methods Appl. Mech. Engrg* **196**, 3049–3060.

G. Fishman (1996), *Monte Carlo: Concepts, Algorithms, and Applications*, Springer Series in Operations Research and Financial Engineering, Springer.

R. FitzHugh (1961), 'Impulses and physiological states in theoretical models of nerve membrane', *Biophys. J.* **1**, 445–466.

L. Formaggia and C. Vergara (2012), 'Prescription of general defective boundary conditions in fluid-dynamics', *Milan J. Math.* **80**, 333–350.

L. Formaggia, J.-F. Gerbeau, F. Nobile and A. Quarteroni (2001), 'On the coupling of 3D and 1D Navier–Stokes equations for flow problems in compliant vessels', *Comput. Methods Appl. Mech. Engrg* **191**, 561–582.

L. Formaggia, J.-F. Gerbeau, F. Nobile and A. Quarteroni (2002), 'Numerical treatment of defective boundary conditions for the Navier–Stokes equation', *SIAM J. Numer. Anal.* **40**, 376–401.

L. Formaggia, D. Lamponi and A. Quarteroni (2003), 'One-dimensional models for blood flow in arteries', *J. Engrg Math.* **47**, 251–276.

L. Formaggia, A. Moura and F. Nobile (2007), 'On the stability of the coupling of 3D and 1D fluid–structure interaction models for blood flow simulations', *M2AN Math. Model. Numer. Anal.* **41**, 743–769.

L. Formaggia, F. Nobile, A. Quarteroni and A. Veneziani (1999), 'Multiscale modelling of the circulatory system: A preliminary analysis', *Comput. Vis. Sci.* **2**, 75–83.

L. Formaggia, A. Quarteroni and A. Veneziani, eds (2009*a*), *Cardiovascular Mathematics: Modeling and Simulation of the Circulatory System*, Springer.

L. Formaggia, A. Quarteroni and C. Vergara (2013), 'On the physical consistency between three-dimensional and one-dimensional models in haemodynamics', *J. Comput. Phys.* **244**, 97–112.

L. Formaggia, A. Veneziani and C. Vergara (2008), 'A new approach to numerical solution of defective boundary value problems in incompressible fluid dynamics', *SIAM J. Numer. Anal.* **46**, 2769–2794.

L. Formaggia, A. Veneziani and C. Vergara (2009*b*), 'Flow rate boundary problems for an incompressible fluid in deformable domains: Formulations and solution methods', *Comput. Methods Appl. Mech. Engrg* **199**, 677–688.

M. Fornefett, K. Rohr and H. Stiehl (2001), 'Radial basis functions with compact support for elastic registration of medical images', *Image Vision Comput.* **19**, 87–96.

C. Forster, W. Wall and E. Ramm (2007), 'Artificial added mass instabilities in sequential staggered coupling of nonlinear structures and incompressible viscous flow', *Comput. Methods Appl. Mech. Engrg* **196**, 1278–1293.

D. Forti and L. Dede' (2015), 'Semi-implicit BDF time discretization of the Navier–Stokes equations with VMS-LES modeling in a high performance computing framework', *Comput. Fluids* **117**, 168–182.

A. Frangi, W. Niessen, R. Hoogeveen, T. van Walsum and M. Viergever (1999), 'Model-based quantitation of 3-D magnetic resonance angiographic images', *IEEE Trans. Med. Imaging* **18**, 946–956.

D. Frazier, W. Krassowska, P. Chen, P. Wolf, N. Danieley, W. Smith and R. Ideker (1988), 'Transmural activations and stimulus potentials in three-dimensional anisotropic canine myocardium', *Circ. Res.* **63**, 135–146.

T. Fritz, C. Wieners, G. Seemann, H. Steen and O. Dossel (2014), 'Simulation of the contraction of the ventricles in a human heart model including atria and pericardium', *Biomech. Model. Mechanobiol.* **13**, 627–641.

Y. Fung (1993), *Biomechanics: Mechanical Properties of Living Tissues*, Springer.

Y. C. Fung, K. Fronek and P. Patitucci (1979), 'Pseudoelasticity of arteries and the choice of its mathematical expression', *Amer. J. Physiol.* **237**, H620–H631.

K. Galvin and H. Lee (2013), 'Analysis and approximation of the cross model for quasi-Newtonian flows with defective boundary conditions', *Appl. Math. Comput.* **222**, 244–254.

K. Galvin, H. Lee and L. Rebholz (2012), 'Approximation of viscoelastic flows with defective boundary conditions', *J. Non-Newton. Fluid Mech.* **169/170**, 104–113.

M. Gee, U. Kuttler and W. Wall (2011), 'Truly monolithic algebraic multigrid for fluid–structure interaction', *Int. J. Numer. Methods Engrg* **85**, 987–1016.

R. van der Geest, E. Jansen, V. Buller and J. Reiber (1994), Automated detection of left ventricular epi- and endocardial contours in short-axis MR images. In *Computers in Cardiology 1994*, IEEE, pp. 33–36.

B. Geiger (1993), Three-dimensional modeling of human organs and its application to diagnosis and surgical planning. Technical Report RR-2105, INRIA.

S. Geneser, R. Kirby and R. MacLeod (2008), 'Application of stochastic finite element methods to study the sensitivity of ECG forward modeling to organ conductivity', *IEEE Trans. Biomed. Engrg* **55**, 31–40.

L. Gerardo-Giorda, L. Mirabella, F. Nobile, M. Perego and A. Veneziani (2009), 'A model-based block-triangular preconditioner for the bidomain system in electrocardiology', *J. Comput. Phys.* **228**, 3625–3639.

L. Gerardo-Giorda, F. Nobile and C. Vergara (2010), 'Analysis and optimization of Robin–Robin partitioned procedures in fluid–structure interaction problems', *SIAM J. Numer. Anal.* **48**, 2091–2116.

J.-F. Gerbeau, D. Lombardi and E. Schenone (2015), 'Reduced order model in cardiac electrophysiology with approximated lax pairs', *Adv. Comput. Math.* **41**, 1103–1130.

P. Gervasio, F. Saleri and A. Veneziani (2006), 'Algebraic fractional-step schemes with spectral methods for the incompressible Navier–Stokes equations', *J. Comput. Phys.* **214**, 347–365.

R. Ghanem and P. Spanos (2003), *Stochastic Finite Elements: A Spectral Approach*, revised edition, Dover. Reprint of the Springer 1991 edition.

G. Gigante and C. Vergara (2015), 'Analysis and optimization of the generalized Schwarz method for elliptic problems with application to fluid–structure interaction', *Numer. Math.* **131**, 369–404.

M. Giles (2015), Multilevel Monte Carlo methods. In *Acta Numerica*, Vol. 24, Cambridge University Press, pp. 259–328.

S. Giordana, S. Sherwin, J. Peiró, D. Doorly, J. Crane, K. Lee, N. Cheshire and C. Caro (2005), 'Local and global geometric influence on steady flow in distal anastomoses of peripheral bypass grafts', *J. Biomech. Engrg* **127**, 1087–1098.

V. Girault and P. Raviart (1986), *Finite Element Methods for Navier–Stokes Equations*, Springer.

S. Glagov, C. Zarins, D. Giddens and D. Ku (1988), 'Hemodynamics and atherosclerosis: Insights and perspectives gained from studies of human arteries', *Arch. Pathol. Lab. Med.* **112**, 1018–1031.

R. Glowinski, T. Pan and J. Periaux (1997), 'A Lagrange multiplier/fictitious domain method for the numerical simulation of incompressible viscous flow around moving rigid bodies (I): Case where the rigid body motions are known *a priori*', *CR Acad. Sci.* I Math. **324**, 361–369.

R. Gnyaneshwar, R. Kumar and K. Balakrishnan (2002), 'Dynamic analysis of the aortic valve using a finite element model', *Ann. Thorac. Surg.* **73**, 1122–1129.

S. Goktepe and E. Kuhl (2010), 'Electromechanics of the heart: A unified approach to the strongly coupled excitation-contraction problem', *Comput. Mech.* **45**, 227–243.

A. Goshtasby and D. Turner (1995), 'Segmentation of cardiac cine MR images for extraction of right and left ventricular chambers', *IEEE Trans. Med. Imaging* **14**, 56–64.

C. Grandmont (1998), Analyse mathématique et numérique de quelques problèmes d'interaction fluide-structure. PhD thesis, Laboratoire d'Analyse Numérique de Paris VI.

B. Griffith, R. Hornung, D. McQueen and C. Peskin (2007), 'An adaptive, formally second order accurate version of the immersed boundary method', *J. Comput. Phys.* **223**, 10–419.

B. Griffith, X. Luo, D. McQueen and C. Peskin (2009), 'Simulating the fluid dynamics of natural and prosthetic heart valves using the immersed boundary method', *Int. J. Appl. Mech.* **1**, 137–176.

L. Grinberg, E. Cheever, T. Anor, J. Madsen and G. Karniadakis (2010), 'Modeling blood flow circulation in intracranial arterial networks: A comparative 3D/1D simulation study', *Ann. Biomed. Engrg* **39**, 297–309.

L. Grinberg, A. Yakhot and G. Karniadakis (2009), 'Analyzing transient turbulence in a stenosed carotid artery by proper orthogonal decomposition', *Ann. Biomed. Engrg* **37**, 2200–2217.

J. Guccione, A. McCulloch and L. Waldman (1991), 'Passive material properties of intact ventricular myocardium determined from a cylindrical model', *J. Biomech. Engrg* **113**, 42–55.

B. Guerciotti, C. Vergara, L. Azzimonti, L. Forzenigo, A. Buora, P. Biondetti and M. Domanin (2015), 'Computational study of the fluid-dynamics in carotids before and after endarterectomy', *J. Biomech.* **195**, 2088–2099.

B. Guerciotti, C. Vergara, S. Ippolito, A. Quarteroni, C. Antona and R. Scrofani (2017), 'Computational study of the risk of restenosis in coronary bypasses', *Biomech. Model. Mechanobiol.* **16**, 313–332.

J. Guermond and L. Quartapelle (1998), 'On the approximation of the unsteady Navier–Stokes equations by finite element projection methods', *Numer. Math.* **80**, 207–238.

J. Guermond and J. Shen (2003), 'Velocity-correction projection methods for incompressible flows', *SIAM J. Numer. Anal.* **41**, 112–134.

J. Guermond and J. Shen (2006), 'An overview of projection methods for incompressible flows', *Comput. Methods Appl. Mech. Engrg* **195**, 6011–6045.

J. Guermond, P. Minev and J. Shen (2005), 'Error analysis of pressure-correction schemes for the time-dependent Stokes equations with open boundary conditions', *SIAM J. Numer. Anal.* **43**, 239–258.

J. Guermond, P. Minev and J. Shen (2006), 'An overview of projection methods for incompressible flows', *Comput. Methods Appl. Mech. Engrg* **195**, 6011–6045.

G. Guidoboni, R. Glowinski, N. Cavallini and S. Canic (2009), 'Stable loosely-coupled-type algorithm for fluid–structure interaction in blood flow', *J. Comput. Phys.* **228**, 6916–6937.

O. Gultekin, G. Sommer and G. Holzapfel (2016), 'An orthotropic viscoelastic model for the passive myocardium: Continuum basis and numerical treatment', *Comput. Methods Biomech. Biomed. Engrg* **19**, 1647–1664.

M. Gunzburger (2003), *Perspectives in Flow Control and Optimization*, Advances in Design and Control, SIAM.

M. Gunzburger, C. Webster and G. Zhang (2014), Stochastic finite element methods for partial differential equations with random input data. In *Acta Numerica*, Vol. 23, Cambridge University Press, pp. 521–650.

A. Gupta, L. von Kurowski, A. Singh, D. Geiger, C. Liang, M. Chiu, L. Adler, M. Haacke and D. Wilson (1993), Cardiac MR image segmentation using deformable models. In *Computers in Cardiology 1993*, IEEE, pp. 747–750.

V. Gurev, P. Pathmanathan, J. Fattebert, H. Wen, J. Magerlein, R. Gray, D. Richards and J. Rice (2015), 'A high-resolution computational model of the deforming human heart', *Biomech. Model. Mechanobiol.* **14**, 132–140.

C. Haggerty, L. Mirabella, M. Restrepo, D. de Zélicourt, J. Rossignac, F. Sotiropoulos, T. Spray, K. Kanter, M. Fogel and A. Yoganathan (2013), Patient-specific surgery planning for the fontan procedure. In *Computer Models in Biomechanics: From Nano to Macro* (G. Holzapfel and E. Kuhl, eds), Springer, pp. 217–228.

A. Hansbo and P. Hansbo (2002), 'An unfitted finite element method, based on Nitsche's method, for elliptic interface problems', *Comput. Methods Appl. Mech. Engrg* **191**, 5537–5552.

A. Hansbo, P. Hansbo and M. Larson (2003), 'A finite element method on composite grids based on Nitsche's method', *ESAIM: Math. Model. Numer. Anal.* **37**, 495–514.

I. Hariton, G. de Botton, T. Gasser and G. Holzapfel (2006), 'Stress-driven collagen fiber remodeling in arterial walls', *Biomech. Model. Mechanobiol.* **3**, 163–175.

J. de Hart, F. Baaijens, G. Peters and P. Schreurs (2003), 'A computational fluid–structure interaction analysis of a fiber-reinforced stentless aortic valve', *J. Biomech.* **36**, 699–712.

J. de Hart, G. Peters, P. Schreurs and F. Baaijens (2000), 'A two-dimensional fluid–structure interaction model of the aortic value', *J. Biomech.* **33**, 1079–1088.

H. Haruguchi and S. Teraoka (2003), 'Intimal hyperplasia and hemodynamic factors in arterial bypass and arteriovenous grafts: A review', *J. Artif. Organs* **6**, 227–235.

X. He, D. Ku and J. Moore Jr (1993), 'Simple calculation of the velocity profiles for pulsatile flow in a blood vessel using Mathematica', *Ann. Biomed. Engrg* **21**, 45–49.

M. Heil (2004), 'An efficient solver for the fully coupled solution of large-displacement fluid–structure interaction problems', *Comput. Methods Appl. Mech. Engrg* **193**, 1–23.

J. Hesthaven, G. Rozza and B. Stamm (2016), *Certified Reduced Basis Methods for Parametrized Partial Differential Equations*, Springer Briefs in Mathematics, Springer.

J. Heywood, R. Rannacher and S. Turek (1996), 'Artificial boundaries and flux and pressure conditions for the incompressible Navier–Stokes equations', *Int. J. Numer. Methods Fluids* **22**, 325–352.

D. Hilgemann and D. Noble (1987), 'Excitation–contraction coupling and extracellular calcium transients in rabbit atrium: Reconstruction of basic cellular mechanisms', *Proc. Royal Soc. London B* **230**, 163–205.

B. Hillen, H. Hoogstraten and L. Post (1986), 'A wave propagation model of blood flow in large vessels using an approximate velocity profile function', *J. Biomech.* **19**, 187–194.

M. Hintermüller, K. Kunisch, Y. Spasov and S. Volkwein (2004), 'Dynamical systems-based optimal control of incompressible fluids', *Int. J. Numer. Methods Fluids* **46**, 345–359.

M. Hinze, R. Pinnau, M. Ulbrich and S. Ulbrich (2009), *Optimization with PDE Constraints*, Vol. 23 of Mathematical Modelling: Theory and Applications, Springer.

C. Hirt, A. Amsden and J. Cook (1974), 'An arbitrary Lagrangian Eulerian computing method for all flow speeds', *J. Comput. Phys.* **69**, 277–324.

A. Hodgkin and A. Huxley (1952), 'A quantitative description of membrane current and its application to conduction and excitation in nerve', *J. Physiol.* **117**, 500–544.

G. Holzapfel (2000), *Nonlinear Solid Mechanics: A Continuum Approach for Engineering*, Wiley.

G. Holzapfel and T. Gasser (2001), 'A viscoelastic model for fiber-reinforced composites at finite strains: Continuum basis, computational aspects, and applications', *Comput. Methods Appl. Mech. Engrg* **190**, 4379–4403.

G. Holzapfel and R. Ogden (2009), 'Constitutive modelling of passive myocardium: A structurally based framework for material characterization', *Philos. Trans. Royal Soc. A* **367**, 3445–3475.

G. Holzapfel and R. Ogden (2010), 'Constitutive modelling of arteries', *Proc. Royal Soc. Lond. A: Math. Phys. Engrg Sci.* **466**, 1551–1596.

G. Holzapfel, T. Gasser and R. Ogden (2000), 'A new constitutive framework for arterial wall mechanics and a comparative study of material models', *J. Elast.* **61**, 1–48.

E. Hopf (1951), 'Über die Anfangswertaufgabe für die hydrodynamischen Grundgliechungen', *Math. Nachrichten* **4**, 213–231.

I. Hoteit, D. Pham and J. Blum (2002), 'A simplified reduced order Kalman filtering and application to altimetric data assimilation in Tropical Pacific', *J. Marine Systems* **36**, 101–127.

P. Houtemaker and F. Zhang (2016), 'Review of the ensemble Kalman filter for atmospheric data assimilation', *Monthly Weather Rev.* **122**, 4489–4532.

M. Hsu, D. Kamensky, Y. Bazilevs, M. Sacks and T. Hughes (2014), 'Fluid–structure interaction analysis of bioprosthetic heart valves: Significance of arterial wall deformation', *Comput. Mech.* **54**, 1055–1071.

W. Huberts, C. de Jonge, W. van der Linden, M. Inda, K. Passera, J. Tordoir, F. van de Vosse and E. Bosboom (2013*a*), 'A sensitivity analysis of a personalized pulse wave propagation model for arteriovenous fistula surgery, B: Identification of possible generic model parameters', *Med. Engrg Phys.* **35**, 827–837.

W. Huberts, C. de Jonge, W. van der Linden, M. Inda, J. Tordoir, F. van de Vosse and E. Bosboom (2013*b*), 'A sensitivity analysis of a personalized pulse wave propagation model for arteriovenous fistula surgery, A: Identification of most influential model parameters', *Med. Engrg Phys.* **35**, 810–826.

T. Hughes (1974), A study of the one-dimensional theory of arterial pulse propagation. PhD thesis, University of California, Berkeley.

T. Hughes (1995), 'Multiscale phenomena: Green's function, the Dirichlet-to-Neumann formulation, subgrid scale models, bubbles and the origins of stabilized formulations', *Comput. Methods Appl. Mech. Engrg* **127**, 387–401.

T. Hughes and J. Lubliner (1973), 'On the one-dimensional theory of blood flow in the larger vessels', *Math. Biosci.* **18**, 161–170.

T. Hughes, L. Mazzei and K. Jansen (2000), 'Large eddy simulation and the variational multiscale method', *Comput. Vis. Sci.* **3**, 47–59.

G. Huiskamp (1998), 'Simulation of depolarization in a membrane- equations-based model of the anisotropic ventricle', *IEEE Trans. Biomed. Engrg* **45**, 847–855.

J. Humpherys, P. Redd and J. West (2012), 'A fresh look at the Kalman filter', *SIAM Review* **54**, 801–823.

J. Humphrey and F. Yin (1987), 'On constitutive relations and finite deformations of passive cardiac tissue, I: A pseudostrain-energy function', *J. Biomech. Engrg* **109**, 298–304.

P. Hunter, M. Nash and G. Sands (1997), Computational electromechanics of the heart. In *Computational Biology of the Heart* (A. Panfilov and A. Holden, eds), Wiley-Blackwell, pp. 345–407.

P. Iaizzo (2009), *Handbook of Cardiac Anatomy, Physiology, and Devices*, Springer.

K. Ide, P. Courtier, M. Ghil and A. Lorenc (1997), 'Unified notation for data assimilation: Operational, sequential and variational', *J. Met. Soc. Japan* **75**, 181–189.

M. Iglesias, K. Law and A. Stuart (2013), 'Ensemble Kalman methods for inverse problems', *Inverse Problems* **29**, 045001.

T. Ijiri, T. Ashihara, T. Yamaguchi, K. Takayama, T. Igarashi, T. Shimada, T. Namba, R. Haraguchi and K. Nakazawa (2008), 'A procedural method for modeling the Purkinje fibers of the heart', *J. Physiol. Sci.* **58**, 90–100.

J. Isaksen, Y. Bazilevs, T. Kvamsdal, Y. Zhang, J. Kaspersen, K. Waterloo, B. Romner and T. Ingebrigtsen (2008), 'Determination of wall tension in cerebral artery aneurysms by numerical simulation', *Stroke* **39**, 3172–3178.

R. Johnstone, E. Chang, R. Bardenet, T. de Boer, D. Gavaghan, P. Pathmanathan, R. Clayton and G. Mirams (2016), 'Uncertainty and variability in models of the cardiac action potential: Can we build trustworthy models?', *J. Molec. Cell. Cardiol.* **96**, 49–62.

S. Julier and J. Uhlmann (2004), 'Unscented filtering and nonlinear estimation', *Proc. IEEE* **92**, 401–422.

S. Julier, J. Uhlmann and H. Durrant-Whyte (1995), A new approach for filtering nonlinear systems. In *Proceedings of the 1995 American Control Conference*, Vol. 3, pp. 1628–1632.

S. Julier, J. Uhlmann and H. Durrant-Whyte (2000), 'A new method for the nonlinear transformation of means and covariance in filters and estimators', *IEEE Trans. Automat. Control* **45**, 477–482.

M. Juntunen and R. Stenberg (2009), 'Nitsche's method for general boundary conditions', *Math. Comp.* **78**, 1353–1374.

J. Kaipio and E. Somersalo (2005), *Statistical and Computational Inverse Problems*, Vol. 160 of Applied Mathematical Sciences, Springer.

R. Kalman (1960), 'A new approach to linear filtering and prediction problems', *ASME. J. Basic Engrg* **82**, 35–45.

G. Karniadakis, M. Israeli and S. Orszag (1991), 'High order splitting methods for the incompressible Navier–Stokes equations', *J. Comput. Phys.* **59**, 414–443.

J. Keener (1991), 'An eikonal-curvature equation for action potential propagation in myocardium', *J. Math. Biol.* **29**, 629–651.

J. Keener and K. Bogar (1998), 'A numerical method for the solution of the bidomain equations in cardiac tissue', *Chaos* **8**, 234–241.

S. Kefayati, D. Holdsworth and T. Poepping (2014), 'Turbulence intensity measurements using particle image velocimetry in diseased carotid artery models: Effect of stenosis severity, plaque eccentricity, and ulceration', *J. Biomech.* **47**, 253–263.

R. Keldermann, M. Nash and A. Panfilov (2009), 'Modeling cardiac mechano-electrical feedback using reaction-diffusion-mechanics systems', *Physica D: Nonlinear Phenomena* **238**, 1000–1007.

C. Kelley (1999), *Iterative Methods for Optimization*, SIAM.

D. Kelly, K. Law and A. Stuart (2014), 'Well-posedness and accuracy of the ensemble Kalman filter in discrete and continuous time', *Nonlinearity* **27**, 2579.

M. Kennedy and A. O'Hagan (2000), 'Predicting the output from a complex computer code when fast approximations are available', *Biometrika* **87**, 1–13.

B. Kerckoffs, O. Faris, P. Boveenderd, F. Prinzen, K. Smits and T. Arts (2003), 'Timing of depolarization and contraction in the paced canine left ventricle', *J. Cardiovasc. Electrophysiol.* **14**, S188–S195.

R. Keynton, M. Evancho, R. Sims, N. Rodway, A. Gobin and S. Rittgers (2001), 'Intimal hyperplasia and wall shear in arterial bypass graft distal anastomoses: An *in vivo* model study', *J. Biomech. Engrg* **123**, 464–473.

S. Khalafvand, L. Zhong and E. Ng (2014), 'Three-dimensional CFD/MRI modeling reveals that ventricular surgical restoration improves ventricular function by modifying intraventricular blood flow', *Int. J. Numer. Methods Biomed. Engrg* **30**, 1044–1056.

H. Kim, J. Lu, M. Sacks and K. Chandran (2008), 'Dynamic simulation of bioprosthetic heart valves using a stress resultant shell model', *J. Biomech.* **36**, 262–275.

H. Kim, I. Vignon-Clementel, C. Figueroa, J. Ladisa, K. Jansen, J. Feinstein and C. Taylor (2009), 'On coupling a lumped parameter heart model and a three-dimensional finite element aorta model', *Ann. Biomed. Engrg* **37**, 2153–2169.

C. Kleinstreuer (2006), *Biofluid Dynamics: Principles and Selected Applications*, CRC Press.

P. Kohl and F. Sachs (2001), 'Mechanoelectric feedback in cardiac cells', *Philos. Trans. Royal Soc. A* **359**, 1173–1185.

P. Kohl, P. Hunter and D. Noble (1999), 'Stretch-induced changes in heart rate and rhythm: Clinical observations, experiments and mathematical models', *Prog. Biophys. Molec. Biol.* **71**, 91–138.

E. Konukoglu, J. Relan, U. Cilingir, B. Menze, P. Chinchapatnam, A. Jadidi, H. Cochet, M. Hocini, H. Delingette, P. Jaïs, M. Haïssaguerre, N. Ayache and M. Sermesant (2011), 'Efficient probabilistic model personalization integrating uncertainty on data and parameters: Application to eikonal-diffusion models in cardiac electrophysiology', *Prog. Biophys. Molec. Biol.* **107**, 134–146.

T. Korakianitis and Y. Shi (2006*a*), 'A concentrated parameter model for the human cardiovascular system including heart valve dynamics and atrioventricular interaction', *Med. Engrg Phys.* **28**, 613–628.

T. Korakianitis and Y. Shi (2006*b*), 'Numerical simulation of cardiovascular dynamics with healthy and diseased heart valves', *J. Biomech.* **39**, 1964–1982.

P. Koutsourelakis (2009), 'Accurate uncertainty quantification using inaccurate models', *SIAM J. Sci. Comput.* **31**, 3274–3300.

A. Krishnamurthy, C. Villongco, J. Chuang, L. Frank, V. Nigam, E. Belezzuoli, P. Stark, D. Krummen, S. Narayan, J. Omens, A. McCulloch and R. Kerckhoffs (2013), 'Patient-specific models of cardiac biomechanics', *J. Comput. Phys.* **244**, 4–21.

R. Kufahl and M. Clark (1985), 'A circle of Willis simulation using distensible vessels and pulsatile flow', *J. Biomech. Engrg* **107**, 112–122.

E. Kuhl and G. Holzapfel (2006), 'A continuum model for remodeling in living structures', *J. Mater. Sci.* **21**, 8811–8823.

K. Kunisch and A. Rund (2015), 'Time optimal control of the monodomain model in cardiac electrophysiology', *IMA J. Appl. Math.* **80**, 1664–1683.

K. Kunisch and B. Vexler (2007), 'Optimal vortex reduction for instationary flows based on translation invariant cost functionals', *SIAM J. Control Optim.* **46**, 1368–1397.

K. Kunzelman and R. Cochran (1990), 'Mechanical properties of basal and marginal mitral valve chordae tendineae', *ASAIO Trans.* **36**, M405.

K. Kunzelman, R. Cochran, C. Chuong, W. Ring, Verrier and R. Eberhart (1993), 'Finite element analysis of the mitral valve', *J. Heart Valve Disease* **2**, 326–340.

U. Kuttler and W. Wall (2008), 'Fixed-point fluid–structure interaction solvers with dynamic relaxation', *Comput. Mech.* **43**, 61–72.

U. Kuttler, M. Gee, C. Forster, A. Comerford and W. Wall (2010), 'Coupling strategies for biomedical fluid–structure interaction problems', *Int. J. Numer. Methods Biomed. Engrg* **26**, 305–321.

A. Laadhari and A. Quarteroni (2016), 'Numerical modeling of heart valves using resistive Eulerian surfaces', *Int. J. Numer. Methods Biomed. Engrg* **32**, e02743.

R. Lal, B. Mohammadi and F. Nicoud (2016), 'Data assimilation for identification of cardiovascular network characteristics', *Int. J. Numer. Methods Biomed. Engrg.* https://doi.org/10.1002/cnm.2824

R. Lancellotti, C. Vergara, L. Valdettaro, S. Bose and A. Quarteroni (2015), Large eddy simulations for blood fluid-dynamics in real stenotic carotids. MOX-Report 63-2015, Department of Mathematics, Politecnico di Milano.

S. Land, S. Niederer, J. Aronsen, E. Espe, L. Zhang, W. Louch, I. Sjaastad, O. Sejersted and N. Smith (2012), 'An analysis of deformation-dependent electromechanical coupling in the mouse heart', *J. Physiol.* **590**, 4553–4569.

T. Lassila, A. Manzoni, A. Quarteroni and G. Rozza (2013*a*), 'Boundary control and shape optimization for the robust design of bypass anastomoses under uncertainty', *ESAIM Math. Model. Numer. Anal.* **47**, 1107–1131.

T. Lassila, A. Manzoni, A. Quarteroni and G. Rozza (2013*b*), 'A reduced computational and geometrical framework for inverse problems in haemodynamics', *Int. J. Numer. Methods Biomed. Engrg* **29**, 741–776.

T. Lassila, A. Quarteroni and G. Rozza (2012), 'A reduced basis model with parametric coupling for fluid–structure interaction problems', *SIAM J. Sci. Comput.* **34**, A1187–A1213.

K. Law, A. Stuart and K. Zygalakis (2015), *Data Assimilation: A Mathematical Introduction*, Vol. 62 of Texts in Applied Mathematics, Springer.

T. Le and F. Sotiropoulos (2013), 'Fluid–structure interaction of an aortic heart valve prosthesis driven by an animated anatomic left ventricle', *J. Comput. Phys.* **224**, 41–62.

F.-X. Le Dimet and O. Talagrand (1986), 'Variational algorithms for analysis and assimilation of meteorological observations: Theoretical aspects', *Tellus A* **38**, 97–110.

O. Le Maître and O. Knio (2010), *Spectral Methods for Uncertainty Quantification: With Applications to Computational Fluid Dynamics*, Springer.

H. Lee (2011), 'Optimal control for quasi-Newtonian flows with defective boundary conditions', *Comput. Methods Appl. Mech. Engrg* **200**, 2498–2506.

J. Lee, A. Cookson, I. Roy, E. Kerfoot, L. Asner, G. Vigueras, T. Sochi, S. Deparis, C. Michler, N. Smith and D. Nordsletten (2016), 'Multiphysics computational modeling in CHeart', *SIAM J. Sci. Comput.* **38**, C150–C178.

S. Lee, S. Lee, P. Fischer, H. Bassiouny and F. Loth (2008), 'Direct numerical simulation of transitional flow in a stenosed carotid bifurcation', *J. Biomech.* **41**, 2551–2561.

M. Legato (1973), 'Ultrastructure of the atrial, ventricular, and Purkinje cell, with special reference to the genesis of arrhythmias', *Circulation* **47**, 178–189.

C. Leguy, A. Bosboom, A. Belloum, A. Hoeks and F. van de Vosse (2011), 'Global sensitivity analysis of a wave propagation model for arm arteries', *Med. Engrg Phys.* **33**, 1008–1016.

M. Lei, J. Archie and C. Kleinstreuer (1997), 'Computational design of a bypass graft that minimizes wall shear stress gradients in the region of the distal anastomosis', *J. Vasc. Surg.* **25**, 637–646.

J. Leiva, P. Blanco and G. Buscaglia (2011), 'Partitioned analysis for dimensionally-heterogeneous hydraulic networks', *Multiscale Model. Simul.* **9**, 872–903.

J. Leray (1934), 'Sur le mouvement d'un liquide visqueux emplissant l'espace', *Acta Math.* **63**, 193–248.

D. Lesage, E. Angelini, I. Bloch and G. Funka-Lea (2009), 'A review of 3D vessel lumen segmentation techniques: Models, features and extraction schemes', *Med. Image Anal.* **13**, 819–845.

A. Leuprecht, K. Perktold, M. Prosi, T. Berk, W. Trubel and H. Schima (2002), 'Numerical study of hemodynamics and wall mechanics in distal end-to-side anastomoses of bypass grafts', *J. Biomech.* **2**, 225–236.

R. LeVeque (1992), *Numerical Methods for Conservation Laws*, Vol. 132 of Lectures in Mathematics, Birkhäuser.

C. Li and C. Vuik (2004), 'Eigenvalue analysis of the simple preconditioning for incompressible flow', *Numer. Linear Algebra Appl.* **11**, 511–523.

D. Li and A. Robertson (2013), 'A structural multi-mechanism damage model for cerebral arterial tissue', *J. Biomech. Engrg* **131**, 101013.

F. Liang and H. Liu (2006), 'Simulation of hemodynamic responses to the valsalva maneuver: An integrative computational model of the cardiovascular system and the autonomic nervous system', *J. Physiol. Sci.* **1**, 45–65.

J. Lions and G. Prodi (1959), 'Un théorème d'existence et d'unicité dans les équations de Navier–Stokes en dimension 2', *CR Acad. Sci. Paris* **248**, 3519–3521.

Y. Liu, C. Charles, M. Gracia, H. Gregersen and G. S. Kassab (2007), 'Surrounding tissues affect the passive mechanics of the vessel wall: Theory and experiment', *Amer. J. Physiol.: Heart Circ. Physiol.* **293**, H3290–H3300.

D. Lombardi (2014), 'Inverse problems in 1D hemodynamics on systemic networks: A sequential approach', *Int. J. Numer. Methods Biomed. Engrg* **30**, 160–179.

M. Lorenzo-Valdés, G. Sanchez-Ortiz, A. Elkington, R. Mohiaddin and D. Rueckert (2004), 'Segmentation of 4D cardiac MR images using a probabilistic atlas and the EM algorithm', *Med. Image Anal.* **8**, 255–265.

F. Loth, P. Fischer and H. Bassiouny (2008), 'Blood flow in end-to-side anastomoses', *Annu. Rev. Fluid. Mech.* **40**, 367–393.

C. Luo and Y. Rudy (1991), 'A model of the ventricular cardiac action potential: Depolarization, repolarization, and their interaction', *Circ. Res.* **68**, 1501–1526.

C. Luo and Y. Rudy (1994*a*), 'A dynamic model of the cardiac ventricular action potential, I: Simulations of ionic currents and concentration changes', *Circ. Res.* **74**, 1071–1096.

C. Luo and Y. Rudy (1994*b*), 'A dynamic model of the cardiac ventricular action potential, II: Afterdepolarizations, triggered activity, and potentiation', *Circ. Res.* **74**, 1097–1113.

M. Lykaser and B. Nielsen (2006), 'Towards a level set framework for infarction modeling: An inverse problem', *Int. J. Numer. Anal. Model.* **3**, 377–394.

M. MacLachlan, B. Nielsen, M. Lysaker and A. Tveito (2006), 'Computing the size and location of myocardial ischemia using measurements of ST-segment shift', *IEEE Trans. Biomed. Engrg* **53**, 1024–1031.

Y. Maday (2009), Analysis of coupled models for fluid–structure interaction of internal flows. Chapter 8 in *Cardiovascular Mathematics* (A. Quarteroni, L. Formaggia and A. Veneziani), Springer, pp. 279–306.

A. Mahmoud, A. El-Barkouky, H. Farag, J. Graham and A. Farag (2013), A non-invasive method for measuring blood flow rate in superficial veins from a single thermal image. In *2013 IEEE Conference on Computer Vision and Pattern Recognition Workshops (CVPRW)*, IEEE, pp. 354–359.

A. Malossi (2012), Partitioned solution of geometrical multiscale problems for the cardiovascular system: Models, algorithms, and applications. PhD thesis, Ecole Polytechnique Fédérale de Lausanne. Thesis 5453.

A. Malossi, P. Blanco and S. Deparis (2012), 'A two-level time step technique for the partitioned solution of one-dimensional arterial networks', *Comput. Methods Appl. Mech. Engrg* **237–240**, 212–226.

A. Malossi, P. Blanco, P. Crosetto, S. Deparis and A. Quarteroni (2013), 'Implicit coupling of one-dimensional and three-dimensional blood flow models with compliant vessels', *Multiscale Model. Simul.* **11**, 474–506.

T. Mansi, I. Voigt, B. Georgescu, X. Zheng, E. Mengue, M. Hackl, R. Ionasec, T. Noack, J. Seeburger and D. Comaniciu (2012), 'An integrated framework for finite-element modeling of mitral valve biomechanics from medical images: Application to MitralClip intervention planning', *Med. Image Anal.* **16**, 1330–1346.

A. Manzoni (2012), Reduced models for optimal control, shape optimization and inverse problems in haemodynamics. PhD thesis, Ecole Polytechnique Fédérale de Lausanne.

A. Manzoni, T. Lassila, A. Quarteroni and G. Rozza (2014), A reduced-order strategy for solving inverse Bayesian shape identification problems in physiological flows. In *Modeling, Simulation and Optimization of Complex Processes, HPSC 2012: Proceedings of the Fifth International Conference on High Performance Scientific Computing* (G. Bock, P. Hoang, R. Rannacher and P. Schlöder, eds), Springer, pp. 145–155.

A. Manzoni, S. Pagani and T. Lassila (2016), 'Accurate solution of Bayesian inverse uncertainty quantification problems combining reduced basis methods and reduction error models', *SIAM/ASA J. Uncert. Quant.* **4**, 380–412.

A. Manzoni, A. Quarteroni and G. Rozza (2012*a*), 'Model reduction techniques for fast blood flow simulation in parametrized geometries', *Int. J. Numer. Methods Biomed. Engrg* **28**, 604–625.

A. Manzoni, A. Quarteroni and G. Rozza (2012*b*), 'Shape optimization of cardiovascular geometries by reduced basis methods and free-form deformation techniques', *Int. J. Numer. Methods Fluids* **70**, 646–670.

S. Marchesseau, H. Delingette, M. Sermesant and N. Ayache (2012), 'Fast parameter calibration of a cardiac electromechanical model from medical images based on the unscented transform', *Biomech. Model. Mechanobiol.* **12**, 815–831.

S. Marchesseau, H. Delingette, M. Sermesant, R. Cabrera-Lozoya, C. Tobon-Gomez, P. Moireau, R. Figueras i Ventura, K. Lekadir, A. Hernandez, M. Garreau, E. Donal, C. Leclercq, S. Duckett, K. Rhode, C. Rinaldi, A. Frangi, R. Razavi, D. Chapelle and N. Ayache (2013), 'Personalization of a cardiac electromechanical model using reduced order unscented Kalman filtering from regional volumes', *Med. Image Anal.* **17**, 816–829.

K. Mardal, B. Nielsen, X. Cai and A. Tveito (2007), 'Semi-implicit time-discretization schemes for the bidomain modelan order optimal solver for the discretized bidomain equations', *Numer. Linear Algebra Appl.* **14**, 83–98.

K. Margaris and R. Black (2012), 'Modelling the lymphatic system: Challenges and opportunities', *J. Royal Soc. Interface* **69**, 601–612.

G. Marom (2015), 'Numerical methods for fluid–structure interaction models of aortic valves', *Arch. Comput. Methods Engrg* **22**, 595–620.

A. Marsden (2014), 'Optimization in cardiovascular modeling', *Annu. Rev. Fluid Mech.* **46**, 519–546.

A. Marsden, J. Feinstein and C. Taylor (2008), 'A computational framework for derivative-free optimization of cardiovascular geometries', *Comput. Methods Appl. Mech. Engrg* **197**, 1890–1905.

V. Martin, F. Clément, A. Decoene and J.-F. Gerbeau (2005), 'Parameter identification for a one-dimensional blood flow model', *ESAIM Proc.* **14**, 174–200.

K. Matthys, J. Alastruey, J. Peiró, A. Khir, P. Segers, P. Verdonck, K. Parker and S. Sherwin (2007), 'Pulse wave propagation in a model human arterial network: Assessment of 1-D numerical simulations against *in vitro* measurements', *J. Biomech.* **40**, 3476–3486.

B. Maury (2013), *The Respiratory System in Equations*, Springer.

K. May-Newman and F. Yin (1998), 'A constitutive law for mitral valve tissue', *J. Biomech. Engrg* **120**, 38–47.

K. May-Newman, C. Lam and F. Yin (2009), 'A hyperelastic constitutive law for aortic valve tissue', *J. Biomech. Engrg* **131**, 081009.

W. Merryman, H. Huang, F. Schoen and M. Sacks (2006), 'The effects of cellular contraction on aortic valve leaflet flexural stiffness', *J. Biomech.* **39**, 88–96.

F. Migliavacca and G. Dubini (2005), 'Computational modeling of vascular anastomoses', *Biomech. Model. Mechanobiol.* **3**, 235–250.

F. Migliavacca, R. Balossino, G. Pennati, G. Dubini, T. Hsia, M. de Leval and E. Bove (2006), 'Multiscale modelling in biofluidynamics: Application to reconstructive paediatric cardiac surgery', *J. Biomech.* **39**, 1010–1020.

R. Mittal and G. Iaccarino (2005), 'Immersed boundary methods', *Annu. Rev. Fluid Mech.* **37**, 239–261.

R. Mittal, J. Seo, V. Vedula, Y. Choi, H. Liu, H. Huang, S. Jain, L. Younes, T. Abraham and R. George (2016), 'Computational modeling of cardiac hemodynamics: Current status and future outlook', *J. Comput. Phys.* **305**, 1065–1082.

P. Moireau and D. Chapelle (2011), 'Reduced-order unscented Kalman filtering with application to parameter identification in large-dimensional systems', *ESAIM Control Optim. Calc. Var.* **17**, 380–405.

P. Moireau, C. Bertoglio, N. Xiao, C. Figueroa, C. Taylor, D. Chapelle and J.-F. Gerbeau (2013), 'Sequential identification of boundary support parameters in a fluid–structure vascular model using patient image data', *Biomech. Model. Mechanobiol.* **12**, 475–496.

P. Moireau, D. Chapelle and P. Le Tallec (2008), 'Joint state and parameter estimation for distributed mechanical systems', *Comput. Methods Appl. Mech. Engrg* **197**, 659–677.

P. Moireau, D. Chapelle and P. Le Tallec (2009), 'Filtering for distributed mechanical systems using position measurements: perspectives in medical imaging', *Inverse Problems* **25**, 035010.

P. Moireau, N. Xiao, M. Astorino, C. Figueroa, D. Chapelle, C. Taylor and J.-F. Gerbeau (2012), 'External tissue support and fluid–structure simulation in blood flows', *Biomech. Model. Mechanobiol.* **11**, 1–18.

J. Moore, D. Steinman and C. Ethier (1997), 'Computational blood flow modelling: Errors associated with reconstructing finite element models from magnetic resonance images', *J. Biomech.* **31**, 179–184.

H. Moradkhani, S. Sorooshian, H. Gupta and P. Houser (2005), 'Dual state-parameter estimation of hydrological models using ensemble Kalman filter', *Adv. Water Resour.* **28**, 135–147.

U. Morbiducci, R. Ponzini, G. Rizzo, M. Cadioli, A. Esposito, F. De Cobelli, A. Del Maschio, F. Montevecchi and A. Redaelli (2009), '*In vivo* quantification of helical blood flow in human aorta by time-resolved three-dimensional cine phase contrast magnetic resonance imaging', *Ann. Biomed. Engrg* **37**, 516–531.

D. Mozaffarian, E. Benjamin, A. Go, D. Arnett, M. Blaha, M. Cushman, S. Das, S. de Ferranti, J.-P. Després, H. Fullerton, V. Howard, M. Huffman, C. Isasi, M. Jiménez, S. Judd, B. Kissela, J. Lichtman, L. Lisabeth, S. Liu, R. Mackey, D. Magid, D. McGuire, E. Mohler, C. Moy, P. Muntner, M. Mussolino, K. Nasir, R. Neumar, G. Nichol, L. Palaniappan, D. Pandey, M. Reeves, C. Rodriguez, W. Rosamond, P. Sorlie, J. Stein, A. Towfighi, T. Turan, S. Virani, D. Woo, R. Yeh and M. Turner (2015), 'Heart disease and stroke statistics: 2016 update', *Circulation* **133**, 447–454.

J. Muller, O. Sahni, X. Lia, K. Jansen, M. Shephard and C. Taylor (2005), 'Anisotropic adaptive finite element method for modelling blood flow', *Comput. Methods Biomech. Biomed. Engrg* **8**, 295–305.

L. Muller and E. Toro (2013), 'Well-balanced high-order solver for blood flow in networks of vessels with variable properties', *Int. J. Numer Methods Biomed. Engrg* **29**, 1388–1411.

L. Muller and E. Toro (2014), 'A global multiscale mathematical model for the human circulation with emphasis on the venous system', *Int. J. Numer. Methods Biomed. Engrg* **30**, 681–725.

M. Munteanu, L. Pavarino and S. Scacchi (2009), 'A scalable Newton–Krylov–Schwarz method for the bidomain reaction-diffusion system', *SIAM J. Sci. Comput.* **31**, 3861–3883.

M. Murillo and X. Cai (2004), 'A fully implicit parallel algorithm for simulating the non-linear electrical activity of the heart', *Numer. Linear Algebra Appl.* **2/3**, 261–277.

A. Muszkiewicz, O. Britton, P. Gemmell, E. Passini, C. Sánchez, X. Zhou, A. Carusi, T. Quinn, K. Burrage, A. Bueno-Orovio and B. Rodriguez (2016), 'Variability in cardiac electrophysiology: Using experimentally-calibrated populations of models to move beyond the single virtual physiological human paradigm', *Prog. Biophys. Molec. Biol.* **120**, 115–127.

C. Nagaiah, K. Kunisch and G. Plank (2011), 'Numerical solution for optimal control of the reaction-diffusion equations in cardiac electrophysiology', *Comput. Optim. Appl.* **49**, 149–178.

C. Nagaiah, K. Kunisch and G. Plank (2013*a*), 'On boundary stimulation and optimal boundary control of the bidomain equations', *Math. Biosci.* **245**, 206–215.

C. Nagaiah, K. Kunisch and G. Plank (2013*b*), 'Optimal control approach to termination of re-entry waves in cardiac electrophysiology', *J. Math. Biol.* **67**, 1–30.

C. Nagaiah, K. Kunisch and G. Plank (2016), 'PDE constrained optimization of electrical defibrillation in a 3D ventricular slice geometry', *Int. J. Numer. Methods Biomed. Engrg* **32**, e02742.

A. Nagler, C. Bertoglio, M. Gee and W. Wall (2013), Personalization of cardiac fiber orientations from image data using the unscented Kalman filter. In *Functional Imaging and Modeling of the Heart* (S. Ourselin, D. Rueckert and N. Smith, eds), Vol. 7945 of Lecture Notes in Computer Science, Springer, pp. 132–140.

A. Nagler, C. Bertoglio, C. Stoeck, S. Kozerke and W. Wall (2015), Cardiac fibers estimation from arbitrarily spaced diffusion weighted MRI. In *Functional Imaging and Modeling of the Heart: FIMH 2015* (H. van Assen *et al.*, eds), Vol. 9126 of Lecture Notes in Computer Science, Springer.

M. Nash and A. Panfilov (2004), 'Electromechanical model of excitable tissue to study reentrant cardiac arrhythmias', *Prog. Biophys. Molec. Biol.* **2/3**, 501–522.

F. Negri (2016), Efficient reduction techniques for the simulation and optimization of parametrized systems: Analysis and applications. PhD thesis, Ecole Polytechnique Fédérale de Lausanne.

M. Nestola, E. Faggiano, C. Vergara, R. Lancellotti, S. Ippolito, C. Antona, A. Quarteroni and R. Scrofani (2017), 'Computational comparison of aortic root stresses in presence of stentless and stented aortic valve bio-prostheses', *Comput. Methods Biomech. Biomed. Engrg* **20**, 171–181.

N. Newmark (1959), 'A method of computation for structural dynamics', *J. Engrg Mech.* **85**, 67–94.

N. Nichols (2010), Mathematical concepts of data assimilation. In *Data Assimilation: Making Sense of Observations* (W. Lahoz, B. Khattatov and R. Menard, eds), Springer, pp. 13–39.

W. Nichols and M. O'Rourke, eds (2005), *McDonald's Blood Flow in Arteries*, Hodder Arnold.

S. Niederer and N. Smith (2007), 'A mathematical model of the slow force response to stretch in rat ventricular myocytes', *Biophys. J.* **92**, 4030–4044.

S. Niederer and N. Smith (2008), 'An improved numerical method for strong coupling of excitation and contraction models in the heart', *Prog. Biophys. Molec. Biol.* **96**, 90–111.

S. Niederer, P. Hunter and N. Smith (2006), 'A quantitative analysis of cardiac myocyte relaxation: A simulation study', *Biophysical Journal* **90**, 1697–1722.

S. Niederer, E. Kerfoot, A. Benson, M. Bernabeu, O. Bernus, C. Bradley, E. Cherry, R. Clayton, F. Fenton, A. Garny *et al.* (2011), 'Verification of cardiac tissue electrophysiology simulators using an N-version benchmark', *Phil. Trans. R. Soc. A* **369**, 4331–4351.

B. Nielsen, X. Cai and M. Lykaser (2007*a*), 'On the possibility for computing the transmembrane potential in the heart with a one shot method: An inverse problem', *Math. Biosciences* **210**, 523–553.

B. Nielsen, M. Lykaser and A. Tveito (2007*b*), 'On the use of the resting potential and level set methods for identifying ischemic heart disease: An inverse problem', *J. Comput. Phys.* **220**, 772–790.

J. Nitsche (1970/71), 'Über ein Variationsprinzip zur lösung von Dirichlet-Problemen bei verwendung von Teilräumen, die keinen Randbedingungen unterworfen sind', *Abhandlungen aus dem Mathematischen Seminar der Universität Hamburg* **36**, 9–15.

F. Nobile (2001), Numerical approximation of fluid–structure interaction problems with application to haemodynamics. PhD thesis, Ecole Polytechnique Fédérale de Lausanne. Thesis 2458.

F. Nobile and C. Vergara (2008), 'An effective fluid–structure interaction formulation for vascular dynamics by generalized Robin conditions', *SIAM J. Sci. Comput.* **30**, 731–763.

F. Nobile and C. Vergara (2012), 'Partitioned algorithms for fluid–structure interaction problems in haemodynamics', *Milan J. Math.* **80**, 443–467.

F. Nobile, M. Pozzoli and C. Vergara (2013), 'Time accurate partitioned algorithms for the solution of fluid–structure interaction problems in haemodynamics', *Comput. Fluids* **86**, 470–482.

F. Nobile, M. Pozzoli and C. Vergara (2014), 'Inexact accurate partitioned algorithms for fluid–structure interaction problems with finite elasticity in haemodynamics', *J. Comput. Phys.* **273**, 598–617.

F. Nobile, R. Tempone and C. Webster (2008), 'A sparse grid stochastic collocation method for partial differential equations with random input data', *SIAM J. Numer. Anal.* **46**, 2309–2345.

J. Nocedal (1992), Theory of algorithms for unconstrained optimization. In *Acta Numerica*, Vol. 1, Cambridge University Press, pp. 199–242.

D. Nordsletten, M. McCormick, P. Kilner, P. Hunter, D. Kayand and N. Smith (2011*a*), 'Fluid–solid coupling for the investigation of diastolic and systolic human left ventricular function', *Int. J. Numer. Methods Biomed. Engrg* **27**, 1017–1039.

D. Nordsletten, S. Niederer, M. Nash, P. Hunter and N. Smith (2011*b*), 'Coupling multi-physics models to cardiac mechanics', *Prog. Biophys. Molec. Biol.* **104**, 77–88.

A. Oberai, N. Gokhale and G. Feijóo (2003), 'Solution of inverse problems in elasticity imaging using the adjoint method', *Inverse Problems* **19**, 297–313.

T. O'Donnell, M. Jolly and A. Gupta (1998), 'A cooperative framework for segmentation using 2D active contours and 3D hybrid models as applied to branching cylindrical structures', *Proc. IEEE Int. Conf. Computer Vision* pp. 454–459.

M. Olufsen, C. Peskin, W. Kim, E. Pedersen, A. Nadim and J. Larsen (2000), 'Numerical simulation and experimental validation of blood flow in arteries with structured-tree outflow conditions', *Ann. Biomed. Engrg* **28**, 1281–1299.

S. Orszag, M. Israeli and M. Deville (1986), 'Boundary conditions for incompressible flows', *J. Sci. Comput.* **1**, 75–111.

H. Osnes and J. Sundnes (2012), 'Uncertainty analysis of ventricular mechanics using the probabilistic collocation method', *IEEE Trans. Biomed. Engrg* **59**, 2171–2179.

A. Owida, H. Do and Y. Morsi (2012), 'Numerical analysis of coronary artery bypass grafts: An over view', *Comput. Methods Programs Biomed.* **108**, 689–705.

M. Padala, M. Sacks, S. Liou, K. Balachandran, Z. He and A. Yoganathan (2010), 'Mechanics of the mitral valve strut chordae insertion region', *J. Biomech. Engrg* **132**, 081004.

S. Pagani (2016), Reduced order models for inverse problems and uncertainty quantification in cardiac electrophysiology. PhD thesis, Mathematical Models and Methods in Engineering, Department of Mathematics, Politecnico di Milano.

S. Pagani, A. Manzoni and A. Quarteroni (2016), A reduced basis ensemble Kalman filter for state/parameter identification in large-scale nonlinear dynamical systems. MOX-Report 24-2016, Department of Mathematics, Politecnico di Milano.

S. Palamara, C. Vergara, D. Catanzariti, E. Faggiano, M. Centonze, C. Pangrazzi, M. Maines and A. Quarteroni (2014), 'Computational generation of the Purkinje network driven by clinical measurements: The case of pathological propagations.', *Int. J. Numer. Meth. Biomed. Engrg* **30**, 1558–1577.

S. Palamara, C. Vergara, E. Faggiano and F. Nobile (2015), 'An effective algorithm for the generation of patient-specific Purkinje networks in computational electrocardiology', *J. Comput. Phys.* **283**, 495–517.

A. Panfilov (1999), 'Three-dimensional organization of electrical turbulence in the heart', *Phys. Rev. E* **59**, R6251–R6254.

S. Pant, B. Fabrèges, J.-F. Gerbeau and I. Vignon-Clementel (2014), 'A methodological paradigm for patient-specific multi-scale CFD simulations: From clinical measurements to parameter estimates for individual analysis', *Int. J. Numer. Meth. Biomed. Engrg* **30**, 1614–1648.

G. Papadakis (2009), 'Coupling 3D and 1D fluid–structure-interaction models for wave propagation in flexible vessels using a finite volume pressure-correction scheme', *Comm. Numer. Methods Engrg* **25**, 533–551.

S. Patankar and D. Spalding (1972), 'A calculation procedure for heat, mass and momentum transfer in three-dimensional parabolic flows', *Int. J. Heat Mass Transfer* **15**, 1787–1806.

P. Pathmanathan, S. Chapman, D. Gavaghan and J. Whiteley (2010), 'Cardiac electromechanics: The effect of contraction model on the mathematical problem and accuracy of the numerical scheme', *Quart. J. Mech. Appl. Math.* **63**, 375–399.

P. Pathmanathan, G. Mirams, J. Southern and J. Whiteley (2011), 'The significant effect of the choice of ionic current integration method in cardiac electrophysiological simulations', *Int. J. Numer. Methods Biomed. Engrg* **27**, 1751–1770.

L. Pavarino and S. Scacchi (2008), 'Multilevel additive Schwarz preconditioners for the bidomain reaction-diffusion system', *SIAM J. Sci. Comput* **31**, 420–443.

L. Pavarino, S. Scacchi and S. Zampini (2015), 'Newton–Krylov–BDDC solvers for nonlinear cardiac mechanics', *Comput. Methods Appl. Mech. Engrg* **295**, 562–580.

T. Pedley (1980), *The Fluid Mechanics of Large Blood Vessels*, Cambridge Monographs on Mechanics and Applied Mathematics, Cambridge University Press.

J. Peiró and A. Veneziani (2009), Reduced models of the cardiovascular system. Chapter 10 in *Cardiovascular Mathematics* (A. Quarteroni, L. Formaggia and A. Veneziani, eds), Springer, pp. 347–394.

J. Peiró, L. Formaggia, M. Gazzola, A. Radaelli and V. Rigamonti (2007), 'Shape reconstruction from medical images and quality mesh generation via implicit surfaces', *Int. J. Numer. Methods Fluids* **53**, 1339–1360.

M. Pennacchio and V. Simoncini (2002), 'Efficient algebraic solution of reaction-diffusion systems for the cardiac excitation process', *J. Comput. Appl. Math.* **145**, 49–70.

P. Perdikaris and G. Karniadakis (2015), 'Model inversion via multi-fidelity Bayesian optimization: A new paradigm for parameter estimation in haemodynamics, and beyond', *J. Royal Soc. Interface* **13**, 20151107.

M. Perego, A. Veneziani and C. Vergara (2011), 'A variational approach for estimating the compliance of the cardiovascular tissue: An inverse fluid–structure interaction problem', *SIAM J. Sci. Comput.* **33**, 1181–1211.

K. Perktold and D. Hilbert (1986), 'Numerical simulation of pulsatile flow in a carotid bifurcation model', *J. Biomed. Engrg* **8**, 193–199.

K. Perktold, E. Thurner and T. Kenner (1994), 'Flow and stress characteristics in rigid walled and compliant carotid artery bifurcation models', *Med. Biol. Engrg Comput.* **32**, 19–26.

P. Perona and J. Malik (1990), 'Scale-space and edge detection using anisotropic diffusion', *IEEE Trans. Pattern Anal. Mach. Intel.* **12**, 629–639.

C. Peskin (1972), 'Flow patterns around heart valves: A numerical method', *J. Comput. Phys.* **10**, 252–271.

C. Peskin (1989), 'Fiber architecture of the left ventricular wall: An asymptotic analysis', *Comm. Pure Appl. Math.* **42**, 1126–1131.

C. Peskin (2002), The immersed boundary method. In *Acta Numerica*, Vol. 11, Cambridge University Press, pp. 479–517.

C. Petitjean and J.-N. Dacher (2011), 'A review of segmentation methods in short axis cardiac MR images', *Med. Image Anal.* **15**, 169–184.

S. Pezzuto (2013), Mechanics of the heart: Constitutive issues and numerical experiments. PhD thesis, Department of Mathematics, Politecnico di Milano.

D. Pham (2001), 'Stochastic methods for sequential data assimilation in strongly nonlinear systems', *Monthly Weather Rev.* **129**, 1194–1207.

D. Pham, C. Xu and J. Prince (2000), 'Current methods in medical image segmentation', *Annu. Rev. Biomed. Engrg* **2**, 315–337.

M. Piccinelli, C. Vergara, L. Antiga, L. Forzenigo, P. Biondetti and M. Domanin (2013), 'Impact of hemodynamics on lumen boundary displacements in abdominal aortic aneurysms by means of dynamic computed tomography and computational fluid dynamics', *Biomech. Model. Mechanobiol.* **12**, 1263–1276.

C. Pierre (2012), 'Preconditioning the bidomain model with almost linear complexity', *J. Comput. Phys.* **231**, 82–97.

S. Piperno and C. Farhat (2001), 'Partitioned prodecures for the transient solution of coupled aeroelastic problems, II: Energy transfer analysis and three-dimensional applications', *Comput. Methods Appl. Mech. Engrg* **190**, 3147–3170.

G. Plank, M. Liebmann, R. dos Santos, E. Vigmond and G. Haase (2007), 'Algebraic multigrid preconditioner for the cardiac bidomain model', *IEEE Trans. Biomed. Engrg* **54**, 585–596.

R. Ponzini, C. Vergara, A. Redaelli and A. Veneziani (2006), 'Reliable CFD-based estimation of flow rate in haemodynamics measures', *Ultrasound Med. Biol.* **32**, 1545–1555.

R. Ponzini, C. Vergara, G. Rizzo, A. Veneziani, A. Roghi, A. Vanzulli, O. Parodi and A. Redaelli (2010), 'Womersley number-based estimates of blood flow rate in Doppler analysis: *In vivo* validation by means of phase contrast magnetic resonance imaging', *IEEE Trans. Biomed. Engrg* **57**, 1807–1815.

A. Porpora, P. Zunino, C. Vergara and M. Piccinelli (2012), 'Numerical treatment of boundary conditions to replace lateral branches in haemodynamics', *Int. J. Numer. Methods Biomed. Engrg* **28**, 1165–1183.

M. Potse, B. Dubé, J. Richer, A. Vinet and R. Gulrajani (2006), 'A comparison of monodomain and bidomain reaction-diffusion models for action potential propagation in the human heart', *IEEE Trans. Biomed. Engrg* **53**, 2425–2435.

S. Pravdin, V. Berdyshev, A. Panfilov, L. Katsnelson, O. Solovyova and V. Markhasin (1989), 'Mathematical model of the anatomy and fibre orientation field of the left ventricle of the heart', *Biomed. Engrg OnLine* **12**, 54.

G. Prodi (1962), 'Teoremi di tipo locale per il sistema di Navier–Stokes e stabilitá delle soluzioni stazionarie', *Rendiconti del Seminario Matematico della Universitá di Padova* **32**, 374–397.

V. Prot, B. Skallerud and G. Holzapfel (2007), 'Transversely isotropic membrane shells with application to mitral valve mechanics. constitutive modelling and finite element implementation', *Int. J. Numer. Methods Engrg* **71**, 987–1008.

A. Pullan, L. Cheng, M. Nash, C. Bradley and D. Paterson (2001), 'Noninvasive electrical imaging of the heart: Theory and model development', *Ann. Biomed. Engrg* **29**, 817–836.

A. Pullan, L. Cheng, M. Nash, A. Ghodrati, R. MacLeod and D. Brooks (2010), The inverse problem of electrocardiography. In *Comprehensive Electrocardiology* (P. Macfarlane *et al.*, eds), Springer, pp. 299–344.

S. Puwal and B. Roth (2007), 'Forward Euler stability of the bidomain model of cardiac tissue', *IEEE Trans. Biomed. Engrg* **5**, 951–953.

Z. Qu and A. Garfinkel (1998), 'An advanced algorithm for solving partial differential equation in cardiac conduction', *IEEE Trans. Biomed. Engrg* **46**, 1166–1168.

A. Quarteroni, ed. (2015), *Modeling the Heart and the Circulatory System*, Springer.

A. Quarteroni and L. Formaggia (2004), Mathematical modelling and numerical simulation of the cardiovascular system. In Handbook of Numerical Analysis, Vol. 12: *Computational Models for the Human Body*, Elsevier, pp. 3–127.

A. Quarteroni and G. Rozza (2003), 'Optimal control and shape optimization of aorto-coronaric bypass anastomoses', *Math. Models Methods Appl. Sci.* **13**, 1801–1823.

A. Quarteroni and G. Rozza, eds (2014), *Reduced Order Methods for Modeling and Computational Reduction*, Vol. 9 of Modeling, Simulation and Applications, Springer.

A. Quarteroni and A. Valli (1994), *Numerical Approximation of Partial Differential Equations*, Springer.

A. Quarteroni and A. Veneziani (1997), Modeling and simulation of blood flow problems. In *Computational Science for the 21st Century* (M.-O. Bristeau *et al.*, eds), Wiley.

A. Quarteroni and A. Veneziani (2003), 'Analysis of a geometrical multiscale model based on the coupling of ODE and PDE for blood flow simulations', *Multiscale Model. Simul.* **1**, 173–195.

A. Quarteroni, T. Lassila, S. Rossi and R. Ruiz-Baier (2017), 'Integrated heart-coupling multiscale and multiphysics models for the simulation of the cardiac function', *Comput. Methods Appl. Mech. Engrg* **314**, 345–407.

A. Quarteroni, A. Manzoni and F. Negri (2016), *Reduced Basis Methods for Partial Differential Equations. An Introduction*, Vol. 92 of Unitext, Springer.

A. Quarteroni, S. Ragni and A. Veneziani (2001), 'Coupling between lumped and distributed models for blood flow problems', *Comput. Vis. Sci.* **4**, 111–124.

A. Quarteroni, R. Sacco and F. Saleri (2000*a*), *Numerical Mathematics*, Springer.

A. Quarteroni, F. Saleri and A. Veneziani (1999), 'Analysis of the Yosida method for the incompressible Navier–Stokes equations', *J. Math. Pures Appl.* **78**, 473–503.

A. Quarteroni, F. Saleri and A. Veneziani (2000*b*), 'Factorization methods for the numerical approximation of Navier–Stokes equations', *Comput. Methods Appl. Mech. Engrg* **188**, 505–526.

A. Quarteroni, M. Tuveri and A. Veneziani (2000*c*), 'Computational vascular fluid dynamics: Problems, models and methods', *Comput. Vis. Sci.* **2**, 163–197.

A. Quarteroni, A. Veneziani and C. Vergara (2016*c*), 'Geometric multiscale modeling of the cardiovascular system, between theory and practice', *Comput. Methods Appl. Mech. Engrg* **302**, 193–252.

G. Querzoli, S. Fortini and A. Cenedese (2010), 'Effect of the prosthetic mitral valve on vortex dynamics and turbulence of the left ventricular flow', *Phys. Fluids* **22**, 041901.

M. Raghavan and D. Vorp (2000), 'Towards a biomechanical tool to evaluate rupture potential of abdominal aortic aneurysm: Identification of a finite strain constitutive model and evaluation of its applicability', *J. Biomech.* **33**, 475–482.

R. Rannacher (1992), *On Chorin's Projection Method for Incompressible Navier–Stokes Equations*, Vol. 1530 of Lecture Notes in Mathematics, Springer, pp. 167–183.

S. Raya and J. Udupa (1990), 'Shape-based interpolation of multidimensional objects', *IEEE Trans. Med. Imaging* **9**, 32–42.

V. Rayz, S. Berger and D. Saloner (2007), 'Transitional flows in arterial fluid dynamics', *Comput. Methods Appl. Mech. Engrg* **196**, 3043–3048.

T. Rees, H. Dollar and A. Wathen (2010), 'Optimal solvers for PDE-constrained optimization', *SIAM J. Sci. Comput.* **32**, 271–298.

J. Relan, P. Chinchapatnam, M. Sermesant, K. Rhode, M. Ginks, H. Delingette, C. Rinaldi, R. Razavi and N. Ayache (2011), 'Coupled personalization of cardiac electrophysiology models for prediction of ischaemic ventricular tachycardia', *Interface Focus* **1**, 396–407.

P. Reymond, F. Merenda, F. Perren, D. Rufenacht and N. Stergiopulos (2009), 'Validation of a one-dimensional model of the systemic arterial tree', *Amer. J. Physiol.: Heart Circ. Physiol.* **297**, H208–H222.

C. Robert and G. Casella (2004), *Monte Carlo Statistical Methods*, second edition, Springer.

A. Robertson, A. Sequeira and R. Owens (2009), Rheological models for blood. Chapter 6 in *Cardiovascular Mathematics* (L. Formaggia, A. Quarteroni and A. Veneziani, eds), Springer, pp. 211–241.

D. Robertson, J. Yuan, G. Wang and M. Vannier (1997), 'Total hip prosthesis metal-artifact suppression using iterative deblurring reconstruction', *J. Comput. Assist. Tomogr.* **21**, 293–298.

J. Rogers and A. McCulloch (1994), 'A collocation-Galerkin finite element model of cardiac action potential propagation', *IEEE Trans. Biomed. Engrg* **41**, 743–757.

S. Rossi (2014), Anisotropic modeling of cardiac mechanical activation. PhD thesis, Ecole Polytechnique Fédérale de Lausanne.

S. Rossi, T. Lassila, R. Ruiz-Baier, A. Sequeira and A. Quarteroni (2014), 'Thermodynamically consistent orthotropic activation model capturing ventricular systolic wall thickening in cardiac electromechanics', *Europ. J. Mech. A: Solids* **48**, 129–142.

B. Roth (1991), 'Action potential propagation in a thick strand of cardiac muscle', *Circ. Res.* **68**, 162–173.

B. Roth (1997), 'Electrical conductivity values used with the bidomain model of cardiac tissue', *IEEE Trans. Biomed. Engrg* **44**, 326–328.

O. Rousseau (2010), Geometrical modeling of the heart. PhD thesis, Université d'Ottawa.

Y. Rudy and B. Messinger-Rapport (1988), 'The inverse problem in electrocardiography: Solutions in terms of epicardial potentials', *Crit. Rev. Biomed. Engrg* **16**, 215–268.

Y. Rudy and J. Silva (2006), 'Computational biology in the study of cardiac ion channels and cell electrophysiology', *Quart. Rev. Biophys.* **39**, 57–116.

A. Saltelli, M. Ratto, T. Andres, F. Campolongo, J. Cariboni, D. Gatelli, M. Salsana and S. Tarantola (2008), *Global Sensitivity Analysis: The Primer*, Wiley.

S. Sankaran and A. Marsden (2010), 'The impact of uncertainty on shape optimization of idealized bypass graft models in unsteady flow', *Phys. Fluids* **22**, 121902.

S. Sankaran and A. Marsden (2011), 'A stochastic collocation method for uncertainty quantification and propagation in cardiovascular simulations', *J. Biomech. Engrg* **133**, 031001.

S. Sankaran, L. Grady and C. Taylor (2015), 'Impact of geometric uncertainty on hemodynamic simulations using machine learning', *Comput. Methods Appl. Mech. Engrg* **297**, 167–190.

S. Sankaran, H. Kim, G. Choi and C. Taylor (2016), 'Uncertainty quantification in coronary blood flow simulations: Impact of geometry, boundary conditions and blood viscosity', *J. Biomech.* **49**, 2540–2547.

R. dos Santos, G. Plank, S. Bauer and E. Vigmond (2005), Preconditioning techniques for the bidomain equations. In *Domain Decomposition Methods in Science and Engineering* (R. Kornhuber *et al.*, eds), Vol. 40 of Lecture Notes in Computational Science and Engineering, Springer, pp. 571–580.

S. Särkkä (2013), *Bayesian Filtering and Smoothing*, Cambridge University Press.

S. Savader, G. Lund and F. Osterman (1997), 'Volumetric evaluation of blood flow in normal renal arteries with a Doppler flow wire: A feasibility study', *J. Vasc. Intervent. Radiol.* **8**, 209–214.

I. Sazonov, S. Yeo, R. Bevan, X. Xie, R. van Loon and P. Nithiarasu (2011), 'Modelling pipeline for subject-specific arterial blood flow: A review', *Int. J. Numer. Methods Biomed. Engrg* **27**, 1868–1910.

D. Schiavazzi, G. Arbia, C. Baker, A. Hlavacek, T. Hsia, A. Marsden, I. Vignon-Clementel and the Modeling Of Congenital Hearts Alliance (MOCHA) investigators (2016), 'Uncertainty quantification in virtual surgery hemodynamics predictions for single ventricle palliation', *Int. J. Numer. Methods Biomed. Engrg* **32**, 02737.

R. Schilling, N. Peters and D. Davies (1998), 'Simultaneous endocardial mapping in the human left ventricle using a noncontact catheter comparison of contact and reconstructed electrograms during sinus rhythm', *Circulation* **98**, 887–898.

R. Sebastian, V. Zimmerman, D. Romero and A. Frangi (2011), 'Construction of a computational anatomical model of the perpheral cardiac conduction system', *IEEE Trans. Biomed. Engrg* **58**, 90–100.

J. Seo, V. Vedula, T. Abraham, A. Lardo, F. Dawoud, H. Luo and R. Mittal (2014), 'Effect of the mitral valve on diastolic flow patterns', *Phys. Fluids* **26**, 121901.

M. Sermesant, R. Chabiniok, P. Chinchapatnam, T. Mansi, F. Billet, P. Moireau, J. Peyrat, K. Wong, J. Relan, K. Rhode, M. Ginks, P. Lambiase, H. Delingette, M. Sorine, C. Rinaldi, D. Chapelle, R. Razavi and N. Ayache (2012), 'Patient-specific electromechanical models of the heart for the prediction of pacing acute effects in CRT: A preliminary clinical validation', *Med. Image Anal.* **16**, 201–215.

M. Sermesant, P. Moireau, O. Camara, J. Sainte-Marie, R. Andriantsimiavona, R. Cimrman, D. Hill, D. Chapelle and R. Razavi (2006), 'Cardiac function estimation from MRI using a heart model and data assimilation: Advances and difficulties', *Med. Image Anal.* **10**, 642–656.

J. Sethian (1999), *Level Set Methods and Fast Marching Methods*, Cambridge University Press.

S. Sherwin, L. Formaggia, J. Peiró and V. Franke (2003*a*), 'Computational modelling of 1D blood flow with variable mechanical properties and its application to the simulation of wave propagation in the human arterial system', *Int. J. Numer. Methods Fluids* **43**, 673–700.

S. Sherwin, V. Franke, J. Peiró and K. Parker (2003*b*), 'One-dimensional modelling of a vascular network in space-time variables', *J. Engrg Math.* **47**, 217–259.

D. Simon (2006), *Optimal State Estimation: Kalman, H-Infinity, and Nonlinear Approaches*, Wiley.

N. Smith, D. Nickerson, E. Crampin and P. Hunter (2004), Multiscale computational modelling of the heart. In *Acta Numerica*, Vol. 13, Cambridge University Press, pp. 371–431.

R. Spilker and C. Taylor (2010), 'Tuning multidomain hemodynamic simulations to match physiological measurements', *Ann. Biomed. Engrg* **38**, 2635–2648.

T. Stankovičová, V. Bito, F. Heinzel, K. Mubagwa and K. Sipido (2003), 'Isolation and morphology of single Purkinje cells from the porcine heart', *Gen. Physiol. Biophys.* **22**, 329–340.

B. Steele, J. Wan, J. Ku, T. Hughes and C. Taylor (2003), '*In vivo* validation of a one-dimensional finite-element method for predicting blood flow in cardiovascular bypass grafts', *IEEE Trans. Biomed. Engrg* **50**, 649–656.

D. Steinman, J. Thomas, H. Ladak, J. Milner, B. Rutt and J. Spence (2001), 'Reconstruction of carotid bifurcation hemodynamics and wall thickness using computational fluid dynamics and MRI', *Magnet. Reson. Med.* **47**, 149–159.

N. Stergiopulos, B. Westerhof and N. Westerhof (1999), 'Total arterial inertance as the fourth element of the windkessel model', *Amer. J. Physiol.: Heart Circ. Physiol.* **276**, H81–H88.

N. Stergiopulos, B. Westerhof, J. Meister and N. Westerhof (1996), The four-element windkessel model. In *Bridging Disciplines for Biomedicine: Proceedings of the 18th Annual International Conference of the IEEE*, Vol. 4, pp. 1715–1716.

J. Stijnen, J. de Hart, P. Bovendeerd and F. van de Vosse (2004), 'Evaluation of a fictitious domain method for predicting dynamic response of mechanical heart valves', *Int. J. Numer. Methods Fluids* **19**, 835–850.

J. Stroud, S. Berger and D. Saloner (2002), 'Numerical analysis of flow through a severely stenotic carotid artery bifurcation', *J. Biomech. Engrg* **124**, 9–20.

A. Stuart (2010), Inverse problems: A Bayesian perspective. In *Acta Numerica*, Vol. 19, Cambridge University Press, pp. 451–559.

T. Sullivan (2015), *Introduction to Uncertainty Quantification*, Vol. 63 of Texts in Applied Mathematics, Springer.

W. Sun, B. Starly, J. Nam and A. Darling (2005), 'Bio-CAD modeling and its applications in computer-aided tissue engineering', *Comput. Aided Design* **11**, 1097–1114.

G. Sundaram, K. Balakrishnan and R. Kumar (2015), 'Aortic valve dynamics using a fluid structure interaction model: The physiology of opening and closing', *J. Biomech.* **48**, 1737–1744.

J. Sundnes, G. Lines and A. Tveito (2005), 'An operator splitting method for solving the bidomain equations coupled to a volume conductor model for the torso', *Math. Biosci.* **2**, 233–248.

J. Sundnes, S. Wall, H. Osnes, T. Thorvaldsen and A. McCulloch (2014), 'Improved discretisation and linearisation of active tension in strongly coupled cardiac electro-mechanics simulations', *Comput. Methods Biomech. Biomed. Engrg* **6**, 604–615.

E. Swim and P. Seshaiyer (2006), 'A nonconforming finite element method for fluid–structure interaction problems', *Comput. Methods Appl. Mech. Engrg* **195**, 2088–2099.

A. Tagliabue, L. Dede' and A. Quarteroni (2015), Fluid dynamics of an idealized left ventricle: The extended Nitsche's method for the treatment of heart valves as mixed time varying boundary conditions. MOX-Report 61-2015, Department of Mathematics, Politecnico di Milano.

H. Talbot, S. Cotin, R. Razavi, C. Rinaldi and H. Delingette (2015), Personalization of cardiac electrophysiology model using the unscented Kalman filtering. In *Computer Assisted Radiology and Surgery: CARS 2015*.

A. Tarantola (2004), *Inverse Problem Theory and Methods for Model Parameter Estimation*, SIAM.

C. Taylor and C. Figueroa (2009), 'Patient-specific modeling of cardiovascular mechanics', *Annu. Rev. Biomed. Engrg* **11**, 109–134.

C. Taylor, T. Hughes and C. Zarins (1996), 'Finite element analysis of pulsatile flow in the abdominal aorta under resting and exercise conditions', *Amer. Soc. Mech. Engineers, Bioengineering Division* **33**, 81–82.

C. Taylor, T. Hughes and C. Zarins (1998), 'Finite element modeling of blood flow in arteries', *Comput. Methods Appl. Mech. Engrg* **158**, 155–196.

R. Temam (1969), 'Sur l'approximation de la solution des équations de Navier–Stokes par la méthode des pas fractionaires (I)', *Arch. Rat. Mech. Anal.* **32**, 135–153.

T. Tezduyar, S. Sathe, T. Cragin, B. Nanna, B. Conklin, J. Pausewang and M. Schwaab (2007), 'Modelling of fluid–structure interactions with the space–time finite elements: Arterial fluid mechanics', *Int. J. Numer. Methods Fluids* **54**, 901–922.

J. Thompson, B. Soni and N. Weatherill, eds (1999), *Handbook of Grid Generation*, CRC Press.

L. Timmermans, P. Minev and F. van de Vosse (1996), 'An approximate projection scheme for incompressible flow using spectral elements', *Int. J. Numer. Methods Fluids* **22**, 673–688.

K. Tomlinson, P. Hunter and A. Pullan (2002), 'A finite element method for an eikonal equation model of myocardial excitation wavefront propagation', *SIAM J. Appl. Math.* **1**, 324–350.

E. Toro (2016), 'Brain venous haemodynamics, neurological diseases and mathematical modelling: A review', *Appl. Math. Comput.* **272**, 542–579.

N. Trayanova (2006), 'Defibrillation of the heart: Insights into mechanisms from modelling studies', *Exp. Physiol.* **91**, 323–337.

N. Trayanova, W. Li, J. Eason and P. Kohl (2004), 'Effect of stretch-activated channels on defibrillation efficacy', *Heart Rhythm* **1**, 67–77.

P. Trenhago, L. Fernandes, L. Müller, P. Blanco and R. Feijóo (2016), 'An integrated mathematical model of the cardiovascular and respiratory systems', *Int. J. Numer. Methods Biomed. Engrg* **32**, e02736.

F. Tröltzsch (2010), *Optimal Control of Partial Differential Equations: Theory, Methods and Applications*, Vol. 112 of Graduate Studies in Mathematics, AMS.

C. Tu and C. Peskin (1992), 'Stability and instability in the computation of flows with moving immersed boundaries: A comparison of three methods', *SIAM J. Sci. Statist. Comput.* **6**, 1361–1376.

S. Turek (1999), *Efficient Solvers for Incompressible Flow Problems*, Vol. 6 of Lecture Notes in Computational Science and Engineering, Springer.

K. ten Tusscher and A. Panfilov (2006), 'Cell model for efficient simulation of wave propagation in human ventricular tissue under normal and pathological conditions', *Phys. Med. Biol.* **51**, 6141–6156.

M. Unser (1999), 'Splines: A perfect fit for signal and image processing', *IEEE Trans. Signal Process. Mag.* **16**, 22–38.

T. Usyk, I. LeGrice and A. McCulloch (2002), 'Computational model of three-dimensional cardiac electromechanics', *Comput. Vis. Sci.* **4**, 249–257.

A. Veneziani (1998*a*), Boundary conditions for blood flow problems. In *Proceedings of ENUMATH* (R. Rannacher *et al.*, eds), World Scientific.

A. Veneziani (1998*b*), Mathematical and numerical modeling of blood flow problems. PhD thesis, University of Milan.

A. Veneziani (2003), 'Block factorized preconditioners for high-order accurate in time approximation of the Navier–Stokes equations', *Numer. Methods Partial Differ. Equations* **19**, 487–510.

A. Veneziani and C. Vergara (2005), 'Flow rate defective boundary conditions in haemodynamics simulations', *Int. J. Numer. Methods Fluids* **47**, 803–816.

A. Veneziani and C. Vergara (2007), 'An approximate method for solving incompressible Navier–Stokes problems with flow rate conditions', *Comput. Methods Appl. Mech. Engrg* **196**, 1685–1700.

A. Veneziani and C. Vergara (2013), 'Inverse problems in cardiovascular mathematics: Toward patient-specific data assimilation and optimization', *Int. J. Numer. Methods Biomed. Engrg* **29**, 723–725.

A. Veneziani and U. Villa (2013), 'ALADINS: An ALgebraic splitting time ADaptive solver for the Incompressible Navier–Stokes equations', *J. Comput. Phys.* **238**, 359–375.

C. Vergara (2011), 'Nitsche's method for defective boundary value problems in incompressibile fluid-dynamics', *J. Sci. Comput.* **46**, 100–123.

C. Vergara, M. Lange, S. Palamara, T. Lassila, A. Frangi and A. Quarteroni (2016), 'A coupled 3D–1D numerical monodomain solver for cardiac electrical activation in the myocardium with detailed Purkinje network', *J. Comput. Phys.* **308**, 218–238.

C. Vergara, S. Palamara, D. Catanzariti, F. Nobile, E. Faggiano, C. Pangrazzi, M. Centonze, M. Maines, A. Quarteroni and G. Vergara (2014), 'Patient-specific generation of the Purkinje network driven by clinical measurements of a normal propagation', *Med. Biol. Engrg Comput.* **52**, 813–826.

C. Vergara, R. Ponzini, A. Veneziani, A. Redaelli, D. Neglia and O. Parodi (2010), 'Womersley number-based estimation of flow rate with Doppler ultrasound: Sensitivity analysis and first clinical application', *Comput. Methods Programs Biomed.* **98**, 151–160.

C. Vergara, F. Viscardi, L. Antiga and G. Luciani (2012), 'Influence of bicuspid valve geometry on ascending aortic fluid-dynamics: A parametric study', *Artificial Organs* **36**, 368–378.

J. A. Vierendeels, K. Riemslagh, E. Dick and P. Verdonck (2000), 'Computer simulation of intraventricular flow and pressure gradients during diastole', *J. Biomech. Engrg* **6**, 667–674.

E. Vigmond, F. Aguel and N. Trayanova (2002), 'Computational techniques for solving the bidomain equations in three dimensions', *IEEE Trans. Biomed. Engrg* **49**, 1260–1269.

E. Vigmond and C. Clements (2007), 'Construction of a computer model to investigate sawtooth effects in the Purkinje system', *IEEE Trans. Biomed. Engrg* **54**, 389–399.

E. Vigmond, R. dos Santos, A. Prassl, M. Deo and G. Plank (2008), 'Solvers for the cardiac bidomain equations', *Prog. Biophys. Molec. Biol.* **96**, 3–18.

I. Vignon-Clementel, C. Figueroa, K. Jansen and C. Taylor (2006), 'Outflow boundary conditions for three-dimensional finite element modeling of blood flow and pressure waves in arteries', *Comput. Methods Appl. Mech. Engrg* **195**, 3776–3996.

N. Virag, V. Jacquemet, C. Henriquez, S. Zozor, O. Blanc, J.-M. Vesin, E. Pruvot and L. Kappenberger (2002), 'Study of atrial arrhythmias in a computer model based on magnetic resonance images of human atria', *Chaos* **12**, 754–763.

F. Viscardi, C. Vergara, L. Antiga, S. Merelli, A. Veneziani, G. Puppini, G. Faggian, A. Mazzucco and G. Luciani (2010), 'Comparative finite element model analysis of ascending aortic flow in bicuspid and tricuspid aortic valve', *Artificial Organs* **34**, 1114–1120.

H. U. Voss, J. Timmer and J. Kurths (2004), 'Nonlinear dynamical system identification from uncertain and indirect measurements', *Int. J. Bifurc. Chaos* **14**, 1905–1933.

J. Voss (2013), *An Introduction to Statistical Computing: A Simulation-Based Approach*, Wiley.

J. Vossoughi, R. Vaishnav and D. Patel (1980), Compressibility of the myocardial tissue. In *Advances in Bioengineering 1980: Papers Presented at the Winter*

Annual Meeting of the American Society of Mechanical Engineers (V. C. Mow, ed.), ASME, pp. 45–48.

E. Votta, T. Le, M. Stevanella, L. Fusinic, E. Caiani, A. Redaelli and F. Sotiropoulos (2013), 'Toward patient-specific simulations of cardiac valves: State-of-the-art and future directions', *J. Biomech.* **46**, 217–228.

E. Votta, F. Maisano, S. Bolling, O. Alfieri, F. Montevecchi and A. Redaelli (2007), 'The geoform disease-specific annuloplasty system: A finite element study', *Ann. Thorac. Surg.* **84**, 92–101.

G. Waiter, F. McKiddie, T. Redpath, S. Semple and R. Trent (1999), 'Determination of normal regional left ventricular function from cine-MR images using a semi-automated edge detection method', *Magnet. Reson. Imaging* **17**, 99–107.

W. Wall, L. Wiechert, A. Comerford and S. Rausch (2010), 'Towards a comprehensive computational model for the respiratory system', *Int. J. Numer. Methods Biomed. Engrg* **26**, 807–827.

M. Wallman, N. Smith and B. Rodriguez (2014), 'Computational methods to reduce uncertainty in the estimation of cardiac conduction properties from electroanatomical recordings', *Med. Image Anal.* **18**, 228–240.

D. Wang, R. Kirby, R. MacLeod and C. Johnson (2013), 'Inverse electrocardiographic source localization of ischemia: An optimization framework and finite element solution', *J. Comput. Phys.* **250**, 403–424.

K. Wang, R. Dutton and C. Taylor (1999), 'Improving geometric model construction for blood flow modeling', *IEEE Engrg Med. Biol. Mag.* **18**, 33–39.

L. Wang, K. L. Wong, H. Zhang, H. Liu and P. Shi (2011), 'Noninvasive computational imaging of cardiac electrophysiology for 3-D infarct', *IEEE Trans. Biomed. Engrg* **58**, 1033–1043.

L. Wang, H. Zhang, K. Wong and P. Shi (2009), A reduced-rank square root filtering framework for noninvasive functional imaging of volumetric cardiac electrical activity. In *2009 IEEE International Conference on Acoustics, Speech and Signal Processing*, pp. 533–536.

H. Watanabe, T. Hisada, S. Sugiura, J. Okada and H. Fukunari (2002), 'Computer simulation of blood flow, left ventricular wall motion and their interrelationship by fluid–structure interaction finite element method', *JSME Int. J. C: Mechanical Systems, Machine Elements and Manufacturing* **45**, 1003–1012.

D. Watson (1981), 'Computing the n-dimensional Delaunay tessellation with application to Voronoi polytopes', *Comput. J.* **24**, 167–172.

N. Weatherill and O. Hassan (1994), 'Efficient three-dimensional Delaunay triangulation with automatic point creation and imposed boundary constraints', *Int. J. Numer. Methods Engrg* **37**, 2005–2039.

E. Weinberg and M. Kaazempur-Mofrad (2006), 'A large-strain finite element formulation for biological tissues with application to mitral valve leaflet tissue mechanics', *J. Biomech.* **39**, 1557–1561.

J. Wenk, L. Ge, Z. Zhang, M. Soleimani, D. Potter, A. Wallace, E. Tseng, M. Ratcliffe and J. Guccione (2013), 'Mechanics of the mitral valve strut chordae insertion region', *Comput. Methods Biomech. Biomed. Engrg* **16**, 807–818.

N. Westerhof, J. Lankhaar and B. Westerhof (2009), 'The arterial windkessel', *Med. Biol. Engrg Comput.* **47**, 131–141.

J. Wong and E. Kuhl (2014), 'Generating fibre orientation maps in human heart models using Poisson interpolation', *Comput. Methods Biomech. Biomed. Engrg* **17**, 1217–1226.

J. Wong, S. Goktepe and E. Kuhl (2013), 'Computational modeling of chemo-electro-mechanical coupling: A novel implicit monolithic finite element approach', *Int. J. Numer. Methods Biomed. Engrg* **29**, 1104–1133.

J. Xi, P. Lamata, J. Lee, P. Moireau, D. Chapelle and N. Smith (2011), 'Myocardial transversely isotropic material parameter estimation from *in-silico* measurements based on a reduced-order unscented Kalman filter', *J. Mech. Behav. Biomed. Materials* **4**, 1090–1102.

F. Xie, Z. Qu, J. Yang, A. Baher, J. Weiss and A. Garfinkel (2004), 'A simulation study of the effects of cardiac anatomy in ventricular fibrillation', *J. Clin. Invest.* **113**, 686–693.

F. Xiong and C. Chong (2008), 'A parametric numerical investigation on haemodynamics in distal coronary anastomoses', *Med. Engrg Phys.* **30**, 311–320.

D. Xiu and J. Hesthaven (2005), 'High-order collocation methods for differential equations with random inputs', *SIAM J. Sci. Comput.* **27**, 1118–1139.

D. Xiu and G. Karniadakis (2002*a*), 'Modeling uncertainty in steady state diffusion problems via generalized polynomial chaos', *Comput. Methods Appl. Mech. Engrg* **191**, 4927–4948.

D. Xiu and G. Karniadakis (2002*b*), 'The Wiener–Askey polynomial chaos for stochastic differential equations', *SIAM J. Sci. Comput.* **24**, 619–644.

D. Xiu and S. Sherwin (2007), 'Parametric uncertainty analysis of pulse wave propagation in a model of a human arterial network', *J. Comput. Phys.* **226**, 1385–1407.

Y. Yamashita (1982), 'Theoretical studies on the inverse problem in electrocardiography and the uniqueness of the solution', *IEEE Trans. Biomed. Engrg* **BME-29**, 719–725.

K. Yanagihara, A. Noma and H. Irisawa (1980), 'Reconstruction of sino-atrial node pacemaker potential based on the voltage clamp experiments', *Japan. J. Physiol* **30**, 841–857.

H. Yang and A. Veneziani (2015), 'Estimation of cardiac conductivities in ventricular tissue by a variational approach', *Inverse Problems* **31**, 115001.

P. Yim, J. Cebral, R. Mullick, H. Marcos and P. Choyke (2001), 'Vessel surface reconstruction with a tubular deformable model', *IEEE Trans. Med. Imaging* **20**, 1411–1421.

M. Yin, X. Luo, T. Wang and P. Watton (2009), 'Effects of flow vortex on a chorded mitral valve in the left ventricle', *Int. J. Numer. Methods Biomed. Engrg* **26**, 381–404.

H. Younis, M. Kaazempur-Mofrad, R. Chan, A. Isasi, D. Hinton, A. Chau, L. Kim and R. Kamm (2004), 'Hemodynamics and wall mechanics in human carotid bifurcation and its consequences for atherogenesis: Investigation of inter-individual variation', *Biomech. Model. Mechanobiol.* **3**, 17–32.

Y. Yu, H. Baek and G. Karniadakis (2013), 'Generalized fictitious methods for fluid–structure interactions: Analysis and simulations', *J. Comput. Phys.* **245**, 317–346.

S. Zhao, X. Xu, A. Hughes, S. Thom, A. Stanton, B. Ariff and Q. Long (2000), 'Blood flow and vessel mechanics in a physiologically realistic model of a human carotid arterial bifurcation', *J. Biomech.* **33**, 975–984.

F. Zhu and J. Tian (2003), 'Modified fast marching and level set method for medical image segmentation', *J. X-Ray Sci. Technol.* **11**, 193–204.

O. Zienkiewicz and R. Taylor (2005), *The Finite Element Method for Solid and Structural Mechanics*, sixth edition, Butterworth-Heinemann.

S. Zonca, L. Formaggia and C. Vergara (2016), An unfitted formulation for the interaction of an incompressible fluid with a thick structure via an XFEM/DG approach. MOX-Report 35-2016, Department of Mathematics, Politecnico di Milano.

P. Zunino (2009), 'Numerical approximation of incompressible flows with net flux defective boundary conditions by means of penalty techniques', *Comput. Methods Appl. Mech. Engrg* **198**, 3026–3038.

Acta Numerica (2017), pp. 591–721
doi:10.1017/S0962492917000083

Algebraic multigrid methods

Jinchao Xu and Ludmil Zikatanov
Center for Computational Mathematics and Applications,
Department of Mathematics,
The Pennsylvania State University,
University Park, PA 16802, USA
E-mail: jinchao@psu.edu, ludmil@psu.edu

This paper provides an overview of AMG methods for solving large-scale systems of equations, such as those from discretizations of partial differential equations. AMG is often understood as the acronym of 'algebraic multigrid', but it can also be understood as 'abstract multigrid'. Indeed, we demonstrate in this paper how and why an algebraic multigrid method can be better understood at a more abstract level. In the literature, there are many different algebraic multigrid methods that have been developed from different perspectives. In this paper we try to develop a unified framework and theory that can be used to derive and analyse different algebraic multigrid methods in a coherent manner. Given a smoother R for a matrix A, such as Gauss–Seidel or Jacobi, we prove that the optimal coarse space of dimension n_c is the span of the eigenvectors corresponding to the first n_c eigenvectors $\bar{R}A$ (with $\bar{R} = R + R^T - R^T AR$). We also prove that this optimal coarse space can be obtained via a constrained trace-minimization problem for a matrix associated with $\bar{R}A$, and demonstrate that coarse spaces of most existing AMG methods can be viewed as approximate solutions of this trace-minimization problem. Furthermore, we provide a general approach to the construction of quasi-optimal coarse spaces, and we prove that under appropriate assumptions the resulting two-level AMG method for the underlying linear system converges uniformly with respect to the size of the problem, the coefficient variation and the anisotropy. Our theory applies to most existing multigrid methods, including the standard geometric multigrid method, classical AMG, energy-minimization AMG, unsmoothed and smoothed aggregation AMG and spectral AMGe.

CONTENTS

1. Introduction

Multigrid methods are among the most efficient numerical methods for solving large-scale systems of equations – linear and non-linear alike – arising from the discretization of partial differential equations. These types of methods can be viewed as an acceleration of traditional iterative methods based on local relaxation, such as Gauss–Seidel and Jacobi methods. For linear systems arising from finite element or finite difference discretization of elliptic boundary value problems, local relaxation methods were observed to converge very rapidly for the high-frequency part of the solution. The low-frequency part of the solution, although slow to converge, corresponds to a relatively smooth part of the function that can be well approximated on a coarser grid. The main idea behind such multigrid methods is to project the error obtained after applying a few iterations of local relaxation methods onto a coarser grid. The projected error equations have two characteristics. First, the resulting system is smaller. Second, part of the slowly converging low-frequency error on a finer grid becomes relatively high-frequency on the coarser grid, and these frequencies can be further corrected via a local relaxation method, but this time on the coarse grid. By repeating this process via ever coarser grids, a multilevel iterative process is obtained. Such algorithms have been proved to have uniform convergence with nearly optimal complexity for a large class of linear algebraic systems arising from the discretization of partial differential equations, especially elliptic boundary problems of second and fourth order. One main component of this type of multilevel algorithm is a hierarchy of geometric grids, typically a sequence of nested grids obtained by successive refinement. The resulting algorithms are known as geometric multigrid (GMG) methods.

Despite their extraordinary efficiency, however, GMG methods have limitations. They depend on a hierarchy of geometric grids, which is often not readily available, and it can be argued that the range of applicability of GMG methods is therefore limited.

Algebraic multigrid (AMG) methods were designed in an attempt to address such limitations. They were proposed as a means to generalize geometric multigrid methods for systems of equations that share properties with discretized PDEs, such as the Laplacian equation, but potentially have unstructured grids in the underlying discretization. The first AMG algorithm in Brandt, McCormick and Ruge (1982) was a method developed under the assumption that such a problem was being solved. The AMG algorithm was later generalized using several heuristics to extend its applicability to more general problems and matrices. As a result, a variety of AMG methods have been developed in the past three decades, and they have been successfully applied to many practical problems.

In this paper we give an overview of AMG methods from a theoretical viewpoint. AMG methods have been developed via a combination of certain theoretical considerations and heuristic arguments, and many AMG methods work, with varying efficiency, for different applications. We find it very hard to give a coherent picture of the state of the art of AMG methods by simply choosing to make a comprehensive list of existing algorithms without digging into their theoretical foundations. Unfortunately, a good theoretical understanding of how and why these methods work is still seriously lacking.

In preparing this paper, we have undertaken the task of making a thorough investigation into the design and analysis of AMG from a theoretical point of view. While there are many bits and pieces of ideas diffused in the literature, we have managed to re-examine most of the existing results and 're-invent the wheel' in trying to deliver a coherent theoretical description. To do this, we have developed several tools for the design and analysis of AMG.

With very few exceptions, most AMG algorithms have targeted the solution of symmetric positive definite (SPD) systems. In this paper we choose to present our studies for a slightly larger class of problems, namely symmetric positive semi-definite (SPSD) systems. This approach is not only more inclusive, but more importantly, the SPSD class of linear systems can be viewed as more intrinsic to AMG ideas. For example, while the standard discretizations of the Laplacian operator with homogeneous Dirichlet boundary conditions results in an SPD system, the design of AMG may be better understood by using local problems (defined on subdomains) with homogeneous Neumann boundary conditions, yielding SPSD sub-systems.

In short, in this paper we consider AMG techniques for solving the linear system of equations

$$Au = f, \tag{1.1}$$

where A is a given SPSD operator or sparse matrix, and the problem is posed in a vector space of large dimension. The starting point of an AMG procedure is first to choose a smoother, which is often taken to be some local relaxation iterative method such as pointwise Jacobi or Gauss–Seidel, or, more generally, overlapping Schwarz methods. The use of pointwise smoothers seems to encompass most of the efforts in the literature to construct and implement most (if not all) AMG methods. More general smoothers based on overlapping Schwarz methods are necessary for some problems, but we shall not study them in detail in this paper. In any event, any chosen smoother is expected to converge well only on certain components of the solution, which will be known as *algebraic high frequencies* with respect to the given smoother. With the smoother fixed, the main task of an AMG method is then to identify a sequence of coarse spaces that would complement this smoother well. Roughly speaking, an ideal sequence of coarse spaces is such that any vector (namely the solution to (1.1) for any f) in the finest space can be well represented by a linear combination of all the algebraic high frequencies on all coarse spaces. As a result, the AMG method would converge well for problem (1.1).

It is hard to translate the above multilevel setting into a concise mathematical statement. Instead we will focus on first answering this question for a two-level setting. For a two-level setting, the first theoretical question is as follows: 'Given a smoother, say R, what is the optimal coarse space of given dimension such that the resulting AMG has the best convergence rate?' This question will be thoroughly addressed in Section 5. As it turns out, the optimal coarse space will consist of the eigenvectors corresponding to the lower end of the spectrum of a matrix such as RA. While our two-level theory is theoretically pleasing, it does not offer a practical solution, as finding these eigenvectors would be too expensive. Thus, the task of the AMG design is to find a good but inexpensive approximation of this *algebraic low-frequency* eigenspace that will still result in an AMG algorithm with desirable convergence properties. We call such an approximation of an optimal coarse space a 'quasi-optimal' coarse space.

In the design of all these AMG algorithms, one key component is the coarsening of spaces via the graph associated with the matrix A. The two main strategies are based on independent sets and on aggregation. We will present several approaches to the construction of quasi-optimal coarse spaces. We would especially advocate two approaches that have sound theoretical foundations. The first approach is outlined in Section 5 and later on in Section 11. We will first prove that the optimal coarse space from the theory in Section 5 can be characterized, in a mathematically equivalent manner, by solutions to a trace-minimization problem. In a functional setting, such a trace-minimization can be interpreted as minimization of the sum of the energy norms of a set of coarse basis functions. These precise equivalents give very clear guidance as to how some AMG methods can be constructed. One practical approach based on such a theorem is to look for energy-minimization basis functions among locally supported function classes. The resulting algorithms are known as energy-min AMG. Other AMG methods, such as classical AMG and aggregation-based AMG, can be viewed as approximations to energy-min AMG. The second approach is outlined in Section 6. The main idea is to construct a quasi-optimal coarse space by piecing together the low-end eigenspaces of some appropriately defined local operators or matrices. This approach can be used to provide quasi-optimal coarse spaces for various AMG methods, including the standard geometric multigrid method, classical AMG, energy-minimization AMG, unsmoothed and smoothed aggregation AMG, and spectral AMGe. As a simple example, this method relies on the fact that an $n_c = Jm$-dimensional low-end eigenspace of an operator can be well approximated by gluing together J-pieces of m-dimensional low-end eigenspaces, for some carefully chosen local operators. Here m is a very small integer, for example, $m = 1$ for the Laplacian operator and $m = 3$ (resp. $m = 6$) for the two-dimensional (resp. three-dimensional) linear elasticity operator. This important property of eigenspaces is closely related to Weyl's lemma on the asymptotic behaviour of eigenvalues for elliptic boundary value problems, discussed in Section 2.3.

Most AMG methods are designed in terms of the adjacency graph of the coefficient matrix of a given linear algebraic system. In Section 7 we give a brief description of graph theory and the adjacency graph of a sparse matrix. One highlight of Section 7 is the concept of M-matrix relatives. This simple tool is instrumental in the design and analysis of the classical AMG method.

One important step in the design of most AMG methods is to zero out some entries of the coefficient matrix A by using the concept of connection strength to construct a filtered matrix $\tilde{A}$, which is equivalent to omitting weakly connected edges in the adjacency graph $\mathcal{G}(A)$ to obtain $\mathcal{G}(\tilde{A})$. Several possible strength functions are introduced in Section 8 to describe the connection strength. In Section 9, the graph $\mathcal{G}(\tilde{A})$ is then coarsened by either keeping a maximal independent set (MIS) as a coarse vertex set $\mathcal{C}$ and then dropping the rest of grid, or using aggregation/agglomeration. In Section 9 we also give some technical details of the construction of a coarse space by using degrees of freedom.

By using the aforementioned general approaches and theoretical techniques, we then motivate and present a number of AMG methods. Some of the highlights of the paper are outlined below.

We first give an overview of GMG and its relationship with AMG in Section 10. After describing some details of a typical GMG method for linear finite element matrix, we argue that the geometric information used in defining a GMG is essentially the graph information of the underlying finite element grid (without using other geometric information such as coordinates of the grid points). This is a strong indication that at least some GMG methods can be realized in a purely algebraic fashion, using only the stiffness matrix, an algebraic smoother, and the adjacency graph of the stiffness matrix. On the other hand, we prove that a GMG method can also be formally obtained via our general AMG approach, presented in Section 6. Furthermore, we use the example of AMGe in Section 10.4 to demonstrate that geometric information on the grid can be used effectively to construct a geometry-based AMG.

In Section 11 we give a detailed account of AMG methods based on energy-minimization. We first present our new theory that the optimal coarse space shown in the two-level theory in Section 5 can actually be obtained by trace-minimization (Section 11.1), after which we derive the energy-min AMG method by seeking a set of locally supported coarse basis functions for energy-minimization.

Classical AMG, as the first class of AMG algorithms studied in the literature, will be presented in Section 12. We derive and analyse this type of method using the framework in Section 6 together with the notion of M-matrix relatives introduced in Section 7.2. We further discuss how a classical AMG method can be viewed as an approximation of the energy-min AMG method.

Aggregation-based AMG will be presented in Section 13. Again we derive and analyse this method using the framework in Section 6. One remarkable feature of aggregation AMG methods is their ease at preserving multidimensional near-null spaces, such as rigid body modes in linear elasticity.

To demonstrate how an AMG method addresses possible heterogeneous properties in a given problem, we devote Section 14 to showing how classical AMG is designed to address the difficulties arising from discretized elliptic problems with strong discontinuous jumps, or anisotropy in the coefficients of the underlying PDE.

In Section 15 we outline a class of AMG methods that attempt to choose the coarse spaces in a bootstrap and adaptive fashion. This class of AMG algorithms does not fall within the theoretical frameworks presented in this paper, but they provide a practical approach to generalizing many existing AMG techniques to more general problems. Finally we make some concluding remarks in Section 16.

We conclude these introductory remarks with a brief summary of the acronyms used for the various AMG algorithms reviewed in this paper.

Aggregation-based AMG:	
– unsmoothed aggregation	UA-AMG
– smoothed aggregation	SA-AMG
Bootstrap and adaptive AMG:	
– classical	αAMG
– smoothed aggregation	αSA-AMG
– bootstrap AMG	BAMG
Element-based AMG	AMGe
Spectral AMGe	ρAMGe

2. Model problems and discretization

While AMG has found applications to a wide range of linear algebraic systems, its development has been mainly motivated by the solution of systems arising from the discretization of partial differential equations by finite element, finite difference or other numerical methods. In this section we will discuss a model of second-order elliptic boundary problems, their finite difference and finite element discretization, and relevant properties of the underlying differential operators and their discretization.

2.1. Model elliptic PDE operators

We consider the boundary value problems

$$\mathcal{L}u = -\nabla \cdot \alpha(x)\nabla u = f, \quad x \in \Omega, \tag{2.1}$$

where $\alpha : \Omega \mapsto \mathbb{R}^{d\times d}$ is an SPD matrix function satisfying

$$\alpha_0\|\xi\|^2 \leq \xi^T\alpha(x)\xi \leq \alpha_1\|\xi\|^2, \quad \xi \in \mathbb{R}^d, \tag{2.2}$$

for some positive constants α_0 and α_1. Here $d = 1, 2, 3$, and $\Omega \subset \mathbb{R}^d$ is a bounded domain with boundary $\Gamma = \partial\Omega$.

A variational formulation for (2.1) is as follows: find $u \in V$ such that

$$a(u, v) = (f, v), \quad \text{for all } v \in V. \tag{2.3}$$

Here

$$a(u, v) = \int_\Omega (\alpha(x)\nabla u) \cdot \nabla v, \quad (f, v) = \int_\Omega fv,$$

and V is a Sobolev space that can be chosen to address different boundary conditions accompanying equation (2.1). One case is that of mixed boundary conditions

$$\begin{aligned} u = 0, \quad & x \in \Gamma_D, \\ (\alpha \nabla u) \cdot n = 0, \quad & x \in \Gamma_N, \end{aligned} \tag{2.4}$$

where $\Gamma = \Gamma_D \cup \Gamma_N$. The pure Dirichlet problem is when $\Gamma_D = \Gamma$, while the pure Neumann problem is when $\Gamma_N = \Gamma$. We thus have V as

$$V = \begin{cases} H^1(\Omega) = \{v \in L^2(\Omega) : \partial_i v \in L^2(\Omega), i = 1 : d\}, \\ H^1_D(\Omega) = \{v \in H^1(\Omega) : v|_{\Gamma_D} = 0\}. \end{cases} \tag{2.5}$$

When we consider a pure Dirichlet problem, $\Gamma_D = \Gamma$, we denote the space by $V = H^1_0(\Omega)$. In addition, for pure Neumann boundary conditions, the following condition is added to ensure the existence of the solution to (2.3):

$$\int_\Omega f = 0. \tag{2.6}$$

One of the most commonly used model problems is when

$$\alpha(x) = 1, \quad x \in \Omega, \tag{2.7}$$

which corresponds to the Poisson equation

$$-\Delta u = f. \tag{2.8}$$

This simple problem provides a good representative model for isotropic problems.

There are two other cases that are of special interest. The first case is when α is a scalar and has discontinuous jumps such as

$$\alpha(x) = \begin{cases} \epsilon & x \in \Omega_1, \\ 1 & x \in \Omega_2. \end{cases} \tag{2.9}$$

The second case is when α is a diagonal matrix such as (for $d = 2$)

$$\alpha(x) = \begin{pmatrix} 1 & 0 \\ 0 & \epsilon \end{pmatrix}, \tag{2.10}$$

which corresponds to the operator

$$-u_{xx} - \epsilon u_{yy} = f. \tag{2.11}$$

In both cases we assume that ϵ is sufficiently small to investigate the robustness of algorithms with respect to discontinuous jumps and anisotropy.

2.2. *Examples of finite difference and finite element discretizations*

As an illustrative example, we consider a finite difference discretization of the Poisson equation (2.8) with pure Dirichlet boundary conditions on the

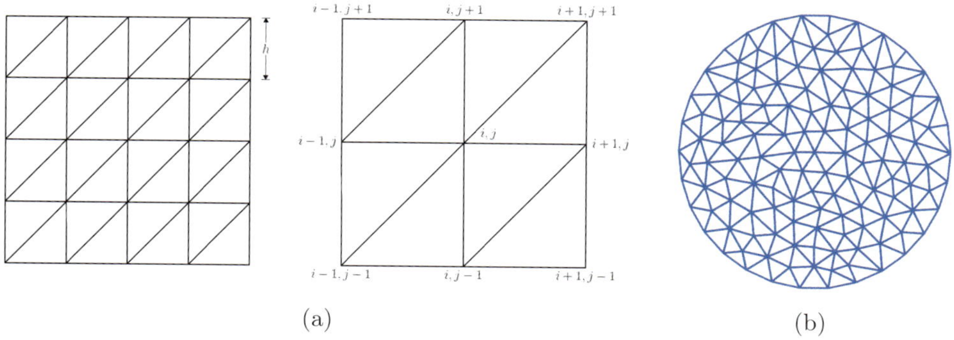

Figure 2.1. (a) Regular (uniform) triangulations for the unit square, and (b) unstructured mesh approximating the unit disk.

unit square $\Omega = (0, 1) \times (0, 1)$. We consider a uniform triangulation of Ω (see Figure 2.1(a)), and we set

$$(x_i, y_j) = \left(\frac{i}{n+1}, \frac{j}{n+1}\right), \quad u_{i,j} \approx u(x_i, y_j), \quad (i, j = 0, \dots, n+1).$$

We use the standard central difference approximation to the Laplacian operator

$$(-\Delta u)(x_i, y_j) \approx \frac{4u_{i,j} - u_{i+1,j} - u_{i-1,j} - u_{i,j+1} - u_{i,j-1}}{h^2}.$$

The finite difference scheme is then given by

$$4u_{i,j} - (u_{i+1,j} + u_{i-1,j} + u_{i,j+1} + u_{i,j-1}) = h^2 f_{i,j}, \tag{2.12}$$

where

$$f_{i,j} = f(x_i, y_j) \tag{2.13}$$

and $u_{i,j} \approx u(x_i, y_j)$. The approximations $u_{i,j}$ are found by solving a linear system. We order the points (x_i, y_j) lexicographically, and for $k = 1, \dots, n^2$ we have

$$k = (j-1)n + i, \quad x_k^h = (x_i, y_j), \quad \mu_k = u_{i,j}, \quad 1 \le i, j \le n, \tag{2.14}$$

We can then write (2.12) as

$$A\mu = b, \tag{2.15}$$

where

$$A = \mathrm{tridiag}(-I, B, -I) \quad \text{and} \quad B = \mathrm{tridiag}(-1, 4, -1). \tag{2.16}$$

A slightly different scheme is obtained using more of the neighbouring points of (x_i, y_j). We can build an approximation using 8 points $(x_{i\pm1}, y_{j\pm1})$ together with the 'centre' point (x_i, y_j). As a result we have the following

9-point finite difference scheme:

$$\begin{aligned} 8\mu_{i,j} - \mu_{i-1,j} - \mu_{i+1,j} - \mu_{i,j-1} - \mu_{i,j+1} \\ - \mu_{i-1,j-1} - \mu_{i+1,j-1} - \mu_{i-1,j+1} - \mu_{i+1,j+1} = 2h^2 f_{i,j}. \end{aligned} \tag{2.17}$$

Again, if we order (x_i, y_j) lexicographically, then (2.17) is the linear system (2.15) corresponding to the 9-point finite difference discretization of the Laplace equation, where

$$A = \text{tridiag}(-C, B, -C), \quad B = \text{tridiag}(-1, 8, -1), \ C = \text{tridiag}(1, 1, 1). \tag{2.18}$$

We now give an example of finite element discretization. Given a triangulation $\mathcal{T}_h$ for Ω, such as that given in Figure 2.1, let $V_h \subset V$ be a finite element space consisting of piecewise linear (or higher-order) polynomials with respect to the triangulation $\mathcal{T}_h$. The finite element approximation of the variational problem (2.3) is: find $u_h \in V_h$ such that

$$a(u_h, v_h) = (f, v_h), \quad \text{for all } v_h \in V_h. \tag{2.19}$$

Assume $\{\phi_i\}_{i=1}^N$ is the nodal basis of V_h, namely $\phi_i(x_j) = \delta_{ij}$ for any nodes x_j. We write $u_h(x) = \sum_{j=1}^N \mu_j \phi_j(x)$. Equation (2.19) is then equivalent to

$$\sum_{j=1}^N \mu_j a(\phi_j, \phi_i) = (f, \phi_i), \quad j = 1, 2, \dots, N,$$

which is a linear system of equations:

$$A\mu = b, \quad (A)_{ij} = a(\phi_j, \phi_i) \quad \text{and} \quad (b)_i = (f, \phi_i). \tag{2.20}$$

The matrix A is known as the stiffness matrix of the nodal basis $\{\phi_i\}_{i=1}^N$.

For $d = 2$ and the special uniform triangulation shown in Figure 2.1, this stiffness matrix for the Laplacian operator turns out to be exactly the one given by (2.12). This special case is an example of the close relationship between finite difference and finite element methods.

We note that the finite element method is based on the variational formulation (2.3), whereas the finite difference method is not. In the development of the AMG method, however, the variational method is also used to derive coarse level equations for finite difference methods.

For any $T \in \mathcal{T}_h$, we define

$$\overline{h}_T = \text{diam}\,(T), \quad h_T = |T|^{1/d}, \quad \underline{h}_T = 2\sup\{r > 0 : B(x, r) \subset T \text{ for } x \in T\}. \tag{2.21}$$

We say that the mesh $\mathcal{T}_h$ is *shape-regular* if there exists a uniformly bounded constant $\sigma \geq 1$ such that

$$\underline{h}_T \leq h_T \leq \overline{h}_T \leq \sigma \underline{h}_T, \quad \text{for all } T \in \mathcal{T}_h. \tag{2.22}$$

We call σ the *shape-regularity constant.*

Let $h = \max_{T\in\mathcal{T}_h} \overline{h}_T$, with $\overline{h}_T$ defined as in (2.21). We say that the mesh $\mathcal{T}_h$ is *quasi-uniform* if there exists a uniformly bounded constant $C > 0$ such that

$$\frac{h}{\underline{h}_T} \le C. \tag{2.23}$$

2.3. Spectral properties

We now discuss the spectral properties of the partial differential operator $\mathcal{L}$ given in (2.1).

We recall the well-known Courant–Fischer min-max principle (Courant and Hilbert 1924) for eigenvalues of symmetric matrices.

Theorem 2.1. Let T be an $n \times n$ symmetric matrix with respect to $(\cdot,\cdot)_*$, and let $\{\lambda_j, \zeta_j\}$ be its eigenpairs with $\lambda_1 \le \lambda_2 \le \cdots \le \lambda_n$. Then

$$\lambda_k = \min_{\dim W = k} \max_{x\in W, x\neq 0} \frac{(Tx,x)_*}{(x,x)_*}, \tag{2.24}$$

where the minimum is achieved if

$$W = \operatorname{span}\{\zeta_j : j = 1 : k\}, \tag{2.25}$$

and

$$\lambda_k = \max_{\dim W = n-k+1} \min_{x\in W, x\neq 0} \frac{(Tx,x)_*}{(x,x)_*}, \tag{2.26}$$

where the maximum is achieved if

$$W = \operatorname{span}\{\zeta_j : j = k : n\}. \tag{2.27}$$

Next, we recall Theorem 1 in Fan (1949), which is known as the Ky–Fan trace-minimization principle.

Theorem 2.2. We assume T is symmetric with respect to $(\cdot,\cdot)_*$, and $\{\lambda_j, \zeta_j\}$ are its eigenpairs with $\lambda_1 \le \lambda_2 \le \cdots \le \lambda_n$. Then

$$\min_{P\in\mathbb{R}^{n\times k}, P^*P=I} \operatorname{trace}(P^*TP) = \sum_{j=1}^{k} \lambda_j.$$

Furthermore, the minimum is achieved when

$$\operatorname{range}(P) = \operatorname{span}\{\zeta_j\}_{j=1}^{k} \quad \text{and} \quad P^*P = I.$$

Here $P^* \in \mathbb{R}^{k\times n}$ is the adjoint of P with respect to the $(\cdot,\cdot)_*$-inner product, namely

$$(P^*u, v) = (u, Pv)_*, \quad \text{for all } u \in \mathbb{R}^n,\ v \in \mathbb{R}^k.$$

Finally, following Xu (1992), we use the notation $a \lesssim b$ to represent the existence of a generic positive constant C, which is independent of salient parameters, such as problem size or anisotropic ratio, and such that $a \le Cb$. Furthermore, we write $a \eqsim b$ if and only if $a \lesssim b$ and $b \lesssim a$.

Theorem 2.3. The PDE operator $\mathcal{L}$ has a complete set of eigenfunctions (φ_k) and non-negative eigenvalues

$$0 \le \lambda_1 \le \lambda_2 \le \cdots$$

such that

$$\mathcal{L}\varphi_k = \lambda_k \varphi_k, \quad k = 1, 2, 3 \ldots,$$

where

(a) $\lim_{k\to\infty} \lambda_k = \infty$,

(b) (φ_i) forms an orthonormal basis of V as well as for $L^2(\Omega)$.

Furthermore,

(c) for the pure Neumann problem, $\lambda_1 = 0$, and φ_1 is the constant function,

(d) for the pure Dirichlet problem, $\lambda_1 > 0$ is simple, and φ_1 does not change sign.

We have the well-known Weyl estimate for the asymptotic behaviour of the Laplacian operator (Weyl 1911, Weyl 1912, Reed and Simon 1978).

Lemma 2.4 (Weyl's law). Assume that Ω is 'contented', which means that Ω can be approximated by unions of cubes in $\mathbb{R}^d$; see Reed and Simon (1978, p. 271) for the exact definition. For the homogeneous Dirichlet boundary condition, the eigenvalues of the pure Laplacian operator satisfy

$$\lim_{k\to\infty} \frac{\lambda_k}{k^{2/d}} = w_\Omega, \quad \text{with } w_\Omega = \frac{(2\pi)^2}{[\omega_d \operatorname{Vol}(\Omega)]^{2/d}}, \tag{2.28}$$

where ω_d is the volume of the unit ball in $\mathbb{R}^d$, and the eigenvalues of the operator $\mathcal{L}$ given in (2.1) satisfy

$$\lambda_k \eqsim k^{2/d}, \quad \text{for all } k \ge 1. \tag{2.29}$$

Next, we extend Weyl's law to discretized PDE operators. The following theorem gives a discrete version of Weyl's law for finite element discretization. Further details of this result can be found in Xu, Zhang and Zikatanov (2016*b*).

Theorem 2.5. Let $V_h \subset H_0^1(\Omega)$ be a family of finite element spaces on a quasi-uniform mesh with $\dim V_h = N$. Consider the discretized operator of (2.1),

$$\mathcal{L}_h : V_h \mapsto V_h, \quad (\mathcal{L}_h u, v) = a(u, v), \quad \text{for all } u, v \in V_h,$$

and its eigenvalues,

$$\lambda_{h,1} \leq \lambda_{h,2} \leq \cdots \leq \lambda_{h,N}.$$

Then, for all $1 \leq k \leq N$, there exists a constant $C_w > 0$ independent of k such that we have the estimates

$$\lambda_k \leq \lambda_{h,k} \leq C_w \lambda_k \tag{2.30}$$

and

$$\lambda_{h,k} \eqsim k^{2/d}. \tag{2.31}$$

2.4. Properties of finite element matrices

The main algebraic property for the stiffness matrices given by (2.20) is that it is sparse with $O(N)$ non-zeros, symmetric positive definite (for both Dirichlet and mixed boundary conditions) and semi-definite for pure Neumann boundary conditions. Its eigenvalues satisfy a discrete Weyl's law.

For simplicity, we will only consider pure Dirichlet boundary conditions in the rest of this section.

Lemma 2.6. The stiffness matrix A given by (2.20) has the following properties.

(a) The condition number of A, defined by the ratio of the extreme eigenvalues of A,

$$\kappa(A) = \frac{\lambda_{\max}(A)}{\lambda_{\min}(A)},$$

satisfies

$$\kappa(A) \eqsim h^{-2}.$$

Furthermore,

$$\lambda_{\min}(A) \eqsim h^2 \quad \text{and} \quad \lambda_{\max}(A) \eqsim 1.$$

(b) The discrete version of Weyl's law holds:

$$\lambda_k(A) \eqsim \left(\frac{k}{N}\right)^{2/d}.$$

We next discuss some more refined spectral properties of finite element stiffness matrices from uniform grids for the unit square domain $\Omega = (0,1) \times (0,1)$ for $d = 2$. We begin with the Poisson equation. It is easy to derive a closed-form solution of the eigenpairs of A given by (2.15), and we have

$$\lambda_{kl}(A) = 4\left(\sin^2 \frac{k\pi}{2(n+1)} + \sin^2 \frac{l\pi}{2(n+1)}\right) \tag{2.32}$$

and

$$\phi_{ij}^{kl} = \sin \frac{ki\pi}{n+1} \sin \frac{lj\pi}{n+1}, \quad 1 \le i \le n,\ 1 \le j \le n. \tag{2.33}$$

Now consider the important case of anisotropic problem (2.11). We order the vertices of the triangulation lexicographically, and, as before, denote them by $\{(ih, jh)\}_{i,j=0}^{n}$. The stiffness matrix is then

$$A = \text{tridiag}(-I, B, -I) \quad \text{with } B = \text{tridiag}(-\epsilon, 2(1+\epsilon), -\epsilon). \tag{2.34}$$

Obviously,

$$A = I \otimes B + C \otimes I \quad \text{with } C = \text{tridiag}(-1, 0, -1),$$

and it is easily verified that

$$\lambda_i(B) = 2(1+\epsilon) - 2\epsilon \cos \frac{i\pi}{(n+1)}, \quad \lambda_j(C) = -2\cos \frac{j\pi}{(n+1)}, \quad 1 \le i,j \le N.$$

which leads to the expression

$$\lambda_{ij}(A) = 4\epsilon \sin^2 \frac{i\pi}{2(n+1)} + 4\sin^2 \frac{j\pi}{2(n+1)}$$

for the eigenvalues, and

$$\phi_{ij}^{k\ell} = \sin \frac{ki\pi}{n+1} \sin \frac{\ell j\pi}{n+1}$$

for the corresponding eigenvectors.

3. Linear vector spaces and duals

In this paper we will mainly consider linear systems of equations of the form

$$Au = f. \tag{3.1}$$

Here

$$A : V \mapsto V', \tag{3.2}$$

$$f \in V',$$

where V is a finite-dimensional linear vector space and V' is the dual of V. If we use the notation $\langle \cdot, \cdot \rangle$ to denote the pairing between V' and V, we can write (3.1) in a variational form: find $u \in V$ such that

$$a(u, v) = \langle f, v \rangle, \quad \text{for all } v \in V, \tag{3.3}$$

where

$$a(u, v) = \langle Au, v \rangle. \tag{3.4}$$

3.1. Dual and inner product

For convenience of exposition, we will assume that V is equipped with an inner product $(\cdot,\cdot)$. By the Riesz representation theorem, for any $f \in V'$, there is a unique $u \in V$ such that

$$(u, v) = \langle f, v \rangle, \quad \text{for all } v \in V. \tag{3.5}$$

It is via the Riesz representation theorem that we identify the dual space V' with V. In the rest of this paper, for convenience, we will always assume that $V' = V$ for any finite-dimensional vector space V. As a result, we have

$$V'' = (V')' = V' = V. \tag{3.6}$$

Thanks to the identification $V' = V$ via (3.5), the identities in (3.6) are clear.

We would like to point out that, in an abstract discussion of all iterative methods for problem (3.1), it suffices to use the abstract dual pairing $\langle \cdot,\cdot \rangle$ without having to introduce an inner product $(\cdot,\cdot)$ on V. However, we find that using an inner product is convenient for exposition, as we shall see later. We further point out that we will not use an inner product to identify $V' = V$ for any infinite-dimensional vector space in this paper.

If $\{\phi_i\}_{i=1}^N$ is a basis of V, we will always choose a basis $\{\psi_i\}_{i=1}^N$ of V' that is dual to the basis of V, namely

$$(\psi_j, \phi_i) = \delta_{ij}, \quad 1 \le i,\ j \le N. \tag{3.7}$$

Such a dual basis will be used only for theoretical considerations and not for the actual implementation of any algorithms.

We will consider two kinds of linear vector spaces. The first kind is $V = \mathbb{R}^n$ and the inner product is just the dot product

$$(u, v)_{\ell^2} = \sum_{i=1}^{n} u_i v_i, \quad \text{for all } u = (u_i),\ v = (v_i) \in \mathbb{R}^n,$$

A canonical basis of $\mathbb{R}^N$ is formed by the column vectors of the identity matrix, $\{e_i\}_{i=1}^N$. It is easy to see that the dual basis of $(\mathbb{R}^N)' = \mathbb{R}^N$ is just the original basis $\{e_i\}_{i=1}^N$ itself.

The second kind is a finite-dimensional functional subspace of $L^2(\Omega)$ for a given domain $\Omega \subset \mathbb{R}^d$ $(1 \le d \le 3)$, equipped with the L^2-inner product

$$(u, v) = \int_\Omega u(x)v(x).$$

One commonly used linear vector space is a finite element space V_h, and the nodal basis functions $\{\phi_i\}_{i=1}^N$ are often used as a basis. In this case, the dual basis functions $\{\psi_i\}_{i=1}^N$ of V' are no longer the original nodal basis functions but rather a set of functions (which are usually globally supported) that satisfy (3.7). This set of dual basis functions is usually needed to derive the matrix representation of operators between various spaces and their duals, but they are not needed for the actual implementation of relevant algorithms.

For a linear operator

$$L : V \mapsto V, \tag{3.8}$$

its adjoint,

$$L' : V \mapsto V, \tag{3.9}$$

is defined as follows:

$$(L'u, v) = (u, Lv), \quad u, v \in V. \tag{3.10}$$

Since V plays the role of both V and its dual V', (3.8) and (3.9) can have four different meanings:

(i) if $L : V \mapsto V$, then $L' : V' \mapsto V'$,

(ii) if $L : V \mapsto V'$, then $L' : V \mapsto V'$,

(iii) if $L : V' \mapsto V$, then $L' : V' \mapsto V$,

(iv) if $L : V' \mapsto V'$, then $L' : V \mapsto V$.

Thanks to the identification we made between V' and V via (3.5), definition (3.10) is applicable to all four cases.

If $V = \mathbb{R}^n$ and $(u, v) = (u, v)_{\ell^2}$, then $L' = L^T$, namely the matrix transpose. We say that an operator $A : V \mapsto V'$ is symmetric positive definite (SPD) if

$$A' = A, \quad (Av, v) > 0, \quad \text{for all } v \in V \setminus \{0\}.$$

When A is SPD, it defines another inner product $(\cdot, \cdot)_A$ on V,

$$(u, v)_A = (Au, v), \quad u, v \in V,$$

and a corresponding norm,

$$\|v\|_A = (v, v)_A^{1/2}, \quad v \in V.$$

We use the superscript '*' for the adjoint operator with respect to $(\cdot, \cdot)_A$, that is,

$$(Bu, v)_A = (u, B^*v)_A.$$

It is easy to see that

$$(BA)^* = B'A, \tag{3.11}$$

and $(BA)^* = BA$ if and only if $B' = B$.

3.2. Matrix representation

Let $V_c \subset V$ be a subspace and consider the inclusion operator $\imath_c : V_c \mapsto V$. Assume that $\{\phi_i^c\}_{i=1}^{n_c}$ and $\{\phi_i\}_{i=1}^{n}$ are basis functions of V_c and V respectively, and the matrix representation of $\imath_c$ is the matrix

$$P : \mathbb{R}^{n_c} \mapsto \mathbb{R}^n \text{ satisfying } (\phi_1^c, \dots, \phi_{n_c}^c) = (\phi_1, \dots, \phi_n)P. \tag{3.12}$$

Equation (3.12) is shorthand for the expansion of the basis in V_c via the basis in V:

$$\phi_k^c = \sum_{j=1}^{n} p_{jk}\phi_j, \quad P = (p_{jk}), \quad k = 1, \dots, n_c, \ j = 1, \dots, n. \tag{3.13}$$

What is the matrix representation of $\imath_c' : V' \mapsto V_c'$? Although we have $V' = V$ and $V_c' = V_c$, we need to use dual bases $\{\psi_i^c\}_{i=1}^{n_c} \subset V_c'$ and $\{\psi_i\}_{i=1}^{n} \subset V'$ respectively. With respect to these dual bases, the matrix representation of $\imath_c'$ is simply P^T (the transpose of P), since it is easy to verify that

$$(\imath_c'\psi_1, \dots, \imath_c'\psi_n) = (\psi_1^c, \dots, \psi_{n_c}^c)P^T.$$

Now consider a linear operator

$$A : V \mapsto V. \tag{3.14}$$

There are two different ways to get a matrix representation of A, because V plays two roles here: V is V itself, and $V = V'$. For the first case, we use the same basis $\{\phi_i\}$ for V as the domain of A and V as the range of A. In this case, the matrix representation of A is the matrix

$$\hat{A} \in \mathbb{R}^{n\times n} \text{ satisfying } (A\phi_1, \dots, A\phi_n) = (\phi_1, \dots, \phi_n)\hat{A}. \tag{3.15}$$

In the second case, we use the basis $\{\phi_i\}$ for V as the domain space of A, but use the dual basis $\{\psi_i\}$ for $V' = V$ as the range space of A. In this case, the matrix representation of A is the matrix

$$\tilde{A} \in \mathbb{R}^{n\times n} \text{ satisfying } (A\phi_1, \dots, A\phi_n) = (\psi_1, \dots, \psi_n)\tilde{A}. \tag{3.16}$$

It is easy to see that

$$\tilde{A} = ((A\phi_j, \phi_i)) \tag{3.17}$$

and

$$\tilde{A} = M\hat{A}, \quad M = ((\phi_j, \phi_i)). \tag{3.18}$$

The matrix $\tilde{A}$ in (3.17) is often called the stiffness matrix of A and the matrix M in (3.18) is called the mass matrix.

In the early multigrid literature, a discrete inner product equivalent to the L^2-inner product was often introduced for finite element spaces so that the corresponding mass matrix becomes diagonal. However, if we view the underlying finite element operator (3.14) in a slightly different way,

$$A : V \mapsto V', \tag{3.19}$$

we will then see easily that the introduction of the discrete L^2-inner product is not necessary.

If $V = \mathbb{R}^n$ and we choose the canonical basis $\{e_i\}$ for V, we would not encounter the mass matrix problem as in the functional space case, since in this case $\{e_i\}$ is also the dual basis of V'. This is certainly convenient, but the convenience tends to mask some subtle but important difference between various vectors and matrices in a given problem and the objects (functions) that they represent.

Given a matrix $A \in \mathbb{R}^{n\times n}$, we can view it as either

$$A : \mathbb{R}^n \mapsto \mathbb{R}^n \tag{3.20}$$

or

$$A : \mathbb{R}^n \mapsto (\mathbb{R}^n)'. \tag{3.21}$$

As it turns out, when A is obtained from the discretization of partial differential equations, (3.21) is more informative than (3.20). Hence we write a matrix equation

$$Ax = b. \tag{3.22}$$

It is sometimes helpful to regard x and b as living in two 'different' spaces,

$$x \in \mathbb{R}^n \quad \text{and} \quad b \in (\mathbb{R}^n)'. \tag{3.23}$$

3.3. Eigenvalues and eigenvectors

Let us briefly discuss eigenvalues and eigenvectors for the symmetric operator $T : V \mapsto V$. If (λ, ϕ) is an eigenpair of T,

$$T\phi = \lambda\phi,$$

then it is easy to see that $(\lambda, \tilde{\phi})$ is an eigenpair of the matrix representation $\tilde{T}$ of T:

$$\tilde{T}\tilde{\phi} = \lambda\tilde{\phi}.$$

Here $\tilde{\phi} \in \mathbb{R}^n$ is the vector representation of ϕ:

$$\phi = (\phi_1, \ldots, \phi_n)\tilde{\phi}.$$

For an operator A defined in (3.14), we need to be cautious when talking about eigenvalues of A. Although we identify $V' = V$ via (3.5), V' and V play two different roles, so A is essentially a mapping between two 'different' spaces V and V', and the spectrum of A should be defined carefully. However, if we consider a symmetric operator

$$R : V' \mapsto V, \tag{3.24}$$

then $RA : V \mapsto V$ is an operator that is symmetric with respect to the A-inner product. In this case, if (λ, ϕ) is an eigenpair of RA, then $(\lambda, \tilde{\phi})$ is an eigenpair of $\tilde{R}\tilde{A}$ (*i.e.* equal to the matrix representation of RA).

We consider the trivial identification operator

$$\jmath : V' \mapsto V \text{ such that } \jmath\psi_i = \psi_i, \text{ for all } i,$$

namely $\jmath v = v$ for all $v \in V' = V$. It is easy to see that the matrix representation of $\jmath$ is the inverse of the mass matrix $M = ((\phi_j, \phi_i))$, namely

$$\tilde{\jmath} = M^{-1}.$$

Using this identification operator, the operator $\jmath A : V \mapsto V$ is a symmetric operator from V to V. We can then talk about its spectrum. For example, if (λ, ϕ) is an eigenpair of $\jmath A$, then $(\lambda, \tilde{\phi})$ satisfies

$$\tilde{A}\tilde{\phi} = \lambda M \tilde{\phi}.$$

This is often the generalized eigenvalue problem appearing in finite element analysis.

Although, for all $v \in V$, $\jmath Av = Av$ because of the identification introduced above, $\jmath A$ and A are, strictly speaking, two different operators: they have two different ranges and their matrix representations are different.

The discussions above, although simple, may seem a little confusing at first glance, but an unambiguous understanding and clarification of these concepts and the underlying subtleties will be helpful for the presentation of algebraic multigrid methods in the rest of this paper. For more detailed discussions on relevant topics, we refer to Xu (2016).

3.4. Bibliographical notes

For a general reading on the basic linear algebra materials used here, we refer to Halmos (1974) and Xu (1992, 2016). In particular, for a more detailed discussion related to dual spaces and matrix representations, we refer to Xu (1992, 2016).

4. Basic iterative methods

We now consider linear iterative methods for solving (2.20). We will focus on two of the most commonly used algorithms, namely the Jacobi and Gauss–Seidel methods. Let us first give a brief introduction to linear iterative methods in a more general setting. Recall the basic problem under consideration. Given a finite-dimensional vector space V equipped with an inner product $(\cdot, \cdot)$, we consider

$$Au = f, \tag{4.1}$$

where $A : V \mapsto V'$ is symmetric positive definite (SPD) and V' is the dual of V. As mentioned in Section 3, we will identify $V' = V$ via an inner product $(\cdot, \cdot)$.

4.1. Basic iterative methods

A general linear iterative method for solving (4.1) can be written as follows. Given $u^0 \in V$, we define

$$u^m = u^{m-1} + B(f - Au^{m-1}), \quad m = 1, 2, \ldots, \tag{4.2}$$

where $B : V' \mapsto V$ is a linear operator which can be thought of as an approximate inverse of A.

Sometimes it is more desirable for the iterator B to be symmetric. If B is not symmetric, there is a natural way to symmetrize it. Consider the following iteration:

$$\begin{cases} u^{m-1/2} = u^{m-1} + B(f - Au^{m-1}), \\ \quad\;\; u^m = u^{m-1/2} + B'(f - Au^{m-1/2}). \end{cases} \tag{4.3}$$

The symmetrized iteration (4.3) can be written as

$$u^m = u^{m-1} + \bar{B}(f - Au^{m-1}), \quad m = 1, 2, \ldots, \tag{4.4}$$

where

$$\bar{B} = B' + B - B'AB, \tag{4.5}$$

which satisfies

$$I - \bar{B}A = (I - BA)^*(I - BA). \tag{4.6}$$

Obviously, $\rho(I - \bar{B}A) < 0 \iff \bar{B} > 0 \iff G \equiv (B')^{-1} + B^{-1} - A > 0$.

Theorem 4.1. The following results hold.

(a) Equation (4.3) converges if and only if $G > 0$, which implies that (4.2) converges. Furthermore,

$$\|I - BA\|_A^2 = \lambda_{\max}(I - \bar{B}A) = 1 - \left(\sup_{\|v\|_A = 1} (\bar{B}^{-1}v, v) \right)^{-1}. \tag{4.7}$$

(b) If $B' = B$, $G > 0$ if and only if (4.2) converges and, with $\eta = \lambda_{\min}(G)$,

$$\frac{2\eta}{\eta + 1}(Bv, v) \le (\bar{B}v, v) \le 2(Bv, v), \quad v \in V. \tag{4.8}$$

Remark 4.2. As an exercise, the reader may generalize Theorem 4.1 to the case when A is SPSD.

Algorithm 1 Modified Jacobi method

$$\text{For } i = 1 : n, \quad x_i^m \leftarrow x_i^{m-1} + \omega a_{ii}^{-1}\Big(b_i - \sum_{j=1}^{n} a_{ij} x_j^{m-1}\Big).$$

Algorithm 2 Modified Gauss–Seidel method

$$\text{For } i = 1 : n, \quad x_i^m \leftarrow x_i^{m-1} + \omega a_{ii}^{-1}\Big(b_i - \sum_{j=1}^{i-1} a_{ij} x_j^{m} - \sum_{j=i}^{n} a_{ij} x_j^{m-1}\Big).$$

4.2. Jacobi and Gauss–Seidel methods

For $A = (a_{ij}) \in \mathbb{R}^{n\times n}$, we write

$$A = D + L + U,$$

where D is the diagonal of A, and L and U are the strict lower and upper triangular parts of A respectively.

Given $\omega > 0$, the (modified) Jacobi method can be written as (4.2) with

$$B = \omega D^{-1} = (\omega^{-1} D)^{-1},$$

as shown in Algorithm 1. The (modified) Gauss–Seidel method can be written as (4.2) with

$$B = (\omega^{-1} D + L)^{-1},$$

as shown in Algorithm 2.

The next result, which follows easily from Theorem 4.1, is well known.

Theorem 4.3. The modified Jacobi method converges if and only

$$0 < \omega < \frac{2}{\rho(D^{-1}A)}, \tag{4.9}$$

and the modified Gauss–Seidel method converges if and only if

$$0 < \omega < 2. \tag{4.10}$$

In practice, it is often easy to choose ω to satisfy (4.9) so that the modified Jacobi method is guaranteed to converge. In the rest of this paper, we may always assume that such a choice of ω is made. For the Gauss–Seidel method, we will always choose $\omega = 1$ (optimal successive over-relaxation is not generally used in multigrid methods). The Jacobi and Gauss–Seidel methods together with their convergence theory can be extended to block matrices in a straightforward fashion.

4.3. The subspace correction method

We consider a sequence of spaces $V_1, \ldots, V_J$. These spaces, which will be known as *auxiliary spaces*, are not necessarily subspaces of V, but each of them is related to the original space V by a linear operator

$$\Pi_k : V_k \mapsto V. \tag{4.11}$$

Our very basic assumption is that the following decomposition holds:

$$V = \sum_{i=1}^{J} \Pi_i V_i. \tag{4.12}$$

This means that for any $v \in V$ there exists $v_i \in V_i$ (which may not be unique) such that

$$v = \sum_{i=1}^{J} \Pi_i v_i. \tag{4.13}$$

Furthermore, we assume that each V_i is equipped with an energy inner product $a_i(\cdot, \cdot)$. We define

$$A_i : V_i \mapsto V_i'$$

by

$$(A_i u_i, v_i) = a_i(u_i, v_i), \quad u_i,\ v_i \in V_i.$$

Let $\Pi_i' : V' \mapsto V_i'$ be the adjoint of Π_i:

$$(\Pi_i' f, v_i) = (f, \Pi_i v_i), \quad f \in V',\ v_i \in V_i.$$

Let $P_i = \Pi_i^* : V \mapsto V_i$ be the adjoint of Π_i with respect to the A-inner products:

$$(P_i u, v_i)_{A_i} = (u, \Pi_i v_i)_A, \quad u \in V,\ v_i \in V_i.$$

The following identity holds:

$$\Pi_i' A = A_i P_i. \tag{4.14}$$

If u is the solution of (4.1), by (4.14), we have

$$A_i u_i = f_i, \tag{4.15}$$

where

$$u_i = P_i u, \quad f_i = \Pi_i' f.$$

This equation may be regarded as the restriction of (4.1) to V_i. We assume that each such A_i has an approximate inverse or preconditioner:

$$R_i : V_i' \mapsto V_i. \tag{4.16}$$

Algorithm 3 Successive subspace correction method

Given $u^0 \in V$, for any $m = 1, 2, \ldots,$

1: $v \leftarrow u^{m-1}$

2: $v \leftarrow v + \Pi_i R_i \Pi_i'(f - Av)$, $i = 1, 2, \ldots, J$,

3: $u^m \leftarrow v$

The *parallel subspace correction* (PSC for short) method is (4.2) with $B = B_{\mathrm{psc}}$ given by

$$B_{\mathrm{psc}} = \sum_{i=1}^{J} \Pi_i R_i \Pi_i'. \tag{4.17}$$

The *successive subspace correction* (SSC for short) method is defined in Algorithm 3.

The algorithm is equivalent to (4.2) with $B = B_{\mathrm{ssc}}$ given by

$$I - B_{\mathrm{ssc}} A = (I - T_J)(I - T_{J-1}) \cdots (I - T_1), \tag{4.18}$$

where

$$T_i = \Pi_i R_i \Pi_i' A = \Pi_i R_i A_i P_i. \tag{4.19}$$

Theorem 4.4. Assume that all R_k are SPD. Then

$$(B_{\mathrm{psc}}^{-1} v, v) = \min_{\sum_i \Pi_i v_i = v} \sum_{k=1}^{J} (R_k^{-1} v_k, v_k), \tag{4.20}$$

with the unique minimizer given by

$$v_k^* = R_k \Pi_k' B_{\mathrm{psc}}^{-1} v. \tag{4.21}$$

Theorem 4.5. Under the assumptions given above, we obtain the identity

$$\begin{aligned} \|I - B_{\mathrm{ssc}} A\|_A^2 &= \|(I - T_J)(I - T_{J-1}) \cdots (I - T_1)\|_A^2 \\ &= 1 - \frac{1}{1 + c_0} \qquad (4.22) \\ &= 1 - \frac{1}{c_1}. \qquad (4.23) \end{aligned}$$

Here

$$c_0 = \sup_{\|v\|_A = 1} c_0(v), \quad c_1 = \sup_{\|v\|_A = 1} c_1(v) = 1 + c_0,$$

and, with

$$w_i = (I - T_i^{-1}) \Pi_i v_i + \sum_{j=i+1}^{J} \Pi_j v_j,$$

we obtain

$$c_0(v) = \inf_{\sum_i \Pi_i v_i = v} \sum_{i=1}^{J} (T_i \overline{T}_i^{-1} T_i^* w_i, w_i)_A, \tag{4.24}$$

and

$$\begin{aligned} c_1(v) &= (\overline{B}_{\rm ssc}^{-1} v, v) \\ &= \inf_{\sum_i \Pi_i v_i = v} (\overline{T}_i^{-1} (\overline{T}_i T_i^{-1} \Pi_i v_i + T_i^* w_i), (\overline{T}_i T_i^{-1} \Pi_i v_i + T_i^* w_i))_A. \end{aligned} \tag{4.25}$$

In particular, if $R_i = A_i^{-1}$, then

$$c_0(v) = \inf_{\sum_i \Pi_i v_i = v} \sum_{i=1}^{J} \left\| P_i \sum_{j=i+1}^{J} \Pi_j v_j \right\|_{A_i}^2 \tag{4.26}$$

and

$$c_1(v) = \inf_{\sum_i \Pi_i v_i = v} \sum_{i=1}^{J} \left\| P_i \sum_{j=i}^{J} \Pi_j v_j \right\|_{A_i}^2 \tag{4.27}$$

Lemma 4.6. If $R_k = A_k^{-1}$ for all k, and V_k are subspaces of V, then

$$\frac{1}{4}(B_{\rm psc}^{-1} v, v) \le (\bar{B}_{\rm ssc}^{-1} v, v) \le c^*(B_{\rm psc}^{-1} v, v), \quad v \in V, \tag{4.28}$$

where

$$c^* = \max_{1 \le k \le M} [N(k)]^2 \text{ and } N(k) = \big\{ j \in \{1, \ldots, J\} : V_j \cap V_k \neq \{0\} \big\}.$$

Proof. Given $v = \sum_{i=1}^{J} v_i$, with $v_i \in V_i$, it follows that

$$\begin{aligned} \|v\|_A^2 &= \sum_{k,j=1}^{J} (v_k, v_j)_A = \sum_{k=1}^{J} (v_k, v_k)_A + 2 \sum_{j>k}^{J} (v_k, v_j)_A \\ &= -\sum_{k=1}^{J} (v_k, v_k)_A + 2 \sum_{j \ge k}^{J} (v_k, v_j)_A. \end{aligned}$$

Thus

$$\begin{aligned} \sum_{k=1}^{J} \|v_k\|_A^2 &\le 2 \sum_{k=1}^{J} \Big(v_k, \sum_{j=k}^{J} v_j \Big)_A = 2 \sum_{k=1}^{J} \Big(v_k, P_k \sum_{j=k}^{J} v_j \Big)_A \\ &\le 2 \Big(\sum_{k=1}^{J} \|P_k \sum_{j=k}^{J} v_j\|_A^2 \Big)^{1/2} \Big(\sum_{k=1}^{J} \|v_k\|_A^2 \Big)^{1/2} \end{aligned}$$

Consequently,

$$\sum_{k=1}^{J} \|v_k\|_A^2 \leq 4 \sum_{k=1}^{J} \left\| P_k \sum_{j=k}^{J} v_j \right\|_A^2 .$$

By (4.20), (4.27) and (4.25), we have

$$(B_{\mathrm{psc}}^{-1} v, v) \leq 4 c_1(v) = 4(\bar{B}_{\mathrm{ssc}}^{-1} v, v).$$

The upper bound also follows easily. From $\|P_k\|_A = 1$ and the Schwarz inequality, we obtain

$$\begin{aligned} \sum_{k=1}^{J} \left\| P_k \sum_{j=k}^{J} v_j \right\|_A^2 &= \sum_{k=1}^{J} \left\| P_k \sum_{j \in N(k); j \geq k} v_j \right\|_A^2 \\ &\leq \sum_{k=1}^{J} \left\| \sum_{j \in N(k); j \geq k} v_j \right\|_A^2 \leq \sum_{k=1}^{J} N(k) \sum_{j \in N(k); j \geq k} \|v_j\|_A^2 \\ &\leq \sqrt{c^*} \sum_{k=1}^{J} \sum_{j \in N(k); j \geq k} \|v_j\|_A^2 \leq c^* \sum_{k=1}^{J} \|v_k\|_A^2 . \end{aligned}$$

The proof is concluded by taking the infimum over all decompositions on both sides and applying (4.27). □

Remark 4.7. We would like to point that the estimate in (4.28) holds for anisotropic and jump coefficient problems, and the constant c^* only depends on the 'topology' of the overlaps between the subspaces and does not depend on other ingredients and properties.

The Jacobi and Gauss–Seidel method can be interpreted as PSC and SSC based on the decomposition

$$\mathbb{R}^n = \sum_{i=1}^{n} \operatorname{span}\{e_i\},$$

with exact subspace solves such that

$$B_{\mathrm{psc}} = D^{-1}, \quad B_{\mathrm{ssc}} = (D+L)^{-1}, \quad \bar{B}_{\mathrm{ssc}} = (D+U)^{-1} D (D+L)^{-1}.$$

By Theorem 4.5,

$$c_0 = \sup_{\|v\|_A = 1} (D^{-1} U v, U v), \quad c_1 = \sup_{\|v\|_A = 1} (D^{-1}(D+U) v, (D+U) v).$$

By Lemma 4.6,

$$\frac{1}{4}(Dv, v) \leq (D^{-1}(D+U)v, (D+U)v) \leq c_*(Dv, v), \quad v \in \mathbb{R}^n. \tag{4.29}$$

We note that

$$c_1 = \sup_{v \in V} \frac{(\bar{B}_{\text{ssc}}^{-1} v, v)}{\|v\|_A^2} \le \sigma \sup_{v \in V} \frac{\|v\|_D^2}{\|v\|_A^2},$$

where

$$\sigma = \sup_{v \in V} \frac{\|v\|^2_{\bar{B}_{\text{ssc}}^{-1}}}{(Dv, v)}. \tag{4.30}$$

In the above presentation, most results are for SPD problems. We would like to point out that all of these results can be extended to a more general class of problems, namely symmetric positive semi-definite problems. When A is a matrix, we further assume that all the diagonals of A are non-zero and hence positive. When the subspace correction method is used for a more general symmetric positive semi-definite operator A, we further assume that each A_i is SPD. We shall use the acronym SPSD to denote matrices or operators that satisfy the aforementioned properties.

4.4. Bibliographical notes

The general notion of subspace corrections by means of space decompositions was described by Xu (1992), based on Bramble, Pasciak, Wang and Xu (1991*b*, 1991*c*). It is an abstract point of view encompassing the theory and practice of a large class of iterative algorithms such as multigrid and domain decomposition methods. In the past two decades a great deal of effort has been put into the investigation of theoretical and practical issues related to these methods. General results, applicable in many cases, in the theory of additive and multiplicative methods in Hilbert space can be found in Xu and Zikatanov (2002). For a literature review and basic results we refer the reader to the monographs and survey articles by Hackbusch (1985), Bramble (1993), Vassilevski (2008), Xu (1989, 1997), Xu and Zou (1998), Yserentant (1993), Toselli and Widlund (2005), Griebel and Oswald (1995) and Smith, Bjørstad and Gropp (1996). For detailed studies of classical iterative methods, we refer to the monographs by Young (1971), Hackbusch (1994), Varga (2000) and Saad (2003).

Note that in this section we have considered SPSD matrices, and according to Lee, Wu, Xu and Zikatanov (2007, 2008) and Ayuso de Dios, Brezzi, Marini, Xu and Zikatanov (2014), all the results in this section are valid for SPSD problems with semi-norms. Relations between the auxiliary space method and subspace correction methods is drawn in Chen (2011). In the classical multigrid literature (Brandt *et al.* 1982, Brandt, McCormick and Ruge 1985, Ruge and Stüben 1987, Trottenberg, Oosterlee and Schüller 2001), the notions of algebraically smooth (low) frequencies and algebraic high frequencies play an important role. They are also instrumental in the design of new AMG methods. As indicated by the convergence estimates, for a given smoother, the desirable coarse spaces should capture or well approximate the lower end of the spectrum of the relaxed matrix $\bar{R}A$ or $D^{-1}A$. This is usually referred to as *near-null space* (Treister and Yavneh 2015, Lai and Olson 2011, Xu 2009, Brezina *et al.* 2006).

5. Abstract multigrid methods and two-level theory

In this section we will present algebraic multigrid methods in an abstract setting. The acronym AMG for 'algebraic multigrid' can also be used to stand for 'abstract multigrid'.

Our focus will be on two-level methods. In view of algorithmic design, the extension of two-level to multilevel is straightforward: a general multilevel V-cycle algorithm can be obtained by recursively applying a two-level algorithm. However, the extension of a two-level convergence theory to a multilevel case can be highly non-trivial.

We will only consider SPSD problems as described in Section 4. As is done in most of the literature, the design principle of an AMG is to optimize the choice of coarse spaces with a given smoother. The most commonly used smoothers are the Gauss–Seidel method and the (modified or scaled) Jacobi method. As these smoothers are convergent as iterative methods, the resulting AMG method is always convergent. The task of our AMG convergence theory is to make sure that this convergence is also fast. In particular, for systems arising from the discretization of partial differential equations, we hope that our AMG method converges uniformly with respect to the size of the problem and/or some crucial parameters from the underlying PDEs. We sometimes speak of this as 'uniformly convergent' or 'uniform convergence'.

As it turns out, we are often able to establish such uniform convergence for two-level AMG, but very rarely can we extend a uniform convergence result to the multilevel case. For second-order elliptic boundary value problems, multilevel convergence is very well understood for geometric multigrid methods. However, a rigorous multilevel convergence theory for AMG without using geometric information is still very much an open problem.

We will mainly focus on two-level convergence theory for AMG methods in this section and also in the rest of the paper.

5.1. *A two-level method*

A two-level method typically consists of the following components:

(i) a smoother, $R : V' \mapsto V$,

(ii) a coarse space V_c, which may or may not be a subspace of V, linked to V via a prolongation operator,

$$P : V_c \mapsto V,$$

(iii) a coarse space solver, $B_c : V_c' \mapsto V_c$.

In the discussion below we need the inner product

$$(u, v)_{\bar{R}^{-1}} = (\overline{T}^{-1} u, v)_A = (\bar{R}^{-1} u, v), \quad \overline{T} = \overline{R}A \tag{5.1}$$

and the accompanying norm $\|\cdot\|_{\bar{R}^{-1}}$. Here we recall that the definition of $\bar{R}$ is analogous to that in (4.5).

We always assume that $\bar{R}$ is SPD, and hence the smoother R is always convergent. Furthermore,

$$\|v\|_A^2 \le \|v\|_{\bar{R}^{-1}}^2. \tag{5.2}$$

The restriction of (4.1) is then

$$A_c u_c = f_c, \tag{5.3}$$

where

$$A_c = P'AP, \quad f_c = P'f.$$

The coarse space solver B_c is often chosen to be the exact solver, namely $B_c = A_c^{-1}$, for analysis, but in a multilevel setting, B_c is recursively defined and it is an approximation to A_c^{-1}. We distinguish two different cases in choosing B_c:

$$\text{it is an } \textit{exact} \text{ two-level method if } B_c = A_c^{-1}, \tag{5.4}$$

$$\text{it is an } \textit{inexact} \text{ two-level method if } B_c \ne A_c^{-1}. \tag{5.5}$$

When A is semi-definite, we use N to denote the kernel of A and we always assume that $N \subset V_c$. Let

$$W := N^{\perp} \quad \text{and} \quad W_c := V_c \cap W, \tag{5.6}$$

where the orthogonality is understood with respect to the $(\cdot,\cdot)_{\bar{R}^{-1}}$-inner product. Let $Q_1 : V \mapsto W$ be the orthogonal projection with respect to the $(\cdot,\cdot)_{\bar{R}^{-1}}$-inner product

$$(Q_1 v, w)_{\bar{R}^{-1}} = (v, w)_{\bar{R}^{-1}}, \quad \text{for all } v \in V, w \in W. \tag{5.7}$$

Here A_c is semi-definite on V_c but invertible on W_c. We denote the restriction of A_c on W_c by $\hat{A}_c$, and define the pseudo-inverse of A_c by

$$A_c^{\dagger} := Q_1' \hat{A}_c^{-1} Q_1. \tag{5.8}$$

With a slight abuse of notation, we will still use A_c^{-1} to denote the pseudo-inverse of A_c, namely

$$A_c^{-1} = A_c^{\dagger}.$$

We will use similar notation for the pseudo-inverse of other relevant singular operators and matrices in the rest of the paper.

Algorithm 4 Two-level multigrid method

Given $g \in V'$, the action Bg is defined via the following two steps.

1: Coarse grid correction: $w = PB_cP'g$.

2: Post-smoothing: $Bg := w + R(g - Aw)$.

We choose to define an AMG algorithm in terms of an operator $B : V' \mapsto V$, which can be considered as an approximate inverse or a preconditioner of A. A typical two-level MG method is shown in Algorithm 4.

In the rest of this section, we take $B_c = A_c^{-1}$.

There are usually two different (and mathematically equivalent) ways to choose V_c, V, P and R. The first, known as the operator version, is such that

$$V_c \subset V.$$

In this case $P = \imath_c$, where

$$\imath_c : V_c \mapsto V \tag{5.9}$$

is the natural inclusion of V_c into V. In the application to finite element discretization for second-order elliptic boundary value problems, V_c and V are just the finite element subspaces of $H^1(\Omega)$. This type of notation is convenient for analysis. However, this is not an algorithm that can be directly used for implementation.

The second, known as the matrix version, is such that

$$V_c = \mathbb{R}^{n_c} \quad \text{and} \quad V = \mathbb{R}^n,$$

and

$$P : \mathbb{R}^{n_c} \mapsto \mathbb{R}^n \tag{5.10}$$

is the prolongation matrix.

These two different notations are related via the use of basis functions of $\{\phi_i^c\}_{i=1}^{n_c} \subset V_c$ and $\{\phi_i\}_{i=1}^{n} \subset V$. As noted earlier, the prolongation matrix P given in (3.12)–(3.13) is simply the matrix representation of $\imath_c$ given in (5.9), and we have

$$(\phi_i^c, \ldots, \phi_{n_c}^c) = (\phi_i, \ldots, \phi_n)P. \tag{5.11}$$

The following observation is clear.

Observation 5.1. Finding a coarse space $V_c \subset V$ is equivalent to finding a prolongation matrix P in (5.10).

Lemma 5.2. The error propagation operator for two-level AMG operator $E = I - BA$ is

$$E = (I - RA)(I - \Pi_c), \tag{5.12}$$

where $\Pi_c = \imath_c A_c^{-1} \imath_c' A$, which is the $(\cdot,\cdot)_A$ orthogonal projection on V_c, in matrix notation $\Pi_c = P A_c^{-1} P^T A$.

5.2. *An optimal two-level AMG theory*

The aim of the AMG method is to balance the interplay between smoother R and the coarse space V_c. Most existing AMG methods first fix a smoother, which is often given by the Jacobi or Gauss–Seidel methods (or their combinations and variations), and then optimize the choice of coarse spaces. This is the approach we will mainly discuss in this paper. However, we also comment on a different approach by first fixing the coarse space and then trying to optimize the choice of the smoother. It is also possible to try to make an optimal choice of smoother and coarse space simultaneously, but we will not address this approach here.

Let $Q_c : V \mapsto V_c$ be the orthogonal projection with respect to the $(\cdot,\cdot)_{\bar{R}^{-1}}$-inner product

$$(Q_c u, v_c)_{\bar{R}^{-1}} = (u, v_c)_{\bar{R}^{-1}}, \quad \text{for all } v_c \in V_c. \tag{5.13}$$

By the definition of W and W_c, we have that $(\cdot,\cdot)_A$ is an inner product on W, $\|\cdot\|_A$ is a norm on W, and the projection $\Pi_c : V \mapsto W_c$ is well defined:

$$(\Pi_c u, v_c)_A = (u, v_c)_A, \quad \text{for all } u \in V, \ v_c \in W_c. \tag{5.14}$$

The two-level convergence rate is obtained in the following theorem.

Theorem 5.3. Assume that $N \subset V_c$. The convergence rate of an exact two-level AMG is given by

$$\|E\|_A^2 = 1 - \frac{1}{K(V_c)}, \tag{5.15}$$

where

$$K(V_c) = \max_{v\in W} \frac{\|(I - Q_c)v\|_{\bar{R}^{-1}}^2}{\|v\|_A^2} = \max_{v\in W} \min_{v_c\in W_c} \frac{\|v - v_c\|_{\bar{R}^{-1}}^2}{\|v\|_A^2}. \tag{5.16}$$

Proof. We note that

$$\|(I - T)v\|_A^2 = ((I - \bar{T})v, v)_A, \quad \text{for all } v \in V.$$

Then we have

$$\begin{aligned}\|E\|_A^2 &= \max_{w\in W} \frac{\|(I-T)(I-\Pi_c)w\|_A^2}{\|w\|_A^2} \\ &= \max_{w\in W} \frac{((I-\overline{T})(I-\Pi_c)w, (I-\Pi_c)w)_A}{\|w\|_A^2} \\ &= 1 - \min_{w\in W} \frac{(\overline{T}(I-\Pi_c)w, (I-\Pi_c)w)_A}{\|w\|_A^2}\end{aligned}$$

$$
\begin{aligned}
&= 1 - \min_{w \in W} \frac{(Q_1 \overline{T}(I - \Pi_c)w, (I - \Pi_c)w)_A}{\|(I - \Pi_c)w\|_A^2 + \|\Pi_c w\|_A^2} \\
&= 1 - \min_{v \in W_c^{\perp_A}} \frac{(Q_1 \overline{T} v, v)_A}{\|v\|_A^2} \\
&= 1 - \min_{v \in W_c^{\perp_A}} \frac{((I - \Pi_c) Q_1 \overline{T} v, v)_A}{\|v\|_A^2} \\
&= 1 - \lambda_{\min}(X),
\end{aligned}
$$

where

$$X = (I - \Pi_c) Q_1 \overline{T} : W_c^{\perp_A} \mapsto W_c^{\perp_A},$$

and it is easy to see that X is self-adjoint with respect to $(\cdot, \cdot)_A$.

One key observation is that the inverse of X can be explicitly written as

$$Z = (Q_1 \overline{T})^{-1}(I - Q_c),$$

since by definition, we have, for any $u, v \in V$,

$$
\begin{aligned}
(\Pi_c Z u, v)_A &= ((Q_1 \overline{T})^{-1}(I - Q_c)u, \Pi_c v)_A = (\overline{T}(Q_1 \overline{T})^{-1}(I - Q_c)u, \Pi_c v)_{\bar{R}^{-1}} \\
&= (Q_1 \overline{T}(Q_1 \overline{T})^{-1}(I - Q_c)u, \Pi_c v)_{\bar{R}^{-1}} = ((I - Q_c)u, \Pi_c v)_{\bar{R}^{-1}} = 0,
\end{aligned}
$$

which implies $\Pi_c Z = 0$. Thus we have

$$Z : W_c^{\perp_A} \mapsto W_c^{\perp_A},$$

and furthermore

$$XZ = (I - \Pi_c)(I - Q_c) = I - \Pi_c = I \text{ on } W_c^{\perp_A}.$$

Consequently

$$\lambda_{\min}(X) = \frac{1}{\lambda_{\max}(Z)}.$$

Finally,

$$
\begin{aligned}
\lambda_{\max}(Z) &= \max_{v \in W_c^{\perp_A}} \frac{((Q_1 \overline{T})^{-1}(I - Q_c)v, v)_A}{(v, v)_A} \\
&= \max_{v \in W_c^{\perp_A}} \frac{(\overline{T}(Q_1 \overline{T})^{-1}(I - Q_c)v, v)_{\bar{R}^{-1}}}{(v, v)_A} \\
&= \max_{v \in W_c^{\perp_A}} \frac{(Q_1 \overline{T}(Q_1 \overline{T})^{-1}(I - Q_c)v, v)_{\bar{R}^{-1}}}{(v, v)_A} \\
&= \max_{v \in W_c^{\perp_A}} \frac{((I - Q_c)v, v)_{\bar{R}^{-1}}}{(v, v)_A} \\
&= \max_{v \in W_c^{\perp_A}} \frac{\|(I - Q_c)v\|_{\bar{R}^{-1}}^2}{(v, v)_A} = K(V_c).
\end{aligned}
$$

The last identity holds because $I - Q_c = (I - Q_c)(I - \Pi_c)$, and we can then take the maximum over all $v \in W$. This completes the proof. $\square$

Theorem 5.3 can be stated as follows using the matrix representation introduced in Section 3.

Theorem 5.4. Assume that $P \in \mathbb{R}^{n\times n_c}$ and $N(\tilde{A}) \subset \text{Range}(P)$. The convergence rate of an exact two-level AMG is given by

$$\|E\|_A^2 = 1 - \frac{1}{\tilde{K}(P)}, \tag{5.17}$$

where

$$\tilde{K}(P) = \max_{v\in\mathbb{R}^n} \min_{v_c\in\mathbb{R}^{n_c}} \frac{\|v - Pv_c\|^2_{\tilde{\bar{R}}^{-1}}}{\|v\|^2_{\tilde{A}}}. \tag{5.18}$$

Remark 5.5. The result of Theorem 5.4 can be viewed as follows. We note that $\overline{T}^{-1}(I - Q_c)$ is a self-adjoint operator with respect to the A-inner product. Hence, we immediately have

$$K(V_c) = \|\overline{T}^{-1}(I - Q_c)\|_A = \|(\overline{R}A)^{-1}(I - Q_c)\|_A.$$

In the case of two subspaces, we have the following theorem giving the precise convergence rate of the corresponding SSC method.

Theorem 5.6. Let $\{\mu_j, \zeta_j\}_{j=1}^n$ be the eigenpairs of $\bar{T} = \bar{R}A$, and assume that $\{\zeta_j\}$ are orthogonal with respect to $(\cdot,\cdot)_{\bar{R}^{-1}}$. The convergence rate $\|E(V_c)\|_A$ is minimal for the coarse space

$$V_c^{\text{opt}} = \text{span}\{\zeta_j\}_{j=1}^{n_c} \in \mathop{\arg\min}_{\dim V_c = n_c, N\subset V_c} K(V_c). \tag{5.19}$$

In this case,

$$\|E\|_A^2 = 1 - \mu_{n_c+1}. \tag{5.20}$$

Proof. By Theorem 5.3, we just need to maximize $1/(K(V_c))$. For any $v \in V_c^{\perp}$, where $\perp$ is with respect to the $\bar{R}^{-1}$-inner product, we have

$$\min_{v_c\in V_c} \|v - v_c\|^2_{\bar{R}^{-1}} = \|v\|^2_{\bar{R}^{-1}}. \tag{5.21}$$

Then it follows that

$$\frac{1}{K(V_c)} = \min_{v\in V} \max_{v_c\in V_c} \frac{\|v\|_A^2}{\|v - v_c\|^2_{\bar{R}^{-1}}} \leq \min_{v\in V_c^{\perp}} \max_{v_c\in V_c} \frac{\|v\|_A^2}{\|v - v_c\|^2_{\bar{R}^{-1}}} = \min_{v\in V_c^{\perp}} \frac{\|v\|_A^2}{\|v\|^2_{\bar{R}^{-1}}}.$$

By the min-max principle (Theorem 2.1), we have

$$\max_{\dim V_c = n_c} \frac{1}{K(V_c)} \leq \max_{\dim V_c = n_c} \min_{v\in V_c^{\perp}} \frac{\|v\|_A^2}{\|v\|^2_{\bar{R}^{-1}}} = \mu_{n_c+1}.$$

On the other hand, if we choose $V_c^{\text{opt}} = \text{span}\{\zeta_j\}_{j=1}^{n_c}$, it is easy to compute that $K(V_c^{\text{opt}}) = 1/(\mu_{n_c+1})$. So we have

$$\max_{\dim V_c = n_c} \frac{1}{K(V_c)} = \mu_{n_c+1},$$

with optimal coarse space

$$V_c^{\text{opt}} = \text{span}\{\zeta_j\}_{j=1}^{n_c}. \qquad \square$$

Using the matrix representation introduced in Section 3, we state the matrix version of Theorem 5.6 below. For simplicity, with abuse of notation, we still use A to denote the matrix representation of operator A.

Theorem 5.7. Let $\{\mu_j, \zeta_j\}_{j=1}^{n}$ be the eigenpairs of $\bar{T} = \bar{R}A$. Let us also assume that $\{\zeta_j\}$ are orthogonal with respect to $(\cdot,\cdot)_{\bar{R}^{-1}}$. The convergence rate $\|E(P)\|_A$ is minimal for P such that

$$\text{Range}(P) = \text{Range}(P^{\text{opt}}), \tag{5.22}$$

where

$$P^{\text{opt}} = (\zeta_1, \ldots, \zeta_{n_c}). \tag{5.23}$$

In this case,

$$\|E\|_A^2 = 1 - \mu_{n_c+1}. \tag{5.24}$$

The following theorem is important in motivating most AMG algorithms.

Theorem 5.8. Given $\eta > 0$, let $\mathcal{X}_\eta$ be defined as

$$\mathcal{X}_\eta = \left\{P \in \mathbb{R}^{n\times n_c} : (Pv, Pv)_{\bar{R}^{-1}} \geq \eta(v,v), \;\; v \in \mathbb{R}^{n_c}\right\}. \tag{5.25}$$

Then, with P^{opt} given by (5.23), we have $P \in \arg\min_{Q\in\mathcal{X}_\eta} \text{trace}(Q^TAQ)$ if

$$P \in \mathcal{X}_\eta \quad \text{and} \quad \text{Range}(P) = \text{Range}(P^{\text{opt}})).$$

Since the eigenvalues of $\bar{R}A$ are expensive to compute, the practical value of Theorem 5.6 is limited. However, it provides useful guidance in the design of practical AMG methods.

For finite element discretizations we can use Weyl's law combined with Theorem 5.6 to prove an estimate for the convergence rate of a two-grid method with an optimal coarse space.

Corollary 5.9. Let the assumptions of the discrete Weyl's law (Theorem 2.5) hold, and let the smoother $\bar{R}$ be spectrally equivalent to the diagonal of the stiffness matrix A. Let $\gamma > 0$ be such that $\gamma n \leq n_c < n$. Then, for the optimal coarse space, we have the estimate

$$\mu_{n_c+1} \geq \delta_0 \quad \text{and} \quad \|E\|_A^2 \leq 1 - \delta_0,$$

where $\delta_0 \in (0,1)$ only depends on γ and the constants γ_0 and γ_1 in (2.31).

Proof. By the assumptions in Theorem 2.5 and the fact that $\bar{R}$ is spectrally equivalent to the diagonal of A, we have

$$|w|_1^2 \eqsim \|w\|_A^2, \quad \|w\|_0^2 \eqsim h^2\|w\|_{\bar{R}^{-1}}^2. \tag{5.26}$$

Further, Lemma 2.4 and Theorem 2.5 then show that

$$\mu_{n_c+1}(\bar{R}A) = \min_{\substack{W\subset V_h \\ \dim W=k}} \max_{\substack{w\in W \\ \|w\|_{\bar{R}^{-1}}\neq 0}} \frac{\|w\|_A^2}{\|w\|_{\bar{R}^{-1}}^2} \eqsim h^2\lambda_{n_c+1} = O\left(\left(\frac{n_c h^d}{\mathrm{Vol}(\Omega)}\right)^{2/d}\right). \tag{5.27}$$

The desired result follows immediately from Theorem 5.6 because $\mathrm{Vol}(\Omega) \eqsim h^d n$ and $\gamma n \le n_c < n$, which gives $\mu_{n_c+1} \eqsim 1$. □

Remark 5.10. Since the coarse space which minimizes the convergence rate is also the coarse space which minimizes $K(V_c)$, as a corollary we have the inequality

$$K(V_c) = \frac{1}{1-\|E\|_A^2} \ge \frac{1}{\mu_{n_c+1}},$$

or

$$\|E\|_A^2 \ge 1 - \mu_{n_c+1}.$$

Theorem 5.3 provides an explicit estimate for the convergence of a two-level method in terms of $K(V_c)$. For a given method, a smaller bound on $K(V_c)$ means a faster convergence rate. In particular, the two-level AMG method is uniformly convergent if $K(V_c)$ is uniformly bounded with respect to mesh parameters.

5.3. Quasi-optimal theories

We now look at the necessary and sufficient condition for uniform convergence of a two-level method as proved in Section 5.2 (see Theorem 5.3):

$$\min_{v_c\in V_c} \|v - v_c\|_{\bar{R}^{-1}} \le K(V_c)\|v\|_A^2. \tag{5.28}$$

Here $K(V_c)$ is the smallest constant for which (5.28) holds for all $v \in V$. The space V_c which minimizes $K(V_c)$ is V_c^{opt}. A similar argument was used in the proof of Theorem 5.6 in Section 5.2. Here we generalize the result to semi-definite A.

For a given smoother R, one basic strategy in the design of AMG is to find a coarse space such that $K(V_c)$ is as small as possible in practice. There are many cases, however, in which the operator $\bar{R}^{-1}$ in the definition of $K(V_c)$ is difficult to work with.

One commonly used approach is to replace $\bar{R}^{-1}$ with a simpler but spectrally equivalent SPD operator. More specifically, we assume that $D : V \mapsto$

V' is an SPD operator such that

$$c_D(Dv,v) \le (\bar{R}^{-1}v,v) \le c^D(Dv,v), \quad \text{for all } v \in V. \tag{5.29}$$

That is,

$$c_D\|v\|_D^2 \le \|v\|_{\bar{R}^{-1}}^2 \le c^D\|v\|_D^2, \quad \text{for all } v \in V, \tag{5.30}$$

where

$$(u,v)_D = (Du,v), \quad \|v\|_D^2 = (v,v)_D.$$

Examples of such equivalent norms for Schwarz smoothers are given in (4.20) and (4.27). As a rule, the norm defined by $\bar{R}$ corresponding to the symmetric Gauss–Seidel method, that is, R defined by the pointwise Gauss–Seidel method, can be replaced with the norm defined by the diagonal of A (*i.e.*, by the Jacobi method, which, while not always convergent as a relaxation, provides an equivalent norm).

In terms of this operator D, we introduce the quantity

$$K(V_c,D) = \max_v \frac{\|v - Q_D v\|_D^2}{\|v\|_A^2} = \max_v \min_{v_c \in V_c} \frac{\|v - v_c\|_D^2}{\|v\|_A^2}, \tag{5.31}$$

where $Q_D : V \mapsto V_c$ is the $(u,v)_D$-orthogonal projection.

By (5.16), (5.31) and (5.30), we have

$$c_D K(V_c,D) \le K(V_c) \le c^D K(V_c,D). \tag{5.32}$$

Theorem 5.11. The two-level algorithm satisfies

$$1 - \frac{1}{c_D K(V_c,D)} \le \|E\|_A^2 \le 1 - \frac{1}{c^D K(V_c,D)} \le 1 - \frac{1}{c^D C}, \tag{5.33}$$

where C is any upper bound of $K(V_c,D)$, namely

$$\min_{w \in V_c} \|v - w\|_D^2 \le C\|v\|_A^2, \quad \text{for all } v \in V. \tag{5.34}$$

The proof of Theorem 5.11 is straightforward and indicates that, if c_D and c^D are 'uniform' constants, the convergence rate of the two-level method is 'uniformly' dictated by the quantity $K(V_c,D)$.

We say that V_c is quasi-optimal if the following inequality holds:

$$\min_{w \in V_c} \|v - w\|_D^2 \le \gamma \mu_{n_c+1}^{-1} \|v\|_A^2, \quad \text{for all } v \in V, \tag{5.35}$$

with a constant $\gamma > 0$ independent of the size of the problem.

The construction of an approximation to the optimal coarse space V_c^{opt} that is used in most AMG algorithms relies on two operators A_M and D_M which satisfy

$$c_1\|v\|_D^2 \le \|v\|_{D_M}^2, \quad \|v\|_{A_M}^2 \le c_2\|v\|_A^2, \quad \text{for all } v \in V, \tag{5.36}$$

with constants c_1 and c_2 independent of the problem size. Here, on the right-hand side, we have a seminorm $\|\cdot\|_{A_M}$, because sometimes A_M is only semi-definite. We point out that here A_M and D_M are analogues of $\underset{\sim}{A}_W$ and $\underset{\sim}{D}$ defined in (6.6) and (6.7), respectively, in the general framework of Section 6; the assumptions in (5.36) are analogous to Assumption 6.4 which we made in the general AMG framework in Section 6.

Theorem 5.12. If D_M and A_M satisfy (5.36), and V_c is a coarse space such that

$$\min_{w\in V_c} \|v-w\|_{D_M}^2 \leq \gamma\mu_{n_c+1}^{-1}\|v\|_{A_M}^2, \quad \text{for all } v\in V, \tag{5.37}$$

then the following hold.

(a) We have

$$\min_{w\in V_c} \|v-w\|_{\overline{R}^{-1}}^2 \leq \frac{c^D}{c_D}\frac{c_2}{c_1}\gamma\mu_{n_c+1}^{-1}\|v\|_A^2, \quad \text{for all } v\in V. \tag{5.38}$$

(b) The corresponding two-level AMG algorithm satisfies

$$\|I-BA\|_A^2 \leq 1-\frac{c_D}{c^D}\frac{c_1}{c_2}\frac{1}{\gamma}\,\mu_{n_c+1}. \tag{5.39}$$

5.4. Algebraically high and low frequencies

In geometric MG, algebraically smooth error is also smooth in the usual geometric sense. However, in AMG settings, smooth error can be geometrically non-smooth. In order to make this distinction, we use the term *algebraically smooth error* when we refer to the error in the AMG setting that is not damped (eliminated) by the smoother R. In general, good interpretation of the algebraically smooth error leads to an efficient and robust AMG algorithm. Careful characterization of the algebraically smooth error is needed, since in this case we can try to construct a coarser level which captures these error components well.

Here is a more formal definition of an algebraically smooth error.

Definition 5.13. Let $R: V\mapsto V$ be a smoothing operator such that its symmetrization $\overline{R}=R+R^T-R^TAR$ is positive definite. Given $\varepsilon\in(0,1)$, we say that the vector v is algebraically ε-smooth (or v is an *ε-algebraic low frequency*) with respect to A if

$$\|v\|_A^2 \leq \varepsilon\|v\|_{\overline{R}^{-1}}^2. \tag{5.40}$$

The set of algebraically smooth vectors will be denoted by

$$\mathcal{L}_\varepsilon = \{v : \|v\|_A^2 \leq \varepsilon\|v\|_{\overline{R}^{-1}}^2\}. \tag{5.41}$$

We point out that this is a set of vectors (a ball, or rather an ellipsoid) and not a linear vector space in general. It is then clear that the elements of this set need to be approximated well by elements from the coarse space.

The rationale of Definition 5.13 can be seen from the following simple result (it is always true that $(\bar{R}Av, v) \le \|v\|_A^2$).

Lemma 5.14. Any vector $v \in V$ that satisfies

$$(\overline{R}A\boldsymbol{v}, \boldsymbol{v})_A \le \varepsilon \|v\|_A^2 \tag{5.42}$$

is ε-algebraically smooth.

Proof. By the Schwarz inequality for the inner product defined by $\overline{R}^{-1}$ and (5.42), we have

$$\|\boldsymbol{v}\|_A^2 = (\overline{R}A\boldsymbol{v}, \overline{R}^{-1}\boldsymbol{v}) \le (\overline{R}A\boldsymbol{v}, A\boldsymbol{v})^{1/2}(\overline{R}^{-1}\boldsymbol{v}, \boldsymbol{v})^{1/2} \le \sqrt{\varepsilon}(\overline{R}^{-1}\boldsymbol{v}, \boldsymbol{v})^{1/2}\|\boldsymbol{v}\|_A.$$

□

We can easily show that this definition is equivalent to saying that the algebraically smooth error components are the components for which the smoother converges slowly. Indeed, inequality (5.42) is clearly equivalent to $((I - \overline{R}A)v, v)_A \ge (1 - \varepsilon)(v, v)_A$, namely

$$\frac{\|S\boldsymbol{v}\|_A^2}{\|\boldsymbol{v}\|_A^2} \ge 1 - \varepsilon, \quad S = I - T \quad \text{and} \quad T = RA. \tag{5.43}$$

The property (5.43) is often referred to as the *smoothing property*.

Remark 5.15. In the classical multigrid literature, an algebraically smooth error is defined as $\boldsymbol{e} \in V$ such that

$$\|\boldsymbol{e}\|_{AD^{-1}A}^2 \le \varepsilon \|\boldsymbol{e}\|_A^2, \tag{5.44}$$

for a small and positive parameter ε which implies

$$\|\boldsymbol{e}\|_A^2 \le \|\boldsymbol{e}\|_D \|\boldsymbol{e}\|_{AD^{-1}A} \le \sqrt{\varepsilon}\|\boldsymbol{e}\|_D \|\boldsymbol{e}\|_A. \tag{5.45}$$

That is,

$$\|\boldsymbol{e}\|_A^2 \le \varepsilon \|\boldsymbol{e}\|_D^2. \tag{5.46}$$

As is clearly seen from Definition 5.13, Lemma 5.14 implies (5.45) with $\overline{R} \approx D^{-1}$, where D is the diagonal of A.

Thanks to (7.9), we have the following result.

Lemma 5.16. If e is algebraically smooth, that is, e satisfies (5.46), then

$$\|\boldsymbol{e}\|_{\tilde{A}}^2 \lesssim \varepsilon \|\boldsymbol{e}\|_{\tilde{D}}^2. \tag{5.47}$$

Namely, e is also algebraically smooth with respect to $\tilde{A}$, the M-matrix relative of A.

On the other hand, note that by the definition of $\|\cdot\|_{\bar{R}^{-1}}$, we always have $\|v\|_{\bar{R}^{-1}} \ge C\|v\|_A$, with constant C independent of the parameters of interest. Drawing from analogy with the geometric multigrid method, we introduce the notion of algebraic high frequency as follows.

Definition 5.17. Given $\delta \in (0,1]$, we call $v \in V$ a *δ-algebraic high frequency* if

$$\|v\|_A^2 \geq \delta \|v\|_{\bar{R}^{-1}}^2.$$

The set of algebraically high-frequency vectors will be denoted by

$$\mathcal{H}_\delta = \{v : \|v\|_A^2 \geq \delta \|v\|_{\bar{R}^{-1}}^2\}. \tag{5.48}$$

The concept of algebraic high and low frequencies will be used in the two-level AMG theory in Section 5.3, and also in the design of classical AMG in Section 12.1.

Lemma 5.18. Let (ϕ_i, μ_i) be all the eigenpairs for $\bar{R}A$, namely $\bar{R}A\phi_i = \mu_i\phi_i$. Then

$$\operatorname{span}\{\phi_i : \mu_i \leq \varepsilon\} \subset \mathcal{L}_\varepsilon$$

and

$$\operatorname{span}\{\phi_i : \mu_i \geq \delta\} \subset \mathcal{H}_\delta.$$

We now introduce the notion of *near-null space.*

Definition 5.19 (near-null space). For sufficiently small $\varepsilon \in (0,1)$, we shall say that $\operatorname{span}\{\phi_i : \mu_i \leq \varepsilon\}$ is an ε-near-null space of $\bar{R}A$.

5.5. *Smoothing properties of Jacobi and Gauss–Seidel methods*

The essence of multigrid methods is that simple iterative methods such as Jacobi and Gauss–Seidel methods have a special property known as the *smoothing property.* As an illustration, we apply the Gauss–Seidel method to

$$A\mu = b,$$

with A given by (2.16) for isotropic problems and (2.34) for anisotropic problems, respectively. We first choose μ randomly (as shown in Figure 5.1) and then compute $A\mu$ for both (2.16) and (2.34) to compute right-hand sides $b = A\mu$ respectively. We then apply the Gauss–Seidel method to both equations with initial guess $\mu^0 = 0$ (see Figure 5.2).

Note that, for A given by (2.34) when $\epsilon \ll 1$, we have

$$\lambda_{11} < \lambda_{21} < \cdots < \lambda_{N1} < \lambda_{ij}, \quad i \geq 1, j \geq 2.$$

The corresponding eigenfunctions, which can be viewed as 'algebraic low frequencies', can be highly oscillatory in the x-direction.

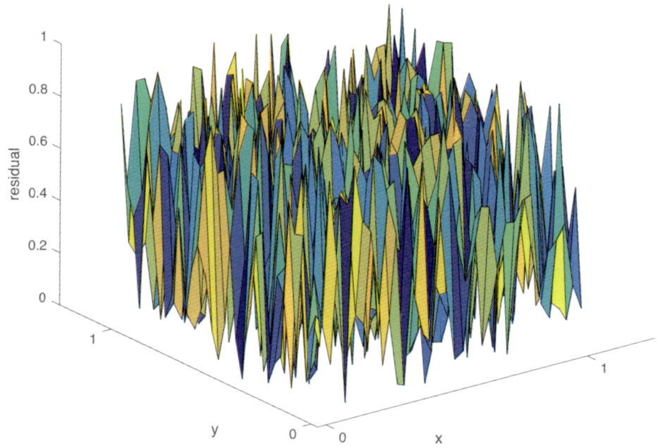

Figure 5.1. Initial error for both (2.16) and (2.34).

As an illustration of the difference between algebraic high/low frequencies and geometric high/low frequencies, we consider the linear system given by (2.34) for anisotropic problems. Clearly, A can be written as

$$A = \epsilon I \otimes M + M \otimes I \quad \text{with } M = \text{tridiag}(-1, 2, -1).$$

We define the vector $\mu \in R^N$ as

$$\mu = x \otimes y, \quad \text{with } x = \mathbf{1}_n, \quad \text{and} \quad y = (1\ 0\ 1\ 0 \cdots 1\ 0\ 1)^T \in R^n.$$

Then it is easy to compute that

$$Mx = \begin{pmatrix} 1 \\ 0 \\ 0 \\ \vdots \\ 0 \\ 1 \end{pmatrix} \quad \text{and} \quad My = \begin{pmatrix} 2 \\ -2 \\ 2 \\ \vdots \\ -2 \\ 2 \end{pmatrix}.$$

We have

$$A\mu = \epsilon(I \otimes M)(x \otimes y) + (M \otimes I)(x \otimes y) = \epsilon(x \otimes My) + (Mx \otimes y)$$

and

$$\begin{aligned} \|\mu\|_A^2 = \mu^T A\mu &= \epsilon(x \otimes y)^T(x \otimes My) + (x \otimes y)^T(Mx \otimes y) \\ &= \epsilon(x^T x) \otimes (y^T My) + (x^T Mx) \otimes (y^T y) = \epsilon n(n+1) + n + 1. \end{aligned}$$

Letting D be the diagonal of A, we then have

$$\|\mu\|_D^2 = 2(1+\epsilon)\mu^T\mu = (1+\epsilon)n(n+1).$$

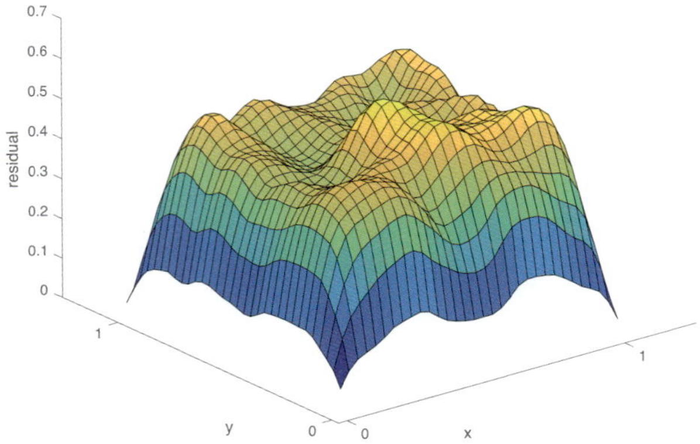

(a)

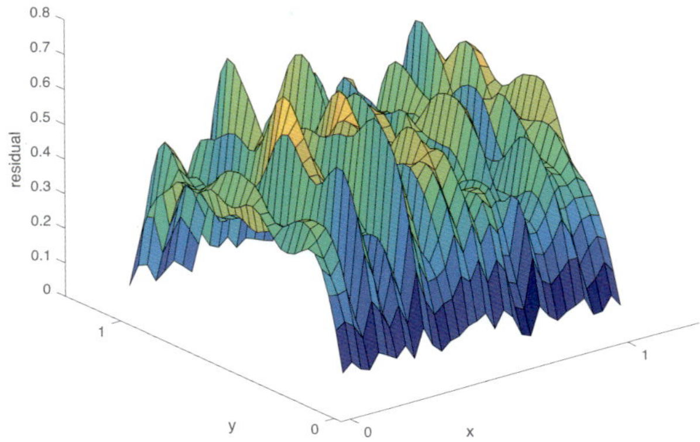

(b)

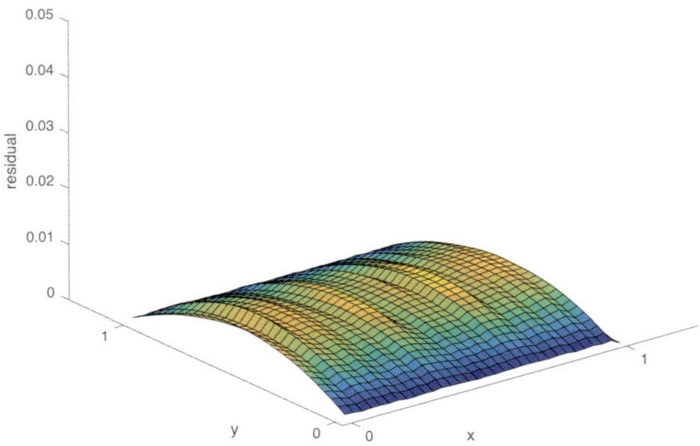

(c)

Figure 5.2. (a) Error after applying five Gauss–Seidel iterations to (2.16). (b) Error after applying five Gauss–Seidel iterations to (2.34). (c) Error after applying one block Gauss–Seidel iteration to (2.34).

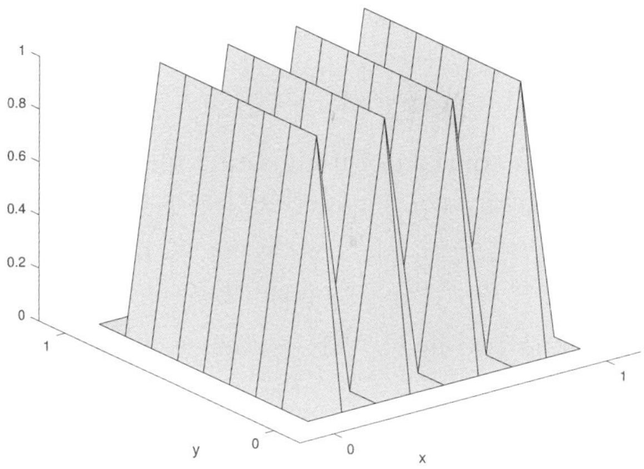

Figure 5.3. Graph of function defined in (5.49). This function is highly oscillatory in the x-direction.

This shows that

$$\frac{\|\mu\|_A^2}{\|\mu\|_D^2} = \frac{\epsilon}{1+\epsilon} + \frac{1}{(1+\epsilon)(n+1)},$$

which implies that μ is an algebraic low frequency if ϵ is sufficiently small.

On the other hand, if we denote the nodal basis functions corresponding to the uniform finite element mesh by $\{\phi_{ij} : 1 \leq i, j \leq n\}$, that is, ϕ_{ij} is a piecewise linear function such that

$$\phi_{ij}(kh, lh) = \delta_{ik}\delta_{jl},$$

then we define

$$\mathbf{\Phi} = (\phi_{11}, \phi_{12}, \ldots, \phi_{1n}, \phi_{21}, \ldots, \phi_{nn}).$$

If we consider the finite element function corresponding to μ, namely, the function defined by

$$u = \mathbf{\Phi}\mu = \sum_{i=1}^{(n+1)/2} \sum_{j=1}^{n} \phi_{2i-1,j}, \tag{5.49}$$

then, in the geometric point of view, this function is highly oscillatory in the x-direction, which is a geometric high frequency (see Figure 5.3).

5.6. Convergence theory in view of algebraic high and low frequencies

We next present a convergence theory based on algebraic high and low frequencies. We first prove the following lemma.

Lemma 5.20. Let $V_c \subset V$ be such that we obtain the 'stable decomposition'

$$V = V_c + V_{hf}$$

for some $V_{hf} \subset V$ which consists of δ-algebraic high frequencies (see Definition 5.17). Namely, for any $v \in V$, there exists $v_c \in V_c$ and $v_{hf} \in V_{hf}$ such that

$$v = v_c + v_{hf}, \quad \|v_{hf}\|_A^2 \leq c_1 \|v\|_A^2.$$

Then the corresponding two-level AMG satisfies

$$\|E\|_A \leq 1 - \frac{\delta}{c_1}.$$

Proof. It follows that

$$\inf_{w_c \in V_c} \|v - w_c\|_{\bar{R}^{-1}}^2 \leq \|v_{hf}\|_{\bar{R}^{-1}}^2 \leq \delta^{-1} \|v_{hf}\|_A^2 \leq \frac{c_1}{\delta} \|v\|_A^2.$$

As a result,

$$K(V_c) \leq \frac{c_1}{\delta},$$

and finally we have

$$\|E\|_A = 1 - \frac{1}{K(V_c)} \leq 1 - \frac{\delta}{c_1}.$$ □

Corollary 5.21. If V_{hf} consists of δ-algebraic high frequencies, then for the coarse space V_c given by

$$V_c = \text{Range}(I - P_{hf}), \tag{5.50}$$

where $P_{hf} : V \mapsto V_{hf}$ is the A-orthogonal projection, we obtain

$$\|E\|_A \leq 1 - \delta.$$

5.7. Bibliographical notes

One of the first results on two-level convergence of AMG methods can be found in papers by Brandt *et al.* (1982) and Ruge and Stüben (1987). There has been much research on extending the MG theory to algebraic settings (Maitre and Musy 1983, Bank and Douglas 1985, Mandel 1988) and the algebraic variational approach to the two-level MG theory (McCormick 1985, McCormick 1984, McCormick and Ruge 1982).

For two-grid convergence, sharper results, including two-sided bounds, are given by Zikatanov (2008) and also considered by Falgout and Vassilevski (2004) and Falgout, Vassilevski and Zikatanov (2005). These two-level results are more or less a direct consequence of the abstract theory provided by Bramble, Pasciak, Wang and Xu (1991*c*), Xu (1992) and Xu and Zikatanov

(2002). A survey of these and other related results can be found in a recent article by MacLachlan and Olson (2014). The two approaches included in this section were recently developed by Xu, Zhang and Zikatanov (2016a, 2016c).

Theorem 5.3 can be found in Zikatanov (2008), and can be viewed as a consequence of the XZ identity (Xu and Zikatanov 2002) in the special case of two subspaces from the general framework of the subspace correction method. The original proof of this theorem in Zikatanov (2008) was based on the XZ identity. The proof here is new and more direct.

Multilevel results are difficult to establish in general algebraic settings, and most of them are either based on unrealistic assumptions or they use geometric grids to prove convergence. We refer to Vaněk, Mandel and Brezina (1996) and Brezina and Vassilevski (2011) for results in this direction. Rigorous multilevel results for finite element equations can be derived using the auxiliary space framework, which was developed by Xu (1996) for quasi-uniform meshes. More recently, Chen, Nochetto and Xu (2012) showed that multilevel convergence results for adaptively refined grids are optimal. A multilevel convergence result on shape-regular grids using AMG based on quad-tree coarsening (in two dimensions) and oct-tree coarsening (in three dimensions) is shown in Grasedyck, Wang and Xu (2015).

Finally, we point out that the notation used in parts of this section originates in Bank and Dupont (1980), Bramble and Pasciak (1987) and Bramble, Pasciak and Xu (1990), and is convenient for the analysis, especially when finite element equations are considered.

6. A general approach to constructing coarse spaces

In this section we describe an abstract framework for constructing coarse spaces by using the notion of space decomposition and subspace corrections.

Let us first introduce some technical results that will be used later as analytic tools.

Lemma 6.1. Let $\underset{\sim}{V}$ and V be two vector spaces and let $\Pi : \underset{\sim}{V} \mapsto V$ be a surjective map. Let $\underset{\sim}{B} : \underset{\sim}{V}' \mapsto \underset{\sim}{V}$ be an SPD operator. Then $B := \Pi \underset{\sim}{B} \Pi'$ is also SPD. Furthermore,

$$(B^{-1}v, v) = \min_{\Pi \underset{\sim}{v} = v} \langle \underset{\sim}{B}^{-1} \underset{\sim}{v}, \underset{\sim}{v} \rangle, \tag{6.1}$$

with the unique minimizer given by

$$\underset{\sim}{v}^* = \underset{\sim}{B} \Pi' B^{-1} v. \tag{6.2}$$

Lemma 6.2. Assume the following two conditions are satisfied for Π.

(a) For all $\underset{\sim}{v} \in \underset{\sim}{V}$,

$$\|\Pi \underset{\sim}{v}\|_A \le \tilde{\mu}_1 \|\underset{\sim}{v}\|_{\underset{\sim}{B}^{-1}}. \tag{6.3}$$

(b) For any $v \in V$, there exists $\underset{\sim}{v} \in \underset{\sim}{V}$ such that $\Pi \underset{\sim}{v} = v$ and

$$\|\underset{\sim}{v}\|_{\underset{\sim}{B}^{-1}} \leq \tilde{\mu}_0 \|v\|_A. \tag{6.4}$$

Then

$$\kappa(BA) \leq \left(\frac{\tilde{\mu}_1}{\tilde{\mu}_0}\right)^2.$$

A direct consequence of Lemma 6.2 is the following result.

Theorem 6.3 (fictitious space lemma). Assume the following two conditions are satisfied for Π. First,

$$\|\Pi \underset{\sim}{v}\|_A \leq \mu_1 \|\underset{\sim}{v}\|_{\underset{\sim}{A}}, \quad \text{for all } \underset{\sim}{v} \in \underset{\sim}{V}$$

Second, for any $v \in V$, there exists $\underset{\sim}{v} \in \underset{\sim}{V}$ such that $\Pi \underset{\sim}{v} = v$ and

$$\|\underset{\sim}{v}\|_{\underset{\sim}{A}} \leq \mu_0 \|v\|_A.$$

Then $\kappa(\Pi) \leq \mu_1/\mu_0$ and, under the assumptions of Lemma 6.1,

$$\kappa(BA) \leq \left(\frac{\mu_1}{\mu_0}\right)^2 \kappa(\underset{\sim}{B}\underset{\sim}{A}).$$

We assume there exists a sequence of spaces $V_1, V_2, \ldots, V_J$, which are not necessarily subspaces of V, but each of them is related to the original space V by a linear operator

$$\Pi_j : V_j \mapsto V. \tag{6.5}$$

We assume that V can be written as a sum of subspaces, and (4.12) and (4.13) hold.

Let

$$\underset{\sim}{W} = V_1 \times V_2 \times \cdots \times V_J,$$

with the inner product

$$(\underset{\sim}{u}, \underset{\sim}{v}) = \sum_{i=1}^{J} (u_i, v_i),$$

where $\underset{\sim}{u} = (u_1, \ldots, u_J)^T$ and $\underset{\sim}{v} = (v_1, \ldots, v_J)^T$. More generally, for $\underset{\sim}{f} = (f_1, \ldots, f_J)^T \in \underset{\sim}{V}'$ with $f_i \in V_i'$, we can define

$$(\underset{\sim}{f}, \underset{\sim}{v}) = \sum_{i=1}^{J} (f_i, v_i).$$

We now define $\Pi_W : \underset{\sim}{W} \mapsto V$ by

$$\Pi_W \underset{\sim}{u} = \sum_{i=1}^{J} \Pi_i u_i, \quad \text{for all } \underset{\sim}{u} = (u_1, \ldots, u_J)^T \in \underset{\sim}{W}.$$

Formally, we can write

$$\Pi_W = (\Pi_1, \dots, \Pi_J) \quad \text{and} \quad \Pi_W' = \begin{pmatrix} \Pi_1' \\ \vdots \\ \Pi_J' \end{pmatrix}.$$

We assume there is an operator $A_j : V_j \mapsto V_j'$ which is symmetric positive semi-definite for each j, and define $\underset{\sim}{A}_W : \underset{\sim}{W} \mapsto \underset{\sim}{W}'$ as follows:

$$\underset{\sim}{A}_W := \mathrm{diag}(A_1, A_2, \dots, A_J). \tag{6.6}$$

For each j, we assume there is a symmetric positive definite operator $D_j : V_j \mapsto V_j'$, and define $\underset{\sim}{D} : \underset{\sim}{W} \mapsto \underset{\sim}{W}'$ as follows:

$$\underset{\sim}{D} := \mathrm{diag}(D_1, D_2, \dots, D_J). \tag{6.7}$$

We associate a coarse space V_j^c, $V_j^c \subset V_j$, with each of the spaces V_j, and consider the corresponding orthogonal projection $Q_j : V_j \mapsto V_j^c$ with respect to $(\cdot,\cdot)_{D_j}$. We define $\underset{\sim}{Q} : \underset{\sim}{W} \mapsto \underset{\sim}{W}'$ by

$$\underset{\sim}{Q} := \mathrm{diag}(Q_1, Q_2, \dots, Q_J). \tag{6.8}$$

Assumption 6.4.

(a) The following inequality holds for all $\underset{\sim}{w} \in \underset{\sim}{W}$:

$$\|\Pi_W \underset{\sim}{w}\|_D^2 \le C_{p,2} \|\underset{\sim}{w}\|_{\underset{\sim}{D}}^2, \tag{6.9}$$

for some positive constant $C_{p,2}$.

(b) For each $w \in V$, there exists a $\underset{\sim}{w} \in \underset{\sim}{W}$ such that $w = \Pi_W \underset{\sim}{w}$ and

$$\|\underset{\sim}{w}\|_{\underset{\sim}{A}_W}^2 \le C_{p,1} \|w\|_A^2, \tag{6.10}$$

with a positive constant $C_{p,1}$ independent of w.

(c) For all j,

$$N(A_j) \subset V_j^c. \tag{6.11}$$

Remark 6.5. Assumption 6.4 implies that

$$w \in N(A) \Rightarrow \underset{\sim}{w} \in N(A_1) \times \cdots \times N(A_J).$$

We define the global coarse space V_c by

$$V_c := \sum_{j=1}^{J} \Pi_j V_j^c. \tag{6.12}$$

Further, for each coarse space V_j^c, we define

$$\mu_j^{-1}(V_j^c) := \max_{v_j \in V_j} \min_{v_j^c \in V_j^c} \frac{\|v_j - v_j^c\|_{D_j}^2}{\|v_j\|_{A_j}^2} \tag{6.13}$$

and

$$\mu_c = \min_{1\le j\le J} \mu_j(V_j^c), \tag{6.14}$$

which is finite, thanks to Assumption 6.4(c) (*i.e.* (6.11)).

By the two-level convergence theory, if D_j provides a convergent smoother, then $(1-\mu_j(V_j^c))$ is the convergence rate for the two-level AMG method for V_j with coarse space V_j^c. The next theorem gives an estimate for the convergence of the two-level method in terms of the constants from Assumption 6.4 and μ_c.

Theorem 6.6. If Assumption 6.4 holds, then for each $v \in V$ we have the error estimate

$$\min_{v_c\in V_c} \|v - v_c\|_D^2 \le C_{p,1}C_{p,2}\mu_c^{-1}\|v\|_A^2. \tag{6.15}$$

Proof. By Assumption 6.4, for each $v \in V$, there exists $\underset{\sim}{v} \in \underset{\sim}{V}$ such that

$$v = \Pi_W \underset{\sim}{v}, \tag{6.16}$$

and (6.10) is satisfied.

By the definition of μ_c, we have

$$\|\underset{\sim}{v} - \underset{\sim}{Q}\underset{\sim}{v}\|_{\underset{\sim}{D}}^2 \le \mu_c^{-1}\|\underset{\sim}{v}\|_{\underset{\sim}{A}_W}^2. \tag{6.17}$$

We let $v_c = \Pi_W \underset{\sim}{Q}\underset{\sim}{v}$. Then $v_c \in V_c$, and by Assumption 6.4 we have

$$\begin{aligned}\|v - v_c\|_D^2 &= \|\Pi_W(\underset{\sim}{v} - \underset{\sim}{Q}\underset{\sim}{v})\|_D^2 \le C_{p,2}\|\underset{\sim}{v} - \underset{\sim}{Q}\underset{\sim}{v}\|_{\underset{\sim}{D}}^2 \\ &\le C_{p,2}\mu_c^{-1}\|\underset{\sim}{v}\|_{\underset{\sim}{A}_W}^2 \le C_{p,1}C_{p,2}\mu_c^{-1}\|v\|_A^2.\end{aligned}$$

□

We define another product space

$$\underset{\sim}{V} := V_c \times V_1 \times V_2 \times \cdots \times V_J, \tag{6.18}$$

and we set $\Pi_c : V_c \mapsto V$ to be the natural inclusion from V_c to V. Then we define $\Pi : \underset{\sim}{V} \mapsto V$ by

$$\Pi := (\Pi_c\ \Pi_1\ \Pi_2\ \cdots\ \Pi_J) \tag{6.19}$$

and $\underset{\sim}{A} : \underset{\sim}{V} \mapsto \underset{\sim}{V}'$ by

$$\underset{\sim}{A}: = \begin{pmatrix} A_c & & & \\ & A_1 & & \\ & & \ddots & \\ & & & A_J \end{pmatrix}, \tag{6.20}$$

where $A_c : V_c \mapsto V_c'$ is given by

$$A_c := \Pi_c' A P i_c \tag{6.21}$$

and $\underset{\sim}{B} : \underset{\sim}{V} \mapsto \underset{\sim}{V}'$ is given by

$$\underset{\sim}{B} := \begin{pmatrix} A_c^{-1} & & & & \\ & D_1^{-1} & & & \\ & & D_2^{-1} & & \\ & & & \ddots & \\ & & & & D_J^{-1} \end{pmatrix}. \tag{6.22}$$

We introduce the additive preconditioner $\widehat{B}$,

$$\widehat{B} := \Pi \underset{\sim}{B} \Pi' = \Pi_c A_c^{-1} \Pi_c' + \sum_{j=1}^{J} \Pi_j D_j^{-1} \Pi_j', \tag{6.23}$$

and we have the following results.

Lemma 6.7. If Assumption 6.4 holds, then for any $v \in V$ there exists $\underset{\sim}{v} \in \underset{\sim}{V}$ such that (6.4) holds, that is,

$$\|\underset{\sim}{v}\|_{\underset{\sim}{B}^{-1}} \leq \tilde{\mu}_0 \|v\|_A,$$

where $\tilde{\mu}_0$ is a constant depending on $C_{p,1}$, $C_{p,2}$, μ_c and c^D.

Lemma 6.8. If Assumption 6.9 holds, then (6.3) holds with constant $\tilde{\mu}_1$ depending on $C_{p,2}$ and c^D.

By directly applying Lemma 6.2, we immediately have the following result.

Theorem 6.9. If Assumption 6.4 holds, then

$$\kappa(\widehat{B}A) \leq \left(\frac{\tilde{\mu}_1}{\tilde{\mu}_0}\right)^2. \tag{6.24}$$

The following two-level convergence result is an application of the convergence theorem (Theorem 5.3) with the error estimate in Theorem 6.6.

Theorem 6.10. If Assumption 6.4 holds, then the two-level AMG method with coarse space defined in (6.12) converges with a rate

$$\|E\|_A^2 \leq 1 - \frac{\mu_c}{C_{p,1} C_{p,2} c^D}.$$

6.1. Bibliographical notes

Matsokin and Nepomnyashchikh (1985) first proved the fictitious space lemma. The work of Xu (1996) on the auxiliary space method is related. An additive version of Lemma 6.1 can be found in Xu and Zikatanov (2002), and the most general case (including multiplicative preconditioners) is found in Xu (2016).

In later sections we show how the general theory used here can be applied to various AMG algorithms, for example classical AMG (Brandt *et al.* 1982, Ruge and Stüben 1987), smoothed aggregation AMG (Míka and Vaněk 1992*a*, 1992*b*), spectral AMGe (Chartier *et al.* 2003, Efendiev, Galvis and Vassilevski 2011), and other algorithms.

Many of the works in the domain decomposition (DD) literature also use techniques for defining coarse spaces, which, to a large extent, have similar aims to the coarse space constructions for AMG outlined in this section. We refer to Toselli and Widlund (2005), Widlund (1994, 2009), Dohrmann, Klawonn and Widlund (2008), Spillane *et al.* (2014) and the references therein for more details on using local eigenspaces for constructing coarse spaces in DD methods.

7. Graphs and sparse matrices

In this section we give a brief introduction to the basic graph theory that is often used for sparse matrices and also for the study of AMG.

7.1. Sparse matrix and its adjacency graph

An *undirected graph* (or simply a *graph*) $\mathcal{G}$ is a pair $(\mathcal{V}, \mathcal{E})$, where $\mathcal{V}$ is a finite set of points called *vertices*, and $\mathcal{E}$ is a finite set of pairs of vertices, known as *edges*. We often write $\mathcal{V} = \{1, \dots, n\}$ for some fixed n. We will not consider directed graphs in this paper because the graphs corresponding to the symmetric sparse matrices are undirected.

An edge $e \in \mathcal{E}$ is an unordered pair $(j,\, k)$, where $j,\, k \in \mathcal{V}$. The vertices j and k are said to be *adjacent* if $(j, k) \in \mathcal{E}$. A *path* from a vertex j to a vertex k is a sequence $(j_0,\, j_1,\, j_2, \dots,\, j_l)$ of vertices where $j_0 = j$, $j_l = k$, and $(j_i,\, j_{i+1}) \in \mathcal{E}$ for all $i = 0,\, 1, \dots,\, l-1$. A vertex j is *connected* to a vertex k if there is a path from j to k. $\mathcal{G} = (\mathcal{V},\, \mathcal{E})$ is *connected* if every pair of vertices is connected by a path; otherwise it is said to be *disconnected*. A graph $\mathcal{G}_0 = (\mathcal{V}_0, \mathcal{E}_0)$ is called a *subgraph* of $\mathcal{G} = (\mathcal{V}, \mathcal{E})$ if $\mathcal{V}_0 \subset \mathcal{V}$ and $\mathcal{E}_0 \subset \mathcal{E}$.

The *neighbourhood* $N(i)$ is the set of vertices adjacent to the vertex i. The *degree* or *valency* of a vertex is the number of edges that connect to it. These are defined as

$$N(i) = \{j : (i,j) \in \mathcal{E}\}, \quad d_i = |\{j : (i,j) \in \mathcal{E}\}|. \tag{7.1}$$

A *path* connecting two vertices i and j is a sequence of edges $(k_0, k_1), (k_1, k_2), \dots, (k_{m-1}, k_m)$ in $\mathcal{E}$ such that $k_0 = i$ and $k_m = j$. The length of the path is the number of edges in it. The *distance* between two vertices i and j is the length of the shortest path connecting i and j, and we denote it by $\operatorname{dist}(i,j)$. If i, j are not connected, then $\operatorname{dist}(i,j) = \infty$. The diameter of a graph is the largest distance between two vertices, that is, $\operatorname{diam}(\mathcal{G}) = \max_{(i,j)\in\mathcal{E}} \operatorname{dist}(i,j)$. An *independent set* is a set of vertices in which no two

$$A = \begin{pmatrix} * & * & * & * & * & * \\ * & * & * & 0 & 0 & 0 \\ * & * & * & * & * & 0 \\ * & 0 & * & * & * & 0 \\ * & 0 & * & * & * & 0 \\ * & 0 & 0 & 0 & 0 & * \end{pmatrix}$$

(a)

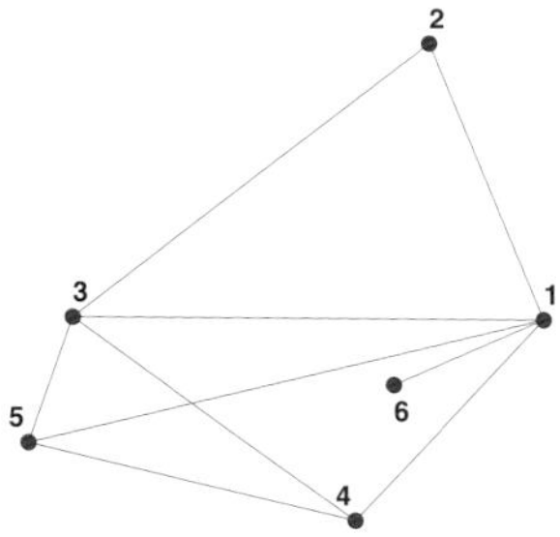

(b)

Figure 7.1. (a) A sparse symmetric matrix, and (b) its associated graph.

are adjacent. A *maximal independent set* is an independent set such that adding any other vertex to the set forces the set to contain an edge.

Given a symmetric matrix $A \in \mathbb{R}^{n\times n}$, the *adjacency graph* of A is an undirected graph, denoted by $\mathcal{G}(A)$, $\mathcal{G} = (\mathcal{V}, \mathcal{E})$ with $\mathcal{V} = \{1, 2, \dots, n\}$. The edges $\mathcal{E}$ are defined as

$$\mathcal{E} = \{(j, k) : a_{jk} \neq 0\}.$$

A matrix A is called *irreducible* if its adjacency graph $\mathcal{G}(A) = (\mathcal{V}, \mathcal{E})$ is connected. Otherwise, A is called *reducible*.

An example of a symmetric matrix is shown in Figure 7.1(a) and the corresponding graph is shown in Figure 7.1(b). Pictorial representation of a graph is often not available, and a graph can be drawn in different ways with different coordinates of the vertices. As a general rule, sparse matrices do not provide any geometric information for the underlying graph and only the combinatorial or topological properties of $\mathcal{G}(A)$.

Given

$$S \subset \{1, \dots, n\} \times \{1, \dots, m\},$$

we define

$$\mathbb{R}_S^{n\times m} = \big\{X = (x_{ij}) \in \mathbb{R}^{n\times m} : x_{ij} = 0 \text{ if } (i, j) \notin S\big\}. \tag{7.2}$$

We say that X has sparsity pattern given by S if and only if $X \in \mathbb{R}_S^{n\times m}$.

Often, the sparsity pattern of a matrix is determined in advance and the set S is determined by a given matrix. For $Y \in \mathbb{R}^{n\times m}$, we let

$$\mathrm{S}(Y) = \big\{(i, j) : y_{ij} \neq 0\big\}.$$

We now consider the graphs associated with finite element or finite difference stiffness matrices. In the case of finite elements, this is the space of finite element functions V_h, and for finite difference discretizations this is the space of *mesh-functions* V_h, which can be identified with $\mathbb{R}^N$.

We assume that we have a finite-dimensional space V_h^N which we use to discretize the Neumann problem. We also have the finite element space for the Dirichlet (or mixed boundary condition) problem, and we assume that the following inclusions hold: $V_h = V_h^D \subset V_h^N$. Equivalently, we have a subspace of the variables which vanish on V_h: for example, the values of the finite element or finite difference solution at the nodes on the Dirichlet boundary vanish.

Let $A^{\mathcal{N}}$ be the matrix corresponding to the finite element or finite difference discretization of the model second-order elliptic equation with Neumann boundary conditions. Clearly we have the identity

$$(A^{\mathcal{N}}u, v) = \sum_{e\in\mathcal{E}} \omega_e \delta_e u \delta_e v. \tag{7.3}$$

Here, the sum is over all edges $\mathcal{E}$ of the graph $\mathcal{G}(A^{\mathcal{N}}) = (\mathcal{V}, \mathcal{E})$, $\delta_e v = v_i - v_j$, if $\mathcal{E} \ni e = (i,j)$, $i < j$. Also, $\omega_e = -(A^{\mathcal{N}})_{ij}$ are the off-diagonal entries of $A^{\mathcal{N}}$. Note that since we consider the Neumann problem, the bilinear form defined by $A^{\mathcal{N}}$ vanishes for u (resp. v) such that $u_i = 1$ (resp. $v_i = 1$) for all i. For both 5-point and 9-point stencils we have that $\omega_e = 1$ for all e.

We consider the stiffness matrix $A^{\mathcal{N}}$, corresponding to the model problem (2.1) with Neumann boundary conditions on a bounded domain $\Omega \subset \mathbb{R}^d$, that is, we have the boundary condition

$$\alpha \nabla u \cdot \boldsymbol{n} = 0 \quad \text{on } \partial\Omega, \tag{7.4}$$

where $\boldsymbol{n}$ is the unit normal vector to $\partial\Omega$ pointing outward. It is easy to derive the stiffness matrices corresponding to the Dirichlet or mixed boundary condition problem: we just restrict the bilinear form defined by $A^{\mathcal{N}}$ to the subspace given by

$$(Au, v) = \sum_{e\in\mathcal{E}} \omega_e \delta_e u \delta_e v, \quad u_j = v_j = 0 \quad x_j \in \Gamma_D. \tag{7.5}$$

Remark 7.1. Similar relations between differential problems with natural (Neumann) and essential (Dirichlet) boundary conditions are seen not only for the model problem considered here, but also for problems on $H(\text{curl})$, $H(\text{div})$, linear elasticity and others.

7.2. M-matrix relatives of finite element stiffness matrices

A symmetric matrix $A \in \mathbb{R}^{n\times n}$ is called an *M-matrix* if it satisfies the following three properties:

$$a_{ii} > 0 \text{ for } i = 1, \dots, n, \tag{7.6}$$

$$a_{ij} \le 0 \text{ for } i \ne j, \; i, \; j = 1, \dots, n, \tag{7.7}$$

$$A \text{ is semi-definite.} \tag{7.8}$$

As the first step in creating a space hierarchy, most AMG algorithms for $Au = f$ with positive semi-definite A use a simple filtering of the entries of A, and construct an M-matrix which is then used to define crucial AMG components.

Definition 7.2 (M-matrix relative). We call a matrix $\widetilde{A}$ an *M-matrix relative* of A if $\widetilde{A}$ is an M-matrix and satisfies the inequalities

$$(v, v)_{\widetilde{A}} \lesssim (v, v)_A \quad \text{and} \quad (v, v)_D \lesssim (v, v)_{\widetilde{D}}, \quad \text{for all } v \in V, \tag{7.9}$$

where $\widetilde{D}$ and D are the diagonals of $\widetilde{A}$ and A respectively.

A few remarks are in order.

(i) We have used the term M-matrix to denote semi-definite matrices, and we are aware that this is not the precise definition. It is, however, much more convenient to refer to M-matrices, and we decided to relax the definition here in the hope that the inaccuracy would pay off by appealing to the reader.

(ii) We point out that the restricted M-matrix relatives are instrumental in the definition of coarse spaces and also in the convergence rate estimates. This is clearly seen later in Section 5.2 where we present the unified two-level theory for AMG.

(iii) Often, the case is that the one-sided inequality in (7.9) is in fact a spectral equivalence.

By definition, we have the following simple but important result.

Lemma 7.3. Let A_+ be an M-matrix relative of A and let D and D_+ be the diagonal matrices of A and A_+, respectively. If $V_c \subset V$ is a subspace, then the estimate

$$\|u - u_c\|_D^2 \lesssim \|u\|_A^2 \tag{7.10}$$

holds for some $u_c \in V_c$, given the estimate

$$\|u - u_c\|_{D_+}^2 \lesssim \|u\|_{A_+}^2. \tag{7.11}$$

This result means that we only need to work on the M-matrix relative of A in order to get the estimate (7.10).

In this section we show how to construct the M-matrix relative to the matrix resulting from a finite element discretization of the model problem (2.1) with linear elements. We first consider an isotropic problem with Neumann boundary condition (7.4) and isotropic $\alpha = a(x)I$. Construction of M-matrix relatives in the case of anisotropic tensor $\alpha(x)$ in (2.1) is postponed to Section 14.2.

In the rest of this section we make the following assumptions on the coefficient and the geometry of Ω.

- The domain $\Omega \subset \mathbb{R}^d$ is partitioned into simplices $\Omega = \cup_{T \in \mathcal{T}_h} T$.

- The coefficient $a(x)$ is a scalar-valued function, and its discontinuities are aligned with the partition $\mathcal{T}_h$.
- We consider the Neumann problem, and hence the bilinear form (2.3) is

$$\int_\Omega a(x)\nabla v \cdot \nabla u = \sum_{(i,j)\in\mathcal{E}} (-a_{ij})\delta_e u \delta_e v = \sum_{e\in\mathcal{E}} \omega_e \delta_e u \delta_e v. \tag{7.12}$$

- It is well known that the off-diagonal entries of the stiffness matrix A are given by

$$\omega_e = -(\phi_j, \phi_i)_A = \sum_{T\supset e} \omega_{e,T},$$

$$\omega_{e,T} = \frac{1}{d(d-1)}\overline{a}_T|\kappa_{e,T}|\cot\alpha_{e,T}, \quad \overline{a}_T = \frac{1}{|T|}\int_T a(x)\,\mathrm{d}x.$$

Here, $e = (i,j)$ is a fixed edge with end-points x_i and x_j, $T \supset e$ is the set of all elements containing e, $|\kappa_{e,T}|$ is the volume of the $(d-2)$-dimensional simplex opposite to e in T, and $\alpha_{e,T}$ is the dihedral angle between the two faces in T not containing e.

- Let $\mathcal{E}$ denote the set of edges in the graph defined by the triangulation and let $\mathcal{E}^-$ be the set of edges where $a_{ij} \geq 0$, $i \neq j$. The set complementary to $\mathcal{E}^-$ is $\mathcal{E}^+ = \mathcal{E} \setminus \mathcal{E}^-$. Then, with $\omega_e = -a_{ij}$ and $\delta_e u = (u_i - u_j)$, $e = (i,j)$, we have

$$\int_\Omega a(x)\nabla v \cdot \nabla u = \sum_{e\in\mathcal{E}^+} \omega_e \delta_e u \delta_e v - \sum_{e\in\mathcal{E}^-} |\omega_e| \delta_e u \delta_e v. \tag{7.13}$$

- We also assume that the partitioning is such that the constant function is the only function in the null space of the bilinear form (7.12). This is, of course, the case when Ω is connected (which is true, as Ω is a domain).

The non-zero off-diagonal entries of A may have either positive or negative sign, and usually $\mathcal{E}^- \neq \emptyset$. The next theorem shows that the stiffness matrix A defined via the bilinear form (7.12) is spectrally equivalent to the matrix A_+ defined as

$$(A_+u, v) = \sum_{e\in\mathcal{E}^+} \omega_e(u_i - u_j)(v_i - v_j). \tag{7.14}$$

Thus, we can ignore any positive off-diagonal entries in A, or equivalently we may drop all ω_e for $e \in \mathcal{E}^-$. Indeed, A_+ is obtained from A by adding to the diagonal all positive off-diagonal elements and setting the corresponding off-diagonal elements to zero. This is a stronger result that we will need later, because it gives spectral equivalence with the M-matrix relative of A_+.

Theorem 7.4. Assume that A is the stiffness matrix corresponding to linear finite element discretization of (2.1) with boundary conditions given by (7.4). Then A_+ is an M-matrix relative of A which is spectrally equivalent to A. The constants of equivalence depend only on the shape-regularity of the mesh. Moreover, the graph corresponding to A_+ is connected.

A simple corollary, which we will use later to prove estimates for the convergence rate, is as follows.

Corollary 7.5. Assume that A is the stiffness matrix for piecewise linear discretization of equation (7.12) and A_+ is the M-matrix relative defined in Theorem 7.4. Then the diagonal D of A and the diagonal D_+ of A_+ are spectrally equivalent.

Proof. For the diagonal elements of A and A_+ we have

$$[D]_j = (\phi_j, \phi_j)_A \eqsim (\phi_j, \phi_j)_{A_+} = [D_+]_j.$$

The equivalences written above follow directly from Lemma 7.4. □

Corollary 7.5 together with Lemma 7.3 provides a theoretical foundation for using M-matrix relatives to design AMG for finite element matrices.

7.3. Bibliographical notes

We have introduced some standard notions from graph theory. For the reader interested in more detailed descriptions, we refer to the classical textbooks by Diestel (2010) and Gibbons (1985) as a general introduction to graph theory, and to Saad (2003) and Varga (2000) for considerations linking graphs, sparse matrices and iterative methods.

Our results on M-matrix relatives are related to the some of the works on preconditioning by Z-matrices and L-matrices (Kraus and Schicho 2006, Kraus 2008). They are implicitly used in most of the AMG literature (Ruge and Stüben 1987) where the classical connection strength definition gives an M-matrix. We point out that the M-matrix property and the existence of an M-matrix relative is often not sufficient to achieve even a two-level uniform convergence of AMG. A typical example is a matrix which has been rescaled and the constant is no longer in the kernel of the discrete operator. In this case, the standard AMG application may fail, and the near kernel needs to be recovered by different means, such as the adaptive AMG processes considered in Section 15 and the references given therein.

8. Connection strength

A central task in AMG is to obtain an appropriate coarse space or prolongation. This process is known as a coarsening process. In a geometric grid or, more generally, the adjacency graph of the stiffness matrix, we need to identify vertices to be deleted from the graph. We need the coarsened graph to still provide a good approximation for algebraically low frequencies.

8.1. Basic idea and strength function

If a subset of vertices on which an algebraically smooth vector, say v, changes very slowly, we only need to keep one degree of freedom to represent v in this subset. In other words, we can either aggregate this subset together or keep one vertex and delete the rest of the vertices in this subset. We say that the vertices in this subset are strongly connected to each other. Connection strength is a concept introduced to identify strongly connected pairs of vertices. Roughly speaking, we say that i and j are strongly connected if $v_i \approx v_j$.

We imagine coarsening the graph in two stages: the first step is to remove some edges and the second step is to remove some vertices. The second step is the goal. Let us examine how the second step is carried out: we either (i) aggregate some neighbouring vertices together or (ii) pick a maximal independent set, denoted by $\mathcal{C}$, in the filtered graph and then remove all the remaining vertices. Using the argument above, (i) each aggregate should only consist of strongly connected vertices, or (ii) each of the deleted vertices should be strongly connected to some point in $\mathcal{C}$. To guarantee either of these two situations, we then have to remove all the weakly connected edges in the first step.

Let us further use some heuristic arguments to motivate how the connection strength should be defined. Let v be an algebraically smooth (5.40), namely

$$\|v\|_A^2 \le \epsilon \|v\|_{\bar{R}^{-1}}^2.$$

Let $u = v/\|v\|_{\bar{R}^{-1}}$. We then have

$$(Au, u) \le \epsilon \Rightarrow \sum_{e=(i,j)\in\mathcal{E}} (-a_{ij})(u_i - u_j)^2 \le \epsilon. \tag{8.1}$$

Thanks to Lemma 5.16, we can assume that A is an M-matrix, namely $-a_{ij} = |a_{ij}|$. From (8.1), we have the following observations.

(i) A larger $|a_{ij}|$ means a smaller $(u_i - u_j)^2$.

(ii) An algebraically smooth error varies more slowly in the direction where $|a_{ij}|$ is larger.

This observation leads to the following definition of connection strength. Given a threshold $\theta > 0$, we say that the vertex j of the adjacency graph is *θ-strongly connected* to vertex i if

$$-a_{ij} \ge \theta \max_{k\neq i} -a_{ik}. \tag{8.2}$$

Note that by the definition above we may have j strongly connected to i while i is not strongly connected to j. As a result, the adjacency graph corresponding to the matrix of strong connections may not be symmetric.

However, our theoretical framework is given in terms of the symmetrized operator $\bar{R}$, regardless of whether the original smoother R is symmetric. This is due to the fact that we used the energy norm, namely the A-norm, to measure the convergence rate, and the resulting convergence rate is given in terms of $\bar{R}$. This is the best convergence theory we have, and we will use this theory to study AMG algorithms. As a result, we will only consider strength functions that are symmetric. A strength function

$$s_c : \mathcal{V} \times \mathcal{V} \mapsto \mathbb{R}_+ \tag{8.3}$$

associated with an SPSD matrix is symmetric if $s_c(i,j) = s_c(j,i)$.

Given a threshold $\theta > 0$, we say that i and j are θ-strongly connected if

$$s_c(i,j) \geq \theta.$$

We then define the strength matrix:

$$S = \sum_{s_c(i,j)\geq\theta} e_i e_j^T. \tag{8.4}$$

Note that S is a Boolean matrix with entries equal to 0 or 1 depending on the connection strength.

Given the non-overlapping decomposition

$$\mathcal{V} = \bigcup_{i=1}^{m} \mathcal{A}_i = \bigcup_{\tilde{\mathcal{A}}\in\mathcal{V}_{\mathcal{A}}} \tilde{\mathcal{A}}, \quad \mathcal{V}_{\mathcal{A}} = (\mathcal{A}_1, \ldots, \mathcal{A}_m), \tag{8.5}$$

we extend the definition of strength function to

$$s_c : \mathcal{V}_{\mathcal{A}} \times \mathcal{V}_{\mathcal{A}} \mapsto \mathbb{R}_+, \tag{8.6}$$

and we assume that s_c is symmetric.

Given a threshold $\theta > 0$, we say that $\mathcal{A}_i$ and $\mathcal{A}_j$ are θ-strongly connected to each other if

$$s_c(\mathcal{A}_i, \mathcal{A}_j) \geq \theta.$$

An example of such a strength function is given in (8.13).

In the AMG literature, a number of heuristics have been proposed for identifying strong connections, particularly when considering discretizations of anisotropic equations. In general, connection strength is a notion that is difficult to address theoretically or to relate to the convergence rate of an algorithm. We refer to the classical papers and monographs mentioned in Section 9.6 for further discussions on related issues. Current trends in AMG development aim to re-evaluate the role of the classical definition of connection strength.

We would finally like to comment that connection strength is used to define the sparsity of P, and it is crucial, for example, in the proof of the convergence of the two-level method for discretizations of elliptic equations with jump coefficients. Choosing the 'right' sparsity of P is crucial, as a denser P would lead to a better approximation from a coarser space, and a sparser P would lead to a less expensive algorithm.

In the rest of the section, we discuss different definitions of connection strength:

(i) classical AMG,

(ii) lean AMG,

(iii) local-optimization based AMG.

8.2. *Classical AMG*

With the above motivation, we define the strength function as follows:

$$\begin{aligned} s_c(i,j) &= \frac{-a_{ij}}{\min\big(\max_{k\neq i}(-a_{ik}), \max_{k\neq j}(-a_{jk})\big)} \\ &= \frac{a_{ij}}{\max\big(\min_{k\neq i} a_{ik}, \min_{k\neq j} a_{jk}\big)}. \end{aligned} \tag{8.7}$$

Definition (8.7) is a symmetrized version of the strength function used in the classical AMG literature (see (8.2)).

The following definition is also commonly used in classical AMG algorithms:

$$s_c(i,j) = \frac{|a_{ij}|}{\min\big(|N(i)|^{-1}\sum_{k\neq i}|a_{ik}|, |N(j)|^{-1}\sum_{k\neq i}|a_{jk}|\big)}. \tag{8.8}$$

Again, this is a symmetrized version of strength functions used in the AMG literature.

Finally, we may also have the following two definitions, which are based on Cauchy–Schwarz for SPSD matrices:

$$s_1(i,j) = \frac{|a_{ij}|}{\sqrt{a_{ii}a_{jj}}} \tag{8.9}$$

and

$$s_2(i,j) = \frac{-2a_{ij}}{a_{ii}+a_{jj}}. \tag{8.10}$$

Note that definition (8.7) (which is mostly associated with classical AMG) ignores all the non-negative entries of the stiffness matrix $A = (a_{ij})$.

8.3. *Local-optimization strength function*

Suppose that each pair of the index $\{i,j\} \subset \{1,\dots,n\}$ is associated with a space V_{ij}, which is not necessarily a subspace of V. Assume now that we have two operators: $A_{ij} : V_{ij} \mapsto V'_{ij}$, which is symmetric positive semi-definite, and $D_{ij} : V_{ij} \mapsto V'_{ij}$, which is symmetric positive definite.

Given a number $k_{ij} < \dim V_{ij}$, we define a coarse space $V^c_{ij} \subset V_{ij}$ as

$$V^c_{ij} := \mathrm{span}\{\zeta^{(k)}_{ij}, k = 1 : k_{ij}\},$$

where $\zeta^{(k)}_{ij}$ is the eigenvector corresponding to the kth smallest eigenvalue of $D^{-1}_{ij} A_{ij}$.

We let $Q_{ij} : V_{ij} \mapsto V^c_{ij}$ denote the orthogonal projection with respect to $(\cdot, \cdot)_{D_{ij}}$. Motivated by (6.13), we define the strength function s_c as follows:

$$s_c(i, j) := \left(\sup_{v \in V_{ij}} \frac{\|(I - Q_{ij})v\|^2_{D_{ij}}}{\|v\|^2_{A_{ij}}} \right)^{-1}. \tag{8.11}$$

A special case of the above definition is introduced in aggregation-based algebraic multigrid (AGMG). Suppose now that we have a set of aggregates $\{\mathcal{A}_1, \dots, \mathcal{A}_J\}$. We fix a pair $\{i, j\} \subset \{1, \dots, J\}$, and define

$$G = \mathcal{A}_i \bigcup \mathcal{A}_j.$$

We then let V_{ij} define the restriction of V to G, that is,

$$V_{ij} := \{v|_G : v \in V\}, \tag{8.12}$$

where

$$v|_G(x) = \begin{cases} v(x) & \text{if } x \in G, \\ 0 & \text{if } x \notin G. \end{cases}$$

We use A_{ij} and D_{ij}, respectively, to denote the restriction of A and D to G. We then choose $k_{ij} = 1$ and define the local coarse space V^c_{ij} by

$$V^C_{ij} = \mathrm{span}\{\zeta_G\}, \quad \zeta_G = \zeta^{(1)}_{ij},$$

and the orthogonal projection $Q_{ij} : V_{ij} \mapsto V^C_{ij}$ with respect to $(\cdot, \cdot)_{D_{ij}}$, namely,

$$Q_{ij} v = \frac{(v, \zeta_G)_{D_{ij}}}{\|\zeta_G\|^2_{D_{ij}}} \zeta_G.$$

The strength function based on aggregation is defined as

$$s_c(i, j) := \left(\sup_{v \in V_{ij}} \frac{\|(I - Q_{ij})v\|^2_{D_{ij}}}{\|v\|^2_{A_{ij}}} \right)^{-1}. \tag{8.13}$$

Another example is choosing $V_{ij} = \mathbb{R}^2$ and

$$A_{ij} = \begin{pmatrix} a_{ii} & a_{ij} \\ a_{ij} & a_{jj} \end{pmatrix}, \quad D_{ij} = \begin{pmatrix} a_{ii} & 0 \\ 0 & a_{jj} \end{pmatrix}.$$

We choose $k_{ij} = 1$ and define the coarse space $V_{ij}^c \subset V_{ij}$ as

$$V_{ij}^c = \mathrm{span}\left\{ \begin{pmatrix} 1 \\ 1 \end{pmatrix} \right\}.$$

By direct computation, the strength function defined in (8.11) is given by

$$s_c(i,j) = \frac{1 - s_1^2}{1 - s_2}, \quad s_1 = \frac{|a_{ij}|}{\sqrt{a_{ii}a_{jj}}}, \quad s_2 = -\frac{2a_{ij}}{a_{ii} + a_{jj}}. \tag{8.14}$$

We note that $s_1 \geq s_2$ and hence

$$s_c(i,j) \leq 1 + s_2 \leq 1 + s_1. \tag{8.15}$$

We point out that the strength function given by (8.14) is obtained by using the theory in Section 6, while the other strength functions such as (8.7) are obtained by heuristic considerations.

8.4. Lean AMG

Instead of using the absolute value of matrix entries as the criterion to determine if two points are strongly coupled, lean AMG uses *affinity* to measure the connection strength, which is based on the following heuristic observation: *Given a vector v, if (i,j) is a strong connected pair of vertices, after several relaxation on v, namely*

$$v \leftarrow (I - RA)^\nu v,$$

the values of v_i and v_j should be close.

In lean AMG, we generate K test vectors. Each test vector is the result of applying ν Gauss–Seidel relaxation sweeps to $Ax = 0$, starting from randomly generated vectors $x^{(1)}, \ldots, x^{(K)} \in \mathbb{R}^n$. Further, we denote

$$X_{n \times K} := \begin{pmatrix} X_1^T \\ \vdots \\ X_n^T \end{pmatrix} = (I - RA)^\nu (x^{(1)} \ \cdots \ x^{(K)}). \tag{8.16}$$

Here X_i^T is the ith row of $X_{n \times K}$. The strength function for lean AMG is then defined as

$$s_c(i,j) := \frac{|(X_i, X_j)|^2}{(X_i, X_i)(X_j, X_j)}. \tag{8.17}$$

8.5. Bibliographical notes

Classical algorithms for determining connection strength can be found in Brandt *et al.* (1982), Stüben (1983), Brandt *et al.* (1985), Ruge and Stüben (1987) and Briggs, Henson and McCormick (2000). The original measure of connection strength given by Ruge (1983, 1985), McCormick and Ruge

(1989) and Brandt *et al.* (1985) is non-symmetric, but for theoretical considerations, which only depend on the symmetrized smoothers, it suffices to use the slightly more restrictive but symmetric versions of connection strength.

Some extensions of these classical algorithms for defining strong connections are based on different measures for connectivity and distance such as *measure of importance* and *algebraic distance*. Details can be found in Ruge and Stüben (1987) and Trottenberg *et al.* (2001, Appendix A).

Connection strength functions have had little theoretical backing in the past. The results developed in this section, such as local Poincaré inequalities and especially the strength functions (8.15), show that such heuristics are reasonable, and their choice can be motivated by theoretical results.

For aggregation AMG, typically a symmetric connection strength function is used, as defined by Vaněk *et al.* (1996). Some recent aggregation algorithms also define connection strength based on sharp theoretical results, and use the local two-level convergence rate (8.11) as a measure (Notay 2010, Napov and Notay 2012).

For aggregations based on matching (aggregates of size 2), the 'heavy edge' matching algorithms in Karypis and Kumar (1998) correspond to a connection strength function selecting aggregates depending on edge weights in the adjacency graph. Some recent works (*e.g.* Livne and Brandt 2012) use connection strength functions based on the size of the entries in the Gram matrix formed by a set of smoothed test vectors.

9. Coarsening strategies

Once the smoother is identified, the central task of an AMG method is to identify a sequence of coarse spaces, in functional terminology, or equivalently, in an algebraic setting, to identify a sequence of prolongation matrices. This procedure is known as 'coarsening'. In this section we will discuss this coarsening procedure.

Roughly speaking, given equation (1.1), namely $Au = f$, on a vector space V, the goal is to find a subspace $V_c \subset V$ such that the solution $u_c \in V_c$ of the 'coarsened' problem

$$A_c u_c = f_c, \quad A_c = \imath_c' A \imath_c, f_c = \imath_c' f \tag{9.1}$$

would provide a good approximation to the original solution $u \in V$. More specifically, the solution of u_c of (9.1) would provide a good approximation to those 'algebraically smooth' components of the error for which the given smoother does not converge well.

9.1. Motivations

In some sense, 'coarsening' is done almost everywhere in numerical analysis. For example, the finite element equation (2.19) can be viewed as a coarsened equation of the original equation (2.3). In this case the finite element space V_h is a coarsened subspace of $V = H_0^1(\Omega)$.

It is therefore informative for us to examine how a finite element space is constructed in general. While there are many different ways to construct finite element spaces, mathematically speaking, the most convenient approach is via the use of 'degrees of freedom', which refers to a basis of a dual space. More specifically, in a finite element discretization, the finite element space V_h is obtained by first specifying the dual space V_h' (the so-called space of degrees of freedom, or variables). For linear finite elements, $V_h' = \{\psi_i : i = 1 : n_h\}$ is such that

$$\psi_i(v) = v(x_i^h).$$

With such variables (nodal evaluation), we then find a dual basis $\{\phi_i : i = 1 : n_h\}$ which are piecewise linear functions such that

$$\psi_i(\phi_j) = \delta_{ij}, \quad 1 \le i, j \le n.$$

The finite element space V_h is then defined by

$$V_h = \text{span}\{\phi_i : i = 1 : n_h\}.$$

In fact, mathematically speaking, all existing finite element spaces V_h can be obtained by first constructing V_h. This is the approach taken in the classical literature on finite element methods: see Ciarlet (2002).

In a similar way to the finite element method, we will therefore focus on techniques for constructing a coarse space $V_c \subset V$ by first identifying its dual basis V_c'. Such an approach is rather abstract, but it turns out to be more intrinsic, more general, and, in fact, more commonly used in the AMG literature, albeit implicitly.

It is interesting to note that we rarely use the word 'coarsening' in the design of a geometric multigrid method. Instead, we use a 'refinement' procedure to define a sequence of nested spaces. As an example, Figure 9.1 shows a uniformly refined triangular grid used for discretization of the Poisson equation with linear finite elements.

In AMG, we do not have the luxury of this hierarchy of spaces given by a geometric refinement. Instead, we carry out a reverse-engineering of the refinement process, namely *coarsening.*

Conceivably, we could also use such a reverse-engineering process to recover geometric multigrid methods by starting from the finest geometric grid, at least for some special cases: for example, if all triangles in the triangulation shown in Figure 9.1 were acute and the graph corresponding to the mesh were also an adjacency graph of the stiffness matrix. It is clear that the coarse grid vertices, that is, the set $\mathcal{C}$ (known from the refinement), is a maximal independent set (MIS) of vertices in the graph corresponding to the refined mesh. In the simplest case, the coarse grid variables (the dual

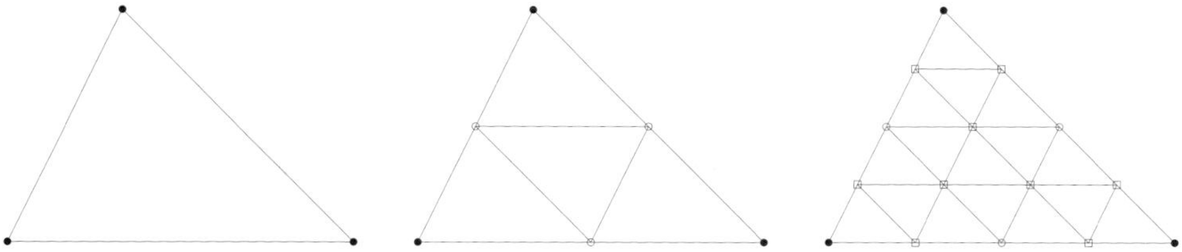

Figure 9.1. Regular refinement of a coarse grid element: •, coarsest level vertices; ◦, first level of refinement; □, second level of refinement.

basis for V_c) precisely correspond to the MIS. This observation is explored later, in Section 9.3.1, when constructing algorithms in the classical AMG framework for a selection of coarse grid vertices via the MIS algorithm (Algorithm 5). For further examples of relations between geometric AMG and GMG coarsening, we refer to Section 10.2.

The above reverse-engineering for GMG gives some hint as to how a coarsening process needs to be done in AMG, but we need to study the process in a broader framework and – more importantly – we will use the degrees of freedom, *i.e.* the dual bases, to obtain coarse spaces. In the above GMG example, each grid point in the geometric grid corresponds to exactly one variable. However, this is not always the case in applications.

9.2. Basic approach

By mimicking the construction of finite element spaces as described above, given a linear algebraic system of equation $Au = f$, we adapt a coarsening strategy consisting of the following steps.

1 We consider the adjacency graph $\mathcal{G}(A)$ of the coefficient matrix A. Based on a certain strength function s_c as described in Section 8, we remove the weakly connected edge in $\mathcal{G}(A)$, that is, we drop certain entries in A to form a filtered matrix $\tilde{A}$.

2 We carry out one of the following two substeps.

   (a) *Classical AMG.* Find a maximal independent set of $\mathcal{G}(\tilde{A})$ to form the set of coarse vertices $\mathcal{C}$. Then remove the rest, namely $\mathcal{V} \setminus \mathcal{C} \equiv \mathcal{F}$.

   (b) *Aggregation AMG.* Agglomerate using some greedy algorithm: pick a point and agglomerate its neighbours and go from there.

3 We obtain a subset of variables obtained from the above steps, *i.e.* $(V')_c$.

4 We use $(V')_c$ to define a high-frequency space $V_{hf} = (V')_c^{(0)}$ by

$$V_{hf} = (V')_c^{(0)} := \{v \in V : \langle g, v\rangle = 0, \text{ for all } g \in (V')_c\}. \tag{9.2}$$

5 Find a tentative coarse space W_c such that

$$V = V_{hf} \oplus W_c. \tag{9.3}$$

6 Apply a certain postprocessing (such as smoothing) to W_c to obtain V_c:

$$V_c = SW_c.$$

7 Using V_c, or equivalently the prolongation P, we form the coarse matrix $A_c = P^T AP$.

8 We then repeat the above steps to A_c in place of A until a desirable coarsest level is reached.

9.2.1. Construction of $(V')_c$
Given $A \in \mathbb{R}^{n\times n}$ and the associated graph $\mathcal{G} = (\mathcal{V}, \mathcal{E})$, we proceed as follows.

(i) Form the two non-overlapping decompositions

$$\mathcal{V} = \mathcal{C} \cup \mathcal{F}, \quad \mathcal{C} = \bigcup_{i=1}^{n_c} \mathcal{A}_i.$$

(ii) Identify $(V')_c = \operatorname{span}\{N_i : i = 1 : n_c\} \subset V'$.

Here are three examples that will be discussed in detail in later sections (see Sections 11, 12 and 13).

Aggregation AMG. $\mathcal{F} = \emptyset$ and

$$N_i(v) = \langle v\rangle_{\mathcal{A}_i} \equiv \frac{1}{|\mathcal{A}_i|}\sum_{j\in\mathcal{A}_i} \psi_j(v) = \frac{1}{|\mathcal{A}_i|}\sum_{j\in\mathcal{A}_i} v_j, \quad i = 1 : n_c. \tag{9.4}$$

Classical AMG. $\mathcal{F} \neq \emptyset$ and

$$\mathcal{A}_i = \{k_i\}, \quad N_i(v) = \psi_{k_i}(v) = v_{k_i}. \tag{9.5}$$

In this case, $\mathcal{C}$ usually consists of disconnected vertices.

Energy-min AMG. $\mathcal{F} \neq \emptyset$ and $\mathcal{A}_i$ are aggregates

$$N_i(v) = \langle v\rangle_{\mathcal{A}_i} \equiv \frac{1}{|\mathcal{A}_i|}\sum_{j\in\mathcal{A}_i} \psi_j(v) = \frac{1}{|\mathcal{A}_i|}\sum_{j\in\mathcal{A}_i} v_j, \quad i = 1 : n_c. \tag{9.6}$$

9.2.2. Construction of V_c
Given coarse grid degrees of freedom $(V')_c \subset V'$, we define V_{hf} as in (9.2). The following lemma shows how to find a subspace W_c, a 'pre-coarse space', such that $V = V_{hf} \oplus W_c$.

Lemma 9.1. If $\phi_{k,c}$, $k = 1, \ldots, n_c$ are elements of V such that the Gram matrix

$$G = (G_{km}) = (N_m(\phi_{k,c}))$$

is non-singular, then we have

$$V = V_{hf} \oplus W_c, \quad W_c = \operatorname{span}\{\phi_{k,c}\}_{k=1}^{n_c}.$$

Proof. We first show that $W_c \cap V_{hf} = \{0\}$. In fact, if

$$v := \sum_{k=1}^{n_c} (\tilde{v})_k \phi_{k,c} \in W_c \cap V_{hf},$$

then we have

$$0 = N_m(v) = \sum_{k=1}^{n_c} N_m(\phi_{k,c}) = \sum_{k=1}^{n_c} (\tilde{v})_k G_{km}, \quad m = 1, \dots, n_c.$$

Hence,

$$G\tilde{v} = 0,$$

and since, by assumption, G is non-singular, we must have $v = 0$. Therefore, $W_c \cap V_{hf} = \{0\}$.

Next, for any $v \in V$, we define $w_c \in W_c$ as

$$w_c = \sum_{k=1}^{n_c} (\widetilde{w}_c)_k \phi_{k,c}, \quad \widetilde{w}_c = G^{-1} \begin{pmatrix} N_1(v) \\ \vdots \\ N_{n_c}(v). \end{pmatrix}.$$

It is immediate to check that

$$N_m(v - w_c) = 0, \quad m = 1, \dots, n_c,$$

and hence $(v - w_c) \in V_{hf}$. This proves that

$$v = w_c + \underbrace{(v - w_c)}_{\in V_{hf}},$$

and completes the proof. □

Lemma 9.1 gives us a way to construct a subspace W_c such that $V = V_{hf} \oplus W_c$.

Lemma 9.2. Assume the coarse grid degrees of freedom are defined as

$$N_k := \sum_{j \in \mathcal{A}_k} \alpha_j \psi_j, \quad k = 1, \dots, n_c,$$

where $\sum_{j \in \mathcal{A}_k} \alpha_j = 1$ and $\{\psi_j\}$ is the dual basis of $\{\phi_j\}$. If $\{\phi_{k,c} : k = 1, \dots, n_c\}$ are elements of V such that the Gram matrix

$$G = (N_l(\phi_{k,c})) = I,$$

then we have

$$\phi_{k,c} = \sum_{j \in \mathcal{A}_k} \phi_j + v_{hf}, \quad v_{hf} \in V_{hf}. \tag{9.7}$$

Proof. We fix a k and consider the subset $W_k \subset V$ such that

$$W_k := \{v \in V : N_k(v) = 1 \text{ and } N_l(v) = 0, \text{ for all } l \neq k\}.$$

Pick any $v_1, v_2 \in W_k$. We have

$$N_l(v_1 - v_2) = N_l(v_1) - N_l(v_2) = 0, \quad \text{for all } l = 1, \ldots, n_c,$$

and this shows that $(v_1 - v_2) \in V_{hf}$.

Furthermore, if we define $v_k^c = \sum_{j \in \mathcal{A}_k} \phi_j$, then

$$N_k(v_k^c) = \sum_{j \in \mathcal{A}_k} \sum_{i \in \mathcal{A}_k} \alpha_j(\psi_j, \phi_i) = \sum_{j \in \mathcal{A}_k} \alpha_j = 1,$$

and $N_l(v_k^c) = 0$ for all $l \neq k$ since $\mathcal{A}_k$ and $\mathcal{A}_l$ have no overlap. We then have $v_k^c \in W_k$, and hence

$$W_k = \{v_k^c + v_{hf} : v_{hf} \in V_{hf}\}.$$

□

Next, we describe how to construct the coarse space V_c using W_c.

Lemma 9.3. Assume that $V = V_{hf} \oplus W_c$ and $\varphi_{1,c}, \ldots, \varphi_{n_c,c}$ is a basis in W_c. Then $\phi_{k,c} = S\varphi_{k,c}$, $k = 1, \ldots, n_c$ are linearly independent if $S : V \mapsto V$ satisfies one of the following conditions:

(a) S maps a linear independent set in W_c into a linear independent set in V,

(b) S is invertible,

(c) $S = I - Q_{hf}$, where $Q_{hf} : V \mapsto V_{hf}$.

As a result, we have

$$V = V_{hf} \oplus V_c, \quad V_c = \text{span}\{S\varphi_{k,c} : k = 1, \ldots, n_c\}.$$

Proof. We only need to prove case (c), as cases (a) and (b) are trivial. If $\phi_{k,c}$ were linearly dependent, there would be a linear combination of $\{\phi_{k,c}\}_{k=1}^{n_c}$ which vanishes. Equivalently, this means that there exists $w_c \in W_c$ such that $(I - Q_{hf})w_c = 0$, which implies that $w_c \in V_{hf}$. Since $V_f \cap W_c = \{0\}$, we have that $w_c = 0$ and the only vanishing linear combination in $\text{span}\{\phi_{k,c}\}_{k=1}^{n_c}$ is the trivial one, and this completes the proof. □

Remark 9.4. **(i)** For smoothed aggregation, $S = I - \omega D^{-1}A$ for some properly chosen ω so that S is non-singular, or maps a special linearly independent set of vectors to a linearly independent set of vectors (see Section 13).

(ii) For classical AMG with the ideal interpolation, $S = I - Q_{hf}$, and Q_{hf} is the A-orthogonal projection (see Section 12).

(iii) For classical AMG with the standard interpolation, $S = I - Q_{hf}$, and Q_{hf} is an approximation to the ideal interpolation (see Section 12).

Algorithm 5 Maximal independent set method

1: **Set** $C \leftarrow \emptyset$, $i \leftarrow 1$.

2: **If** i and all its neighbours are not visited, **then** set $C \leftarrow C \cup \{i\}$ and mark i and all vertices in $N(i)$ as visited.

3: **If** all vertices are visited, **then** output C and stop; **else** set $i \leftarrow i + 1$ and go to 2.

To give a summary of the discussions above, the coarsening algorithms in AMG are methods for determining the coarse grid degrees of freedom, or coarse grid variables. Such algorithms are based on selecting degrees of freedom associated with subsets of vertices in the adjacency graph that correspond to the matrix A, or to the strength matrix S, as is done in geometric coarsening, when the hierarchy of meshes or adjacency graphs is known.

9.3. Two basic coarsening algorithms

In the next two subsections we present typical algorithms for finding the coarse grid degrees of freedom. Each such degree of freedom is associated with a vertex or a subset of a graph. Two types of algorithm are distinguished: classical AMG algorithms pick coarse grid degrees of freedom that correspond to a maximal independent set of vertices in the adjacency graph of the strength matrix, and aggregation-based AMG algorithms use a splitting of the adjacency graph of the strength matrix in connected subgraphs.

9.3.1. A maximal independent set algorithm

Here we present a simple 'greedy' maximal independent set (MIS) algorithm, which has been used in classical AMG algorithms to identify coarse grid degrees of freedom. Given the adjacency graph of the strength matrix, the simple greedy MIS algorithm is shown in Algorithm 5.

Remark 9.5. We note that the MIS algorithm recovers the geometric coarsening if the vertices are visited in such an order that coarse grid vertices are ordered first and all the connections in Figure 9.1 are strong. This is obvious, but nevertheless shows that geometric coarsening can sometimes be recovered by an algebraic algorithm.

Let us point out that for finite element stiffness matrices obtained via adaptive refinement algorithms, the hierarchy of vertices is naturally included in the refinement procedure. For regular refinement this choice of maximal independent set is illustrated in Figure 9.1.

Algorithm 6 Greedy aggregation method

Input: Graph $\mathcal{G}$ with n vertices. **Output:** $\mathcal{V} = \cup_{k=1}^{n_c} \mathcal{V}_k$, and $\mathcal{V}_k \cap \mathcal{V}_j = \emptyset$ when $k \neq j$.

1: Set $n_c \leftarrow 0$ and for $k = 1 : n$ do:

a. If k and all its neighbours have not been visited, then (i) we set $n_c \leftarrow n_c + 1$, (ii) we label with n_c the subgraph whose vertices are k and the neighbours of k, and (iii) we mark k and all its neighbours as visited.

b. If at least one neighbour of k has been visited, we continue the loop over the vertices.

2: Since after this procedure there might be vertices which do not belong to any aggregate (but definitely have a neighbouring aggregate), we add each such vertex to a neighbouring aggregate and we pick the one which has a minimal number of vertices in it.

3: The algorithm ends when all vertices are in a subset.

9.3.2. An aggregation algorithm

The class of algorithms known as aggregation algorithms refers to the splitting of the adjacency graph of the strength matrix into a union of connected subgraphs. Let $\{\mathcal{V}_k\}_{k=1}^{n_c}$ be a non-overlapping splitting of the set of vertices

$$\mathcal{V} = \cup_{k=1}^{n_c} \mathcal{V}_k, \quad \mathcal{V}_j \cap \mathcal{V}_k = \emptyset, \quad \text{for } j \neq k.$$

We then define

$$\mathcal{E}_k = \{(l, m) \in \mathcal{E} : l \in \mathcal{V}_k \text{ and } m \in \mathcal{V}_k\} \tag{9.8}$$

to be the set of edges associated with $\mathcal{V}_k$.

An aggregation can be done in many different ways, some very sophisticated. In general, all combinatorial graph-partitioning algorithms can be used for aggregation. We will not consider such algorithms in detail, however, but provide the basic and most important example of the greedy aggregation algorithm: see Algorithm 6. This algorithm can be applied recursively to provide a multilevel hierarchy of aggregates.

9.3.3. Aggressive coarsening

The extended strong connections and the corresponding strength operator are used to construct coarse spaces of smaller dimension. This procedure is also known as aggressive coarsening. We recall the definition of a path in the graph given in Section 7.1, and all our considerations concern the adjacency graph $\mathcal{G}(S)$ of the strength matrix $S \in \mathbb{R}^{n \times n}$ defined in (8.4). Aggressive coarsening refers to a selection of coarse grid vertices as the independent set in the adjacency graph. Specifically, we choose vertices whose distance is larger than 2 with respect to the strength operator.

Definition 9.6 (strong connection along a path). A vertex i is said to connect strongly to a vertex j along a path of length l if there exists a path $(k_0, k_1, \dots, k_l)$ in $\mathcal{G}(S)$ such that $k_0 = i$, $k_l = j$ and $s_c(k_m, k_{m+1}) \geq \theta$, $m = 0, 1, \dots, l-1$.

The next definition is related to the number of strongly path-connected vertices.

Definition 9.7 ((m, l)-strong connection). For given integers $m > 0$ and $l > 0$, a vertex i is (m, l)-strongly connected to a vertex j if and only if i strongly connects to j along at least m paths of length l (*cf.* Definition 9.6).

An aggressive coarsening algorithm generates a maximal independent set using Algorithm 5 for the graph $\mathcal{G}_{m,l} = (\mathcal{V}, \mathcal{E}_{m,l})$ with a set of vertices $\mathcal{V} = \{1, \dots, n\}$ and a set of edges $\mathcal{E}_{m,l}$ defined as

$$\mathcal{E}_{m,l} := \{(i, j) : i \text{ is } (m, l) \text{ strongly connected to } j\}. \tag{9.9}$$

As is well known (Diestel 2010), $(S^l)_{ij}$ is non-zero if and only if there is a path of length $\leq l$ between i and j. An aggressive coarsening exploits this property, and is an algorithm which selects a set of coarse grid degrees of freedom corresponding to vertices in the graph which are at graph distance larger than l. It uses the adjacency graph $\mathcal{G}(S^l)$ of S^l in place of $\mathcal{G}(S)$ in an aggregation or a maximal independent set algorithm.

As an example, let us consider aggressive coarsening with $l = 2$ and $m = 1$. The set of coarse grid degrees of freedom is obtained by applying the standard MIS algorithm twice. First we find a maximal independent set in $\mathcal{G}(S)$ and obtain a set of coarse grid degrees of freedom C (these are at graph distance at least 2). Then we apply the MIS algorithm for a second time to the graph with vertices the C-points, the edges between them being given by the strength operator corresponding to S^2.

Similarly, for aggregation, an aggressive coarsening corresponds to applying Algorithm 6 recursively, or applying it directly to the graph corresponding to S^l for a given l.

9.4. Adaptive coarsening for classical AMG

An adaptive coarsening algorithm is an algorithm which adaptively chooses the coarse grid degrees of freedom based on a given definition of strength function based on the smoother in a two-grid algorithm. One example of adaptive coarsening follows from the classical compatible relaxation introduced by Brandt. The algorithm takes as input a smoother which leaves the coarse grid variables invariant and only smooths the components in the algebraic high-frequency space $V_{\rm hf}$.

A typical adaptive coarsening algorithm follows the steps given below.

Step 0. Set $k=0$ and choose $(V')_{k,c} \subset V'$, for example, using the MIS or aggregation method introduced in Section 9.3.

Step 1. Define $V_{k,f}$ as the subspace of V which is annihilated by the functionals in $(V')_{k,c}$, namely

$$V_{k,f} = \{v \in V : (g,v) = 0, \text{ for all } g \in (V')_{k,c}\}.$$

Step 2. Let $\imath_{k,f} : V_{k,f} \mapsto V$ be the natural inclusion operator, and compute an estimate $\rho_{k,f}$ of the norm of the smoother on $V_{k,f}$ using

$$\rho_{k,f} \approx \sup_{v \in V_{k,f}} \frac{\|(I - \imath_{k,f} R_{k,f} \imath_{k,f}' A)v\|_A^2}{\|v\|_A^2}.$$

Here, $R_{k,f}$ could be the restriction of the smoother R on $V_{k,f}$, or more generally, any relaxation on $V_{k,f}$.

Step 3. Given a threshold $\delta_f > 0$, if $\rho_{k,f} > \delta_f$, we set $k = k+1$, add more functionals to $(V')_c$ and go to **Step 1**. Otherwise, we set $(V')_c = (V')_{k,c}$, and accordingly $V_f = V_{k,f}$ and stop the iteration.

In **Step 3**, if the stopping criterion is not satisfied, we need to enrich the space $(V')_c$ by extending the set. One example of doing so is introduced in compatible relaxation methods by extending the set $\mathcal{C}$ using the following procedure.

First we randomly choose a vector $v^0 \in V_{k,f}$, and form

$$v = (I - \imath_{k,f} R_{k,f} \imath_{k,f}' A)^\nu v^0 \quad \text{for some } \nu \geq 1.$$

Then, with a given threshold $\theta \in (0,1)$, we let

$$\mathcal{C}_0^1 = \Big\{ i \in \mathcal{F} : |v_i| > \theta \max_k |v_k| \Big\}.$$

and

$$\mathcal{C}_1 = \mathcal{C}_0 \cup \mathrm{MIS}(\mathcal{C}_0^1).$$

Finally, we update $\mathcal{C} \leftarrow \mathcal{C}_1$ and proceed with the next compatible relaxation iteration.

As is clear from the algorithm outlined above, we can use any of the definitions of connection strength to obtain $(V')_{0,c}$ at **Step 0**. When $\rho_{k,f} > \delta_f$, it means that $\mathcal{C}$ obtained from the strength function s_0 is not satisfactory, namely, too coarse. This means that either the threshold for s_0 is too small or s_0 itself is not satisfactory. We could still use s_0, but with a smaller threshold to obtain a $\mathcal{C}_0'$ that is bigger than $\mathcal{C}_0$ but does not necessarily contain $\mathcal{C}_0$. For example, if we initially use the (m,l)-strong connection defined in Definition 9.7, $\mathcal{C}$ can be extended by increasing the value of m or decreasing the value of l. This approach may not be computationally efficient. A more effective approach, as used in compatible relaxation, is to find candidates for additional C-points by examining the filtered matrix

$$A^{(1)} = \{a_{ij} : i,j \in \mathcal{F}_0\},$$

where $\mathcal{F}_0 = \Omega \setminus \mathcal{C}_0$, to get the set of coarse grid degrees of freedom $\mathcal{C}_0^1$ and add them to $\mathcal{C}_0$ to extend the size of $\mathcal{C}$.

9.5. AGMG coarsening: a pairwise aggregation

Aggregation-based algebraic multigrid (AGMG) uses the strength function of (8.13) to form aggregations such that the local convergence rate (see (6.13)) on each aggregate is bounded by a given threshold. The main idea of the algorithm can be explained as follows.

It first splits the index set Ω into aggregates, each of which has at most two elements, namely

$$\Omega = \bigcup_j \mathcal{A}_j^{(0)}, \quad \mathcal{A}_i^{(0)} \bigcap \mathcal{A}_j^{(0)} = \emptyset \quad \text{and } |\mathcal{A}_j^{(0)}| \leq 2. \tag{9.10}$$

This process is done by a greedy algorithm. At each step, the algorithm finds the pair $G = \{i,j\}$ for which the strength function defined in (8.14) is maximal.

Using these pairs as aggregates $\{\mathcal{A}_j^{(0)}\}$, we form the unsmoothed aggregation prolongation P, which is piecewise constant with respect to the aggregates and P has orthogonal columns.

We let $A^{(0)} := A$ and $P^{(0)} := P$. Then $A^{(1)} := (P^{(0)})^T A^{(0)} P^{(0)}$. We then apply the pairwise aggregation algorithm on $A^{(1)}$ and find larger aggregates $\{\mathcal{A}_j^{(1)}\}$. Each $\mathcal{A}_j^{(1)}$ is a union of two pairs in $\{\mathcal{A}_j^{(0)}\}$ which minimize the strength function defined in (8.13). Then we obtain $P^{(1)}$ and $A^{(2)}$. Applying this procedure recursively, we obtain the final aggregates $\mathcal{A}_j$, each of which is a union of several pairs in $\{\mathcal{A}_j^{(0)}\}$.

The pairwise aggregation strategy aims to find the aggregates on which the Poincaré constant $\mu_j(V_j^c)^{-1}$ is bounded, which is introduced in our abstract framework by (6.13). As stated in the abstract convergence theorem, bounding $\mu_j(V_j^c)^{-1}$ will bound the convergence rate of the AMG method.

9.6. Bibliographical notes

The coarsening strategies used in AMG are basically of two types: the first uses connection strength to define a 'strength' graph and then performs the greedy aggregation or MIS algorithms, and the other is based on algorithms such as compatible relaxation coarsening, which uses the smoother to detect slowly converging components. These are heuristic approaches, which work well on a certain class of problems but rarely have a theoretical justification for their efficiency as AMG splitting algorithms.

Classical algorithms for selection of the coarse grid degrees of freedom are found in Brandt *et al.* (1982), Stüben (1983), Brandt *et al.* (1985), Ruge and Stüben (1987), and the MG tutorial of Briggs *et al.* (2000).

Parallel coarse-grid selection algorithms are found in Sterck, Yang and Heys (2006) and, in combination with scalable interpolation algorithms, in De Sterck, Falgout, Nolting and Yang (2008). Coarsening using information about discretization, *i.e.* AMGe, is given in Jones and Vassilevski (2001) and Brezina *et al.* (2001). Spectral AMGe coarsening is considered in detail in Chartier *et al.* (2003). Many of the 'upscaling' and related techniques in homogenization (see Efendiev, Hou and Wu 2000, Hou, Wu and Cai 1999), resemble the coarsening procedures introduced in the classical and modern AMG literature.

More sophisticated maximal independent set (MIS) algorithms for selection of coarse grid degrees of freedom, using different measures for connectivity and distance in the graph corresponding to A, are found in Ruge and Stüben (1987) and Trottenberg *et al.* (2001, Appendix A). Most of these algorithms are refinements of the greedy algorithm given in this section. For parallel versions, we refer to Luby (1986), Cleary, Falgout, Henson and Jones (1998) and Sterck *et al.* (2006) for specific details on parallel and parallel randomized MIS algorithms. Other coarsening schemes that are also suitable for parallel implementation are the coupled and decoupled coarsening schemes (Yang 2006, Henson and Yang 2002).

Regarding the aggregation coarsening methods, we refer to Vakhutinsky, Dudkin and Ryvkin (1979), Blaheta (1986) and Marek (1991) for earlier work on such methods. The greedy aggregation algorithm presented here is found in Vaněk *et al.* (1996). A special class of aggregation coarsening methods based on matching were first employed by Karypis and Kumar (1998) for fast graph partitioning, and later used in several of the AMG methods. One example is the AGMG algorithm described in Section 9.5 and found in Napov and Notay (2012) and Notay (2012). The algorithm given by Kim, Xu and Zikatanov (2003) also uses this coarsening approach. Special matching techniques which optimize matrix invariants were used by D'Ambra and Vassilevski (2013, 2014). The compatible relaxation (CR) algorithm, first introduced by Brandt (2000) and further investigated by Livne (2004), Falgout and Vassilevski (2004) and Brannick and Falgout (2010), is a device that reduces the role of connection strength to only define initial set of coarse grid degrees of freedom and then use the smoother to select additional degrees of freedom. Other coarse/fine degrees of freedom partitioning algorithms are considered in MacLachlan and Saad (2007) from both classical and compatible relaxation perspectives. The aggregation algorithms, which aggregate vertices based on a local measure for two-level convergence (Notay 2010, Napov and Notay 2012, Livne and Brandt 2012), are somewhat different adaptive coarsening algorithms.

10. GMG, AMG and geometry-based AMG

Historically, the algebraic multigrid method was motivated by the geometric multigrid method. In this section we will explore the relationship between these two types of method.

10.1. Geometric multigrid method

We begin our discussion with a simple one-dimensional model problem, namely (2.1) for $d = 1, \Omega = (0,1)$ and $\alpha \equiv 1$ with zero Dirichlet boundary condition. For any integer N, we consider a uniform grid, denoted by $\mathcal{T}_h$, of the interval $[0,1]$ as follows:

$$0 = x_0 < x_1 < \cdots < x_{N+1} = 1, \quad x_j = \frac{j}{N+1}(j = 0 : N+1). \tag{10.1}$$

This partition consists of uniform sub-intervals with length $h = 1/(N+1)$, that is, $\mathcal{T}_h = \bigcup_i \{\tau_i\}$, where $\tau_i = (x_{i-1}, x_i)$ for $d = 1$. Such a uniform partition is shown in Figure 10.1.

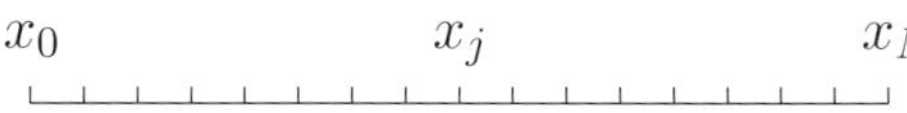

Figure 10.1. One-dimensional uniform grid.

We define a linear finite element space associated with the partition $\mathcal{T}_h$:

$$V_h = \{v : v \text{ is continuous and piecewise linear w.r.t. } \mathcal{T}_h,\ v(0) = v(1) = 0\}. \tag{10.2}$$

Let $V_h = V_h$. Recall from the previous section that the finite element approximation of our model problem is then $u_h \in V_h$, satisfying (2.19). We introduce the operator $A_h : V_h \mapsto V_h$ such that

$$(A_h v_h, w_h) = a(v_h, w_h), \quad v_h, w_h \in V_h.$$

Then the finite element solution u_h satisfies

$$A_h u_h = f_h, \tag{10.3}$$

where $f_h \in V_h$ is the L^2-projection of f: $(f_h, v_h) = (f, v_h)$, $v_h \in V_h$.

To describe a geometric multigrid algorithm, we need to have a multiple level of grids, say $\mathcal{T}_k$, with $k = 1 : J$ and $\mathcal{T}_J = \mathcal{T}_h$ being the finest mesh.

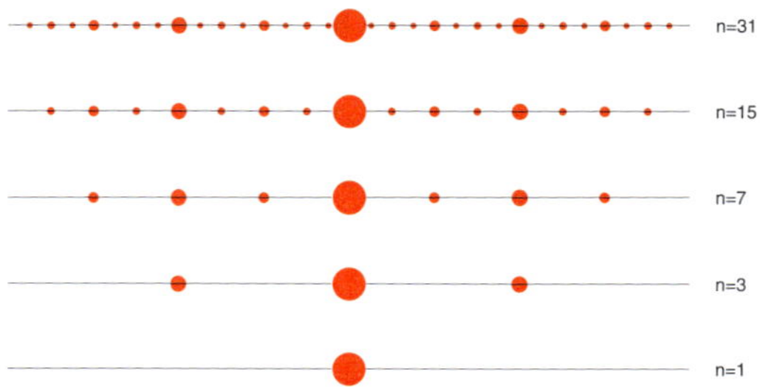

Figure 10.2. Multiple grids in one dimension.

One simple definition of the grid points in $\mathcal{T}_k$ is

$$x_i^k = \frac{i}{2^k}, \quad i = 0, 1, 2, \ldots, N_k + 1, \; k = 1, 2, \ldots, J,$$

where $N_k = 2^k - 1$. Note that $\mathcal{T}_k$ can be viewed as being obtained by adding midpoints of the sub-intervals in $\mathcal{T}_{k-1}$. For each k, the set of above nodes will be denoted by $\mathcal{N}_k$. This is illustrated in Figure 10.2.

For $k = 1 : J$, similar to the finite element space V_h defined as in (10.2), we define the finite element space V_k associated with the grid $\mathcal{T}_k$, to obtain a nested sequence of finite element spaces as follows:

$$V_1 \subset \cdots \subset V_k \subset \cdots \subset V_J. \tag{10.4}$$

The classical V-cycle GMG method simply applies Algorithm 4 recursively with the following setting:

(i) $\mathcal{V}_k = V_k$,

(ii) $A_k : V_k \mapsto V_k$ defined by

$$(A_k u_k, v_k) = a(u_k, v_k), \quad u_k, v_k \in V_k,$$

(iii) $\imath_{k-1}^k : V_{k-1} \mapsto V_k$, the inclusion operator,

(iv) $R_k : V_k \mapsto V_k$ corresponding to a smoother such as the Gauss–Seidel method.

Algebraic setting

Equation (10.3) may be called the operator form of the finite element equation. To get an equation in terms of vectors and matrix, we use the nodal basis functions for V_h:

$$\phi_i(x) = \begin{cases} \dfrac{x - x_{i-1}}{h} & x \in [x_{i-1}, x_i], \\ \dfrac{x_{i+1} - x}{h} & x \in [x_i, x_{i+1}], \\ 0 & \text{elsewhere.} \end{cases} \tag{10.5}$$

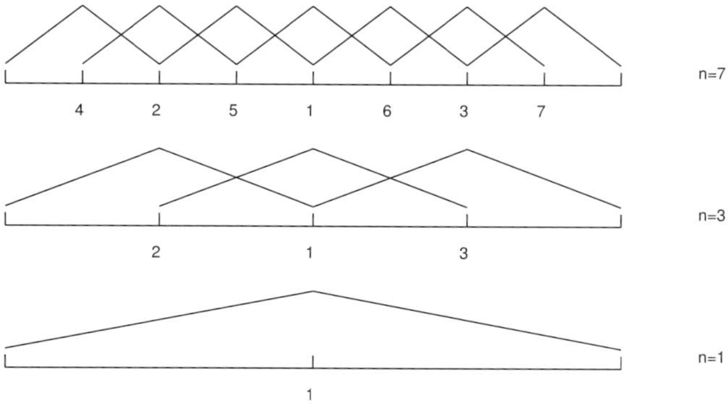

Figure 10.3. One-dimensional nodal basis functions on each level.

On each level, as in (10.5), we introduce a set of nodal basis functions $\{\phi_i^{(k)} : i = 1 : N_k\}$ for the finite element space V_k. This is illustrated in Figure 10.3.

Each $v \in V_k$ can be uniquely written as a linear combination of the basis functions:

$$v = \xi_1 \phi_1^{(k)} + \xi_2 \phi_2^{(k)} + \cdots + \xi_{N_k} \phi_{N_k}^{(k)}. \tag{10.6}$$

This gives an isomorphism from V_k to $\mathbb{R}^{N_k}$, which maps $v \in V_k$ to $\mu \in \mathbb{R}^{N_k}$, that is,

$$v = \xi_1 \phi_1^{(k)} + \xi_2 \phi_2^{(k)} + \cdots + \xi_{N_k} \phi_{N_k}^{(k)} \longrightarrow \mu = \begin{pmatrix} \xi_1 \\ \xi_2 \\ \vdots \\ \xi_{N_k} \end{pmatrix}. \tag{10.7}$$

Here μ is called the *matrix representation* of v. Recall from the discussion in Section 2.2, giving a basis of V_h defined in (10.5), that (2.19) is equivalent to the linear system equations in (2.20).

We introduce the auxiliary space $V_k := \mathbb{R}^{N_k}$. The transition operator P_k^{k+1} from V_k to V_{k+1} is a matrix in $\mathbb{R}^{N_{k+1} \times N_k}$, and satisfies

$$(\phi_1^{(k)} \cdots \phi_{N_k}^{(k)}) = (\phi_1^{(k+1)} \cdots \phi_{N_{k+1}}^{(k+1)}) \imath_k^{k+1}. \tag{10.8}$$

For the special one-dimensional problem we are now considering, for $k = 1, 2, \ldots, J-1$ we have

$$\phi_j^{(k)} = \frac{1}{2} \phi_{2j-1}^{(k+1)} + \phi_{2j}^{(k+1)} + \frac{1}{2} \phi_{2j+1}^{(k+1)}. \tag{10.9}$$

The matrix that encodes this relation is

$$P_k^{k+1} = \begin{pmatrix} \frac{1}{2} & & & \\ 1 & & & \\ \frac{1}{2} & \frac{1}{2} & & \\ & 1 & & \\ & \frac{1}{2} & \frac{1}{2} & \\ & & 1 & \\ & & \frac{1}{2} & \\ & & & \ddots & \frac{1}{2} \\ & & & & 1 \\ & & & & \frac{1}{2} \end{pmatrix}. \tag{10.10}$$

The classical V-cycle AMG method is obtained by simply applying Algorithm 4 recursively with the following setting:

(i) $V_k = \mathbb{R}^{N_k}$,

(ii) $A_k \in \mathbb{R}^{N_k \times N_k} : V_k \mapsto V_k$ defined by
$$(A_k)_{ij} = a(\phi_i^{(k)}, \phi_j^{(k)}), \quad 1 \le i, j \le N_k,$$

(iii) $P_{k-1}^k : V_{k-1} \mapsto V_k$, a matrix in $\mathbb{R}^{N_{k+1} \times N_k}$ defined by (10.8),

(iv) $R_k : V_k \mapsto V_k$ corresponding to a smoother such as the Gauss–Seidel method.

A similar multigrid algorithm can be obtained for problems in two and three dimensions as long as we have a multiple level of grids and the corresponding finite element spaces on each level.

10.2. Obtaining AMG from GMG

The first obstacle to extending GMG to AMG is the geometric information used in GMG. However, a close inspection of GMG reveals that a GMG method only depends on the following two major ingredients:

(i) the stiffness matrix corresponding to the finest grid,

(ii) the prolongation matrix on each level.

Once all the prolongation matrices are given, the stiffness matrices on all coarser levels are given by

$$A_k = P_k^T A P_k, \quad k = J-1, J-2, \ldots, 1, \quad P_k = \prod_{j=k}^{J-1} P_j^{j+1}. \tag{10.11}$$

Of course, a smoother is also needed on each level, but its definition can be considered purely algebraic for the moment.

As the stiffness matrix on the finest grid is always available in any given application, the only thing left is the prolongation matrices. We will now use the example of the linear finite element method to discuss the relationship between the prolongation matrix and geometric information. Two observations are highly relevant.

Observation 1. The prolongation matrix only depends on the natural graph associated with the underlying grid, but not on the coordinates of grid points.

Observation 2. The graph of the underlying grid is very close to the adjacency graph of the stiffness matrix.

Based on the above discussions, roughly speaking, we can essentially recover a geometric multigrid method for the stiffness matrix corresponding to the continuous linear finite element discretization of the Laplace equation by only using the algebraic and graph information provided by the stiffness matrix:

(i) form the adjacency graph $\mathcal{G}(A)$ of the stiffness matrix A,

(ii) coarsen $\mathcal{G}(A)$.

As an illustrative example, let us consider the stiffness matrices corresponding to a discretization of the Laplace equation on a square domain with bilinear elements. It is well known (Ciarlet 2002) that the stiffness matrix in this case is the same as the scaled matrix for the 9-point finite difference stencil (2.18). The corresponding adjacency graph shown in Figure 10.4(c) is denser (has more edges) than the mesh graph shown in Figure 10.4(a). The set of its edges includes the diagonals of each of the squares forming the mesh. For the construction of the prolongation/interpolation matrix, we recall that the prolongation matrix gives the coefficients of the expansion of a coarse grid basis function on a grid of size $2h$ in terms of the finer grid basis on a grid of size h. Locally this matrix looks as follows:

$$[(P_{2h}^{h})^T]_i = \begin{pmatrix} \frac{1}{4} & \frac{1}{2} & \frac{1}{4} \\ \frac{1}{2} & 1 & \frac{1}{2} \\ \frac{1}{4} & \frac{1}{2} & \frac{1}{4} \end{pmatrix}.$$

This matrix is often called a 'prolongation stencil', and it shows the coefficients in the expansion of a coarse grid basis function in a compact form. In the centre we have the coefficient 1, in front of the fine grid basis function associated with a coarse grid vertex. The rest of the entries refer to the coefficients corresponding to the fine grid basis functions in the expansion.

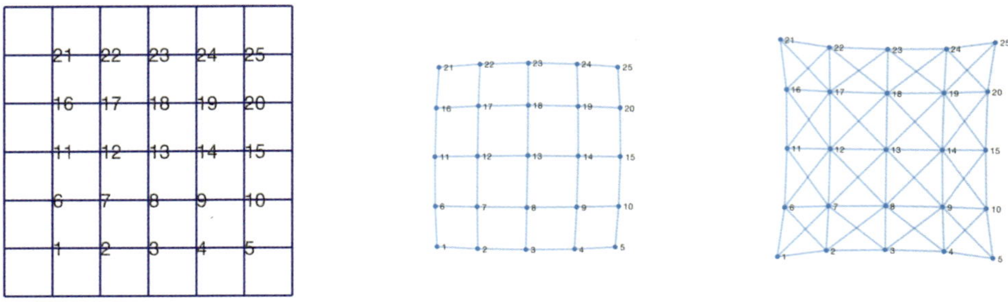

(a) (b) (c)

Figure 10.4. (a) 6×6 uniform grid, (b) graph of the matrix corresponding to the 5-point finite difference stencil, and (c) graph of the matrix corresponding to the 9-point finite difference stencil.

On regularly refined triangular grids we have an analogous situation. We refer to Section 9 for details of the selection of coarse grid degrees of freedom in this case. Prolongation and restriction matrices only depend on the topological structure of this graph. Similar observations led to the development of the AMG: if the geometric coordinates are unknown, different avenues for constructing coarse spaces are needed, leading to different variants of AMG algorithms.

10.3. Obtaining GMG from AMG

In this section we use the unified theory in Section 6 to obtain GMG from AMG. The main ingredients needed are the spaces V_j, operators Π_j, A_j, D_j, and coarse spaces V_j^c.

We now consider constructing a two-level geometric multigrid method for (2.1) (or the variational formulation (2.3)). Suppose we have two grids: a fine grid $\mathcal{T}_h$ and a coarse grid $\mathcal{T}_H$. On each grid we define a linear finite element space V_h and V_H with nodal basis functions $\{\phi_j^h\}$ and $\{\phi_j^H\}$, respectively, and we consider the following partition of the domain Ω:

$$\overline{\Omega} = \bigcup_{j=1}^{J} \overline{\Omega}_j, \quad \text{with } \overline{\Omega}_j = \operatorname{supp}(\phi_j^H). \tag{10.12}$$

We then define V_j as

$$V_j := \{\chi_j v : v \in V_h\}, \tag{10.13}$$

where χ_j is the characteristic function of Ω_j. We note that V_j is not a subspace of $H^1(\Omega)$. The operator $\Pi_j : V_j \mapsto V_h$ is defined as follows:

$$\Pi_j v_j := I_h(\phi_j^H v_j), \quad \text{for all } v_j \in V_j,\ j = 1, \dots, J, \tag{10.14}$$

where I_h is the nodal interpolation operator on the fine grid. Here we note that $\phi_j^H v_j$ is continuous in Ω. We notice that, by definition, $\phi_j^H \chi_j = \phi_j^H$ on Ω, which implies the identities

$$\sum_{j=1}^{J} \Pi_j \chi_j v = I_h\left(\sum_{j=1}^{J} \phi_j^H \chi_j v\right) = I_h\left(\sum_{j=1}^{J} \phi_j^H v\right) = I_h(v) = v,$$

and hence

$$\sum_{j=1}^{J} \Pi_j \chi_j = \mathrm{Id}.$$

The operator $A_j : V_j \mapsto V_j'$ is the local restriction of the bilinear form $a(\cdot,\cdot)$, namely,

$$(A_j u_j, v_j) := a(u_j, v_j)_{\Omega_j} = \int_{\Omega_j} \alpha(x) \nabla u_j \cdot \nabla v_j, \quad u_j, v_j \in V_j. \tag{10.15}$$

The following estimate tells us that A_j satisfies (6.10) with decomposition $v = \sum_{j=1}^{J} \Pi_j v_j$ where $v_j = \chi_j v$:

$$\sum_{j=1}^{m_c} \|v_j\|_{A_j}^2 = \sum_{j=1}^{m_c} a(v,v)_{\Omega_j} \le C_1 \|v\|_A^2,$$

where C_1 depends on the number of overlaps of Ω_j.

We choose $D : V \mapsto V'$ using the fine grid basis functions as follows:

$$(D\phi_i^h, \phi_j^h) := a(\phi_i^h, \phi_j^h)\delta_{ij}, \quad 1 \le i, j \le n, \tag{10.16}$$

and

$$(Du, v) := \sum_{j=1}^{n} a(\phi_j^h, \phi_j^h) u_j v_j, \quad u, v \in V. \tag{10.17}$$

Note that $(D\cdot,\cdot)$ is an inner product on V which induces a norm $\|\cdot\|_D$.

In the above definition, the matrix representation of D is the diagonal of the matrix representation of A. In a similar way, we can define $D_j : V_j \mapsto V_j'$ using the basis functions for V_j.

By a simple scaling argument, we obtain

$$\|v\|_0^2 \eqsim h^d \|v\|_{\ell^2}^2.$$

From (10.17), using inverse inequality, we then have

$$\|v\|_D^2 \lesssim h^{-2} \sum_{j=1}^{n} \|\phi_j^h\|_0^2 v_j^2 \eqsim h^{d-2} \|v\|_{\ell^2}^2 \eqsim h^{-2} \|v\|_0^2.$$

We also have

$$\|I_h v\|_0^2 \eqsim h^d \|I_h v\|_{\ell^2}^2 = h^d \|v\|_{l^2}^2 \eqsim \|v\|_0^2.$$

Assumption 6.9 can then be verified by

$$\begin{aligned}
\Big\|\sum_{j=1}^{J}\Pi_j v_j\Big\|_D^2 &\lesssim h^{-2}\Big\|\sum_{j=1}^{J}\Pi_j v_j\Big\|_0^2 = h^{-2}\Big\|I_h\Big(\sum_{j=1}^{J}\phi_j^H v_j\Big)\Big\|_0^2\\
&\eqsim h^{-2}\Big\|\sum_{j=1}^{J}\phi_j^H v_j\Big\|_0^2 = h^{-2}\sum_{i,j=1}^{J}\int_\Omega \phi_i^H v_i \phi_j^H v_j\\
&\le h^{-2}\sum_{i,j=1}^{J}\|\phi_i^H\|_\infty\|\phi_j^H\|_\infty\int_\Omega v_i v_j \le h^{-2}\sum_{i,j=1}^{J}\int_{\Omega_i\bigcap\Omega_j} v_i v_j\\
&\le C_o\sum_{j=1}^{J}h^{-2}\int_{\Omega_j}|v_j|^2 \lesssim C_o\sum_{j=1}^{J}\|v_j\|_{D_j}^2.
\end{aligned}$$

By the definition of A_j, the kernel of A_j consists of all constant functions in V_j, and we choose V_j^c to be the one-dimensional space of constant functions on Ω_j. Then

$$\mu_j(V_j^c) = \lambda_j^{(2)}, \tag{10.18}$$

where $\lambda_j^{(2)}$ is the second smallest eigenvalue of the operator $D_j^{-1}A_j$.

The global coarse space V_c is defined as in (6.12). Note that in this case it is easy to show, by definition, that the coarse space V_c constructed by (6.12) is in fact identical to V_H, namely

$$V_c = \operatorname{span}\{\phi_j^H : j = 1, \ldots, J\}. \tag{10.19}$$

By Theorem 6.10, the convergence rate of this two-level geometric multigrid method depends on the $\min_j(\lambda_j^{(2)})$. If the Poincaré inequality is true for each V_j, namely

$$\inf_{v_c\in V_j^c}\|v - v_c\|_{D_j}^2 \le c_j\|v\|_{A_j}^2, \quad \text{for all } v \in V_j, \tag{10.20}$$

with c_j a constant, then the two-level geometric multigrid method converges uniformly.

10.4. Spectral AMGe: a geometry-based AMG

We now consider the *element*-based AMG approaches, and also define the ingredients needed to fit these methods into the general results in Section 6. The AMGe methods are less algebraic, as they assume an underlying grid and use element local stiffness matrices to define interpolation operators. In the AMGe setting, it is assumed that we know the decomposition of the $n \times n$ matrix A, that is,

$$A = \sum_{\tau\in\mathcal{T}} \tilde{A}_\tau, \tag{10.21}$$

where $\mathcal{T}$ is the set of finite elements used to discretize the problem and, for each element $\tau \in \mathcal{T}$, $\tilde{A}_\tau$ is the zero-extension of the local stiffness matrix A_τ on τ (which is symmetric positive semi-definite).

To define V_j, D_j, A_j and Π_j in Section 6, corresponding to AMGe, we partition the domain Ω into disjoint subdomains, $\Omega_1, \ldots, \Omega_J$. Each subdomain is an agglomeration of elements, and

$$\bar{\Omega} = \bigcup_{j=1}^{J} \bar{\Omega}_j.$$

For each subdomain Ω_j, we introduce the cutoff operator $\chi_j : V \mapsto V_j$, whose action on $v \in V$ is defined by

$$(\chi_j v)(x) := \begin{cases} v(x) & \text{if } x \in \bar{\Omega}_j, \\ 0 & \text{if } x \notin \bar{\Omega}_j. \end{cases} \tag{10.22}$$

Then we define the space V_j by

$$V_j := \chi_j V. \tag{10.23}$$

Here A_j is defined by summing up all the associated stiffness matrices for elements in Ω_j, namely

$$A_j := \sum_{\tau \subset \Omega_j} \tilde{A}_\tau. \tag{10.24}$$

Clearly, A_j is symmetric positive semi-definite. It is easy to verify that (6.10) holds. In fact, we have the following equations:

$$\sum_{j=1}^{J} \|\chi_j v\|_{A_j}^2 = \sum_{j=1}^{J} a(v,v)_{\Omega_j} = \sum_{\tau \in \mathcal{T}} a(v,v)_\tau = \|v\|_A^2. \tag{10.25}$$

If we denote the diagonal of A by D, then D_j is a diagonal matrix defined by

$$[D_j]_{ii} := \begin{cases} D_{ii} & \text{if } i \in \Omega_j, \\ 0 & \text{if } i \notin \Omega_j. \end{cases} \tag{10.26}$$

The operator Π_j is provided by diagonal matrices defined as

$$[\Pi_j]_{ii} := \begin{cases} [A_j]_{ii}/[A]_{ii} & \text{if } i \in \Omega_j, \\ 0 & \text{if } i \notin \Omega_j. \end{cases} \tag{10.27}$$

Note that $[\Pi_j]_{ii} = 1$ if i is an inner point of Ω_j. Inequality (6.9) is verified in Lemma 12.4.

On each local space V_j, spectral AMGe chooses locally the 'best' coarse space V_j^c, which is the subspace spanned by eigenvectors of $D_j^{-1} A_j$ belonging to its m_j^c smallest eigenvalues. The global coarse space is then defined by (6.12).

By the abstract convergence theorem (Theorem 5.3), the convergence rate of two-level spectral AMGe depends on the minimum of each m_j^c+1 smallest eigenvalue of $D_j^{-1}A_j$ on V_j. More precisely,

$$\|E\|_A \leq 1 - \frac{\mu_c}{C_{p,1}C_{p,2}},$$

with $\mu_c = \min_{1\leq j\leq J} \mu_{m_j^c+1}^{(j)}$.

10.5. Bibliographical notes

The main ideas of the GMG method were first demonstrated in the pioneering works by Fedorenko (1961, 1964) and Bahvalov (1966). Similar ideas, using group relaxation methods, can be traced back to the works of Southwell (1940, 1946). The first description of a truly multigrid method is found in the seminal work by Brandt (1973). Further developments in multilevel methods are those of Brandt (1977) and Hackbusch (1979, 1978). These works have attracted a lot of attention from the computational mathematics and engineering communities. Advances in the convergence analysis of multigrid methods have been made by Nicolaides (1975, 1977), Bank and Dupont (1980), Braess and Hackbusch (1983), Bramble and Pasciak (1987), Bramble, Pasciak and Xu (1990, 1991*a*), Bramble, Pasciak, Wang and Xu (1991*c*) and Xu (1992).

BoxMG (Dendy 1982, 1983) is a method that uses geometrically refined grids and defines interpolation using algebraic techniques. We refer to Dendy (1982, 1983) and de Zeeuw (1990) for details, and also to MacLachlan, Moulton and Chartier (2012) for results on the equivalence between BoxMG and classical AMG.

Element-based AMG approaches are less algebraic, as they assume an underlying grid and use element stiffness matrices to define interpolation operators. Such methods include *plain* AMGe, *element-free* AMGe, *spectral* AMGe and *spectral agglomerate* AMGe, and are developed to improve AMG robustness for finite element based AMG. For results and discussions on different flavours of AMGe we refer to Jones and Vassilevski (2001), Brezina *et al.* (2001), Henson and Vassilevski (2001) and Brezina *et al.* (2006).

11. Energy-min AMG

Here we consider the energy-minimization algorithms for the construction of coarse spaces. While this is not historically the first AMG approach to coarsening, we focus on this technique first, as it can be used to motivate most other AMG algorithms.

11.1. Energy-minimization versus trace-minimization

In the next theorem we add a constraint to Theorem 5.8 and give a relation between the optimal coarse space V_c^{opt} and the energy-minimization. We refer to Section 5.2 for the definition of P^{opt} and $\mathcal{X}_\eta$.

Theorem 11.1 (trace-minimization theorem). Given $\eta > 0$, let $\mathcal{Z}_\eta$ be defined as

$$\mathcal{Z}_\eta = \left\{P \in \mathbb{R}^{n\times n_c} : (Pv, Pv)_{\bar{R}^{-1}} \geq \eta(v,v),\ v \in \mathbb{R}^{n_c} \text{ and } P\mathbf{1} = \sqrt{n_c\eta}\zeta_1\right\}. \tag{11.1}$$

Then $P \in \arg\min_{Q\in\mathcal{Z}_\eta} \operatorname{trace}(Q^T AQ)$ if

$$P \in \mathcal{Z}_\eta \quad \text{and} \quad \operatorname{Range}(P) = \operatorname{Range}(P^{\text{opt}}).$$

Let $\hat{P} = \bar{R}^{-1/2}P$ and define

$$\mathcal{Y}_\eta = \left\{P \in \mathbb{R}^{n\times n_c} : (Pv, Pv) \geq \eta(v,v), v \in \mathbb{R}^{n_c} \text{ and } P\mathbf{1} = \sqrt{n_c\eta}\hat{\zeta}_1\right\}, \tag{11.2}$$

where $\hat{\zeta}_j$ is the eigenvector corresponding to the jth smallest eigenvalue of $\bar{R}^{1/2}A\bar{R}^{1/2}$. It is clear that $\bar{R}^{1/2}A\bar{R}^{1/2}$ and $\bar{R}A$ have the same spectrum. Theorem 11.1 can be written as follows.

Theorem 11.2. Given $\eta > 0$, let $\mathcal{Y}_\eta$ be defined as in (11.2). Then,

$$P \in \arg\min_{Q\in\mathcal{Y}_\eta} \operatorname{trace}(Q^T\bar{R}^{1/2}A\bar{R}^{1/2}Q)$$
$$\text{if } P \in \mathcal{Y}_\eta \text{ and } \operatorname{Range}(P) = \operatorname{Range}(P^{\text{opt}}).$$

Suppose now that we have a bilinear form $a(\cdot,\cdot)$ on V which is symmetric positive semi-definite, and an inner product $(\cdot,\cdot)_{\bar{R}^{-1}}$ on V. Here, for example, the operator $\bar{R}$ is the scaled parallel (resp. successive) subspace correction method corresponding to the splitting of V as

$$V = \sum_{i=1}^{n} \operatorname{span}\{\phi_i\}.$$

In practice, $\bar{R}$ can be a symmetrization of any A-norm convergent smoother on V. Here, for simplicity, we choose $\bar{R} = D^{-1}$.

We now consider a finite element space V with basis functions

$$\{\phi_j : j = 1 : n\}.$$

Let $P = (p_{ij}) \in \mathbb{R}^{n\times n_c}$ be such that

$$(\phi_1^c, \dots, \phi_{n_c}^c) = (\hat{\phi}_1, \dots, \hat{\phi}_n)P, \quad \hat{\phi}_j = \phi_j/\|\phi_j\|_A, \quad j = 1 : n. \tag{11.3}$$

Let $\{(\mu_j, \zeta_j)\}$ denote the eigenpairs of $\bar{R}A$, and $\hat{\zeta}_j = \zeta_j/\|\zeta_j\|_A$. We then define

$$X_\eta = \{(\phi_1^c, \dots, \phi_{n_c}^c) : (\phi_1^c, \dots, \phi_{n_c}^c) \text{ satisfies (11.3) with } P \in \mathcal{Y}_\eta\}. \quad (11.4)$$

We consider the minimization problem

$$\min_{(\phi_1^c,\dots,\phi_{n_c}^c)\in X_\eta} \sum_{j=1}^{n_c} \|\phi_j^c\|_A^2. \quad (11.5)$$

Note that

$$\sum_{j=1}^{n_c} \|\phi_j^c\|_A^2 = \sum_{j=1}^{n_c} a(\phi_j^c, \phi_j^c) = \sum_{j=1}^{n_c} a\Big(\sum_{k=1}^{n} p_{kj}\hat{\phi}_k, \sum_{l=1}^{n} p_{lj}\hat{\phi}_l\Big)$$
$$= \sum_{j=1}^{n_c}\sum_{k=1}^{n}\sum_{l=1}^{n} p_{kj} a(\hat{\phi}_k, \hat{\phi}_l) p_{lj} = \operatorname{trace}(P^T D^{-1/2} \tilde{A} D^{-1/2} P).$$

Then Theorem 11.1 implies that

$$\operatorname{span}\{\zeta_j, 1 \le j \le n_c\} = \operatorname{span}\{\phi_j^0, 1 \le j \le n_c\}, \quad (11.6)$$

where

$$(\phi_1^0, \dots, \phi_{n_c}^0) \in \operatorname*{arg\,min}_{(\phi_1^c,\dots,\phi_{n_c}^c)\in X_\eta} \sum_{j=1}^{n_c} \|\phi_j^c\|_A^2. \quad (11.7)$$

In the next section we use the functional setting, and provide details of the design of energy-minimizing bases.

11.2. Energy-minimization basis for AMG and Schwarz methods

The discussion above motivates the computation of an energy-minimizing basis as the solution of a global optimization problem with constraint. This could be of concern regarding the efficiency of the proposed approach. As we show later in this section, however, this is not a concern because the optimization problem is well conditioned and can be solved efficiently. We also show below that the basis functions solving the energy-minimization problem are locally harmonic within each coarse grid 'element'. This property of the energy-minimizing basis suggests that various 'harmonic extension' techniques, used to define coarse spaces in multigrid methods (see Chan, Xu and Zikatanov 1998, Brezina *et al.* 2001, Jones and Vassilevski 2001) are very closely related to the energy-minimization algorithms. This property also suggests that the energy-minimizing basis may also be used for numerical homogenization for multiscale problems (see Efendiev, Hou and Wu 2000, Hou, Wu and Cai 1999).

We start our description with a given set of subdomains Ω_i with the property that none of the subdomains is fully contained in the union of the rest of them. More precisely, we have

$$\Omega = \bigcup_{i=1}^{n_c} \Omega_i \quad \text{and} \quad \bar{\Omega}_i \bigcap \Big(\bigcup_{j \neq i} \Omega_j \Big)^c \neq \emptyset, \tag{11.8}$$

where the superscript c is the standard set-complement. Equivalently, in the purely algebraic setting, when there is no function space in the background, we may choose Ω_i to be a subset of vertices of the adjacency graph corresponding to a matrix A. The aim is to construct basis functions $\{\phi_i^H\}_{i=1}^{n_c}$ that are in $\mathcal{X}_\eta$, with the following additional restrictions:

$$\text{supp}(\phi_i^H) \subset \bar{\Omega}_i, \quad 1 \leq i \leq n_c.$$

We want the basis functions to have a total minimal energy among all such functions, namely $\{\phi_i^H\}_{i=1}^J$ is the minimizer of

$$\min \sum_{i=1}^{n_c} \|\psi_i\|_A^2 \quad \text{subject to } \psi_i \in V_i \text{ and } (\psi_1, \ldots, \psi_{n_c}) \in \mathcal{X}_\eta. \tag{11.9}$$

Here

$$V_i = \{v \in V_h : \text{supp}(v) \subset \bar{\Omega}_i\}, \quad 1 \leq i \leq n_c. \tag{11.10}$$

Thanks to (11.8), the decomposition (11.11) holds, namely

$$V = \sum_{i=1}^{n_c} V_i. \tag{11.11}$$

Remark 11.3. In AMG, the minimization problem (11.9) written in terms of the prolongation matrices is as follows: find $P \in \mathbb{R}^{n \times n_c}$ such that

$$P = \underset{Y \in \mathbb{R}^{n \times n_c}}{\arg\min}\, \mathcal{F}(Y), \quad Y \mathbf{1}_{n_c} = \mathbf{1}_n, \quad \mathcal{F}(Y) = \text{trace}(Y^T A Y). \tag{11.12}$$

In terms of vectors, local support means that few non-zeros per column (or per row) are allowed in P. We note that the functions $\{\phi_i^H\}$ satisfying the properties mentioned above are linearly independent due to the second assumption in (11.8) and the constraint in (11.12). This linear independence is equivalent to assuming that P is a full rank matrix (*i.e.* $\text{rank}(P) = n_c$).

By the assumption in (11.8), for each j, there exists $k \in \Omega_j$ such that $k \notin \Omega_i$ for all $i \neq j$. We define

$$\mathcal{A}_j = \{k \in \Omega_j : k \notin \Omega_i, i \neq j\}, \quad j = 1 : n_c.$$

Then $\mathcal{A}_j \cap \mathcal{A}_i = \emptyset$ if $i \neq j$. We then define

$$\mathcal{C} = \bigcup_{j=1}^{n_c} \mathcal{A}_j \quad \text{and} \quad \mathcal{F} = \Omega \setminus \mathcal{C}.$$

We define $N_j \in V'$ as follows:

$$N_j(v) = \frac{1}{|\mathcal{A}_j|} \sum_{i \in \mathcal{A}_j} \psi_i(v) = \frac{1}{|\mathcal{A}_j|} \sum_{i \in \mathcal{A}_j} v_i, \quad j = 1 : n_c.$$

Clearly $\{N_j\}_{j=1}^{n_c}$ are linearly independent, and if $\mathrm{supp}(v) \subset \Omega_j$, then $N_i(v) = 0$ for all $i \neq j$.

We define $(V')_c \subset V'$ by

$$(V')_c = \mathrm{span}\{N_j : j = 1 : n_c\},$$

and $V_{hf} \subset V$ by

$$V_{hf} = \{v \in V : (g, v) = 0, \text{ for all } g \in (V')_c\}.$$

If the basis $\{\varphi_j\}_{j=1}^{n_c}$ satisfies

$$\mathrm{supp}(\varphi_j) \subset \Omega_j \quad \text{and} \quad \sum_{j=1}^{n_c} \varphi_j = 1,$$

then we have the Gram matrix

$$G = (G_{ij}) = (N_j(\varphi_i)) = I.$$

By Lemma 9.1, we have

$$V = V_{hf} \oplus W_c, \quad W_c = \mathrm{span}\{\varphi_j : j = 1 : n_c\}.$$

Let us first introduce some notation. We define the restriction A_i of A on each subspace V_i as

$$(A_i u_i, v_i) = (A u_i, v_i), \quad \text{for all } u_i, v_i \in V_i, \ i = 1 : n_c. \tag{11.13}$$

Let $Q_i : V' \mapsto V_i'$ be a projection defined as the adjoint of the natural inclusion $V_i \subset V$:

$$\langle Q_i u', v_i \rangle = \langle u', v_i \rangle \quad \text{for all } v_i \in V_i, \ u' \in V'.$$

We now define the following PSC-type preconditioner (see (4.17)):

$$B = \sum_{i=1}^{n_c} A_i^{-1} \imath_i' = \sum_{i=1}^{n_c} \imath_i A_i^{-1} \imath_i'. \tag{11.14}$$

Thanks to (11.8), it is easy to see that the operator $T : V \mapsto V$ is an isomorphism.

We are now in a position to state and prove the first result in this section.

Theorem 11.4. The minimization problem (11.9) has a unique solution which is given by

$$\phi_i^H = A_i^{-1} Q_i B^{-1} 1, \tag{11.15}$$

satisfying

$$\text{supp}(\phi_i^H) \subset \Omega_i.$$

Proof. This result actually follows directly from Theorem 4.4 with $v = 1$. However, let us give a different proof by finding the critical point of the quadratic functional

$$L = \sum_{i=1}^{n_c} \left(\frac{1}{2} \|\phi_i\|_A^2 - \langle \lambda, \phi_i \rangle \right).$$

Differentiating this functional gives

$$[\partial_{\phi_i} L]\xi_i = (A\phi_i, \xi_i) - \langle \lambda, \xi_i \rangle, \quad \xi_i \in V_i.$$

Hence the the ith component of the critical point (ϕ_i^H) is given by

$$(\phi_i, \xi_i)_A = \langle \lambda, \xi_i \rangle, \quad \text{for all } \xi_i \in V_i, \ i = 1 : n_c. \tag{11.16}$$

From the above equations we obtain that

$$\phi_i^H = A_i^{-1} Q_i \lambda.$$

Summing up leads to

$$\lambda = B^{-1} 1.$$

This gives a derivation of (11.15).

It is obvious that this unique critical point (ϕ_i^H) is indeed the unique global minimizer of (11.9) that has a convex objective functional and a convex constraint. □

We now show that the constructed basis functions are locally discrete A-harmonic. We say that a function $w \in V$ is discrete A-harmonic on a subdomain D if

$$(w, v)_A = 0, \quad \text{for all } v \in V_{h,0}(D) \equiv \{v \in V_h : \text{supp}(v) \subset \bar{D}\}.$$

This property requires us to define the 'subdomains' D on which it holds. Below, we introduce such subdomains in terms of function spaces. Matrix/vector representations of the considerations below are easy to write. To define an analogue of coarse grid elements (an analogue to a finite element coarse grid), we first consider the set of all points in Ω that are interior to all the Ω_i:

$$\omega_0 = \left(\bigcup_{i=1}^{n_c} \partial\Omega_i \right)^c \bigcap \Omega.$$

Given $x \in \omega_0$, define the following function with values in the subsets of $\{1, \ldots, n_c\}$, which is the set of indices of subdomains Ω_i that contain x:

$$I(x) = \{i : x \in \Omega_i\}. \tag{11.17}$$

To rule out any ambiguity, we shall assume that for any $x \in \omega_0$ the set $I(x)$ is ordered in ascending order. We then define

$$K_x = \{y \in \omega_0 : I(y) = I(x)\}. \tag{11.18}$$

That is, K_x is the intersection of all the Ω_i that contain x (see Figure 11.1).

The following simple proposition will lead us to an appropriate definition of coarse grid elements.

Proposition 11.5. For the sets K_x defined in (11.18) we have:

(a) $K_x = K_y$ if and only if $I(x) = I(y)$,

(b) either $K_x \cap K_y = \emptyset$ or $K_x = K_y$, $x \in \omega_0$, $y \in \omega_0$,

(c) there are a finite number m_H of different sets K_x, $x \in \omega_0$.

Proof. The ($\Rightarrow$) direction in (a) follows from the fact that $x \in K_x = K_y$, and hence $I(x) = I(y)$. The other direction follows from the definition of $K_{x,y}$.

To prove (b) let us assume that there exists $z \in \omega_0$, such that $z \in K_x$ and $z \in K_y$. The definition of K_x and K_y then gives that $I(x) = I(y) = I(z)$. By (a), $K_x = K_y$. This proves (b).

The conclusion (c) follows directly from (b). □

Let $\mathcal{T}_H$ denote the finite collection of m_H sets from Proposition 11.5(c). We have

$$\omega_0 = \bigcup_{x \in \omega_0} K_x = \bigcup_{K \in \mathcal{T}_H} K.$$

As it is obvious that $\bar{\omega}_0 = \bar{\Omega}$,

$$\bar{\Omega} = \bar{\omega}_0 = \overline{\bigcup_{K \in \mathcal{T}_H} K} = \bigcup_{K \in \mathcal{T}_H} \bar{K}. \tag{11.19}$$

This means that the collection of $\mathcal{T}_H$ forms a non-overlapping partition of Ω. Each element in $\mathcal{T}_H$ will be called a *coarse grid element.*

Remark 11.6. It is tempting to show how these macro-elements look on an unstructured grid, and in Figure 11.1 we have depicted three such supports together with their intersection. However, let us point out that an essential feature of the technique we present here is that the coarse elements need not be defined explicitly, and they might have a fairly complicated shape.

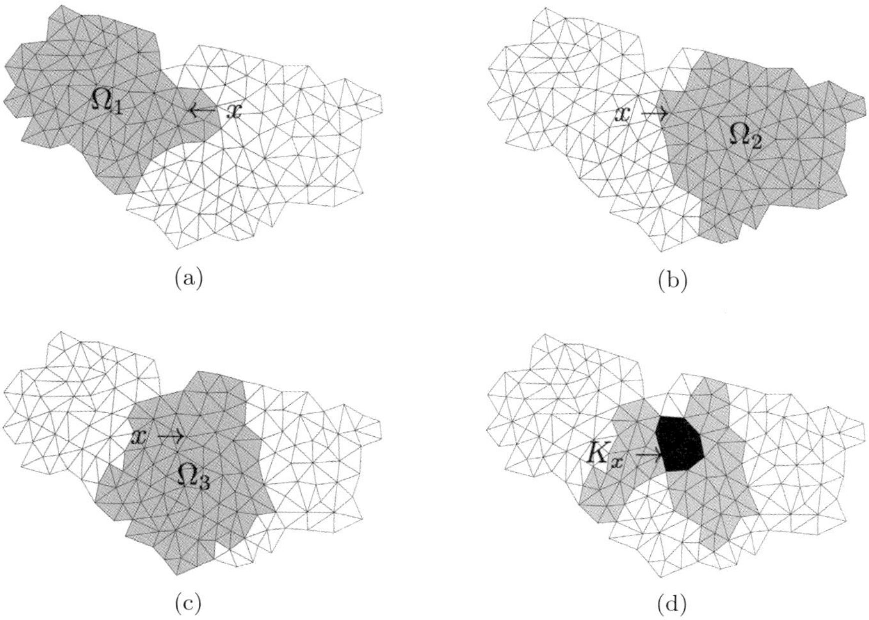

Figure 11.1. (a–c) A triangular grid and the supports of three basis functions. The intersections are plotted in (d): the darker shaded domain corresponds to a coarse element and is the intersection of all three supports; the lighter shaded domains are intersections of two supports; the white area corresponds to no intersection.

Lemma 11.7. Let $\lambda = B^{-1}1$. Assume that for each coarse element $K \in \mathcal{T}_H$ as defined above, we have $(\mathbf{1}, w_K)_A = 0$ for all w_K supported in K. Then

$$\langle \lambda, \xi \rangle = 0, \quad \text{for all } \xi \in V_{h,0}(K).$$

Proof. By definition $K = K_y$ for some $y \in \Omega$. Thus

$$V_{h,0}(K) = \bigcap_{i \in I(y)} V_i \quad \text{and} \quad \sum_{i \in I(y)} \phi_i^H(x) = 1, x \in K.$$

Thus, by (11.16), we have

$$(\phi_i, \xi)_A = \langle \lambda, \xi \rangle, \quad \text{for all } \xi \in V_{h,0}(K),$$

and

$$\langle \lambda, \xi \rangle = \sum_{i \in I(y)} (\phi_i, \xi)_A = (\mathbf{1}, \xi)_A = 0.$$

The desired result then follows. □

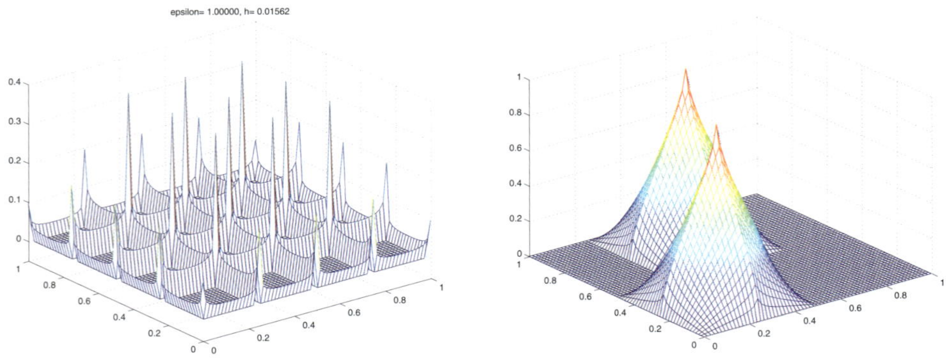

(a) (b)

Figure 11.2. (a) The profile of $\lambda = B^{-1}\mathbf{1}$, and (b) a typical basis function ϕ_i^H.

When the coarsening corresponds to a geometric multigrid and uniform refinement, the lemma shows that $\lambda = B^{-1}1 \in V'$ is a discrete edge δ-function with respect to the coarse elements (*i.e.* λ is supported around ∂K). Figure 11.2 shows an example profile of λ and a basis function ϕ_i^H.

Combining the above result with the identity (11.16), we immediately obtain our second main result in this section.

Theorem 11.8. Each basis function ϕ_i^H is discrete A-harmonic on each coarse element $K \in \mathcal{T}_H$, namely

$$(\phi_i^H, v)_A = 0, \quad v \in V_{h,0}(K). \tag{11.20}$$

In one space dimension ($d = 1$) the above result is rather trivial, and this has in fact already been discussed by Wan, Chan and Smith (1999/2000). In this case, the basis function (ϕ_i^H) is analogous to the generalized finite element basis function in Babuška and Osborn (1983).

The local harmonic properties in all the aforementioned literature are obtained by *construction* from *local* element boundaries. It is interesting to note that the energy-minimizing basis studied here is the result of a more *global* construction, and the local harmonic properties are a by-product of the construction.

11.3. Convergence of energy-minimization AMG

We present a proof of the two-level convergence of AMG based on energy-minimization.

We define cutoff operators χ_j on the subdomains Ω_j introduced in (11.8) as follows:

$$(\chi_j v)(x) := \begin{cases} v(x) & \text{if } x \in \bar{\Omega}_j, \\ 0 & \text{if } x \notin \bar{\Omega}_j. \end{cases}$$

Then we define spaces W_j by

$$W_j := \chi_j V = \chi_j v, \quad \text{for } v \in V.$$

It is easy to verify that

$$\{\chi_j \phi_i : \operatorname{supp} \phi_i \cap \Omega_j \neq \emptyset\} \tag{11.21}$$

forms a basis of W_j.

The operator $\Pi_j : W_j \mapsto V$ is defined as in (10.14), with ϕ_j^H being the solution of the minimization problem (11.9) ((11.15) in Theorem 11.4).

We then have that

$$\sum_{j=1}^{n_c} \Pi_j \chi_j = \mathrm{Id}.$$

In fact, we have

$$\sum_{j=1}^{n_c} \phi_j^H(x) = \sum_{j : x \in \Omega_j} \phi_j(x) = \sum_{j \in I(x)} \phi_j(x) = 1,$$

which implies the identity

$$\sum_{j=1}^{n_c} \Pi_j \chi_j v = I_h \Bigg(\sum_{j=1}^{n_c} \phi_j^H \chi_j v \Bigg) = I_h(v) = v, \text{ and hence } \sum_{j=1}^{n_c} \Pi_j \chi_j = \mathrm{Id}. \tag{11.22}$$

The operator $A_j : W_j \mapsto W_j'$ is the local restriction of the bilinear form $a(\cdot, \cdot)$ as in (10.15).

Inequality (6.10) is satisfied with the decomposition $v = \sum_{j=1}^{n_c} \Pi_j v_j$, where $v_j = \chi_j v \in W_j$:

$$\sum_{j=1}^{m_c} \|v_j\|_{A_j}^2 = \sum_{j=1}^{m_c} a(v, v)_{\Omega_j} \leq C_o \|v\|_A^2,$$

where C_o depends on the number of overlaps of Ω_j.

We choose $D : V \mapsto V'$ using the fine grid basis functions as follows:

$$(D\phi_i^h, \phi_j^h) := (A\phi_i^h, \phi_j^h)\delta_{ij}, \quad 1 \leq i, j \leq n. \tag{11.23}$$

Notice that in the above definition the matrix representation of D is the diagonal of the matrix representation of A. In a similar way, we can define $D_j : W_j \mapsto W_j'$ using basis functions of W_j defined in (11.21), and the matrix representation of D_j is the diagonal of the matrix representation of A_j. It is well known that

$$\|v\|_D^2 \eqsim h^{-2} \|v\|_0^2, \text{ for all } v \in V, \quad \text{and} \quad \|v_j\|_{D_j}^2 \eqsim h^{-2} \|v_j\|_0^2, \text{ for all } v_j \in W_j.$$

Inequality (6.9) can be verified by

$$\begin{aligned}\Big\|\sum_{j=1}^{n_c}\Pi_j v_j\Big\|_D^2 &\lesssim h^{-2}\int_\Omega\Big(\sum_{j=1}^{n_c}\phi_j^H v_j\Big)^2 = h^{-2}\sum_{i,j=1}^{n_c}\int_\Omega \phi_i^H v_i \phi_j^H v_j\\ &\le h^{-2}\sum_{i,j=1}^{n_c}\|\phi_i\|_\infty\|\phi_j\|_\infty\int_\Omega v_i v_j\\ &\le h^{-2}\Big(\max_{1\le j\le n_c}\|\phi_j\|_\infty\Big)^2\sum_{i,j=1}^{n_c}\int_{\Omega_i\bigcap\Omega_j} v_i v_j\\ &\le C_o\Big(\max_{1\le j\le n_c}\|\phi_j\|_\infty\Big)^2\sum_{j=1}^{n_c}h^{-2}\int_{\Omega_j}|v_j|^2\\ &\lesssim C_o\Big(\max_{1\le j\le n_c}\|\phi_j\|_\infty\Big)^2\sum_{j=1}^{n_c}\|v_j\|_{D_j}^2.\end{aligned}$$

We choose local coarse spaces $W_j^c \subset W_j$ to be the space of constant functions on $\bar{\Omega}_j$. Then the global coarse space V_c is defined as

$$V_c := \sum_{j=1}^{n_c}\Pi_j W_j^c.$$

In fact, for this case, V_c is the subspace spanned by $\{\phi_j^H\}$:

$$V_c = \operatorname{span}\{\phi_j^H : j = 1, \ldots, n_c\}. \tag{11.24}$$

We choose the subdomain Ω_j such that the Poincaré inequality holds:

$$\inf_{v_c\in W_j^c}\|v - v_c\|_0^2 \le c d_j^2 |v|_1^2,$$

where d_j is the diameter of Ω_j, and c is a constant independent of the mesh size and the size of Ω_j. Since

$$\|v\|_{D_j}^2 \eqsim h^{-2}\|v\|_0^2, \quad \|v\|_{A_j}^2 \eqsim |v|_1^2,$$

we have

$$\inf_{v_c\in W_j^c}\|v - v_c\|_{D_j}^2 \le c\Big(\frac{d_j}{h}\Big)^2\|v\|_{A_j}^2,$$

Combining the above discussion with Theorem 6.10, we obtain the following result.

Theorem 11.9. The convergence rate of two-level AMG based on energy-minimization is bounded by

$$\|E\|_A \le 1 - \mu, \tag{11.25}$$

where $0 < \mu < 1$ depends only on the size and overlaps of the subdomains Ω_j.

11.4. Bibliographical notes

Energy-minimization seems to encompass several algorithms: the Lagrange equations for this minimization are solved approximately in classical AMG (Section 12), while the functional is approximately minimized in smoothed aggregation (Section 13).

For energy-minimization approaches we refer to the works by Mandel, Brezina and Vaněk (1999), Wan *et al.* (1999/2000), Chan, Xu and Zikatanov (1998), Xu and Zikatanov (2004) and Brannick and Zikatanov (2006). An interesting fact is that for regularly refined grids, given the right supports, the trace (energy) minimization prolongation recovers the coarse basis very closely, although not exactly. The small discrepancies are due to effects from the boundary, and the further in graph distance the coarse grid basis function is, the closer it is to piecewise linear.

The extensive numerical experiments reported in Wan *et al.* (1999/2000), Wan (1998), Mandel *et al.* (1999) and Xu and Zikatanov (2004) show that the energy-minimizing basis leads to uniformly convergent two-grid and multigrid methods for many problems of practical interest, and especially for problems with rough coefficients. These methods also provide a framework for numerical homogenization, and are related to the homogenization methods used by Grauschopf, Griebel and Regler (1997) and the M^3 techniques of Lipnikov, Moulton and Svyatskiy (2011).

The 'smoothed aggregation' approach in the algebraic multigrid method, as well as the theoretical framework, is reported in works by Vaněk, Mandel and Brezina (Vaněk *et al.* 1996, 1998, Mandel *et al.* 1999), who draw an explicit relation between the construction of a basis for the coarse space and the 'energy' of the basis functions. As pointed out by Wan *et al.* (1999/2000) and Mandel *et al.* (1999), this can be viewed as a step (or steps) towards obtaining basis functions minimizing a quadratic (energy) functional, which we considered earlier.

Olson, Schroder and Tuminaro (2011) explore different sparsity patterns, including long-distance energy-minimizing prolongations. Schroder (2012) considers anisotropic problems.

Vassilevski and Zikatanov (2006) consider constrained energy-minimization preserving multiple vectors. In the finite element settings, when the element stiffness matrices are available, local energy-minimization provides coarse spaces which result in uniform two-level convergence (Kolev and Vassilevski 2006).

12. Classical AMG

A coarse space in classical AMG is always viewed as a subspace defined via a prolongation (interpolation) matrix. Its dimension, n_c, is a fraction of the dimension of the finer space. Popular interpolation schemes in classical AMG are *direct, standard* or *multipass* interpolation. The matrix representations of such interpolations are of the form $P = \binom{W}{I}$, with $W \in \mathbb{R}^{n_F \times n_C}$, and they can be viewed as *sparse* approximations to the so-called 'ideal' interpolation with $W = -A_{FF}^{-1} A_{FC}$, which is, in general, a full matrix. Here, the matrix A_{FF} is the block of the matrix corresponding to the F-points and A_{FC} is the block corresponding to the connections between the C-points and F-points. The splitting of vertices of the adjacency graph corresponding to A in subsets F and C is done using one of the coarsening strategies described in Section 9.

12.1. Coarse spaces in classical AMG

The coarsening algorithm in classical AMG uses a splitting of the set of vertices $\{1, 2, \ldots, n\}$ of the graph corresponding to A into two disjoint sets

$$\mathcal{C} \cup \mathcal{F} = \{1, 2, \ldots, n\}, \quad \mathcal{C} \cap \mathcal{F} = \emptyset, \tag{12.1}$$

where $\mathcal{C}$ is a maximal independent set whose independence is with respect to the graph of the strength operator S defined in Section 8.2, and the splitting is referred to as a 'C/F splitting'. A simple greedy C/F splitting algorithm is introduced in Algorithm 5, Section 9.3.1.

We assume that V is equipped with a basis $\{\phi_k\}_{k=1}^n$, and we let $n_f = |\mathcal{F}|$, $n_c = |\mathcal{C}|$ denote the cardinality of the sets forming the C/F splitting. When convenient we assume that

$$\mathcal{F} = \{1, \ldots, n_f\} \quad \text{and} \quad \mathcal{C} = \{n_f + 1, \ldots, n\}. \tag{12.2}$$

In this case, the high-frequency space V_{hf} as defined in (9.2) can be written as

$$V_{hf} = \operatorname{span}\{\phi_j, j \in \mathcal{F}\}. \tag{12.3}$$

Following the procedure in Section 9, we first proceed to identify a tentative coarse space W_c. One easy choice is

$$W_c = \operatorname{span}\{\phi_j, j \in \mathcal{C}\}. \tag{12.4}$$

Obviously the above tentative space satisfies (9.3), namely

$$V = V_{hf} \oplus W_c,$$

but W_c is hardly a low-frequency space. Given a basis function in W_c, ϕ_{j_k}, we filter out its high-frequency component to obtain the following coarse basis function:

$$\phi_{k,c} := (I - \Pi_{hf})\phi_{j_k} = \phi_{j_k} + \sum_{j \notin \mathcal{C}} p_{jk}\phi_j. \tag{12.5}$$

Here P_{hf} is the $(\cdot,\cdot)_A$ orthogonal projection onto V_{hf}. By the definition of Π_{hf},

$$\Pi_{hf}\phi_k = \begin{cases} \phi_k & \text{if } k \in \mathcal{F}, \\ -\sum_{j\in\mathcal{F}} p_{jk}\phi_j & \text{if } k \in \mathcal{C}, \end{cases}$$

where p_{jk} satisfies

$$-\sum_{j\in\mathcal{F}} p_{jk}(\phi_j, \phi_i)_A = (\phi_k, \phi_i)_A, \quad \text{for all } i \in \mathcal{F},\ k \in \mathcal{C}. \tag{12.6}$$

In matrix notation, with the ordering given by (12.2), the matrix form of the equations in (12.6) is

$$A_{FF}W = A_{FC}, \quad \text{where } A = \begin{pmatrix} A_{FF} & A_{FC} \\ A_{FC}^T & A_{CC} \end{pmatrix}. \tag{12.7}$$

Here, $W_{jk} = p_{jk}$, $j \in \mathcal{F}$, $k \in \mathcal{C}$, where p_{jk} are the coefficients given in (12.6).

In view of Lemma 9.3, we have $S = I - \Pi_{hf}$. V_c is the span of the functions in (12.5):

$$V_c = \operatorname{span}\{\phi_{k,c}\}_{k=1}^{n_c} = \operatorname{Range}(I - P_{hf}).$$

We note that

$$(\phi_{1,c}, \dots, \phi_{n_c,c}) = (\phi_1, \dots, \phi_n)\begin{pmatrix} W \\ I \end{pmatrix}.$$

Thus the corresponding prolongation matrix is

$$P = \begin{pmatrix} W \\ I \end{pmatrix}, \quad W = -A_{FF}^{-1}A_{FC}. \tag{12.8}$$

The functions $\{\phi_{k,H}\}$ given in (12.5) form a basis of V_c similar to the geometric multigrid method for Lagrangian finite elements. We denote the prolongation matrix defined in (12.8) by P^{opt}, and refer to it below as the *ideal interpolation.*

The following result can be easily established.

Lemma 12.1. If A satisfies $A\mathbf{1} = 0$, the solution of (12.7) also satisfies the constraint in (11.12), namely, $W\mathbf{1}_{n_c} = \mathbf{1}_{n_F}$.

Next, we introduce the set of the prolongations $\mathcal{P}_C$,

$$\mathcal{P}_C = \left\{P : P = \begin{pmatrix} W \\ I \end{pmatrix}\right\}.$$

and from the definition given in (5.25) in Section 11 we recall that

$$\mathcal{P}_C \subset \eta\mathcal{X},$$

for some constant $\eta > 0$. For example, $\eta \eqsim \max_{1\le j\le n} |D_{jj}|$ if $\bar{R}^{-1} \eqsim D$, where D is the diagonal of A. Equivalently, $P \in \mathcal{P}_C$ means that the coarse grid basis is defined as

$$\phi_{k,H} = \phi_{j_k} + v_f, \quad j_k \in \mathcal{C}, \quad v_f \in V_f, \quad k = 1,\dots,n_c.$$

Clearly, all elements of $\mathcal{P}_{\mathcal{C}}$ are of full rank, and we also have that $P^{\rm opt} \in \mathcal{P}_C$ by definition.

We have the following theorem, showing that the coarse grid matrix corresponding to $V_c^{\rm opt} = \mathrm{Range}(P^{\rm opt})$ has a minimal trace.

Theorem 12.2. If we fix the set of indexes $\mathcal{C}$ and coarse grid degrees of freedom, then for $P^{\rm opt}$ we have

$$P^{\rm opt} = \arg\min_{P\in\mathcal{P}_C} \mathrm{trace}(P^T AP). \tag{12.9}$$

Furthermore, if we denote $A_c = (P^{\rm opt})^T AP^{\rm opt}$, then

$$\|v_c\|_{A_c}^2 = \min\{\|v\|_A^2 : \langle\phi'_{j_k}, v\rangle = v_{c,k},\ k = 1,\dots,n_c\}. \tag{12.10}$$

Proof. The relation (12.9) follows from the simple identities

$$\begin{aligned}
\mathrm{trace}(A_c) &= \sum_{k=1}^{n_c} \|\phi_{k,H}\|_A^2 = \sum_{k=1}^{n_c} \|(I - P_{hf})\phi_{j_k}\|_A^2 \\
&= \sum_{k=1}^{n_c} \min_{v_k\in V_{hf}} \|\phi_{j_k} + v_k\|_A^2 \\
&= \min\left\{\sum_{k=1}^{n_c} \|\phi_{j_k} + v_k\|_A^2 : v_k \in V_{hf},\ k = 1,\dots,n_c\right\} \\
&= \min_{P\in\mathcal{P}_C} \mathrm{trace}(P^T AP).
\end{aligned}$$

To prove (12.10), we note that any $v \in V$ such that $\langle\phi'_{j_k}, v\rangle = v_{c,k}$ can be written as

$$v = w_c + v_f, \quad \text{where } w_c = \sum_{k=1}^{n_c} v_{c,k}\phi_{j_k}.$$

By the definition of P_{hf}, we have $(I - P_{hf})v_f = 0$ and $v_c = (I - P_{hf})(v - v_f) = (I - P_{hf})v$. We then have

$$\|v\|_A^2 = \|P_{hf}v\|_A^2 + \|(I - P_{hf})v\|_A^2 = \|P_{hf}v\|_A^2 + \|v_c\|_{A_c}^2 \ge \|v_c\|_{A_c}^2. \qquad \square$$

Theorem 12.2 shows that the minimizer P satisfies equation (12.7). By Theorem 12.2 it follows that the minimizer of (11.12) and the solution to (12.7) are the same for $A\mathbf{1} = 0$.

12.2. Quasi-optimality of the ideal interpolation

The following two-level convergence result is well known: see *e.g.* MacLachlan, Manteuffel and McCormick (2006).

Theorem 12.3. For P^{opt} defined as the solution to the minimization problem (11.12), or equivalently the solution to (12.7), the two-level AMG method with prolongation P^{opt} converges with a rate $\|E\|_A \le 1-\delta$, where δ is a constant depending only on the maximum degree of the vertices in the graph corresponding to A and the threshold θ in choosing the strong connections.

Proof. According to Corollary 5.21, we only need to verify that V_{hf} consists of δ-algebraic high frequencies as defined in Definition 5.17. Clearly, from the discussion in Section 7.2, we can assume that A is an M-matrix with all connections being strong connections. We consider the graph corresponding to A and recall that the set of coarse grid degrees of freedom is a maximal independent set of vertices in this graph. By definition of connection strength, for any $j \in \{1,\dots,n\}$ and any $i \in N(j)$ we have

$$a_{jj} = \sum_{k\in N_j} -a_{jk} \le |N_j| \max_{k\in N_j}(-a_{jk}) \le -\frac{|N_j|}{\theta} a_{ji}.$$

Next, for any $j \in F$, let $k_j \in \mathcal{C}$ be such that $|a_{k_j,j}| > 0$. Such a k_j exists because the set of C-vertices is a maximal independent set. This choice is not unique, and we just fix one such index k_j for every j. Using the notation from Section 9, and the fact that $v \in V_{hf}$ vanishes at the C-vertices, we obtain

$$\begin{aligned}
\|v\|^2_{\bar{R}^{-1}} &\le c^D \|v\|^2_D = c^D \sum_{j\in F} a_{jj} v_j^2 = c^D \sum_{j\in F} a_{jj}(v_j - v_{k_j})^2 \\
&\le c^D \sum_{j\in F} -\frac{|N_j|}{\theta} a_{j,k_j}(v_j - v_{k_j})^2 \\
&\le \max_j(|N_j|)\frac{c^D}{\theta} \sum_{(i,j)\in\mathcal{E}} -a_{ij}(v_i - v_j)^2 \\
&\le \max_j(|N_j|)\frac{c^D}{\theta}\|v\|_A^2. \qquad \square
\end{aligned}$$

We note that interpolations like P^{opt} (with the possible exception of one-dimensional problems) are not used in practice, because the prolongation matrix P^{opt} could have a lot of fill-in and $(P^{\text{opt}})^T AP^{\text{opt}}$ is dense. However, it is also important to note that sparse approximations to P^{opt} are used in practice. Thus the energy-minimization technique for constructing coarse spaces may be viewed as a motivation for other AMG techniques used to construct approximations of V_c^{opt}. For example, most of the known techniques approximate the minimizer of the functional $\mathcal{F}$ given in equation (11.12): (i) the prolongation matrices constructed in the classical AMG framework approximate the solution to equation (12.7) over the subset of $\mathcal{P}_C$ consisting of sparse matrices, and (ii) the energy-minimization techniques outlined in Section 11.2 and the smoothed aggregation considered in Section 13.3 minimize (approximately) the trace of the coarse grid matrix $\mathcal{F}(P)$ over a set of sparse matrices P.

12.3. Construction of prolongation matrix P

12.3.1. Prolongation

An intuitive idea for finding an approximate solution of the problem (12.7) is to use some basic iterative methods such as the Jacobi method, and then properly rescale the coefficients of W so that it satisfies

$$W\mathbf{1}_{n_c} = \mathbf{1}_{n_f} \tag{12.11}$$

and also fits into a sparsity pattern:

$$W \in \mathbb{R}_S^{n_f \times n_c}. \tag{12.12}$$

Following this idea, we construct the interpolations used in classical AMG (Trottenberg *et al.* 2001, Section A.7) – direct interpolation, standard interpolation and multipass interpolation – in a unified fashion.

Direct interpolation. Here we approximate the solution to (12.7)–(12.12) by one Jacobi iteration with initial guess $W_0 = 0$, namely,

$$W_1 = -D_{FF}^{-1} A_{FC}.$$

In order to satisfy the constraint (12.11), we rescale W_1 to obtain that

$$W = M(-D_{FF}^{-1} A_{FC}) = [\operatorname{diag}(A_{FC}\mathbf{1})]^{-1} A_{FC},$$

where M is a rescaling operator defined by

$$M(Y) = [\operatorname{diag}(Y\mathbf{1})]^{-1} Y. \tag{12.13}$$

Standard interpolation. The construction of the prolongation matrix via standard interpolation can be viewed in several different ways. Probably the most natural way is to view it as a smoothing of the direct interpolation, again followed by rescaling. Indeed, assuming that the smooth error (e_F^T, e_C^T) satisfies

$$A_{FF} e_F + A_{FC} e_C = 0,$$

we approximate this equation via

$$e_F = -D_{FF}^{-1} A_{FC} e_C + D_F^{-1} A_{FF} (D_{FF}^{-1} A_{FC}) e_C \tag{12.14}$$

$$W = (I - D_{FF}^{-1} A_{FF}) W^1, \quad W^1 = -D_{FF}^{-1} A_{FC}, \tag{12.15}$$

using the same notation e_F, e_C for the approximations (see Trottenberg *et al.* 2001). This is equivalent to a Jacobi smoothing iteration applied to the unscaled direct interpolation W^1. Rescaling is also needed for the standard interpolation, and the final formula is

$$W = [\mathrm{diag}((I - D_{FF}^{-1} A_{FF}) W^1 \mathbf{1})]^{-1} (I - D_{FF}^{-1} A_{FF}) W^1. \tag{12.16}$$

The similarity to the smoothed aggregation discussed in Section 13.3 is obvious, as this is indeed a smoothing applied to W^1.

Multipass interpolation. This is an approximation to the solution of (12.7) when the C/F splitting has been constructed by means of aggressive coarsening, namely, using the (m, l)-strong connection defined in Definition 9.7. The set of F is divided into l disjoint subsets $F_1, F_2, \ldots, F_l$ as follows. We first define the distance from a point j to a subset of points C:

$$\mathrm{dist}(j, C) = \min\{\mathrm{dist}(j, i) : i \in C\}.$$

Next, we set

$$F_k = \{j \in \mathcal{F} : \mathrm{dist}(j, \mathcal{C}) = k\}, \quad k = 1, 2, \ldots, l, \quad l = \max\{\mathrm{dist}(j, \mathcal{C}) : j \in F\}.$$

Then A_{FF}, A_{FC} can be written as the following block matrices:

$$A_{FF} = \begin{pmatrix} A_{F_1F_1} & A_{F_1F_2} & \cdots & A_{F_1F_l} \\ A_{F_2F_1} & A_{F_2F_2} & \cdots & A_{F_2F_l} \\ \vdots & \vdots & \ddots & \vdots \\ A_{F_lF_1} & A_{F_lF_2} & \cdots & A_{F_lF_l} \end{pmatrix}, \quad A_{FC} = \begin{pmatrix} A_{F_1C} \\ A_{F_2C} \\ \vdots \\ A_{F_lC} \end{pmatrix}. \tag{12.17}$$

We can also write W block-wise as $W^T = (W_{F_1}^T, W_{F_2}^T, \ldots, W_{F_l}^T)$. Then (12.7) takes the form

$$\begin{cases} A_{F_1F_1} W_{F_1} + A_{F_1F_2} W_{F_2} + \cdots + A_{F_1F_l} W_{F_l} + A_{F_1C} = 0, \\ A_{F_2F_1} W_{F_1} + A_{F_2F_2} W_{F_2} + \cdots + A_{F_2F_l} W_{F_l} + A_{F_2C} = 0, \\ \vdots \\ A_{F_lF_1} W_{F_1} + A_{F_lF_2} W_{F_2} + \cdots + A_{F_lF_l} W_{F_l} + A_{F_lC} = 0. \end{cases} \tag{12.18}$$

To define the entries of W_{F_k} via multipass interpolation, we use the following steps.

1 For $k = 1$, use direct interpolation to approximate W_{F_1}. More precisely, we write the first equation in (12.18) as

$$A_{F_1F_1} W_{F_1} + \hat{A}_{F_1C} = 0,$$

with $\hat{A}_{F_1C} = A_{F_1F_2}W_{F_2} + \cdots + A_{F_1F_l}W_{F_l} + A_{F_1C}$. Then we apply direct interpolation and write W_{F_1} as a function of W_{F_j}, $j > 1$.

2 While $k < l$:

(a) Write the $(k+1)$st equation in (12.18) substituting the expressions for W_{F_m}, $m < (k+1)$ obtained from the previous steps. The $(k+1)$th equation then has the form

$$\hat{A}_{F_{k+1}F_{k+1}}W_{F_{k+1}} + \hat{A}_{F_{k+1}F_{k+2}}W_{F_{k+2}} + \cdots \hat{A}_{F_{k+1}F_l}W_{F_l} + \hat{A}_{F_{k+1}C} = 0. \tag{12.19}$$

(b) Apply direct interpolation to (12.19) to write $W_{F_{k+1}}$ as a function of W_{F_j}, $j > (k+1)$.

(c) Set $k \leftarrow k+1$.

From the derivation and the definitions above, we have the following set inclusions describing the sparsity patterns of the prolongations defined earlier:

$$\mathrm{S}(P) \subset \mathrm{S}\begin{pmatrix} A_{FC} \\ I \end{pmatrix} \qquad \text{(direct interpolation (12.3.1))},$$

$$\mathrm{S}(P) = \mathrm{S}\begin{pmatrix} \widehat{A}_{FC} \\ I \end{pmatrix} \qquad \text{(standard interpolation (12.16))},$$

$$\mathrm{S}(P) = \mathrm{S}\begin{pmatrix} \widehat{A}_{F_1C} \\ \widehat{A}_{F_2C} \\ \vdots \\ \widehat{A}_{F_lC} \\ I \end{pmatrix} \qquad \text{(multipass interpolation)}.$$

12.3.2. Interpolation preserving a given vector

From the above discussions, it is important that the prolongation P preserves some vectors which can represent algebraic smooth errors. Both direct and standard interpolation use a diagonal scaling to make sure that the prolongation preserves constant vectors, which is the kernel of scalar elliptic operators. To generalize this idea, here we introduce αAMG interpolation, which constructs the prolongation matrix by choosing an initial guess that preserves the near-null component $v^{(1)}$, which may not necessarily be constant vectors.

We construct $P \in \mathcal{P}_C$ such that it 'preserves' a given vector v, namely,

$$v_F = Wv_C.$$

To do this, we first pick an initial guess W^0 for W (or P),

$$W^0 = D_v^{-1}A_{FC}, \tag{12.20}$$

where D_v is a diagonal matrix such that the following identity holds:

$$v_F = D_v^{-1} A_{FC} v_C.$$

It is easy to derive that the explicit formula for diagonal entries of D_v is

$$d_{kk} = \frac{\sum_{j\in C} a_{kj} v_j^{(1)}}{v_k^{(1)}}.$$

Then the W in this 'vector-preserving' interpolation is obtained by applying one Jacobi iteration for the linear problem (12.7) with initial guess W^0:

$$W = D^{-1} A_{FC} + D_{FF}^{-1}(-A_{FC} - A_{FF} D^{-1} A_{FC}).$$

A fully detailed description of the construction of a prototype vector v and coarse space interpolating this vector exactly, using αAMG (the classical AMG version of adaptive AMG), can be found in Brezina *et al.* (2006). The ideas, however, were outlined earlier in Brandt *et al.* (1982).

12.4. Classical AMG within the abstract AMG framework

Classical AMG falls within the abstract theory developed in Sections 5 and 6. To do this, we first consider an *M-matrix relative* of A using the adjacency graph corresponding to a strength function. We then use a maximal independent set algorithm to identify C, the set of coarse points, to form a C/F splitting. We further split the set of indices $\Omega = \{1, 2, \dots, n\}$ into subsets $\Omega_1, \Omega_2, \dots, \Omega_J$ so that

$$\Omega = \bigcup_{j=1}^{J} \Omega_j. \tag{12.21}$$

Then, for each $j \in C$, we define

$$\Omega_j := \{j\} \bigcup F_j^s, \quad j = 1, \dots, J. \tag{12.22}$$

where $F_j^s := \mathcal{F} \bigcap s_j$ and s_j is the set of strong neighbours of j. This depends on the definition of connection strength. For example, in the direct interpolation we introduced in the previous section, we simply use the strength connection defined in (8.8); in the standard interpolation we use the (m, l)-strong connection defined in Definition 9.7, with $m = 1$ and $l = 2$.

For each Ω_j we let

$$\Omega_j = \{m_1, m_2, \dots, m_{n_j}\}, \tag{12.23}$$

and let $n_j := |\Omega_j|$, namely, n_j is the cardinality of Ω_j. In accordance with the notation in Section 6, we then define

$$V_j := \mathbb{R}^{n_j}, \tag{12.24}$$

and the associated operator $\Pi_j : V_j \mapsto V$,

$$(\Pi_j v)_i = \begin{cases} p_{m_k,k} v_k & \text{if } i = m_k, \\ 0 & \text{if } i \notin \Omega_j, \end{cases} \tag{12.25}$$

where $p_{m_k,k}$ are given weights. As all the constructions below will be based on the M-matrix relative of A, without loss of generality, we use A to denote the M-matrix relative.

Following Section 6, we introduce the operator $\chi_j : V \mapsto V_j$,

$$(\chi_j v)_i := v_{m_i}, \tag{12.26}$$

which takes as argument a vector v and returns only the portion of it with indices in Ω_j. That is, $\chi_j v$ is a vector in $\mathbb{R}^{n_j}$. It is immediate to verify that

$$\sum_{j=1}^{J} \Pi_j \chi_j = I.$$

To estimate the convergence rate using the theory in Section 5, we need to verify all the items in Assumptions 6.4. To do this, we choose a decomposition $v = \sum_{j=1}^{J} \Pi_j v_j$ with $v_j = \chi_j v$. We further define $C_{p,2}$ as a constant depending on the overlaps in the partition $\{\Omega_j\}_{j=1}^{J}$,

$$C_o = \max_{1 \le j \le J} |\{l : \Omega_l \cap \Omega_j \neq \emptyset\}|. \tag{12.27}$$

The local operators A_j on V_j are defined as

$$(A_j u, v) = \sum_{\substack{e \in \mathcal{E} \\ e \subset \Omega_j}} \omega_e \delta_{j,e} u \delta_{j,e} v. \tag{12.28}$$

Here, $e \subset \Omega_j$ means the two vertices connected by e are in Ω_j, and $\delta_{j,e} u = u_{m_k - m_l}$ for $e = (k, l)$ and ω_e are the weights determined by the off-diagonal elements in the M-matrix relative of A. Note that A_j is a symmetric positive semi-definite matrix because all the weights ω_e are non-negative. Then (6.10) easily verified:

$$\begin{aligned} \sum_{j=1}^{m_c} \|\chi_j v\|_{A_j}^2 &= \sum_{j=1}^{m_c} \sum_{\substack{e \in \mathcal{E} \\ e \subset \Omega_j}} \omega_e (\delta_e v)^2 \le C_o \sum_{e \in \mathcal{E}} \omega_e (\delta_e v)^2 \\ &= C_o \|v\|_{A_M}^2 \le C_o C_M \|v\|_A^2. \end{aligned} \tag{12.29}$$

If D is the diagonal of A, then we set D_j, $j = 1 : J$ to be the restriction of D on Ω_j, namely, in $\mathbb{R}^{n_j \times n_j}$ and

$$(D_j)_{ii} = D_{m_i, m_i}, \quad \text{or equivalently } D_j = \chi_j D_j \chi_j' \tag{12.30}$$

We have the following lemma, which shows (6.9).

Lemma 12.4. For D_j defined in (12.30), the following inequality holds:

$$\left\|\sum_{j=1}^{m_c} \Pi_j w_j\right\|_D^2 \le C_o \sum_{j=1}^{m_c} \|w_j\|_{D_j}^2, \quad \text{for all } w_j \in V_j. \tag{12.31}$$

Proof. Recall from the definition of Π_j that we have

$$\|\Pi_j v\|_D \le \|v\|_{D_j}, \quad \text{for all } v \in V_j. \tag{12.32}$$

Therefore,

$$\begin{aligned}\left\|\sum_{j=1}^{J} \Pi_j v_j\right\|_D^2 &= \left(D\sum_{i=1}^{J} \Pi_i v_i, \sum_{j=1}^{J} \Pi_j v_j\right) = \sum_{i=1}^{J}\sum_{j=1}^{J}(D\Pi_i v_i, \Pi_j v_j) \\ &= \sum_{\substack{1\le i,j\le J\\ \Omega_i\cap\Omega_j\ne\emptyset}} (D\Pi_i v_i, \Pi_j v_j) \le \sum_{\substack{1\le i,j\le J\\ \Omega_i\cap\Omega_j\ne\emptyset}} \frac{\|\Pi_j v_i\|_D^2 + \|\Pi_j v_j\|_D^2}{2} \\ &\le C_o \sum_{j=1}^{J} \|v_j\|_{D_j}^2.\end{aligned}$$

□

We choose the local coarse spaces V_j^c as

$$V_j^c := \operatorname{span}\{\mathbf{1}_{n_j}\}. \tag{12.33}$$

Then, by definition, we have

$$\mu_j(V_j^c) = \frac{1}{\lambda_j^{(2)}}, \tag{12.34}$$

where $\lambda_j^{(2)}$ is the second smallest eigenvalue of the matrix $D_j^{-1}A_j$. The global coarse space V_c is then obtained by (6.12), and is

$$V_c = \operatorname{span}\{P_1, P_2, \dots, P_J\}. \tag{12.35}$$

Finally, by Theorem 5.3, the convergence rate of this two-level geometric multigrid method depends on the $\min_j(\lambda_j^{(2)})$. If the discrete Poincaré inequality is true for each V_j, namely,

$$\inf_{v_c\in V_j^c} \|v - v_c\|_{D_j}^2 \le c_j\|v\|_{A_j}^2, \quad \text{for all } v \in V_j, \tag{12.36}$$

with c_j a constant, then the two-level classical AMG method converges uniformly.

12.5. Bibliographical notes

The classical coarse space definition in AMG was introduced by Brandt *et al.* (1982), and then somewhat improved by Stüben (1983), Brandt *et al.* (1985) and Ruge and Stüben (1987).

Despite its great success in practical applications, classical AMG algorithms still lack solid theoretical justifications beyond the theory for two-level methods. It is important to note that a multigrid method that converges uniformly in the two-level case with an exact coarse grid solver may not converge uniformly in the multilevel case: see Brandt (1986) and Ruge and Stüben (1987). For classical AMG, the early theoretical studies of convergence date back to the 1980s: see Mandel (1988), McCormick (1982), Brandt (1986) and Ruge and Stüben (1987). A survey of Stüben's results can be found in the monograph by Trottenberg *et al.* (2001), who also gives the three classical prolongation constructions (direct, standard and multipass). In all cases, it is crucial to define coarsening and interpolation operators so that the interpolation error is uniformly bounded. The role of the ideal interpolation as the minimizer of an upper bound for the convergence rate was emphasized by Falgout and Vassilevski (2004).

13. Aggregation-based AMG

Aggregation (or agglomeration) refers to an algorithm that splits the set of vertices of the graph of the filtered matrix as a union of non-overlapping subsets (aggregates) (each of which forms a connected sub-graph):

$$\{1,\ldots,n\} = \bigcup_{j=1}^{J} \mathcal{A}_j, \quad \mathcal{A}_i \bigcap \mathcal{A}_j = \emptyset, \ i \neq j. \tag{13.1}$$

Such a partition can be obtained using the algorithms described in Section 9.

If we are solving a finite element system, the partition (13.1) would correspond to a non-overlapping decomposition of Ω,

$$\Omega = \bigcup_{j=1}^{J} \Omega_j, \quad \Omega_i \bigcap \Omega_j = \emptyset, \ i \neq j, \tag{13.2}$$

such that $\mathcal{A}_j$ contains the indices associated with the enumerations of the vertices in the subdomain Ω_j.

We denote the elements in $\mathcal{A}_j$ by

$$\mathcal{A}_j = \{m_1, m_2, \ldots, m_{n_j}\}, \tag{13.3}$$

and let $n_j := |\mathcal{A}_j|$, that is, n_j is the number of elements in $\mathcal{A}_j$.

13.1. Unsmoothed aggregation: preserving one kernel vector

Using the framework introduced in Section 6, we define

$$V_j := \mathbb{R}^{n_j},$$

and the associated operator $\Pi_j : \mathbb{R}^{n_j} \mapsto \mathbb{R}^n$ is the trivial extension of $v \in \mathbb{R}^{n_j}$:

$$(\Pi_j v) = \begin{cases} v_k & i = m_k, \\ 0 & i \notin \mathcal{A}_j, \end{cases} \tag{13.4}$$

$$P = (p_1, p_2, \ldots, p_J), \quad p_j = \Pi_j \mathbf{1}_{n_j}. \tag{13.5}$$

This prolongation matrix obviously satisfies

$$P\mathbf{1}_J = \sum_{j-1}^{J} P_j = \mathbf{1}_n. \tag{13.6}$$

The local coarse space V_j^c is

$$V_j^c := \text{span}\{\mathbf{1}_{n_j}\}.$$

Then the global coarse space V_c is obtained by (6.12). In fact

$$V_c = \text{span}\{p_1, p_2, \ldots, p_J\}.$$

We note that

$$(\phi_{1,c}, \ldots, \phi_{J,c}) = (\phi_1, \ldots, \phi_n)P, \tag{13.7}$$

where

$$\phi_{j,c} = \sum_{k \in \mathcal{A}_j} \phi_k, \quad j = 1, 2, \ldots, J. \tag{13.8}$$

Further, Π_j corresponds to matrix representation of the operator $I_h(\phi_j^H \cdot)$ with the coarse grid basis $\phi_{j,c}$ defined above. In view of Section 10.3 (see especially (10.14)), aggregation-based AMG can also be viewed as a GMG method.

The above procedure gives a full description of the unsmoothed aggregation AMG method. We can use the framework in Section 6 to carry out a two-level convergence analysis. The local matrices A_j are defined in exactly the same way as in the case of classical AMG in Section 12.4. Thus we write A as in (12.4) and then define A_j by (12.28). The matrices D_j are defined as the restriction of the diagonal of A to Ω_j, as in (12.30). Using the same argument as in Section 12.4, Assumption 6.4 is satisfied by (12.29) and Lemma 12.4. Theorem 5.3 can then be applied to prove that the two-level unsmoothed aggregation method has a convergence rate depending only on the local Poincaré constants in (12.36).

13.2. Unsmoothed aggregation: preserving multiple vectors

One great advantage of the aggregation AMG method is that it can be easily extended to the case when the stiffness matrix A has multiple kernel or near-kernel vectors.

To give an illustration of how to construct prolongation preserving more than one vector, we consider two-dimensional elasticity problems. In this case we have three functions, namely the rigid body motion, to preserve

$$u_1 = \begin{pmatrix} 1 \\ 0 \end{pmatrix}, \quad u_2 = \begin{pmatrix} 0 \\ 1 \end{pmatrix}, \quad u_3 = \begin{pmatrix} -y \\ x \end{pmatrix}.$$

The corresponding vectors are

$$\zeta_1 = \begin{pmatrix} 1 \\ 0 \\ 1 \\ 0 \\ \vdots \\ 1 \\ 0 \end{pmatrix}, \quad \zeta_2 = \begin{pmatrix} 0 \\ 1 \\ 0 \\ 1 \\ \vdots \\ 0 \\ 1 \end{pmatrix} \quad \text{and} \quad \zeta_3 = \begin{pmatrix} -y_1 \\ x_1 \\ -y_2 \\ x_2 \\ \vdots \\ -y_n \\ x_n \end{pmatrix} \in \mathbb{R}^{2n}.$$

On each aggregate $\mathcal{A}_j$, we consider

$$\zeta_1^{(j)} = \begin{pmatrix} 1 \\ 0 \\ 1 \\ 0 \\ \vdots \\ 1 \\ 0 \end{pmatrix}, \quad \zeta_2^{(j)} = \begin{pmatrix} 0 \\ 1 \\ 0 \\ 1 \\ \vdots \\ 0 \\ 1 \end{pmatrix} \quad \text{and} \quad \zeta_3^{(j)} = \begin{pmatrix} -y_{m_1} \\ x_{m_1} \\ -y_{m_2} \\ x_{m_2} \\ \vdots \\ -y_{m_{n_j}} \\ x_{m_{n_j}} \end{pmatrix} \in \mathbb{R}^{2n_j}.$$

The prolongation matrix is then given by

$$P = (p_1, p_2, \dots, p_J) \in \mathbb{R}^{(2n)\times(3J)}, \quad P_j = \Pi_j(\zeta_1^{(j)}, \zeta_2^{(j)}, \zeta_3^{(j)}) \in \mathbb{R}^{(2n)\times 3}.$$

This prolongation matrix satisfies

$$P(\mathbf{1}_J \otimes e_j) = \zeta_j, 1 \le j \le 3. \tag{13.9}$$

The rest of the AMG algorithm based on this prolongation matrix is similar to the 1-vector case.

Extension from the 3-vector case, as mentioned above, to arbitrary m-vectors is straightforward. We let $m = \dim(N(A))$, and consider the case when $m \ge 1$. We now assume that we are given m vectors $\{\zeta_j\}_{j=1}^m$ which are linearly independent and $m \ll n$, and we can then proceed to construct a prolongation P such that $P(\mathbf{1}_{n_c} \otimes e_j) = \zeta_j$, $j = 1 : m$.

We would like to point out that for the multiple kernel or near-kernel vector case we often need some geometric information to describe the corresponding vectors. In the finite element case, these vectors should be obtained by taking the canonical interpolation of the corresponding kernel functions of the underlying partial differential operators, and we then split

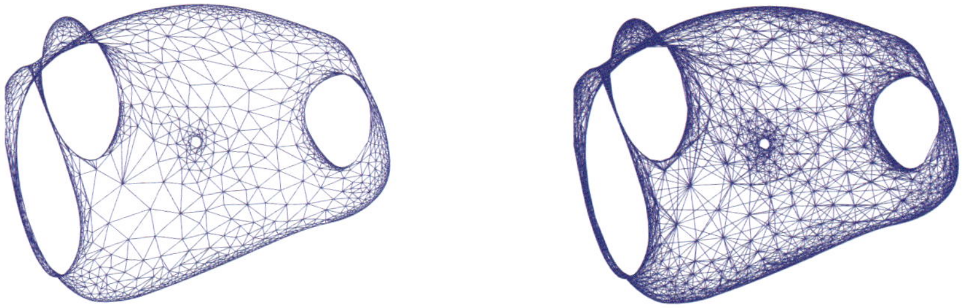

Figure 13.1. Graph of the coarse grid matrix corresponding to the unsmoothed aggregation (a) and the 'denser' graph for the coarse grid matrix obtained by smoothed aggregation (b).

these vectors into different aggregates by using the local degrees of freedom associated with different aggregates.

13.3. Smoothed aggregation

The aggregation procedure described above provides a simple yet efficient AMG method. The resulting method is known as the unsmoothed aggregation AMG (UA-AMG) method. This terminology may be justified by examining the shape of the basis function as defined by (13.8). A typical basis of this form is not smooth. There is a procedure to smooth out these basis function by using a smoother, which is equivalent to a few applications of smoothing on the prolongation matrix (13.5) as follows:

$$P_S = (I - R_s A)^\nu P, \quad \text{for some } \nu \geq 1. \tag{13.10}$$

A typical choice is the scaled Jacobi smoother with $R_S = \omega D^{-1}$. We note that, in view of (13.9), if ζ_j is in the kernel of A, we still have

$$P_S(\mathbf{1}_J \otimes e_j) = \zeta_j, 1 \leq j \leq m.$$

Conceivably, a bigger ν in (13.10) would lead to an AMG algorithm that has a better convergence rate. However, a bigger ν in (13.10) also means a denser P_S and hence a denser A_c, and a more expensive setup for the resulting AMG algorithm. As an example, Figure 13.1 shows the graphs of the coarse grid matrices corresponding to unsmoothed and smoothed aggregation. Clearly the smoothed aggregation graph is denser than the unsmoothed one.

In view of Lemma 9.3, the basis function (13.8) and the resulting UA-AMG corresponds to the use of a tentative coarse space W_c. The smoothed basis functions by means of (13.10) and resulting SA multigrid corresponds to the use of the V_s in Lemma 9.3 with $S = (I - R_s A)^\nu$.

13.4. Bibliographical notes

One of the first aggregation algorithms comes from applications in economics, and is due to Vakhutinsky *et al.* (1979). Subsequently, aggregation methods were developed and used by Blaheta (1986) to solve discrete elliptic problems, and by Marek (1991) for the calculation of stationary points of Markov chains.

A special class of aggregation coarsening methods is based on matching in graphs (also known as pairwise aggregation), and such methods were employed in Kim *et al.* (2003) and for non-symmetric problems in Kim, Xu and Zikatanov (2004). Aggregations using matching in graphs were further used by Notay (2010), Brannick, Chen and Zikatanov (2012), Livne and Brandt (2012) and D'Ambra and Vassilevski (2013, 2014). To improve the scalability of the unsmoothed aggregation, a number of algorithms have been developed: Kim *et al.* (2003) used a variable V-cycle, Notay and Vassilevski (2008) proposed non-linear Krylov subspace acceleration, and Olson *et al.* (2011) designed a procedure for correcting the energy of the coarse-level Galerkin operator.

Aggregation, especially the unsmoothed aggregation, can be used in conjunction with non-linear (variable-step/flexible) preconditioning methods to result in an optimal algorithm. Such non-linear methods are called *algebraic multilevel iteration methods*, and were introduced by Axelsson and Vassilevski (1991). Non-linear multilevel preconditioners were proposed and an additive version of them was first analysed by Axelsson and Vassilevski (1994); see also Golub and Ye (1999/2000), Notay (2000), Saad (2003) and Kraus (2002). In these non-linear multilevel preconditioners, n steps of a preconditioned CG iterative method provide a polynomial approximation of the inverse of the coarse grid matrix. The same idea can be used to define the MG cycles, as shown by Vassilevski (2008) and Notay and Vassilevski (2008). A comprehensive convergence analysis of the non-linear AMLI-cycle multigrid method for symmetric positive definite problems was conducted by Hu, Vassilevski and Xu (2013). Based on classical assumptions for approximation and smoothing properties, the non-linear AMLI-cycle MG method is shown to be uniformly convergent.

The smoothed aggregation AMG method, first developed by Míka and Vaněk (1992*a*, 1992*b*) and later extended by Vaněk, Mandel and Brezina (1996), was motivated by some early work on aggregation-based MG studied by Blaheta (1986, 1988).

A major work on the theory and practice of the SA algorithm is found in Vaněk, Mandel and Brezina (1998). The convergence result proved there assumes the sparsity of the coarse grid matrix and a certain ratio between the number of coarse and fine degrees of freedom. For general sparse matrices, and even for general adapted finite element matrices corresponding to elliptic equations, it is difficult to verify such assumptions, and this is yet to be done.

14. Problems with discontinuous and anisotropic coefficients

A good AMG method should be robust with respect to possible heterogeneity features such discontinuous jumps and anisotropy. These heterogeneities should be detected and properly resolved automatically in an AMG process. Extensive numerical experiments have shown that AMG such as classical AMG and SA-AMG are very robust with respect to these heterogeneities. One main technique used to detect and resolve these heterogeneities is by means of the connection strength (see Section 8.2). In this section we shall use the model problem (2.1) with two special sets of coefficients, (2.9) and (2.10), to discuss how classical AMG works for problems with strong heterogeneities.

14.1. Jump coefficients

In this section we consider the classical AMG method when applied to a problem with heterogeneous (jump) coefficients, namely (2.1) with (2.9). We begin with a discussion of how the connection strength is used to define the sparsity pattern of the prolongation.

The connection strength function was introduced to handle cases such as jump coefficients and anisotropies in the matrices corresponding to discretizations of scalar PDEs. An important observation regarding classical AMG is that the prolongation matrix P, which defines the basis in the coarse space, uses only strong connections. Here we need the *strength operator* $S : V \mapsto V$ defined in (8.4). We now focus on the jump coefficient problem defined in (2.9),

$$\alpha(x) = \begin{cases} \epsilon & x \in \Omega_\epsilon, \\ 1 & x \in \Omega_1, \end{cases}$$

where ϵ is sufficiently small that the graph corresponding to the strength operator has at least two connected components. Directly from the definition of S_i in (8.8), we have the following.

- The graph corresponding to the connection strength matrix S is obtained from the graph corresponding to A by removing all entries a_{ij} corresponding to an edge connecting a vertex from the interior of Ω_ϵ to a vertex in $\overline{\Omega}_1$.
- In other words, in this setting we have a *block lower triangular* S with at least two blocks, in which the first block corresponds to the vertices interior to Ω_ϵ and the second block corresponds to the rest of the vertices.

- The considerations above apply to any configuration of the subsets Ω_ϵ and Ω_1. In fact, they can even be disconnected, forming cross-points and so on.

In Figures 14.1–14.3 we illustrate the connection strength graphs and their connected components for specific coefficients $\alpha(x)$, which is defined as in (2.9) with $\epsilon = 10^{-3}$. The domain Ω_ϵ is a union of elements, and $T \in \Omega_\epsilon$ if and only if $(x_T - 1/2)(y_T - 1/2) < 0$, where (x_T, y_T) is the barycentre of T. The domain Ω_1 is $\Omega_1 = \Omega \setminus \overline{\Omega}_1$. In Figure 14.2, Ω_ϵ and Ω_1 are disconnected in the graph of the strength matrix S. In the graph of S, Ω_ϵ is split into two connected components, denoted by $\Omega_\epsilon^{(1)}$ and $\Omega_\epsilon^{(2)}$; Ω_1 is split into two parts, denoted by $\Omega_1^{(1)}$ and $\Omega_1^{(2)}$, which are connected via only one point, $x_0 = (1/2, 1/2)$. With a reordering of indices, the strength matrix for this case can be written as

$$S = \begin{pmatrix} S_{\epsilon,1} & & & & \\ & S_{\epsilon,2} & & & \\ & & S_{11} & & \\ & & & S_{12} & \\ & & S_{x1} & S_{x2} & S_x \end{pmatrix},$$

where $S_{\epsilon,1}$ and $S_{\epsilon,2}$ are two diagonal blocks corresponding to the two connected components $\Omega_\epsilon^{(1)}$ and $\Omega_\epsilon^{(2)}$, S_{11} and S_{12} are two diagonal blocks corresponding to the two connected components $\Omega_1^{(1)}$ and $\Omega_1^{(2)}$, S_0 is the diagonal entry corresponding to the grid point x_0, and S_{x1} and S_{x2} are low-rank matrices with only a few non-zero entries (two non-zero entries in this example), which contain the connections from x_0 to points in $\Omega_1^{(1)}$ and $\Omega_1^{(2)}$ respectively.

We define an operator $A_S : V \mapsto V'$ using A and S:

$$(A_S u, v) := \sum_{S_{ij} \neq 0} \omega_{ij}(u_i - u_j)(v_i - v_j),$$

where $\omega_{ij} = |a_{ij}|$. We have the following lemma.

Lemma 14.1. For all $v \in V$, we have

$$(A_S v, v) \leq (A_+ v, v) \lesssim (Av, v).$$

For classical AMG, the definition of the prolongation matrix P only uses strong connections, and the coarse space is the union of the coarse spaces in each strongly connected component of the graph of A_+. Note that the classical AMG construction gives the same coarse space for both A and its M-matrix relative A_+. For this reason, in the considerations that follow we write A instead of A_+ and $\mathcal{E}$ instead of $\mathcal{E}_+$, as the coarse space is defined using only entries from A_+.

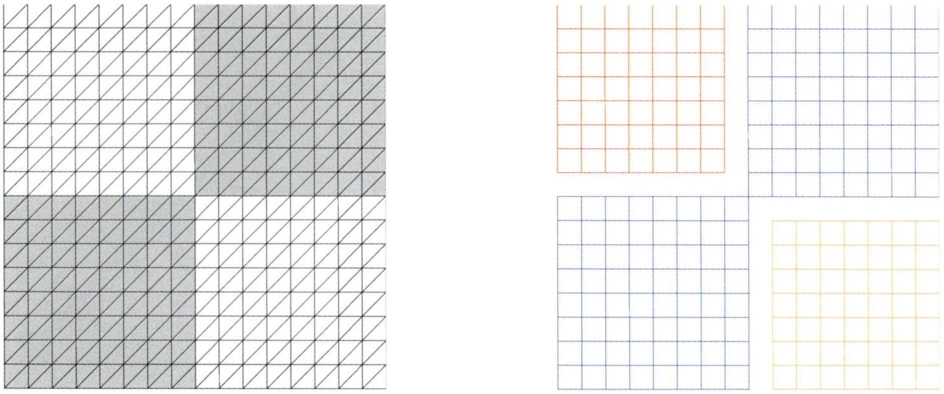

(a) (b)

Figure 14.1. Jump coefficient problem on a uniform mesh. (a) The coefficient is 1 in the shaded region, and 10^{-3} in the blank region. (b) The strongly connected components in the graph correspond to the strength matrix S.

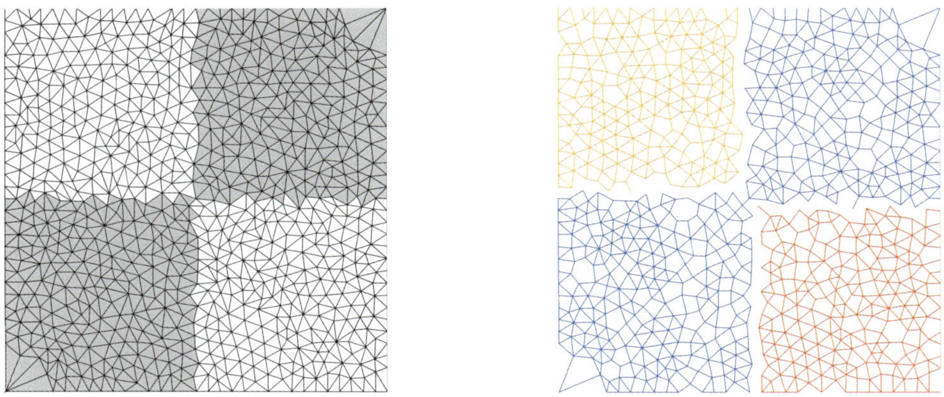

(a) (b)

Figure 14.2. Jump coefficient problem on an unstructured mesh. (a) The coefficient is 1 in the shaded region, and 10^{-3} in the blank region. (b) The strongly connected components in the graph correspond to the strength matrix S.

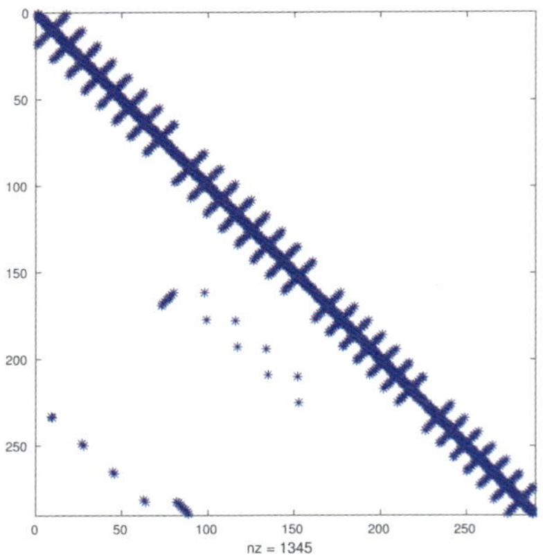

(a)

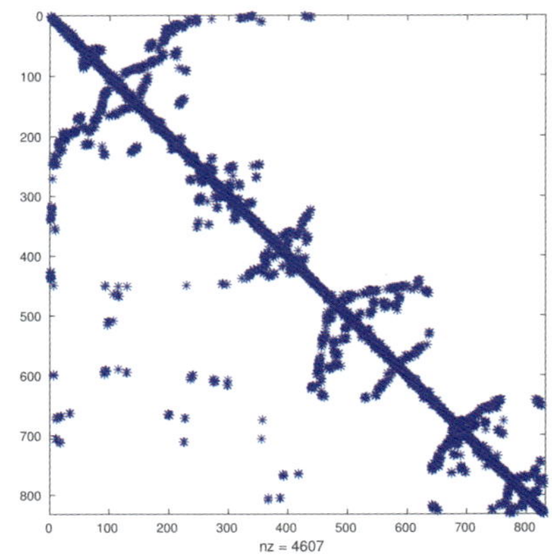

(b)

Figure 14.3. Jump coefficient problem: non-zero pattern of strength matrix S when reordered in block lower triangular form. (a) Structured uniform grid, and (b) unstructured mesh.

We now consider the convergence of classical two-level AMG with the standard interpolation for the jump coefficient problem, and we prove a uniform convergence result for the two-level method. The same result for direct interpolation can be obtained via slight modification of the proof for the standard interpolation case. Before we go through the AMG two-level convergence proof, we first introduce the following result on a connected graph, which can be viewed as a discrete version of the Poincaré inequality.

Lemma 14.2. We consider the following graph Laplacian on a connected undirected graph $\mathcal{G} = (\mathcal{V}, \mathcal{E})$:

$$\langle Au, v\rangle = \frac{1}{2} \sum_{(i,j)\in\mathcal{E}} (u_i - u_j)(v_i - v_j). \tag{14.1}$$

For any $v \in V$, the following estimate is true:

$$\|v - v_c\|_{\ell^2}^2 \leq \mu n^2 d \langle Av, v\rangle, \tag{14.2}$$

where $n = |\mathcal{V}|$ is the size of the graph, $v_c = \sum_{j=1}^n w_j v_j$ is a weighted average of v, $\mu = \sum_{j=1}^n w_j^2$, and d is the diameter of the graph.

We now consider classical AMG coarsening as defined in Section 12.4, and with an abuse of notation we use Ω_j to denote the set of vertices defined via the C/F splitting in (12.21). The next lemma is a spectral equivalence result, showing that the local operators A_j, defined in (12.28), for a shape-regular mesh, are spectrally equivalent to a scaling of the graph Laplacian operators $A_{L,j}$ defined as

$$(A_{L,j}u, v) = \frac{1}{2} \sum_{(i,k)\in\Omega_j} (u_i - u_k)(v_i - v_k). \tag{14.3}$$

Lemma 14.3. With the assumption we made on the shape-regularity of the finite element mesh, the following inequalities hold for A_j defined as in (12.28) using the standard interpolation:

$$c_L h^{d-2}\langle A_{L,j}v_j, v_j\rangle \le (A_j v_j, v_j) \le c^L h^{d-2}\langle A_{L,j}v_j, v_j\rangle, \tag{14.4}$$

where $A_{L,j}$ is a graph Laplacian defined in (14.1) on the graph $\mathcal{G}_j$, h is the mesh size, and c_L, c^L are constants depending on the shape-regularity constant and the threshold θ for the connection strength.

Proof. By definition of connection strength, we have

$$a_{ii} = \sum_{k\in N_k} -a_{ik} \le -\frac{|N_i|}{\theta} a_{ij}.$$

Since A is symmetric, we also have

$$a_{ii} \le -\frac{|N_i|}{\theta} a_{ji}.$$

By the definition of Ω_j in the standard interpolation, for any $i \in \Omega_j \setminus \{j\}$, either $i \in F_j^s$ or there exists a $i \in F_j^s$ such that $i \in F_k^s$. For the latter, (j, k, i) forms a path between j and i going along strong connections. We then have

$$-a_{ik} \ge -\frac{\theta}{|N_k|} a_{kk} \ge -\frac{\theta}{|N_k|} a_{kj} \ge -\frac{\theta^2}{|N_k||N_j|} a_{jj}$$

and

$$a_{jj} \ge -a_{jk} \ge \frac{|N_k|}{\theta} a_{kk} \ge -\left(\frac{|N_k|}{\theta}\right)^2 a_{ik}.$$

Combining the above two inequalities and using the assumption that the mesh is shape-regular, for any $l \in \Omega_j$ that is connected to i we have

$$\sigma_1 a_{ij} \le -a_{il} \le \sigma_2 a_{jj},$$

with constants σ_1 and σ_2 depending on the shape-regularity constant and θ.

Since in the definition of A_j in (12.28), $\omega_e = -a_{ij}/2$ for $e = (i,j)$, we obtain

$$c_1 a_{jj}\langle A_{L,j}v_j, v_j\rangle \le (A_j v_j, v_j) \le c_2 a_{jj}\langle A_{L,j}v_j, v_j\rangle. \tag{14.5}$$

Then, by a scaling argument, $a_{jj} \eqsim h^{d-2}$ and the proof is complete. $\square$

Theorem 14.4. The two-level method using a coarse space V_c defined via classical AMG is uniformly convergent.

Proof. By Theorem 5.3, we only need to show that μ_c is bounded, which can be easily obtained by combining Lemmas 14.2–14.3 with Lemma 7.3. □

14.2. Anisotropic problem

We consider the following problem:

$$\begin{cases} -u_{xx} - \varepsilon u_{yy} = f & \text{in } \Omega, \\ \dfrac{\partial u}{\partial n} = 0 & \text{on } \partial\Omega. \end{cases} \tag{14.6}$$

We discretize the problem using the finite element method on an $n \times n$ uniform triangular grid in $\Omega = (0,1) \times (0,1)$. We order the vertices of the triangulation lexicographically and denote them by $\{(ih, jh)\}_{i,j=0}^{n}$. Then the stiffness matrix is

$$A = \operatorname{diag}(-\varepsilon I, B, -\varepsilon I) \quad \text{with } B = \operatorname{diag}(-1, 2(1+\varepsilon), -1). \tag{14.7}$$

It is immediate to see that, for sufficiently small $0 < \varepsilon \ll 1$, the strength operator S takes the form

$$S = \operatorname{diag}(0, S_B, 0) \quad \text{with } S_B = \operatorname{diag}(1,1,1), \tag{14.8}$$

and the M-matrix relative of A is

$$A_+ = \operatorname{diag}(0, B_+, 0) \quad \text{with } B_+ = \operatorname{diag}(-1, 2, -1). \tag{14.9}$$

If we use uniform coarsening to solve the above linear problem, it has been proved by Yu, Xu and Zikatanov (2013) (see also Zikatanov 2008) that uniform convergence is not achieved when using point relaxation as a smoother and standard coarsening. A fix for this is to use connection strength and coarsen the adjacency graph of the strength operator S. The C/F splitting constructed in this way using the MIS algorithm (Algorithm 5) from Section 9.3.1 results in semi-coarsening (coarsening in only one direction).

We now move on to show uniform convergence of two-level classical AMG for the anisotropic problem (14.6). Recall that we consider a uniform grid in $\mathbb{R}^2$ with lexicographical ordering of the vertices. Then the stiffness matrix is (14.7). We further assume, without loss of generality, that $n = 2m+1$ for some m. The coarse grid points using the strength operator defined in (14.8) are then with coordinates $((2i)h, jh)$, where $i = 1:m$ and $j = 1:n$.

Further, as each coarse grid function is uniquely determined by its values at the coarse points, the function that corresponds to the point $((2i)h, jh)$ for some i and j is defined as the unique piecewise linear function $\phi_{i,j,H}$ satisfying

$$\phi((2i)h, jh) = 1, \quad \phi((2i-1)h, jh) = 1/2, \quad \phi((2i+1)h, jh) = 1/2,$$

and $\phi_{i,j,H}(x) = 0$ at all other grid nodes.

Alternatively we can use the basis functions for bilinear elements, which can be written as tensor products. Let us first define the piecewise linear basis in one dimension:

$$\phi_{j,h}(t) = \begin{cases} \dfrac{(t-(j-2)h)}{h} & t \in ((j-2)h, (j-1)h), \\ \dfrac{(jh-t)}{h} & t \in ((j-1)h, jh), \\ 0 & \text{otherwise.} \end{cases}$$

The basis in V_h is then

$$\phi_{ij,h}(x,y) = \phi_{i,h}(x)\phi_{j,h}(y), \tag{14.10}$$

and for the coarse grid basis we have

$$\phi_{ij,H}(x,y) = \phi_{i,2h}(x)\phi_{j,h}(y). \tag{14.11}$$

The basis functions for the linear elements are formed by piecewise linear interpolation of the bilinear basis.

The subset Ω_{ij} is the support of this basis function, that is, the grid points where $\phi_{ij,H}$ is non-zero. More precisely,

$$\Omega_{ij} = \{((2i-1)h, jh), (2ih, jh), ((2i+1)h, jh)\}. \tag{14.12}$$

Thus Ω_{ij} consists of the coarse grid point $(2ih, jh)$ and its neighbours in the x-direction. Then we define $V_j := \mathbb{R}^3$.

The operator $\Pi_j : V_j \mapsto V$ is defined by the matrix representation of $I_h(\phi_{ij,H}\cdot)$. A_{ij} is defined as in (12.28). In this case

$$A_{ij} = \begin{pmatrix} 1 & -1 & 0 \\ -1 & 2 & -1 \\ 0 & -1 & 1 \end{pmatrix}. \tag{14.13}$$

Hence A_{ij} is symmetric positive semi-definite, and since for any $v_{ij} \in V_{ij}$ we have

$$\begin{aligned} \sum_{i,j} (v_{ij}, v_{ij})_{A_{ij}} &= \sum_{ij} \sum_{\substack{(k,l)\in\mathcal{E} \\ k,l\in\Omega_{ij}}} -a_{kl}(v_k - v_l)^2 \\ &\le 2 \sum_{(k,l)\in\mathcal{E}} -a_{kl}(v_k - v_l)^2 \\ &= 2(v,v)_A, \end{aligned}$$

A_{ij} satisfies (6.10).

We define D_{ij} as in (12.30). As $A_{ij}\mathbf{1} = \mathbf{0}$, the minimum eigenvalue of $D_{ij}^{-1}A_{ij}$ is 0 and the corresponding eigenvector is the constant vector, we choose the local coarse space V_{ij}^c to be the space of all constant vectors in V_{ij}. The corresponding global coarse space is then defined as in (6.12).

Theorem 14.5. The two-level method with coarse space defined above converges uniformly for the anisotropic problem in (14.6), with convergence rate independent of ε and the mesh size h.

Proof. By Theorem 5.3, the convergence rate depends on the second smallest eigenvalue of $D_{ij}^{-1}A_{ij}$, which is 1 for all i and j.

Next, Theorem 5.3 can be applied to this case, and we obtain

$$\|E\|_A \leq 1 - \frac{1}{C}, \tag{14.14}$$

with C independent of ε and the mesh size, which proves the uniform convergence of the AMG method. □

14.3. Bibliographical notes

Fast solvers for problems with heterogeneous and or anisotropic coefficients have been the focus of research for the last three or four decades. The AMG methods are among the preferred solvers due to their robust behaviour with respect to the coefficient variation and independence of the geometry. Standard multilevel methods for these problems do have limitations, as their convergence may deteriorate, as shown by Alcouffe, Brandt, Dendy and Painter (1981). The cause for this in geometric multigrid with standard coarse spaces was discussed by Bramble and Xu (1991) and Xu (1991), and later by Oswald (1999). Attempts to remove the dependence on the size of the coefficient jumps was made by introducing the matrix-dependent prolongations. We refer to Dendy (1982, 1983), Reusken (1993) and de Zeeuw (1990). Several theoretical and numerical results on geometric and algebraic methods for discontinuous coefficients can be found in the survey paper by Chan and Wan (2000) and the references therein. Anisotropic equations and AMG coarsening are considered by Brannick *et al.* (2006).

Finally, in addition to the theory presented here, some partial theoretical results on the convergence of AMG are found in the classical papers on AMG (Ruge and Stüben 1987) and smoothed aggregation (SA) (Vaněk *et al.* 1996). Related works are the frequency filtering and decompositions found in Hackbusch (1989), Wittum (1992), Wagner and Wittum (1997), Weiler and Wittum (1997) and Nägel, Falgout and Wittum (2008).

15. Bootstrap and adaptive AMG

In all the algorithms studied earlier, we assume that the near-null spaces are known in advance. Eliminating such an assumption and extending the range of applicability of optimal multigrid techniques is attempted in the framework of the bootstrap/adaptive AMG methods, which we describe in this section. In summary, the adaptive AMG (αAMG, αSA) and bootstrap AMG (BAMG) algorithms are intended for harder problems for which the

Algorithm 7 Prolongation via least-squares minimization

Input: Matrices $\Psi \in \mathbb{R}^{n\times m}$, $\Psi_c \in \mathbb{R}^{n_c\times n_c}$, a norm $\|\cdot\|$, initial sparsity pattern $\mathrm{S}(P)$ for P; bound on maximum non-zeros per row $s_{\max}$.
Output: Prolongation matrix P such that $P\Psi_c \approx \Psi$.
Set $i \leftarrow 1$.
while $(i \le n)$ **do**
  Find $\widetilde{p}_i \leftarrow \arg\min_{\widetilde{p}_i \in \mathbb{R}^{n_c}_{S_i}} \|\Psi_c^T \widetilde{p}_i - \widetilde{\Psi}_i\|$.
  if $(\|\Psi_c^T \widetilde{p}_i - \widetilde{\Psi}_i\| > \varepsilon)$ **and** $(|S(i)| \le s_{\max})$ **then**
    $S(i) \leftarrow S(i) \cup_{j_c} \{j_c\}$, where j_c are close to i in graph distance.
  else
    $i \leftarrow i+1$
  end if
end while
for $i = 1:n$, $j_c \in S(i)$ **do**
  if $(p_{i,j_c} \le \varepsilon_p)$ **then** Set $p_{i,j_c} \leftarrow 0$ **end if**
end for
return P

standard variants of the classical AMG method or the smoothed aggregation method may converge too slowly. The bootstrap/adaptive algorithms make special choices of coarse spaces. Based on a given smoother, they self-improve until they achieve the desired convergence rate.

15.1. Sparsity of prolongation matrices

Here we give a very short summary of the sparsity patterns of the prolongation matrices defined for energy-minimization AMG in Section 11, classical AMG in Section 12, and aggregation AMG in Section 13. First, for energy-minimizing AMG we have that the sparsity pattern of the prolongation could be prescribed in advance, and in some sense this approach is more general as it can also use the sparsity patterns given below for classical AMG and aggregation AMG approaches. For specific restrictions related to the number of vectors interpolated exactly by the energy-minimizing prolongation, we refer to Xu and Zikatanov (2004) and Vassilevski and Zikatanov (2006). There are also ways to define the prolongation with changing patterns. Such algorithms are useful because they provide a mechanism to control the sparsity pattern of the prolongation. Algorithm 7 was first described by Brandt (2002) and later included in the bootstrap adaptive AMG algorithm designed by Brandt, Brannick, Kahl and Livshits (2011). In the algorithm description, we refer to the graph of the M-matrix relative as defined in Section 9.

Algorithm 8 Prolongation via aggregation

Input: Matrix $\Psi \in \mathbb{R}^{n\times m}$, an aggregation $\cup_{i=1}^{n_c}\mathcal{A}_i = \{1,\dots,n\}$, and a prolongator smoother $S : \mathbb{R}^n \mapsto \mathbb{R}^n$ (for smoothed aggregation (SA)).
Output: Prolongation matrix P such that Range$(P\Psi_c) \approx$ Range(Ψ).
Set $(\Psi_{\mathcal{A}_i})_{kj} = \begin{cases} \Psi_{kj} & k \in \mathcal{A}_i, j = 1:m, \\ 0 & k \notin \mathcal{A}_i, j = 1:m. \end{cases}$
Set $P = (\Psi_{\mathcal{A}_1}, \dots, \Psi_{\mathcal{A}_{n_c}})$.
if (SA) **then** $P \leftarrow (I - R_s A)P$ **end if**
return P

To conclude this section, we point out that the algorithmic details for the construction of *direct*, *standard* and *multipass* interpolations are already described in detail in Section 12.

15.2. Notation

Given a matrix $A \in \mathbb{R}^{n\times n}$ and a relaxation (smoother) R for this matrix, we use an adaptive procedure to construct a sequence of coarse spaces which are characterized by the sequence of prolongation matrices

$$P_j^m : \mathbb{R}^{n_j^m} \mapsto \mathbb{R}^{n_{j+1}^m}.$$

The corresponding V-cycle matrix is denoted by B^m. We introduce the following notation, used throughout this section.

- For a matrix $Y \in \mathbb{R}^{n\times m}$, we set

$$Y = (y_1, \dots, y_m) = \begin{pmatrix} \widetilde{y}_1^T \\ \vdots \\ \widetilde{y}_n^T \end{pmatrix}.$$

  Clearly, $\{y_i\}_{i=1}^m$ are the columns of Y and $\{\widetilde{y}_j^T\}_{j=1}^n$ are the rows of Y.

- $\mathcal{V} = (V_1, \dots, V_J)$ is a multilevel hierarchy of spaces $V_j \subset V_{j+1}$, $j = 1 : (J-1)$ and the fine grid spaces are denoted by V_J.

- We let $\{P_{j-1}^j\}_{j=2}^J$ denote the prolongation matrix from a coarser level $(j-1)$ to finer level j, and $P_j = P_{J-1}^J P_{J-2}^{J-1} \cdots P_j^{j+1}$, $j = 1, \dots, (J-1)$ is the prolongation from level j to level J

- We let $\mathcal{P}_j = \{P_{j-1}^j\}_{j=2}^j$ and $\mathcal{P} = \mathcal{P}_J$ denote the set of all prolongations up to level j. As we have mentioned on several occasions, the set of prolongation matrices $\mathcal{P}$ completely determines the multilevel hierarchy of spaces $\mathcal{V}$.

- A set of test vectors on every level is available:

$$\mathcal{C} = \{\Psi_1, \dots, \Psi_J\}, \quad \Psi_j \in \mathbb{R}^{n_j \times m}.$$

  In an adaptive method, the hierarchy of coarse spaces is constructed so that $\Psi_j \approx P^j_{j-1}\Psi_{j-1}$, or P^j_{j-1} is constructed so that Ψ_j can be well approximated by elements from $\operatorname{Range} P^j_{j-1}$.

- We need the standard V-cycle preconditioner $B(\mathcal{P}_j)$ with a hierarchy of spaces given by $\mathcal{P}_j$, starting at level j.
- A_j is the restriction of the fine grid matrix A on level j, and M_j is the restriction of the 'mass' matrix, defined as $A_j = P_j^T A P_j$, $M_j = P_j^T P_j$.

A generic adaptive AMG algorithm changes the set $\mathcal{P}$, adjusting the hierarchy of spaces $\mathcal{V}$. In general, in the adaptive procedure the number of levels is not known, and in some of the algorithms below we use V_1 as the finest space by mapping the indexes in the notations above as

$$j \leftarrow J - j + 1, \quad j = 1, \dots, J.$$

15.3. A basic adaptive algorithm

We now describe a generic adaptive algorithm for constructing coarse spaces. We use an approach slightly different to that found in the literature, and fit both BAMG and αSA into one framework.

Step 0. Given $A \in \mathbb{R}^{n\times n}$ and the associated graph $\mathcal{G}(A) = (\mathcal{V}, \mathcal{E})$.

Step 1. Initialization.

1: Choose $m_0 \geq 1$, $q \geq 1$, $1 \leq n_0 < n$ and $\delta_0 \in (0, 1)$.

2: $\mathcal{P} \leftarrow \emptyset$.

3: $V_c \leftarrow V = \mathbb{R}^n$; $\mathcal{V} \leftarrow \{V_c\}$.

4: $n_c \leftarrow n$; $m \leftarrow m_0$.

5: $B \leftarrow R$.

6: Randomly pick m test vectors $\mathbf{\Psi}_0 = (\psi_1, \dots \psi_m)$, $\mathbf{\Psi}_0 \in \mathbb{R}^{n\times m}$.

7: $\mathbf{\Psi} \leftarrow \mathbf{\Psi}_0$; $\mathcal{C} \leftarrow \{\mathbf{\Psi}_0\}$.

Step 2. If $n_c \leq n_0$, go to **Step 3**, else do:

1: Make a copy of $\mathbf{\Psi}$: $\hat{\mathbf{\Psi}} \leftarrow \mathbf{\Psi}$. Then compute

$$\mathbf{\Psi} \leftarrow (I - BA)^q \mathbf{\Psi}, \quad \delta = \max_{1\leq i\leq m} \frac{\|\Psi_i\|_A}{\|\hat{\Psi}_i\|_A}. \tag{15.1}$$

Algorithm 9 Approximation of l_e eigenpairs of A

for $j = J, \dots, 1$ **do**
  if $j = J$ (coarsest grid) **then**
    Find $\{(\varphi_J^{(k)}, \lambda_J^{(k)})\}_{k=1}^{l_e}$– the solutions to the eigenvalue problems: $A_J\varphi_J^{(k)} = \lambda_J^{(k)} M_J \varphi_J^{(k)}$, $k = 1, \dots, l_e$;
  else
    for $k = 1, \dots, l_e$ **do**
      Set $v_j^{(k)} = (P_{j-1}^j)^T \varphi_{j-1}^{(k)}$, $\mu_j^{(k)} = \lambda_{j-1}^{(k)}$, and $C_j^{(k)} = A_j - \mu_j^{(k)} M_j$
      Relax on $C_j^{(k)} w = 0$, *i.e.*, $\varphi_j^{(k)} = (I - S_j^{(k)} C_j^{(k)})\, v_j^{(k)}$.
      Set $\lambda_j^{(k)} = \dfrac{\langle A_j\varphi_j^{(k)}, \varphi_j^{(k)}\rangle}{\langle M_j\varphi_j^{(k)}, \varphi_j^{(k)}\rangle}$
    end for
  end if
end for

2: If $\delta \le \delta_0$ then **stop**.

3: Use a coarsening strategy (see Section 9) to update n_c and find a set of coarse grid DOFs $\{N_i(\cdot)\}_{i=1}^{n_c}$. Then set

$$V_c \leftarrow \mathbb{R}^{n_c}, \quad \mathcal{V} \leftarrow \mathcal{V} \cup \{V_c\}.$$

4: Form the 'restriction' of $\boldsymbol{\Psi}$ to the coarse space:

$$\boldsymbol{\Psi}_c \leftarrow \text{Restrict}(\boldsymbol{\Psi}, \{N_i(\cdot)\}_{i=1}^{n_c}).$$

Then set $\mathcal{C} \leftarrow \mathcal{C} \cup \{\boldsymbol{\Psi}_c\}$.

5: Identify a sparsity pattern $\mathrm{S}(P)$ (see Section 15.1; also Sections 12.3 and 13.2).

6: Find a prolongation matrix P using $\boldsymbol{\Psi}_c$ and $\boldsymbol{\Psi}$ by applying Algorithms 7 or 8. Then set

$$\mathcal{P} \leftarrow \mathcal{P} \cup \{P\}, \quad A \leftarrow P^T A P, \quad \boldsymbol{\Psi} \leftarrow \boldsymbol{\Psi}_c, B \leftarrow R_c,$$

where R_c is the relaxation on the coarse grid V_c.

Step 3. Order the spaces in $\mathcal{V}$ increasing with respect to their dimension as

$$\mathcal{V} = \{V_1, V_2, \dots, V_J\} \quad \text{with } V_J = \mathbb{R}^n,$$

and the corresponding prolongation matrices

$$\mathcal{P} = \{P_j^{j+1}\}_{j=1}^{J-1}, \quad P_j^{j+1} : V_j \mapsto V_{j+1}, \quad j = 1, 2, \dots, J-1.$$

Algorithm 10 αSA: adding a test vector ψ to the current set of test vectors Ψ_J

Step 1. Update $\mathbf{\Psi}_J$ by adding ψ as a new column: $\mathbf{\Psi}_J \leftarrow [\mathbf{\Psi}_J, \psi]$.

Step 2. For $l = J, \dots, 3$:

1: Update Ψ_{l-1} by evaluating the coarse grid degrees of freedom $N_j(\psi)$, $j = 1 : n_{l-1}$. This process adds n_{l-1} rows to Ψ_{l-1}.

2: Use Algorithm 8 (SA) to define a prolongation P_{l-1}^l matrix and coarse grid operator A_{l-1}.

3: Update $\mathcal{P}_l \leftarrow \{P_{l-1}^l\} \cup \mathcal{P}_{l-1}$.

4: As the number of rows in Ψ_{l-1} was increased in Step 2.1, we need to change P_{l-2}^{l-1} in $\mathcal{P}_{l-1}$ in order to keep the set $\mathcal{P}$ consistent. We define $\widetilde{P}_{l-2}^{l-1}$ as the 'bridge' prolongation via Algorithm 11, and we set

$$\widetilde{\mathcal{P}}_{l-1} =\leftarrow \{\widetilde{P}_{l-2}^{l-1}\} \cup \mathcal{P}_{l-2}.$$

5: Test the convergence of $B(\widetilde{\mathcal{P}}_{l-1})$ on the newly added test vector ψ_{l-1}:

$$\hat{\psi}_{l-1} \leftarrow \psi_{l-1}, \quad \psi_{l-1} \leftarrow (I - B_{l-1}(\widetilde{\mathcal{P}}_{l-1})A)^q(\psi_{l-1}).$$

If $\left(\dfrac{\|P_{l-1}\psi_{l-1}\|_A^2}{\|P_{l-1}\widehat{\psi}_{l-1}\|_A^2}\right)^{1/q} \le \delta$ then **stop**.

6: Update the coarse representation of $\mathbf{\Psi}_{l-1}$:

$$\mathbf{\Psi}_{l-1} \leftarrow [\hat{\mathbf{\Psi}}^{l-1}, \psi_{l-1}].$$

Set B to be the V-cycle AMG method based on the above coarse spaces and prolongation matrices.

Step 4. Compute δ in (15.1) using current B. If $\delta \le \delta_0$, **stop**. Else, update $\mathcal{C}$, $\mathcal{P}$ and $\mathcal{V}$ by modifying, removing from or adding to the set of test vectors; bootstrap AMG uses Algorithm 9 and adaptive SA uses Algorithm 10. Then go to **Step 3**.

The restriction operator in Step 2.4 is as follows:

$$\mathbf{\Psi}_c = (N_i(\psi_j))_{1\le i\le n_c, 1\le j\le m} \qquad \text{(bootstrap AMG)},$$

$$\mathbf{\Psi}_c = (N_i(\psi_j))_{1\le i\le n_c} \otimes e_j, \quad j = 1 : m \quad \text{(adaptive aggregation)}.$$

In the test vectors all the sets $\mathcal{P}$, $\mathcal{C}$, $\mathcal{V}$ are chosen so that J corresponds to the finest grid.

Algorithm 11 αSA setup: construction of a 'bridging' prolongator

1: Denote the last column of Ψ_{l-1} by ψ_{l-1}, and let $\hat{\Psi}_{l-1}$ consist of all other columns of Ψ_{l-1}.

2: Create a prolongation P_{l-2}^{l-1} using Algorithm 8 with the Ψ_{l-2} by fitting all the vectors in $\hat{\Psi}_{l-1}$.

15.4. Bibliographical notes

The AMG approaches we have considered are aimed at the adaptive choice of coarse spaces and multilevel hierarchies in an AMG algorithm. The majority of adaptive AMG methods known to date use the operator A and aim to capture the worst-case errors. All specific details regarding *bootstrap AMG* (BAMG) are found in Brandt (2002), Brandt *et al.* (2011) and Brandt, Brannick, Kahl and Livshits (2015). Adaptive classical AMG is described in Brezina *et al.* (2006), and adaptive smoothed aggregation AMG is discussed in detail by Brezina *et al.* (2004, 2005). While the adaptive and bootstrap MG processes have been successful in several settings, they are still only a heuristic attempt to overcome serious obstacles to achieving good performance from the classical AMG perspective. Indeed, the costs of achieving added robustness using a bootstrap or adaptive MG algorithm are significant.

Central to adaptive MG is an important distinction between the role of the underlying multigrid algorithm (aggregation or classical AMG) and what additional adaptive elements it should provide. If the idea of self-improving the coarse spaces is poorly implemented, the adaptive and bootstrap multigrid algorithms can easily degenerate into an algorithm with no better convergence properties than a classical Krylov method preconditioned by the MG smoother (Falgout 2004).

The basic ideas on adaptive AMG are outlined in early works on classical AMG (Brandt *et al.* 1982). In fact, the adaptive process of constructing interpolation based on fitting a set of test vectors for badly scaled matrices was introduced by Ruge (1983). Some of the basic ideas of adaptive AMG can also be found in Ruge (1985, 1986), McCormick and Ruge (1989) and Brandt *et al.* (1985). Further advances in adaptive AMG methods, along with new ideas for using eigenvector approximations to guide the adaptive process, were introduced in the bootstrap AMG (BAMG) framework of Brandt (2002). The BAMG method is a self-learning multigrid algorithm that automatically determines the algebraically smooth errors in a given problem, and was further developed by Brandt *et al.* (2011, 2015). Adaptive AMG algorithms were further developed and new algorithms were introduced in some more recent works: αAMG in the framework of classical AMG (Brezina *et al.* 2006), and αSA in the framework of smoothed aggregation AMG (Brezina *et al.* 2004, Brezina *et al.* 2005). Other adaptive multilevel algorithms are the adaptive filtering and the filtering decompositions (Wittum 1992, Wagner and Wittum 1997) and the multilevel multigraph algorithms (Bank and Smith 2002).

These references have more specific details on implementation of BAMG, αAMG and αSA. One improvement on the original BAMG method is that of Manteuffel, McCormick, Park and Ruge (2010), who introduce an *indirect* BAMG (iBAMG) method. Compared to BAMG, the iBAMG method is closer to the spirit of classical AMG, which attempts to collapse unwanted connections on the assumption that the smooth error is locally constant.

Brandt *et al.* (2011) pair BAMG with an adaptive relaxation (Brandt 2000, Kahl 2009) and a multigrid eigensolver (Brezina *et al.* 2008). A combination of the bootstrap cycling scheme (Livshits 2008, Kahl 2009) and then the multigrid eigensolver is used to compute sufficiently accurate sets of test vectors, and adaptive relaxation is used to improve the AMG setup cycle.

In recent years, other adaptive approaches to constructing hierarchy of spaces have been introduced. The classical AMG approach to defining the hierarchy of spaces is considered by Brannick, Frommer, Kahl, MacLachlan and Zikatanov (2010) and MacLachlan *et al.* (2006) in the framework of adaptive reduction algorithms. Adaptive BoxMG was considered by MacLachlan *et al.* (2012). A specialized adaptive approach for Markov chains has been presented by De Sterck, Miller, Treister and Yavneh (2011).

16. Concluding remarks

In this paper we try to give a coherent presentation of a number of algebraic multigrid (AMG) methods. However, this presentation, limited by our current theoretical understanding of AMG algorithms in general, is by no means complete. We choose to include those AMG algorithms that can fit into the theoretical frameworks presented in this paper. One notable exception is the bootstrap and adaptive AMG presented in Section 15. These types of algorithm still lack good theoretical understanding, but they provide an algorithmic framework to use and extend the AMG algorithms presented in Sections 10–13 to a wider range of applications.

There are still many AMG algorithms in the literature that we are not able to include in this paper, for two main reasons. One is that these algorithms cannot be easily cast within our theoretical framework; another is that there are algorithms that the authors are yet to comprehend on a reasonable theoretical level. Examples of algorithms and results that need further investigation and analysis include adaptive filtering, multilevel ILU methods, the multilevel convergence properties of SA-AMG, BAMG, adaptive AMG, and many others.

The AMG methods studied in this paper are obtained by optimizing the choice of coarse spaces based on a given smoother. Indeed, almost all the existing AMG methods follow this strategy. It is possible, however, that an AMG method could also be designed by optimizing the choice of smoother based on a given coarsening strategy. This, in our view, is a subject worthy of further investigation. Theoretically, it would also be possible to optimize both coarsening and smoother simultaneously.

Finally, we would like to note that several AMG software packages have been developed and are currently in use, most notably:[1]

Hypre	http://acts.nersc.gov/hypre/
Trilinos	https://trilinos.org/
Multigraph	http://ccom.ucsd.edu/~reb/software.html
AGMG	http://homepages.ulb.ac.be/~ynotay/AGMG/
PyAMG	http://pyamg.org/
FASP	http://fasp.sourceforge.net/

Acknowledgements

The authors wish to thank Stephen F. McCormick, Xiaozhe Hu, James Brannick and Scott MacLachlan for many helpful discussions and suggestions, and especially Hongxuan Zhang for his help throughout the preparation of this paper.

The work of the first author was supported by US DOE grant DE-SC-0009249, as part of the 'Collaboratory on Mathematics for Mesoscopic Modeling of Materials', and NSF grants DMS-1412005 and DMS-1522615. The work of the second author was supported by NSF grants DMS-1418843 and DMS-1522615.

REFERENCES

R. Alcouffe, A. Brandt, J. Dendy Jr and J. Painter (1981), 'The multigrid method for the diffusion equation with strongly discontinuous coefficients', *SIAM J. Sci. Statist. Comput.* **2**, 430–454.

O. Axelsson and P. Vassilevski (1991), 'A black box generalized conjugate gradient solver with inner iterations and variable-step preconditioning', *SIAM J. Matrix Anal. Appl.* **12**, 625–644.

O. Axelsson and P. Vassilevski (1994), 'Variable-step multilevel preconditioning methods I: Self-adjoint and positive definite elliptic problems', *Numer. Linear Algebra Appl.* **1**, 75–101.

[1] The URLs cited in this work were correct at the time of going to press, but the publisher and the authors make no undertaking that the citations remain live or are accurate or appropriate.

B. Ayuso de Dios, F. Brezzi, L. Marini, J. Xu and L. Zikatanov (2014), 'A simple preconditioner for a discontinuous Galerkin method for the Stokes problem', *J. Sci. Comput.* **58**, 517–547.

I. Babuška and J. Osborn (1983), 'Generalized finite element methods: Their performance and their relation to mixed methods', *SIAM J. Numer. Anal.* **20**, 510–536.

N. Bahvalov (1966), 'Convergence of a relaxation method under natural constraints on an elliptic operator', *Ž. Vyčisl. Mat. i Mat. Fiz.* **6**, 861–883.

R. Bank and C. Douglas (1985), 'Sharp estimates for multigrid rates of convergence with general smoothing and acceleration', *SIAM J. Numer. Anal.* **22**, 617–633.

R. Bank and T. Dupont (1980), Analysis of a two-level scheme for solving finite element equations. Technical report CNA–159, Center for Numerical Analysis, University of Texas, Austin.

R. Bank and R. Smith (2002), 'An algebraic multilevel multigraph algorithm', *SIAM J. Sci. Comput.* **23**, 1572–1592.

R. Blaheta (1986), 'A multilevel method with correction by aggregation for solving discrete elliptic problems', *Aplikace Matematiky* **31**, 365–378.

R. Blaheta (1988), 'Iterative methods for the numerical solution of the boundary value problem of elasticity' (abstract of thesis), *Commentationes Mathematicae Universitatis Carolinae* **29**, 200.

D. Braess and W. Hackbusch (1983), 'A new convergence proof for the multigrid method including the V-cycle', *SIAM J. Numer. Anal.* **20**, 967–975.

J. Bramble (1993), *Multigrid Methods*, Vol. 294 of Pitman Research Notes in Mathematics, Longman Scientific & Technical.

J. Bramble and J. Pasciak (1987), 'New convergence estimates for multigrid algorithms', *Math. Comp.* **49**(180), 311–329.

J. Bramble and J. Xu (1991), 'Some estimates for a weighted L^2 projection', *Math. Comp.* **56**(194), 463–476.

J. Bramble, J. Pasciak and J. Xu (1990), 'Parallel multilevel preconditioners', *Math. Comp.* **55**(191), 1–22.

J. Bramble, J. Pasciak and J. Xu (1991*a*), 'The analysis of multigrid algorithms with nonnested spaces or noninherited quadratic forms', *Math. Comp.* **56**(193), 1–34.

J. Bramble, J. Pasciak, J. Wang and J. Xu (1991*b*), 'Convergence estimates for multigrid algorithms without regularity assumptions', *Math. Comp.* **57**(195), 23–45.

J. Bramble, J. Pasciak, J. Wang and J. Xu (1991*c*), 'Convergence estimates for product iterative methods with applications to domain decomposition', *Math. Comp.* **57**(195), 1–21.

A. Brandt (1973), Multi-level adaptive technique (MLAT) for fast numerical solution to boundary value problems. In *Proc. Third International Conference on Numerical Methods in Fluid Mechanics*, Vol. 18 of Lecture Notes in Physics, Springer, pp. 82–89.

A. Brandt (1977), 'Multi-level adaptive solutions to boundary-value problems', *Math. Comp.* **31**(138), 333–390.

A. Brandt (1986), 'Algebraic multigrid theory: The symmetric case', *Appl. Math. Comput.* **19**, 23–56.

A. Brandt (2000), 'General highly accurate algebraic coarsening', *Electron. Trans. Numer. Anal.* **10**, 1–20.

A. Brandt (2002), Multiscale scientific computation: Review 2001. In *Multiscale and Multiresolution Methods: Theory and Applications* (T. Barth, T. Chan and R. Haimes, eds), Springer, pp. 3–95.

A. Brandt, J. Brannick, K. Kahl and I. Livshits (2011), 'Bootstrap AMG', *SIAM J. Sci. Comput.* **33**, 612–632.

A. Brandt, J. Brannick, K. Kahl and I. Livshits (2015), 'Bootstrap algebraic multigrid: Status report, open problems, and outlook', *Numer. Math. Theory Methods Appl.* **8**, 112–135.

A. Brandt, S. McCormick and J. Ruge (1982), Algebraic multigrid (AMG) for automatic multigrid solutions with application to geodetic computations. Technical report, Institute for Computational Studies, Fort Collins, CO.

A. Brandt, S. McCormick and J. Ruge (1985), Algebraic multigrid (AMG) for sparse matrix equations. In *Sparsity and its Applications: Loughborough 1983*, Cambridge University Press, pp. 257–284.

J. Brannick and R. Falgout (2010), 'Compatible relaxation and coarsening in algebraic multigrid', *SIAM J. Sci. Comput.* **32**, 1393–1416.

J. Brannick and L. Zikatanov (2006), Algebraic multigrid methods based on compatible relaxation and energy minimization. In *Domain Decomposition Methods in Science and Engineering XVI* (O. Widlund and D. Keyes, eds), Springer, pp. 15–26.

J. Brannick, M. Brezina, S. MacLachlan, T. Manteuffel, S. McCormick and J. Ruge (2006), 'An energy-based AMG coarsening strategy', *Numer. Linear Algebra Appl.* **13**, 133–148.

J. Brannick, Y. Chen and L. Zikatanov (2012), 'An algebraic multigrid method for anisotropic elliptic equations based on subgraph matching equations based on subgraph matching', *Numer. Linear Algebra Appl.* **19**, 279–295.

J. Brannick, A. Frommer, K. Kahl, S. MacLachlan and L. Zikatanov (2010), 'Adaptive reduction-based multigrid for nearly singular and highly disordered physical systems', *Electron. Trans. Numer. Anal.* **37**, 276–295.

M. Brezina and P. Vassilevski (2011), Smoothed aggregation spectral element agglomeration AMG: SA-ρAMGe. In *Large-Scale Scientific Computing 2011* (I. Lirkov, S. Margenov and J. Wasniewski, eds), Vol. 7116 of Lecture Notes in Computer Science, Springer, pp. 3–15.

M. Brezina, A. Cleary, R. Falgout, V. Henson, J. Jones, T. Manteuffel, S. McCormick and J. Ruge (2001), 'Algebraic multigrid based on element interpolation (AMGe)', *SIAM J. Sci. Comput.* **22**, 1570–1592.

M. Brezina, R. Falgout, S. MacLachlan, T. Manteuffel, S. McCormick and J. Ruge (2004), 'Adaptive smoothed aggregation (αSA)', *SIAM J. Sci. Comput.* **25**, 1896–1920.

M. Brezina, R. Falgout, S. MacLachlan, T. Manteuffel, S. McCormick and J. Ruge (2005), 'Adaptive smoothed aggregation (αSA) multigrid', *SIAM Rev.* **47**, 317–346.

M. Brezina, R. Falgout, S. MacLachlan, T. Manteuffel, S. McCormick and J. Ruge (2006), 'Adaptive algebraic multigrid', *SIAM J. Sci. Comput.* **27**, 1261–1286.

M. Brezina, T. Manteuffel, S. McCormick, J. Ruge, G. Sanders and P. Vassilevski (2008), 'A generalized eigensolver based on smoothed aggregation (GES-SA) for initializing smoothed aggregation (SA) multigrid', *Numer. Linear Algebra Appl.* **15**, 249–269.

W. Briggs, V. Henson and S. McCormick (2000), *A Multigrid Tutorial*, second edition, SIAM.

T. Chan and W. Wan (2000), 'Robust multigrid methods for nonsmooth coefficient elliptic linear systems', *J. Comput. Appl. Math.* **123**, 323–352.

T. Chan, J. Xu and L. Zikatanov (1998), An agglomeration multigrid method for unstructured grids. In *Domain Decomposition Methods 10*, Vol. 218 of Contemporary Mathematics, AMS, pp. 67–81.

T. Chartier, R. Falgout, V. Henson, J. Jones, T. Manteuffel, S. McCormick, J. Ruge and P. Vassilevski (2003), 'Spectral AMGe (ρAMGe)', *SIAM J. Sci. Comput.* **25**, 1–26.

L. Chen (2011), Deriving the X-Z identity from auxiliary space method. In *Domain Decomposition Methods in Science and Engineering XIX*, Vol. 78 of Lecture Notes in Computational Science and Engineering, Springer, pp. 309–316.

L. Chen, R. Nochetto and J. Xu (2012), 'Optimal multilevel methods for graded bisection grids', *Numer. Math.* **120**, 1–34.

P. Ciarlet (2002), *The Finite Element Method for Elliptic Problems*, Vol. 40 of Classics in Applied Mathematics, SIAM. Reprint of the 1978 original.

A. Cleary, R. Falgout, V. Henson and J. Jones (1998), Coarse-grid selection for parallel algebraic multigrid. In *Solving Irregularly Structured Problems in Parallel: Berkeley, CA, 1998*, Vol. 1457 of Lecture Notes in Computer Science, Springer, pp. 104–115.

R. Courant and D. Hilbert (1924), *Methoden der Mathematischen Physik*, Vol. 1, Julius Springer.

P. D'Ambra and P. Vassilevski (2013), 'Adaptive AMG with coarsening based on compatible weighted matching', *Comput. Vis. Sci.* **16**, 59–76.

P. D'Ambra and P. Vassilevski (2014), αAMG based on weighted matching for systems of elliptic PDEs arising from displacement and mixed methods. Technical report LNLL-CONF-656131, Lawrence Livermore National Laboratory.

H. De Sterck, R. Falgout, J. Nolting and U. Yang (2008), 'Distance-two interpolation for parallel algebraic multigrid', *Numer. Linear Algebra Appl.* **15**, 115–139.

H. De Sterck, K. Miller, E. Treister and I. Yavneh (2011), 'Fast multilevel methods for Markov chains', *Numer. Linear Algebra Appl.* **18**, 961–980.

P. de Zeeuw (1990), 'Matrix-dependent prolongations and restrictions in a blackbox multigrid solver', *J. Comput. Appl. Math.* **33**, 1–27.

J. Dendy Jr (1982), 'Black box multigrid', *J. Comput. Phys.* **48**, 366–386.

J. Dendy Jr (1983), 'Black box multigrid for nonsymmetric problems', *Appl. Math. Comput.* **13**, 261–283.

R. Diestel (2010), *Graph Theory*, fourth edition, Vol. 173 of Graduate Texts in Mathematics, Springer.

C. Dohrmann, A. Klawonn and O. Widlund (2008), A family of energy minimizing coarse spaces for overlapping Schwarz preconditioners. In *Domain Decomposition Methods in Science and Engineering XVII*, Vol. 60 of Lecture Notes in Computational Science and Engineering, Springer, pp. 247–254.

Y. Efendiev, J. Galvis and P. Vassilevski (2011), Spectral element agglomerate algebraic multigrid methods for elliptic problems with high contrast coefficients. In *Domain Decomposition Methods in Science and Engineering XIX* (Y. Huang, R. Kornhuber, O. Widlund and J. Xu, eds), Vol. 78 of Lecture Notes in Computational Science and Engineering, Springer, pp. 407–414.

Y. Efendiev, T. Hou and X.-H. Wu (2000), 'Convergence of a nonconforming multiscale finite element method', *SIAM J. Numer. Anal.* **37**, 888–910.

R. Falgout (2004), A note on the relationship between adaptive AMG and PCG. Technical report UCRL-TR-205838, Lawrence Livermore National Laboratory.

R. Falgout and P. Vassilevski (2004), 'On generalizing the AMG framework', *SIAM J. Numer. Anal.* **42**, 1669–1693. Also available as technical report UCRL-JC-150807, Lawrence Livermore National Laboratory.

R. Falgout, P. Vassilevski and L. Zikatanov (2005), 'On two-grid convergence estimates', *Numer. Linear Algebra Appl.* **12**, 471–494.

K. Fan (1949), 'On a theorem of Weyl concerning eigenvalues of linear transformations I', *Proc. Nat. Acad. Sci. USA* **35**, 652.

R. Fedorenko (1961), 'A relaxation method of solution of elliptic difference equations', *Ž. Vyčisl. Mat. i Mat. Fiz.* **1**, 922–927.

R. Fedorenko (1964), 'On the speed of convergence of an iteration process', *Ž. Vyčisl. Mat. i Mat. Fiz.* **4**, 559–564.

A. Gibbons (1985), *Algorithmic Graph Theory*, Cambridge University Press.

G. Golub and Q. Ye (1999/2000), 'Inexact preconditioned conjugate gradient method with inner–outer iteration', *SIAM J. Sci. Comput.* **21**, 1305–1320.

L. Grasedyck, L. Wang and J. Xu (2015), 'A nearly optimal multigrid method for general unstructured grids', *Numer. Math.* pp. 1–30.

T. Grauschopf, M. Griebel and H. Regler (1997), 'Additive multilevel preconditioners based on bilinear interpolation, matrix-dependent geometric coarsening and algebraic multigrid coarsening for second-order elliptic PDEs', *Appl. Numer. Math.* **23**, 63–95.

M. Griebel and P. Oswald (1995), 'On the abstract theory of additive and multiplicative Schwarz algorithms', *Numer. Math.* **70**, 163–180.

W. Hackbusch (1978), A fast iterative method for solving Poisson's equation in a general region. In *Numerical Treatment of Differential Equations: Proc. Conf., Math. Forschungsinst., Oberwolfach 1976*, Vol. 631 of Lecture Notes in Mathematics, Springer, pp. 51–62. Longer version: Ein Iteratives Verfahren zur Schnellen Auflösung Elliptischer Randwertprobleme, Report 76-12 (1976), Mathematisches Institut der Universität zu Köln.

W. Hackbusch (1979), 'On the computation of approximate eigenvalues and eigenfunctions of elliptic operators by means of a multi-grid method', *SIAM J. Numer. Anal.* **16**, 201–215. Earlier version: Report 77-10, Mathematisches Institut der Universität zu Köln, 1977.

W. Hackbusch (1985), *Multigrid Methods and Applications*, Vol. 4 of Springer Series in Computational Mathematics, Springer.

W. Hackbusch (1989), 'The frequency decomposition multi-grid method I: Application to anisotropic equations', *Numer. Math.* **56**, 229–245.

W. Hackbusch (1994), *Iterative Solution of Large Sparse Systems of Equations*, Vol. 95 of Applied Mathematical Sciences, Springer.

P. Halmos (1974), *Finite-dimensional Vector Spaces*, second edition, Undergraduate Texts in Mathematics, Springer.

V. Henson and P. Vassilevski (2001), 'Element-free AMGe: General algorithms for computing interpolation weights in AMG', *SIAM J. Sci. Comput.* **23**, 629–650.

V. Henson and U. Yang (2002), 'BoomerAMG: A parallel algebraic multigrid solver and preconditioner', *Appl. Numer. Math.* **41**, 155–177.

T. Hou, X.-H. Wu and Z. Cai (1999), 'Convergence of a multiscale finite element method for elliptic problems with rapidly oscillating coefficients', *Math. Comp.* **68**(227), 913–943.

X. Hu, P. Vassilevski and J. Xu (2013), 'Comparative convergence analysis of nonlinear AMLI-cycle multigrid', *SIAM J. Numer. Anal.* **51**, 1349–1369.

J. Jones and P. Vassilevski (2001), 'AMGe based on element agglomeration', *SIAM J. Sci. Comput.* **23**, 109–133.

K. Kahl (2009), Adaptive algebraic multigrid for lattice QCD computations. PhD thesis, Fachbereich Mathematik und Naturwissenschaften, Bergische Universität Wuppertal.

G. Karypis and V. Kumar (1998), 'A fast and high quality multilevel scheme for partitioning irregular graphs', *SIAM J. Sci. Comput.* **20**, 359–392.

H. Kim, J. Xu and L. Zikatanov (2003), 'A multigrid method based on graph matching for convection diffusion equations', *Numer. Linear Algebra Appl.* **10**, 181–195.

H. Kim, J. Xu and L. Zikatanov (2004), 'Uniformly convergent multigrid methods for convection–diffusion problems without any constraint on coarse grids', *Adv. Comput. Math.* **20**, 385–399.

T. Kolev and P. Vassilevski (2006), 'AMG by element agglomeration and constrained energy minimization interpolation', *Numer. Linear Algebra Appl.* **13**, 771–788.

J. Kraus (2002), 'An algebraic preconditioning method for M-matrices: Linear versus non-linear multilevel iteration', *Numer. Linear Algebra Appl.* **9**, 599–618.

J. Kraus (2008), 'Algebraic multigrid based on computational molecules 2: Linear elasticity problems', *SIAM J. Sci. Comput.* **30**, 505–524.

J. Kraus and J. Schicho (2006), 'Algebraic multigrid based on computational molecules 1: Scalar elliptic problems', *Computing* **77**, 57–75.

J. Lai and L. Olson (2011), 'Algebraic multigrid for high-order hierarchical $H(\text{curl})$ finite elements', *SIAM J. Sci. Comput.* **33**, 2888–2902.

Y.-J. Lee, J. Wu, J. Xu and L. Zikatanov (2007), 'Robust subspace correction methods for nearly singular systems', *Math. Models Methods Appl. Sci.* **17**, 1937–1963.

Y.-J. Lee, J. Wu, J. Xu and L. Zikatanov (2008), 'A sharp convergence estimate for the method of subspace corrections for singular systems of equations', *Math. Comp.* **77**(262), 831–850.

K. Lipnikov, J. Moulton and D. Svyatskiy (2011), 'Adaptive strategies in the multilevel multiscale mimetic (M^3) method for two-phase flows in porous media', *Multiscale Model. Simul.* **9**, 991–1016.

O. Livne (2004), 'Coarsening by compatible relaxation', *Numer. Linear Algebra Appl.* **11**, 205–227.

O. Livne and A. Brandt (2012), 'Lean algebraic multigrid (LAMG): Fast graph Laplacian linear solver', *SIAM J. Sci. Comput.* **34**, 499–523.

I. Livshits (2008), 'One-dimensional algorithm for finding eigenbasis of the Schrödinger operator', *SIAM J. Sci. Comput.* **30**, 416–440.

M. Luby (1986), 'A simple parallel algorithm for the maximal independent set problem', *SIAM J. Comput.* **15**, 1036–1053.

S. MacLachlan and L. Olson (2014), 'Theoretical bounds for algebraic multigrid performance: Review and analysis', *Numer. Linear Algebra Appl.* **21**, 194–220.

S. MacLachlan and Y. Saad (2007), 'A greedy strategy for coarse-grid selection', *SIAM J. Sci. Comput.* **29**, 1825–1853.

S. MacLachlan, T. Manteuffel and S. McCormick (2006), 'Adaptive reduction-based AMG', *Numer. Linear Algebra Appl.* **13**, 599–620.

S. MacLachlan, J. Moulton and T. Chartier (2012), 'Robust and adaptive multigrid methods: Comparing structured and algebraic approaches', *Numer. Linear Algebra Appl.* **19**, 389–413.

J.-F. Maitre and F. Musy (1983), 'Méthodes multigrilles: Opérateur associé et estimations du facteur de convergence; le cas du V-cycle', *CR Acad. Sci. Paris Sér. I Math.* **296**, 521–524.

J. Mandel (1988), 'Algebraic study of multigrid methods for symmetric, definite problems', *Appl. Math. Comput.* **25**, 39–56.

J. Mandel, M. Brezina and P. Vaněk (1999), 'Energy optimization of algebraic multigrid bases', *Computing* **62**, 205–228.

T. Manteuffel, S. McCormick, M. Park and J. Ruge (2010), 'Operator-based interpolation for bootstrap algebraic multigrid', *Numerical Linear Algebra Appl.* **17**, 519–537.

I. Marek (1991), Aggregation methods of computing stationary distributions of Markov processes. In *Numerical Treatment of Eigenvalue Problems | Numerische Behandlung von Eigenwertaufgaben*, Vol. 5, Springer, pp. 155–169.

A. Matsokin and S. Nepomnyashchikh (1985), 'The Schwarz alternation method in a subspace', *Izv. Vyssh. Uchebn. Zaved. Mat.* **85**, 61–66.

S. McCormick (1982), 'An algebraic interpretation of multigrid methods', *SIAM J. Numer. Anal.* **19**, 548–560.

S. McCormick (1984), 'Multigrid methods for variational problems: Further results', *SIAM J. Numer. Anal.* **21**, 255–263.

S. McCormick (1985), 'Multigrid methods for variational problems: General theory for the V-cycle', *SIAM J. Numer. Anal.* **22**, 634–643.

S. McCormick and J. Ruge (1982), 'Multigrid methods for variational problems', *SIAM J. Numer. Anal.* **19**, 924–929.

S. McCormick and J. Ruge (1989), Algebraic multigrid methods applied to problems in computational structural mechanics. Technical report 19900060129, NASA.

S. Míka and P. Vaněk (1992*a*), 'Acceleration of convergence of a two level algebraic algorithm by aggregation in smoothing process', *Appl. Math.* **37**, 343–356.

S. Míka and P. Vaněk (1992*b*), 'A modification of the two-level algorithm with overcorrection', *Appl. Math.* **37**, 13–28.

A. Nägel, R. Falgout and G. Wittum (2008), 'Filtering algebraic multigrid and adaptive strategies', *Comput. Vis. Sci.* **11**, 159–167.

A. Napov and Y. Notay (2012), 'An algebraic multigrid method with guaranteed convergence rate', *SIAM J. Sci. Comput.* **34**, A1079–A1109.

R. Nicolaides (1975), 'On multiple grid and related techniques for solving discrete elliptic systems', *J. Comput. Phys.* **19**, 418–431.

R. Nicolaides (1977), 'On the l^2 convergence of an algorithm for solving finite element equations', *Math. Comp.* **31**(140), 892–906.

Y. Notay (2000), 'Flexible conjugate gradients', *SIAM J. Sci. Comput.* **22**, 1444–1460.

Y. Notay (2010), 'An aggregation-based algebraic multigrid method', *Electron. Trans. Numer. Anal.* **37**, 123–146.

Y. Notay (2012), 'Aggregation-based algebraic multigrid for convection–diffusion equations', *SIAM J. Sci. Comput.* **34**, A2288–A2316.

Y. Notay and P. Vassilevski (2008), 'Recursive Krylov-based multigrid cycles', *Numer. Linear Algebra Appl.* **15**, 473–487.

L. Olson, J. Schroder and R. Tuminaro (2011), 'A general interpolation strategy for algebraic multigrid using energy minimization', *SIAM J. Sci. Comput.* **33**, 966–991.

P. Oswald (1999), 'On the robustness of the BPX-preconditioner with respect to jumps in the coefficients', *Math. Comp.* **68**(226), 633–650.

M. Reed and B. Simon (1978), *Methods of Modern Mathematical Physcis*, Vol. IV: *Analysis of Operators*, Elsevier.

A. Reusken (1993), 'Multigrid with matrix-dependent transfer operators for a singular perturbation problem', *Computing* **50**, 199–211.

J. Ruge (1983), Algebraic multigrid (AMG) for geodetic survey problems. In *Preliminary Proc. Internat. Multigrid Conference, Fort Collins, CO.*

J. Ruge (1985), Final report on AMG02. Report, Gesellschaft für Mathematik und Datenverarbeitung, St. Augustin, GMD contract 5110, 022090.

J. Ruge (1986), 'AMG for problems of elasticity', *Appl. Math. Comput.* **19**, 293–309.

J. Ruge and K. Stüben (1987), Algebraic multigrid (AMG). In *Multigrid Methods* (S. McCormick, ed.), Vol. 3 of Frontiers in Applied Mathematics, SIAM, pp. 73–130.

Y. Saad (2003), *Iterative Methods for Sparse Linear Systems*, second edition, SIAM.

J. Schroder (2012), 'Smoothed aggregation solvers for anisotropic diffusion', *Numer. Linear Algebra Appl.* **19**, 296–312.

B. Smith, P. Bjørstad and W. Gropp (1996), *Domain Decomposition: Parallel Multilevel Methods for Elliptic Partial Differential Equations*, Cambridge University Press.

R. Southwell (1940), *Relaxation Methods in Engineering Science: A Treatise on Approximate Computation*, Oxford Engineering Science Series, Oxford University Press.

R. Southwell (1946), *Relaxation Methods in Theoretical Physics*, Clarendon Press, Oxford University Press.

N. Spillane, V. Dolean, P. Hauret, F. Nataf, C. Pechstein and R. Scheichl (2014), 'Abstract robust coarse spaces for systems of PDEs via generalized eigenproblems in the overlaps', *Numer. Math.* **126**, 741–770.

H. Sterck, U. Yang and J. Heys (2006), 'Reducing complexity in parallel algebraic multigrid preconditioners', *SIAM J. Matrix Anal. Appl.* **27**, 1019–1039.

K. Stüben (1983), 'Algebraic multigrid (AMG): Experiences and comparisons', *Appl. Math. Comput.* **13**, 419–451.

A. Toselli and O. Widlund (2005), *Domain Decomposition Methods: Algorithms and Theory*, Vol. 34 of Springer Series in Computational Mathematics, Springer.

E. Treister and I. Yavneh (2015), 'Non-Galerkin multigrid based on sparsified smoothed aggregation', *SIAM J. Sci. Comput.* **37**, A30–A54.

U. Trottenberg, C. Oosterlee and A. Schüller (2001), *Multigrid*, Academic Press.

I. Vakhutinsky, L. Dudkin and A. Ryvkin (1979), 'Iterative aggregation: A new approach to the solution of large-scale problems', *Econometrica* **47**, 821–841.

P. Vaněk, J. Mandel and M. Brezina (1996), 'Algebraic multigrid based on smoothed aggregation for second and fourth order problems', *Computing* **56**, 179–196.

P. Vaněk, J. Mandel and M. Brezina (1998), 'Convergence of algebraic multigrid based on smoothed aggregation', *Computing* **56**, 179–196.

R. Varga (2000), *Matrix Iterative Analysis*, expanded edition, Vol. 27 of Springer Series in Computational Mathematics, Springer.

P. Vassilevski (2008), *Multilevel Block Factorization Preconditioners: Matrix-based Analysis and Algorithms for Solving Finite Element Equations*, Springer.

P. Vassilevski and L. Zikatanov (2006), 'Multiple vector preserving interpolation mappings in algebraic multigrid', *SIAM J. Matrix Anal. Appl.* **27**, 1040–1055.

C. Wagner and G. Wittum (1997), 'Adaptive filtering', *Numer. Math.* **78**, 305–328.

W. Wan (1998), Scalable and multilevel iterative methods. PhD thesis, University of California, Los Angeles.

W. Wan, T. Chan and B. Smith (1999/2000), 'An energy-minimizing interpolation for robust multigrid methods', *SIAM J. Sci. Comput.* **21**, 1632–1649.

W. Weiler and G. Wittum (1997), 'Parallel frequency filtering', *Computing* **58**, 303–316.

H. Weyl (1911), 'Über die asymptotische Verteilung der Eigenwerte', *Nachrichten von der Gesellschaft der Wissenschaften zu Göttingen, Mathematisch–Physikalische Klasse* **1911**, 110–117.

H. Weyl (1912), 'Das asymptotische Verteilungsgesetz der Eigenwerte linearer partieller Differentialgleichungen', *Math. Ann.* **71**, 441–479.

O. Widlund (1994), Exotic coarse spaces for Schwarz methods for lower order and spectral finite elements. In *Domain Decomposition Methods in Scientific and Engineering Computing*, Vol. 180 of Contemporary Mathematics, AMS, pp. 131–136.

O. Widlund (2009), The development of coarse spaces for domain decomposition algorithms. In *Domain Decomposition Methods in Science and Engineering XVIII*, Vol. 70 of Lecture Notes in Computational Science and Engineering, Springer, pp. 241–248.

G. Wittum (1992), *Filternde Zerlegungen: Schnelle Löser für große Gleichungssysteme*, Teubner Skripten zur Numerik, Teubner.

J. Xu (1989), *Theory of Multilevel Methods*, Cornell University.

J. Xu (1991), 'Counterexamples concerning a weighted L^2 projection', *Math. Comp.* **57**(196), 563–568.

J. Xu (1992), 'Iterative methods by space decomposition and subspace correction', *SIAM Rev.* **34**, 581–613.

J. Xu (1996), 'The auxiliary space method and optimal multigrid preconditioning techniques for unstructured grids', *Computing* **56**, 215–235.

J. Xu (1997), An introduction to multilevel methods. In *Wavelets, Multilevel Methods, and Elliptic PDEs* (M. Ainsworth *et al.*, eds), Numerical Mathematics and Scientific Computation, Oxford University Press, pp. 213–302.

J. Xu (2009), Optimal algorithms for discretized partial differential equations. In *ICIAM 07: 6th International Congress on Industrial and Applied Mathematics*, European Mathematical Society, pp. 409–444.

J. Xu (2016), *Multilevel Iterative Methods*. Lecture notes, 1989–2016, The Pennsylvania State University.

J. Xu and L. Zikatanov (2002), 'The method of alternating projections and the method of subspace corrections in Hilbert space', *J. Amer. Math. Soc.* **15**, 573–597.

J. Xu and L. Zikatanov (2004), 'On an energy minimizing basis for algebraic multigrid methods', *Comput. Vis. Sci.* **7**, 121–127.

J. Xu and J. Zou (1998), 'Some nonoverlapping domain decomposition methods', *SIAM Rev.* **40**, 857–914.

J. Xu, H. Zhang and L. Zikatanov (2017*a*), Obtaining optimal coarse spaces for AMG via trace minimization. Submitted to *J. Comput. Math.*

J. Xu, H. Zhang and L. Zikatanov (2017*b*), On the Weyl's law for discretized elliptic operators. Submitted to *Math. Comp.*

J. Xu, H. Zhang and L. Zikatanov (2017*c*), A unified approach to the design and analysis of AMG. Submitted to *Numer. Math.*

U. Yang (2006), *Parallel Algebraic Multigrid Methods: High Performance Preconditioners*, Springer.

D. Young (1971), *Iterative Solution of Large Linear Systems*, Academic Press.

H. Yserentant (1993), Old and new convergence proofs for multigrid methods. In *Acta Numerica*, Vol. 2, Cambridge University Press, pp. 285–326.

G. Yu, J. Xu and L. Zikatanov (2013), 'Analysis of a two-level method for anisotropic diffusion equations on aligned and nonaligned grids', *Numer. Linear Algebra Appl.* **20**, 832–851.

L. Zikatanov (2008), 'Two-sided bounds on the convergence rate of two-level methods', *Numer. Linear Algebra Appl.* **15**, 439–454.